International Series on
MICROPROCESSOR-BASED AND INTELLIGENT SYSTEMS ENGINEERING

VOLUME 10

Editor

Professor S. G. Tzafestas, *National Technical University, Athens, Greece*

Editorial Advisory Board

Professor C. S. Chen, *University of Akron, Ohio, U.S.A.*
Professor T. Fokuda, *Nagoya University, Japan*
Professor F. Harashima, *University of Tokyo, Tokyo, Japan*
Professor G. Schmidt, *Technical University of Munich, Germany*
Professor N. K. Sinha, *McMaster University, Hamilton, Ontario, Canada*
Professor D. Tabak, *George Mason University, Fairfax, Virginia, U.S.A.*
Professor K. Valavanis, *University of Southern Louisiana, Lafayette, U.S.A.*

The titles published in this series are listed at the end of this volume.

Robotic Systems

Advanced Techniques and Applications

edited by

SPYROS G. TZAFESTAS
*National Technical University of Athens,
Department of Electrical and Computer Engineering,
Athens, Greece*

KLUWER ACADEMIC PUBLISHERS
DORDRECHT / BOSTON / LONDON

Library of Congress Cataloging-in-Publication Data

```
Robotic systems : advanced techniques and applications / edited by
  Spyros G. Tzafestas.
       p.    cm. -- (Microprocessor based and intelligent systems
  engineering ; v. 10)
    Includes index.
    ISBN 0-7923-1749-1
    1. Robotics.    I. Tzafestas, S. G.   II. Series.
  TJ211.R548  1992
  670.42'72--dc20                                              92-12521
```

ISBN 0-7923-1749-1

Published by Kluwer Academic Publishers,
P.O. Box 17, 3300 AA Dordrecht, The Netherlands.

Kluwer Academic Publishers incorporates
the publishing programmes of
D. Reidel, Martinus Nijhoff, Dr W. Junk and MTP Press.

Sold and distributed in the U.S.A. and Canada
by Kluwer Academic Publishers,
101 Philip Drive, Norwell, MA 02061, U.S.A.

In all other countries, sold and distributed
by Kluwer Academic Publishers Group,
P.O. Box 322, 3300 AH Dordrecht, The Netherlands.

Printed on acid-free paper

All Rights Reserved
© 1992 Kluwer Academic Publishers
No part of the material protected by this copyright notice may be reproduced or
utilized in any form or by any means, electronic or mechanical,
including photocopying, recording or by any information storage and
retrieval system, without written permission from the copyright owner.

Printed in the Netherlands

CONTENTS

PREFACE — xi

PART 1

ROBOT KINEMATIC AND DYNAMIC ANALYSIS

Direct Kinematics of the 6–4 Fully Parallel Manipulator with Position and Orientation Uncoupled
  C. Innocenti, V. Parenti-Castelli — 3

A General Method for a Closed-Form Solution to Robot Kinematics Problems
  C. Galletti, P. Fanghella — 11

Fast Implementation of Robotic Manipulator Kinematics using Cordic and Systolic Processors
  V.G. Mertzios, S.S. Scarlatos — 19

A Formal Framework for Specifying Robot Kinematics
  V.S. Alagar, T.D. Bui, K. Periyasamy — 27

Non-Geometrical Parameters Identification for Robot Kinematic Calibration by Use of Neural Network Techniques
  J.-M. Renders, J.R. Millan, M. Becquet — 37

The Use of Skew-Symmetric Cartesian Tensors in Describing Orientations and Invariants of Spatial Rotations
  C.A. Balafoutis, R.V. Patel — 45

An Automatic Verification of Simplified Robot Models
  K. Swider — 53

A Comparison Between Theoretically Derived and Experimentally Verified Dynamic Modelling of a SCARA Robot
  T.A. Rieswijk, I. de Haas, G. Honderd, W. Jongkind — 61

Dynamics of Robots Having Supple Bodies
  P. Andre, J.-P. Taillard, P. Mitrouchev, L. Losco — 71

A Finite Elements Model of Two-Joint Robots with Elastic Forearm
  P. Muraca, M. La Cava, A. Ficola — 79

On the Modelling of the Contact Forces of Constrained Robots
  K.A. Tahboub, P.C. Müller   87

PART 2

ROBOT CONTROL

Passivity and Learning Control for Robot Tasks under Geometric Constraints
  S. Arimoto, T. Nariwa   99

A Decentralized Adaptive Controller for Robot Manipulator Trajectory Control
  P. Desai   109

Robust Control of Robot Arms Using the Variable Structure Approach
  A.J.L. Nievergeld, H.H. van de Ven   117

A Comparison of Direct Adaptive Controllers Applied to a Class 1 Manipulator
  D.P. Stoten, S.P. Hodgson   125

An Algorithm for the Nonlinear Adaptive Robot Control Synthesis
  B.M. Novaković   135

Model Reference Adaptive Control of Robotic Manipulators Using a Modified Output Error Method
  R.Y.J. Tam, S.C.A. Thomopoulos   143

Application of Modified Generalized Predictive Controllers to Trajectory Control of a Robot Arm
  S.S. Gençev, Y. Istefanopoulos, T. Ergün   151

Design of Manipulator Trajectories with Minimum Motion Time and Specified Accuracy
  Y. Song, R.M.C. De Keyser   159

The Use of Simulation Systems to Control Manually Operated Remote Manipulators, with Long Pure Time Delays
  M. Griffin, R. Mitchell   167

Path Coordinated Control of Robotic Arms
  S. Arnaltes, J. Pérez Oria   175

Command Matching for a Constraint Robot System
  P.N. Paraskevopoulos, F.N. Koumboulis, K.G. Tzierakis   183

Control of a Flexible Arm for Prescribed Frequency Domain Tolerances: Modelling, Controller Design, Laboratory Experiments
  M. Kelemen, J. Starrenburg, A. Bagchi   193

PART 3

ROBOT PATH PLANNING AND TRACKING

High-Speed Robot Path Planning in Time-Varying Environment Employing a Diffusion Equation Strategy
G. Schmidt, W. Neubauer 207

Path Planning of Transfer Motions for Industrial Robots by Heuristically Controlled Decomposition of the Configuration Space
G. Duelen, C. Willnow 217

Planning and Optimization of Geometrical Trajectories Inside Collision-Free Subspaces with the Aid of High-Order Hermite Splines
T.A. Rieswijk, P. Schalkwijk, G. Honderd 225

A Fast and Efficient Algorithm for the Computation of Path Constrained Time-Optimal Motions
T.A. Rieswijk, M. Sirks, G. Honderd, W. Jongkind 235

Three-Dimension Abstraction of Convex Space Path Planning
P.K. Sinha, P.L. Ho 245

An Approach to Real-Time Flexible Path Planning
A.C. Meng, S. Ntafos, M. Tsoukalas 253

A Path Planning Method for Mobile Robots in a Structured Environment
F.V. Hatzivasiliou, S.G. Tzafestas 261

Minimum-Time Motion Planner for Mobile Robots on Uneven Terrains
A. Liegeois, C. Moignard 271

Mobile Robot Trajectory Planning
P.K. Sinha, A. Benmounah 279

Task Modeling for Planning an Autonomous Mobile Robot
V. Schaeffer, A. Mauboussin 287

An Optimal Solution to the Robot Navigation Planning Problem Based on an Electromagnetic Analogue
V. Petridis, T.D. Tsiboukis 297

Real-Time LQG Robotic Visual Tracking
N.P. Papanikolopoulos, P.K. Khosla 305

Trajectory Tracking for Mobile Robot
S. Delaplace, P. Blazevic, J.G. Fontaine, N. Pons, J. Rabit 313

A Robust Tracking System for Mobile Robot Guidance
  J. Frau, A. Larré, E. Montseny, G. Oliver   321

A Real-Time Multiple Lane Tracker for an Autonomous Road Vehicle
  K.P. Wershofen, V. Graefe   333

PART 4

MOBILE ROBOTS: ARCHITECTURES, PERCEPTION, NAVIGATION AND CONTROL

An Architecture for Intelligent Mobile Robots
  J. Sequeira, J. Sentieiro   343

Software Architecture for an Autonomous Manipulator
  J.B. Thevenon, E.J. Gaussens, P. LePage, F. Arlabosse   351

Automatic Control of Mobility: The VAHM Project
  K. Moumen, A. Pruski   359

A Modular Approach to Mobile Robot Design
  G. Barrall, K. Warwick   367

Environment Representation by a Mobile Robot Using Quadtree Encoding of Range Data
  L. Piotrowski   375

Perception Planning in PANORAMA
  A.M. de Campos, M.M. Matos, P. Fogaça   383

Navigation and Perception Approach of PANORAMA Project
  G. Frappier, P. Lemarquand, T. Van den Bogaert   391

Trajectory Generation for Mobile Robots with Clothoids
  G.M. van der Molen   399

Digital Models for Autonomous Vehicle Terrain-Following
  N. Christou, K. Parthenis, B. Dimitriadis, N. Gouvianakis   407

Navigation of a Mobile Robot
  P. van Turennout, G. Honderd   415

Robot Navigation and Exploration in an Unknown Environment
  R. Malik, S. Prasad   423

The Optimal Next Exploration: Uncertainty Minimization in Mobile Robot Self-Location
  V. Caglioti   431

A Parallel Blackboard Model for Mobile Robotics Control
  M. Occello, C. Chaouiya, M.-C. Thomas   439

PART 5

ROBOT PROGRAMMING AND SENSORY DATA PROCESSING

Graphical Robot Programming: Requirements and Existing Systems
  G. Nikoleris   451

Practical Error Compensation for Use in Off-Line Programming of Robots
  S. Albright, K. Schröer   459

Intelligent Programming of Force-Constrained Cooperating Robots
  G. Duelen, H. Münch, Y. Zhang   469

Sensing Strategies Generation for Monitoring Robot Assembly Programs
  V. Caglioti, M. Danieli, D. Sorrenti   479

Integration of a Constraint Scheme in a Programming Language for Multi-Robots
  D. Duhaut, E. Monacelli   487

Software Architecture and Simulation Tools for Autonomous Mobile Robots
  G.D. van Albada, J.M. Lagerberg, B.J.A. Kröse   495

A Logical Framework of Sensor/Data Fusion
  M.M. Kokar, K.P. Zavoleas   505

Maintaining World Model Consistency by Matching Parametrized Object Models to 3d Measurements
  M. Järviduoma, S. Pieskä, T. Heikkilä   515

PART 6

SOPHISTICATED ROBOTIC SYSTEMS AND APPLICATIONS

A Model of Manned Robotic Systems
  P.H. Wewerinke   525

Design, Construction and Performance of an Antrhopomorphic Robot Head
  P. Mowforth, D. Wilson   535

Multi-fingered Robot Hands
  E.A. Al-Gallaf, A.J. Allen, K. Warwick   543

Multidirectional Pneumatic Force Sensor for Grippers
  R. Caen, S. Colin   551

Grasping in an Unstructured Environment Using a Coordinated Hand Arm Control
    S. Agrawal, R. Bajcsy   559

Effective Integration of Sensors and Industrial Robots by Means of a Versatile Sensor Control Unit
    J. Wahrburg   569

Identification and Evaluation of Hydraulic Actuator Models for a Two-Link Robot Manipulator
    J.-J. Zhou, F. Conrad   577

A Modular Architecture for Controlling Autonomous Agents
    L. Petropoulakis   585

Development of Intelligent Control for Robot Cells Using Knowledge Based Simulation
    Z. Doulgeri, G. D'Alessandro   595

Recent Advances in Robot Grinding
    A. Ikonomopoulos, L. Dritsas   603

A Low Cost Robot Based Integrated Manufacturing System for the Garment Industry
    I. Gibson, P. Bowden, P.M. Taylor, A.J. Wilkinson   611

Vision for Robot Guidance in Automated Butchery
    G. Purnell, K. Khodabandehloo   619

Use of Robots in Surgical Procedures
    P.N. Brett, K. Khodabandehloo   627

    AUTHOR INDEX   637

PREFACE

This book contains a selection of papers presented at the "European Robotics and Intelligent Systems Conference" (EURISCON '91) held in Corfu, Greece (June 23–28, 1991). It is devoted to the analysis, design and applications of robotic systems.

Today's industrial robots have their origin in computerized numerical control (CNC) of machine tools and remote manipulation, and have a geometrical similarity with the human arm. The basic feature of industrial robots that was drawn from CNC is their ability to be programmed and carry out various operations by using and processing stored data. The analogy with the human arm comes from the requirement to perform dexterous human-like actions and motions of modern manufacturing systems and other applications. The new field that emerged from the marriage of numerical control and remote manipulation, i.e. the field of *Robotics*, has now been expanded and involves mobile robots and robotic systems with intelligence. Many new concepts and techniques have been developed in this field which go far beyond the original ones. Actually, *robotics* has now become an interdisciplinary field with continuously increasing interest and expansion. Fields like mechanics, mechanical engineering, electronics, electrical engineering, computer science and engineering, control, vision, sensor technology, manufacturing engineering, and bioengineering are among the principal partners in the design and development of modern industrial and nonindustrial robotic systems.

The papers of the book are grouped in six parts which have some unavoidable overlap and contain a representative but nonexhaustive set of topics within the field. These parts are:

Part 1: Robot kinematic and dynamic analysis
Part 2: Robot control
Part 3: Robot path planning and tracking
Part 4: Mobile robots: Architectures, perception, navigation, and control
Part 5: Robot programming and sensory data processing
Part 6: Sophisticated robotic systems and applications

I am deeply grateful to all colleagues who have contributed to the success of EURISCON'91 through their presentations, active participation and scientific interaction. I believe that their works, put together in the present book, will provide an important reference pool of knowledge which will inspire potential researchers and practitioners to develop robotic systems with much more sophisticated and dexterous capabilities.

Finally, particular thanks are expressed to my University and to the supporting Societies for their encouragement and contribution.

December 1991   Spyros G. Tzafestas

EURISCON '91

CHAIRMAN Spyros G. Tzafestas

ADVISORY COMMITTEE
P. Coiffet (France)
T. Fukuda (Japan), R. Isermann (Germany)
P. Kopacek (Austria), P. MacConaill (EEC, ESPRIT CIM), A. Meystel (U.S.A.)
U. Rembold (Germany), G. Saridis (U.S.A.), M. Singh (U.K.)

TECHNICAL PROGRAM COMMITTEE
P. Albertos (Spain)
P. Borne (France), E. Eloranta (Finland), J.C. Gentina (France)
G. Honderd (Netherlands), T.Jordanides (U.S.A.), K.Khodabandehloo (U.K.)
J. Kontos (Greece), P. Kool (Belgium), G. Papakonstantinou (Greece)
L. Pau (France), A. Sanderson (U.S.A.), G. Schmidt (Germany)
H. Schneider (Germany), E. Siores (Australia), H. Stephanou (U.S.A.)
T. Takamori (Japan), I. Troch (Austria), K. Valavanis (U.S.A.)
A. Villa (Italy), K. Warwick (U.K.)
B. Zeigler (U.S.A.)

ORGANIZING COMMITTEE
S. Tzafestas (Chairman)
C. Athanassiou, P. Dalianis, F. Hatzivassiliou
G. Kapsiotis, A. Triantafyllakis
A. Zagorianos

PART 1
ROBOT KINEMATIC AND DYNAMIC ANALYSIS

DIRECT KINEMATICS OF THE 6-4 FULLY PARALLEL MANIPULATOR WITH POSITION AND ORIENTATION UNCOUPLED

C. Innocenti
DIEM - Facoltà di Ingegneria - Università di Bologna
Viale Risorgimento, 2 - 40136 Bologna - ITALY

V. Parenti-Castelli
Istituto di Ingegneria meccanica - Università di Salerno
84081 Baronissi - Salerno - ITALY

ABSTRACT. This paper presents the direct position analysis in analytical form of the general fully in-parallel actuated six degrees-of-freedom manipulator having geometric uncoupling between position and orientation of the platform. The analysis, performed in two steps, leads first to a 2nd order polynomial equation in one unknown for the position of the platform reference point; then, for each position, an 8th order polynomial equation in one unknown can be solved for the orientation of the platform. Consequently, the mechanism can be assembled at most in 16 configurations. An example is reported for illustrating the new theoretical results.

1. INTRODUCTION

The problem of positioning and orienting a rigid body in spatial motion can be accomplished by means of open and closed kinematic chains [1], often referred to as serial and parallel manipulators, respectively. Serial mechanisms are open chains with all pairs actively controlled, whereas parallel mechanisms consist of one or more closed loops where only some pairs are actuated. Advantages and disadvantages of these two classes of mechanisms have been stressed in the literature and some of their performances compared [2,3]. Higher mechanical stiffness, end effector acceleration and position accuracy are major advantages that make parallel mechanisms competitive with serial ones.
This paper presents the direct position analysis (DPA) in analytical form of a parallel manipulator with position and orientation uncoupled.
The DPA of parallel mechanisms, which aims to find position and orientation of the output controlled link (platform) when a set of actuator displacements is given, turns out to be a difficult problem because non-linear equations are involved, and many solutions are possible. Numerical iterative methods are used, in general, although they prove difficulty to find all solutions. On the contrary, analytical form solution, when achievable, would guarantee to find all mechanism

configurations, providing a deeper understanding of the kinematic behaviour of the mechanism. In addition, more efficient algorithms for platform position control in cartesian space could be implemented.

The uncoupling between position and orientation of the platform is a further desirable feature that simplifies the kinematic equations, and may speed up the control algorithms. Only few papers, however, have proposed manipulators with uncoupled position and orientation, and all of them resort to simplified geometries for obtaining a DPA solution in analytical form. In [4] a hybrid (parallel-serial) manipulator is developed with position-orientation nearly uncoupled. Uncoupled manipulators are presented in [5,6]: in [5] a planar platform with a fully-parallel spherical wrist and a serial subchain for platform position are adopted, and in [6] a simplified geometry is considered.

In this paper the most general arrangement of the fully in-parallel actuated six-degrees-of-freedom manipulator with uncoupled platform position and orientation is presented, and the analytical solution of the manipulator direct position analysis is developed.

Basically, the manipulator (see Fig. 1) consists of two bodies (base and platform) connected by six independent links (legs) with variable lengths. The six legs meet the base at six distinct points, while three legs meet the platform at the same point Q. Hence there are four connection points in the platform. Points on both base and platform are not constrained to lie on a plane. At each leg extremity a spherical pair connection exists, so that a triple spherical pair occurs in Q. The basic geometrical structure of the manipulator is then formed by a fully in-parallel actuated positioning part - the three legs that allocate reference point Q - and by a spherical wrist architecture, having the most general fully-parallel geometry, that can orient the platform after the desired position of Q has been achieved. Because of its 6 and 4 connection point pattern, the manipulator is also denoted as 6-4 manipulator.

The analytical form DPA of the 6-4 manipulator is presented. The DPA can be solved in two distinct steps: first the position of point Q is determined by solving a 2nd order polynomial equation, then the

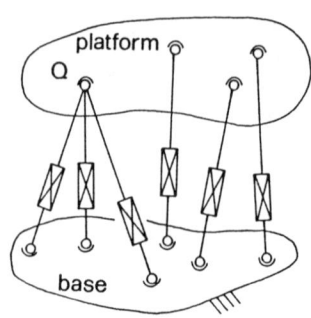

Fig. 1 Sketch of 6-4 manipulator

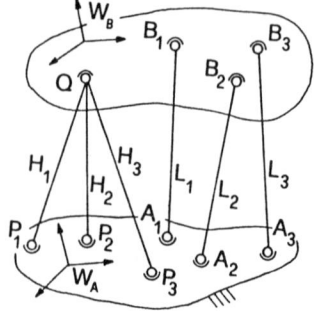

Fig. 2 Sketch of 6-4 structure

orientation of the platform is obtained from the solution of an 8th order polynomial equation. The second step, presented in a previous paper [7], is briefly reviewed here for the sake of completeness.
The analysis, in conclusion, provides 16 solutions in the complex field, i.e., for a given set of actuator displacements, the manipulator can be assembled in 16 real configurations at most.
Finally, a numerical example confirms the new theoretical results.

2. DIRECT POSITION ANALYSIS

Kinematic model. When a set of actuator displacements is given the manipulator becomes a statically determined structure (see Fig. 2), denoted as 6-4 structure, and the position analysis reduces to find all possible closure configurations of the structure.
A careful inspection of Fig. 2 shows that the 6-4 structure can be analyzed by considering first a statically determined substructure, called positioning substructure, which is responsible for the positioning of reference point Q (see Fig. 3). Once the position of point Q has been determined, the orientation of the platform W can be found by referring to the structure of Fig. 4, which represents the most general fully-parallel constraint defining the relative orientation of two bodies having a point in common. Thus position and orientation of the platform can be solved separately.

Determination of point Q position. The geometrical dimensions of the base are given; hence the coordinates of points P_i, i=1,3, in an arbitrary reference coordinate system W_A fixed to the base are known (see Fig. 3). For a given input the leg lengths H_i, i=1,3, are also known.
The closure equations of the positioning substructure are:

$$\begin{cases} (Q-P_1)^2 = H_1^2 & (1.1) \\ (Q-P_2)^2 = H_2^2 & (1.2) \\ (Q-P_3)^2 = H_3^2 & (1.3) \end{cases}$$

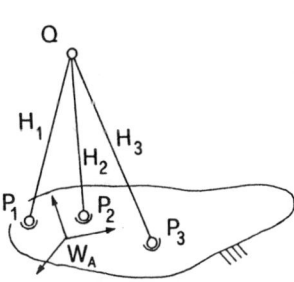
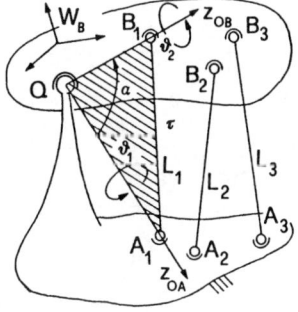

Fig. 3 Positioning substructure   Fig. 4 Orientation structure

Provided the three points P_i, $i=1,3$, are not aligned, the following expression is introduced:

$$(Q-P_1) = \delta \cdot (P_2-P_1) + \mu \cdot (P_3-P_1) + \sigma \cdot (P_2-P_1) \times (P_3-P_1) \qquad (2)$$

where δ, μ, and σ are parameters to be determined. Equations (1), after some algebra, become:

$$\begin{cases} \delta^2(P_2-P_1)^2 + \mu^2(P_3-P_1)^2 + \sigma^2[(P_2-P_1) \times (P_3-P_1)]^2 + \\ \qquad\qquad 2 \cdot \delta \cdot \mu \cdot (P_2-P_1) \cdot (P_3-P_1) = H_1^2 \qquad (3.1) \\ \delta \cdot (P_2-P_1)^2 + \mu \cdot (P_2-P_1) \cdot (P_3-P_1) = [H_1^2 - H_2^2 + (P_2-P_1)^2]/2 \qquad (3.2) \\ \delta \cdot (P_2-P_1) \cdot (P_3-P_1) + \mu \cdot (P_3-P_1)^2 = [H_1^2 - H_3^2 + (P_3-P_1)^2]/2 \qquad (3.3) \end{cases}$$

Indeed the straight substitution of (2) in (1.1) leads to (3.1). Equation (1.2), if written as:

$$[(Q-P_1) - (P_2-P_1)]^2 = H_2^2 \qquad (4)$$

can be rearranged in the following form:

$$2 \cdot (P_2-P_1) \cdot (Q-P_1) - (Q-P_1)^2 - (P_2-P_1)^2 + H_2^2 = 0 \qquad (5)$$

If expression (2) is considered for vector $(Q-P_1)$ in the first term of equation (5) and, from (1.1), H_1^2 is substituted for $(Q-P_1)^2$ in the second term of (5), equation (3.2) can be obtained. In a similar way equation (3.3) can be derived from (1.3).

The last two equations of system (3) are linear in δ and μ; the determinant D of their coefficient matrix is given by the expression:

$$D = [(P_2-P_1) \times (P_3-P_1)]^2 \qquad (6)$$

which is not equal to zero because points P_1, P_2, and P_3 are supposed to be not aligned. Then the unknowns δ and μ can be linearly determined.

By substituting the values of δ and μ in the first equation of system (3), two roots can be found for σ. Hence, according to (2), two possible positions of the point Q are obtained. (If the roots were complex conjugate the manipulator would not have any real closure).

Determination of platform orientation. The geometry of the orientation structure (see Fig. 4) is given. In particular, the lengths L_j of the legs $A_j B_j$, $j=1,3$, are known, and the positions of points Q, A_1, A_2, and A_3 are given in W_A. Moreover, the positions of points Q, B_1, B_2, and B_3 are given in an arbitrary reference system W_B fixed to the platform.

Let W_{OA} be a second reference system fixed to the base with origin in Q and its z axis, namely z_{OA}, directed from Q to A_1; axis x_{OA} can have arbitrary direction on a plane orthogonal to z_{OA}. Let M_A be the 4x4 matrix for coordinate transformation from W_{OA} to W_A. For a given leg length L_1 the triangle $\tau \equiv A_1 B_1 Q$ remains defined. The position of point B_1 with respect to W_{OA} can then be parameterized by means of the rotation θ_1 of τ about z_{OA} axis. Angle θ_1 is measured counterclockwise from the half-plane (x_{OA}, z_{OA}), x_{OA} positive.

Yet, let W_{OB} be a second reference system fixed to the platform with origin in Q and axis z_{OB} directed from Q to B_1; axis x_{OB} can have arbitrary direction on a plane orthogonal to z_{OB}. 4x4 matrix M_B allows coordinate transformation from W_B to W_{OB}. If the position of point B_1 is considered as known in W_{OA}, then the position of the platform in W_{OA} is defined by the angle θ_2 which measures the counterclockwise rotation about axis z_{OB} to superimpose triangle τ on the half-plane (x_{OB}, z_{OB}), x_{OB} positive.

Then angles θ_1 and θ_2 uniquely parameterize the position of the platform with respect to the base. The values of θ_1 and θ_2 corresponding to the closures of the structure can be determined by imposing that distances A_2B_2 and A_3B_3 equal the leg lengths L_2 and L_3 respectively.

To this respect, let consider the 3x3 rotation matrix R for the coordinate transformation from system W_{OB} to system W_{OA}:

$$R = \begin{vmatrix} c_1c_2 - u \cdot s_1s_2 & -c_1s_2 - u \cdot s_1c_2 & v \cdot s_1 \\ s_1c_2 + u \cdot c_1s_2 & -s_1s_2 + u \cdot c_1c_2 & -v \cdot c_1 \\ v \cdot s_2 & v \cdot c_2 & u \end{vmatrix} \qquad (7)$$

written according to the Denavit-Hartenberg notation, where $c_j = \cos\theta_j$, $s_j = \sin\theta_j$, $u = \cos\alpha$, $v = \sin\alpha$, and α is the angle between the axes z_{OA} and z_{OB}. With reference to Fig. 4, it can be found:

$$\cos\alpha = \frac{(A_1-Q)^2 + (B_1-Q)^2 - L_1^2}{2 \cdot |A_1-Q| \cdot |B_1-Q|} \quad ; \qquad \sin\alpha = (1-\cos^2\alpha)^{\frac{1}{2}} \qquad (8)$$

where all right hand side quantities of (8) can be determined.

The closure equations for the orientation structure, written in reference system W_{OA}, are:

$$(R \cdot \underline{B}_{20} - \underline{A}_{20})^2 = L_2^2; \qquad (R \cdot \underline{B}_{30} - \underline{A}_{30})^2 = L_3^2 \qquad (9)$$

where $\underline{A}_{jo}$ is the position vector of point A_j in W_{OA} and $\underline{B}_{jo}$ is the position vector of point B_j in W_{OB}.

By developing equations (9) it can be obtained:

$$\underline{A}_{20}^T \cdot R \cdot \underline{B}_{20} = (\underline{A}_{20}^2 + \underline{B}_{20}^2 - L_2^2)/2; \qquad \underline{A}_{30}^T \cdot R \cdot \underline{B}_{30} = (\underline{A}_{30}^2 + \underline{B}_{30}^2 - L_3^2)/2 \qquad (10)$$

Each equation in system (10) is cosine-sine linear in θ_1 and θ_2; i.e., any term, if dependent on θ_j (j=1,2), contains either $\cos\theta_j$ or $\sin\theta_j$. By substituting in (10) the relations:

$$\cos\theta_j = (1-t_j^2)/(1+t_j^2); \qquad \sin\theta_j = 2 \cdot t_j/(1+t_j^2) \qquad (11)$$

where $t_j = \tan(\theta_j/2)$, and after rationalizing, it follows:

$$\sum_{h,k=0,2} a_{hk} \cdot t_1^h \cdot t_2^k = 0; \qquad \sum_{h,k=0,2} b_{hk} \cdot t_1^h \cdot t_2^k = 0 \qquad (12)$$

where a_{hk} and b_{hk} are constant coefficients that can be directly determined from the first and second equation of system (10), respectively.

In order to solve system (12) one of the two unknowns, for instance t_2, must be eliminated. The dependence of equations (12) on t_2 can be outlined by writing (12) as follows:

$$E \cdot t_2^2 + F \cdot t_2 + G = 0; \qquad M \cdot t_2^2 + N \cdot t_2 + S = 0 \qquad (13)$$

where E,F,G,M,N,S are second order polynomials in t_1; their analytic expressions can be found straightly by comparison with equations (12).

The elimination of the unknown t_2 from system (13) can be worked out by finding the eliminant V of equations (13). The equation V=0 is the condition under which the closure equations (12) have the same solution for t_2. The equation V=0 resulted [7]:

$$(E \cdot S - G \cdot M)^2 + (E \cdot N - F \cdot M) \cdot (G \cdot N - F \cdot S) = 0 \qquad (14)$$

Equation (14) represents an 8th order algebraic equation in t_1, and can be written in the form:

$$\sum_{j=0,8} w_j \cdot t_1^j = 0 \qquad (15)$$

where the coefficients w_j, j=0,8, depend only on the link geometry of the 6-4 structure.

Equation (15) has eight solutions in the complex field. Let t_{1k}, k=1,8, be the generic root of (15); then, the polynomials E, F, G, M, N, S assume a definite value, and system (13) can be linearly solved in the unknowns t_2^2 and t_2. In particular the solution t_{2k} for t_2, together with the value t_{1k} for t_1, is a solution of system (12). Moreover, by means of (11), matrix R is determined that, together with constant matrices M_A and M_B, defines the coordinates of points B_j, j=1,3, in system W_A. In conclusion, each of the eight roots of equation (15) leads, in the complex field, to a closure of the orientation structure.

DPA of 6-4 manipulator. Since for every position of point Q eight different platform orientations can be found, the DPA of the 6-4 manipulator admits sixteen solutions in the complex field.

3. NUMERICAL EXAMPLE

The DPA of a 6-4 manipulator with the structure represented in Fig. 2 is developed. The manipulator geometric data, in arbitrary length unit, are: P_1=(2.2, -1.5, 8.6); P_2=(-3.3, 0.7, 5.1); P_3=(5.21, 16.08, -21.42); A_1=(1.6, -10.0, 4.9); A_2=(0.7, 0.9, 1.6); A_3=(-4.5, -5.6, 2.8) in reference system W_A, and Q=(1.5, 2.4, -3.8); B_1=(6.1, -6.6, 5.3); B_2= (-0.9, 4.0, -7.1); B_3=(1.7, -7.0, 4.3); in reference system W_B. The leg lengths are: H_1=13.1; H_2=10.0; H_3=22.15; L_1=8.0; L_2=10.0; L_3=6.1.

Two position vectors (in W_A) of point Q are determined: Q_{A1}=(1.1551, 4.3435, -3.0778), Q_{A2}=(1.1953, 5.9060, -2.1588). Then for each of them eight closures of the orientation structure have been found.

Table 1. B_1, B_2 and B_3 x,y,z coordinates (with real and imaginary parts) in reference system W_A, for all solutions.

Solution No.	x	y	z
	\multicolumn{3}{c}{Solutions for $Q_A=Q_{A1}$}		

Solution No.	x	y	z
1	(1.7177E+00, 0.0000E+00) (1.7572E+00, 0.0000E+00) (1.2903E+00, 0.0000E+00)	(-2.8363E+00, 0.0000E+00) (5.3188E+00, 0.0000E+00) (-5.3264E+00, 0.0000E+00)	(8.4592E+00, 0.0000E+00) (-7.3082E+00, 0.0000E+00) (4.6993E+00, 0.0000E+00)
2	(6.5180E-01, 0.0000E+00) (2.3717E+00, 0.0000E+00) (1.2970E+00, 0.0000E+00)	(-2.8847E+00, 0.0000E+00) (5.4746E+00, 0.0000E+00) (-5.3412E+00, 0.0000E+00)	(8.4317E+00, 0.0000E+00) (-7.1338E+00, 0.0000E+00) (4.6808E+00, 0.0000E+00)
3	(1.4629E-01, 0.0000E+00) (-1.0225E-02, 0.0000E+00) (-3.7537E+00, 0.0000E+00)	(-2.9374E+00, 0.0000E+00) (6.0462E+00, 0.0000E+00) (-1.1524E+00, 0.0000E+00)	(8.3651E+00, 0.0000E+00) (-6.9447E+00, 0.0000E+00) (6.9076E+00, 0.0000E+00)
4	(-4.5897E+00, 0.0000E+00) (1.6364E+00, 0.0000E+00) (-5.3664E+00, 0.0000E+00)	(-4.9336E+00, 0.0000E+00) (8.1117E+00, 0.0000E+00) (-6.3341E-01, 0.0000E+00)	(5.0402E+00, 0.0000E+00) (-5.2640E+00, 0.0000E+00) (6.2340E+00, 0.0000E+00)
5	(-4.8377E+00, 0.0000E+00) (3.0839E+00, 0.0000E+00) (-5.7586E+00, 0.0000E+00)	(-7.1196E+00, 0.0000E+00) (7.1705E+00, 0.0000E+00) (-5.9620E+00, 0.0000E+00)	(1.1238E+00, 0.0000E+00) (-5.8160E+00, 0.0000E+00) (-3.1578E+00, 0.0000E+00)
6	(-3.9784E+00, 0.0000E+00) (1.9429E+00, 0.0000E+00) (-6.6892E+00, 0.0000E+00)	(-7.9769E+00, 0.0000E+00) (8.0509E+00, 0.0000E+00) (-5.2710E+00, 0.0000E+00)	(-4.6550E-01, 0.0000E+00) (-5.2790E+00, 0.0000E+00) (-2.8841E+00, 0.0000E+00)
7	(2.5984E+01, 2.4309E+01) (-1.8236E+01,-4.7354E+00) (-3.1061E-01, 4.3026E+01)	(-1.7504E+01, 1.2277E+01) (1.1474E+01,-1.5639E+01) (-3.0109E+01,-1.6244E+01)	(-1.9265E+01, 2.0717E+01) (-4.7222E+00,-1.1973E+01) (-3.8766E+01, 1.3915E+01)
8	(2.5984E+01,-2.4309E+01) (-1.8236E+01, 4.7354E+00) (-3.1061E-01,-4.3026E+01)	(-1.7504E+01,-1.2277E+01) (1.1474E+01, 1.5639E+01) (-3.0109E+01, 1.6244E+01)	(-1.9265E+01,-2.0717E+01) (-4.7222E+00, 1.1973E+01) (-3.8766E+01,-1.3915E+01)

Solution No.	x	y	z
	\multicolumn{3}{c}{Solutions for $Q_A=Q_{A2}$}		
9	(4.0373E+00, 0.0000E+00) (-7.1279E-01, 0.0000E+00) (-2.6134E-01, 0.0000E+00)	(-2.9420E+00, 0.0000E+00) (7.5229E+00, 0.0000E+00) (-3.0151E+00, 0.0000E+00)	(7.7712E+00, 0.0000E+00) (-5.7581E+00, 0.0000E+00) (6.3443E+00, 0.0000E+00)
10	(4.7194E+00, 0.0000E+00) (-5.8468E-01, 0.0000E+00) (7.3856E-01, 0.0000E+00)	(-3.0789E+00, 0.0000E+00) (7.4324E+00, 0.0000E+00) (-3.9057E+00, 0.0000E+00)	(7.4236E+00, 0.0000E+00) (-5.8618E+00, 0.0000E+00) (5.4262E+00, 0.0000E+00)
11	(-4.2440E+00, 0.0000E+00) (4.2382E+00, 0.0000E+00) (-1.0573E+00, 0.0000E+00)	(-6.1449E+00, 0.0000E+00) (8.8092E+00, 0.0000E+00) (-6.2979E+00, 0.0000E+00)	(1.0288E+00, 0.0000E+00) (-3.3925E+00, 0.0000E+00) (-2.1871E+00, 0.0000E+00)
12	(3.3675E+00, 0.0000E+00) (-8.4199E-01, 0.0000E+00) (-9.8432E-01, 0.0000E+00)	(-7.5155E+00, 0.0000E+00) (9.7263E+00, 0.0000E+00) (-6.3111E+00, 0.0000E+00)	(-2.4962E+00, 0.0000E+00) (-2.8405E+00, 0.0000E+00) (-2.1340E+00, 0.0000E+00)
13	(-4.4731E+00,-4.2549E-01) (4.2024E+00,-8.8608E-01) (1.2160E+00,-1.2971E+00)	(-4.6084E+00,-5.7212E-01) (9.4565E+00, 8.5949E-01) (-4.0023E+00, 1.3986E+00)	(4.5042E+00,-1.2648E+00) (-2.5352E+00, 1.0279E+00) (5.7504E+00, 1.7555E+00)
14	(-4.4731E+00, 4.2549E-01) (4.2024E+00, 8.8608E-01) (1.2160E+00, 1.2971E+00)	(-4.6084E+00, 5.7212E-01) (9.4565E+00,-8.5949E-01) (-4.0023E+00,-1.3986E+00)	(4.5042E+00, 1.2648E+00) (-2.5352E+00,-1.0279E+00) (5.7504E+00,-1.7555E+00)
15	(3.5204E+01, 5.0746E+01) (-2.7825E+01,-2.7934E+01) (-2.9381E+01, 4.3522E+01)	(-2.5197E+01, 1.4695E+01) (2.7251E+01,-1.5824E+01) (-1.1566E+01,-2.9405E+01)	(-4.4164E+01, 3.0205E+01) (1.6944E+01,-2.4755E+01) (-4.6941E+01,-1.8243E+01)
16	(3.5204E+01,-5.0746E+01) (-2.7825E+01, 2.7934E+01) (-2.9381E+01,-4.3522E+01)	(-2.5197E+01,-1.4695E+01) (2.7251E+01, 1.5824E+01) (-1.1566E+01, 2.9405E+01)	(-4.4164E+01,-3.0205E+01) (1.6944E+01, 2.4755E+01) (-4.6941E+01, 1.8243E+01)

Sixteen closures in the complex field are thus determined. Table 1 shows the numerical results in terms of the coordinates of points B_j (j=1,3) in the base reference system W_A. It can be easily verified that, for each closure, the distance between all pairs of points connected by a leg equals the corresponding leg length.

4. CONCLUSIONS

The paper presents the direct position analysis in analytical form of the general six d.o.f. fully in-parallel actuated manipulator with platform position and orientation uncoupled. The analysis can be performed in two steps: first, a 2nd order algebraic equation in one unknown is solved for the platform reference point position; then, for each of the two positions, an 8th order algebraic equation in only one unknown is solved for platform orientation. Thus, 16 solutions are possible in the complex field, i.e., 16 real closures of the 6-4 manipulator can be found at most. An example confirms the new theoretical results.

Acknowledgements

The financial support of the Italian Ministry of Education (MPI-40% and MPI-60%) is gratefully acknowledged.

5. REFERENCES

1 HUNT K.J., "Structural Kinematics of in-Parallel-Actuated Robot-Arms", Trans. ASME, J. Mech. Trans. Autom. Design., Vol. 105, 1983, pp. 705-712.
2 WILLIAMS II R. L., REINHOLTZ C. F., "Forward Dynamic Analysis and Power Requirement Comparison of Parallel Robotic Mechanisms", The 1988 ASME Design Technology Conferences-20th Biennial Mechanisms Conference, September 25-28, Kissimmee, Florida, pp. 71-78.
3 McCLOY D., "Some Comparisons of Serial-Driven and Parallel-Driven Manipulators", Robotica, Vol. 8, 1990, pp.355-362.
4 THORNTON G. S., "The GEC Tetrabot - A New Serial-Parallel Assembly Robot", Proceedings 1988 IEEE Int. Conference on Robotics and Automation, April 24-29, 1988, Philadelphia, Pennsylvania, pp.437-439.
5 ZHANG C., SONG S., "Kinematics of Parallel Manipulators with a Positional Subchain", 1990 21th ASME Mechanisms Conference, 16-19 Sept., 1990 Chicago, IL, DE-Vol. 25, pp. 271-278.
6 NANUA P., WALDRON K. J., "Direct Kinematic Solution of a Special Parallel Robot Structure", Proc. 8th Int. Symposium on Theory and Practice of Robots and Manipulators, Ro.man.sy '90, July 2-6, 1990, Cracow, Poland, pp. 131-139.
7 INNOCENTI C., PARENTI CASTELLI V., "Closed Form Direct Position Analysis for the Most General Fully-Parallel Spherical Wrist", Report N. 67 DIEM, Bologna, August 1990.

A GENERAL METHOD FOR A CLOSED-FORM SOLUTION TO ROBOT KINEMATICS PROBLEMS

Carlo Galletti and Pietro Fanghella
Istituto di Meccanica Applicata alle Macchine
via Opera Pia 15a
16145 Genova - ITALY

ABSTRACT
This work deals with the applications of the mechanism-composition principle and of displacement-group theory to the kinematics of multiloop robot mechanisms. It is shown that they can give a unified approach to two typical problems of robot kinematics, i.e. mobility analysis and direct and inverse position analyses.
Starting from the description of a multiloop kinematic chain in terms of its kinematic pairs, link shapes, link-pair connections, and driving constraints, the robot mechanism is decomposed into single-loop determined chains. For each chain, all constraints between any two links (i.e., group, connectivity and invariant geometric properties) are found. Then, the information about the constraint is used to obtain a set of independent closure equations for the chain considered.

1. INTRODUCTION
Multiloop kinematic chains are employed in almost all real robot applications in order to add mechanism actuators to the functional robot structure. In fact, either the basic functional structure of a robot arm is embedded in the actuating one (with only a minor exception for directly driven mechanisms) in order to realize a complete kinematic structure that allows the motion control, or the end-effector is actuated in parallel (platforms).
On the basis of structural criteria, several authors [1,2] identified some properties of multiloop spatial mechanisms, and studied how kinematic structures can be modified by transfer of drives. Their works dealt specifically with topological issues.
The general problem of producing a mathematical model for determining the positions of robot joints and bodies, when direct or inverse kinematics is required, involves writing and solving a simultaneous set of nonlinear congruence equations
$$\mathbf{f}(\mathbf{x},\mathbf{q}) = 0 \qquad (1)$$
where $\mathbf{x}$ is a suitable vector of robot-body positions, and $\mathbf{q}$ is a vector defining the values of the driving constraints. The number of components of each vector depends on the mobilities and structures of the complete robot and of its subassemblies.

This problem is a rather complex one. Different subassemblies of the robot mechanism satisfy different general constraints, and may appear overconstrained if this fact is not taken into account. As a consequence, the number of components of the vectors **f** and **x**, the structure of the system of equations, and the techniques for its solution are different for each subassembly.

2. APPROACHES TO MULTILOOP PROBLEMS

It is easy to show that general approaches are not well suited to analyzing multiloop robot mechanisms. It is well known that the actual architectures of many industrial robots are not of the general type but are based on some specific choices, and that their loops use particular link assemblies that can simplify synthesis and analysis problems. As a consequence, a large number of practical cases can be solved through the systematic serial/parallel application of simple algorithms.

The convenience of fully exploiting this possibility is stressed in the current literature: both "particular" [3,4] and "theoretical" [5] approaches have been suggested; the oldest ones are based on intuitive reasoning, but stronger systematic methods for solving the problem have been proposed in the past few years.

Using these methods, it is possible to realize a "robot compiler" that processes the data describing a robot mechanism, and produces automatically an explicit robot model and suitable algorithms to perform the kinematic analyses of countless robot systems of practical interest.

To design such a compiler a composition approach to mechanism kinematics has been adopted.

Extending the classical principle of mechanism composition [6,7], we state that multiloop robot mechanisms of any complexity can be obtained by sequential additions of certain link arrangements to pairs of links whose relative displacements are known data (driving constraints). The kinematic chain defined by such arrangements and by driving constraints can be (and often is) a single-loop determined chain (SLDC). According to the mechanism composition principle, the position of any SLDC is independent of the positions of all successively added SLDCs.

Then, the solution process of a given robot structure consists in recognizing, organizing, and solving all the SLDCs of the robot considered. The modelling and solution of each SLDC can be accomplished separately through the analysis of its mobility properties and kinematic constraints [5]. Once all the SLDCs of a given robot have been recognized and their congruence equations have been obtained and solved, the solution to the complete set (1) is reached.

In order to describe the kinematic properties of SLDCs and to obtain explicit expressions for their closure equations, we have adopted the method of displacement groups [8,9].

Ten displacement groups exist, which are subgroups of the Euclidean space D; six of them (revolute R, prismatic P, cylindrical C, planar F, spherical S, and helical H) correspond to lower kinematic pairs; the other four (planar translation P2, spatial translation P3, spatial translation and rotation in a given direction RP3, and helical displacement in a given direction with a given pitch HP2) cannot be achieved by a single link pair. Each displacement group is defined by invariant geo-

metric properties.
The relative motion of two links in a displacement group GR can be obtained by connecting them with an appropriate sequence of lower pairs. Fig. 1 shows a kinematic chain connecting two links, 1 and 4, whose relative displacements are contained in the group RP3; the unit vector **v**, representing the direction of rotation, is the invariant geometric property of the group. If the connectivity of the terminal links is equal to the group dimension and no passive mobility arises, the sequence of links and pairs realizes a single-loop determined chain in the group GR. For instance, the well known 5 kinds of planar dyads are the complete set of SLDCs in the plane; a chain made up of three intersecting revolutes is the only spherical SLDC; the chain R-S-S2 is a spatial dyad.
In the following, any constraint on the motion between two links of a mechanism will be regarded as a kinematic constraint (KC). A kinematic constraint coinciding with a displacement group or with a proper subset of a group, inherits the invariant geometric properties of that group.

3. MAIN TASKS OF THE ROBOT COMPILER
According to the approach outlined in the previous section, the problem of the kinematic analysis of a multiloop robot mechanism can be solved by accomplishing two main tasks.
1) The first task (decomposition into SLDCs) consists in deriving the hierarchy of SLDCs from the robot-mechanism description.
After this task, the robot is described by:
- the driving constraints;
- a set of SLDCs, each represented by a sequence of pairs and links;
- hierarchical connections among the SLDCs.
2) The second task lies in the generation of a set of closure equations (derived from the information about the constraints) for each SLDC.
In its implicit form, each congruence equation is completely defined by:
- the definition of a single-loop chain as a sequence of constraints;
- a metric relation between the properties of two constraints selected from the sequence. The set of implicit closure equations has to be combined with the rigid transformations defining the relative positions of the selected constraints. These transformations depend on the link dimensions and on the variables of the remaining constraints in the SLDC. As a result, we obtain the explicit form of each congruence equation, given by its type (e.g., linear, quadratic, etc.) and symbolic coefficients.
Explicit congruence equations and specific solution algorithms can be expressed in a programming language, like Pascal or FORTRAN. As the hierarchical connections among SLDCs determine the input/output information available to and from any SLDC, the same programming language can be used to code such connections.

4. MECHANISM DECOMPOSITION
The starting data for the decomposition task are the geometric link shapes (i.e., the relative positions of the pairs on each link), the types of pairs, and the connections between links and pairs. Moreover, a list of generalized (independent) coordinates for the robot must be chosen as input data in order to define the driving constraints. Such coor-

dinates can be the pairs actuated by motors, the position coordinates of the end-effector, or any feasible combination of them. In this way, no distinction is to be made between direct and inverse kinematics.

All these data are used to find at least one single-loop closed chain, with zero mobility, and with one or more driving constraint(s), in a displacement group. If this chain exists, it is the first SLDC.

As, by definition, the positions of all the bodies of an SLDC are determined by the known driving constraints, all the couples of links in the SLDC can be regarded as driving constraints for any other SLDC to be found. Therefore, a new list of driving constraints can be created by adding the couples of links of the found SLDC to the previously determined ones. Starting from this augmented list, it is possible to search for another SLDC, whose couples of links will in turn become new driving constraints. The search and updating processes are repeated until the starting robot mechanism is completely decomposed into SLDCs. These SLDCs are naturally organized into a hierarchy, according to the order in which they have been found.

Fig. 2-a shows the search process: an algorithm for searching for minimum-mobility closed paths containing at least one driving constraint is used to find a trial chain. A mobility analysis of the trial chain is performed in order to find the KCs between all its links and the number of its degrees of freedom. To this end, a particular algorithm uses a set of predefined rules for composition and intersection of kinematic constraints (Fig. 2-b). The rules allow one to perform binary operations between kinematic constraints by using the invariant geometric properties of the corresponding displacement groups. The result of each operation is computed in accordance with the geometric relations between the known geometric properties of the starting groups.

By exhaustive serial use of the composition and intersection operations, one can evaluate all true KCs between links, link connectivities, and the number of degrees of freedom of the trial chain. When the number of degrees of freedom is equal to 0 the chain is an SLDC.

5. GENERATION OF CLOSURE EQUATIONS

The geometric relations between the invariant geometric properties (IPs) of any couple of displacement groups, GR1 and GR2, are used to generate closure equations for a given SLDC. These relations express connections between the IPs of the two groups; such connections hold in the group obtained by composition of GR1 and GR2.

As an example, Table 1 presents the geometric relations between two revolutes whose IPs are represented by points (A) and unit vectors ($\mathbf{v}$) (see Fig. 3). Column 1 of the Table gives all the groups obtained by composing two revolute groups. The geometric relations are divided into three categories, listed in the columns of the Table. Column 2 gives the composition relations that specify the composed groups. Column 3 gives the metric relations (valid for each composed group), which depend on some pair variables of the SLDC; therefore, they can be used to obtain the closure equations for the SLDC, in the group considered. Column 4 lists the relations that are independent of all pair variables of the SLDC: they are used to obtain offset relations that must be satisfied by the links of the SLDC in order to allow a correct closure at every position.

Similar Tables are available for all other couples of groups [10].
For a given SLDC, two constraints, KC1 and KC2, and the respective
groups, GR1 and GR2, are considered. The metric relations MRs between
the groups GR1 and GR2 are expressed through the patterns p1 and p2 connecting the two KCs. These relations are reference-independent scalar
functions of the link dimensions, the driving constraints, and the unknown pair variables, belonging to both patterns, and can be equated,
thus obtaining the following independent closure equations

$$\mathbf{MR}(p1) = \mathbf{MR}(p2) \qquad (2)$$

which are independent off the pair variables of KC1 and KC2. Many equivalent arrangements for the same single-loop chain are feasible; each of
them corresponds to a distinct sequence of links and kinematic constraints, and can result in a set of congruence equations.
It is possible to define a solvability condition for eq. (2): all the KC
combinations must be found for which the number of independent closures
is equal to the number of unknown variables [5]. In many cases of practical interest for robotics, one equation in one unknown can be obtained.
A suitable procedure for generating the closure equations for an SLDC is
the following (see Fig.4):
- by using the composition algorithms described in the previous section,
all KC combinations of the SLDC are found;
- the group of the SLDC is identified;
- all entries in the table of metric relations corresponding to the
group of the SLDC and to any couple of KCs, are analyzed: the entries
fulfilling the solvability condition are marked. The set of corresponding metric relations with a minimum number of unknowns is chosen;
- the offset conditions are also defined to allow for a further check;
- the relative positions of the invariant properties of the KCs originating the metric relations are expressed through sequences of rigid
transformations along the two patterns, p1 and p2, connecting the KCs.
Combining the transformation sequences with the considered metric relations yields an independent subset of congruence equations for the SLDC:
the unknown variables of such equations are the pair variables outside
the selected constraints;
- the pairs whose variables can be computed by using the obtained closure equations are flagged as "solved", and the procedure is repeated
until all pairs of the SLDC are "solved".

6. EXAMPLE

The first closure equation is derived for the inverse kinematic problem
of a typical robot arm, shown in Fig. 5-a.
Two alternative constraint sequences, obtained by composing the original
robot pairs, are presented in Figs 5-b and 5-c. The first sequence is:
revolute R1, plane π, and sphere K1; the second is: sphere K2, revolute
R3, and sphere K1. In the first case, the constraints K1 and π are considered. A systematic Table giving metric relations [5] shows that one
metric relation (i.e., the distance between point K1 and plane π) holds
between the groups S and F; this metric relation is used to write the
following closure equation:

$$\text{distance}(\pi, K1)_{p1} = \text{distance}(\pi, K1)_{p2} \qquad (3)$$

where p1 and p2 are the patterns describing the subchains connecting the constraints π and K1 (Fig. 5-b). The left-hand side of equation (3) depends on the known transformation between the frame and the end-effector and on the unknown variable of the pair R1; the right-hand side is computed from the link dimensions. Equation (3) can be solved in closed form. In a similar way, the constraint sequence in Fig. 5-c can be used to find an equation depending on the unknown variable of the pair R3.

REFERENCES
[1] Earl C.F., Rooney J. *Trans. ASME, J. of Mechanism, Transmission, and Automation in Design*, 105, 1983, pp 15-22
[2] Hunt K.H. *Trans. ASME, J. of Mechanism, Transmission, and Automation in Design*, 105, 1983, pp 705-712
[3] Hunt K.H. *Mech.& Mach. Theory*, 21, 1986, pp 481-487
[4] Hiller M., Woerlne C. *Proc. 7th Conf. Th. Mach. Mech.*, Sevilla, 1987, pp 215-221
[5] Fanghella P., Galletti C. *Mech.& Mach. Theory*, 24, 1989, pp 383-394
[6] Galletti C. *Mech. Mach. Theory*, 21, 1986, pp 385-391
[7] Fanghella P., Galletti, C. *The Int. J. of Robotics Research*, 6, 1990, pp 19-24
[8] Herve' J. *Mech.& Mach. Theory*, 13, 1978, pp 437-450
[9] Fanghella P. *Mech.& Mach. Theory*, 23, 1988, pp 171-183
[10] Querci C. *Master Thesis*, IMAGE, 1981 (in Italian)

composed group	composition relations	metric relations	offset relations
D	–	$\mathbf{v}_a \cdot \mathbf{v}_b$	–
	(skew axes)	$(A_a-A_b)\cdot \mathbf{v}_a \times \mathbf{v}_b$	
		$(A_a-A_b)\cdot \mathbf{v}_a$	
		$(A_a-A_b)\cdot \mathbf{v}_b$	
RP3	$\mathbf{v}_a \times \mathbf{v}_b = 0$	$(A_a-A_b)\cdot \mathbf{v}_a$	–
	(parallel axes)	$[(A_a-A_b)\times \mathbf{v}_a]^2$	
F	$\mathbf{v}_a \times \mathbf{v}_b = 0$	$[(A_a-A_b)\times \mathbf{v}_a]^2$	$(A_a-A_b)\cdot \mathbf{v}_a$
S	$(A_a-A_b)\cdot \mathbf{v}_a \times \mathbf{v}_b = 0$	$\mathbf{v}_a \cdot \mathbf{v}_b$	$[(A_a-A_{ab})\cdot \mathbf{v}_a]$
	(axes intersecting at point A_{ab})		
C	$\mathbf{v}_a \times \mathbf{v}_b = 0$	$(A_a-A_b)\cdot \mathbf{v}_a$	–
	$(A_a-A_b)\times \mathbf{v}_a = 0$		
	(coincident axes)		
R	$\mathbf{v}_a \times \mathbf{v}_b = 0$	–	$(A_a-A_b)\cdot \mathbf{v}_a$
	$(A_a-A_b)\times \mathbf{v}_a = 0$		

Table 1 Geometric relations between the IPs of two revolute groups

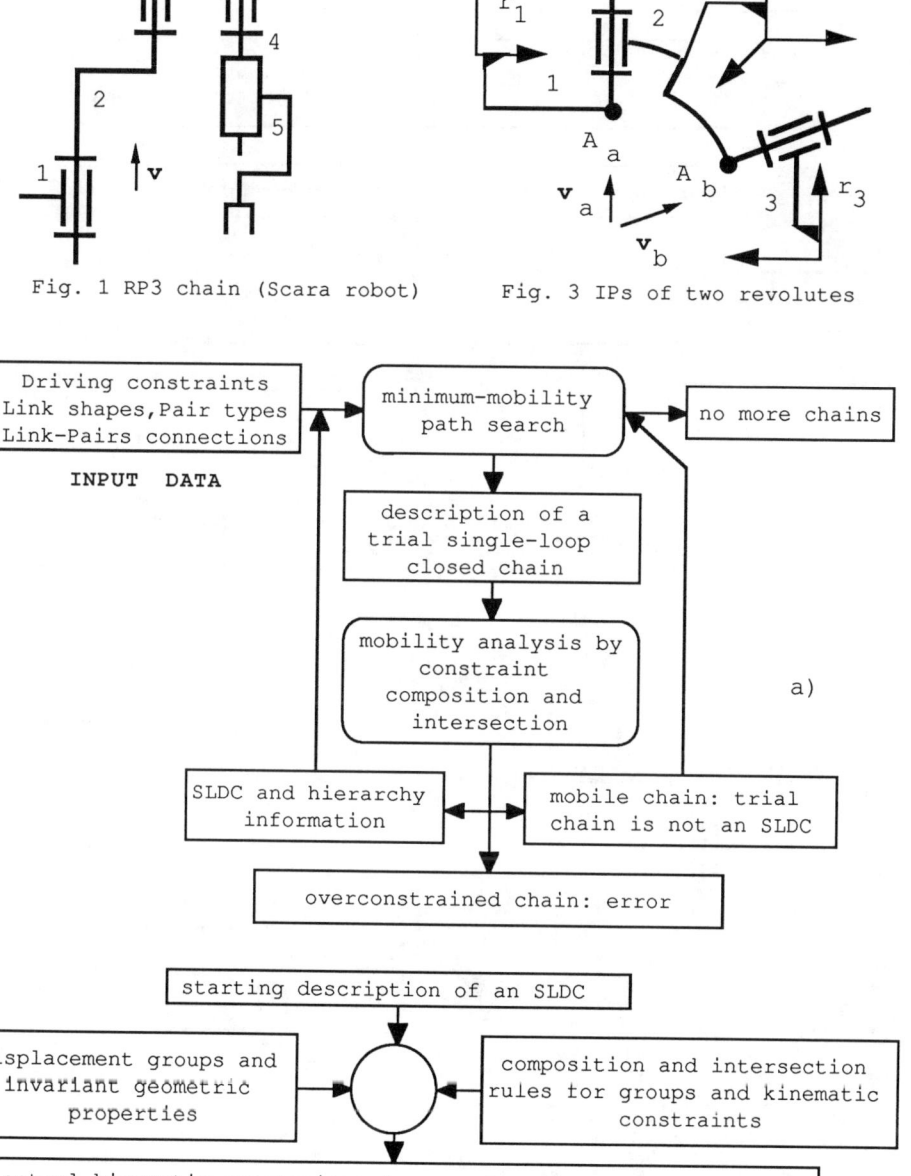

Fig. 1 RP3 chain (Scara robot)

Fig. 3 IPs of two revolutes

Fig. 2 Search for SLDCs

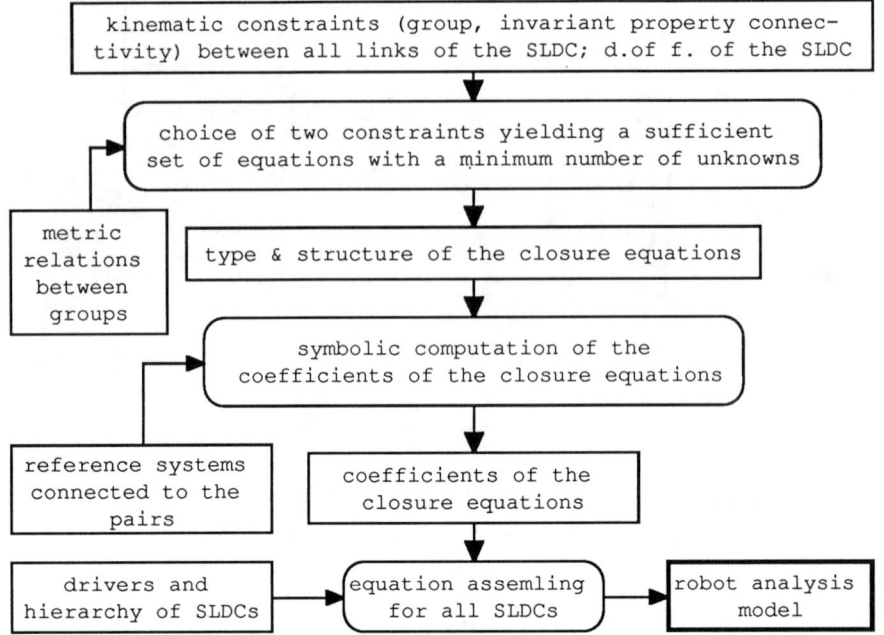

Fig. 4 Generation of a robot kinematic model

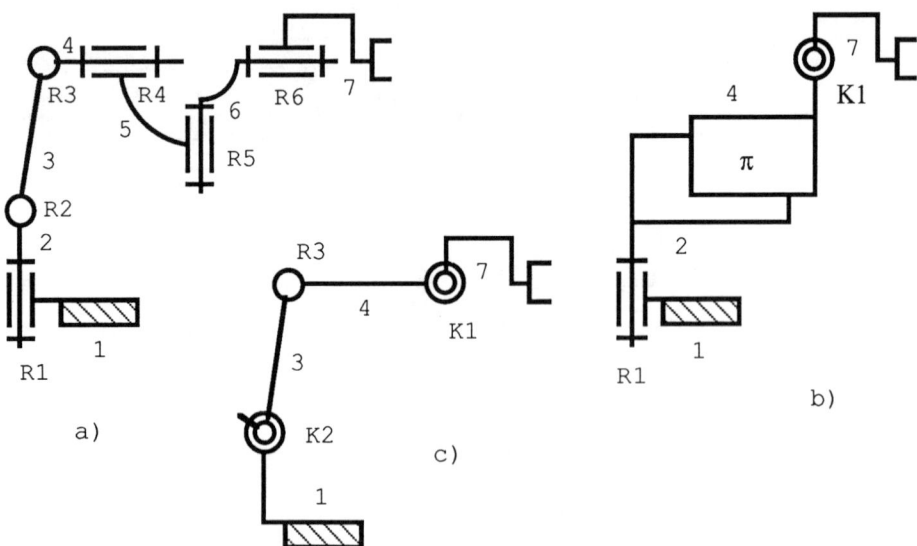

Fig. 5 Equivalent constraints in a robot arm

FAST IMPLEMENTATION OF ROBOTIC MANIPULATOR KINEMATICS USING CORDIC AND SYSTOLIC PROCESSORS

Vassilios (Basil) G. Mertzios and Stylianos S. Scarlatos

Department of Electrical Engineering, Democritus University of Thrace, 67 100 Xanthi, Greece

Abstract

This paper introduces the use of CORDIC processors for the fast implementation of the kinematics of robotic manipulators. Each rotation and translation of the links is implemented independently. Rotation processing elements incorporate one CORDIC rotor. The overall systolic architecture is derived by cascading the rotation and translation processing elements and pipelining the array. The use of CORDIC processors results in simple, hardware-efficient, fast, and VLSI suitable implementations of the kinematic equations appropriate for run-time operation.

I. INTRODUCTION

The kinematics of robotic manipulators involve intense trigonometric computations [1], [2], [9]. This kind of computations, when carried out via software or conventional hardware, consumes considerable time and imposes a burden for the use of sophisticated design algorithms, such as adaptive and stochastic schemes.

The CORDIC (Coordinate Rotation Digital Computer) algorithm is an efficient method for computing two-dimensional vector generalized rotations by simple shift and add operations [3]. The applications of CORDIC rotors include orthogonal filters [5], [7] and methods with complex arithmetic, e.g., complex filtering [4]. The CORDIC algorithm in comparison with other relevant methods, is fast, hardware-efficient, and suitable for VLSI implementation.

For the ever-increasing demands for high-speed and massive computations, VLSI technology is a cost-effective and efficient solution. The potential of VLSI technology is fully exploited by concurrent array processors (APs) for the implementation of modular, regular, and locally communicative structures [6]. Two dominant types of the special-purpose VLSI APs are the systolic APs, which are synchronous, and their data flow counterparts, the wavefront APs.

In order to achieve a systolic implementation of robot kinematics involving CORDIC-type processing elements (PEs), the *transformation matrix* between the first and the last link is adopted in the chain-product form composed of matrices that correspond to elementary transformations, i.e., rotations and translations. In this particular case, the CORDIC algorithm for circular rotations can readily implement the rotations of the links. The link translations, as it will be shown in the sequel, can be implemented using add-type PEs.

II. MANIPULATOR KINEMATICS

The forward kinematics problem consists in the determination of the position and orientation of the end-effector or the tool with respect to a reference coordinate system (e.g., the base coordinate system), given the link parameters.

For the description of the translational and rotational relationships between neighbor links, the Denavit-Hartenberg (D-H) representation is adopted [1],[2]. For the *frame* (i.e., the link coordinate system) assignment and the definition of the link parameters (*link length* a_i, *link offset* d_i, *link twist* θ_i, and *joint angle* α_i), see [2]. The general *homogeneous transformation matrix* for neighbor frames, i and i-1, is adopted in its composite form:

$$^{i-1}T_i = T_{x,\alpha} T_{x,a} T_{z,\theta} T_{z,d}$$

$$= \begin{bmatrix} 1 & 0 & 0 & 0 \\ 0 & C\alpha_i & -S\alpha_i & 0 \\ 0 & S\alpha_i & C\alpha_i & 0 \\ 0 & 0 & 0 & 1 \end{bmatrix} \begin{bmatrix} 1 & 0 & 0 & a_i \\ 0 & 1 & 0 & 0 \\ 0 & 0 & 1 & 0 \\ 0 & 0 & 0 & 1 \end{bmatrix} \begin{bmatrix} C\theta_i & -S\theta_i & 0 & 0 \\ S\theta_i & C\theta_i & 0 & 0 \\ 0 & 0 & 1 & 0 \\ 0 & 0 & 0 & 1 \end{bmatrix} \begin{bmatrix} 1 & 0 & 0 & 0 \\ 0 & 1 & 0 & 0 \\ 0 & 0 & 1 & d_i \\ 0 & 0 & 0 & 1 \end{bmatrix} \quad (1)$$

By notation, $Cx = \cos x$ and $Sx = \sin x$. The first and third matrices correspond to rotations by the *link twist* α_i and the *joint angle* θ_i [2]. The second and forth matrices correspond to translations by the *link length* a_i and the *link offset* d_i.

There are two types of joints, *revolute* and *prismatic*. Revolute joints realize rotations, while prismatic joints realize translations. In practice, the transformation matrix (1) is composed of a product of at most three matrices; a variable term corresponding to the type of joint, and constant terms corresponding to the different configuration in space of the two adjacent frames.

The homogeneous transformation *matrix* ^0T_i, which specifies the position and orientation of frame i with respect to the base frame 0, is the chain product of successive transformation matrices for neighbor frames:

$$^0T_i = \prod_{j=1}^{i} {}^{j-1}T_j \qquad (2)$$

III. THE CORDIC ALGORITHM

The CORDIC algorithm is an efficient, bit-recursive method for computing two-dimensional vector generalized rotations by simple shift and add operations [3].

For coordinate systems, parametrized by a quantity m, in which the norm R, and angle φ, of a vector $\mathbf{x} = (x_0, y_0)$ are defined by

$$R = \sqrt{x_0^2 + m y_0^2}$$

$$\varphi = m \tan^{-1}\left[\frac{y_0 \sqrt{m}}{x_0}\right],$$

the iterative CORDIC algorithm is given by

$$\mathbf{x}_{i+1} = \begin{bmatrix} 1 & -\mu_i \delta_i \\ \mu_i \delta_i & 1 \end{bmatrix} \mathbf{x}_i \qquad (3)$$

$$z_{i+1} = z_i - a_i$$

where

$$i = \frac{1}{m} \tan^{-1}\left[\delta_i \sqrt{m}\right],$$

$\{\delta_i\}$ is a set of arbitrary constants, and z_0 is the rotation angle of $\mathbf{x}_0$. These equations specify the rotation of $\mathbf{x}_i$ through an angle $\mu_i a_i$ to $\mathbf{x}_{i+1}$, where $\mu_i = \pm 1$ is the direction of rotation.

For circular rotations (m=1) the CORDIC algorithm implements Givens rotations normalized by a constant. A simplified version of the CORDIC rotor exhibiting output scaling is shown in Fig. 1.

For selection of sequences $\{\delta_i\}$ and $\{a_i\}$, output scaling, and convergence of the method see [4].

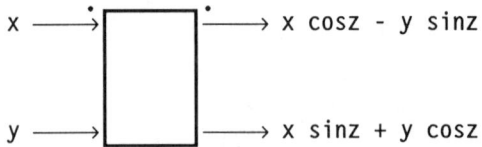

Figure 1. A CORDIC rotor realizing Givens rotations of a vector **x**=(x,y) about angle z.

IV. IMPLEMENTATION

A. Adjacent Links

A link rotation is now easily implemented using a CORDIC processor. Specifically, a rotation of a three-dimensional rotation or position vector **v** about a rotation axis, corresponding to a basis vector, is implemented by:

(i). broadcasting the component of the vector corresponding to the rotation axis, and

(ii). furnishing into the CORDIC module the other two components.

For example, the rotation about axis **z** is shown in Fig. 2 using a Rotation PE (R-PE). The z-component v_z, of the three-dimensional input vector **v**, is not affected.

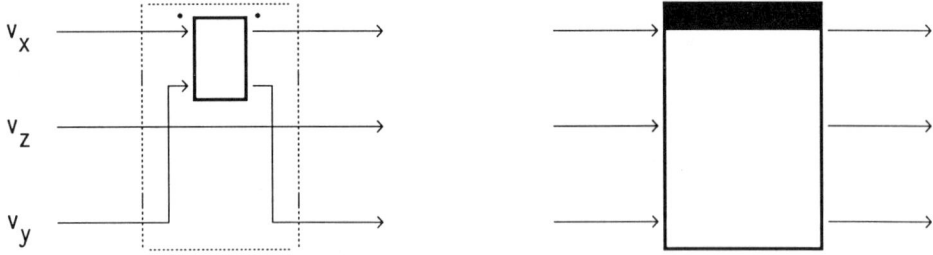

Figure 2. (a) Rotation PE (R-PE) about axis **z**. (b) Symbol of R-PE.

Due to internal structure of the translation matrix, a link translation requires just an addition. Specifically, a translation of a three-dimensional position vector **p** along a translation axis, corresponding to a basis vector, is implemented by:

(i). furnishing into an adder the component of the position vector corresponding to the translation axis, and

(ii). broadcasting the other two components.

The Translation PE (T-PE) of Fig. 3 can effectively represent translations. The critical path corresponds to the vector component which is translated, i.e., p_z.

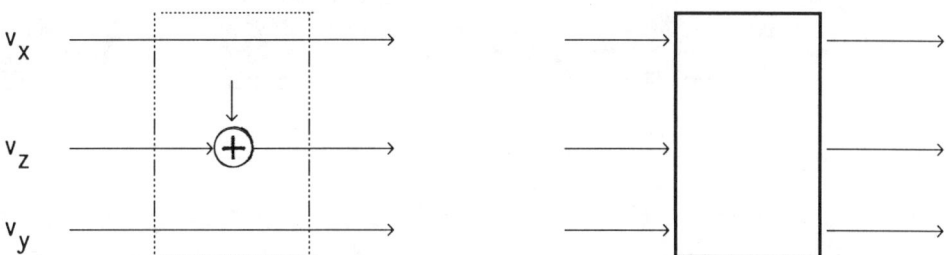

Figure 3. (a) Translation PE (T-PE) along axis z. (b) Symbol of T-PE.

The transformation matrix for neighbor frames of (1) is implemented in Fig. 4. The first PE designated by a lower dashed line corresponds to the a mirror R-PE. Note that in practical robots the hardware is considerably decreased, since at most three T-PEs and/or R-PEs are required, and the fixed value of a joint parameter yields a simplified transformation matrix (e.g., $a_i = 0°$ for frame i).

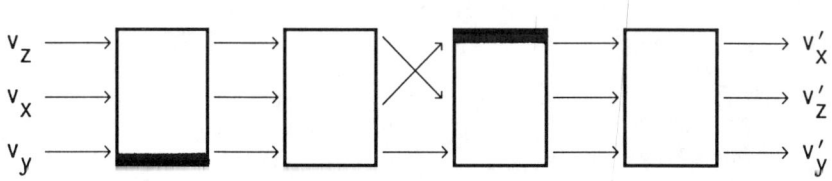

Figure 4. Implementation of transformation between adjacent frames.

The second and the third PE are lattice interconnected, in order to commute the transformation axis **x** by **z**.

Note that the translation quantity is fed into the adder of a T-PE every four systolic clock cycles, in order to interact with a position vector; otherwise, it is zero. In contrast, R-PEs operate in each systolic clock cycle, since rotations affect also translation vectors (observe the form of the multiplication of two homogeneous trasformation matrices).

B. Overall Manipulator

The kinematics of the overall robotic manipulator can be implemenented by further cascading the parts corresponding to the transformations of all adjacent frames.

The various CORDIC-type and add-type stages, implemented by R-PEs and T-PEs respectively, communicate via lattice interconnections, in order to match elementary transformations about different axes.

In the sequel, implementation of the kinematics of two of the most popular manipulators, the PUMA 560 and the Stanford robot, will be discussed. The Puma 560 exhibits a revolute configuration, while the Stanford robot exhibits a spherical configuration.

B1. PUMA 560

The D-H parameter table for the PUMA 560 is shown in Table 1. According to the framework presented, we start from the first link. Since $a_1=0°$, no rotation is performed on the input vector, thus the CORDIC-type stage of a_1 is ommited. Furthermore, rotations about -90° or 90° is possible to be implemented by means of just an inverter to the appropriate terminal, which is not done, in order to maintain a modular and regular structure.

The generation of a systolic linear array for the kinematics of the whole manipulator is straightforward. Since only feedforward paths exist, the array is pipelined by assigning to all interconnections a delay equal to the latency of the slowest stage, i.e., the CORDIC-type stage [6], [8]. The systolic array for computing the kinematics of PUMA 560 is shown in Fig. 5.

B2. Stanford robot

The D-H parameter table for the Stanford robot is shown in Table 2. By following the above described framework, the systolic array shown in Fig. 6 is generated.

V. CONCLUSIONS

The use of CORDIC processors for the implementation of robotic manipulator kinematics results in simple, hardware-efficient, and fast architectures of concurrent array type, which are appropriate for run-time operation. Within a CORDIC processor, its adders can be pipelined. Moreover, the whole scheme can be bit-level pipelined, thus, resulting

in very high throughput rates.
The idea can be extended towards the implementation of inverse kinematics, dynamics, and control robotic problems.

Link i D-H parameters	1	2	3	4	5	6
α_i	0°	-90°	0°	-90°	-90°	+90°
a_i	0	0	a_3	0	0	0
θ_i	θ_1	θ_2	θ_3	θ_4	θ_5	θ_6
d_i	d_1	0	d_3	d_4	0	d_6

Table 1. D-H parameter table for the PUMA 560.

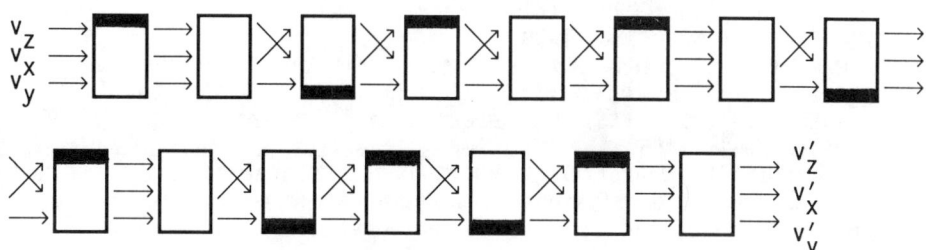

Figure 5. Systolic array for the kinematics of PUMA 560. The wright side of the upper array communicates with the left side of the lower array.

Link i D-H parameters	1	2	3	4	5	6
α_i	0°	-90°	+90°	0°	-90°	+90°
a_i	0	0	0	0	0	0
θ_i	θ_1	θ_2	0°	θ_4	θ_5	θ_6
d_i	d_1	d_2	d_3	0	0	d_6

Table 2. D-H parameter table for the Stanford robot.

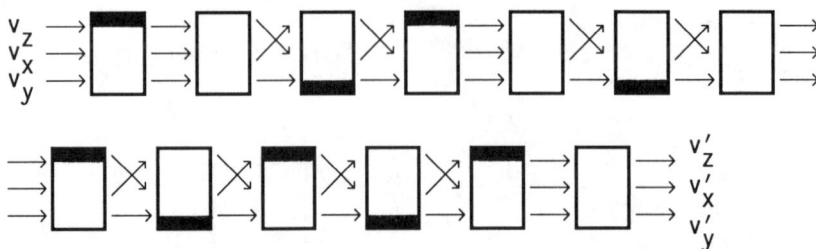

Figure 6. Systolic array for the kinematics of Stanford robot. The wright side of the upper array communicates with the left side of the lower array.

REFERENCES

[1] K.S. Fu, R.C. Gonzalez, and C.S.G. Lee, *Robotics: Control, Sensing, Vision, and Intelligence*, McGraw-Hill, 1987.
[2] J.J. Craig, *Introduction to Robotics: Mechanics and Control*, Addison-Wesley, 1986.
[3] J.E. Volder, "The CORDIC trigonometric computing technique," *Trans. IRE*, EC-8, pp. 330-334, 1959.
[4] A.M. Ahmed, "Alternative arithmetic unit architectures for VLSI digital signal processors," in S.Y. Kung, H.J. Whitehouse, and T. Kailath (editors), *VLSI and Modern Signal Processing*, pp. 277-303, Prentice Hall, 1985.
[5] P. Dewilde, E. Deprettere, and R. Nouta, "Parallel and pipelined VLSI implementation of signal processing algorithms," in S.Y. Kung, H.J. Whitehouse, and T. Kailath (editors), *VLSI and Modern Signal Processing*, pp. 257-276, Prentice Hall, 1985.
[6] S.Y. Kung, *VLSI Array Processors*, Englewood Cliffs, NJ: Prentice Hall, 1987.
[7] A. Fettweis, "Design of Orthogonal and Related Digital Filters by Network-Theory Approach," *Archiv fur Elektronik und Ubertragungstechnik*, vol. 44, pp. 65-74, March/April 1990.
[8] H.V. Jagadish, R.G. Mathews, T. Kailath, and J.A. Newkirk, "A study of pipelining in computing arrays," *IEEE Trans. Comput.*, vol. 35, No. 5, pp. 431-440, May 1986.
[9] W.A. Wolovich, *Robotics: Basic Analysis and Design*, CBS College Publishing, 1987.

A Formal Framework for Specifying Robot Kinematics

V.S. Alagar, T.D. Bui and K. Periyasamy
Department of Computer Science, Concordia University
Montreal, Quebec, CANADA H3G 1M8

1. Introduction

An *intelligent robot* is a physical machine endowed with computational mechanisms to plan, choose and execute actions and reason about the consequences of such choices. Intelligent robots are autonomous and they are required in environments where human interaction is hazardous or impossible. The computational mechanisms that make a robot intelligent are mainly software packages consisting of a variety of complex programs whose inputs and outputs are not just mathematical entities but physical objects. Assuring correctness of the software used for intelligent robots is mandatory because online error recovery is almost impossible. Consequently, there is a need for a formal offline framework to specify the structure and properties of robots and their application domains so that (1) a static analysis can be conducted to reason about the behavior of the robotic system to be built and (2) the specifications can be transformed into robots implementing the specified tasks. This paper describes one component of our on-going research in this area : formal specification supporting a rigorous analysis and a correct synthesis of robotic agents.

A formal framework introduced in this paper is founded on mathematical principles and hence promotes abstraction and supports proofs. A robot is abstracted away from its physical characteristics and its particular physical environments. An abstract robot will be called an *agent*, a mathematical object endowed with operations whose manifestations in the real-world will drive a robot into its actions. Devoid of irrelevant details, an agent represents, in general, a class of real-life robots and the behavior of an agent permeates through this class of real-life robots. In this paper, the term 'robot' is used to mean its 'agent'.

The formal specification language used in this paper is VDM (Vienna Development Method), a method based on type theory and first order logic. In addition, VDM contains a number of basic constructs for building complex data types and structuring specifications. In particular, it enables a system to be viewed as a state machine, promotes top-down refinement and permits a *calculational* style of formal reasoning.

As remarked earlier, our goal has been to formalize the structure and architecture of intelligent robots. Given the vastness and diversity of ideas one has to gather and bring them to bear on a problem of this magnitude and complexity for specification purposes, we have been very selective in the initial choice of subdomains. Brady [3] has recently remarked that the automation of industrial processes such as mechanical parts assembly using robots is an important open problem. The validation of mechanical assembly is the answer to the question "given solids S_1, S_2 and S_3, whether the solids S_1 and S_2 can be assembled in *some correct way* to produce S_3?". The formal specification given by us in [1] leads to an answer to this question in a formal calculational manner. Formal verification of assembly process requires a mathematical definition of assembly and logical tools for proving that at every stage of subassembly the definition is respected. A full treatment of this problem appears in [2]. The most fundamental aspect that supports assembly is "robot kinematics" and for the purpose of brief illustration we have chosen "inverse kinematics problem".

2. Vienna Development Method (VDM)

VDM is based on the mathematical primitives *sets, lists, maps and trees*. In addition, it also provides facilities for the user to define new data types which are aggregations of the basic types. Operations on these data types are specified as a collection of predicates, grouped into two major categories, namely *pre-conditions* and *post-conditions*. The predicates are combined using the logical connectives **and** ($\wedge$), **or** ($\vee$) and **not** ($\sim$). The pre-condition for an operation is a logical formula stating the system constraints and it must be satisfied before the operation is invoked; that is, if any one of the predicates in the pre-condition fails, the operation fails. Consequently the post-condition specifies the constraints that are to be satisfied after the operation successfully terminates. The set of properties of the system that are to be satisfied at every instant is called *invariant*, and is expressed as a set of constraints. Some other useful notations used are : "$\exists$!" denotes "there exists exactly one", " $\triangleq$ " defines the left side in terms of the right side and "**let ... tel**" clause is the usual mathematical substitution mechanism.

A specification which does not require an explicit reference to a global variable, is written in a purely *functional style*. Specifications in Section 3 illustrate this style. However, when a specification requires an explicit reference to one or more global variables, which incidently define the *state* of the system, it is written in *model-based style*. Specifications in Section 4 are model-based. Another major difference between these two styles is the presence of **ext** clause in the model-based style that lists all the global variables accessed in that operation. We use a number of *auxiliary functions* in the specifications. These are, in fact, modules of a complex specification. Auxiliary functions represent actions returning some values as opposed to the constraints which always return logical values. An exhaustive list of auxiliary functions arising in robot kinematics and their semantics can be found in

[2,7].

3. Specification for Rigid Solids

A robot consists of several links joined together to form a chain, each link being a rigid solid. Robot kinematics is the study of the positional information and the associated transformations of these links in 3-dimensional space. Hence any formal framework for robotics must include a formal study of rigid solids. In this section, we provide the definition of a rigid solid and two primitive operations, namely *translation* and *rotation*. Theorems and their proofs can be found in [2].

A rigid solid in space is defined by its shape and its position and orientation with respect to a global coordinate frame. Hence the type definition of a rigid solid can be stated in VDM as

Solid   = Structure
Structure   :: POSI-ORIE : Transformation, SHAPE : Primitive | Composite

The equality of two vectors u and v belonging to the coordinate frames T_u and T_v is formally specified as below :

Vector-equal : Vec-Frame-Pair × Vec-Frame-Pair → Boolean
post-Vector-equal (A,B,b) $\triangleq$
b' = **let** u = VECTOR (A), v = VECTOR (B),
    T_u = FRAME (A), T_v = FRAME (B) **in**
(norm (u) = norm (v)) ∧ (angle-X (u,T_u) = angle-X (v,T_v)) ∧
(angle-Y (u,T_u) = angle-Y (v,T_v)) ∧ (angle-Z (u,T_u) = angle-Z (v,T_v)) **tel**

Using this specification, next we specify a rigid solid.

Rigid : Solid → Boolean
post-Rigid (S,b) $\triangleq$
b' = **let** O = position (POSI-ORIE (S)) **in**
  ($\forall$ p ∈ Point) ((p ≠ O) ∧ (on (p, S)) ⇒ ($\forall$ T ∈ Transformation)
    (($\exists$! p' ∈ Point)
      ((p' = transform-point (T,p)) ∧ (O' = transform-point (T,O)) ∧
        (**let** u_T_u = const-vec-frame (point-vector (O',p'), POSI-ORIE (S))),
          v_T_v = const-vec-frame (point-vector (O,p), POSI-ORIE (S'))) **in**
        vector-equal (u_T_u,v_T_v) **tel**)))) **tel**

Translation of an object is that transformation which defines the positional change in the object without change in its orientation. Formally,

Translation : Solid × Axis–Rep → Solid
pre-Translation (S, A) $\triangleq$ rigid (S)
post-Translation (S, A, S') $\triangleq$ (rigid (S')) ∧
  (**let** O = position (POSI-ORIE (S)), O' = position (POSI-ORIE (S')) **in**

(parallel (const-line (O, O'), A)) ∧ (**let** T = image (S,S') **in**
(T ≠ NIL) ∧ (∀ p ∈ Point) ((p ≠ O) ∧ (on (p, S)) ⇒
parallel (const-line (O, p), const-line (O', transform-point (T,p)))) **tel**) **tel**)

The above specification for translation is quite abstract; see [2] for an extended specification for translation through a given distance.

Rotation changes the orientation of an object continuously in such a way that every point on the object describes a circular path. Formally,

Rotation : Solid × Axis–Rep → Solid
pre-Rotation (S,A) $\triangleq$ rigid (S)
post-Rotation (S,A,S') $\triangleq$ (rigid (S') ∧ (**let** T = image (S,S') **in**
(T ≠ NIL) ∧ (∀ p ∈ Point) (on (p,S) ⇒ (p' = transform-point (T, p)) ∧
(**let** circ = circle-pt-axis (p,A) **in**
lie-on-circle (p, circ) ∧ lie-on-circle (p', circ) **tel**)) **tel**)

Circle-pt-axis : Point × Axis–Rep → Circle
post-Circle-pt-axis (p,A,Circ) $\triangleq$
let c = CENTRE (Circ), r = RADIUS (Circ) **in**
c = intersect (normal (p,A), A) ∧ r = distance (p,c) **tel**

See [2] for a specification of rotation through a definite angle.

Prismatic and *revolute* joints are commonly used structures for building robotic manipulators. A joint connects two rigid solids in which one of them is fixed (with respect to the joint) and the other moves along/about an axis defined within the joint; see Figure 1.

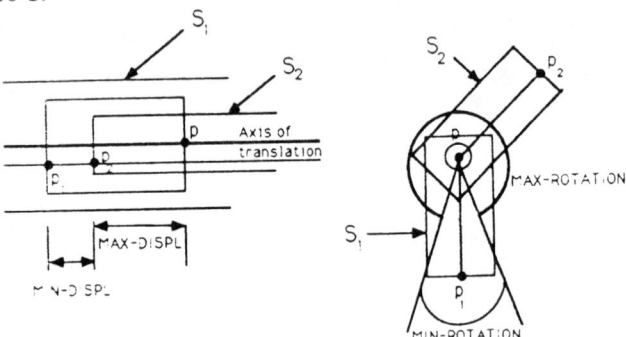

Figure 1 : Prismatic and Revolute Joints

Type Definitions
Prisjoint   :: S1 : Solid, S2 : Solid, AXIS-OF-MOVE : Axis–Rep
              MIN-DISPL : Dist–Rep, MAX-DISPL : Dist–Rep
Revoljoint   :: S1 : Solid, S2 : Solid, AXIS-OF-ROTATION : Axis–Rep
              MIN-ROTATION : Angle–Rep, MAX-ROTATION : Angle–Rep

Type Invariants : The type invariant for revolute joint is given below; see [2] for the type invariant for prismatic joint.

inv-Revoljoint (jn) $\triangleq$ ($\exists$ p $\in$ Point)
  ((on (p, S1)) $\wedge$ (on (p, S2)) $\wedge$ (lie-on-line (p, AXIS-OF-ROTATION (jn))) $\wedge$
    ($\exists$ p_1, p_2 $\in$ Point)
      ((on (p_1, S1)) $\wedge$ (on (p_2, S2)) $\wedge$
        (perpendicular (const-line (p, p_1), AXIS-OF-ROTATION (jn))) $\wedge$
        (perpendicular (const-line (p, p_2), AXIS-OF-ROTATION (jn))) $\wedge$
        (let θ = angle (const-line (p, p_1), const-line (p, p_2)) in
          (MIN-ROTATION (jn) $\leq$ θ MAX-ROTATION (jn)) $\wedge$
          (θ − MIN-ROTATION (jn) > $\Rightarrow$
            on (p_2, rotate-angle (S2, AXIS-OF-ROTATION (jn), θ))) tel)))

4. Formalism of Inverse Kinematics

The problem of inverse kinematics can be informally stated as follows : *"given the positional change in the end-effector, determine the positional changes on all the links of the robot"*. We do not provide procedures for finding all possible solutions to this problem; rather, we characterize these solutions so that an offline formal verification of an algorithm computing these solutions can be carried out.

4.1. Formal Definition of a Robot

In order to simplify our discussions, we consider only one general purpose multi-link robot. One end of the first link is fixed at a known location in the workspace. Two consecutive links are joined by only one type of joint, either Prismatic or Revolute. The end-effector is attached to the last link of the robot and is treated differently from the links. The type definitions given below show the robot environment in **bold face**. The robot consists of a list of joints, a list of links and an end-effector (gripper). Each link has a unique identification number for reference. Its structure is defined by the variable 'GEOMETRY'. We have added another variable 'WRIST' in gripper definition to indicate the base of the end-effector. This type definition is extendable to the case when tools and fingers are attached to the wrist in the context of more general manipulations.

State ::
BASE-COORD : Transformation; **ROBOT-ARM** : Manipulator
Manipulator :: LINKS : Armtype-list, JOINTS : Jointtype-list
                GRIPPER : Grippertype
Armtype   :: LINKID : ID–Rep, GEOMETRY : Structure
Grippertype :: WRIST : Wristtype
Wristtype   :: GEOMETRY : Structure
Jointtype   = Prisjoint | Revoljoint

See [2] for type invariants of the robot structure.

4.2. Specification for Inverse Kinematics

The four problems in inverse kinematics are the following :
Let P_w and Q_w denote two spatial positions. (1) The wrist has same orientations

in these positions. (a) Determine the valid configurations of the links to obtain Q_w from P_w using only *translations*. (b) Determine the valid configurations of the links to obtain Q_w from P_w using only *rotations*. (2) The wrist has different orientations in these positions. (a) Determine the valid configurations of the links to obtain Q_w from P_w using only *rotations*. (b) Determine the valid configurations of the links to obtain Q_w from P_w using a combination of *translations* and *rotations*.

A complete specification for all the four problems can be found in [2]. Specifications for 2(b) can be given for two cases - interleaved activation (only one link is activated at a time) and concurrent activation (more than one link can be activated simultaneously). Below, we give specification for concurrent activation; specification for interleaved activation can be derived from this by restricting the cardinality of the subset of links activated at each instant to be 1.

An Informal Description of the Solution : Let Pjn and Rjn be the set of prismatic and revolute joints such that (**card** Pjn = l) and (**card** Rjn = k), Pjn $\cup$ Rjn = Jns, the set of joints. Between P_w and Q_w, there exist a non-empty list of intermediate positions Plist such that size (Plist) = N, N > 0, Plist(1) = P_w and Plist(N) = Q_w. For all t, $1 \le t \le$ (N-1), Plist(t+1) can be obtained only from Plist(t) by applying translation on some subset (possibly empty) of prismatic joints and/or rotation on some subset (possibly empty) of revolute joints simultaneously.

Let SPjn with cardinality n and SRjn with cardinality m denote arbitrary subsets of prismatic and revolute joints, chosen for activation. Let T be the change in orientation of the wrist, $\vec{d}$ be the linear displacement between Plist(t+1) and Plist(t), T_x, T_y and T_z be its X,Y,Z components, ϕ_i be the angle of rotation applied to the second link of SRjn$_i$, T_i be the change in orientation of the second link due to ϕ_i and T_{ix}, T_{iy} and T_{iz} be the X,Y,Z components respectively of T_i, $1 \le i \le m$. Let $\vec{d_j}$ be the displacement vector applied to the second link of the prismatic joint SPjn$_j$, $1 \le j \le n$. The following conditions are to be satisfied for moving the wrist from Plist(t) to Plist(t+1).
$\sum_{i=1}^{m} T_{ix} = T_x$, $\sum_{i=1}^{m} T_{iy} = T_y$ and $\sum_{i=1}^{m} T_{iz} = T_z$.
$\sum_{j=1}^{n} \vec{d_j} = \vec{d_l}$ and $\vec{d} = \vec{d_l} + \vec{d_\phi}$ where $\vec{d_\phi}$ is the net displacement vector due to rotations of m links of SRjn.

Specification
MOVE-WRIST-OTRS (DESTINATION : Transformation)
(* Move the wrist to 'DESTINATION' with change in orientation, by a combination of translations and rotations applied to the links. More than one link may be activated simultaneously. *)

ext ROBOT-ARM : **wr** Manipulator, BASE-COORD : **rd** Transformation
Pre
  let jns = JOINTS(robot-arm), grip = GRIPPER(robot-arm),
    wrs = WRIST(grip) **in**
    ($\exists$ Pjn, Rjn $\in$ Jointtype–list)

$((\forall\ i\ \{1\cdot\cdot\ \textbf{len}\ \text{Pjn}\})\ (\text{Pjn}(i) \in \text{Prisjoint}) \wedge$
$(\forall\ j\ \{1\cdot\cdot\ \textbf{len}\ \text{Rjn}\})\ (\text{Rjn}(j) \in \text{Reveljoint}) \wedge$
$((\textbf{elems}\ \text{Pjn} \cup \textbf{elems}\ \text{Rjn}) = \textbf{elems}\ \text{jns}) \wedge$
$(\exists\ \text{Plist} \in \text{Transformation–list})$
  $((\textbf{len}\ \text{Plist} > 0) \wedge$
  $(\text{position}(\text{Plist}(1)) = \text{position}(\text{POSI-ORIE}(\text{GEOMETRY}(\text{wrs})))) \wedge$
  $(\text{orientation}(\text{Plist}(1)) = \text{orientation}(\text{POSI-ORIE}(\text{GEOMETRY}(\text{wrs})))) \wedge$
  $(\text{position}(\text{Plist}(\textbf{len}\ \text{Plist})) = \text{position}(\text{destination})) \wedge$
  $(\text{orientation}(\text{Plist}(\textbf{len}\ \text{Plist})) = \text{orientation}(\text{destination})) \wedge$
  $(\forall\ t \in \{2\cdot\cdot\ \textbf{len}\ \text{Plist}\})$
  $(\text{POSI-ORIE}\ (\text{wrs}) = \text{Plist}(t) \Rightarrow$
    (* dvec is the vector displacement and T is the change in
    orientation between successive positions. *)
    let dvec = vector $(\text{position}(\text{Plist}(t)), \text{position}(\text{Plist}(t+1)))$,
        T = change-in-orientation $(\text{orientation}(\text{Plist}(t)), \text{orientation}(\text{Plist}(t+1)))$,
        T_x = X-component $(T, \text{base-coord})$, T_y = Y-component $(T, \text{base-coord})$,
        T_z = Z-component $(T, \text{base-coord})$ **in**
    $(\exists\ \text{SPjn}, \text{SRjn} \in \text{Jointtype-list})$
      $((\forall\ i \in \{1\cdot\cdot\ \textbf{len}\ \text{SPjn}\})\ (\text{Spjn}(i) \in \text{Prisjoint}) \wedge$
      $(\forall\ j \in \{1\cdot\cdot\ \textbf{len}\ \text{SRjn}\})\ (\text{SRjn}(j) \in \text{Reveljoint}) \wedge$
      $(\textbf{len}\ \text{SPjn} \leq \textbf{len}\ \text{Pjn}) \wedge (\textbf{len}\ \text{SRjn} \leq \textbf{len}\ \text{Rjn}) \wedge$
      $(\textbf{len}\ \text{SPjn} + \textbf{len}\ \text{SRjn} \neq 0) \wedge$
      (* dlist is the list of vector displacements on prismatic joints.
      ϕlist is the list of angles, imposed on revolute joints,
      and Tlist is the corresponding list of orientations of the links of
      revolute joints. *)
      $(\exists\ \text{dlist} \in \text{Vectortype-list}, \phi\text{list} \in \text{Angle–Rep-list},$
        $\text{Tlist} \in \text{Transformation-list})\ ((\textbf{len}\ \text{dlist} = \textbf{len}\ \text{SPjn}) \wedge$
        $(\textbf{len}\ \phi\text{list} = \textbf{len}\ \text{Tlist}) \wedge (\textbf{len}\ \phi\text{list} = \textbf{len}\ \text{SRjn}) \wedge$
        $(\forall\ i \in \{1\cdot\cdot\ \textbf{len}\ \text{SPjn}\})\ (\text{norm}\ (\text{dlist}(i)) \leq \text{MAX-DISPL}\ (\text{SPjn}(i)))) \wedge$
        $(\forall\ j \in \{1\cdot\cdot\ \textbf{len}\ \text{SRjn}\})\ (\phi\text{list}(j) \leq \text{MAX-ROTATION}\ (\text{SRjn}(j)))) \wedge$
        $(\textbf{let}\ T_{old}$ = orientation (POSI-ORIE (GEOMETRY (LINK2(SRjn(j))))),
            T_{new} = orientation (POSI-ORIE (rotate-angle (GEOMETRY (LINK2
              (SRjn(j))), AXIS-OF-ROTATION (SRjn(j)), $\phi\text{list}(j)$))) **in**
        Tlist(j) = change-in-orientation (T_{old}, T_{new}) **tel**) $\wedge$
      (* $\overrightarrow{dvec} = \overrightarrow{dlvec} + \phi vec$, ϕvec is net displacement due to rotations $\phi list$. *)
      (**let** dlvec $\in$ Vectortype, ϕvec $\in$ Vectortype-list **in**
        (dlvec = vector-sum (dlist) $\wedge$ (**len** ϕvec = **len** Tlist) $\wedge$
        $(\forall\ j \in \{1\cdot\cdot\ \textbf{len}\ \text{Tlist}\})$
          (**let** P_{old} = position (POSI-ORIE (GEOMETRY (LINK2(SRjn(j))))),
              P_{new} = position (POSI-ORIE (rotate-angle (GEOMETRY (LINK2
                (SRjn(j))), AXIS-OF-ROTATION (SRjn(j)), $\phi\text{list}(j)$))) **in**

$\phi\text{vec}(j) = \text{vector}(P_{old}, P_{new})$ **tel**) ∧
(dϕvec = vector-sum (ϕvec)) ∧
(**let** dvec_B = const-vec-frame (dvec, base-coord),
    dl_ϕ_B = const-vec-frame (vector-sum (dlvec, dϕvec), base-coord) **in**
    vector-equal (dvec_B, dl_ϕ_B) **tel**) **tel**) ∧
(∃ Xlist, Ylist, Zlist ∈ Real-list)
((**len** Xlist = **len** Tlist) ∧ (**len** Ylist = **len** Tlist) ∧
len Zlist = **len** Tlist) ∧ (∀ j ∈ {1 ·· **len** Tlist})
((Xlist(j) = X-component (Tlist(j), base-coord)) ∧
(Ylist(j) = Y-component (Tlist(j), base-coord)) ∧
(Zlist(j) = Z-component (Tlist(j), base-coord)) ∧
(Real-sum (Xlist) = T_x) ∧ (Real-sum (Ylist) = T_y) ∧
(Real-sum (Zlist) = T_z))))) **tel**)) **tel**)) **tel**

Post
let jns = JOINTS(robot-arm), grip = GRIPPER(robot-arm),
    wrs = WRIST(grip) **in**
(∃ Pjn, Rjn ∈ Jointtype–list)
((∀ i {1·· **len** Pjn}) (Pjn(i) ∈ Prisjoint) ∧
(∀ j {1·· **len** Rjn}) (Rjn(j) ∈ Revoljoint) ∧
((elems Pjn ∪ elems Rjn) = elems jns) ∧
(∃ dlist ∈ Dist–Rep-list, ϕlist ∈ Angle–Rep-list)
((**len** dlist = **len** Pjn) ∧ (**len** ϕlist = **len** Rjn) ∧
(∀ i ∈ {1 ·· **len** Pjn})
    (translate-link (LINKID (LINK2 (Pjn(i))), dlist(i))) ∧
(∀ j ∈ {1 ·· **len** Rjn})
    (rotate-link (LINKID (LINK2 (Rjn(j))), ϕlist(j))))))
⇔ POSI-ORIE (wrs)' = destination **tel**

5. Conclusion

There have been only a few attempts [4] to formalize the notion of an integrated architecture for an intelligent robot. Our approach is different from these and can be extended to various problem domains in robotics. From the specifications in Section 3, robots may be generated automatically and the specification in Section 4 can be the basis for a formal verification of the software implementing kinematics. Forces, cognitive mechanisms, sensors and control aspects can be specified independently and then combined for a particular class of application. These are some of our goals to pursue.

For simplicity at the logical level, we have considered every joint to have only one degree of freedom. This is consistent with the mechanical design considerations being practised [6]. It is only rarely that the joints with n degrees of freedom are used in practice. Such joints can be accommodated in our formalism by letting n joints, each with one degree of freedom, connecting links of negligible length. That is, the structural integrity of a joint with n degrees of freedom has been modeled without loss of generality in our formalism. We also remark that the behavior of the controller for the joint with n degrees of freedom is also faithfully captured in our formalism. Controller that activates only one degree of freedom at a time corresponds to the interleaved specification model whereas the controller which activates more than one degree of freedom at a time corresponds to the concurrent model. Interleaved activation of links obtain successive positions due to a single translation or a single rotation. Consequently, the path traversed by the wrist is a collection of piecewise line segments and circular arcs. The converse is also true. However, if the trajectory is not composed of piecewise line segments and circular arcs, then an interleaved activation of links cannot realize the trajectory. In this situation, a concurrent activation of links becomes necessary.

6. References

1. V.S. Alagar, T.D. Bui and K. Periyasamy, "Formal Specifications for Regularized Boolean Operations in Solid Modeling", submitted for publication to *Science of Computer Programming*.

2. V.S. Alagar, T.D. Bui and K. Periyasamy, "Formal Specifications for Robot Kinematics", *Technical-Report–CSD-90-5*, Dept. of Computer Science, Concordia University, Montreal, Canada.

3. Michael Brady, *Robotics Science*, The MIT Press, 1989.

4. A.D. Christiansen, "A Framework for Specifying Robotic Agents", *Technical Report, CMU-CS-89-155*, Department of Computer Science, Carnegie-Mellon University, Pittsburgh, 1989.

5. B. Cohen, W.T. Harwood and M.I. Jackson, *Specification of Complex Systems*, Addison-Wesley Publishing Company, 1986.

6. J.J. Craig, *Introduction to Robotics*, Addison-Wesley Publishing Company, 1989.

7. K. Periyasamy, V.S. Alagar and T.D. Bui, "Specification for Geometric Primitives", *Technical Report–CSD-90-4*, Dept. of Computer Science, Concordia University, Montreal, Canada.

NON-GEOMETRICAL PARAMETERS IDENTIFICATION FOR ROBOT KINEMATIC CALIBRATION BY USE OF NEURAL NETWORK TECHNIQUES

Jean-Michel RENDERS*, José del R. MILLAN** and Marc BECQUET**

* Laboratoire d'Automatique, Université Libre de Bruxelles,
 50 Av. Roosevelt, B-1050 Bruxelles, BELGIUM
**Institute for Systems Engineering and Informatics, Commission of the European Communities, Joint Research Center, I-21020 Ispra, ITALY

Abstract: *This paper presents a new technique for the calibration of robots based on a neural network approach for the identification of non-geometrical errors. Identification of geometrical errors is not a problem any more since several methods have been presented recently. The remaining problem is the identification of the non-geometrical errors. Non-geometrical errors modeling is a very complex and heavy process. The originality of this paper is the use of a neural network approach avoiding explicit modeling of this kind of errors. Simulations have been carried out on a robot with 6 degrees of freedom. Finally, two compensation algorithms are presented, based on the improved knowledge of the model: the first one is based on the construction of false target, the second one compensates directly into the joint space.*

I. INTRODUCTION

One of the most important breakthroughs in robot technology is the off-line programming process in CAD system that makes possible the general design and layout of a robotic cell and also the robot task simulation. The most important output of this simulation is a task file that can be downloaded into the robot control system to perform the desired motion. The major problem is the difference between the *nominal geometry* of a robot, based on the robot design specifications, and the real geometry of the same robot affected by manufacturing tolerances, mounting errors during robot link assembly and kinematic model simplification in the control unit. Nominal models used are simple and are based on several assumptions, such as parallelism or othogonality of the axes of the joints. These errors are known as *geometrical* ones. Differences between nominal and real kinematic models also arise from *non-geometrical errors*: link and joint flexibility, backlash, gear runout, etc.

Robot kinematic calibration consists of identifying a more accurate geometrical relationship between joint encoder readings and the actual position of the end-effector; then the robot positioning software is changed according to this identified relationship. Kinematic calibration usually involves four steps: modeling, measurement, identification and correction (see Roth[1] for a general overview on robot calibration).

Models of geometrical position error have been widely studied. A linearized geometrical error model has been established by Veitshegger[2]; it was used for *a priori* accuracy analysis. An[3] used Veitschegger's linearized error model for geometrical parameter identification. Renders[4] developed an algorithm for estimating geometrical parameter errors, taking into account the following characteristics: sensor accuracy, resolver resolution, backlash in transmission units, manufacturing tolerances and position/orientation errors of the measuring device; these limits were used for building a global error model and a maximum likelihood estimator.

Details about possible measuring devices for calibration purposes can be found in Payannet[5] (potentiometric transducers), Stone[6] (ultrasonic systems), An[3] (infrared 3D measurement), Renders[4] (linear motion tracking system with magnetostrictive and magnetic sensors), Whitney[7] and Judd[8] (theodolites and stereo triangulation).

Modeling and identification of non-geometrical errors are not so common in the field of robot calibration. There are only a few works dealing with this problem: Withney[7] uses a model including non-geometrical errors and a least squares numerical search algorithm on theodolite measurements, but the error model is completely non-linear; moreover, the definition of parameters to be identified is rather unusual and related to measuring instruments. Judd[8], on the other hand, considers mainly non-geometrical errors (gear train errors, joint and link flexibility, etc.) and proposes error models which can be used for identification with a common least squares procedure; unfortunately, the quality of the identification results strongly depends on the presumed structure of the model.

In this work, we will assume that the nominal model is already tuned to compensate for the geometrical error effects; then, the real position reached by the end-effector is obtained by adding to the nominal model output the contribution of a *residual error function* containing all the non-geometrical effects. We use a *neural network* in order to identify this error function, without the need of an explicit modeling stage, while taking advantage of the potential computational benefits expected from this new approach (parallel computation, adaptation to large number of parameters, natural fault tolerance, natural robustness, generalization capacity, etc.). A *supervised learning process*, based on the *backpropagation algorithm*, is applied to develop the network.

The following section briefly describes the main sources of positioning error in robot kinematics and introduces the residual error function. Section III develop the neural network architecture and the training stage methodology. Then, different ways to use the neural network's results for correction and compensation are presented in section IV. Results of simulations on a robot with 3 degrees of freedom illustrating the performance of the network for calibration purposes are shown in section V. Finally, section VI contains some conclusions about the work.

II. POSITIONING ERROR SOURCES

a. Geometrical errors

A classical choice for the geometrical parameters used to describe the robot geometry is the Denavit and Hartenberg parameters; for kinematic identification and calibration, it is usually necessary to introduce an extra parameter (the link twist, β_i) in the case of successive nominally parallel rotational joints (Veitschegger[2]).

Nominal values of those geometrical parameters can be different from the real ones in the robot system, because manufacturing and assembly tolerances affect the kinematic model of the robot; moreover, the electrical zeroes of the joint encoders do not generally coincide with the mechanical zeroes of the joints themselves. Usually, geometrical errors on geometrical parameters are responsible for about 90% of the total error, while the remaining 10% are due to non-geometrical errors. As previously mentioned, in the following we consider that the nominal geometric model is already tuned, so that only non-geometrical errors will be identified and compensated for.

b. Non-geometrical errors

The sources of non-geometrical errors lie with:
-joint flexibility,
-link flexibility,
-structural deformation of the base,
-gear transmission error (runout and orientation error),

-backlash in gear transmission,
-temperature effect.

Link flexibility is usually less than *joint flexibility* (below 5%); the flexibility of the transmission unit can be considered as a torsional spring (or linear spring for prismatic joints) inserted between the output of the gear box (assumed as rigid) and the link drived by the transmission.

Gear runout error occurs when the actual center of rotation does not lie at the center of the pitch circle. It produces a cyclic error depending on the angle of rotation of the gear and the reduction ratio. Orientation error is a second important error existing in gear trains; it appears when the two axis of the gears are not exactly parallel: the contact line of one gear with respect to the other is no longer circular, but ellipsoidal; this error is cyclic with a period that is half a revolution. Details on this kind of errors are explained in Judd[8].

Backlash is probably one of the most difficult error source to identify. We will assume that the backlash is closed; it is true if the torque applied by the actuator is larger than the friction torque; then, following the direction of the applied torque, the gear will be at one or the other extreme of the dead zone, even if this torque is due to gravity in static condition. Otherwise, backlash should be considered as a noise (with roughly uniform distribution) perturbating the accuracy of the mechanism.

The effect of *temperature* is to expand the robot mechanical structure. Given the materials used for robot links and their thermal expansion coefficient, the error due to temperature is responsible for 0.1% of the total error. This means that for the present purpose it can be ignored.

c. Residual error function

Let us call θ the vector of the *joint encoder angles*; these angles are measured directly on the rotor shaft of each motor, prior to backlash, gear runout and flexibility effects. Let us call f_n, the *nominal geometrical model*, that gives the *nominal vector of the end effector position/orientation* x_n as a function of θ:

$$x_n = f_n(q) \qquad (1)$$

Let us call x_r the *vector of the real position/orientation of the end effector*, which differs from x_n because of non-geometrical errors, and f_r the function of x_n and θ which produces x_r:

$$x_r = f_r(x_n, \theta) = f_r(f_n(\theta), \theta) \qquad (2)$$

Introducing the *residual error function* $\varepsilon(\theta)$ defined by:

$$\varepsilon(\theta) = f_r(f_n(\theta), \theta) - f_n(\theta) \qquad (3)$$

we have the following relationship:

$$x_r = x_n + \varepsilon(\theta) \qquad (4)$$

The objective consists then of trying to identify (or to approximate) this residual error function $\varepsilon(\theta)$ having a set of measurement data (θ, x_r). Note that this function also depends on the payload carried by the robot, because of the gravity torque and the flexibility. In our work, we have chosen to perform this identification task by training a neural network with supervised learning.

III. NEURAL NETWORK: ARCHITECTURE AND TRAINING

An artificial neural network consists of many simple processing elements called *units* that interact using weighted *connections*. Each unit has a *state* or *activity level* that is determined by the input received from other units in the network.

The *short-term* knowledge of a neural network is usually encoded by the states of the units. All the *long-term* knowledge is codified by the weights. Thus, learning amounts to modify the weights in a way that allows the networks to construct the necessary internal representations of their environments in order to solve the task they face.

Recently, it has been proved that *multilayer feedforward networks* with at least one hidden layer of *sigmoid units* (see below) and using a learning algorithm of the *backpropagation* type [9] can approximate any square-integrable function to any desired degree of accuracy, provided sufficient hidden units are available [10]. Nevertheless, there are no guarantees that the solution can be found. In addition, the optimal number of weights is not known in advance, what could result in a network larger than required which, consequently, arbitrarily approximates the training data but generalizes poorly. The learning algorithm we have used addresses simultaneously these two topics.

In a multilayer feedforward network, units are arranged into layers in such a way that a given unit can only send its output to units located in layers above. In this manner, a network is made of an input layer (bottom), one or more hidden layers (middle) and an output layer (top). The learning phase is aimed at reducing the difference between the actual output of the network and the desired one.

An input unit simply forwards the information it receives from outside. Two are the kinds of input signals to our network: either a joint angle, a real number in the interval $[-\pi,\pi]$, or a payload, a positive real number.

A hidden unit computes its output by applying a sigmoid function to its activation level:

$$o_i = \frac{2}{1+e^{-\beta(\sum_j w_{ij}o_j + \sigma_i)}} - 1$$

where o_i is the output of the unit i, σ_i is a bias, j ranges over all the units sending their outputs to the unit i, w_{ij} is the weight associated to the connection from the unit j to the unit i, and β is a positive constant (0.4 in the simulations reported here). We have chosen a symmetrical sigmoid in order to avoid rapid saturation with extreme inputs that makes learning even more difficult.

An output unit produces as output its activation level.

The learning algorithm we have used is characterized by two processes. The first of them is the rule for updating the weights. The weights are modified according to a variant of the basic backpropagation algorithm:

$$\Delta w_{ij} = -\eta_i \frac{\delta E}{\delta w_{ij}}$$

where η_i is the learning rate associated to the unit i and E is the error function to be minimized. Three are the modifications we have introduced. First, the weights are not updated after the gradients have been accumulated over the whole training set, but after the presentation of each pattern. This version, known as *stochastic* or *on-line* version, allows to exploit redundancy in the training data (as it is our case. See Section V, where it is described how the training data were obtained).

Second, the learning rate is specific for each unit:

$$\eta_i = \frac{\alpha}{\rho_i}$$

where α is a constant and ρ_i is the number of input connections to the unit i [11].

Third, the value of α decreases as learning proceeds until reaching a sufficiently small value. At this moment the learning process has converged (successfully or not). Nevertheless, we have observed that if, at this moment, the value of α is risen and the process of incrementally reducing α is repeated, then the weight configuration is better than before. A possible explanation is that when the learning rate is increased, the weight configuration is perturbed sufficiently as to climb the walls of the basin of attraction it was in. Experiments confirm this explanation since the error increases when the learning rate increases. In the simulations, we have iterated the cycle "set α to its initial value (0.2) --- decrease gradually α up to a final value (0.001)" until the network does not improve significatively.

The second feature of our learning algorithm is that the network is built incrementally. A new hidden unit is added (either in an existen hidden layer or in a new one) when the current network cannot reduce any more the error. By modifying the network architecture, the shape of the weight space is also changed, what could remove the local minimum the network was trapped in. Experiments confirm this fact: after adding a new hidden unit, the performance of the network always improves. In addition, this technique could allow to build the network with the smallest number of weights required to solve the task at hand. Consequently, it is expected to obtain a network with good generalization abilities. Again, the experiments we have carried out confirm this hypothesis.

Finally, it is worth noting that we have observed that the performance of a network built incrementally is always better than the performance of an equivalent network trained from the beginning with the final architecture.

IV. POSITIONING ERROR COMPENSATION AND TRAJECTORY CORRECTION

In this section, we will present two ways to compensate for the effects of non-geometrical errors, using the output $\hat{\varepsilon}(\theta)$ of the neural network, which is nothing else that the estimate of the residual error function $\varepsilon(\theta)$.

a. Action on the target: task space correction

The idea is to define a *false target* in such a way that the end effector will reach a position closer to the desired one (Payannet[5]).
Let us call:

x_d, the desired position/orientation

x_r, the real position/orientation of the robot

x_m, the modified position/orientation (false target)

f_n, the nominal direct geometrical model

f_n^{-1}, the nominal inverse geometrical model (implemented in the robot control unit).

If we present to control unit of the robot the modified target given by:

$$x_m = x_d - \hat{\varepsilon}(f_n^{-1}(x_d)) \qquad (5)$$

the robot will reach the following position/orientation:

$$x_r = x_m + \varepsilon(f_n^{-1}(x_m)) \qquad \text{(using eq. 5)}$$

$$= x_d + \varepsilon(f_n^{-1}(x_m)) - \hat{\varepsilon}(f_n^{-1}(x_d)) \qquad (6)$$

$$\cong x_d \quad \text{since, with a very good approximation, } \varepsilon(f_n^{-1}(x_m)) \cong \hat{\varepsilon}(f_n^{-1}(x_d)).$$

Then, the control unit computes:

$$\theta_m = f_n^{-1}(x_m) \qquad (7)$$

This type of algorithm requires 2 nominal inverse model calls and one neural network computation.

b. Correction in the joint space

If the control unit allows to control the robot directly in the joint space, then we can use the following algorithm to compensate for the positioning error (which is nothing else that the differential version of the preceding one):

$$\theta_m = f_n^{-1}(x_d) - J_n^{-1} \cdot \hat{\varepsilon}(f_n^{-1}(x_d)) \qquad (8)$$

where J_n is the *Jacobian matrix of the robot*, computed in $\theta_d = f_n^{-1}(x_d)$.
Indeed, using eq. 8, the robot will reach the position/orientation given by:

$$x_r = f_n(\theta_m) + \varepsilon(\theta_m) \qquad \text{(using eq. 5)}$$

$$= x_d - J_n \cdot J_n^{-1} \cdot \hat{\varepsilon}(f_n^{-1}(x_d)) + \varepsilon(\theta_m) \text{ (developing } f_n \text{ up to the first order)} \qquad (9)$$

$$\cong x_d \quad \text{since, with a very good approximation, } \varepsilon(\theta_m) \cong \hat{\varepsilon}(f_n^{-1}(x_d)).$$

Note that both algorithms use the same approximation, but the second one uses one nominal inverse model, one inverse Jacobian (which could be sometimes easier to compute than the inverse model) and one neural network computation.

V. CASE STUDY - RESULTS OF SIMULATION

Simulations were performed on a robot with 6 degrees of freedom; the end effector is defined as the intersection point of the three last axis (see Fig.1). Only the end effector position (x,y,z) was considered for measurement and correction, so that only the first 3 joints have to be considered. To simulate the real behavior of the manipulator, we have introduced backlash, gear runout and joint flexibility.

For the *training stage* of the neural network, 80 measurement lines, all parallel to the y_o-axis, were selected. These lines are inside the largest cube contained in the spherical workspace of this robot. Each line has a length of 0.4 m with 9 different measurement points located on it. During the training, the end effector is placed on each of the 720 points and the difference between the real position and the nominal one is measured. For each point, two different payloads are considered (M=0kg and M=1kg). The total number of measures for the training set is therefore 1440. Each element of the training set consists of an input pattern specifying the angular configuration and payload of the robot and of an output pattern giving the positioning error. Once the training stage is finished, a *validation phase* is performed in order to test and to quantify the performances of the neural network. For this, we have introduced several validation lines parallel to the training lines and inside the reference cube, and other lines with payloads different from the training ones to assess the *interpolation* and *generalization capabilities* of the network. The number of elements used for the interpolation and generalization test are 64 and 36, respectively.

Figure 2 (dashed line) gives a typical positioning error profile along a "demonstration" line, perpendicular to the training lines. Effects of runout (oscillations) and backlash (discontinuity) are clearly observed.

Before any correction, the *mean positioning error* (r.m.s.) measured over the training set is about 1.0mm. Note that the mean value of the error (r.m.s.) is defined by :

$$\bar{e} = \sqrt{\frac{1}{N} \sum_{m=1}^{N} \sum_{i=1}^{3} e_{m,i}^2}$$

where N is the number of measurement points and $e_{m,i}$ is the ith component of the positioning error at the mth point.

The procedure described in section III was applied for developing a good neural network architecture. The resulting structure was composed of two hidden layers with respectively 16 and 4 units. After learning, the *mean estimation error* (difference between the real positioning error and the output of the network) was of the order of 0.2mm, both on the training lines and on the validation lines, as illustrated by Table 1 (second column). For these cases, the network shows effectively good interpolation and generalization capabilities. Figure 2 (continous line) gives the estimated positioning error along the "demonstration" line; it can be observed that the network has considered runout effects as noise and has smoothed over the real profile. The introduction of auxiliary inputs (such as $\cos(n\theta_i)$ and $\sin(n\theta_i)$, where n is the reduction ratio of the gears) is likely to help the effective learning of this kind of effect.

Table 1 - Mean Errors for Training Set and Test Set

	Positioning Error Without Correction	Estimation Error	Positioning Error After Correction
Training Set (1440 points)	1.10 mm	0.22 mm	0.24 mm
Test Set, Generalization (36 points)	1.08 mm	0.22 mm	0.24 mm
Test Set, Interpolation (64 points)	1.07 mm	0.23 mm	0.25 mm

Finally, we have implemented the two compensation techniques presented in section IV. Both have been tested on the training lines and the validation lines, and give very similar results, as expected from eq. 6 and eq. 9. Figure 3 presents the positioning error before and after correction along the demonstration line. The third column of table 1 gives the r.m.s. positioning error after compensation following the false target technique. Comparison between column 1 and 3 shows that an improvement of a factor 5 in the accuracy can be achieved by the present technique. On the other hand, comparison between column 2 and 3 shows that the remaining positioning error after correction is mainly due to the estimation error of the neural network.

VI. CONCLUSION

A new technique for the calibration of robot based on a neural network approach for the identification of non-geometrical errors was presented. After learning of the network and correction by a false target technique, the positioning error can be reduced up to 20% of the initial error. Although the neural network seems to have sufficiently good generalization and interpolation capabilities, it fails to discover and to learn high-frequency effects such as runout. Further work must focus on this problem (for example, by considering the introduction of auxiliary inputs to the network) and address real experiments on a physical robot.

REFERENCES

[1] Z.S. Roth, B.W. Mooring and B. Ravani (1987). An Overview of Robot Calibration, *IEEE Journal of Robotics and Automation*, **RA-3**, 377-384.
[2] W.K. Veitschegger and C.H. Wu (1986). Robot accuracy analysis based on kinematics, *IEEE Journal of Robotics and Automation*, **RA-2**, 171-180.
[3] C.H. An, C.G. Atkeson and J.M. Hollerbach (1988). *Model-based control of a robot manipulator*, MIT Press, Cambridge, MA, Chapter 3, 49-64.
[4] J.M. Renders, E. Rossignol, M. Becquet and R. Hanus (1991). Kinematic Calibration and Geometrical Parameter Identification for Robots, to appear in *IEEE Journal of Robotics and Automation*.
[5] D. Payannet, M.J. Aldon and A. Liégeois (1985). Identification and Compensation of mechanical errors for industrial robots, *Proc. 15th ISIR*, 857-864.
[6] H.W. Stone and A.S. Sanderson (1987). A prototype arm signature identification system,

Proc. IEEE Int. Conf. Robotics and Automation, 175-182.
[7] D.E. Whitney, C.A. Lozinski and J.M. Rourke (1986). Industrial robot forward calibration method and results, *ASME Journal of Dynamic Systems, Measurements and Control,* **108**, 1-8.
[8] R.P. Judd and A.B. Knasinski (1987). A technique to calibrate industrial robots with experimental verification, *Proc. IEEE Int. Conf. Robotics and Automation,* 351-357.
[9] D.E. Rumelhart, G.E. Hinton and R.J. Williams (1986). Learning representations by back-propagating errors, *Nature,* **323**, 533-536.
[10] R. Hecht-Nielsen (1990). Theory of the backpropagation neural network, *Proc. Int. Joint Conf. on Neural Networks,Vol. I,*593-611.
[11] D.C. Plaut, S.J. Nowlan and G.E. Hinton (1986). Experiments on learning by back propagation, Technical Report CMU-CS-86-126, Carnegie-Mellon University.

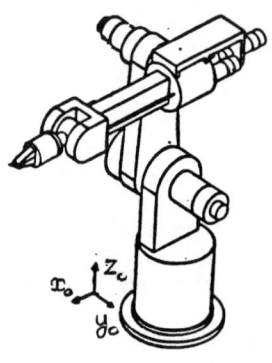

Fig.1. ULB-Manipulator (6 d.o.f.)

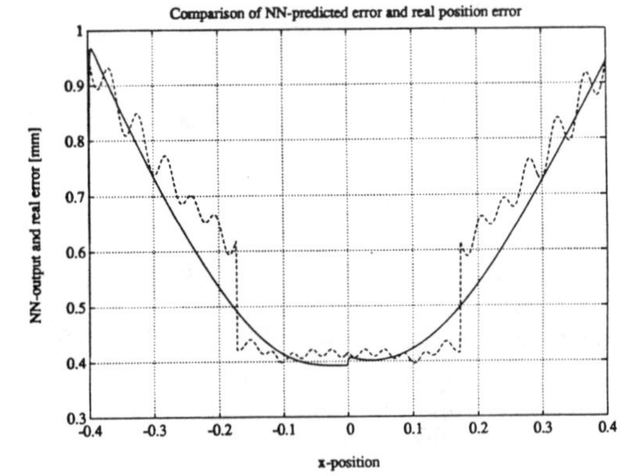

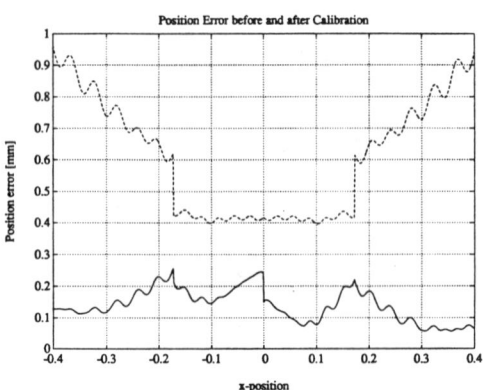

Fig.2. Comparison of NN-predicted error and real position error

Fig.3. Positioning error before and after calibration.

THE USE OF SKEW-SYMMETRIC CARTESIAN TENSORS IN DESCRIBING ORIENTATIONS AND INVARIANTS OF SPATIAL ROTATIONS[†]

C. A. Balafoutis and R. V. Patel
Electrical and Computer Engineering Department
Concordia University, Montreal, Quebec
CANADA H3G 1M8

ABSTRACT

In this paper, we provide a brief analysis of second-order skew-symmetric Cartesian tensors and present some of their applications in the analysis of spatial rotations. In particular by exploring various relationships between second-order skew-symmetric Cartesian tensors and their vector invariants, we provide a number of important tensor identities which enable us to manipulate effectively (and thus simplify) other complex tensor equations. Also, based on relatively oriented skew-symmetric second-order Cartesian tensors, we provide an analysis for the orientations of spatial rotations and derive some important formulations for their axes and the angles of rotation.

1. INTRODUCTION

One of the most interesting and elusive ideas in geometry and linear algebra is that of *orientation*. In geometry, we handle the concept of orientation by using "arrows". In linear algebra, orientation can be defined in a number of ways. For example, we can use a definite order of the unit vectors of a basis, we can use the sign (plus or minus) of a determinant function, or we can even use expressions from nature language such as: *positive* or *negative, clockwise* or *counterclockwise*, and *right-handed* or *left-handed*. In particular, the expressions "right-handed" or "left-handed" are universally used to express the orientation of spatial rotations. For spatial rotations, this is very convenient since the algebra of their classical descriptions (such as real 3×3 *orthogonal* matrices or *quaternions*) does not distinguish between the two possible orientations of a spatial rotation [1-5]. However, the use of expressions from nature language impose restriction on the analysis of spatial rotations since these expressions are not part of a mathematical language. Obviously, to avoid using these expressions we must be able to describe the orientation of a spatial rotation in a mathematical (geometric or algebraic) manner. Such a description will be useful in many practical applications in fields such as robotics, computer vision, computer graphics, and other engineering disciplines where efficient means of representing and applying *oriented* spatial rotations is necessary.

[†] This research was supported by NSERC of Canada under Grant OGP001345 and by the Institute of Robotics and Intelligent Systems (IRIS) under the Networks of Centres of Excellence Program of the Government of Canada.

Mathematically, in an abstract algebraic approach, we can define [2] orientation by using a determinant function and factor spaces. However, although this approach allows one to define oriented angles in a plane, it has limitations when it is extended into three dimensional (3-D) space. Alternatively, in a geometric approach, we can use multi-vectors and their geometric algebra [3]. In this approach, oriented rotations are defined simply as multi-vectors, called *spinors*. In this approach, we face the limitation that the action of a spinor on a vector is the same as that of a unit quaternion. Therefore, in applications it is not possible to take into consideration the orientation of spinors. In this paper, based on second order Cartesian tensor analysis, we provide new physical insight into the orientation of 3-D Euclidean space and spatial rotations. In particular, we use relatively oriented skew-symmetric Cartesian tensors to define orientation in 3-D Euclidean space, and based on this, we analyze the orientation of spatial rotations. Also, using Cartesian tensor analysis, we provide some new formulations for the axis and angle of a spatial rotation tensor.

The layout of this paper is as follows. In Section 2, we review some relevant facts from Cartesian tensor analysis. In Section 3, we define a number of algebraic tensor identities and in Section 4, we use Cartesian tensor analysis for the study of spatial rotations.

2. Second Order Cartesian Tensors

2.1 Basic Tensor Operations

As is well known [6-9], the set of second-order Cartesian tensors, together with the algebraic operations of tensor addition and scalar multiplication, constitutes a vector space over a scalar field. However, besides the addition (or subtraction) of tensors and their scalar multiplication, the following algebraic operations are well defined [6,7].

Tensor product: Let $\mathbf{T}$ and $\mathbf{S}$ be two second-order tensors whose components, referred to an orthogonal Cartesian coordinate system, are t_{ij} and s_{kl}. Then the 3^4 scalars

$$u_{ijkl} = t_{ij}s_{kl} \quad \Leftrightarrow \quad \mathbf{U} = \mathbf{T} \otimes \mathbf{S} \tag{2.1}$$

form the components of a tensor $\mathbf{U}$, say, of order 4 which is called the tensor product of $\mathbf{T}$ and $\mathbf{S}$.

Dot product: Let T and S be coordinate matrices, relative to the same coordinate system, of second-order tensors $\mathbf{T}$ and $\mathbf{S}$, respectively. Then the equation

$$U_{ij} = T_{il}S_{lj} \quad \Leftrightarrow \quad \mathbf{U} = \mathbf{T} \cdot \mathbf{S} \equiv \mathbf{TS} \tag{2.2}$$

defines the components of the second-order tensor $\mathbf{U}$ which we call the dot product or multiplication of $\mathbf{T}$ and $\mathbf{S}$.

Double dot or inner product: The double dot product of two second-order tensors $\mathbf{T}$ and $\mathbf{S}$ is defined in terms of the trace operator by the equation

$$\mathbf{T} : \mathbf{S} = tr(\mathbf{T} \cdot \mathbf{S}) \tag{2.3}$$

Left and right dot product: Let $\mathbf{T}$ be a second-order tensor and $\mathbf{v}$ be a vector, with coordinate matrices T and v relative to coordinate systems $\{\mathbf{e}_i \otimes \mathbf{e}_j\}$ and $\{\mathbf{e}_i\}$ respectively. Then the equations

$$u_i = T_{ij}v_j \quad \text{and} \quad w_i = v_j T_{ji} \quad \Leftrightarrow \quad \mathbf{u} = \mathbf{T} \cdot \mathbf{v} \text{ and } \mathbf{w} = \mathbf{v} \cdot \mathbf{T} = \mathbf{T}^T \cdot \mathbf{v} \tag{2.4}$$

define, respectively, the right dot product (or post-multiplication) and the left dot product (or pre-multiplication) of $\mathbf{T}$ and $\mathbf{v}$.

2.2 Vector Invariants, Dual Vectors and Dual Tensors

As is well known [1,6,7], every Cartesian tensor **T**, defined on a 3-D Euclidean space, has a *vector invariant* which is defined component-wise by the equation

$$t_i = \frac{1}{2}\varepsilon_{ijk} t_{kj} \tag{2.5}$$

where ε_{ijk} and t_{kj} are the components of the *Levi-Civita* tensor $\mathcal{E}$ and the tensor **T** respectively. We denote the vector invariant **t** of a tensor **T** by writing

$$\mathbf{t} = vect(\mathbf{T}) \equiv vect(\mathbf{T}_-) \tag{2.6}$$

where, $vect(\cdot)$ denotes the *tensor valued vector operator* which is defined by (2.5) and $\mathbf{T}_-$ is the skew-symmetric component in the Cartesian decomposition of **T**.

Observe that the *vect* operator is not 1–1. However, it is easy to see that if we restrict the domain of the *vect* operator on the linear space of the skew-symmetric tensors we get a 1–1 operator. Based on this, we can give the following definition [6,7].

Definition 2.1: The *right (left) dual operator* is a 1–1 *tensor-valued tensor operator* which has the following property: when this operator is evaluated at a tensor of order one (i.e., a vector) we get a skew-symmetric tensor of order two, and when it is evaluated at a second-order skew-symmetric tensor we get a tensor of order one. We define the action of the right (left) dual operator on a vector or a skew-symmetric tensor, component-wise, using the following 1–1 correspondence

$$\begin{bmatrix} u_1 \\ u_2 \\ u_3 \end{bmatrix} \Leftrightarrow \begin{bmatrix} 0 & -u_3 & u_2 \\ u_3 & 0 & -u_1 \\ -u_2 & u_1 & 0 \end{bmatrix} \quad \left(\begin{bmatrix} u_1 \\ u_2 \\ u_3 \end{bmatrix} \Leftrightarrow \begin{bmatrix} 0 & u_3 & -u_2 \\ -u_3 & 0 & u_1 \\ u_2 & -u_1 & 0 \end{bmatrix} \right) \tag{2.7}$$

Symbolically, we denote the action of the right (left) dual operator on a vector or a skew-symmetric tensor by writing

$$\tilde{\mathbf{u}} \triangleq dual(\mathbf{u}) \quad (\overline{\mathbf{u}} \triangleq (\mathbf{u})dual) \tag{2.8a}$$

and

$$\mathbf{u} \triangleq dual(\tilde{\mathbf{u}}) \quad (\mathbf{u} \triangleq (\overline{\mathbf{u}})dual) \tag{2.8b}$$

Remark 2.1: As is well known [9], the components of a skew-symmetric tensor change sign when the "handedness" of the coordinate system is changed, from right-handed to left-handed, say. This is also true for other quantities whose orientation or sense is established by *convention*, such as the familiar *axial vectors*. Therefore, it is obvious that the dual tensor and its vector invariant are closely related to the vector cross product operation which produces axial vectors in three dimensional vector analysis. In particular, it is easy to see that the following equalities are true.

$$\mathbf{w} = \mathbf{u} \times \mathbf{v} = \tilde{\mathbf{u}} \cdot \mathbf{v} \tag{2.9}$$

An important consequence of this equation is that the *Lie algebra* which is introduced in E^3 by the vector cross product operation [10] is equivalent to the Lie algebra of 3×3 skew-symmetric tensors. In particular, it is not difficult to see that these two Lie algebras are *isomorphic*, with the isomorphism provided by the equation

$$dual(\mathbf{u} \times \mathbf{v}) = dual(\tilde{\mathbf{u}}\mathbf{v}) = \tilde{\mathbf{u}}\tilde{\mathbf{v}} - \tilde{\mathbf{v}}\tilde{\mathbf{u}} = [\tilde{\mathbf{u}}, \tilde{\mathbf{v}}] \tag{2.10}$$

3. CARTESIAN TENSORS IDENTITIES

By exploring the relationships between second-order Cartesian tensors and their vector invariants we can prove [7] the following identities.

$$\tilde{u}v = -\tilde{v}u = -v\cdot\tilde{u} = v\cdot\tilde{u} = \tilde{v}\cdot u = -\tilde{v}\cdot u \tag{3.1}$$

$$\tilde{u}u = u\tilde{u} = 0 \tag{3.2}$$

$$\tilde{u}\tilde{v} = v\otimes u - u\cdot v\,1 \tag{3.3}$$

$$v\otimes u = \tilde{u}\tilde{v} + (v\cdot u)1 \tag{3.4}$$

$$dual(\tilde{u}\cdot v) = [\tilde{u},\tilde{v}] = \tilde{u}\tilde{v} - \tilde{v}\tilde{u} = v\otimes u - u\otimes v \tag{3.5}$$

$$dual(\tilde{a}\tilde{b}v) = [\tilde{a},[\tilde{b},\tilde{v}]] = \tilde{b}(a\cdot v) - \tilde{v}(a\cdot b) = 0 \iff a\perp b, a\perp v \tag{3.6}$$

$$tr[\tilde{u}\tilde{v}] \equiv \tilde{u}:\tilde{v} = -2u\cdot v, \tag{3.7}$$

Now, using these identities, we prove the following propositions.

Proposition 3.1: The dual tensors $\tilde{u}$ and $\tilde{v}$ satisfies the following equations.

$$a)\quad \tilde{v}\tilde{u}\tilde{u} + \tilde{u}\tilde{u}\tilde{v} = -(u\cdot u)\tilde{v} - (v\cdot u)\tilde{u} \tag{3.8}$$

$$b)\quad \tilde{v}\tilde{v}\tilde{u}\tilde{u} - \tilde{u}\tilde{u}\tilde{v}\tilde{v} = (u\cdot v)[\tilde{u},\tilde{v}] \tag{3.9}$$

$$c)\quad \tilde{u}\tilde{v}\tilde{u} = -(v\cdot u)\tilde{u} \tag{3.10}$$

Proof: Using equations (3.1) and (3.3), the left-hand side of (3.8) can be transferred into the right-hand side as follows:

$$\tilde{v}\tilde{u}\tilde{u} + \tilde{u}\tilde{u}\tilde{v} = \tilde{v}(u\otimes u - u\cdot u\,1) + \tilde{u}(v\otimes u - v\cdot u\,1) = (\tilde{v}u)\otimes u - (u\cdot u)\tilde{v} + (\tilde{u}v)\otimes u - (v\cdot u)\tilde{u}$$

$$= -(\tilde{u}v)\otimes u - (u\cdot u)\tilde{v} + (\tilde{u}v)\otimes u - (v\cdot u)\tilde{u} = -(u\cdot u)\tilde{v} - (v\cdot u)\tilde{u}$$

To prove equation (3.9), we pre- and post-multiply equation (3.8) by $\tilde{v}$ and subtract one of the resulting equations from the other. Then, after some straightforward manipulations we obtain equation (3.9). Equation (3.10) results from the following manipulations

$$\tilde{u}\tilde{v}\tilde{u} = \tilde{u}(u\otimes v - (v\cdot u)1) = (\tilde{u}u)\otimes\tilde{v} - (v\cdot u)\tilde{u} = -(v\cdot u)\tilde{u} \qquad \square$$

Proposition 3.2: The vector cross product $a\times b$ is related to the vector invariant of the tensor $\tilde{a}\tilde{b}$, $a\otimes b$ and $[\tilde{a},\tilde{b}]$ by the following equations

$$a\times b = 2vect(\tilde{a}\tilde{b}) \tag{3.11}$$

$$= 2vect(b\otimes a) \tag{3.12}$$

$$= vect[\tilde{a},\tilde{b}] \tag{3.13a}$$

$$= dual[\tilde{a},\tilde{b}] \tag{3.13b}$$

Proof: For the vector invariant of the tensor $\tilde{a}\tilde{b}$ we have

$$vect(\tilde{a}\tilde{b}) = \frac{1}{2}dual\left\{[\tilde{a}\tilde{b} - (\tilde{a}\tilde{b})^T]\right\} = \frac{1}{2}dual\left\{[\tilde{a}\tilde{b} - \tilde{b}\tilde{a}]\right\} = \frac{1}{2}dual\left[dual(\tilde{a}b)\right] \tag{3.14}$$

Now, since the dual operation is the inverse of itself, equations (3.14) and (2.9) imply (3.11). Equation (3.12) follows from (3.11) and (3.3) and, finally, equation (3.13) follows from (3.11) and (3.5). $\qquad\square$

4. APPLICATIONS OF CARTESIAN TENSOR ANALYSIS TO SPATIAL ROTATIONS

4.1 Oriented Three-Dimensional Euclidean Spaces

It is well known from spatial intuition that the 3-D Euclidean space E^3 admits two possible orientations, namely, the positive and the negative, and also that *relative* orientations can be defined between its various subspaces. Mathematically, in an abstract algebraic approach, we can define [2] orientation by using a determinant function and factor spaces. Alternatively, in a geometric approach, we can use multi-vectors and their geometric algebra [3] or we can use a more traditional vector approach relying on a basis $\{e_1, e_2, e_3\}$ of E^3 and the equations

$$e_1 \times e_2 = \pm e_3 \quad \text{and} \quad e_1 \cdot e_2 \times e_3 = \pm 1 \tag{4.1}$$

We shall demonstrate here that orientation can also be described by using Cartesian tensor analysis. As is well known [6], Cartesian tensor analysis can be used to describe planes which are oriented relative to their normal vectors. In particular, as has been demonstrated in [6], given a vector $\mathbf{u}$ we can use its right dual tensor $\tilde{\mathbf{u}}$ to denote the plane which is normal to $\mathbf{u}$ and has been oriented relative to $\mathbf{u}$ with the positive or right-handed orientation. Similarly, the left dual tensor $\bar{\mathbf{u}}$ can be used to describe the plane which is normal to $\mathbf{u}$ and has the negative or left-handed orientation relative to $\mathbf{u}$. Now, observe that a vector $\mathbf{u}$ and its normal plane define the space E^3 which can be assumed to be oriented when the normal plane is oriented relative to $\mathbf{u}$. Thus, based on this vector-tensor approach, we can describe mathematically the orientation of E^3 in a parametric form by the equation

$$e_1 \cdot \tilde{e}_2 \cdot e_3 = +1 \quad \text{and} \quad e_1 \cdot \bar{e}_2 \cdot e_3 = -1 \tag{4.2}$$

or, in a nonparametric form, by considering the following oriented direct sum decompositions

$$E^3_{rh} = \mathbf{u} \oplus \tilde{\mathbf{u}} \quad \text{and} \quad E^3_{lh} = \mathbf{u} \oplus \bar{\mathbf{u}} \tag{4.3}$$

where E^3_{rh} and E^3_{lh} denote, respectively, right- and left-handed oriented spaces.

Remark 4.1 : Observe that the 3-D Euclidean space can be divided into two *half* spaces and in each one of them we can introduce both the right-handed and the left-handed orientations. Also, observe that based on Definition 2.1, we can say that the skew-symmetric tensor $\bar{\mathbf{u}}$ ($\tilde{\mathbf{u}}$) is positively (negatively) oriented relative to the negative vector $-\mathbf{u}$. Therefore, consistently with spatial intuition, we can define two equivalent right-handed or left-handed orientations in E^3 based on skew-symmetric Cartesian tensors.

4.2 On the Orientation of Spatial Rotation Tensors

As is well known [5], there is no 1-1 global description of a spatial rotation tensor $\mathbf{R}$. Popular descriptions of $\mathbf{R}$, such as those based on 3×3 *orthogonal matrices* or *unit quaternions* are "unoriented" 2-1 descriptions [1-5]. From the foregoing, it is obvious that for a 1-1 description of $\mathbf{R}$ we need to introduce some form of constraints on $\mathbf{R}$. There are two equivalent approaches for imposing these constraints. One is to characterize the orientation of $\mathbf{R}$ and the other is to impose limitations on the domain of the angle or the orientation of the unit vector along the axis of rotation.

Following the first approach, to distinguish between the two possible descriptions of $\mathbf{R}$, we introduce [4] the "active" or the "passive" interpretation of $\mathbf{R}$ or, equivalently, we use expressions from natural language such as "clockwise" or "counterclockwise". However, since the algebra which describes $\mathbf{R}$ remains the same no matter which point of view is followed, the distinction between the two orientations is not clear. In order to have a better physical insight, which may lead to a more clear description of orientation, we shall use the idea of relatively oriented skew-symmetric tensors. Let us consider the following equation

$$\mathbf{R} \equiv \mathbf{R}(\mathbf{u}, \theta) = \mathbf{u} \otimes \mathbf{u} + \left[1 - \mathbf{u} \otimes \mathbf{u}\right] cos(\theta) + \tilde{\mathbf{u}} sin(\theta), \quad (4.4)$$

which defines [1] a spatial rotation $\mathbf{R}$ in terms of its axis and the angle of rotation. Now, observe that $cos(\theta) = cos(2\pi-\theta)$ and $sin(\theta) = -sin(2\pi-\theta)$, which implies the equality $\tilde{\mathbf{u}} sin(\theta) = \tilde{\mathbf{u}} sin(2\pi-\theta)$. Therefore, we have $\mathbf{R}(\mathbf{u}, \theta) = \mathbf{R}(-\mathbf{u}, 2\pi-\theta)$ i.e., that equation (4.4) provides a 2–1 description for $\mathbf{R}$. From the foregoing, $\mathbf{R}$ alone does not provide all the necessary information needed for a unique description. Now, let us analyze the action of $\mathbf{R}$ on a vector $\mathbf{v}$ by considering the product $\mathbf{R} \cdot \mathbf{v}$. Observe that since the tensors $\mathbf{u} \otimes \mathbf{u}$ and $1 - \mathbf{u} \otimes \mathbf{u}$ are projection tensors, the orientation of the resulting rotation depends only on the product $sin(\theta)\tilde{\mathbf{u}} \cdot \mathbf{v}$. This implies that the orientation of a rotation tensor $\mathbf{R}$ is associated exclusively with the skew-symmetric part of $\mathbf{R}$. Therefore, the problem of describing an oriented rotation is similar to that of describing an oriented skew-symmetric Cartesian tensor. Now, based on equation (2.9) and relatively oriented skew-symmetric tensors we have the following observations.

Remark 4.2: The dot product of relatively oriented skew-symmetric tensors with vectors can be used to define *oriented rotations* in 3-D Euclidean space. In particular, since for skew-symmetric tensors every vector is orthogonal to its image-vector, the product $\tilde{\mathbf{u}} \cdot \mathbf{v}$ involves a 90° right-handed rotation and the product $\mathbf{v} \cdot \tilde{\mathbf{u}}$ involves a 90° left-handed rotation.

From this remark it follows that for a complete and clear characterization of the orientation of $\mathbf{R}$ we need to define two things: the relative orientation of its skew-symmetric part (right- or left-handed) and the *order* with which the tensor $\mathbf{R}$ acts on a vector $\mathbf{v}$. Thus, without specifying the order of the dot product between the tensor $\mathbf{R}$ and the vector $\mathbf{v}$ the description of the orientation of $\mathbf{R}$ is not complete.

In the second approach, to resolve the 2–1 ambiguity in the description of a rotation tensor $\mathbf{R}$, we usually define [1] the orientation of the unit vector $\mathbf{u}$ so that the angle θ will lie in a restricted domain. Thus, for example, if we assume that the angle of rotation θ is restricted in the interval $0 \le \theta < \pi$, then we have

$$sin(\theta) = \| vect(\mathbf{R}) \| = \| \mathbf{u} sin(\theta) \| \quad \text{and} \quad cos(\theta) = \frac{1}{2}\left[tr[\mathbf{R}] - 1\right] \quad (4.5)$$

which allows for a unique characterization of $\mathbf{R}$ in this restricted domain of θ. However, there are many applications, specially in robotics, where the vector $\mathbf{u}$ is known, say, from sensor based information and the domain of θ is larger that $[0, \pi)$. In applications of this nature, it is obviously desirable to be able to work with no restrictions on the angle θ. Therefore, as an alternative, we provide the following lemma which removes the constraints from the angle of rotation θ.

Lemma 4.1: If the vector $\mathbf{u}$ along the axis of a right-handed rotation tensor $\mathbf{R}$ is given, then the angle of rotation θ for $\theta \in [0, 2\pi)$ can be computed by using the equations

$$sin(\theta) = -\frac{1}{2\|\mathbf{u}\|}\mathbf{R} : \tilde{\mathbf{u}} \quad \text{and} \quad cos(\theta) = \frac{1}{2}\left[tr[\mathbf{R}] - 1\right] \quad (4.6)$$

Proof: The equation which defines the $cos(\theta)$ is well known [1]. To prove the equation for $sin(\theta)$ we proceed as follows. Using equation (3.4), equation (4.4) can be written as

$$\mathbf{R} = 1 + \tilde{\mathbf{u}}\tilde{\mathbf{u}}(1 - cos(\theta)) + \tilde{\mathbf{u}} sin(\theta) \quad (4.7)$$

from where, by using the definition of the double dot product (i.e., eq. (2.3)), we have

$$\mathbf{R}:\tilde{\mathbf{u}} = tr[\mathbf{R}\cdot\tilde{\mathbf{u}}]$$
$$= tr[\tilde{\mathbf{u}}\cdot\tilde{\mathbf{u}}]\sin(\theta) \quad \text{(by (3.10))}$$
$$= -2\|\mathbf{u}\|\sin(\theta) \quad \text{(by (3.7))}$$

which then gives equation (4.6). □

4.3 Some Remarks on the Axis and the Angle of a Rotation Tensor

As another demonstration of Cartesian tensor analysis in the study of spatial rotations, we provide formulations which define the axis of rotation when the rotation tensor is defined based on two well known decompositions.

First, as is well known [1], a rotation tensor $\mathbf{R}$ can be defined as the product of two *reflections* $\mathbf{H}_1$ and $\mathbf{H}_2$. It is also known [1] that a reflection $\mathbf{H}_i$ about a plane perpendicular to the unit vector $\mathbf{h}_i$ satisfies the tensor equation

$$\mathbf{H}_i = 1 - 2\mathbf{h}_i \otimes \mathbf{h}_i \quad (4.8)$$

Therefore, when a rotation tensor $\mathbf{R}$ is defined as the product of two reflections, say $\mathbf{H}_1$ and $\mathbf{H}_2$, we can write the following equation

$$\mathbf{R} = \mathbf{H}_1\mathbf{H}_2 = 1 - 2\mathbf{h}_1\otimes\mathbf{h}_1 - 2\mathbf{h}_2\otimes\mathbf{h}_2 + 4(\mathbf{h}_1\cdot\mathbf{h}_2)\mathbf{h}_1\otimes\mathbf{h}_2 \quad (4.9)$$

where $\mathbf{h}_1$ and $\mathbf{h}_2$ are unit vectors with an angle $\phi \in [0,\pi)$ between them. In this case, we can state the following lemma.

Lemma 4.2 : If a rotation tensor $\mathbf{R}$ satisfies equation (4.9) then its angle θ and the unit vector along its axis satisfy the equations

$$\theta = 2\phi \quad \text{and} \quad \mathbf{u} = \lambda \mathbf{h}_2 \times \mathbf{h}_1 \quad (4.10)$$

respectively, where λ is a scalar.

Proof: Let $\mathbf{r}$ be the vector invariant of $\mathbf{R}$, i.e., let $\mathbf{r} = vect(\mathbf{R})$. Then, since the vector invariant of a symmetric tensor is zero, we have from (4.9)

$$\mathbf{r} = 4(\mathbf{h}_1\cdot\mathbf{h}_2)vect(\mathbf{h}_1\otimes\mathbf{h}_2)$$
$$= 2(\mathbf{h}_1\cdot\mathbf{h}_2)\mathbf{h}_2\times\mathbf{h}_1 \quad \text{(by eq. (3.13))} \quad (4.11)$$

Now, for unit vectors $\mathbf{h}_1$ and $\mathbf{h}_2$ with angle ϕ between them, we can write

$$\mathbf{h}_1\cdot\mathbf{h}_2 = \cos(\phi) \quad \text{and} \quad \mathbf{h}_2\times\mathbf{h}_1 = \mathbf{v}\sin(\phi) \quad (4.12)$$

where $\mathbf{v}$ is a unit vector perpendicular to both $\mathbf{h}_1$ and $\mathbf{h}_2$. Therefore, using equations (4,4), (4.11) and (4.12), we can write

$$\mathbf{u}\sin(\theta) - 2\cos(\phi)\sin(\phi)\mathbf{v} - \sin(2\phi)\mathbf{v}$$

which implies that (4.10) must be true. □

Another important decomposition of a rotation tensor $\mathbf{R}$, which has eigenvalues different from -1, is provided [1] by *Cayley's* formula

$$\mathbf{R} = (1-\mathbf{S})(1+\mathbf{S})^{-1} = (1-\tilde{\mathbf{s}})(1+\tilde{\mathbf{s}})^{-1} \quad (4.13)$$

where $\mathbf{S} \equiv \tilde{\mathbf{s}}$ is a skew-symmetric tensor. Based on this decomposition we can state the following lemma.

Lemma 4.3: Let a rotation tensor **R** satisfy equation (4.13) then its axis of rotation satisfies the equation

$$\mathbf{r} = -\frac{2}{1 + \|\mathbf{s}\|^2}\mathbf{s} \qquad (4.14)$$

i.e., the axis of rotation is parallel to the dual vector of the skew-symmetric tensor **S**.

Proof: Based on the expressions

$$\mathbf{R} + \mathbf{1} = (1 - \tilde{\mathbf{s}})(1 + \tilde{\mathbf{s}})^{-1} + \mathbf{1} = (1 - \tilde{\mathbf{s}})(1 + \tilde{\mathbf{s}})^{-1} + (1 + \tilde{\mathbf{s}})(1 + \tilde{\mathbf{s}})^{-1} = 2(1 + \tilde{\mathbf{s}})^{-1}$$

and the fact (which can be easily verified) that $(1 + \tilde{\mathbf{s}})^{-1} = 1 - \dfrac{\tilde{\mathbf{s}} - \tilde{\mathbf{s}}^2}{1 + \|\mathbf{s}\|^2}$, we can write $\mathbf{R} = 1 - \dfrac{2(\tilde{\mathbf{s}} - \tilde{\mathbf{s}}^2)}{1 + \|\mathbf{s}\|^2}$. From this expression for **R**, equation (4.14) is easily obtained since $\tilde{\mathbf{s}}^2$ is a symmetric tensor. □

5. CONCLUSIONS

In this paper, by exploring relationships between second-order skew-symmetric Cartesian tensors and their vector invariants, we have stated and proved a number of important tensor identities. These identities enable one to manipulate effectively (and thus simplify) complex tensor equations. Also, by using the *dual* operators and a geometric characterization of second-order skew-symmetric Cartesian tensors, the orientation of spatial rotations has been analyzed and new formulations for the angle and axis of rotation have been derived. These formulations and the characterization of oriented rotations are currently under investigation in relation to robot kinematics (cyclic motion of kinematically redundant manipulators).

6. REFERENCES

[1] J. Angeles, *Rational Kinematics*, Springer-Verlag, New York, 1988.
[2] W. H. Greub, *Linear Algebra*, Springer-Verlag, New York, 1967.
[3] D. Hestenes, *New Foundations of Classical Mechanics*, D. Reidel Publishing Company, Dordrecht, Holland, 1986.
[4] H. Goldstein, *Classical Mechanics*, Reading, MA:, Addison Wesley, 1980.
[5] J. Stuelpnagel, "On the Parametrization of the Three-Dimensional Rotation Group", *SIAM REVIEW*, Vol. 6, No. 4, pp. 422-430, October 1964.
[6] C. A. Balafoutis, and R. V. Patel, "A Cartesian Tensor Methodology for the Study of Classical Newtonian Dynamics, Part I : Cartesian Tensor Analysis," in *Proc. 10th Symposium on Engineering Applications of Mechanics*, pp. 55-60, Kingston, Ontario, May 27-30, 1990.
[7] C. A. Balafoutis and R. V. Patel, *Dynamic Analysis of Robot Manipulators: A Cartesian Tensor Approach,* Kluwer Academic Publishers, Boston, MA, 1991.
[8] A. M. Goodbody, *Cartesian Tensors: With Applications to Mechanics, Fluid Mechanics and Elasticity*, Ellis Horwood, England, 1982.
[9] A. Lichnerowicz, *Elements of Tensor Calculus*, Methuen, London, 1962.
[10] R. Gilmore, *Lie Groups, Lie Algebras, and Some of Their Applications*, John Wiley & Sons, New York, 1974.

AN AUTOMATIC VERIFICATION OF SIMPLIFIED ROBOT MODELS

Krzysztof SWIDER

Department of Electrical Engineering
Technical University Rzeszów, Poland

Abstract - The computer program ALMORO has been implemented by the author in 1987 to generate symbolically the forward solution and the complete Lagrangian dynamic robot model. The derived *closed-form* of dynamic equations could become even extremally complex - especially for more complicated manipulators. At the same time model-based controllers require the *on-line* evaluation of the robot dynamics in the control law. This becomes computationally expensive and time-consuming task and often dictates the use of numerically simplified models. It also turned out that the simplification procedure applied without accounting some specific model properties (configuration dependence, inter-relationships etc.) can violate the fundamental physical principles and destroy the structure of the controller. The objective of the present paper is to demonstrate a method of model simplifications based on systematic approach to simplification process. The complete closed-form robot models are simplified automatically using a set of *symbolically generated simplification strategies*. The author has expanded the capabilities of ALMORO to include the operations necessary for systematic simplification procedure.

1. INTRODUCTION

The complexity of dynamic equations of robot manipulator motion causes a lot of interest in robot dynamics formulation methods [2,5,8]. The model applications using *on-line* evaluation of dynamic coefficients have led to development of efficient computational algorithms [2,5,7].

Sometimes we are interested in obtaining the better insight into the structure of manipulator. We can get it writing *closed-form* of dynamic equations with general terms - inertia-matrix, vector of centifugal and

Coriolis forces and gravity vector - separated. Moreover, it turned out that for *kinematically* and *dynamically simple* manipulators ([1]) the computational scheme based on closed-form equations is about 2 times more efficient then general Newton-Euler method. The formulation of closed-form equation however requires a fair amount of human effort. Therefore symbolic computations are often applied for automatic generation of robot models [3,6,9].

The general structure of robot dynamics space equations for robot manipulator with N DOF and symbolically generated models will be outlined in section 2. From control engineering point of view we try to obtain the *simplest* and *sufficiently adequate* model. The model simplicity means just reducing the number of necessary arithmetic operations in dynamics evaluation scheme. The model adequacy will be considered as (a) compliance with some modelling tolerance and (b) preserving fundamental physical principles of manipulator model. The sample procedure for numerical simplification of closed-form model will be discussed in section 3. The thesis of the paper - the *simplification strategies generation* will be developed in sections 4 and 5. The results are finally summarized in section 6.

2. ALGEBRAIC ROBOT MODELLING

The basic equation for the closed-form dynamic model of an open chain robot with N rigid links (DOF) is

$$D(q)\ddot{q} + C(q,\dot{q}) + G(q) = F(t) \quad (1)$$

It is the system of N coupled, non-linear second-order differential equations whose parameters are functions of the instantenous configuration in state space of generalized joint coordinates q and velocities $\dot{q}$. $F(t)$ is N-vector of actuating (motor) joint forces (torques). The detailed specification of inertial coefficient (NxN)-matrix $D(q)$, cetrifugal and Coriolis forces N-vector $C(q,\dot{q})$ and gravitational forces N-vector has been widely presented in robotic literature [6,8]. Their physical and mathematical properties are demonstrated in [13].

The formulation of closed-form robot dynamics equations is known as an arduous and time-consuming task which is prone to errors. A number of computer programs for symbolic generation of dynamic robot models has been worked out to free the engineer from routine and mundane tasks [3,6]. ALMORO - implemented by the author [9] generates symbolically the forward solution and complete robot dynamics model using Lagrangian formulation.

3. SIMPLIFIED ROBOT MODELS

The on-line evaluation of the computer model coefficients becomes computationally expensive and time consuming task. This often dictates the use of numerically simplified models in the controllers [12]. Lets consider a sample simplification procedure [14] used in [10].

Denoting the inertia-matrix closed-form coefficient as $d(q_1,\ldots,q_N)$ we have

$$d(q_1,\ldots,q_N) = \sum_{i=1}^{N} k_i \alpha_i(q_1,\ldots,q_N) \qquad (2)$$

where k_i are constants depending on the geometrical and mass manipulator parameters as well as the range of translational join coordinates. Since each α_i is a product of (i) some sines and cosines (and their powers) and of (ii) some normalized translational join coordinates, hence

$$|\alpha_i(q_1,\ldots,q_N)| \le 1 \quad \forall i \qquad (3)$$

Ordering terms in (2) so that

$$|k_1| \le |k_2| \le \ldots \le |k_M| = k_{max} \qquad (4)$$

we proceed with simplification algorithm as follows

1° Write inertial coefficient in (2) as

$$d(q_1,\ldots,q_N) = k_{max} \sum_{i=1}^{M} p_i \alpha_i(q_1,\ldots,q_N),$$

where $p_i \equiv k_i / k_{max}$, therefore $|p_i| \leq 1$, $\forall i$.

2° Define the modelling tolerance ε, $(0<\varepsilon<1)$.
3° Find the greatest integer $L \leq M$ for which
$$\sum_{i=1}^{L} |p_i| \leq \varepsilon$$

4° Define β_i as:
$$\begin{cases} p_i & \text{if } i>L \\ 0 & \text{if } i \leq L \end{cases}$$

5° Evaluate the simplified coefficient as:
$$\bar{d}(q_1,\ldots,q_N) = k_{max} \sum_{k=1}^{M} \beta_i \alpha_i (q_1,\ldots,q_N).$$

As prooved in [10] $k_{max} \varepsilon$ is the upper bound of simplification error.

The simplification algorithm 1°- 5° applied for sample manipulators [10,11] can destroy *positive-definiteness* - the fundamental property of inertia-matrix. It leads to significant problems in manipulator simulation, controllers design and trajetory planning. The numerical simplifications which ignore classical mechanics leads to models that do not characterize any real manipulator. This can violate fundamental physical principles and destroy the structure of the controller. The special attention, therefore, must be paid to development of simplification procedures preserving the fundamental properties of robot models.

Presenting the thesis of the paper a systematic approach to simplification process will be proposed in the next two sections.

4. SYSTEMATIC SIMPLIFICATION PROCEDURE

The numerical simplifications of inertia-matrix coefficients making by straightforward application of $1^\circ - 5^\circ$ lead to posible losing of matrix positive definiteness. The terms for removing from some matrix coefficient are determined according to a simple rule. The same rule is applied to simplify every individual coefficient without regarding the consequences for a model as a whole. Let us notice, however, that for some modelling tolerance ε there is not only one possible combination of the terms to neglect in coefficient under simplification. Let TE be a set of all the terms in some coefficient. By applying simplification algorithm for a given ε, TE is divided into two subsets:

TR - with the terms for removing from the coefficient and

TL - with the terms which should be left in the coefficient (TL=TE-TR).

In the case of algorithm in section 3 the members of TR are defined permanently for all simplified matrix elements as

$$T_i = k_i \alpha_i (q_1, \ldots, q_N), \quad i=1, 2, \ldots, L.$$

If the simplification destroys inertia matrix positive-definiteness, simplification process fails and the method must be assumed as not applicable. The simplification procedure can be intuitively generalized defining TR as

(i) any combination of terms

$$T_{i_j} = k_{i_j} \times \alpha_{i_j} (q_1, \ldots, q_N), \quad \text{where } 1 \leq i_j \leq M \quad \text{and} \quad \sum_{s=1}^{NR} |p_{i_s}| \leq \varepsilon$$

(NR - number of coefficients to remove) or

(ii) an empty set $\emptyset$ (no terms are neglected).

Assume, that NA is a number of inertia matrix elements affected by the simplification procedure. For any such element e_l (l=1...NA) a set containing admissible combinations of neglected terms can be generated:

$$M_l = \{\emptyset, TR_1, TR_2, \ldots\}$$

Moreover, the same should be done independently for each element e_l (l= 1...NA), and than all the M_l should be combined together as

$M_1 \times M_2 \times \ldots \times M_{NA}$ (x - denotes Cartesian product operation)
composing a set of possible *simplification strategies*.

In order to perform effectively the arduous and time-consuming operations, symbolic computations are incorporated in ALMORO (a) for automatic generation of simplification strategies and (b) to proceed with simplification process according to some strategy. The numerical evaluation of simplified coefficients should be than applied to verify the impact of simplifications on the fundamental properties of model parameters.

5. EXAMPLE

In this section the systematic simplification procedure wil be applied to generate a set of simplification strategies for inertia matrix of 3 DOF laboratory robot [4]. The manipulator kinematic structure is specified in Table 1 (standard robotic terminology is used [6,8]).

Table 1.

link	variable	θ	α	a	d
1	θ_1	θ_1	$90°$	0	0
2	θ_2	θ_2	0	0.3	0
3	θ_3	θ_3	0	0.3	0

The ALMORO generated inertia matrix coefficients are

$d_{11} = 2.849 - 1.067\sin^2\theta_2 + 0.321\cos\theta_2\cos(\theta_2 + \theta_3)$
$d_{21} = -0.093\sin\theta_2 + 0.008\sin(\theta_2 + \theta_3)$
$d_{31} = 0.008\sin(\theta_2 + \theta_3)$
$d_{22} = 4.078 + 0.446\cos\theta_3$
$d_{32} = 1.681 + 0.223\cos\theta_3$
$d_{33} = 1.681$

For $\varepsilon=0.4$ the following coefficients are affected by the simplification procedure

d_{32}: $M_1 = \{ \emptyset, \{0.223\cos\theta_3\} \}$

d_{22}: $M_2 = \{ \emptyset, \{0.446\cos\theta_3\} \}$
d_{21}: $M_3 = \{ \emptyset, \{0.008\sin(\theta_2+ \theta_3)\} \}$
d_{11}: $M_4 = \{ \emptyset, \{0.321\cos\theta_2\cos(\theta_2+ \theta_3)\} \{-1.067\sin^2\theta_2\} \}$

Than 24 simplification strategies are generated by Cartesian product $M_1 \times M_2 \times M_3 \times M_4$

Conserving space in the paper, we list only their representatives:

Strategy 1: $\{ \emptyset, \emptyset, \emptyset, \emptyset \}$ *(no simplifications)*

Strategy 2: $\{ \{0.223\cos\theta_3\}, \emptyset, \emptyset, \emptyset \}$ *(the term $0.223\cos\theta_3$ should be neglected in d_{32})*

...   ...

Strategy 24: $\{ \{0.223\cos\theta_3\}, \{0.446\cos\theta_3\},$ *(the terms specified should*
$\{0.008\sin(\theta_2+ \theta_3)\},$ *be neglected in d_{32}, d_{22},*
$\{0.321\cos\theta_2\cos(\theta_2+ \theta_3)\} \}$ *d_{21} and d_{11} respectively)*

6. CONCLUSIONS

In the paper robot dynamics and symbolically generated robot models have been reviewed. The formulation of closed-form dynamic models is a routine and time-consuming task - recently often performed by computer programs for algebraic robot modeling. The numerically simplified models for real-time dynamics evaluation in modern controllers have been also discussed. The deficiences of numerical simplification process cause the necessity of development of simplification procedures preserving fundamental properties of model structure. A sample simplification procedure known from robotic literature has been expanded in order to enable systematic verification of simplifications. The necessary operations are performed by computer program AIMORO. Some capabilities of the program are demonstrated for a real manipulator.

This work was supported by univerity research theme U2691/BW "Automatic Generation of Robot Dynamics Symbolic Models" at Technical University Rzeszow (Poland).

REFERENCES

[1] John J. Craig: Introduction to Robotics. Addison-Wesley Publishing Company, 1986.

[2] John M. Hollerbach: A Recursive Lagranian Formulation of Manipulator Dynamics and a Comparative Study of Dynamics Formulation Complexity. IEEE Trans. Syst. Man Cybern., Vol SMC-10, No 11, Nov 1980.

[3] A. Izaguirre, R. Paul: Automatic generation of the dynamic equations using a LISP program. 1986 Int. Conf. on Robotics and Automation, Apr. 7-10, San Francisco.

[4] R. Leniowski: "Robot laboratoryjny KRĘPY - model matematyczny i program symulacyjny". Tech. report, Department of Electr. Engin., Techn. University Rzeszów, CPBR 7.1 "Roboty przemysłowe", ss. 1-26, 1987.

[5] J. Y. S. Luh, M. W. Walker, R. P. C. Paul: On-line Computational Scheme for Mechanical Manipulators, ASME Journal of Dynamic Systems, Measurement and Control, vol 102, June 1980, pp. 69-76.

[6] J. J. Murray, C. P. Neuman: ARM: An algebaic robot robot dynamic modelling program. In Proc. 1-st Int. IEEE Conf. Robotics, R. P. Paul Ed., Atlanta, GA, Mar 13-15, 1984, pp. 103-114.

[7] —————— : Organizing Customized Robot Dynamics Algorithms for Efficient Numerical Evaluation. IEEE Trans. Syst. Man. Cybern., Vol. 18, No 1, Jan-Feb 1988.

[8] R. P. Paul: Robot Manipulators : Mathematics, Programming and Control. Cambridge, MA, MIT, 1981.

[9] K. Świder: Symboliczne generowanie modeli prostych manipulatorów. Tech. report Department of Electr. Engineering, Technical Univ. Rzeszow CPBR 7.1 "Roboty przemysłowe" pp. 163-184, 1987.

[10] K. M. W. Tang and V. D. Tourassis: Mathematical Deficiencies of Numerically Simplified Dynamic Robot Models. IEEE Trans. on Autom. Control, Vol. 34, No 10, Oct. 1989.

[11] —————— : Systematic simplification of dynamic robot models," in Proc. Midwest Symp. Circuits Syst., Syracuse, NY, Aug. 17-18, 1987, pp. 1031-1034.

[12] V. D. Tourassis: Principles and design of model-based controllers, Int. Journal of Control, vol. 47, pp. 1267-1275, May 1988.

[13] V. D. Tourassis and C. P. Neuman, "Properties and Structure of dynamic robot models for control engineering applications," Mech. Machine Theory, vol. 20, pp. 27-40, Jan. 1985.

[14] M. Vukobratović and N. Kirćanski, Real-Time Dynamics of Manipulation Robots. New York: Springer-Verlag, 1985.

A COMPARISON BETWEEN THEORETICALLY DERIVED AND EXPERIMENTALLY VERIFIED DYNAMIC MODELLING OF A SCARA-ROBOT

T.A. Rieswijk, J. de Haas, G. Honderd, W. Jongkind

Delft University of Technology,
Faculty of Electrical Engineering, Control Laboratory.
Mekelweg 4, P.O. Box 5031, 2600 GA Delft. The Netherlands.

1. INTRODUCTION

To achieve optimal use of the capabilities of a robot it has become increasingly important to include the dynamical behaviour of the robot in control strategies. In this case a dynamic model is needed for collision-free motion planning and control of two cooperating robots in an assembly cell. The two robots are of different types, an anthropomorphic type and a SCARA type. In this paper the derivation and identification of the dynamical model for the industrial Bosch SR800 TurboScara robot is described. The robot has 4 degrees of freedom of which the first two are the most important. Both these degrees consist of joints that are actuated with DC-motors and use harmonic-drives for speed reduction. In recent literature [1],[2] it is shown that the use of harmonic-drives can introduce considerable flexibility. Analyses of the actuator response after a step input shows flexibility in the first joint. This flexibility will be incorporated in the model. The model has been identified using black-box identification and parameter optimization. With the black-box model orders and rough parameter values are estimated. With a parameter optimization method the individual parameters, like link inertia, are estimated through minimizing the difference between measured and simulated time response. The use of time responses enables the individual estimation of parameters with an influence on a specific part of the respons.

2. MODELLING

The robot that is modelled is a four degree industrial robot of the SCARA-type. The dynamic model for this robot is derived in two stages. The first stage consists of the derivation of the dynamical equations with the Newton-Euler method. The second stage consists of the modelling of the actuator dynamics. Both actuators are composed of a DC-motor and a harmonic-drive for speed reduction.

2.1 Dynamical equations

The dynamical equations describe the motion of the links as a result of applied forces and torques on the joints. The two best known methods to derive the closed form of

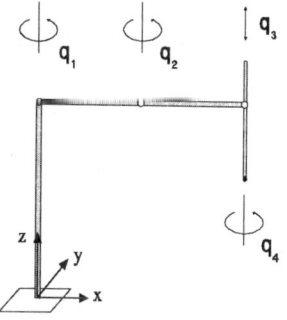

Figure 2.1 Schematic representation of the SCARA-robot.

these dynamical equations are the Newton-Euler method and the Lagrange method. In this case the Newton-Euler method has been used to derive the dynamic model for the first three degrees of freedom. The fourth axis only rotates the gripper and is left out of the model. In figure 2.1 a schematic representation of the robot is given. The first two links are assumed to have a symmetric mass distribution in the y- and z-direction. The third link is assumed to have a symmetric mass distribution in the x- and y-direction.

$$\underline{T} = M(\underline{q})\underline{\ddot{q}} + \underline{C}(\underline{q},\underline{\dot{q}}) + \underline{G}(\underline{q}) \tag{2.1}$$

The dynamic equations of the robot are given in the form of eq. 2.1. This equation gives the vector $\underline{T}$ of joint forces and torques as a result of joint positions, velocities and accelerations. In the equation $\underline{q}$ is the vector of joint positions, M is the moment of inertia matrix, $\underline{C}$ is the vector with Centrifugal and Coriolis effects and $\underline{G}$ is a vector with the gravitational effects. When the closed form of the dynamical equations is calculated the following vectors and matrices are derived :

$$M(\underline{q}) = \begin{bmatrix} I_1 + 2I_2\cos(q_2) & I_3 + I_2\cos(q_2) & 0 \\ I_3 + I_2\cos(q_2) & I_3 & 0 \\ 0 & 0 & I_4 \end{bmatrix}, \underline{C}(\underline{q},\underline{\dot{q}}) = \begin{bmatrix} -2I_2\sin(q_2)\dot{q}_1\dot{q}_2 - I_2\sin(q_2)\dot{q}_2^2 \\ I_2\sin(q_2)\dot{q}_1^2 \\ 0 \end{bmatrix}, \underline{G}(\underline{q}) = \begin{bmatrix} 0 \\ 0 \\ -gm_3 \end{bmatrix}.$$

The terms I_1 to I_4 are given in appendix A. From the moment of inertia matrix $M(\underline{q})$ it can be seen that the third link is completely decoupled from the first two links. Furthermore the first two links are not effected by gravitational forces. Because there is no dynamic coupling between the first two links and the third link, the modelling will concentrate on the first two degrees of freedom.

2.2 Dynamics of the actuators

The actuators consist of armature controlled DC-motors and reducers in the form of harmonic-drives. The well known block diagram for a DC-motor is presented in figure 2.2. In this diagram K_T is the torque constant, R_a and L_a are the armature resistance and inductance respectively, J_m is the rotor moment of inertia and V_m is the viscous friction of the rotor. With the disturbance torque T_d the stiction and Coulomb friction are taken into account. Because of it's small value, L_a will further be neglected.

As mentioned earlier, harmonic-drives are known to introduce a flexibility that can not be neglected. It has been shown [2] that this flexibility can be modelled as a torsional

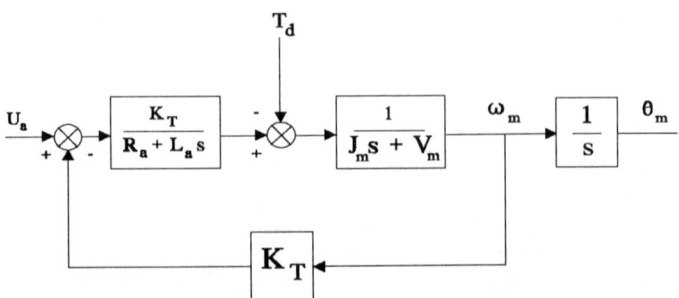

Figure 2.2 *Block diagram of an armature controlled DC-motor.*

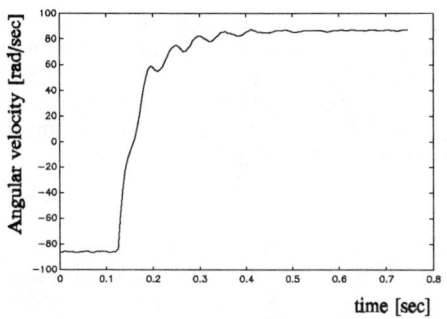

Figure 2.3 Step response of first axis. **Figure 2.4** Step response of second axis.

spring between the actuator and the link. The dynamical properties of the spring are described with the spring constant K and the spring damping D according to eq. 2.2. In this equation Δq is the difference between the positions of the two ends of the spring.

$$T_{spring} = K\Delta q + D\frac{d\Delta q}{dt} \tag{2.2}$$

When looking at the responses of both actuators after a step input (fig. 2.3 and 2.4) it can be seen that only in the first joint flexibility is present. The second joint can be accurately modelled with a rigid gear. When the spring is incorporated into the model of the DC-motor and the dynamic coupling is added, the complete model of figure 2.5 is achieved. For the first link the reaction torque delivered by the spring is fed back to the motor through the gear. The dynamics of the link that is driven with the torque of the spring are described by the dynamic equations that have been derived before.

3. IDENTIFICATION

Identification of the model has been done in two phases. The first phase involves black-box modelling with the prediction error method described by Ljung [3]. With the estimated transfer functions it's possible to deduce the order of the system and get a rough estimate of the parameters of the model obtained in the previous section. The second phase consists of a parameter optimization method. An error criterium is minimized using the 'pattern search' method of Hooke and Jeeves. With this method it is possible to get a good estimate of the individual parameters. First a short explanation is given of the method of measurement that has been used for the identification.

3.1 Measurements

To identify the model of the previous section, measurements have been done of the response of the motors with a step input. Both motors axis are provided with incremental encoders of 4320 increments per revolution, which are attached before the gears and are sampled at a rate of 2000 Hz. At the same time a 12 bit AD converter measures the input voltage to one of the motors with the same sample-rate. The measurement system can record up to 16383 time samples. From the position measurement the speed of the motor can be derived by simple differentiation. With gear ratios of 1:128 and 1:100 for motor 1 and 2 respectively an accuracy of $2.3 \cdot 10^{-2}$ rad/sec and $2.9 \cdot 10^{-2}$ rad/sec is achieved. The resulting signal has been filtered with a first order 100 Hz low-pass filter to reduce

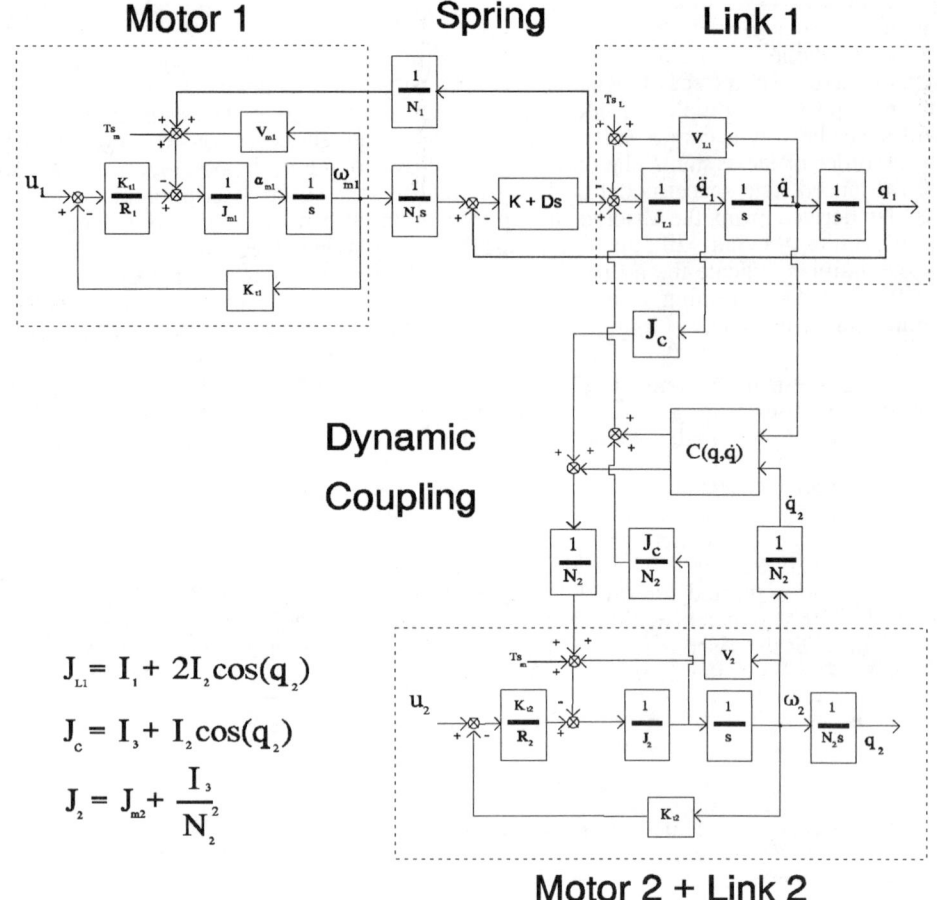

Figure 2.5 *Complete model of the first two joints of the SCARA-robot.*

quantization noise. The step input has been generated directly with a power supply without an additional controller. The power supply has been used to generate steps of ±5 to ±25V. The input of 25V is respectively a quarter and a half of the maximum input voltage allowed for motors 1 and 2. For the black-box estimation the dynamic coupling has been reduced by short circuiting the remaining motor. Also the second link has been placed at an angle of 130 degrees with the first link to minimize the coupling effects.

3.2 Black-box model

To obtain information about the order and the rough values of the parameters black-box modelling is used. The transfer functions of the Output Error structure are estimated according to the prediction error method [3]. They are estimated with the measured input and output (voltage and motor speed) for a voltage step on the motor. The discrete functions are transformed to continuous functions to obtain information about the rough parameter values.

The prediction error method is used to estimate the transfer functions of the two separate axes. The effect of dynamic coupling between both links has been reduced as much as possible, complete decoupling however is not possible. The estimate therefore will never be very accurate, still it will be accurate enough to get a good approximation of the order of the system. The order is derived by looking at the observability gramian of a balanced state space model of the system. The state space description of the model can be derived from the discrete transfer function. When the model is balanced the observability gramian will equal the controllability gramian. The values on the diagonal of the gramian indicate the influence of every state on the output. By estimating transfer functions with increasing order the gramian will indicate when no further relevant improvement is made.

In table 3.1 the gramians for the transfer functions from the first through the fifth order of the first axis are given. From the elements of the gramians it can be concluded that only the first three states have a relevant influence. From the calculated gramians in table 3.2 it follows that the second axis can be modelled with a first order transfer function. The difference in influence between the states is not as large as for the first axis. A possible explanation for this can be the influence of the dynamic coupling.

By deriving the algebraic transfer functions for input voltage to motor speed from the model in figure 2.5, the rough parameter values can be calculated. They follow from equating the coefficients of the derived and estimated transfer functions. When the damping D is neglected for simplicity, eq. 3.1 and 3.2 are derived. The values that are found for the parameters will be used as initial values for the parameter identification. The values for the viscous friction (V_{ml}, V_{LI} and V_2) are adjusted because of the lack of Coulomb friction in the linear black-box model.

$$\frac{\omega_{ml}}{U_1} = \frac{\frac{K_{TI}}{R_1}(J_{LI}s^2 + V_{LI}s + K)}{J_{ml}J_{LI}s^3 + \left[J_mV_{LI} + J_{LI}\left(V_{ml} + \frac{K_{TI}^2}{R_1}\right)\right]s^2 + \left[V_{LI}\left(V_{ml} + \frac{K_{TI}^2}{R_1}\right) + K\left(J_{ml} + \frac{J_{LI}}{N_1^2}\right)\right]s^1 + K\left(V_{ml} + \frac{K_{TI}^2}{R_1} + \frac{V_{LI}}{N_1^2}\right)} \quad (3.1)$$

$$\frac{\omega_2}{U_2} = \frac{K_{T2}}{J_2R_2s + (V_2R_2 + K_{T2}^2)} \quad (3.2)$$

Order for the gramian	States				
	1	2	3	4	5
1	1.76	-		-	-
2	1.60	0.18	-	-	-
3	1.72	1.21	1.17	-	-
4	1.66	1.25	1.23	0.09	-
5	1.66	1.22	1.20	0.07	0.03

Table 3.1 Gramians for the first joint.

Order for the gramian	States		
	1	2	3
1	3.20	-	-
2	2.99	0.23	-
3	2.99	0.23	0.04

Table 3.2 Gramians for the second joint.

3.3 Parameter-identification

The next step in identifying the model is the derivation of individual parameters of the model. This derivation is done in three steps. The first step is determining parameters that allow (almost) direct measurement because of there physical nature. The second and third step consist of the estimation through parameter optimization. The optimization minimizes a criterium which penalizes the difference between measured velocity response and the response of the model using the 'pattern search' method of Hooke and Jeeves. This method allows the optimization of a number of parameters at the same time. It is implemented in the simulation software packet PSI/e [4]. The criterium that is used is given as 3.3. The total time T during which the error is integrated depends on the part of the response which is influenced by the parameters. For the optimization in step three for example the time T is taken as in figure 3.1.

$$criterium = \int_0^T |\omega_{motor} - \omega_{model}| dt \qquad (3.3)$$

Directly measurable parameters

The torque constant K_T, the armature resistance R_a and stiction and Coulomb friction are the only directly measurable parameters of the model. K_T can be determined by measuring the voltage across the motor armature and the velocity of the motor axis while moving the motor axis manually. R_a can then be determined by applying a voltage V_a to a temporary blocked motor and measuring the current I_a.

The stiction and Coulomb friction can roughly be determined by applying a triangular input current. The torque can be derived by measuring the current I_a when the motor starts (stiction) and stops (Coulomb friction). These two values are time and position variant, the values given in table 3.3 are therefore the means of several measurements. The values of K_T and R_a are also given in table 3.3.

Individually optimized parameters

The parameter optimization in the second step is used to optimize single parameters. The purpose is to individually optimize as many parameters as possible. This separate optimization is performed to avoid drift of mutually compensating parameters when they are optimized.

There are five parameters that are optimized individually. These parameters have a corresponding part of the response of a step input with a dominant influence of that parameter. The parameters that are optimized are the total moment of inertia of the second link J_2, the moment of inertia of the first motor axis J_{ml} and the viscous frictions V_{ml}, V_{L1} and V_2. These last parameters have a very nonlinear behaviour and are therefore newly optimized for every measured step response. The friction is optimized on the steady state part of the response. V_{ml} is fixed at 10^{-3} Nm/(rad/sec) based on the black-box estimation and a number of optimizations. When the viscous friction V_2 is estimated and dynamic coupling is minimized, the moment of inertia J_2 of the

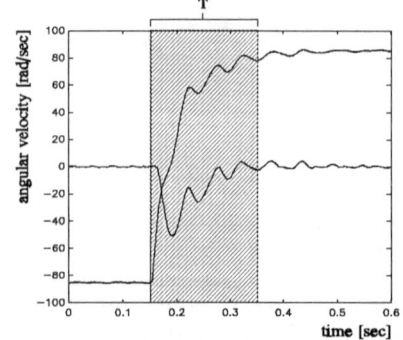

Figure 3.1 The time T over which the criterium is minimized in step 3.

second link is the only unknown parameter that influences the transient part of the step response.

The remaining parameter that can be optimized individually is the moment of inertia J_{ml} of the first motor axis. The reason why this parameter can be individually optimized is the presence of elasticity between the axis and the link. In the first moments of a step response the axis will move independently of the link until the spring is wound up and the link also starts to move. The optimization of this parameter is done for the first 30 ms of the step response. The remaining parameters of the model should be roughly known to assure a good estimate of J_{ml}. The values found for J_2 and J_{ml} are given in table 3.3.

	Motor 1	Motor 2
Torque constant K_T	0.240 Nm/A	0.099 Nm/A
Armature resistance R_a	1.45 Ω	1.87 Ω
Stiction	0.14 Nm	0.11 Nm
Coulomb friction	0.11 Nm	0.09 Nm
Mom. Iner. J_2	0.42 10^{-3} kgm²	
Mom. Iner. J_{ml}	1.075 10^{-3} kgm²	

Table 3.3 *Individual estimated motor parameters.*

Jointly optimized parameters

The remaining unknown parameters are the spring characteristics K and D, the link inertia J_{L1} and the coupling inertia J_c. These parameters have a combined influence on the transient part of the step response of the first motor axis and, through J_c, also an influence on the response of the second axis. Therefore they will be optimized together using the criterium 3.4.

$$criterium = \int_0^T |\omega_{motor} - \omega_{model}|_{axis1} dt + \int_0^T |\omega_{motor} - \omega_{model}|_{axis2} dt \qquad (3.4)$$

The criterium is calculated for T as indicated in figure 3.1 The parameters are optimized for three different positions of the second link : $q_2=0°$, $q_2=60°$, $q_2=90°$. The expected values for J_{L1} will then be respectively I_1+2I_2, I_1+I_2 and I_1. For J_c they will be respectively I_3+I_2, $I_3+\frac{1}{2}I_2$ and I_3. The results of the optimizations for the three different positions are given in table 3.4. Beside the mentioned parameters the viscous friction V_2 is also optimized for the different responses. When we look at the values for the spring constant K we see a difference between the first two positions and the last. This difference is caused by mutual compensation between estimated parameters. The oscillating frequency ω_n depends on the reciprocal of J_{L1} and K. From a number of different optimizations a good value for K appears to be $8.2 \cdot 10^4$ Nm/rad. With this value of K, J_{L1} and J_c are estimated again with the results given in table 3.5.

Adaptive estimation of viscous friction

Because of the nonlinear and time variant behaviour of the viscous friction an adaptive estimating algorithm has been implemented. With this algorithm the viscous frictions V_2 and V_{L1} are estimated using an observer based on stable Lyapunov rules. With this kind of estimation the time-variant and nonlinear behaviour can be studied. The viscous

q_2	J_{L1} [kgm²]	K [Nm/rad]	J_c [kgm²]	D [Nm/(rad/s)]
0°	10.9	$8.2 \cdot 10^4$	3.2	21
60°	9.0	$8.0 \cdot 10^4$	2.4	0
90°	7.0	$7.6 \cdot 10^4$	1.5	23

Table 3.4 *Jointly optimized parameters for three positions of link 2.*

friction appears to increase during a period with a constant input voltage. The variation in V_2 with a constant input voltage can amount to 25%. The viscous friction is to complex to be modelled, an adaptive controller seems therefore a necessary requirement.

q_2	J_{L1} [kg m^2]	J_c [kg m^2]
0°	10.8	3.3
60°	9.1	2.3
90°	7.5	1.5

4. VALIDATION

Table 3.5 Optimized J_{L1} and J_c for fixed $K=8.2\ 10^4$

The model that has been derived in the previous chapter will now be validated. To this purpose a known load is placed in the end-effector of the robot. The measurements with this load are compared to the calculated effect with the model. The effect on the model of a load in the end-effector can be calculated by increasing the mass m_3 by the mass of the load. With a load of 4.2 kg the following adjustments to the terms J_c, J_{L1} and J_2 have to be made (with $L_1=0.445$ m, $L_2=0.355$ m, $^{C}I_{load_{zz}}=0$, $N_2=100$) :

$$J_2' = J_2 + \frac{1}{N_2^2}(L_2^2 m_{load} + {}^C I_{load_{zz}}) = J_2 + 0.53 \cdot 10^{-4} \quad \text{kgm}^2$$

$$J_{L1}' = J_{L1} + 2m_{load}L_1L_2\cos(q_2) + m_{load}(L_1^2+L_2^2) + {}^C I_{load_{zz}} = J_{L1} + 1.33 \cdot \cos(q_2) + 1.36 \quad \text{kgm}^2$$

$$J_c' = J_c + m_{load}L_1L_2\cos(q_2) + L_2^2 m_{load} + {}^C I_{load_{zz}} = J_c + 0.66 \cdot \cos(q_2) + 0.53 \quad \text{kgm}^2$$

The response of the loaded robot can now be simulated with the model. To get a reasonable response from the model the viscous frictions are optimized on the steady state part of the responses. The two responses for a 25V/-25V step are given in figure 4.1 and 4.2. When we compare robot- and model-response, the frequency of the oscillation seems to correspond quit good. The amplitude however is much more damped for the real response then for the simulated one. To check this the spring damping D is optimized individually for the response. The value found is indeed larger then the previously determined value (see fig 4.3). To validate the other parameters of the model a new series of joint optimizations is performed for the three positions as described before. The results are given in table 4.1 together with the values as calculated according to the model. In this table the good agreement between the estimated and calculated parameters can be seen. The spring damping D apparently increases with the load. The value found for J_2 is $0.47 \cdot 10^{-3}$ kgm^2, which again agrees with the predicted value.

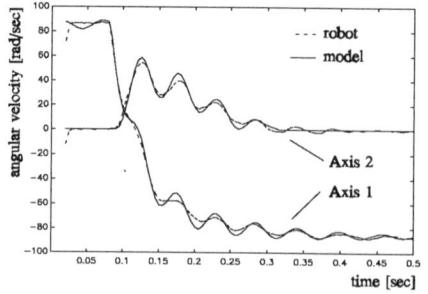
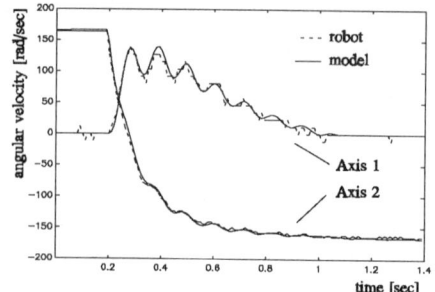

Figure 4.1 Measured and simulated response of step on 1st axis of robot with load (D=10).

Figure 4.2 Measured and simulated response of step on 2nd axis of robot with load.

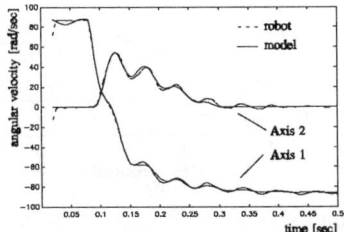

q_2	J_{L1} [kg m²]		J_c [kg m²]		D
	Est.	Calc.	Est.	Calc.	[Nm/(rad/s)]
0°	13.5	13.5	4.4	4.5	107.5
60°	11.2	11.1	3.3	3.2	72.4
90°	8.9	8.9	1.9	2.0	88.0

Figure 4.3 *Measured and simulated response of step on 1st axis of robot with load (D=80).*

Table 4.1 *Calculated and estimated values for J_{L1} and J_c.*

5. CONCLUSION

In this paper a method is presented to identify a robot model including the flexibility introduced by harmonic-drives. Through order estimation of the transfer functions it is shown that flexibility can accurately be modelled with a third order model. With the optimization method it is possible to estimate individual link parameters which can directly be used in control strategies and collision-free motion planning. It is also shown that the influence of load changes on the inertial parameters can be predicted with the dynamical equations. Friction appears to be a very nonlinear and position- and time-variant parameter. An adaptive controller seems to be a reasonable solution to this problem. Further research will concentrate on modelling of the third link which also shows flexibility, developing a controller to reduce vibration and the implementation of time-optimal collision-free motion planning.

Literature

[1] Good, M.C., Sweet, L.M., Strobel, K.L., *'Dynamic Models for Control System Design of Integrated Robot and Drive Systems.'* Trans. ASME, Journal of Dynamical Systems, Measurement and Control. Vol. 107, march 1985, p. 53-59.
[2] Spong, M.W., *'Modelling and Control of Elastic Joint Robots.'* Trans. of ASME, Journ. of Dyn. Sys., Measurement and Control. Vol. 109, dec. 1987, p.310-319
[3] Ljung, L., *System Identification - Theory for the User*. Prentice-Hall, Englewood Cliffs, 1987.
[4] Bosch, P.P.J. van den, Butler, H., Soeterboek, A.R.M., *Modelling and Simulation with PSI/e*. Pijnacker, 1990. BOZA Automatiserings BV.

Appendix A

$$I_1 = m_1 P_{C1_x}^2 + m_2(P_{C2_x}^2 + L_1^2) + m_3(L_1^2 + L_2^2) + {}^C I_{zz_1} + {}^C I_{zz_2} + {}^C I_{zz_3},$$
$$I_2 = m_2 L_1 P_{C2_x} + m_3 L_1 L_2,$$
$$I_3 = m_2 P_{C2_x}^2 + m_3 L_2^2 + {}^C I_{zz_2} + {}^C I_{zz_3},$$
$$I_4 = m_3. \tag{A.1}$$

m_i — Mass link i $\qquad L_i$ — Length link i

P_{Ci} — Vector from joint i to centre of mass link i

${}^C I_{zz_i}$ — Moment of inertia with respect to z-axis

DYNAMICS OF ROBOTS HAVING SUPPLE BODIES

P.ANDRE professor   J-P.TAILLARD professor
P.MITROUCHEV PHD researcher   L.LOSCO professor
Laboratoire D'Automatique de Besançon. URA CNRS 822. 15, Impasse des Saint
Martin.25000 BESANCON Cedex. FRANCE. ☏ 81 88 53 44 - Fax. 81 88 65 02

Abstract
We present a method for the dynamic modelling of articulated open chain robots taking into account the elastic deformations undergone by the links. We have used Newton-Euler's formalism for a simple robot in rotary movement. So we have studied how to calculate the own pulsations of the robot's mechanical structure. The robot's segments have been modelized using Euler-Bernoulli beams and taking into account the stiffness of the articulations as well as masses and inertias including those of the load. The dynamic stiffness method has been chosen for the frequency determination. In order to test our model, a software has been developed under DEC VAX-VMS, providing the values of natural frequencies in a low frequency spectrum for any given position of the robot. A vibration test with a sine excitation has been performed on an " AID-5V " robot. The results of automatic modelling have proved to be consistent with those obtained by experimental tests. The software modelling tools that we have developed are under integration into the "EUCLID"-IS 210 CAD system running under VAX-VMS.

Keywords: Robotics, Dynamic modelling, Supple robots, Vibration test

1.INTRODUCTION
Generally, putting the problems of the dynamics of solids into equation involves the Newton-Euler equations [RAK 85, BOD 89], the principle of virtual powers or d'Alembert formalism [YAZ 88], Hamilton principle-using Lagrange's equations [BAR 88,LUC 89, SER 88, NIC 89, YUH 87]...
Considering these different method, we have chosen the Newton-Euler equations. A problem that we now are facing is the choice of the frame against which we measure the displacement. Several solutions are being used [BRI 87a]
* <u>Rigid-body Movement (of U.L.M. type)</u>: For each link we use as a reference the position it would occupy in its movement if it was considered rigid [BRI 87b].
* <u>Movable Intermediate Frame</u>: The displacements are measured with respect to the position that the segment would have if it was considered rigid and situated at the end of the previous deformed one. The axis x_i is tangent to the middle line of segment S_i in O_i. It seems better to use the movable intermediate frame because in the plan, the solid will have 3 degrees of

freedom instead of 6. Basing ourselves on this frame, we have obtained a kinematic model of flexible-segment robots [TAI 98, LOS 90].

The aim of this study is to take into account the <u>elastic deformation of the structure</u> in the <u>dynamic modelling</u> of robots, for the purpose of correcting the control which initially considers this system to be rigid. This is a necessity in the case of structures for which the inertial characteristics are important concerning the stiffness characteristics, and for which the dynamics imposed by the control and the own modes of the servo loops are likely to excite the natural modes of the structure. In recent years, this problem has been the subject of many works. At modelling level, the modal approach and the discretization by finite elements [CHE 87b] are often used. In the works of Chretien [CHR 85] an original method called "fictitious" joinings method is put forward. It consists in discretizing the potential energy of a flexible segment by introducing suitably-selected and situated springs.

2.PRESENTATION OF MODEL, NEWTON-EULER EQUATIONS

Let us consider a link as a beam embedded at its origin O into a mobile support, its other extremity L(M) being loaded and free to move in the plan. The frame R_O (OoXoYo) is the Galilean frame and R_1 ($O_1X_1Y_1$) is the mobile frame, attached to the support, and to which the beam's deformation will be referred. Let's consider an element around P between x and dx (see fig.1) with: u(x,t) and w(x,t) - elastic longitudinal and deflexion deformation of the body respectively, $\phi(x,t)=dw/dx=w'$- angular elastic displacement of body, it represents the rotation between the tangents to the mean line at the origin and point P. Using the general Mechanism theorem of the inertia center, we obtain:

$$\rho S \underline{\Gamma}(P) = \underline{T}' + \underline{F} \qquad (1)$$

with: $\underline{\Gamma}(P)$ -the absolute acceleration of P point, $\underline{T}'$- derivate we the respect to x of the internal effort $\underline{T}=[T_x \quad T_y]^T$, $\underline{F}=[F_x \quad F_y]^T$- external effort.

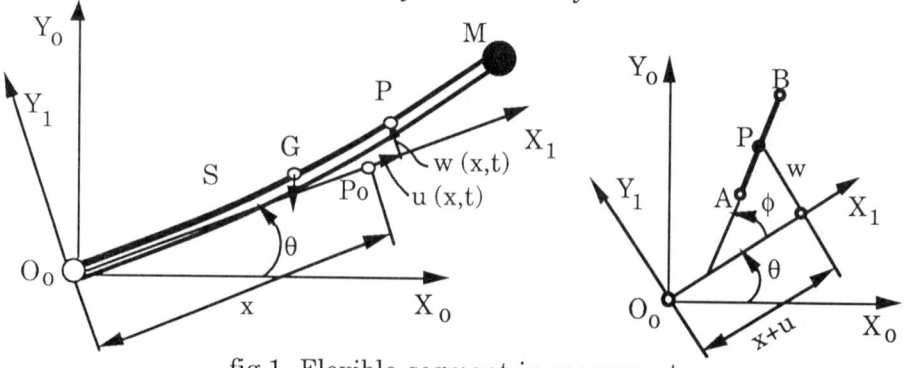

fig.1 Flexible segment in movement.

The theorem of the angular momentum on P, midlle of AB, gives:

$\mathcal{M}'+T_y=0$, $\mathcal{M}=EIw''$ and $T_x=ESu'$ \hfill (2), (3), (4)

Putting back equations (2) and (4) in equation (1) and taking into account equation (3), we get:

$$\begin{vmatrix} \rho S\Gamma_x = ESu'' + F_x \\ \rho S\Gamma_y = -EIw'''' + F_y \end{vmatrix} \quad (5)$$

with: E - Young's modulus, S - area, ρ - mass per volume unit, supposed to be constant and I area moment of inertia.

Let's note $\underline{PoP}/R = [u \ w \ 0]^T$ the vector of elastic displacements of a current point P of the segment in the frame R_1 attached to the link end; $\underline{\Omega}=[0 \ 0 \ \dot{\theta}]^T$ - instantaneous rotation of the frame R_1 with respect to the ground frame R_O.

The position of point P in the frame R_1 is:

$\underline{OoP} = \underline{OoPo} + \underline{PoP} = [x+u \ w \ 0]^T$

The absolute speed of point P in this frame is:

$$\underline{V}(P) = \begin{vmatrix} \dot{u} - \dot{\theta}w \\ \dot{w} + \dot{\theta}(x+u) \end{vmatrix} \quad (6)$$

The absolute acceleration of P point in the frame Ro (R_1) is given by:

$$\underline{\Gamma}(P) = \begin{vmatrix} -\dot{\theta}^2 x + \ddot{u} - \ddot{\theta}w - 2\dot{\theta}\dot{w} - \dot{\theta}^2 u \\ \ddot{\theta}x + \ddot{w} + \ddot{\theta}u + 2\dot{\theta}\dot{u} - \dot{\theta}^2 w \end{vmatrix} \quad (7)$$

Replacing equations (7) in the equation (1) we get:

$$\rho S[-\dot{\theta}^2 x + \ddot{u} - \ddot{\theta}w - 2\dot{\theta}\dot{w} - \dot{\theta}^2 u] - ESu'' = F_x \quad (8)$$
$$\rho S[\ddot{\theta}x + \ddot{w} + \ddot{\theta}u + 2\dot{\theta}\dot{u} - \dot{\theta}^2 w] + EIw'''' = F_y \quad (9)$$

with boundary conditions: 1) in point O, $\begin{vmatrix} x=0, w(0)=w'(0)=0 \\ u(0)=0 \end{vmatrix}$

Let M is the given charge;   2) in point L free extremity: x=L,

$$\begin{vmatrix} M[-\dot{\theta}^2 L + \ddot{u}(L,t) - \ddot{\theta}w(L,t) - 2\dot{\theta}\dot{w} - \dot{\theta}^2 u(L,t)] = -ESu'(L,t) + F_x(L) \\ M[\ddot{\theta}L + \ddot{w}(L,t) + \ddot{\theta}u(L,t) + 2\dot{\theta}\dot{u}(L,t) - \dot{\theta}^2 w(L,t)] = EIw'''(L,t) + F_y(L) \\ \mathcal{M}(L) = \mathcal{M}_{ext}(L) \end{vmatrix}$$

If $\mathcal{M}_{ext}(L)=0$, then $\mathcal{M}(L)=0$ then $w''(L)=0$

Particular case, considering the low frequences case u<<w, u is insignificant (u=0), so equation (9) becomes:

$$EIw'''' + \rho S\ddot{w} - \rho S\dot{\theta}^2 w = F_y - \rho S\ddot{\theta}x \quad (10)$$

For $\theta=0$, (R is fixed, static case)

$$EIw'''' + \rho S\ddot{w} = F_y$$

In conclusion, if we consider a RR...R type robot having n links, for each link, we have the equation (10) with the boundary conditions depending on the type of segment: - embedding w=0, w'=0; - rotary joint w=0, $\mathcal{M}$ =0, w''=0 and - free extremity $\mathcal{M}$ =0, T_y=0, w''=0, w'''=0

3. EULER'S EQUATION

In the following part, we will consider the case where u=0 and where $F_x=F_y=0$ (horizontal arm). Let us apply the in O angular momentum theorem to the total system. The in O angular momentum is by definition:

$$\underline{K_o} = \int_0^L \underline{O_oP} \wedge \underline{V}(P) dm + M\underline{O_oL} \wedge \underline{V}(P) = \int_0^L (x\dot{w}+x^2\dot{\theta}) dm + M(L\dot{w}(L)+L^2\dot{\theta}) \quad (11)$$

with: $dm = \rho S dx$ - for a uniform link (girder) with a constant section, an m mass and $w^2 \cong 0$. We will present the equation (11) as the sum of two terms:

a) in rigid link movement:
$$\underline{K}_{or} = (mL^2/3 + ML^2)\dot{\theta} \quad (12)$$

b) in elastic displacements:
$$\underline{K}_{of} = \int_0^L x\dot{w} dm + ML\dot{w}(L,t) \quad (13)$$

where: w - is the derivate with respect to time of w(x,t). The theorem of angular momentum in O is:
$$\underline{\dot{K}}_{or} + \underline{\dot{K}}_{of} = \underline{C} \quad (14)$$

where: $\underline{C}$ the actions torque in O (connection, outside actions, activators...) The latter equation shows that if one doesnot change the $\underline{C}$ torc of actions in O in relation to the movement supposed to by rigid, the introduction of $\underline{\dot{K}}_{of}$ related to elastic distorsion creates a modification of $\underline{\dot{K}}_{or}$ hence of angle θ, and this variation of θ is of the same nature as the w distorsion.

4. DYNAMIC STUDY OF THE LINK BY MODAL ANALYSIS

Let us suppose that the rigid movement θ is nil. We have just explained that flexibility causes θ to vary in the same way as w, so much so that one can ignore $\dot{\theta}^2$. The equation of the rotating (9) movement of the flexible link, on the basis of the derivation of Euler's equation is [THO 81, RAK 85]:

$$EIw'''' + \rho S\ddot{w} + \rho Sx\ddot{\theta} = 0 \quad (15)$$

And moreover:
$$(mL^2/3 + ML^2)\ddot{\theta} + \rho S\int_0^L \ddot{w}x dx + ML\ddot{w}(L,t) = 0 \quad (16)$$

with boundary conditions:
$$\begin{cases} w(0,t) = w'(0,t) = 0 \\ EIw''(L,t) = 0 \\ -EIw'''(L,t) + M\ddot{w}(L,t) = 0 \end{cases} \quad (17)$$

One can easily verify that $w=-x\theta$ is a particular solution of equations (15) and (16) and one can state:
$$w=-x\theta+F(x,t) \qquad (18)$$
with $F(x,t)$ any function of x and t, that verifies the following equation:
$$EIF''''+\rho S\ddot{F}=0 \qquad (19)$$

The latter equation is immediately solved by modal analysis, F being an infinite sum whose form is:
$$F(x,t)=\sum_{i=1}^{\infty}\Phi_i(x)q_i(t) \qquad (20)$$

Thus one obtains $\ddot{q}=-\omega^2 q$ and $q=a\cos\omega t+b\sin\omega t$, ω being the specific vibration pattern of the system.
$$\Phi''''-\rho S\omega^2\Phi/EI=0 \qquad (21)$$

Let us state $\alpha^4=\rho S\omega^2/EI$, then we will find
$$w(x,t)=-x\theta+(a\cos\omega t+b\sin\omega t)(a_1 Ch\alpha x+a_2 Sh\alpha x+a_3 \cos\alpha x+a_4 \sin\alpha x) \qquad (22)$$

All the constants a_1, a_2, a_3 and a_4 as well as the specific pulsation ω are determined by the boundary conditions (17) that is to say:

$w(0,t)=0$, thus $a_1=-a_3$,

$w'(0,t)=0$, thus $\theta=\alpha(a_2+a_4)(a\cos\omega t+b\sin\omega t)$

From the (16) equation, one gets:
$$\rho S\int_0^L \Phi(x)x\,dx+ML\Phi(L)=0 \qquad (23)$$

which gives an equation in a_1, a_2 and a_4.

For $w''(L,t)=0$: $a_1(Ch\alpha L+\cos\alpha L)+a_2 Sh\alpha L-a_4\sin\alpha L=0$ which gives once again an equation in a_1, a_2 and a_4.

We can thus calculate the other constants a_1, a_2, and a_4 according to α.

Finally $-EIw'''(L,t)+M\ddot{w}(L,t)=0$ thus:

$[a_1(Sh\alpha L-\sin\alpha L)+a_2 Ch\alpha L-a_4\cos\alpha L]+$

$\qquad +(M\alpha/\rho S)[a_1(Ch\alpha L-\cos\alpha L)+a_2 Sh\alpha L+a_4\sin\alpha L]=0$ - is the equation of the specific patterns.

Thus, the model we are trying to determine is defined by.

$w(x,t)=(a\cos\omega t+b\sin\omega t)[-\alpha(a_2+a_4)x+a_1(Ch\alpha x-\cos\alpha x)+a_2 Sh\alpha x+a_4\sin\alpha x]$

$\theta(t)=\alpha(a_2+a_4)(a\cos\omega t+b\sin\omega t)$

A approch method allows us to obtain $w(x,t)$ easily. One chooses $W_1(x)$ the solution of $EIw''''+\rho S\ddot{w}=0$ as a function, limiting oneself to the first vibration pattern ω_1; W_1 will have to fit the equation (17). Setting down w(x,t)=W_1(x)q_1(t). Transferring in equations (15) and (16), then multiplying (15) by W_1(x) and integrating between 0 and L, one obtains:

$$\begin{bmatrix} \frac{mL^2}{3}+ML^2 & \rho S\int_0^L W_1(x)x\,dx+MLW_1(L) \\ \rho S\int_0^L W_1(x)x\,dx & \rho SL \end{bmatrix} \begin{bmatrix} \ddot{\theta} \\ \ddot{q}_1 \end{bmatrix} + \begin{bmatrix} 0 & 0 \\ 0 & EI\alpha^4 L \end{bmatrix} \begin{bmatrix} \theta \\ q_1 \end{bmatrix} = \begin{bmatrix} 0 \\ 0 \end{bmatrix}$$

The generalised variables θ and q are obtained by integration.

5. EXPERIMENTAL TEST.

In the study the dynamic vibrating behaviour of a robot, the models of vibrating behaviour for each flexible link and the definition of an assembly process for these models are required to obtain the whole model.
The mechanical structure of the robot is presented by five bi-dimensional segments of Euler-Bernoulli (fig. 2). An approach which allows us to build accurate models of a continuous milieu in low frequencies is the <u>modal analysis</u> method. The position of the sensors in the robot structure AID-5V is given in fig.2

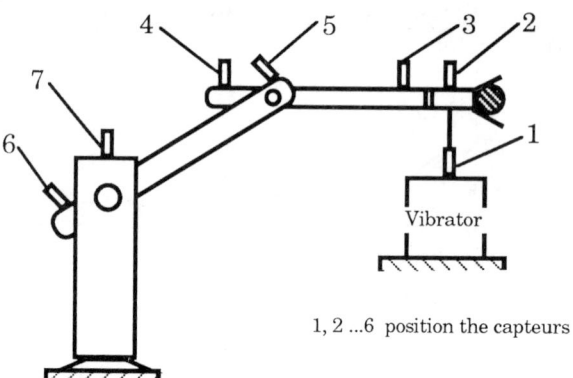

1, 2 ...6 position the capteurs

fig.2 Robot AID - 5V and the points in position the capteurs.

The natural frequencies are obtained by the dynamic stiffness method, where each segment is characterized by its matrix called "dynamic stiffness" matrix. The matricial relationships between the generalized forces and the generalised displacements can be written in a matricial form.

$$\underline{f_j}=[K(\omega_j)]\underline{a_j}$$

with: $\underline{a_j}$ - vector of the generalized displacements of segment S_j and $\underline{f_j}$ - vector of the generalized stresses.

The assembly of dynamic stiffness matrices consists in arranging the five dynamic stiffness matrices into a big (nxn) matrix. Writing that the determinant of this matrix equals zero leads to the frequency equation, which can be solved by a dichotomy method, det($[K(\omega_j)]$)=0 with: $[K(\omega_j)]$ - nxn) dynamic stiffness matrix, n=17- number of degrees of freedom which are dependent of the discretized structure.

The structure is excited by a sine signal. For the analysis method, we have chosen the programmed linear lissage available at the Applied Mechanics Laboratory, University of Besançon on computer MV-15000, which makes a frequency scanning with a given step and automatically treats the results of measures. The fig. 3 gives the phase and amplitude versus frequency of the transfer function at point 3 (cf. fig. 2). On the figure, the peaks correspond to the natural frequencies. A software on VAX-VMS has been developed for the purpose of automatically obtaining the natural frequencies for the whole structure, as well as for each link the longitudinal and transversal natural frequencies.

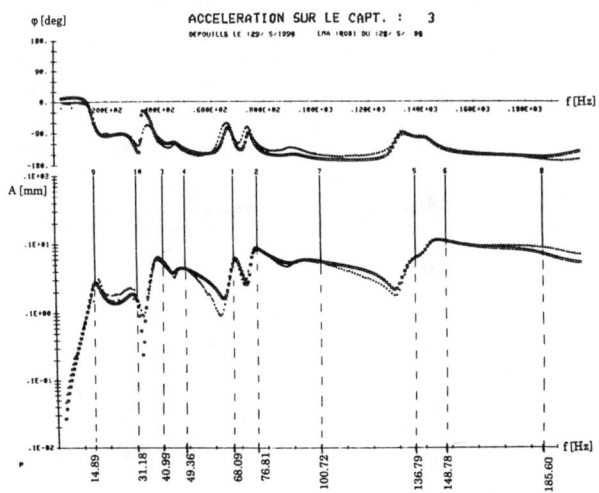

fig.3 The phase and amplitude versus frequency at point 3.

CONCLUSION

We have put forward a method of dynamic modelling of open-loop chain manipulators taking into account elastic displacements of the links by means of the Newton-Euler methods. If one is only interested in behaviour of the links (i.e. considering the joints to be rigid), the Newton-Euler method we have used has the advantage of avoiding the calculation of partial and time derivatives of the potential and kinetic energies, thus leading to simpler calculation thon, for instance. In addition, this method provides a very easy link with modal analysis. The problem of the determination of the natural frequencies of the robot's mechanical structure has been outlined. The dynamical stiffness method we have chosen is well suited to one programming of the study of its own frequencies, from the robot structure Based upon this method we have developed a program (on VAX-VMS) which derives the specific modes of the structure from the C.A.D. model of the robot. The experimental results confirm the calculated results with acceptable accuracy.

REFERENCES

[BAR 88] E. Barbieri and U. Ozguner, "Unconstrained and constrained mode expention for a flexible slewing link", Jurn. dyn. syst., meas., control. (USA), 1988, 110, no 4, pp 416-421.

[BOD 89] A. Bodner, "Adding flexibility to the links of a rigid-link dynamic model of an articulated robot.", - Robotica, GBR, (1989), 7, part.2, pp. 165-168.

[BRI 87 a] J.-N.Bricout, J.C Debus, P. Micheau " Modélisation par éléments finis du comportement dynamique de manipulateurs déformables," Colloque Recherche et transfert de technologie en mécanique, Lille, 16-17 Décembre, 1987.

[BRI 87 b] J.-N. Bricout, "Contribution à la modélisation du comportement dynamique de manipulateurs déformables", Thèse de docteur en mécanique, Lille, Juin. 1987.

[CHE 87 b] P. Chedmail, P. Glumineau, J.C. Bardiaux, "Plane flexible robot modelisation and application to the control of an elastic arm", Towards Third Generation Robotics Proc. of the 3rd International Conference Advanced Robotics, ICAR'87, 13-15 Octobre 1987, Versailles, France, pp.525-536.

[CHR 85] J.-P. Chretien, M. Delpech and A. Louhmadi, "Modeling and simulation of distributed flexibilty in a spaceborn manipulator", 10th IFAC Symposium on Automatic Control in Space. Toulouse, France, 1985, pp.268-277.

[LOS 90] L. Losco, P. Andre, P. Mitrouchev and J.-P. Taillard, "Kinematic modelling of robots having supple bodies", Proceeding Thirteenth IASTED International Symposium Robotics and Manufacturing, Santa - Barbara, California, U.S.A., November 13 - 15 1990, ACTA PRESS Anaheim *Calgaty* ZURICH, pp. 30-33

[LUC 89] A. DE Luca, P. Lucibello and G. Ulivi, "Invertion techniques for trajectory contre of flexible robot arms", Jurn. rob. syst., USA, (1989), 6, no 4, pp. 325-344.

[NIC 89] S. Nicosia, P. Tomei and A. Tornambe, "Hamiltonian description and dynami control of flexible robot", Jurn. rob. syst., USA, (1989), 6, no 4, pp. 345-361.

[RAK 85] F. Rakhsha and A. Goldenberg, "Dynamics modelling of a single-link flexibl robot",Proceeding of IEEE International Conference on Robotic and Automation, S Louis, 1985, pp. 984-989.

[SER 88] M. Serna and E. Bayo, "A simple and efficient computational approach for th forward dynamics of elastic robots", Jurn. rob. syst., USA, (1988), 6, no 4, pp. 363-382.

[TAI 89] J.-P. Taillard, L. Losco, P. Mitrouchev, "Génération automatique de modèles de systèmes mécaniques articulés de solides rigides et déformables," 2ème Congrè StruCoMe'89, 14-16 Novemre, 1989, Palais de Congrès, Paris, France, pp 329-337.

[THO 81] W.T. Thomson, "Theory of Vibration with Application," 2^{-nd} Edition. Prentice Hall, Inc., 1981

[YAZ 88] A. Yazman,"Modélisation des robots fléxibles par les Bond-Graphs, application l'analyse de leurs performances dynamiques", Thèse de doctorat en Automatique Paris 11, 1988.

A FINITE-ELEMENTS MODEL OF TWO-JOINT ROBOTS WITH ELASTIC FOREARM

P. Muraca (*), M. La Cava, A. Ficola (+)
(*) Systems Department, University of Calabria
  87036 Arcavacata (CS), Italy
(+) Institute of Electronics, University of Perugia
  Via Cairoli 24, 06100 Perugia, Italy.

ABSTRACT
  The finite-elements technique is applied in modelling robots with flexible forearm. The forearm is divided into a number of elemental beams; each element consists of a rigid pendulum with a spring and a mass at its extremities. This model has a particular structure that can be easily used in simulations and implemented on a standard PC by using 4th-order Runge-Kutta method.
  The comparison with the eigenfunction expansion model shows that the modes of the elastic forearm are achieved with a satisfactory accuracy. Moreover the proposed model yields the progressive wave solution of the Bernoulli-Euler equation for the flexible link, which is not possible to put into evidence with the eigenfunction expansion model.

1 - INTRODUCTION

The behaviour of an elastic beam subjected to inertia and external forces is described by the following Euler - Bernoulli equation:

$$w'''' + K\, d^2w/dt^2 = 0 \tag{1a}$$
$$K = E\, J_s\, L\, /M \tag{1b}$$

Its solution is a class of functions

$$w = f(x,t) \tag{2}$$

where the function depends on the boundary conditions, x and w are respectively the coordinate of the axis of the beam and the deflection. A general closed form solution like (2) is practically impossible to be found, specially if the boundary conditions are functions of time; thus a particular solution of the form

$$w(x,t) = \pi(x)\, \Phi(t) \tag{3}$$

is searched for [1],[2],[3],[4]. The hypothesis of separability into a function of time and a function of space leads to the modal expansion model.
This approach allows to determine a model which put into evidence the eigenvalues of the elastic beam; on the other hand the progressive solution

$$w(x,t) = F(x + v\, t) \tag{4}$$

is not kept into account. This solution is particularly important in

transient phenomena and in the case of external variable forces.
Another drawback of the modal expansion model consists in the fact that during transients the force between the elastic beam and the rigid link is not correctly represented. This is the case of robots with elastic forearm.
If a control law is searched for, the above mentioned drawbacks have negligible consequences, because only the eigenfunctions determine the stability of the system; on the other hand, if a simulation is requested a more realistic model, which keeps into account solution (4) too, is useful.
In this paper we propose a model based on the finite elements technique. The model is obtained by dividing the elastic beam into a number of elemental beams; each of them has a suitable structure and they are conveniently linked each others; the conditions to be satisfied are:
- the kinetic and elastic energies of the whole beam is equal to the sum of those of their parts;
- the eigenvalues of the model coincide with those determined with (3)
- the model is easily implementable.

Several models can be obtained, depending on the structure of the elemental beam. The choice is in general characterized by two conflicting requirements:
- the model must be easily implementable;
- the modes must be represented with accuracy.
    The efficiency of the proposed model to satisfactorily describe the robot dynamic behaviour is investigated by comparing its eigenvalues with those of the eigenfuntion expansion model. This comparison is worked out considering also a variable payload mass.
    Finally, several simulation tests have been performed. They put clearly in evidence the contribution of the progressive solution of the Euler-Bernoulli equation during rapid transients.

2 - THE PROPOSED MODEL
The element of the elastic beam is modelled as represented in Fig. 1

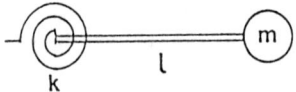

Fig. 1 - The elemental beam

The finite elements decomposition is applied only to the elastic forearm. The elastic beam is composed of a number of mathematical pendulums connected each others by springs.
The following parametres are defined:
    m is the mass of the pendulum
    l is its length
    k is the stiffness of the spring.

Referring to the elastic beam and the rigid link the following parameters are considered:
- Lo length of the rigid link
- Mo mass of the rigid link
- Jo momentum of inertia of the rigid link
- N   number of parts
- L   length of the elastic forearm
- J   momentum of inertia of the elastic beam with respect to the joint with the rigid arm;
- M   mass of the forearm
- Ml payload

If the forearm is a prismatic beam the following relations can be written

$\Sigma l = L \Rightarrow l = L/N$ (5a)
$\Sigma m = M \Rightarrow m = M/N$ (5b)
$J = M L^2/12 + Jm = m \times l^2 (1 + 2^2 + \ldots + N^2) + Jn$ (5c)
$k = E Js/l$ (5d)

being
Jm the momentum of inertia of the motor of the flexible link,
Jo a suitable value to obtain the dynamic equivalence.

Relations (5a-c) represent the fact that in the rigid motion of the elastic beam, its kinetic energy is equal to the sum of those of their elements; equation (5d) imposes that the elastic potential energy of the beam under static load is equal to sum of that of the springs which link the parts.

To determine the equation of motion of the above defined mechanical system, the Lagrange approach is considered.

The following lagrangian coordinates are defined with reference to Fig.2
- θ is the angle between the rigid link and the orizzontal axis;
- ϕ_i angle between the axis of the element i and the orizzontal one;
- $\phi_i - \phi_{i-1}$ angle between element i and i-1;

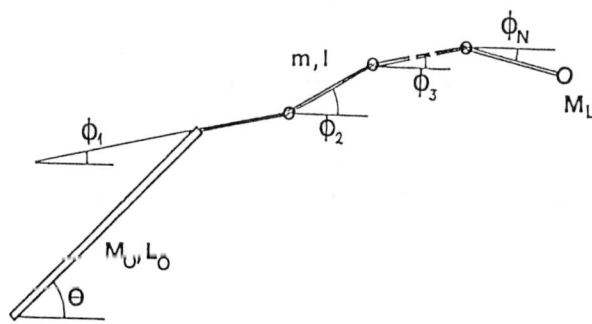

Fig. 2 - Lagrangian coordinates

The kinetic energy of element "p" is

$$T_p = m/2 \ [\ (L_o \ \dot\theta \ )^2 + l^2 \sum_{i=1}^{p} \dot\phi_i^2 + 2 \ l \ L_o \ \dot\theta \sum_{i=1}^{p} \dot\phi_i \cos(\theta-\phi_i) +$$

$$+ \ 2 \ l^2 \sum_{i=1}^{p} \sum_{j=i+1}^{p} \dot\phi_i \ \dot\phi_j \cos(\phi_i-\phi_j)] \tag{6}$$

The potential energy is

$$U_1 = k/2 \ (\phi_0^* + \theta - \phi_1)^2 = k/2 \ (\phi_0 - \phi_1)^2 \qquad U_i = k/2 \ (\phi_i - \phi_{i-1})^2 \tag{7}$$

$$T = \sum_{p=1}^{N} T_p \qquad U = \sum_{i=1}^{N} U_i \tag{8a-b}$$

The lagrangian is
$$L = T - U \tag{9}$$
and the equations of motion are given by

$$\frac{d}{dt} \left(\frac{\partial L}{\partial \dot q_i} - \frac{\partial L}{\partial q_i} \right) = u_i \qquad q_i = [\theta, \phi_0^*, \ldots, \phi_N] \tag{10}$$

$$\phi_0 = \phi_0^* + \theta$$

In matrix notation
$$B \ q'' + K \ q + C = u(t) \tag{11}$$
where
  B is the matrix of inertia
  K is the elastic matrix
  C is the Coriolis term.
  u are the torques.
Damping is not considered.
Being the inverse of B not easily calculable, a linearization can be performed; only the elastic part of the model is linearized assuming that
$$\phi_i - \phi_{i-1} \approx 0 \tag{12}$$
The following semi-linearized equations yield:

if $q_i = \theta$
$$(J_o + J_m + mm \ L^2) \ \theta'' + m \sum [L_o \ \theta'' + L_o \ l \sum \phi''_i \cos(\theta-\phi_i) \ ]$$
$$+ \ Ml \ [\ L_o^2 \ \theta'' + L_o \ l \sum \phi''_i \cos(\theta-\phi_i)] \ + \ k \ (\phi_0 - \phi_1) = u_\theta \tag{13a}$$

if $q_i = \phi_0$
$$J_m \ \phi''_0 + k \ (\phi_0 - \phi_i) = u_\phi \tag{13b}$$

if $q_i = \phi_i$, $i = 1..N$
$$m \sum [L_o \ l \ \theta'' \cos(\theta-\phi_i) + l^2 \sum \phi''_i] + M \ [L_o \ l \ \theta'' \cos(\theta-\phi_N) +$$
$$+ \ l^2 \sum \phi''_i] + k \ (2 \ \phi_i - \phi_{i-1} - \phi_{i+1}) = 0$$
$$\phi_{N+1} = 0 \tag{13c}$$

where sums are performed for $i = 1..N$

The matrices are

$$B = \begin{bmatrix} a & 0 & b(N+c) & b(N-1+c) & b(N-2+c) & . & b(2+c) & b(1+c) \\ 0 & J & 0 & 0 & 0 & . & 0 & 0 \\ b(N+c) & 0 & d(N+c) & d(N-1+c) & d(N-2+c) & . & d(2+c) & d(1+c) \\ b(N-1+c) & 0 & d(N-1+c) & d(N-1+c) & d(N-2+c) & . & d(2+c) & d(1+c) \\ . & . & . & . & . & . & . & . \\ . & . & . & . & . & . & . & . \\ b(2+c) & 0 & d(2+c) & d(2+c) & d(2+c) & . & d(2+c) & d(1+c) \\ b(1+c) & 0 & d(1+c) & d(1+c) & d(1+c) & . & d(1+c) & d(1+c) \end{bmatrix}$$

(14a)

where
$a = J + (M + Ml) Lo^2$
$b = m\ Lo\ l\ \cos(\theta-\phi)$
$c = Ml/m$
$d = m\ l^2$

$$K = \begin{bmatrix} 2 & -1 & 0 & 0 & 0 & . & . & 0 & 0 & 0 \\ -1 & 2 & -1 & 0 & 0 & . & . & 0 & 0 & 0 \\ 0 & -1 & 2 & -1 & 0 & . & . & 0 & 0 & 0 \\ 0 & 0 & -1 & 2 & -1 & . & . & 0 & 0 & 0 \\ . & . & . & . & . & . & . & . & . & . \\ 0 & 0 & 0 & 0 & 0 & . & . & -1 & 2 & -1 \\ 0 & 0 & 0 & 0 & 0 & . & . & 0 & -1 & 1 \end{bmatrix} k$$

(14b)

As can be easily observed the matrices are structured in such a way that increasing their order does not request to calculate them from Lagrangian. The inverse of B is easy to write, having the following structure:

$$B1 = \begin{bmatrix} -d\,s & 0 & b\,s & 0 & 0 & 0 & . & 0 & 0 & 0 \\ 0 & 1/J & 0 & 0 & 0 & 0 & . & 0 & 0 & 0 \\ b\,s & 0 & t\,s & -1/d & 0 & 0 & . & 0 & 0 & 0 \\ 0 & 0 & -1/d & 2/d & -1/d & 0 & . & 0 & 0 & 0 \\ 0 & 0 & 0 & -1/d & 2/d & -1/d & 0 & 0 & 0 \\ . & . & . & . & . & . & . & . & . \\ . & . & . & . & . & . & 2/d & -1/d & 0 \\ 0 & 0 & 0 & 0 & 0 & 0 & . & -1/d & 2/d & -1/d \\ 0 & 0 & 0 & 0 & 0 & 0 & . & 0 & -1/d & (c+2)/[d(c+1)] \end{bmatrix}$$

where (14c)
$s = 1/[b^2(N+c) - a\,d]$
$t = b^2(N-1+c) - a\,d$

3 - EIGENVALUES OF THE ELASTIC BEAM

The eigenvalues are determined referring to submatrices K*, B1* the entries of which are rows 3÷N and columns 3÷N of matrices (14b) and (14c), by solving

$$\|\tau^2 - B1^* K^*\| = 0 \tag{15}$$

In the following Table 1 the eigenvalues computed with (3) and those of the proposed model are reported; four values of the payload are

considered; the proposed model has order N such that N/2 = 11÷14 (i.e. the elastic beam consists of N = 11÷14 elements);
The following parametres are assumed:
    M = 0.425 kg
    L = 1.05 m
    E Js = 2.043696 N m²

Table 1				
EIGENVALUES OF THE PROPOSED MODEL				
	M1 = 0	M1=0.156	M1=0.312	M1=0.624 kg
N = 11; k = 67051.43				
1	6.88	4.42	3.50	2.65
2	42.53	33.79	32.26	31.29
3	116.19	100.16	98.30	97.21
4	219.23	198.31	196.49	195.47
5	344.46	321.50	319.87	318.98
6	482.54	460.44	459.09	458.36
N = 12; k = 94964.89				
1	6.93	4.44	3.52	2.66
2	42.93	34.05	32.51	31.53
3	117.74	101.23	99.33	98.22
4	223.46	201.41	199.52	198.47
5	353.91	328.90	327.16	326.21
6	500.84	475.63	474.13	473.32
N = 13; k = 130801.12				
1	6.97	4.47	3.54	2.66
2	43.27	34.28	32.72	31.73
3	119.02	102.12	100.19	99.07
4	226.95	203.97	202.02	200.93
5	361.66	334.93	333.10	332.10
6	515.88	487.98	486.36	485.48
N = 14; k = 175934.17				
1	7.01	4.62	3.70	2.67
2	43.56	34.45	32.89	31.91
3	120.11	102.88	100.93	99.79
4	229.86	206.10	204.10	202.99
5	368.12	339.92	338.02	336.98
6	528.38	498.17	496.44	495.51
EXACT EIGENVALUES				
	M1 = 0	M1=0.156	M1=0.312	M1=0.624 kg
1	7.17	4.55	3.60	2.73
2	44.91	35.24	33.64	32.63
3	125.75	106.43	104.35	103.16
4	246.41	217.33	215.06	213.81
5	407.34	368.35	365.97	364.69
6	608.49	559.54	557.09	555.79

As can be easily observed the eigenvalues of the proposed model tend to the exact values as N increases.
Errors arise from the fact that the dynamic equivalence has been imposed only for the first mode, which the whole elastic energy is assigned to.
The proposed model has been simulated on a PC with the 4th order Runge-Kutta method. The program is written in Pascal language.
In the following Figs. 3a,b,c the motion of the end point $w(1)$ is represented for various kinds of input torques $u\emptyset$. A comparison is made in the same figures with the model obtained by the eigenfunction expansion technique.
Simulation results show that the proposed model put into evidence some oscillations due to the wave which travels through the elastic beam.

CONCLUSIONS

A model of manipulators with elastic forearm is presented. It has been developed to simulate the behaviour of the elastic beam under variable forces and during transients. The main advantages of that model are that it is easily implementable and increasing its dimension does not require hard calculations, because of its simple structure. Recursive formulae can be obtained for the entries of the model matrices by using algebraic manipulation languages, running also on a standard personal computer.

A better reconstruction of the eigenvalues can be performed assuming more sophisticated schemes for the elemental beam.

REFERENCES

[1] Y. Sakawa, F. Matsumi and S. Fukushima. "Modelling and feedback control of a flexible arm". Journ. of robotics systems, vol.2, n.4, pp. 453-472 (1985).

[2] G. Hastings and W. Book. "Verification of a linear dynamic model for flexible robotic manipulators". Proc. of the IEEE Int. Conf. on Robotics and Automation, pp. 1024-1029, S.Francisco, CA, April 1986.

[3] E. Barbieri and U.Ozguner. "Unconstrained and constrained mode expansions for a flexible slewing link". ASME Trans. Journ. of Dynamic Systems, Measurement and Control, December 1988.

[4] P.Lucibello, F.Nicolò, R.Pimpinelli. "Automatic symbolic modelling of robots with deformable links". IFAC Int. Symp. on theory of robots, Wien, 1986.

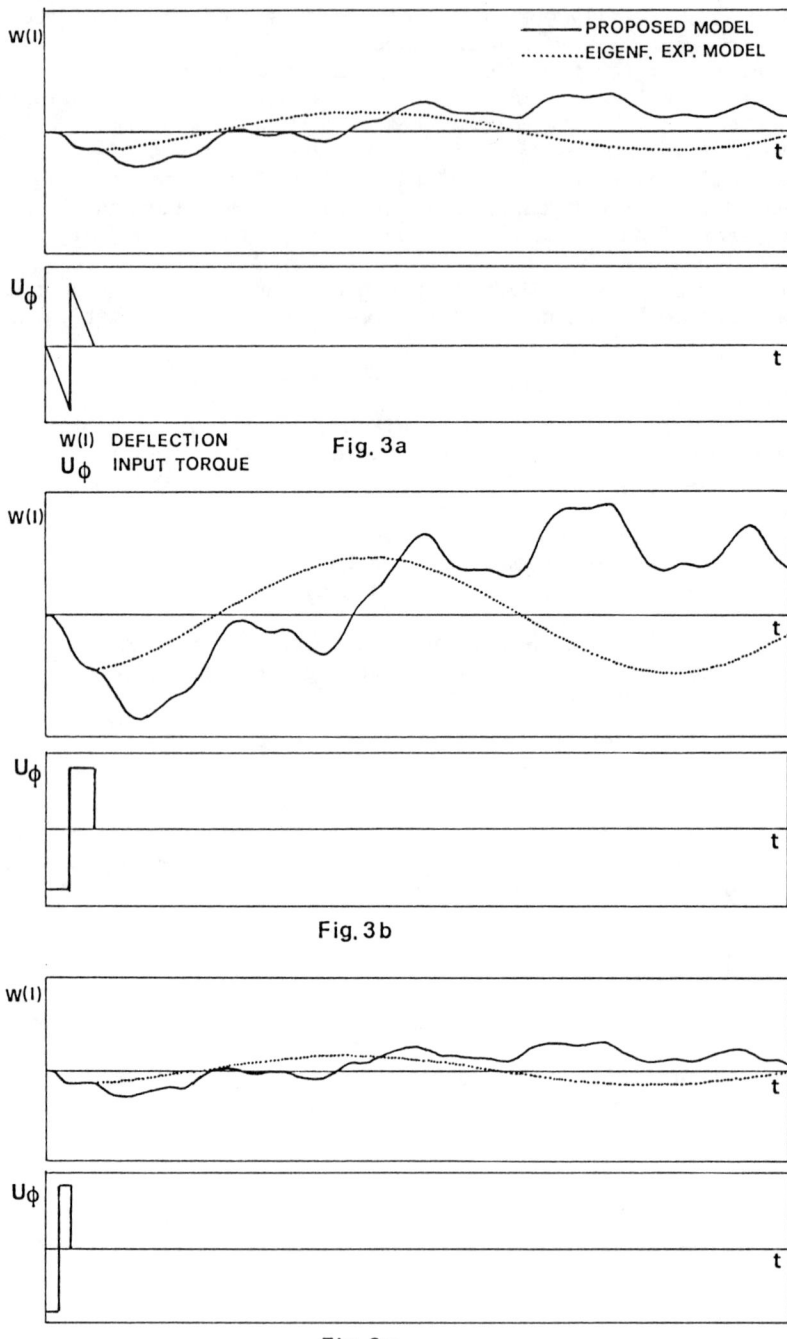

W(l) DEFLECTION
U_φ INPUT TORQUE

Fig. 3a

Fig. 3b

Fig. 3c

On the Modeling of the Contact Forces of Constrained Robots

K. A. Tahboub* and P. C. Müller

Safety Control Engineering, Wuppertal University
Gauß Str. 20, D-5600 Wuppertal 1, Germany

* On leave from the Hebron Technical Engineering College, Palestine.

Abstract. In this paper, the modeling of the contact forces between the end effector of a robot and an external body is considered; the descriptor model is examined in depth. It is shown that the descriptor model can handle the same tasks coverd by the classical approaches if the constraints are carefully and practically defined. The task oriented descriptor modeling method is presented, this method introduces the task space as the starting point in finding the kinematic constraints which cover a large class of tasks including machining and assembling. The constraint equations can be then transformed to the desired space or frame; while it is assured that the rank condition can be always easily checked.

Keywords. Robot; Kinematic Constraints; Geometric Constraints; Contact Forces; Descriptor Model; Assembling; Machining.

1 Introduction and Motivations

There are numerous applications where the end effector of a robot manipulator comes in contact with a rigid external object. These applications include the use of manipulators for carrying out assembling and machining tasks. Driving a screw, inserting a peg into a hole, and turning a crank are representative examples of assembling tasks. In the machining tasks the tool, which is hold by the end effector, cuts parts of the external object, as it is the case in the deburring, the grinding, and the broaching for example.

In both cases, the end effector (which may include a gripper holding an object or a tool holder containing a tool) contacts the external environment, and so the contact forces have to be monitored and controlled. However, there are profound differences between the two applications. In the assembling, in the absence of friction forces, the contact forces are passive; they are the necessary forces to maintain the contact; the force control is required in this case to overcome the positional inaccuracies resulting from elasticities in the joints and in the links, and from the modeling errors for example. Whereas, in the machining case, the normal force for example, takes a desired value depending on the specific application, on the tool, and on the material. The second difference is that the assembling tasks can be generally discussed in a stationary frame, while this is not always possible for general machining tasks since the path of the end effector depends on the specific task to be done.

There have been a lot of research publications which have dealt with constrained robots. Basically, there are two appraoches to handle this subject; the first approach which forms the majority do not focus on the role of the constraints in defining the task [6], [9], [17], and [18]. While the constraints in the second approach play a more important and explicit role [1] ,[5], and [12],. This paper concentrates on the second approach with the premise that there is a need to pay more attention to the practical considerations of finding the constraint equations. A great help can be seen in the gathered experiences arising from developing the methods of the first approach. In this paper, we present some ideas and concepts with the hope to reach a suitable theoretical and practical framework to handle these cases.

2 Modeling Constrained Robots

The traditional robot model used for position and force control in the classical approach in the joint space is

$$\mathbf{M}(\mathbf{q})\ddot{\mathbf{q}} + \mathbf{C}(\mathbf{q},\dot{\mathbf{q}}) = \mathbf{\Gamma} + \mathbf{J}_0^T(\mathbf{q}) \begin{bmatrix} \mathbf{F}_s \\ \mathbf{\Gamma}_s \end{bmatrix}, \qquad (1)$$

where $\mathbf{M}(\mathbf{q})$ is the square mass matrix, $\mathbf{C}(\mathbf{q},\dot{\mathbf{q}})$ represents the Coriolis, centrifugal and gravity force vector, $\mathbf{q} \in \mathbf{R}^n$ is the vector of joint coordinates, $\mathbf{\Gamma} \in \mathbf{R}^n$ is the vector of the generalized input control forces at the joints, $\mathbf{F}_s \in \mathbf{R}^3$ and $\mathbf{\Gamma}_s \in \mathbf{R}^3$ are the force and moment vectors acting on the end effector mapped to the joint space through the transpose of the basic Jacobian matrix $\mathbf{J}_0(\mathbf{q})$. It is important to note here that $\mathbf{F}_s$ and $\mathbf{\Gamma}_s$ are any external forces and moments acting on the end effector; they can result from touching external objects, from an extra attached weight, or from pulling or pushing the end effector in some manner. The basic Jacobian expresses the relation between the end effector's translational $\mathbf{V}$ and rotational $\mathbf{\Omega}$ velocities written in the reference Cartesian frame, and the joint velocities $\dot{\mathbf{q}}$

$$\begin{bmatrix} \mathbf{V} \\ \mathbf{\Omega} \end{bmatrix} = \mathbf{J}_0(\mathbf{q})\dot{\mathbf{q}}. \qquad (2)$$

The use of the virtual work concept [4] to map the end effector forces from the frame of refernce to the joint space, interprets the appearance of the Jacobian in (1).

2.1 Descriptor models

McClamroch [12] proposed a method to model constrained robots, which can be considered as the base in modeling the constrained robot as a differential algebraic system.

Let $\mathbf{P} \in \mathbf{R}^n$ denote the position vector of the end effector, in terms of the frame of reference. Let the constraints on the end effector be given as

$$\mathbf{\Phi}(\mathbf{P}) = \mathbf{0}, \qquad (3)$$

where $\mathbf{\Phi} : \mathbf{R}^n \to \mathbf{R}^m$ is twice continuously differentiable. The relation between the robot coordinates $\mathbf{q}$ and the end effector position $\mathbf{P}$ can be expressed as

$$\mathbf{P} = \mathbf{H}(\mathbf{q}), \qquad (4)$$

where $\mathbf{H} : \mathbf{R}^n \to \mathbf{R}^n$ is invertible and twice continuously differentiable. Let the Jacobian matrix be defined as

$$\mathbf{D}(\mathbf{P}) = \frac{\partial \mathbf{\Phi}(\mathbf{P})}{\partial \mathbf{P}}. \qquad (5)$$

Then the dynamic model of the constrained robot can be written as

$$\mathbf{M}(\mathbf{q})\ddot{\mathbf{q}} + \mathbf{C}(\mathbf{q}, \dot{\mathbf{q}}) = \mathbf{\Gamma} + \mathbf{J}_o^T(\mathbf{q})\mathbf{D}^T(\mathbf{P})\boldsymbol{\lambda}, \qquad (6)$$

where $\boldsymbol{\lambda} \in \mathbf{R}^m$ are Lagrange multipliers.

This work can be considered as a theoretical framework for handling constrained robots as a descriptor system. Some applications and practical considerations as well as some controller designs following this work can be found in many publications [1], [2], [3], [7], [8], [10], [11], [13], [14], [15], and [16]. The common features of the mentioned publications can be summerized as follow:

1. The robot model was written always in the joint space.

2. Geometric constraints were considered, and it was assumed that the constraint equations are given.

3. It is difficult to check Grübler condition which requieres that the rank of the Jacobian matrix $\mathbf{D}$ is m, this condition is necessary for independent constraint force vector $\boldsymbol{\lambda}$.

4. Little attention was given to constraints imposed on the orientation of the end effector.

5. Simple applications were shown as examples, the proposed methods do not guarantee practical solutions in the case of more complicated applications.

In the sequel, we discuss the above points in more depth and present ideas and suggestions which are thought to lead to better chances for the descriptor model in the constrained robot field.

2.2 The Cartesian model

Since, as seen from (3), the constraints are imposed on the end effector and as a result on the joints, and since we are interested in controlling and regulating the motion and contact forces of the end effector in the first place, it is more suitable to write the robot model (1) in Cartesian space, as the frame of reference, as

$$\mathbf{M_x(q)} \begin{bmatrix} \dot{\mathbf{V}} \\ \dot{\mathbf{\Omega}} \end{bmatrix} + \mathbf{C_x(q, \dot{q})} = \mathbf{\Gamma} + \mathbf{J_0^T(q)} \begin{bmatrix} \mathbf{F_s} \\ \mathbf{\Gamma_s} \end{bmatrix}, \qquad (7)$$

where

$$\mathbf{M_x(q)} = \mathbf{M(q)J_0^{-1}} \qquad (8)$$

and

$$\mathbf{C_x(q, \dot{q})} = \mathbf{C(q, \dot{q})} - \mathbf{M_x(q)\dot{J}_0(q)\dot{q}}. \qquad (9)$$

This art of modeling cancels the necessity of transforming the constraint equations from the frame of reference as given in (3) to the joint space through transformation (4), and at the same time provides an effective means to describe the dynamic behaviour of the end effector and to integrate the position and the force control.

2.3 Geometric versus kinematic constraints

The geometric constraints $\mathbf{\Phi(P)}$ describe the surface, with which the robot has to maintain contact; they are refered to as geometric since they describe only the geometry of the surface without explicitly giving information about the path of the end effector on this surface. In fact, what is needed is a description of the constraints as a function of the time, we refer to such constraints as kinematic constraints; they are given as

$$\mathbf{A(q)} \begin{bmatrix} \mathbf{V} \\ \mathbf{\Omega} \end{bmatrix} = \mathbf{0}, \qquad (10)$$

where $\mathbf{A(q)}$ is some matrix of dimension $m \times n$ and function of the joint vector $\mathbf{q}$, and $\mathbf{0} \in \mathbf{R}^m$ is the zero vector. Although, we handle holonomic constraints here and thus equation (10) is integratable, it is more helpful to leave (10) without integration as will be seen later.

In contrary to the geometric constraints, it is possible to describe many robot applications through the kinematic constraints, especially in the assembling and machining fields, where a specific path of the end effector on the constraining surface is desired, as in the case of cutting a certain shape at the surface of a cylinder. Moreover, sometimes it is not necessary to model the shape of the mating surfaces, as in the case of inserting a peg into a hole, where the shape of the hole and the peg are not important to consider, but in order to fulfil the task it is important to know in which direction does the motion take place and in which it is constrained.

2.4 The natural and artificial constraints

A common starting point of most of the classical force / position control methods is analysing the task by writing a set of natural and artificial constraints [9], then the orthogonality between the motion and force directions can be clearly observed. The natural and artificial constraints for the peg into the hole task are given in Figure 1 as an example.

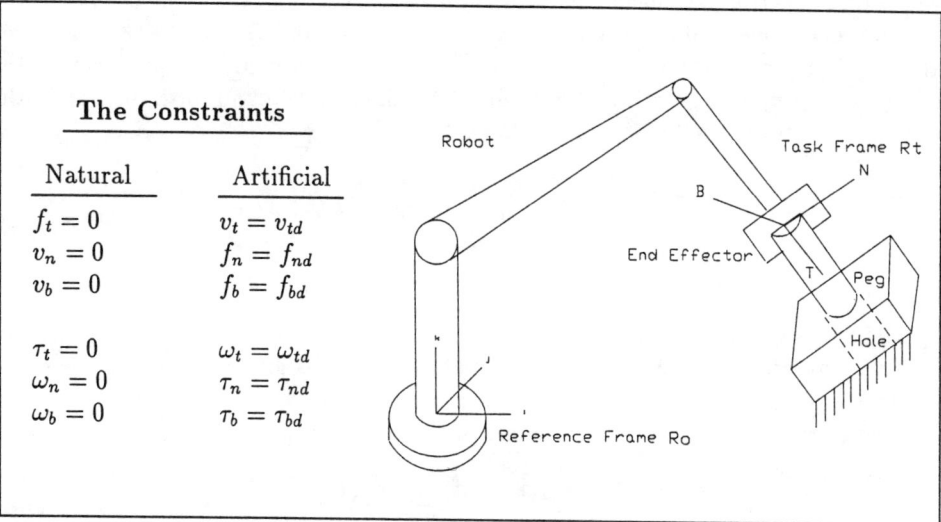

The Constraints

Natural	Artificial
$f_t = 0$	$v_t = v_{td}$
$v_n = 0$	$f_n = f_{nd}$
$v_b = 0$	$f_b = f_{bd}$
$\tau_t = 0$	$\omega_t = \omega_{td}$
$\omega_n = 0$	$\tau_n = \tau_{nd}$
$\omega_b = 0$	$\tau_b = \tau_{bd}$

Fig. 1: The peg into the hole task with the constraints.

The natural constraints are of great importance to develop the constraint equations necessary for the descriptor model; they are imposed by the nature of the task. For example, no force, in the absence of the friction forces, can be exerted in the t direction, no velocity components are allowed in the n or in the b directions; moreover no rotational velocity components are allowed about the n or the b axes to maintain the same orientation for the peg and the hole. The artificial constraints are set as desired in accordance with the natural constraints, the subscript d stands for 'desired' in the table. In the sequel we formulate the task oriented descriptor model starting from the natural constraints imposed on the velocity of the end effector as explained in the given example.

3 The Task Oriented Descriptor Model

We introduce here the task frame as the frame which, along and around its axes, the natural constraints are given; in the above example it corresponds to $\mathbf{R}_t$ as shown in Fig. 1. It can be interpreted as the reference frame rotated by the orthogonal

rotation matrix
$$R_t = \begin{bmatrix} T & N & B \end{bmatrix}. \tag{11}$$

The kinematic constraint equations in this frame are given by
$$\Phi_t(\dot{P}_t) = S \begin{bmatrix} V_t \\ \Omega_t \end{bmatrix} = 0, \tag{12}$$

where the subscript t stands for the task frame, $S \in \mathbf{R}^{m \times n}$ of rank m is occupied with zeros except m elements which are occupied with 1, V_t and Ω_t are the translational and rotational velocity vectors of the end effector written in the task frame. In the given example, where there are four constraints on the 2nd, 3rd, 5th, and 6th degrees of freedom, matrix $S \in \mathbf{R}^{4 \times 6}$ looks like

$$S = \begin{bmatrix} 0 & 1 & 0 & 0 & 0 & 0 \\ 0 & 0 & 1 & 0 & 0 & 0 \\ 0 & 0 & 0 & 0 & 1 & 0 \\ 0 & 0 & 0 & 0 & 0 & 1 \end{bmatrix}. \tag{13}$$

Here, it is guaranteed that the rank of the constraint equations is m, in contrary to the geometric constraints in the joint space, where it is difficult to check the rank condition.

The constraint forces are accordingly given by
$$\begin{bmatrix} F_{st} \\ \Gamma_{st} \end{bmatrix} = \frac{\partial \Phi_t^T(\dot{P}_t)}{\partial \dot{P}_t} \lambda, \tag{14}$$

which is
$$\begin{bmatrix} F_{st} \\ \Gamma_{st} \end{bmatrix} = S^T \lambda. \tag{15}$$

3.1 The descriptor model in the frame of reference

The kinematic constraint equations can be found in the frame of reference by substituting for the velocity vector in (12) by
$$\begin{bmatrix} V_t \\ \Omega_t \end{bmatrix} = \begin{bmatrix} R_t^T & 0 \\ 0 & R_t^T \end{bmatrix} \begin{bmatrix} V \\ \Omega \end{bmatrix}, \tag{16}$$

and thus
$$\Phi(P, \dot{P}) = S \begin{bmatrix} R_t^T & 0 \\ 0 & R_t^T \end{bmatrix} \begin{bmatrix} V \\ \Omega \end{bmatrix} = 0. \tag{17}$$

The constrained forces are found similarly to be
$$\begin{bmatrix} F_s \\ \Gamma_s \end{bmatrix} = \begin{bmatrix} R_t & 0 \\ 0 & R_t \end{bmatrix} S^T \lambda. \tag{18}$$

The derivation of the descriptor model in the frame of reference is thus completed. Equations (7) and (18) represent the differential part whereas equation (17) represents the algebraic part, it is to be noted that the constraint equations (17) are independent and of rank m, since the transformation matrix $\mathbf{R}_t$ is nonsingular.

3.2 The descriptor model in the joint space

Though it is preferable, as explained in section 2.2, to analyse, simulate, and control constrained robots using the Cartesian model (7), we give here the kinematic constraint equations in the joint space for the completeness of the discussion.

The kinematic constraint equations can be found by substituting for the velocity vector in (17) by

$$\begin{bmatrix} \mathbf{V} \\ \Omega \end{bmatrix} = \mathbf{J}_0(\mathbf{q})\dot{\mathbf{q}}, \qquad (19)$$

and thus

$$\Phi(\mathbf{q}, \dot{\mathbf{q}}) = \mathbf{S} \begin{bmatrix} \mathbf{R}_t^T & 0 \\ 0 & \mathbf{R}_t^T \end{bmatrix} \mathbf{J}_0(\mathbf{q})\dot{\mathbf{q}} = 0. \qquad (20)$$

and similarly the constraint forces in the joint space Γ_{con} are given by

$$\Gamma_{con} = \mathbf{J}_0^T(\mathbf{q}) \begin{bmatrix} \mathbf{R}_t & 0 \\ 0 & \mathbf{R}_t \end{bmatrix} \mathbf{S}^T \lambda, \qquad (21)$$

In this space, as well, the constraint equations (20) are independent and of rank m, if the basic Jacobian $\mathbf{J}_0(\mathbf{q})$ is not singular at the given configuration $\mathbf{q}$.

4 Discussion and Conclusions

In this paper, we have presented a short review on modeling the contact forces of constrained robots; we have discussed two main groupes: the classical approach which integrates the constraints in the dynamical model structure directly, and the descriptor model approach which is consisted of a differential equation describing the dynamic behaviour of the robot and an algebraic equations set describing the constraints explicitly. It is noted that, not much interchanges between the two approaches related to the constraints have been made, so the premise of this paper was to reconsider the descriptor model approach trying to benefit from the well developed concepts in the classical approach.

The common features of the descriptor model methods were reviewed in section 2.1, and a treatment of these features has been made in the following sections.

The term kinematic constraints has been used to replace the geometric constraints; the kinematic constraints are derived in the task space as a list of natural constraints and are then transformed to the space of interest. This method guarantees a wide range of applications with practical significance; an example of an assembling task was given, it was shown in a previous work [19] how to use the same method for general machining tasks. Moreover, it is easy to prove the independence of the constraint equations in all of the considered tasks and frames.

It is hoped, that this work has fulfilled its aims, in giving the descriptor model approach a better chance to handle constrained robots in a practical way.

5 Acknowledgment

The authors would like to thank the members of the Differential Algebraic Systems Group: Mr. Ming Hou, Mr. Thorsten Schmidt, and Mr. Rolf Schüpphaus, for the helpful discussions.

References

[1] Cai, L., and A. A. Goldenberg (1988). General dynamic model for analysis and simulation of robots in contact tasks. IFAC Sympos. Robot Control, pp. 61.1-6.

[2] Cai, L., and A. A. Goldenberg (1989). An approach to force and position control of robot manipulators. IEEE Int. Conf. Robotics and Automation, pp. 86-91.

[3] Cai, L., and A. A. Goldenberg (1989). A new approach to force and position control of robot manipulators. IEEE Int. Conf. on Control and Applications, pp. RA-3-5.1-7.

[4] Craig, J. J. (1986). Introduction to robotics, mechanics and control. Addison-Wesley, Reading, Msddschusets, pp. 152-153.

[5] Hemami, H., and B. F. Wyman (1979). Modeling and control of constrained dynamic systems with applications to biped locomotion in the frontal plane. IEEE Trans. on Automatic control, Vol. AC-24, pp. 526-535.

[6] Hogan, N. (1985). Impedence Control: an approach to manipulations: part I-theory; part II-implementation; part III-applications. ASME J. Dynamics Systems, Measurement and Control, Vol. 107, pp. 1-24.

[7] Huang, H. P., and N. H. McClamroch (1988). Time-optimal control for a robotic contour following problem. IEEE J. Robotics and Automation, Vol. 4, No. 2, pp. 140-149.

[8] Huang, H. P., and M. Lin (1990). Variable structure control of constrained dynamic systems. IEEE Int. Conf. Robotics and Automation, pp. 1362-1367.

[9] Mason, M. T. (1981). Compliance and force control for computer controlled manipulators. IEEE Trans. Systems, Man and Cybernetics, Vol. SMC-11, pp. 418-432.

[10] McClamroch, N. H. (1986). Singular systems of differential equations as dynamic models for constrained robots. IEEE Int. Conf. Robotics and Automation, pp. 21-28.

[11] McClamroch, N. H. (1990). A singular perturbation approach to modeling and control of manipulators constrained by a stiff environment. IEEE Conf. on Decision and Control.

[12] McClamroch, N. H., and H. P. Huang (1985). Dynamics of a closed chain manipulator. American control conference, pp. 50-54.

[13] McClamroch, N. H., and D. Wang (1988). Feedback stabilization and tracking of constrained robots. IEEE Trans. Automatic Control, Vol. 33, No. 5, pp. 419-426.

[14] McClamroch, N. H., and D. Wang (1990). Linear feedback control of position and contact force for a nonlinear constrained mechanism. Transactions of ASME, Vol. 112, pp. 640-645.

[15] Krishnan, H., and N. H. McClamroch (1990). A new approach to position and contact force regulation in constrainred robot systems. IEEE Int. Conf. Robotics and Automation, pp. 1344-1349.

[16] Mills, J., and A. A. Goldenberg (1989). Force and position control of manipulators during constrained motion tasks. IEEE Trans. Robotics and Automation, Vol. 5, No. 1, pp. 30-46.

[17] Raibert, M. H., and J. J. Craig (1982). Hybrid position/force control of manipulators. ASME J. of Dynamics Systems, Measurment and Control, Vol. 102, pp. 126-133.

[18] Salisbury, J. K. (1980). Active stiffness control of a manipulator in Cartesian coordinates. IEEE Conf. on Decision and Control, pp. 95-100.

[19] Tahboub, K. A., and P. C. Müller (1991) A reduced dynamic model for constrained robots in the task frame. Robotersysteme, Vol. 7, pp. 49-52.

PART 2
ROBOT CONTROL

Passivity and Learning Control for Robot Tasks under Geometric Constraints

Suguru ARIMOTO and Tomohide NANIWA

Faculty of Engineering, University of Tokyo,
Bunkyo-ku, Tokyo, 113 Japan

SUMMARY

Learning control is a new approach to the problem of skill refinement for robotic manipulators. It is considered to be a mathematical model of motor program learning for skilled motions in the central nervous system.

This paper proposes a class of learning control algorithms for bettering operation of the robot arm under a geometrical end-point constraint. At each trial, the command input torque is modified by present joint velocity errors deviated from the desired velocity trajectory. It is shown that motion trajectories approach the desired one in the sense of squared integral norm provided the local feedback loop consists of both position and velocity feedbacks plus a feedback term of the error force vector between the reaction force and desired force on the end-point constrained surface. It is explored that various passivity properties of residual error dynamics of the manipulator play a crucial role in the proof of convergence of both position and velocity trajectories.

1. INTRODUCTION

Learning control is a new approach for the control problem of skill refinement. We humans are able to acquire the skill via a long series of repeated exercises. Motivated by this observation, much literature concerned with learning control techniques for robotic systems has accumulated very recently[1-7], mainly during the past several years. However, most of the papers were so far concerned with only a class of tasks described in terms of joint-trajectory tracking. In other words, for a desired motion given to a robot arm in which its end-point is free to move in the three-dimensional space, the arm repeats exercises to reduce the trajectory tracking errors gradually and eventually can learn to exactly trace the given desired motion.

However, there is a variety of tasks for robotic systems that must be described in terms of end-point constraint. Writing with a pen is such an example, since the tip of the pen must move in touch with a given sheet of paper fixed on a table. In such a case, not only a desired end-point trajectory but also a desired time-evolution of contact force acting on the surface along the end-point trajectory must be specified. This is called traditionally a hybrid (position and force) control problem in the field of robotics.

This paper proposes a simple learning control algorithms, which is effective for a class of tasks with geometrical end-point constraints. Under the assumptions that the contact force can be measured via a force sensor and the local servo loop consists of a feedback of the error force vector being evaluated at joints in addition to the ordinary feedback of both position and velocity signals, it is shown that the learning control algorithm assures the convergence of motion trajectories to the desired motion in the sense of squared integral function norm. In the proof, various passivity properties of residual error dynamics of the robot arm are vital the same as in the case of ordinary joint-trajectory tracking with free end-point[5-7].

2. PASSIVITY PROPERTIES OF ROBOT DYNAMICS

A class of serial-rink manipulators with all revolute-type joints is considered in the paper. First we discuss the passivity of the dynamics of such a manipulator with free end-point, which can be described in terms of joint coordinates vector $q = (q^1, \cdots, q^n)^T$ in the following way:

$$L(q) \triangleq (H_0 + H(q))\ddot{q} + (B_0 + \frac{1}{2}\dot{H}(q))\dot{q} + S(q,\dot{q})\dot{q} + g(q) = u \quad (1)$$

where H denotes an inertia matrix, H_0 a positive diagonal matrix representing inertial terms of internal load distribution of actuators, $g(q)$ a vector of gravity terms, u a vector of input torques generated at servo actuators, B_0 a positive definite matrix representing damping factors and coefficients of electro-motive forces. It is well known that the inertia matrix $H(q)$ is symmetric and positive definite and, moreover, each entry H_{ij} of H is constant or a trigonometric function of components of joint vector q. Hence, $H(q)$ and any of partial derivatives of $H(q)$ with respect to q^i are uniformly Lipschitz continuous in q. The term $S(q,\dot{q})\dot{q}$ expresses

$$S(q,\dot{q})\dot{q} = \frac{1}{2}\dot{H}(q)\dot{q} - \frac{\partial}{\partial q}\left\{\frac{1}{2}\dot{q}^T H(q)\dot{q}\right\}. \quad (2)$$

and hence the ij-entry of S can be represented by

$$S_{ij} = \frac{\partial}{\partial q_j}\left(\sum_{k=1}^{n}\dot{q}_k H_{ik}\right) - \frac{\partial}{\partial q_i}\left(\sum_{k=1}^{n}\dot{q}_k H_{jk}\right) \quad (3)$$

which implies the skew-symmetry of $S(q,\dot{q})$, i.e., $r^T S(q,\dot{q})r = 0$, in general. The passivity of robot dynamics follows directly from the skew-symmetry of S.

Property 1 Motion equation (1) implies the passivity of velocity vector $\dot{q}$ with respect to torque input u, i.e.,

$$\int_0^t \dot{q}^T(\tau)u(\tau)d\tau \geq \gamma \quad (4)$$

for any $t \geq 0$ and a fixed constant γ depending only on the initial state $x(0) (= (q(0), \dot{q}(0)))$.

In fact, we see that

$$\int_0^t \dot{q}^T(\tau)u(\tau)d\tau = \int_0^t \dot{q}^T(\tau)L(q(\tau))d\tau$$

$$= \int_0^t \left[\frac{d}{dt} \left\{ \frac{1}{2} \dot{q}^T (H_0 + H(q)) \dot{q} + G(q) \right\} + \dot{q}^T B_0 \dot{q} \right] d\tau$$

$$= \int_0^t \dot{q}^T B_0 \dot{q} d\tau + V(t) - V(0), \tag{5}$$

where $V(t)$ is defined as

$$V(t) = \frac{1}{2} \dot{q}^T(t) \{H_0 + H(q(t))\} \dot{q}(t) + G(q(t)) \tag{6}$$

and $G(q)$ denotes the potential function induced by the gravity force, i.e., $g(q) = (\partial G / \partial q^1, \cdots, \partial G / \partial q^n)^T$. Since the constant term of potential is arbitrary, it is reasonable to assume that $\min_q G(q) = 0$. Then, $V(t) \geq 0$ and therefore eq.(6) implies

$$\int_0^t \dot{q}^T(t) u(t) dt \geq -V(0) = \gamma. \tag{7}$$

which proves Property 1.

The passivity of robot dynamics is quite natural as well as the passivity of electrical lumped-parameter circuits. In fact, the left-hand side of eq.(4) implies the total work done by torques generated by actuators during the time interval $[0, T]$. This quantity is reasonably equivalent to the increase or decrease of total energy, $V(t) - V(0)$, plus the energy consumption during $[0, T]$ as shown in eq.(5).

Now we consider the robot dynamics in the case that the end-effector is in touch with a surface as shown in Fig. 1. Suppose that the surface is described by a scalar equation, $\varphi(x^1, x^2, x^3) = 0$, where $x = (x^1, x^2, x^3)^T$ denotes the cartesian coordinates (task coordinates) fixed at the inertial reference frame, and the contact friction arises in the direction of $-\dot{x}$ with the magnitude $\xi(\|\dot{x}\|)$ where $\xi(\alpha)$ is a positive function of α. Then the dynamics is expressed by the form

$$L(q) = J_\varphi^T(q) f - \xi(\|\dot{x}\|) J_x^T(q) \dot{x} + v, \tag{8}$$

where f is the magnitude of the contact force as shown in Fig. 1 and $J_\varphi(q)$ and $J_x(q)$ denote the $1 \times n$ Jacobian matrix of φ and the $3 \times n$ Jacobian matrix of x, respectively, with respect to joint vector q, that is,

$$J_\varphi(q) = \left(\sum_{i=1}^3 \frac{\partial \varphi}{\partial x^i} \frac{\partial x^i}{\partial q^1}, \cdots, \sum_{i=1}^3 \frac{\partial \varphi}{\partial x^i} \frac{\partial x^i}{\partial q^n} \right) = \frac{\partial \varphi}{\partial x} \frac{\partial x}{\partial q} = \frac{\partial \varphi}{\partial x} J_x. \tag{9}$$

Similarly to Property 1, it is possible to state the following:

Property 2 As long as the end-point of the manipulator is constrained on the surface $\varphi(x) = 0$, the passivity condition of robot dynamics described by eq.(8) is satisfied, i.e., it holds that

$$\int_0^t \dot{q}^T(\tau) v(\tau) d\tau \geq \gamma \tag{10}$$

where γ is a constant depending on only the initial state.

To see this, first note that the geometrical constraint described by eq.(11) implies $J_\varphi(q) \dot{q} = 0$, which in fact follows directly from $d\varphi/dt = 0$. Then, by taking the inner product of $\dot{q}(\tau)$ with eq.(8), it is possible to obtain the same conclusion as in Property 1. This verifies eq.(10).

3. P-TYPE LEARNING CONTROL

For a desired motion trajectory which is given in terms of joint velocity $\dot{q}_d(t)$ over $t \in [0,T]$, the learning control law is described in the following recursive form:

$$u_{k+1}(t) = u_k(t) - \Phi \dot{r}_k(t) \tag{11}$$

where r_k denotes the residual error defined by

$$r_k(t) = q_k(t) - q_d(t) \tag{12}$$

and Φ denotes a positive definite constant gain matrix. The recursive form of eq.(11) is called the P-type learning algorithm since a Proportional term of the velocity error is used in modification of the input torque. Differently from a D-type learning algorithm[1-2] in which the derivative of the velocity error signal is used, a certain extended concept of passivity concerning the residual error dynamics played a crucial role in the proof[5-7] of uniform boundedness and convergence of the motion trajectories during repetitive learning. To gain an insight into this, note that subtraction of the ideal input $u_d(t)$ realizing the desired trajectory $q_d(t)$ from both sides of eq.(11) yields

$$\Delta u_{k+1}(t) = \Delta u_k(t) - \Phi \dot{r}_k(t) \tag{13}$$

where

$$\Delta u_i = u_i - u_d. \tag{14}$$

Then, it follows from eq.(13) that

$$\Delta u_{k+1}^T \Phi^{-1} \Delta u_{k+1} = \Delta u_k^T \Phi \Delta u_k - 2\dot{r}_k^T \Delta u_k + \dot{r}_k^T \Phi \dot{r}_k. \tag{15}$$

Hence, if there are positive constants $\lambda > 0$ (not so large) and $\beta > 0$ such that

$$\int_0^t e^{-\lambda \tau} \dot{r}_k^T(\tau) \Delta u_k(\tau) d\tau \geq \frac{1+\beta}{2} \int_0^t e^{-\lambda \tau} \dot{r}_k^T(\tau) \Phi \dot{r}_k(\tau) d\tau, \tag{16}$$

then it follows from eq.(15) that

$$\int_0^T e^{-\lambda t} \Delta u_{k+1}^T(t) \Phi^{-1} \Delta u_{k+1}(t) dt$$
$$\leq \int_0^T e^{-\lambda t} \Delta u_k^T(t) \Phi^{-1} \Delta u_k(t) dt - \beta \int_0^T e^{-\lambda t} \dot{r}_k^T(t) \Phi \dot{r}_k(t) dt. \tag{17}$$

This means that the squared integral norm of the input error signal decreases with repetition of exercises as long as the squared integral of velocity error dynamics does not vanish.

As in the previous literature[5-7], the inequality

$$\int_0^t e^{-\lambda \tau} r^T(\tau) \Delta u(\tau) d\tau \geq \gamma \tag{18}$$

is called the condition of exponential passivity concerning the residual dynamics. Clearly the exponential passivity is a weaker and more relaxed condition than the ordinary passivity discussed in the previous section. However, the inequality of (16) is stronger and hence more restricted in comparison with the exponential passivity. Therefor, it is reasonable to call the inequality (16) with $\beta > 0$ and $\lambda > 0$ the exponential passivity with a specified quadratic margin. In the next section, we will show that such a stronger condition of exponential passivity is valid for the residual error dynamics of the manipulator under the end-point constraint provided that the inner servo loop is properly composed of a force feedback in addition to the ordinary position and velocity feedback.

4. PASSIVITY OF RESIDUAL DYNAMICS

Now suppose that a desired end-point trajectory $x_d(t)$ and a desired time-evolution $f_d(t)$ of the magnitude of the contact force are defined on $t \in [0,T]$ and given to the manipulator. We reasonably assume that $x_d(t)$ satisfies $\varphi(x_d(t)) = 0$ and the contact force directs inside the contact surface, that is, $f_d(t) > 0$, for all $t \in [0,T]$. If the degree of freedom of the manipulator is greater than three ($n > 3$), then there is a possibility of existence of many $q_d(t)$ that may satisfy $\varphi(x(q_d)) = 0$ for all $t \in [0,T]$, where $x(q) = (x^1(q), x^2(q), x^3(q))$ denotes the coordinates transformation from joint coordinates q to cartesian coordinates x. In the present paper we assume that in that case ($n > 3$) the posture of ($n-3$) end-components of q is specified and hence there exists a unique $q_d(t)$ satisfying $x(q_d) = x_d$. However, we neither need the computation data of q_d on the basis of inverse kinematics nor use it in the close-loop servo. As a matter of course we need on-line measurement data on the reaction force $f(t)$ caused by the contact between the end-point and the surface. Thus we consider a servo-loop for the manipulator in the following way:

$$v = -Aq - B_1\dot{q} - J_\varphi^T(q)(f - f_d) + u. \tag{19}$$

The third term of the right hand side refers to the force feedback through the transpose of the Jacobian matrix and the fourth term u refers to the feedforward input that must be determined through learning. Substitution of eq.(19) into eq.(8) yields

$$L(q) + B_1\dot{q} + Aq = J_\varphi^T(q)f_d - \xi(\|\dot{q}\|)J_x^T(q)\dot{x} + u \tag{20}$$

where we implicitly assume that $\dot{x}_d$ (and hence $\dot{q}_d$) is differentiable and both $\ddot{x}_d$ (and hence $\ddot{q}_d$) and f_d are piecewise continuous. Since eq.(20) is invertible from output $\dot{q}$ to input u, it is possible to assume the existence of an ideal input u_d that realizes the desired output $\dot{q}_d$, i.e.,

$$u_d = L(q_d) + B_1\dot{q}_d + Aq_d - J_\varphi^T(q_d)f_d + \xi(\|\dot{x}\|)J_r^T(q_d)\dot{x}_d. \tag{21}$$

Subtracting this equation from eq.(20) yields

$$(H_0 + H(q_d + r))\ddot{r} + (B + \frac{1}{2}\dot{H}(q_d + r))\dot{r} + S(q_d + r, \dot{q}_d + \dot{r})\dot{r} + Ar + h = \Delta u \tag{22}$$

where

$$\Delta u = u - u_d, \qquad r = q - q_d, \qquad B = B_0 + B_1,$$

$$h = h(r, \dot{r}) = \{H(q_d + r) - H(q_d)\} \ddot{q}_d + \frac{1}{2}\{\dot{H}(q_d + r) - \dot{H}(q_d)\} \dot{q}_d$$
$$+ \{S(q_d + r, \dot{q}_d + \dot{r}) - S(q_d, \dot{q}_d)\dot{q}_d\} + g(q_d + r) - g(q_d)$$
$$- \{J_\varphi(q_d + r) - J_\varphi(q_d)\}^T f_d + \xi(\|\dot{q}_d + \dot{r}\|)\dot{q} - \xi(\|\dot{q}_d\|)\dot{q}_d. \quad (23)$$

Note that every entry of $H(q)$ is constant or a sinusoidal function of components of q and every entry of $S(q, \dot{q})$ is linear in $\dot{q}$. Therefore, h is linear in $\dot{r}$ and hence can be rewritten into the following form:

$$h = E(f_d, q_d, \dot{q}_d, \ddot{q}_d)r + F(q_d, \dot{q}_d, r)\dot{r} + \bar{h}(f_d, q_d, \dot{q}_d, \ddot{q}_d, r, \dot{r}) \quad (24)$$

where all linear terms of $\dot{r}$ in h in eq.(28) are firstly recast into the second term of the right hand side and hence the remaining terms in the right hand side become irrelevant to $\dot{r}$. In detail,

$$F(q_d, \dot{q}_d, r)\dot{r} = \frac{1}{2}\left\{\sum_{i=1}^{n} H^i(q_d + r)\dot{r}^i\right\} \dot{q}_d$$
$$+ \xi(\|\dot{q}_d + \dot{r}\|)\dot{r} + \{S(q_d + r, \dot{q}_d + \dot{r}) - S(q_d + r, \dot{q}_d)\} \dot{q}_d \quad (25)$$

where $H^i = \partial H/\partial r^i$, and

$$E(f_d, q_d, \dot{q}_d, \ddot{q}_d)r + \bar{h}(f_d, q_d, \dot{q}_d, \ddot{q}_d, r, \dot{r}) = g(q_d + r) - g(q_d)$$
$$+ \frac{1}{2}\left[\sum_{i=1}^{n}\{H^i(q_d + r) - H^i(q_d)\} \dot{q}_d^i\right] \dot{q}_d$$
$$+ \{H(q_d + r) - H(q_d)\} \ddot{q}_d + \{S(q_d + r, \dot{q}_d) - S(q_d, \dot{q}_d)\} \dot{q}_d$$
$$- \{J_\varphi(q_d + r) - J_\varphi(q_d)\}^T f_d + \{\xi(\|\dot{q}_d + \dot{r}\|) - \xi(\|\dot{x}_d\|)\} \dot{q}_d. \quad (26)$$

Note again that all H, S, g, and J_φ are periodic in r and thereby all entries and components of F and $\bar{h}$ are bounded provided all components of $\ddot{q}_d$ and f_d are piecewise continuous and hence bounded. In addition, we assume that there exists a constant $\bar{\rho} > 0$ such that the friction in the direction $-\dot{x}$ satisfies $|\xi(\|\dot{x}\|) - \xi(\|\dot{x}_d\|)| < \bar{\rho}\|\dot{x}\|$. According to these observations, we see that there exist constants $\rho_0 > 0$ and $\rho_1 > 0$ such that

$$|\dot{r}^T h| \leq |\dot{r}^T E r| + |\dot{r}^T F \dot{r}| + |\dot{r}^T \bar{h}| \leq \rho_0 r^T r + \rho_1 \dot{r}^T \dot{r} \quad (27)$$

for any r and $\dot{r}$.

We are now in a position to show the exponential passivity with a quadratic margin for the residual dynamics described by eq.(22).

Property 3. As long as the end-point of the manipulator is constrained on the surface $\varphi(x) = 0$, the exponential passivity of the residual robot dynamics of eq.(22) is satisfied, i.e., it holds that

$$\int_0^t e^{-\lambda \tau} \dot{r}^T(\tau) u(\tau) d\tau \geq \frac{1+\beta}{2} \int_0^t e^{-\lambda \tau} \dot{r}^T(\tau) \Phi \dot{r}(\tau) d\tau + \gamma \quad (28)$$

with $\lambda > 0$ (not so large), and $\beta > 1$, where γ depends on only the initial state $(r(0), \dot{r}(0))$.

To prove this, we observe that

$$\int_0^t e^{-\lambda \tau} \dot{r}^T(\tau) \Delta u(\tau) d\tau$$
$$= \frac{1}{2} \int_0^t \frac{d}{d\tau} \left[e^{-\lambda \tau} \left\{ \dot{r}^T (H_0 + H(q_d + r)) \dot{r} + r^T A r \right\} \right] d\tau$$
$$+ \frac{1}{2} \int_0^t \lambda e^{-\lambda \tau} \left\{ \dot{r}^T (H_0 + H(q_d + r)) \dot{r} + r^T A r \right\} d\tau$$
$$+ \int_0^t e^{-\lambda \tau} \left\{ \dot{r}^T B \dot{r} + \dot{r}^T h(r, \dot{r}) \right\} d\tau$$
$$\geq e^{-\lambda t} U(r(t), \dot{r}(t)) - U(r(0), \dot{r}(0))$$
$$- \int_0^t e^{-\lambda \tau} \left[\lambda U(r(\tau), \dot{r}(\tau)) + \dot{r}^T(\tau) B \dot{r}(\tau) \right] d\tau$$
$$- \int_0^t e^{-\lambda \tau} \left\{ \rho_0 r^T(\tau) r(\tau) + \rho_1 \dot{r}^T(\tau) \dot{r}(\tau) \right\} d\tau \qquad (29)$$

where

$$U(r, \dot{r}) = \frac{1}{2} \left\{ \dot{r}^T (H_0 + H(q_d + r)) \dot{r} + r^T A r \right\}. \qquad (30)$$

Next it is convenient to define a scalar function

$$W(\lambda; r, \dot{r}) = \lambda U(r, \dot{r}) + \dot{r}^T B \dot{R} - \rho_0 r^T r - \rho_1 \dot{r}^T \dot{R} - \frac{1+\beta}{2} \dot{r}^T \Phi \dot{r}$$
$$= \frac{1}{2} r^T (\lambda A - 2\rho_0) r + \frac{1}{2} \dot{r}^T \left[\lambda \left\{ H_0 + H(q_d + r) \right\} + 2B - 2\rho_1 I - (1+\beta) \Phi \right] \dot{r}$$

which becomes positive definite in r and $\dot{r}$ with an appropriate choice for positive λ. From this it follows that

$$\int_0^t e^{-\lambda \tau} \dot{r}^T \Delta u_d d\tau \geq e^{-\lambda t} U(r(t), \dot{r}(t)) - U(r(0), \dot{r}(0))$$
$$+ \int_0^t e^{-\lambda \tau} W(\lambda; r(\tau), \dot{r}(\tau)) d\tau + \frac{1+\beta}{2} \int_0^t e^{-\lambda \tau} \dot{r}^T \Phi \dot{r} d\tau$$
$$\geq -U(r(0), \dot{r}(0)) + \frac{1+\beta}{2} \int_0^t e^{-\lambda \tau} \dot{r}^T(\tau) \Phi \dot{r}(\tau) d\tau \qquad (31)$$

which proves Property 3.

It is important to remark that if $\beta > 0$ and Φ satisfies $2B \geq (1+\beta)\Phi > 0$, then the choice for λ depends on neither γ nor $\beta \Phi$. In other words, λ can be chosen in such a way that

$$\lambda = \max \left\{ 2\rho_0 \left\| A^{-1} \right\|, 2\rho_1 \max_q \left\| \{H_0 + H(q)\}^{-1} \right\| \right\}. \qquad (32)$$

Here $\|X\|$ denotes the spectre radius of matrix X. Another important remark is that the above exponential passivity is valid even when at some instant the end-point may leave or repeat to touch and leave may times the surface $\varphi(x) = 0$ during $[0, T]$, provided that the tip of the end-effector does not slip on the surface.

5. CONVERGENCE OF TRAJECTORIES

We now return to the problem of convergence of trajectories when the P-type learning control law described by eq.(11) is applied for the robot arm whose dynamics is represented by eq.(20). We assume at the first stage that the arm is reinitialized perfectly the same at the beginning of exercise in every trial, that is

$$q_k(0) = q_d(0), \quad \dot{q}_k(0) = \dot{q}_d(0) \quad \text{for } k = 0, 1, 2, \cdots. \tag{33}$$

Then, γ in eq.(28) vanishes since $U(r_k(0), \dot{r}_k(0))$ in eq.(31) vanishes and therefore we see the validity of inequality (17) which can be written in the form

$$\|\Delta u_{k+1}\|_{\Phi^{-1}}^2 \leq \|\Delta u_k\|_{\Phi^{-1}}^2 - \beta \|\dot{r}_k\|_\Phi^2. \tag{34}$$

Since the norm $\|\Delta u_k\|_{\Phi^{-1}}$ is bounded and monotonously decreasing in k as long as $\|\dot{r}_k\|_\Phi \neq 0$, velocity error norms $\|\dot{r}_k\|_\Phi$ converge to zero. Since $q_k \to q_d$ as $k \to \infty$ implies $x_k \to q_k$ as $k \to \infty$, we conclude as follows:

Theorem 1 When the learning control law of eq.(11) is applied for the manipulator with the geometrical end-point constraint, motion trajectories $x_k(t)$ converge to the given desired one $x_d(t)$ as $k \to \infty$, provided that the manipulator dynamics is subject to eq.(20) and the perfect reinitialization is satisfied.

6. CONCLUDING REMARKS

A recursive algorithm of P-type learning control is proposed for bettering robot tasks with geometric end-point constraints. A new concept of exponential passivity with a quadratic margin is introduced for a class of residual error dynamics of robots with or without geometric end-point constraints. A rigorous proof of the convergence of motion trajectories to the desired motion is presented on the basis of passivity properties concerning residual error dynamics of robots.

We finally remark that the convergence of force trajectories $f_k(t)$ to the desired one $f_d(t)$ can not be assured by the argument given above, for the force $J_\varphi^T f$ effective at joints and caused by the reactive force at the end-point contact is compensated by the force feedback and therefore the residual error force term $\Delta f_k = f_k - f_d$ dose not appear explicitly in eq.(22). When 1) reinitialization is not perfect, 2) the force feedback is not completed owing to incorrect measurement of the contact force, and 3) there are unknown fluctuations to some extent to be included in the dynamics, we can obtain

$$\|\Delta u_{k+1}\|_{\Phi^{-1}}^2 \leq \|\Delta u_k\|_{\Phi^{-1}}^2 - \beta \|\dot{r}_k\|_\Phi^2 + \varepsilon \tag{35}$$

instead of eq.(34), where $\varepsilon > 0$ depends on all errors cited above. In such a case, introduction of a forgetting factor into the recursive learning becomes in effect likely as shown[7-8] in the case of free end-point. The detailed discussions of this problem will be presented elsewhere in a future.

REFERENCES

1. S.Arimoto, S, Kawamura, and F. Miyazaki "Bettering operation of robots by learning" *Journal of Robotic Systems* **1**, 123–140, (1984).

2. ibid: Can mechanical robots learn by themselves?, in *"Robotics Research" The Second International Symposium*, H. Hanafusa & H. Inoue, Eds., MIT Press, Cambridge, Massachusetts, 127–134, (1985).
3. J.J. Craig "Adaptive control of manipulators through repeated trials" *Proc. of 1984 American Control Conference*, San Diego, California, (1984).
4. P. Bondi, G. Casalino, and L. Gambardella "On the iterative learning control theory for robotic manipulators" *IEEE J. of Robotics and Automation 4*, 14–22, (1988).
5. S. Arimoto "Mathematical theory of learning with applications to robot control" *Proc. of 4th Yale Workshop on Applications of Adaptive Systems Theory*, Yale University, New Haven, Connecticut, (1985).
6. S. Kawamura, F. Miyazaki, and S. Arimoto "Realization of robot motion based on a learning method" *IEEE Trans. on Systems, Man and Cybernetics SMC-18*, 126–134, (1988).
7. S. Arimoto "Learning control theory for robotic motion" *International Journal of Adaptive Control and Signal Processing 4*, 543–564, (1990).
8. S. Arimoto "Learning for skill refinement in robotic systems" *IEICE (Institute of Electronics, Information and Communication Engineers) Transactions*, Vol.*E74*, No.2, 235–243, (1991).

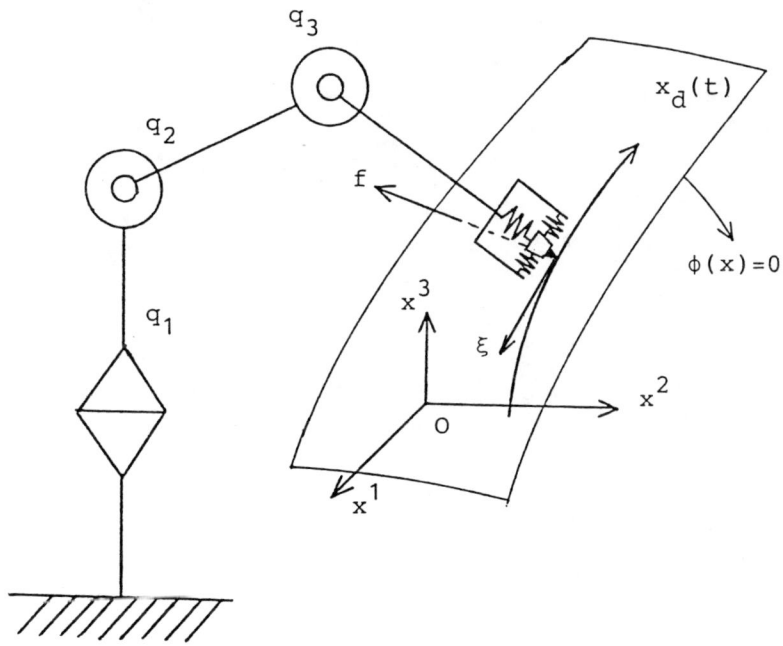

Fig. 1 Hybrid (position/force) Control under geometric constraint

A DECENTRALIZED ADAPTIVE CONTROLLER FOR ROBOT MANIPULATOR TRAJECTORY CONTROL

Premal Desai

Division of Science, Engineering and Technology

Penn State University at Erie

Erie, PA 16563

U. S. A.

Abstract

A new form of decentralized adaptive control scheme for multi-joint rigid robot manipulators has been proposed in this paper. The control objective is accurate trajectory tracking. The control law is based on the independent joint control and self-tuning control philosophies. The resulting controller is simple in structure, computationally fast and amenable to parallel processing implementation.

1. Introduction

Conventional controllers for motion control of industrial robots are based on independent joint control schemes, where each joint is controlled by a position servo-loop with constant prespecified gains. Such control schemes are suitable for simple pick-and-place tasks where only point-to-point motion is of interest. However, these controllers are often inadequate for accurate tracking of fast and complex trajectories under diverse payload conditions. In view of these limitations, researchers have explored two major design philosophies for developing versatile trajectory controllers for industrial robot manipulators.

The first design methodology may be referred to as the model-based methods such as the Computed Torque Method [1]. This technique requires precise knowledge of the manipulator dynamic model in terms of its parameters as well as its structure and results in a complex, computationally intensive centralized controller. The performance of such a control system is highly sensitive to the accuracy of the robot dynamic model [2]. Since the manipulator dynamic model contains time-varying parameters (such as joint friction, motor parameters and payloads) which are often difficult to measure accurately, modelling errors and payload variations adversely affect the manipulator performance.

The second design methodology commonly known as the performance - based methods such as the adaptive control schemes [3], do not use mathematically complex models. These methods are free of some of the problems associated with the model-based techniques. Most of the existing adaptive controllers for robot manipulator motion control application are *centralized* in structure and computationally burdensome [4-7]. These adaptive control schemes are based on both the Model Reference Adaptive Control (MRAC) [8-9] as well as the Self-Tuning Control (STC) [10-12] philosophies. Recently, several researchers have proposed *decentralized* adaptive controllers [13-17].

In this paper, a new form decentralized adaptive controller for advanced trajectory control of robot manipulators has been proposed. This control technique is based on the adaptive independent joint control philosophy and yields a decentralized state-space self-tuning controller. Some of the important features of the new decentralized self-tuning controller must be noted. First, due to its adaptive structure, precise knowledge of the manipulator dynamic model in terms of the parameter values and payload configuration is not required. Second, due to its decentralized nature, the control scheme is computationally simple and is amenable to parallel processing implementation with a distributed processing architecture. Finally, in this work, the trajectory control problem is formulated in decentralized control context at the outset; unlike the centralized approaches mentioned earlier.

2. Problem Formulation

In this paper, we consider a n-link robot manipulator model based on the Langrangian dynamics, which may be given as:

$$M(q)\ddot{q} + C(q,\dot{q}) + G(q) + H(\dot{q}) = \tau \tag{1}$$

where q is an n-dimensional vector of joint displacements, $M(q)$ is an n-dimensional symmetric positive-definite manipulator inertia matrix, $C(q,\dot{q})$ is an n-dimensional vector of centrifugal and Coriolis torques, $G(q)$ is an n-vector of gravitational torques, $H(\dot{q})$ is an n-vector of frictional torques and τ is an n-vector of applied joint torques. It is important to note that the matrices H and G are not independent. Thus, a robot arm model is highly non-linear. The manipulator trajectory control problem is to devise a control scheme which generates the joint torque vector τ such that joint displacement vector q and the joint velocity vector $\dot{q}$ track the desired trajectory vector q_r and the reference velocity vector $\dot{q}_r$ respectively; where q_r and $\dot{q}_r$ are arbitrary functions of time.

3. A Decentralized Adaptive Controller

In order to develop a decentralized control scheme, it is convenient to view each joint as a subsystem of the entire manipulator system, where such systems are interconnected by *coupling torques* which include the inertial coupling terms as well as the Coriolis, centrifugal, friction and gravity terms in equation (1). It is now possible to represent the manipulator dynamic model by a collection of n second-order nonlinear scalar differential equations

$$m_{ii}(q)\ddot{q}_i + \left[\sum_{j=1,j\neq i}^{n} m_{ij}(q)\ddot{q}_j\right] + c_i(q,\dot{q}) + g_i(q) + h_i(\dot{q}) = \tau_i \qquad (2)$$

where $i = 1, \ldots, n$

The subscript i in equation (2) refers to ith joint, $m_{ii}(q)$ is the varying effective inertia seen at the ith joint which is always positive due to the positive-definiteness of M. It is then possible to express equation (2) as

$$m_{ii}(q)\ddot{q}_i + d_i(q,\dot{q},\ddot{q}) = \tau_i \qquad (3)$$

such that $i = 1, \ldots, n$. and

$$d_i(q,\dot{q},\ddot{q}) = \left[\sum_{j=1,j\neq i}^{n} m_{ij}(q)\ddot{q}_j\right] + c_i(q,\dot{q}) + g_i(q) + h_i(\dot{q}). \qquad (4)$$

The term d_i may be treated as the disturbance torque by ith joint controller and contains the gravity, friction, Coriolis and centrifugal torques for i th joint as well as the inertial coupling from the other joints. In other words, d_i represents the coupling between ith subsystem and the remaining subsystems.

In order to derive the decentralized self-tuning control law, we rewrite equation (3) in the following state-space form:

$$\dot{x}(t) = Ax(t) + Bu(t) + D + w(t) \tag{5}$$

$$y(t) = Fx(t) + v(t) \tag{6}$$

where $x(t)$ is the 2n-dimensional state vector given as: $x(t) = \begin{pmatrix} q(t) \\ \dot{q}(t) \end{pmatrix}$
such that $q(t)$ and $\dot{q}(t)$ are the joint displacement and joint velocity respectively. $u(t)$ is the m-dimensional input torque vector and $y(t)$ is the p-dimensional output vector. $w(t)$ and $v(t)$ are zero-mean white gaussian noises with covariances $Q(t)$ and $R(t)$ respectively and represent the modelling and measurement errors. The matrices A, B, D and F may be given by the following relations: $A = \begin{pmatrix} 0 & 1 \\ 0 & 0 \end{pmatrix}$, $B = \begin{pmatrix} 0 \\ m_{ii}^{-1} \end{pmatrix}$, $D = \begin{pmatrix} 0 \\ -m_{ii}^{-1}d_i \end{pmatrix}$ and assuming that the joint displacements are the only measurements available; we have $F = (1 \ \ 0)$.

It is possible to discretize the state space model in equations (5) and (6) at a sampling interval T_s, to obtain the following discretized linear state space model given as:

$$x(k+1) = \Phi x(k) + \Psi u(k) + \Lambda + \Gamma w(k) \tag{7}$$

$$y(k) = \Omega x(k) + v(k) \tag{8}$$

Next, we define the state error vector $x_e(k)$ as:

$$x_e(k) = x(k) - x_d(k) \tag{9}$$

where $x_d(k) = \begin{pmatrix} q_r \\ \dot{q}_r \end{pmatrix}$.

We also assume that the desired state $x_d(k)$ is the solution of the deterministic form of the state equation

$$x_d(k+1) = \Phi x_d(k) + \Psi r(k), \qquad x_d(0) = 0 \tag{10}$$

where $r(k)$ is a slowly varying reference input.

We then consider the minimum state error variance (MSEV) performance index $J(k)$ given as:

$$J(k) = E\left\{ \begin{pmatrix} x_e(k+1) \\ u^1(k) \end{pmatrix}^T \begin{pmatrix} W & S \\ S^T & U \end{pmatrix} \begin{pmatrix} x_e(k+1) \\ u^1(k) \end{pmatrix} \Big/ Y(k-1) \right\} \quad (11)$$

where W and S are known non-negative definite weighting matrices and U is a known positive definite matrix. The control objective is to drive the state tracking error $x_e(k)$, to the origin of the state-space asymptotically such that

$$lim_{k \to \infty} E\left\{ x_e(k) \Big/ Y(k-1) \right\} = 0 \quad (12)$$

while minimizing the performance index in equation (11) under the constraints of equations (7) and (8). Note that $u^1(k) = u(k) - r(k)$.

It is possible to realize the solution of this stochastic control problem in the form [18]:

$$u(k) = u(k-1) + [r(k) - r(k-1)] + \hat{L}_x(k)[\hat{x}_e(k) - \hat{x}_e(k-1)] + \hat{L}_y(k)[y_d(k) - \hat{y}(k-1)] \quad (13)$$

$$\hat{L}_x(k) = \hat{N}_1(k)\Pi_{11} + \hat{N}_2(k)\Pi_{21} \quad (14)$$

$$-\hat{L}_y(k) = \hat{N}_1(k)\Pi_{12} + \hat{N}_2(k)\Pi_{22} \quad (15)$$

$\hat{N}_1(k)$ and $\hat{N}_2(k)$ in the above equations may be defined as:

$$(\hat{N}_1(k) \quad \hat{N}_2(k)) = -[\hat{\Psi}_a^T W_a \hat{\Psi}_a + U_a]^{-1}[\hat{\Psi}_a W_a \hat{\Phi}_a + S_a^T \hat{\Phi}_u] \quad (16)$$

and $\Pi_{i,j}$ for $i,j = 1,2$ may be defined as:

$$\begin{pmatrix} \Pi_{11} & \Pi_{12} \\ \Pi_{21} & \Pi_{22} \end{pmatrix} = \begin{pmatrix} \hat{\Phi} - I & \hat{\Psi} \\ \hat{F} & 0 \end{pmatrix}^{-1} \quad (17)$$

The subscript a stands for the augmented version of a given matrix. The augmented matrices are defined by the following equations:

$$\Phi_a = \begin{pmatrix} \Phi & \Psi \\ 0 & I \end{pmatrix}, \Psi_a = \begin{pmatrix} 0 \\ I \end{pmatrix}, \Gamma_a = \begin{pmatrix} \Gamma \\ 0 \end{pmatrix}, W_a = \begin{pmatrix} W & S \\ S^T & U \end{pmatrix}, S_a = \begin{pmatrix} S_{1p} \\ S_{2p} \end{pmatrix}$$

and $U_a = U_p$.

It is important to note that the augmented weighting matrices S_a and U_a need to be chosen by the user in the same fashion as the other weighting matrices W, S and U. Further, the derivation of the control law (13) is mathematically involved and the reader is referred to an earlier work by the author [18] for details.

Remark(1):

The decentralized control law given in equation (13) is in its incremental form. It involves proportional + integral (P + I) feedback of the joint position error and proportional feedforward of the desired output which translates into proportional + derivative (P + D) feedforward of the desired position.

Remark (2):

The superscript $\hat{}$ in the control law (13) represents the estimated values of the corresponding matrices and vectors. Thus, it is necessary to estimate the robot parameter matrices as well as the state vector $x(k)$ in order to implement the control law. In our earlier work [18], we have used the *Pseudo Linear Regression algorithm* for combined parameter as well as state estimation. In this work, we propose to use the same algorithm to estimate the states and the parameters of the robot dynamic model. The pseudo linear regression algorithm is summerized below.

4. Conclusion

A decentralized adaptive control has been proposed in this paper for the robot manipulator trajectory control problem. The control law is in its discrete time, incremental form and involves P + I feedback of the joint position and P + D feedforward of the desired joint position. It is based on the self-tuning control philosopy and minimization of a single stage performance index that includes controller energy and a quadratic in the position control error. The control law proposed here bares close similarity to the decentralized control law proposed by Seraji [15], which uses the MRAC philosophy and has been given in its continuous time form. It requires P + D feedback of the joint position and P + D^2 feedforward of the desired joint position.

Simulation results are available, where the performance of the propsed decentralized MSEV P + I controller has been studied. They are omitted here due to space limitations.

5. References:

1. Bejczy A. K., "**Robot Arm Dynamics and Control**, Tech. Memo. 33-669, Jet Propulsion Lab., Cal. Inst. Technology, Pasadena, 1974.

2. Gilbert E. G. and Ha I. J., **An Approach to Nonlinear Feedback control with Applications to Robotics**, Proc. of the IEEE Conf. on Decision and Control, San Antonio, 1983.

3. Vukobratovic M. and Kircanski N., **Decentralized Adaptive Control for A Class of Large-Scale Mechanical Systems**, Proceedings of the IFAC/IFORS Symposium on Large-Scale Systems, Warsaw, 1983.

4. Anex R. P. and Hubbard M., **Modeling and Adaptive Control of A Mechanical Manipulator**, ASME Journal of Dynamic Systems, Measurements and Control, p. 106, 1984.

5. Koivo A. J. and Guo T., **Adaptive Linear Controller for Robot Manipulators**, IEEE Transactions on Automatic Control, vol. AC-28(2),p. 162, 1983.

6. Lee C. S. G. and Chung M. J., **An Adaptive Control Strategy for Mechanical Manipulators**, IEEE Transactions on Automatic Control, vol. AC-29(9),p. 837, 1984.

7. Luo G. L. and Saridis G. N., **L-Q Design of PID Controllers fo Robot Arms**, IEEE Journal of Robotics and Automation, vol. RA-1(2), p.57-65, 1985.

8. Lee C. S. G. et al., **An Approach of Adaptive Control for Robot Manipulators**, Journal of Robotic Systems, vol. 1, 1984.

9. Dubowsky S. and Desforges T., "**The application of model- referenced adaptive control to robotic manipulators**", ASME Journal of Systems, Measurement and Control, vol. 101, p. 193, 1979.

10. Chung C. H. and Leininger G., **Adaptive Self-Tuning Control of Manipulators in Task Coordinate System**, International Conf. on Robotics, Atlanta, 1984.

11. Koivo A. J., **Force-Velocity Control with Self-Tuning for Robot Manipulators**, IEEE Conf. on Robotics and Automation, San Francisco, 1986.

12. Desai P. and Abbas F., **A Minimum State Error Variance Controller Using an Extended Kalman Filter for Robot Manipulator Trajectory Control**, Proc. of the International Conf. on Automation, Robotics and Computer Vision, Singapore, p.611-615, 1990.

13. Sundareshan M. K. and Koenig M. A., **Decentralized Model Reference Adaptive Control of Robot Manipulators**, Proc. of the American Control Conference, p.44-49, 1985.

14. Al-Abass F. and Ozguner U., **Decentralized Model Reference Adaptive System using a Variable Structure Control**, Proc. of the IEEE Conf. on Decision and Control, p.1473-1478, 1985.

15. Seraji H., **Decentralized Adaptive Control of Manipulators :Theory, Simulation and Experimentation**, IEEE Trans. on Robotics and Automation, vol.5(2), p.183-201, 1989.

16. Hsia T. C. and Gao L. S., **Robot Manipulator Control Using Decentralizd Linear Time-Invariant Time-Delayed Joint Controllers**, Proc. of the IEEE International Conf. on Robotics and Automation, p.2070-2075, 1990.

17. Morgan R. G. and Ozguner U., **A Decentralized Variable Structure Control Algorithm for Robot Manipulators**, IEEE Journal of Robotics and Automation, vol. RA-1(2), p.57-65, 1985.

18. Mahalanabis A. K. and Desai P.,**A Robust Self-Tuned Controller Based on State-space Approach**, Proc. of the IFAC Workshop on Robust Adaptive Control, New Castle, Australia, p. 129-134, 1988.

ROBUST CONTROL OF ROBOT ARMS USING THE VARIABLE STRUCTURE APPROACH

A.J.L. Nievergeld and H.H. van de Ven
Eindhoven University of Technology
Department of Electrical Engineering
Eindhoven, The Netherlands

ABSTRACT

The implementation of a sliding mode controller on an existing control system for a robot is presented. Sliding mode is a control method that belongs to variable structure systems. According to a suitable switching logic the structure is changed, i.e. it switches from one continuous function to another continuous function of the state. In sliding mode the state follows the switching surface and the dynamics of the robot have no influence on the tracking performance. Much attention is paid to reducing the mechanical stress caused by the switching in the control signal.
 A model of the robot is used in designing the controller. This model describes the dynamic behaviour of the first two links. With the help of this model, the control signal is calculated. The model does not describe the behaviour of the robot exactly so the calculated control is not sufficient to move the robot accurately to the desired position.
Therefore, the calculated signal is corrected by a discontinuous signal which compensates the influence of the unmodelled dynamics, disturbances and variations in the dynamics caused by load variations. This correction guarantees the existence of sliding mode and reduces the chattering of the control signal.
 The controller is implemented on an ASEA robot with five links. The first two links are controlled by the sliding mode algorithm and the third link is held in position by a PD-controller. The fourth and fifth link do not move with respect to the third link.

1   INTRODUCTION

Industrial robots will become a more important component of automated flexible manufacturing systems as their speed, accuracy and capabilities increase. The design and implementation of control strategies for such robots presents interesting problems. The dynamic equation of each link of a robot with several degrees-of-freedom represents a coupled non-linear and time-varying system [1].
Many control schemes are proposed for positional control of a single joint, with compensation for the unwanted dynamic effects, such as gravitational torques, coriolis and centrifugal torques [2].
A great deal of effort is now being concentrated on the develop-

ment of adaptive control systems. In particular, the so-called Model Reference Adaptive Control (MRAC) receives much attention [3,4]. In this paper, the designed robust controller for the control of joint angles is based on the variable structure systems (VSS) theory [5,6,7]. In variable structure systems, a discontinuous control law is obtained because switching can take place from one kind of trajectory to another in the state space at any instant. The switching occurs when the trajectories cross a certain chosen surface. From any arbitrary initial condition the trajectory of the system should be steered towards the switching surface and subsequently a switching function keeps the trajectory on it. The trajectory slides on this surface to the equilibrium state. This new type of motion is termed "sliding motion". In sliding mode, the system remains insensitive to parameter variations and external disturbances. The design of a variable structure control using sliding mode does not require accurate modeling, it is sufficient to know the bounds of the parameters.

2. Variable structure control

Variable structure control is characterized by discontinuous control, which switches from one 'structure' of a system to another 'structure' upon reaching a set of predetermined switching surfaces in state space [5]. When the state slides on these surfaces to the equilibrium state, the system is in so-called sliding mode. Consider a system of order n, which is linear with respect to the control and given by the following state equation

$$\underline{\dot{x}} = \underline{f}(\underline{x},t) + B(\underline{x},t)\underline{u} \quad (1)$$

where $\underline{x}, \underline{f} \in \mathbb{R}^n$, $\underline{u} \in \mathbb{R}^m$ and $B \in \mathbb{R}^{n \times m}$

The sliding mode control input has the following form [5]

$$u_i(\underline{x},t) = \begin{cases} u_i^+(\underline{x},t) & \text{if } s_i(\underline{x}) > 0 \\ u_i^-(\underline{x},t) & \text{if } s_i(\underline{x}) < 0 \end{cases} \quad (2)$$

In this equation is $u_i(\underline{x},t)$ the ith component of $\underline{u}$ and $s_i(\underline{x})$ the ith switching surface. The switching surfaces are described by

$$\underline{s}(\underline{x}) = \underline{0} \quad \text{with} \quad \underline{s} \in \mathbb{R}^m \quad (3)$$

To find the equations of ideal sliding mode, a technique named the 'equivalent control $\underline{u}_{eq}$', is used [5,7]. This method involves selecting a continuous control by setting the time derivative of the vector $\underline{s}(\underline{x})$ equal to zero ((4) and (5)).

$$\underline{\dot{s}}(\underline{x}) = \frac{\partial \underline{s}}{\partial \underline{x}^T} \underline{\dot{x}} = J\underline{\dot{x}} = J(\underline{f}(\underline{x},t) + B(\underline{x},t)\underline{u}_{eq}) = \underline{0} \quad (4)$$

$$\underline{u}_{eq} = -(JB)^{-1} J\underline{f} \quad (5)$$

The det(JB) is assumed to be different from zero which is true in most mechanical processes.

When the system is in sliding mode, the control can be considered as a function of a low frequency, or average component, and a (theoretical) infinite frequency component ((2), [5]). The behaviour of the system depends primarily on the average rather than on the high frequency component. In real sliding mode, disturbances cause the state to move in some Δ-vicinity of the discontinuous surface instead of following it, and cause the control to chatter at a finite frequency. The equivalent control is close to the low frequency component in real sliding mode. Thus, the physical meaning of the equivalent control is the control that causes the state to move towards the intersection of the switching surfaces and subsequently, in the ideal situation, maintain the state on it.

In ideal sliding mode, the dynamic behaviour of the system is described by (1) and (5).

$$\dot{\underline{x}} = \underline{f} - B(JB)^{-1}J\underline{f} \qquad (6)$$

and invariant to parameter variation.
In the non-ideal situation the control is presented in the following way

$$\underline{u} = \underline{u}_{eq} + \Delta\underline{u} \qquad (7)$$

with $\Delta u_i = \begin{cases} \Delta u_i^+ & \text{if } s_i(\underline{x}) > 0 \\ \Delta u_i^- & \text{if } s_i(\underline{x}) < 0 \end{cases} \qquad (8)$

The use of $\underline{u}_{eq}$ reduces the amplitude of the switching control considerably compared with equation (2) [7], which reduces the mechanical stress.

Sliding mode occurs when all of the trajectories are pointed to the switching surfaces. The necessary condition for sliding mode, is for SISO systems, rather easy [5]

$$\lim_{s \to 0} s\dot{s} < 0 \qquad (9)$$

In the case of MIMO systems, (9) is not immediately applicable. The occurrence of sliding mode in systems with coupled dynamics cannot be described by an easy equation. But by using a proper transformation, the n-dimensional MIMO system can be transformed in a canonical system and considered as n SISO systems. The occurrence of sliding mode for the n SISO systems is then described by (9).
By making a proper choice of $\Delta\underline{u}$, the decoupling is possible [7]

$$\Delta\underline{u} = -(JB)^{-1}\Delta\underline{\tau} \qquad (10)$$

where $\Delta\underline{\tau}$ is the sliding mode input corresponding to (2) and described by

$$\Delta\underline{\tau} = \underline{w}.\text{sgn}(\underline{s}) \qquad (11)$$

where $\underline{w}$ is a weighting vector ($w_i > 0$).
However, in most cases $(JB)^{-1}$ is not obtained so easily. In practical controller design, an estimation $\overline{(JB)}^{-1}$ is performed for $(JB)^{-1}$, and its accuracy is dependent on system model and the calculating capacity of the control system. If the following assumption is satisfied, existence condition (9) is applicable in the same case where accurate $(JB)^{-1}$ is obtained in advance [7].

$$(JB)\overline{(JB)}^{-1} = I_n + \delta I \qquad (12)$$
with $(\delta I)_{ij} \ll 1 \qquad i,j = 1,\ldots,n$

and I_n the nxn unity matrix.
In most mechanical systems the coupling between the outputs is small, therefore assumption (12) is valid for designing the controller and $(\overline{JB})^{-1}$ can be used.

The existence condition of sliding mode (9) means:
$$s_i(\underline{x}) = 0 \text{ and } \dot{s}_i(\underline{x}) = 0 \qquad (13)$$
In sliding mode the system dynamics are only described by the parameters of the switching surfaces. Thus, external disturbances and variation of the system parameters have no influence on the performance of the controlled system. The dynamics of this system can freely be chosen giving the entries of the matrix J proper values (> 0).

The main drawback of sliding mode strategy is that the control signal is discontinuous and chatters at a theoretically infinite frequency (11). This signal can not be used to drive a plant and has to be filtered. Replacing the discontinuous function $sgn(s_i)$ in (11) by a continuous function $s_i/(|s_i|+ \varepsilon_i)$, takes care of this filtering. The ε_i is a small positive constant which determines the breakpoint' of the filter. Consequently, (9) does not apply when the state is on the switching surface and so it deviates from the surface.

3. Controller design

The dynamics of a n-joint manipulator are represented by:
$$M(\underline{\theta})\underline{\ddot{\theta}} + \underline{f}(\underline{\theta},\underline{\dot{\theta}}) + V\underline{\dot{\theta}} + \underline{g}(\underline{\theta}) = \underline{u} \qquad (14)$$
where $M(\underline{\theta})$ is the nxn symmetric inertia matrix, $\underline{f}(\underline{\theta},\underline{\dot{\theta}})$ the n-dimensional vector representing the coriolis and the centrifugal torque, $V\underline{\dot{\theta}}$ the n-dimensional vector representing the viscous damping torque and $\underline{g}(\underline{\theta})$ the n-dimensional vector representing the torque due to the gravity. $\underline{u}$ is the n-dimensional input vector, $\underline{\theta}^T$, $\underline{\dot{\theta}}^T$, $\underline{\ddot{\theta}}^T$ are the n-dimensional angular position, velocity and acceleration vector, respectively. The state vector $\underline{x}$ is defined by $\underline{x}^T = (\underline{\theta} - \underline{\theta}_d, \underline{\dot{\theta}} - \underline{\dot{\theta}}_d) = (\underline{e},\underline{v})$, where $\underline{\theta}_d^T$ is the desired path and $\underline{v} = \underline{\dot{e}}$.

The switching surfaces s_i are described by
$$s_i(e_i,v_i) = c_i e_i + v_i \text{ with } c_i > 0 \text{ and } i = 1,\ldots,n \qquad (15)$$

The $-c_i$ are the eigenvalues of the system projected on the ith switching plane. They have to be situated in the left half-plane of the complex plane.

The equivalent control $\underline{u}_{eq}$ is calculated by transforming (14) in
$$\underline{\ddot{\theta}} = M(\underline{\theta})^{-1}(\underline{u} - \underline{f}(\underline{\theta},\underline{\dot{\theta}}) - V\underline{\dot{\theta}} - \underline{g}(\underline{\theta})) \qquad (16)$$
From (16) and $\underline{\dot{s}} = \underline{0}$ follows
$$\underline{u}_{eq} = \underline{f}(\underline{\theta},\underline{\dot{\theta}}) + V\underline{\dot{\theta}}) + \underline{g}(\underline{\theta}) + M(\underline{\theta})(\underline{\ddot{\theta}}_d - C\underline{v}) \qquad (17)$$
The equivalent control given by (17) does not account for unmodelled dynamics, disturbances and payload variations so it is not sufficient to move the state towards the switching surfaces and direct it along the intersection of the surfaces to the origin of the state space. The actual control (7) consists of

the equivalent control and a correction to this signal, which
causes the state to move towards the switching surfaces and
ensures that the state is maintained on it.

In robot control is $\Delta\underline{u}$ in (7) $\Delta\underline{u} = -\overline{M(\underline{\theta})}\Delta\underline{\tau}$

The following assumption is imposed on the disturbances d_i
(including also unmodelled dynamics and payload variations)

$$|d_i(\underline{\theta},\underline{\dot{\theta}},\underline{\ddot{\theta}})| < \alpha_i |e_i| + \beta_i |v_i| + \gamma_i \tag{18}$$

The control input $\Delta\tau_i$ has the following form

$$\Delta\tau_i = (\varphi_i |e_i| + \xi_i |v_i| + \kappa_i)\,\text{sgn}(s_i) \tag{19}$$

The condition that sliding mode occurs is given by equation (9)
and is used to calculate the φ_i, ξ_i and κ_i.
The following conditions apply [7]

$$\begin{aligned}(1 + \delta I_{ii})\varphi_i &> \alpha_i \\ (1 + \delta I_{ii})\xi_i &> \beta_i \\ (1 + \delta I_{ii})\kappa_i &> \gamma_i\end{aligned} \tag{20}$$

4. Derivation of control parameters

In order to derive the control algorithm, a mathematical model of
the robot is necessary. The derived sliding mode algorithm must
be able to control the first two links of an ASEA IRb6 robot.
The third link is held in position by a PD-controller. The
fourth and fifth link do not move with respect to the third
link. Corresponding to (14) the dynamics of the first two links
are described by (21):

$$\begin{bmatrix} u_1 \\ u_2 \end{bmatrix} = \begin{bmatrix} a_{11} & 0 \\ 0 & a_{22} \end{bmatrix} \begin{bmatrix} \ddot{\theta}_1 \\ \ddot{\theta}_2 \end{bmatrix} + \begin{bmatrix} 120.8 \\ 148.2 \end{bmatrix} \begin{bmatrix} \dot{\theta}_1 \\ \dot{\theta}_2 \end{bmatrix}$$

$$+ \begin{bmatrix} 0.19\dot{\theta}_1\dot{\theta}_2 \sin(2\theta_2) \\ -0.19\dot{\theta}_1^2 \sin(2\theta_2) \end{bmatrix} + \begin{bmatrix} 0 \\ -39.5 + 47.8\sin\theta_2 \end{bmatrix} \tag{21}$$

where $a_{11} = 11.4 - 3.7\sin\theta_2 + 3.5\sin^2\theta_2$; $a_{22} = 10.7 - 3.8\sin\theta_2$

The equivalent control is given by

$$\underline{u}_{cq} = v\underline{\dot{\theta}} + q(\underline{\theta}) + M(\underline{\theta})\,(\underline{\ddot{\theta}}_d - C\underline{v}) \tag{22}$$

The coriolis- and centrifugal torque are also considered to be
disturbances.
The 'disturbances' $f(\underline{\theta},\underline{\dot{\theta}})$ are given by (18) with

$$\alpha_1 = 0;\quad \alpha_2 = 0;\quad \beta_2 = 0;\quad \beta_1 = |1 - \frac{0.2}{a_{11}}\dot{\theta}_2\sin(2\theta_2)|$$

$$\gamma_1 = |1 - \frac{0.2}{a_{11}}\dot{\theta}_2\sin(2\theta_2)\dot{\theta}_{d1}|;\quad \gamma_2 = |1\,\frac{0.2}{a_{22}}\dot{\theta}_1^2\sin(2\theta_2)|$$

The 'correction' to the equivalent control $\Delta \underline{u}$ has the following form

$$\Delta u_i = - a_{ii}(\varphi_i |e_i| + \xi_i |v_i| + \kappa_i) \frac{s_i}{|s_i| + \varepsilon_i} \qquad (23)$$

where $\varphi_i > \max(\alpha_i)$, $\xi_i > \max(\beta_i)$ and $\kappa_i > \max(\gamma_i)$, for $i = 1, 2$.

5. Experimental results

In order to determine the performance of the controller, the robot is stimulated by a step function. In the criterion used, the position error is considered to be more important in determining the quality of the controller than the velocity error. During the measurements, it turned out that the control chatters with a large amplitude, in spite of the low-pass filtering. This is caused by the calculation of the velocity error by differentiating the position error. This velocity error is used to calculate the $\underline{u}_{eq}$, which is not filtered. The factor $M(\theta)(\ddot{\theta}_d - C\underline{v})$ makes a low contribution to the average value of the equivalent control, because it consists only of 'spikes'. Therefore it is left out of the calculation of $\underline{u}_{eq}$. The control parameters are chosen as:

$\varphi_1 = 250$; $\varphi_2 = 300$; $\xi_1 = \xi_2 = 100$; $\kappa_1 = 0.2$; $\kappa_2 = 0.3$;
$c_1 = c_2 = 20$; $\varepsilon_1 = \varepsilon_2 = 0.1$

The exprimental results of the two links are given in figures 1a - d, respectively. In ideal sliding mode, the systems behaviour is that of a first order linear system, with a time constant $1/c$. Figure 1a shows that the response of the first link approximates that ideal situation. Figure 1c shows the trajectory in the phase-plane and we see that the state does not follow the switching line exactly but stays in its vicinity. This is caused by the replacing of sgn(s) by $s/(|s| + \varepsilon)$. The response of the second link (fig. 1b) does not follow the ideal situation as we see from the phase-plane in figure 1d.

Next the tracking performance is tested by forcing the links to follow a path. The path is simulated by a saw-tooth function. The first measurments are done with only one link being moved at the same time. The position error of the first link is given in figure 2a. The average error is about 0.2 degrees.
The control is given in figure 2c. The amplitude of the switching control component Δu_1 is small in accordance with the amplitude of the equivalent control. This means that the model describes the dynamics of this link very well. The dynamic behaviour of the second link is not entirely described by the model. According to figure 2b, the tracking performance is not as good as that of the first link. The 'correction' to the equivalent control Δu_2 is larger (figure 2d).

In figures 3a and b the position error of the two links is shown when they move at the same time. Comparing figures 2a and 3a and 2c and 3b shows that the error has not increased so the links are decoupled.

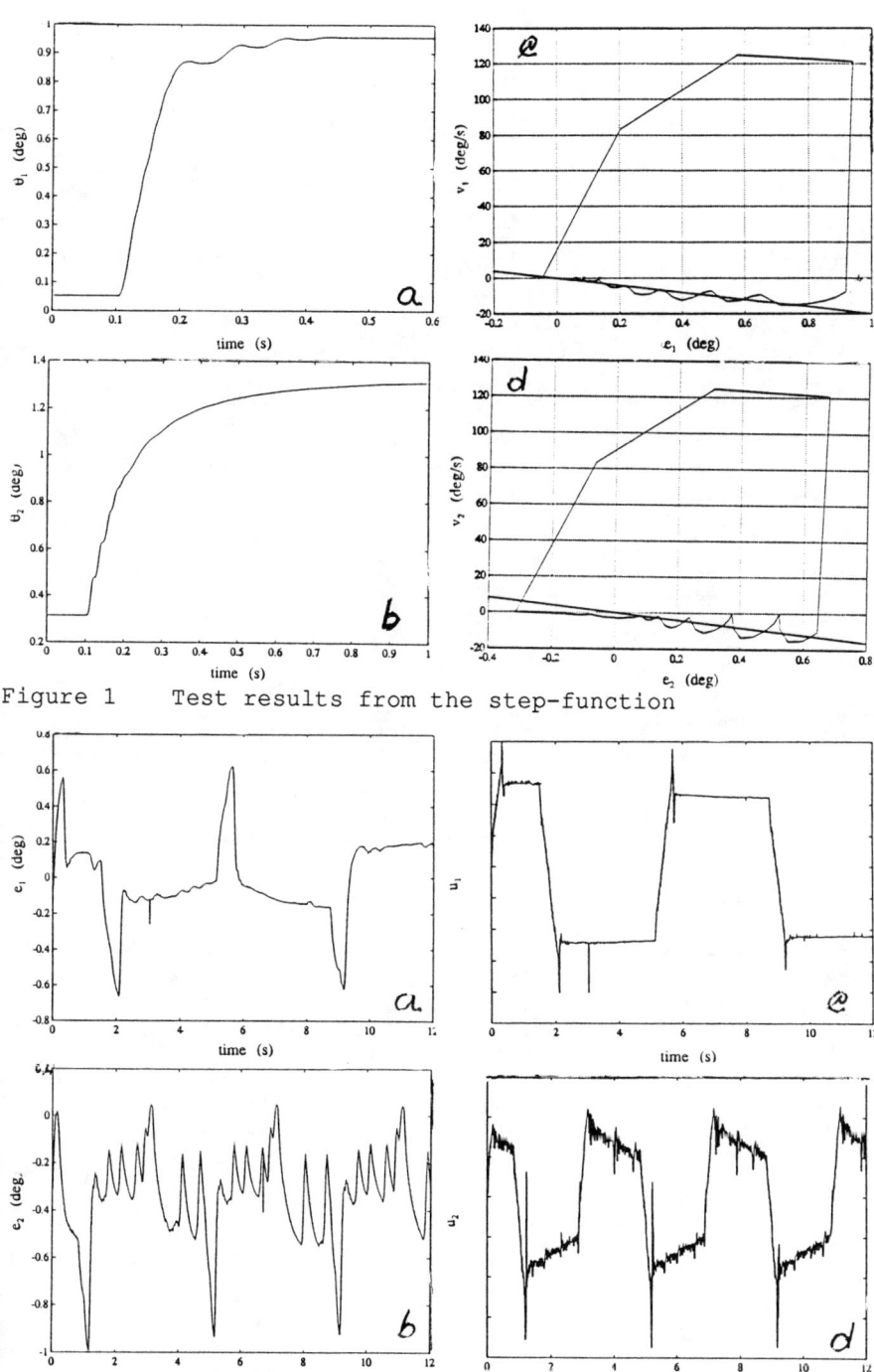

Figure 1 Test results from the step-function

Figure 2 Test results from a saw-tooth function

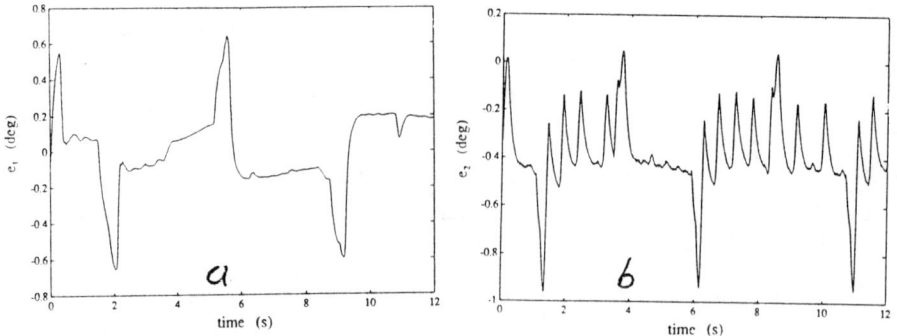

Figure 3   Tracking results: two links move together

CONCLUSIONS

A sliding mode control algorithm for two links of an industrial robot manipulator is proposed. The sliding mode control suppresses the nonlinear dynamic interactions of the manipulator joints. Tracking performance of the first link is good and the dynamics of the link are almost entirely determined by the controller. The dynamics of the second link are still affected by robot parameters. Tuning of the control parameters will improve the performances of the second link. The chattering in the control signal is small and causes little mechanical stress.

REFERENCES

[1] B.C. McInnes and C.K.F. Liu; Kinematics and Dynamics in Robotics: A Tutorial Based upon Classical Concepts of Vectorial Mechanics. IEEE Journal of Robotics and Automation, vol. RA-2, no. 4, pp. 181-187 (1986).

[2] J.Y.S. Luk; Conventional Controller Design for Industrial Robots- a Tutorial. IEEE Trans. on Systems, Man, and Cybernetics, vol. SMC-13, no. 3, pp. 210-228 (1983).

[3] Y.D. Landau, Adaptive Control: The Model Reference Approach, Marcel Dekker Inc., New York (1979).

[4] S. Nicosia and P. Tomei; Model Reference Adaptive Control Algorithms for Industrial Robots. Automatica, vol. 20, no. 5, pp. 635-644 (1984).

[5] V.I. Utkin, Sliding Mode and their Application to Variable Structure Systems. Mir Publishers, Moscow, (1978)

[6] K.K.D. Young; Controller Design for a Manipulator Using Theory of Variable Structure Systems. IEEE Trans. on Systems, Man, and Cybernetics, vol. SMC-8, no. 2, pp. 101-109 (1978).

[7] F. Harashima, H. Hashimoto and K. Maruyamai; Practical Robust Control of Robot Arm Using Variable Structure System. Proc. Robotics and Automation Conf., San Francisco, pp. 532-539 (1986).

A Comparison of Direct Adaptive Controllers Applied to a Class 1 Manipulator

D P Stoten   Reader, Department of Mechanical Engineering, Bristol University, Bristol BS8 1TR, UK.

S P Hodgson, Research Student, Department of Mechanical Engineering, Bristol University, Bristol BS8 1TR, UK.

Summary

This paper compares the results of implementing three well-known adaptive algorithms on the motion-control of a planar revolute axis manipulator. The main emphasis is on the experimental results, and the relative efficacy of each algorithm. However, we also draw comparisons on the structure of each algorithm, and the corresponding implications for discrete-time implementations. The main conclusion is that the decentralised version of the Minimal Control Synthesis (MCS) algorithm of Stoten and Benchoubane [1,2] resulted in excellent model-following, with relative noise insensitivity and minimum flops per sampling interval. MCS is classed as a direct adaptive controller[3]. The two other direct adaptive controllers of Seraji [4] and Oh and Jamshidi [5] did not perform as well as MCS: model-following was not so close, noise was more of a problem and the implemented forms were more complicated.

1. Introduction

In recent years several adaptive algorithms have been developed for general dynamic systems control, and more specifically, manipulator control. These may be classified into direct and indirect adaptive control schemes [3].
Direct adaptive controllers compute the controller gains without recourse to plant parameter estimation. The Minimal Control Synthesis (MCS) algorithm of Stoten and Benchoubane [1,2] was developed for general dynamic control problems, and it has been successfully applied to various manipulator problems [8,9]. The algorithms of Seraji [4] and Oh and Jamshidi [5] were developed specifically for the problems associated with manipulator control.
Indirect adaptive controllers require on-line plant parameter identification in order to compute the requisite control signals via a procedure of inverse dynamic compensation (the 'computed-torque' method). Good estimates are required for the initial conditions of the parameter estimates: this implies that an off-line system identification exercise is also necessary. Examples of this type of algorithm are given by Craig [6], Slotine and Li [7] and Ortega and Spong [10]. The purpose of this work is to compare the results of implementing the above direct adaptive controllers on a single task, for a given manipulator. To the authors' knowledge, this type of comparison has not been attempted before. In a subsequent work, we will also compare the results of implementing indirect adaptive control algorithms.

2. The Control Algorithms

Each algorithm is presented in summary form: we have attempted to adopt a unified notation (which does not necessarily correspond with the original description). All of the algorithms are based on a state-space description of the manipulator dynamics (see Section 3), and they all have a decentralised structure.

2.1 The MCS Algorithm of Stoten and Benchoubane [1,2]   See Fig 1.

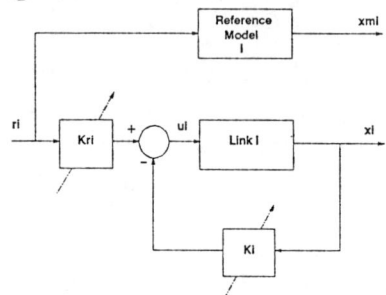

Fig 1   *The MCS Algorithm of Stoten and Benchoubane*

The control equation for each degree of freedom (i) is:
$$u_i = K_i x_i + K_{ri} r_i$$
where $x_i = [\text{position} \quad \text{velocity}]^T$. The adaptive gains K_i and K_{ri} have arbitrary initial conditions (often zero), and are renewed according to a 'proportional-plus-integral' law:
$$\theta_i = (\alpha y_e w)_i + \partial/\partial t (\beta y_e w)_i$$
where $\theta_i^T = [K_i \quad K_{ri}]$, $y_{ei} = C_{ei} x_{ei}$, $C_{ei} = \text{even rows}(P_i)$, $w_i^T = [x_i^T \quad r_i]$

$x_{ei} = x_{mi} - x_i$, $\dot{x}_{mi} = A_{mi} x_{mi} + B_{mi} r_i$, $P_i A_{mi} + A_{mi}^T P_i = -Q_i$, $Q_i > 0$

The scalars $\{\alpha, \beta\}_i$ are chosen to be positive, whilst the reference model parameters $\{A_m, B_m\}_i$ have desirable characteristics [8]. This law ensures that the state error (x_{ei}) - the difference between the reference model state (x_{mi}) and the plant state (x_i) - is hyperstable according to the theorem of Popov [11]. Furthermore, the equations can be readily transformed into discrete form (for digital controller implementation) using the principle of zero order hold discrete equivalence.

2.2 The Algorithm of Seraji [4]   See Fig 2.

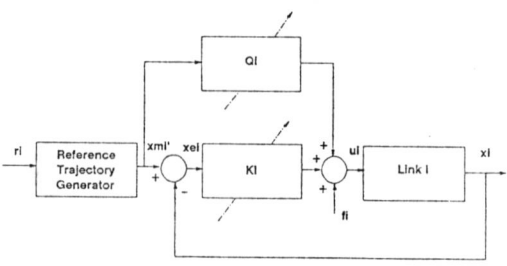

Fig 2   *The Algorithm of Seraji*

This control algorithm was developed specifically for manipulator problems (the MCS algorithm is for general dynamic system control); it is augmented by an 'auxiliary' signal f_i:
$$u_i = K_i x_{ei} + Q_i x_{mi} + f_i$$

where $Q_i = [q_p \ q_v \ q_a]_i$, $f_i = (\alpha \ y_e)_i + \partial/\partial t \ (\beta \ y_e)_i$

$x_{mi}' = [\text{position velocity acceleration}]^T$ of reference trajectory i

The adaptive law for the gains K_i and Q_i is of a similar form to that given for θ_i in MCS. However, θ_i and w_i are now defined by:

$$\theta_i^T = [K_i \ Q_i] \quad \text{and} \quad w_i^T = [x_{ei}^T \ x_{mi}'^T]$$

All other terms in the algorithm are similar to those in MCS.

2.3 The Algorithm of Oh and Jamshidi [5]   See Fig 3.

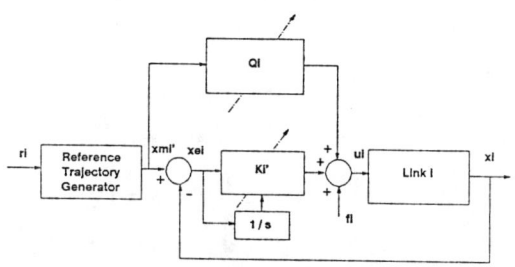

Fig 3 The Algorithm of Oh and Jamshidi

This algorithm is structurally similar to that of Seraji; the control equation is:

$$u_i = K_i' x_{ei}' + Q_i x_{mi}' + f_i$$

where $K_i' = [k_{Ii} \ K_i]$, $x_{ei}'^T = [x_{eIi} \ x_{ei}^T]$, $\dot{x}_{eIi} = x_{ei1}$

Now θ_i and w_i are defined as:

$$\theta_i^T = [K_i' \ Q_i] \quad \text{and} \quad w_i^T = [x_{ei}'^T \ x_{mi}'^T]$$

3. The Manipulator.

A planar, revolute, two-axis machine has been developed by one of the authors (SPH) for investigations into manipulator motion and force control (Fig 4).

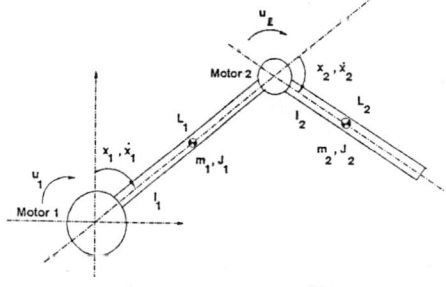

Fig 4 Hodgson's Class 1 Manipulator

The manipulator belongs to Class 1 of Coiffet [12], and its equation of motion can be written in state-space form as:

$$\begin{bmatrix} \dot{x}_1 \\ \ddot{x}_1 \\ \dot{x}_2 \\ \ddot{x}_2 \end{bmatrix} = \begin{bmatrix} 0 & 1 & 0 & 0 \\ 0 & -c_1/J_1^* & 0 & c_2/J_1^* \\ 0 & 0 & 0 & 1 \\ 0 & c_1/J_1^* & 0 & -c_2(1/J_2^*+1/J_{x2}) \end{bmatrix} \begin{bmatrix} x_1 \\ \dot{x}_1 \\ x_2 \\ \dot{x}_2 \end{bmatrix} + \begin{bmatrix} 0 & 0 \\ k_1/J_1^* & -k_2/J_1^* \\ 0 & 0 \\ -k_1/J_1^* & k_2(1/J_2^*+1/J_{x2}) \end{bmatrix} \begin{bmatrix} u_1 \\ u_2 \end{bmatrix} + \begin{bmatrix} 0 \\ d_1 \\ 0 \\ d_2 \end{bmatrix}$$

or $\dot{x} = Ax + Bu + d$

In the above expression, the inertia terms J_1^* and J_{x2} are derived from the respective link principal moments of inertia J_1 and J_2 (about the principal axes perpendicular to the page) as follows:

$$J_1^* = J_1 + m_1 l_1^2 + m_2 L_1^2 \quad \text{and} \quad J_{x2} = J_2 + m_2 l_2^2$$

The non-linear 'disturbance' terms d_1 and d_2 are given by

$d_1 = (f_2 - f_1)/J_1^*$ $\qquad d_2 = -d_1 - f_2/J_{x2}$

$f_1 = (2m_2 L_1 l_2 \cos x_2)\ddot{x}_1 + (m_2 L_1 l_2 \cos x_2)\ddot{x}_2 - (m_2 L_1 l_2 \sin x_2)\dot{x}_2^2$
$\quad -(2m_2 L_1 l_2 \sin x_2)\dot{x}_1\dot{x}_2 - ((m_1 l_1 + m_2 L_1)\sin x_1 + m_2 l_2 \sin(x_1 + x_2))g$

$f_2 = (m_2 L_1 l_2 \cos x_2)\ddot{x}_1 + (m_2 l_2 L_1 \sin x_2)\dot{x}_1^2 - (m_2 l_2 \sin(x_1 + x_2))g$

The datum for the state vector $x = [x_1\ \dot{x}_1\ x_2\ \dot{x}_2]^T$ is $x = 0$, i.e. the links are stationary and vertical. Nominal values were found to be:
$m_1 = 2.53$ kg, $l_1 = 0.201$ m, $L_1 = 0.3$ m, $J_1 = 0.606$ kgm^2, $J_1^* = 0.762$ kgm^2
$m_2 = 0.596$ kg, $l_2 = 0.202$ m, $L_2 = 0.3$ m, $J_2 = 0.268$ kgm^2, $J_{x2} = 0.292$ kgm^2
$c_1 = 9.33$ Nms rad^{-1}, $k_1 = 2.79$ NmV^{-1}, $c_2 = 1.81$ Nms rad^{-1}, $k_2 = 1.74$ NmV^{-1}

so that the {A,B} matrices were derived as:

$$A = \begin{bmatrix} 0 & 1 & 0 & 0 \\ 0 & -12.2 & 0 & 2.38 \\ 0 & 0 & 0 & 1 \\ 0 & 12.2 & 0 & -8.57 \end{bmatrix} ; \quad B = \begin{bmatrix} 0 & 0 \\ 3.66 & -2.28 \\ 0 & 0 \\ -3.66 & 8.24 \end{bmatrix}$$

whilst the transducer outputs (potentiometers and tachometers) were scaled so that 1.0 V ≡ 1.0 rad and 1.0 V ≡ 1.0 rad s^{-1} (respectively).

4 Results and Discussion

The control task, described below, does not test the algorithms over all conceivable manipulator trajectories. However, based upon the results from previous investigations [9], we can say that the task *does* provide valuable insights into the feasibility of the control algorithms, when implemented on a real machine. None of the algorithms was modified in order to overcome problems introduced by noise and unmodelled, high-order, plant dynamics.

Results generated by the controllers are presented graphically, with brief comments in each case. Control programmes were implemented as compiled QuickBasic code, with control signal wind-up protection. This protection was vital during periods of control signal saturation, otherwise overshoots could occur when the signals became unsaturated, due to unjustified changes (*eg* increases) in the gains. The controller hardware was a 20 MHz Dell 310 PC machine (80386/80387 processors), equipped with 12-bit d/a and a/d converters. Furthermore, all the fixed parameters in the control algorithms $\{\alpha, \beta\}_i$ were selected in order to achieve maximum equitability of adaptive effort.

Reference trajectories in all cases were equivalent to the critically - damped response to a square wave of frequency ~ 0.15 Hz and amplitude ~ 0.6 rads. The settling - time of the reference trajectory, following a constant demand, was ~ 1.0 s, with zero error in the steady - state. This was the standard test signal applied (simultaneously) to both links of the manipulator. In addition, at a time t ~ 23 s after each test commenced, a sudden increase was made in the gain of motor amplifier 1 (*ie* to the value of k_1). This element of the programme tested each algorithm's capability in overcoming the effect of a large parameter change in the plant dynamics. In a previous study [9] we found that this change could drive a linear controller to the verge of instability.

The results are given in Figs 5-7; the order of these figures corresponds to the presentation of the algorithms in Section 2. The most obvious conclusion is that each algorithm has generated controlled responses that follow the reference trajectories quite well (see the graphs entitled 'references & states'). However, closer inspection of Figs 5-7 reveals that the DMCS (D = decentralised) algorithm produced the most accurate trajectory - following. In descending order, the Oh/Jamshidi and Seraji algorithms produced the next best results.

However, link responses are not the only criterion on which to base a judgment of controller performance: we also consider the quality of the control signals, gain/parameter magnitudes and convergence, and complexity of design and implementation

It is vital that noise is not propagated via the control signals: the consequences of noise problems are actuator / mechanism failures and reduced steady - state tracking performance. Thus, examination of the graphs entitled 'controls' shows that the DMCS and Oh/Jamshidi algorithms resulted in the least noisy (and possibly the smoothest) of all the control signals. The controls from the Seraji algorithm were more noisy.

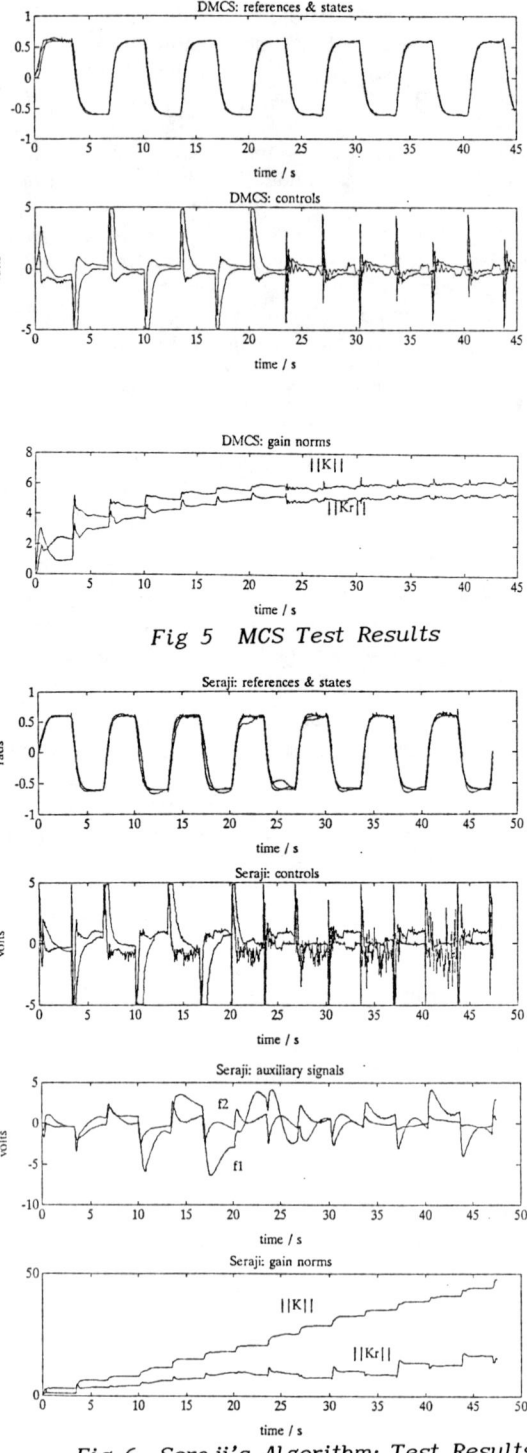

Fig 5 MCS Test Results

Fig 6 Seraji's Algorithm: Test Results

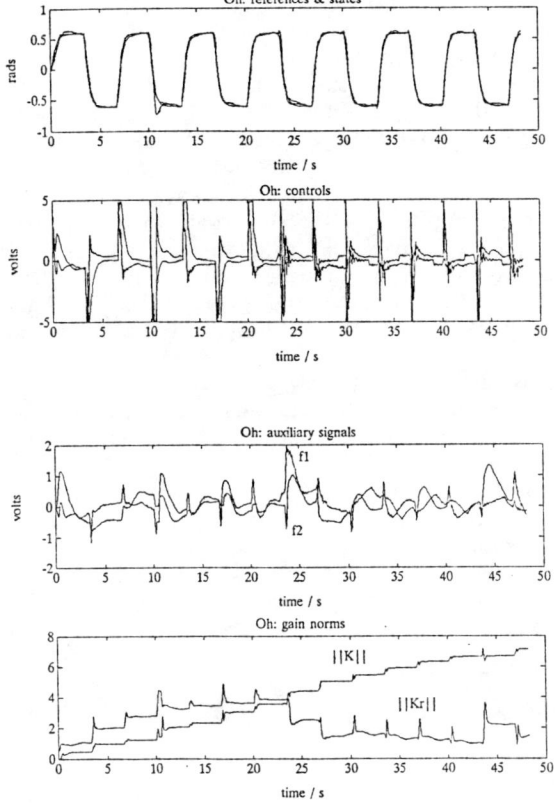

Fig 7   Oh and Jamshidi's Algorithm: Test Results

An investigation of the characteristics of the adaptive gains leads to a judgment on the controllers' abilities to cope with unknowns in the plant dynamics. For example, unnecessarily excessive gain magnitudes lead to noise propagation and to further gain wind-up. We see that the DMCS and Oh/Jamshidi algorithms produced the best gain norm trajectories, in terms of magnitudes and convergence rates. Unfortunately, the gains in Seraji's controller did not converge at all - even after an extended test of several minutes; the amplitudes of the gain norms were becoming excessive, as were the attendant noise problems. One possible explanation for this phenomenon is that, compared with the DMCS algorithm, certain terms in Seraji's regressor matrix are subject to low signal:noise ratios.

Reliable indexes of adaptive controller complexity (with respect to design and implementation) are the accuracy of the plant model required, the number of gains to be updated and the minimum length of the sampling interval (for given hardware and software). Other indexes can be used. In all of these respects, DMCS again comes at the top of our list. Like the other direct adaptive controllers, DMCS requires no plant parameter estimates: this eases the design problem considerably in comparison with the design of indirect controllers. Moreover, DMCS requires only 6 gains to be updated on our two-axis problem. Compare this with 10 gains for Seraji and 12 gains for Oh/Jamshidi. Excluding the a/d and d/a conversion times, the value of the minimum sampling interval obtainable for DMCS was 0.53 ms, compared to 0.79 ms for Seraji and 0.87 ms for Oh/Jamshidi. It is notable that the given time for DMCS will increase *linearly* with the number of degrees of freedom in the manipulator.

Other comments on the results now follow.
The auxiliary signals within the Seraji and Oh/Jamshidi algorithms can be removed without significant deterioration in their respective controller performances (results not included here).
Our transducers consisted of potentiometers and permanent magnet tachometers: their signals were relatively noise free, with rms noise levels of ~ 0.01 V ($\cong 0.5°$ on link positions).
Unlike all the other algorithms, DMCS does not use an explicit 'front-end' trajectory planner. However, this can be accommodated by inversion of the reference model dynamics to generate a reference input, r_i, which guarantees a required reference output, x_{mi}:

$$r_i = B_{mi}^{\dagger} (\dot{x}_{mi} - A_{mi} x_{mi})$$

where $B_{mi}^{\dagger}$, the pseudo-inverse of $B_{mi}^{\dagger}$, will always exist.

5. Conclusions

This paper compares the implementation of three direct adaptive control algorithms on a Class 1 manipulator:
  The (Decentralised) Minimal Control Synthesis (DMCS) algorithm of Stoten and Benchoubane.
  The algorithm of Seraji.
  The algorithm of Oh and Jamshidi.
It was evident that DMCS resulted in a closed-loop system which gave excellent trajectory-following, with good noise suppression and the most efficient design/implementation exercise.
The algorithms of Oh/Jamshidi and Seraji resulted in the least accurate responses (in relative terms). Also, the algorithm of Seraji suffered from poor convergence problems, with excessive wind-up of the controller gains and attendant noise problems.

6. References

1. **Stoten, D.P. and H. Benchoubane** Robustness of a minimal controller synthesis algorithm, *Int.J. Control*, 1990, Vol 51, No 4, 851-861.
2. **Stoten, D.P. and H. Benchoubane** Empirical studies of an MRAC algorithm with minimal controller synthesis, *Int.J. Control*, 1990, Vol 51, No 4, 823-849.
3. **Sastry, S. and M. Bodson** *Adaptive Control: Stability, Convergence, and Robustness*, Prentice Hall, London, 1989.
4. **Seraji, H.** Decentralised adaptive control of manipulators: theory, simulation, and experimentation, *IEEE Trans. Robotics and Auto.*, 1989, Vol 5, No 2, 183-201.
5. **Oh, B.J. and M. Jamshidi** Decentralised adaptive control of robot manipulators, *Journ. Robotic Manipulators*, 1989, Vol 6, No 4, 461-483.
6. **Craig, J.J.** *Adaptive Control of Mechanical Manipulators*, Addison-Wesley, Reading(MA), 1988.
7. **Slotine, J-J.E. and W. Li** On the adaptive control of robot manipulators, *The Int. Journ. Robotics Research*, 1987, Vol 6, No 3, 49-59.
8. **Stoten, D.P.** *Model Reference Adaptive Control of Manipulators*, Research Studies Press Ltd., Taunton, England, 1990.
9. **Stoten, D.P. and S.P. Hodgson** Comparative implementation studies of the minimal control synthesis algorithm on a class 1 manipulator, accepted for publication in *I Mech E Proceedings - Part I: Systems and Control*, 1991.
10. **Ortega, R. and M.W. Spong** Adaptive motion control of rigid robots: a tutorial, *Automatica*, 1989, Vol 25, No 6, 877-888.
11. **Popov, V.M.** *Hyperstability of Control Systems*, Springer-Verlag, Berlin, 1973
12. **Coiffet, P.** *Robot Technology: Volume 1 - Modelling and Control*, Kogan-Page, London, 1983.

AN ALGORITHM FOR THE NONLINEAR ADAPTIVE ROBOT CONTROL SYNTHESIS

BRANKO M. NOVAKOVIĆ
FSB-University of Zagreb, Salajeva 1
POB 194, YU-41000 Zagreb, Yugoslavia

Abstract - This paper presents an unified approach to the nonlinear adaptive control synthesis of industrial robots. Following the ideas of the external linearization, the control problem of nonlinear industrial robot model (IRM) has been reduced to optimal control one of the equivalent linear IRM, with control and state constraints. Introducing the tracking error model of a full robot model (manipulator + actuators), the nonlinear control law has been derived, which guarantees the linear behaviour of the nonlinear IRM in the closed loop system. An unified direct adaptation procedure of the position and velocity regulator gains, has also been proposed, where desired stage of the relative and exponential stability have been preserved. In this sense a simple control synthesis procedure has been built.

Keywords - Robots; nonlinear systems; external linearization; optimal control; adaptive control; time-domain analysis; state-space.

INTRODUCTION

There exists to-day a great number of papers deal with adaptive control synthesis of the industrial robots. The basic procedures of the adaptive control problem solution have been described by Aström and Wittenmark [1], then by Vukobratović, Stokić, and Kirćanski [2], and by Bodson and Sastry [3]. Generally speaking, there exist the two different schemes of the adaptive control [4]; direct and indirect adaptive control. Some of the recent results on the direct adaptive control of robot manipulators may be found in [5]-[8], and the references therein. These works only show asymptotic stability results. One of the recent papers on indirect adaptive control of robot manipulators is [9], where exponential stability results for indirect adaptive controllers have been presented. Adaptive control of robot manipulators with only a few unknown robot parameters, the focus of which is on parameter estimation, has been presented in [10], [11] and [12], where in [11] the on-line estimation procedure of pseudo-physical parameters of robot manipulators has been used. The well known model-reference adaptive control is applied on the linear robot manipulator model in [13]. Another approach to adaptive control of manipulator is based on the full nonlinear manipulator model and on the inverse dynamics algorithms in [14] and [15]. The algorithms for rapid continuous-time adaptive control and Lyapunov theory stability investigation have been presented in [16], while a new exponentially stable direct adaptive control law for robot manipulator motion control has been proposed in [17].

In this paper an unified approach to the nonlinear adaptive control synthesis of industrial robots, has been presented. Starting with a full robot model (manipulator + actuators), a nonlinear robot tracking error model has been derived. Following the ideas of the external linearization in [20], and [22], and using the state-space approach, an unified direct adaptation procedures of the position and velocity regulator gains has successfully been applied. As results of this approach the position and velocity regulator gains become the time-varying parameters, and the high (desired) stage of the relative stability, in the case of state and control constraints, has been obtained. Moreover, the exponential stability of the full robot model, in the closed loop system, is preserved.

This paper is organized as follows: the next or the second section presents the full robot model, the main results are given in the third section, and the corresponding conclusions have been pointed out by the last section.

THE FULL ROBOT MODEL

Mathematical model of the dynamic motion of the manipulator with n-degrees of freedom, incorporating all relevant mechanical parameters, is given by Paul [18] as a system of complex nonlinear differential equations, in the standard form:

$$M(q)\ddot{q} + N(q,\dot{q}) = P(t), \qquad (1)$$

where $q(t)$ is a real n-vector of generalized (joint) coordinates, $M(q)$ is a real (nxn) - inertial matrix, $N(q,\dot{q})$ is a real n-vector of centrifugal, Coriolis and gravity forces, and $P(t)$ is a real n-vector of joint controls (torques and/or forces). The great number of papers, that deal with the feedback or adaptive robot control synthesis, have only been based on the manipulator model (1). From the real control synthesis problem of view, this approach is of the academic importance only, since the real control variables are not torques and/or forces, and the dynamics of actuators has not been included in the model (1). As it is well known, the actuator input variables are real robot control variables. In this sense the full robot model (manipulator + actuators) has to be built, where the dynamics of robot actuators has to be included in it. Following this idea, and introducing the corresponding actuator models, in this paper, the full robot model has been derived. For this purpose we can start with the second-order linear actuator models with linear connections between manipulator and actuators:

$$\dot{x}_i = x_{n+i}, \quad \dot{x}_{n+i} = a_i x_{n+i} + b_i u_i + c_i P_i,$$
$$x_i = \alpha_i q_i, \quad x_{n+i} = \alpha_i \dot{q}_i, \quad \dot{P}_i = \beta_i P_i, \qquad (2)$$

where $i = 1, 2, \ldots, n$, $q_i \in q$, and $P_i \in P$ from (1), x_i and x_{n+i} are actuator state variables, u_i is a real control variable, P_i is an actuator torque (or force), $\underline{a}_i$, $\underline{b}_i$, and $\underline{c}_i$ are actuator parameters, and α_i and β_i are corresponding constant connections between manipulator and actuators. The actuator models (2) can be translated into the following compact form:

$$P(t) = \underline{C}^{-1}(\alpha\ddot{q}-\underline{A}\dot{q}-\underline{B}U), \qquad (3)$$

where matrices and vectors from (3) are given by the next form:

$$\underline{A}=\text{diag }(\alpha_i \underline{a}_i), \underline{B} = \text{diag }(\underline{b}_i), \underline{C}= \text{diag }(\beta_i \underline{c}_i),$$

$$\alpha= \text{diag }(\alpha_i), \ P = (P_1..P_n)^T, \ U = (u_1..u_n)^T, \qquad (4)$$

and $i = 1,2,\ldots, n$.

After substitution of vector P(t) from (3) into the equation (1), the full robot model, including the dynamics of manipulator and actuators, has been derived:

$$E(q)\ddot{q} + F(q,\dot{q}) = U(t) \qquad (5)$$

where U(t) is the real robot control vector. The real (nxn) matrix $E(q)$, and the real n-vector $F(q,\dot{q})$ are given by the equations:

$$E(q)=\underline{B}^{-1}[\alpha-\underline{C}M(q)], \quad F(q,\dot{q})=-\underline{B}^{-1}[\underline{A}\dot{q}+\underline{C}N(q,\dot{q})]. \qquad (6)$$

It has to be noticed, the model (5) has the analogical structural form as the model (1). In this sense, the same control synthesis methods for the vectors P(t) and U(t) can be used. Thus, the main result, or the main conclusion of this approach is: the all control synthesis algorithms, developed by using the model (1), can be applied on the model (5) just using the substitution of the triplet (M,N,P) by the triplet (E,F,U). The procedures of the full robot model derivation, in the case of the third-order linear actuator models with linear or nonlinear connections between manipulator and actuators, will be presented in the next paper.

THE MAIN RESULTS

In order to develop an unified direct adaptation procedure of the position and velocity regulator gains, and starting with (5), a nonlinear tracking error model of industrial robot has been derived:

$$\ddot{e}(t) = r(t) - E^{-1}(q) [U(t) - F(q,\dot{q})], \qquad (7)$$

where a vector of tracking error, e, and a vector of desired acceleration, r, are given by the following equations:

$$e=q_w-q, \dot{e}=\dot{q}_w-\dot{q}, \ddot{e} = \ddot{q}_w-\ddot{q}, \ r(t)=E^{-1}(q_w)[U_w(t)-F(q_w,\dot{q}_w)]. \qquad (8)$$

In the equations (8) a subscript w characterizes the desired or nominal system variables of full robot model. Some procedures of the nominal control synthesis of a robot model are given by Vukobratović and Stokić [19], and by Novaković [21], [22]. Following the ideas of the external linearization (Isidori and Ruberti [20]; Novaković [22]), and using substitution:

$$v(t) = r(t) - E^{-1}(q)[U(t)- F(q,\dot{q})], \qquad (9)$$

the nonlinear tracking error model (7) can be translated into the corresponding linear one:

$$\ddot{e}(t) = v(t), v(t) = (v_1 v_2 .. v_n)^T, \qquad (10)$$

where $v(t)$ is a control (input) vector of the system (10). Selecting phase variables as state variables:

$$Z_I = (z_1 z_2 ... z_n)^T = (e_1 e_2 ... e_n)^T = e,$$
$$Z_{II} = (z_{n+1} z_{n+2} ... z_{n+n})^T = (\dot{e}_1 \dot{e}_2 ... \dot{e}_n)^T = \dot{e}, \qquad (11)$$

the corresponding state-space model has been obtained:

$$\begin{bmatrix} \dot{Z}_I \\ \dot{Z}_{II} \end{bmatrix} = \begin{bmatrix} 0 & I \\ 0 & 0 \end{bmatrix} \begin{bmatrix} Z_I \\ Z_{II} \end{bmatrix} + \begin{bmatrix} 0 \\ I \end{bmatrix} v(t), \qquad (12)$$

where I is (nxn) - unit matrax. This state-space model can be rewritten as the corresponding model of subsystems:

$$\dot{Z}_i(t) = A_i Z_i(t) + B_i v_i(t), Z_i = (z_i z_{n+i})^T,$$
$$A_i = \begin{bmatrix} 0 & 1 \\ 0 & 0 \end{bmatrix}, B_i = \begin{bmatrix} 0 \\ 1 \end{bmatrix}, i = 1, 2, ..., n. \qquad (13)$$

In accordance with (13) the special linear quadratic criterion, with exponential stability characteristic, can be used:

$$J_i = \int_0^\infty (\exp(2\sigma_i t)) (Z_i^T Q_i Z_i + v_i g_i v_i) dt,$$
$$Q_i = \text{diag}(d_{i1}, d_{i2}) = S_i^T S_i, S_i = \text{diag}(d_{i1}^{1/2}, d_{i2}^{1/2}), \qquad (14)$$
$$Q_i \geq 0, \quad g_i > 0, i = 1, 2, ..., n,$$

where σ_i is a positive control parameter which has to be adapted, d_{i1} and d_{i2} are state - weights, and g_i is a control weight. In order to connect the optimal control approach with the pole-placement method, the following selection of the state and control weights has been applied:

$$d_{i1} = g_i (p_{i1} p_{i2} + 2\sigma_i (p_{i1} + p_{i2}) + 4\sigma_i^2) p_{i1} p_{i2},$$
$$d_{i2} = g_i (p_{i1}^2 + p_{i2}^2 + 2\sigma_i (p_{i1} + p_{i2})), \qquad (15)$$
$$p_{i1} = p_{0i1} - \sigma_i, p_{i2} = p_{0i2} - \sigma_i,$$

where p_{i1} and p_{i2} are the desired, real and stable poles or eigenvalus of i-th subsystem in the closed loop system, and p_{0i1} and p_{0i2} are the corresponding real and stable initial eigenvalus ($\sigma_i = 0$).

As the result of the proposed optimization procedure, which guarantees the desired pole - placement (15) in the closed loop system, is the corresponding control $v_i(t)$:

$$v_i(t) = -g_i^{-1} B_i^T H_i Z_i = -K_i Z_i = -(k_i \; k_{n+i}) Z_i,$$

$$k_i = g_i^{-1} h_{112} = P_{11} P_{12} = (P_{011} - \sigma_i)(P_{012} - \sigma_i), \quad (16)$$

$$k_{n+i} = g_i^{-1} h_{122} = -(P_{11} + P_{12}) = 2\sigma_i - (P_{011} + P_{012}), \quad i=1,2,\ldots,n,$$

where K_i is a state-regulator of i-th subsystem, k_i and k_{n+i} are position and velocity gains, h_{112} and h_{122} are elements of matrix H_i, which is the real positive definite solution of the Riccati equation:

$$H_i B_i g_i^{-1} B_i^T - H_i(A_i + \sigma_i I) - (A_i + \sigma_i I)^T H_i - Q_i = 0, \quad (17)$$

where $i = 1, 2, \ldots, n$, and I is the (2x2)- unit matrix. Using (9) and (16), the nonlinear control law of the full robot model (5) has been derived:

$$U(t) = E(q)[r(t) + K_I Z_I + K_{II} Z_{II}] + F(q, \dot{q}) \quad (18)$$

where:

$$K_I = \text{diag}(k_i), \quad K_{II} = \text{diag}(k_{n+i}), \quad i=1,2,\ldots,n, \quad (19)$$

and where K_I and K_{II} are the position and velocity regulators of the full robot model (5), respectively. By applying this nonlinear control, the linear behaviour of the nonlinear full robot model (5), in the closed loop system, has been obtained:

$$\ddot{q}(t) = r(t) + K_I Z_I + K_{II} Z_{II}. \quad (20)$$

In order to achieve the high (desired) stage of the relative stability in the case of nonlinear control (16), (18) and (19), the control synthesis free-parameters σ_i, from (16), have to be adapt in the region:

$$0 \leq \sigma_i(t) \leq \sigma_{i\max}, \quad i=1,2,\ldots,n, \quad (21)$$

where $\sigma_{i\max} > 0$. In this case the desired poles, or eigenvalus, of the closed loop system (20) are retained in the stable area:

$$(P_{0ij} - \sigma_{i\max}) \leq P_{ij}(t) \leq P_{0ij}, \quad (22)$$

$$P_{0ij} < 0, \; i=1,2,\ldots,n; \; j=1,2.$$

The above mentioned conditions (21) and (22) can be realized, among the others, by using one of the adaptation schemes. In this sense the following unified direct adaptation procedure of the control synthesis free-parameters $\sigma_i(t)$ can be applied;

if. $|z_i(t)| > z_{i\max}$, or/and $|z_{n+i}(t)| > z_{n+i\max}$,

or/and $|u_i(t)| > u_{i\max}$, then $\sigma_i(t)=0$,

else:

$$\sigma_i(t) = \frac{\sigma_{i\max}}{3}\left[3 - \gamma_i \frac{|z_i(t)|}{z_{i\max}} - \varphi_i \frac{|z_{n+i}(t)|}{z_{n+i\max}} - \xi_i \frac{|u_i(t)|}{u_{i\max}}\right], \quad (23)$$

where $i = 1, 2, \ldots, n$, and γ_i, φ_i and ξ_i are the weighting parameters of the previous algorithm, with constraints:

$$0 \leq \gamma_i, \varphi_i, \xi_i \leq 1, \quad (24)$$
$$z_{i\,max} > 0, \quad z_{n+i\,max} > 0, \quad u_{i\,max} > 0, i=1,2,\ldots,n.$$

According to (16), and (23), the position and velocity regulator gains become the time-varying parameters, which have to be calculated by the equations:

$$k_i(t) = p_{i1}(t)p_{i2}(t) = (p_{oi1} - \sigma_i(t))(p_{oi2} - \sigma_i(t)),$$
$$k_{n+i}(t) = -(p_{i1}(t) + p_{i2}(t)) = 2\sigma_i(t) - (p_{oi1} + p_{oi2}), \quad (25)$$
$$K_I(t) = \text{diag}(k_i(t)), K_{II}(t) = \text{diag}(k_{n+i}(t)), i=1,2,\ldots,n,$$

where $\sigma_i(t)$ is from (23). The control synthesis parameters, p_{oi1}, p_{oi2}, $\sigma_{i\,max}$, γ_i, φ_i and ξ_i should be selected in order to obtain the practical closed loop stability, and the desired exponential stability of the full robot model (including (5), (18), and (23) to (25)), taking into account the state and control constraints:

$$|z_i(t)| \leq z_{i\,max} \leq z_{ic}, \quad i=1,2,\ldots,n,$$
$$|z_{n+i}(t)| \leq z_{n+i\,max} \leq z_{n+ic}, \quad (26)$$
$$|u_i(t)| \leq |u_{wi}(t) + \Delta u_i(t)| \leq u_{i\,max} \leq u_{ic},$$

where $u_{wi}(t)$ is the nominal control of i-th subsystem, $\Delta u_i(t)$ is the corresponding increase (decrease) of control variable in the closed loop system, and subscript c characterizes the constraints of the full robot model variables, and can substitute the subscrypt max in (23). In addition to this $z_{i\,max}$ and $z_{n+i\,max}$ can be calculated by the equations:

$$z_{i\,max} = \left| \left[a_i e^{\lambda_{i1}(t_i^* - t_o)} + b_i e^{\lambda_{i2}(t_i^* - t_o)} \right] / c_i \right|,$$
$$z_{n+i\,max} = \left| \left[\lambda_{i1} a_i e^{\lambda_{i1}(t_i^{**} - t_o)} + \lambda_{i2} b_i e^{\lambda_{i2}(t_i^{**} - t_o)} \right] / c_i \right|, \quad (27)$$

where:

$$a_i = \lambda_{i2} z_i(t_o) - z_{n+i}(t_o), \quad b_i = z_{n+i}(t_o) - \lambda_{i1} z_i(t_o)$$
$$t_i^* = \frac{1}{c_i} \ln\left(-\frac{a_i \lambda_{i1}}{b_i \lambda_{i2}}\right) + t_o, \quad c_i = \lambda_{i2} - \lambda_{i1}, \quad (28)$$
$$t_i^{**} = \frac{1}{c_i} \ln\left(-\frac{a_i}{b_i}\left(\frac{\lambda_{i1}}{\lambda_{i2}}\right)^2\right) + t_o, \quad \lambda_{i1} \neq \lambda_{i2},$$
$$\lambda_{i1} = p_{oi1} - \sigma_{i\,max}, \quad \lambda_{i2} = p_{oi2} - \sigma_{i\,max}, \quad i=1,2,\ldots,n.$$

As results of application of nonlinear direct adaptation procedures (23) on the full robot model (5), the high (desired) stage of the relative stability, in the case of state and control constraints, has been obtained, and an exponential stability has been preserved. Besides, the sampling period in the adaptive control case can be greater, than in the nonadaptive one, without loss of the desired stage of the relative and exponential stability. The corresponding simulated example, based on the RRTR-robot structure [23], has been discussed at the conference.

CONCLUSIONS

In this paper an unified approach to the nonlinear adaptive control synthesis of industrial robots has been presented. Starting with the mathematical dynamic models of manipulator and actuators and using the linear connections between them, the full robot model hes been obtained. Following the ideas of the external linearization, and using the state-space approach, the tracking error model of industrial robot has been built, and the corresponding nonlinear control law has been derived. An unified direct adaptation procedure of the position and velocity regulator gains has also been proposed. As results of this approach, the high (desired) stage of the relative stability, including the state and control constraints, has been obtained, where an exponential stability is preserved. Besides, the sampling period in the adaptive control case can be greater, than in the nonadaptive one, without loss of the desired stage of the relative and exponential stability.

REFERENCES

[1] K.J. Aström and B. Wittenmark, "Adaptive Control", Reading, MA: Addison-Wesley, 1989.
[2] M.Vukobratović, D. Stokić, and Kirćanski, "Adaptive and nonadaptive control of manipulation robots" Springer-Verlag, Berlin, 1985.
[3] M.Bodson and S. Sastry, "Adaptive Control Stability, Convergence, and Robustness", Englewood Cliffs: Prentice Hall, 1989.
[4] H. Unbehauen, "Theory and application of adaptive control, Prepr. of 7th Conf. on Dig. Comp. Appl. to Proc. Control, IFAC, IFIP, IMACS, pp. 3-19, Vienna, 1985.
[5] J.J. Craig, P.Hsu, and S.S.Sastry, "Adaptive control of mechanical manipulators", Int. J. Robotics Res., vol.6, pp. 16-28, 1987.
[6] J.J.E. Slotine and W.Li. "On the adaptive control of robot manipulators", "Int. J. Robotics Res., vol. 6, no.3, pp. 49-59, 1987.
[7] N.Sadegh and R. Horowitz, "Stability analysis of an adaptive controller for robotic manipulators", in Proc. IEEE Int. Conf. Robotics Automat. (Raleigh, NC), Mar. 1987., pp. 1223-1239.
[8] D.S. Bayard and J.T. Wen, "New class of control laws for robotic manipulators-Part 2. Adptive case", Int. J. Contr. vol. 47, no. 5, pp. 1387-1406, 1988.

[9] J. J. E. Slotine and W. Li, "Indirect adaptive robot control", in Proc. IEEE Int. Conf. Robotics Automat. (Philadelphia, PA), Apr. 1988, pp. 704-709.
[10] C. S. G. Lee and M. J. Chung, "An adaptive control strategy for mecanical manipulators", IEEE Trans. Automat. Contr., vol. AC-29, pp. 837-840, 1984.
[11] G. Campion, A. M. Guillaume, and G. Bastin- Greco, "Adaptive external linearization feedback control for flexible link manipulators: Robustness analysis" Prepr. IFAC/IFIP/IMACS Intern. Symp. on Theory of robots, pp. 37-40, OPWZ, Vienna, Dec. 3-5, 1986.
[12] R. L. Leal and C. C. De Wit, "Passivity based adaptive control for mechanical manipulators using LS-type estimation", IEEE Trans. Autom. Contr., vol 35, no. 12, pp. 1363-1365, 1990.
[13] S. Dubowsky and D. T. Desforges, "The application of model-reference adaptive control to robotic manipulators", ASME J. Dyn. Syst. Meas. Contr., vol. 101, 1979.
[14] J. J. Craig, P. Hsu, and S. S. Sastry, "Adaptive control of mechanical manipulators", in IEEE Int. Conf. on Robotics and Automation (San Francisco, CA), 1986.
[15] P. Hsu, M. Bodson, S. Sastry, and B. Paden, "Adaptive identification and control for manipulators without using joint accelerations", in IEEE Int. Conf. on Rolbotics and Automation (Raleigh, NC), 1987.
[16] R. Johansson, "Adaptive Control of Robot Manipulator Motion", IEEE Trans. on robot and Autom., vol. 6, no. 4, pp. 483-490, 1990.
[17] N. Sadegh and R. Horowitz, "An exponentially stable adaptive control law for robot manipulators" IEEE Trans. on Robot. and Autom., vol. 6, no. 4, pp. 491-496, 1990.
[18] R. Paul, "Robot manipulators: Mathematics, programming, and control; The computer control of robot manipulators", The MIT Press, London, 1982.
[19] M. Vukobratović, and D. Stokić, "Control of manipulation robots: Theory and application", Springer-Verlag, Vol. 2, Berlin, 1982.
[20] A. Isidori and A. Ruberti, "On the synthesis of linear input-output responses for nonlinear systems" Systems and Control Letters, 4, pp. 17-22, 1984.
[21] B. M. Novaković, "A time and energy optimal control of industrial robots", in Theory of robots, ed. by P. Kopacek, I. Troch, K. Desoyer, IFAC, Proc. Series, No. 3, pp. 205-210, 1988.
[22] B. M. Novaković, "An external linearization and energy optimal control of industrial robots", in AMSE Conf. Modell. and Simul. Proc. Vol. 3A, pp. 107-118, Karlsruhe, 1987.
[23] M. Crneković, "Developing of simulation robot model of RRTR structurte for control synthesis", Scientific report, M. Sc. degree FSB, University of Zagreb (YU), 1988.

Model Reference Adaptive Control (MRAC) of Robotic Manipulators using A Modified Output Error Method (MOEM)

Ricky Y. J. Tam
Department of Electrical Engineering
Southern Illinois University
Carbondale, IL 62901, USA

Stelios C. A. Thomopoulos
Decision and Control Systems Laboratory
Department of Electrical Engineering
The Pennsylvania State University
University Park, PA 16802, USA

ABSTRACT

Model reference adaptive control of a robotic manipulator using a modified output error method is introduced. In the absence of nonlinearities in an integrated dynamical model of a manipulator and actuators, the defined new regressors and augmented error in the modified output error method result a closed-loop system that is globally asymptotically stable without over-parameterization. Knowledge of the high frequency gain sign of the unknown plant is not required to implement the adaptive controller. Moreover, no requirement of strictly positive realness is imposed on the reference model transfer function. Numerical simulation with the modified output error method in the feedback loop and computed-torque compensator for model nonlinearities of the manipulator has resulted in stability of the overall closed-loop system and output tracking errors converging to zero asymptotically.

INTRODUCTION

Model reference adaptive control (**MRAC**) of a robotic manipulator using a modified output error method (**MOEM**) for the case of structural uncertainty is studied. A correct structural model of the controlled plant consisting of a d.c. servomotor driving a robotic joint is assumed with all uncertainties resulting from incorrect parameter values. Hence, there exists a correct (but unknown) set of values for the parameters such that the controlled plant model matches the actual system [1]. A desired joint trajectory that a robotic joint is required to track is described compactly by a reference model excited by a reference command input. Thus, in this study a correct set of values for the controller parameters is determined such that both output tracking and identification of the unknown controlled plant are achieved simultaneously.

Modifications of the output error method (**OEM**) in **MRAC** has been described in adaptive control literatures [2]÷[5]. The condition of strictly positive realness (**SPR-ness**) of transfer function is imposed in [4]. In addition, knowledge of the sign of the high frequency gain kp of the controlled plant transfer function is required to be known a priori in both [4] and [5].

In contrast, **MOEM** requires only a lower bound on kp for assuring the boundedness of all signals and the uniform boundedness of parameter errors is guaranteed by using the normalized gradient algorithm with projection in the adaptation law for updating the controller parameter estimates [6]. Furthermore, **MOEM** imposes no requirement of **SPR-ness** on the reference model transfer function and its linear error equation enables other more effective identification algorithms, such as least squares, to be used in the adaptation law.

Nonlinearities arising from gravity loading, Coriolis and centrifugal forces in the manipulator dynamics can be treated as an external disturbance load torque input to the joint actuator. Computed-torque compensation is applied for compensating such an external disturbance load torque.

Lastly, graphical results from computer simulations on the performance of **MOEM** with and without computed-torque compensation under various external disturbance loads are presented.

§1. MODELLING OF CONTROLLED PLANT AND REFERENCE MODEL

In this paper, the Laplace transform of any function f(t) in the time t-domain is represented by $\hat{f}(s)$ in the complex frequency s-domain. A block diagram of a d.c. servomotor with negligible armature inductance driving a robotic joint is depicted in Figure 1.1. The torque developed Td at the motor shaft is required to accelerate an inertia (usually consisting of the armature and an external load) and to overcome any viscous damping torque (due to the motion of the armature and external load) and any external disturbance load torque DL due to the presence of gravity loading, Coriolis and centrifugal forces in the manipulator dynamics. From Figure 1.1, the plant transfer function with unknown or slowly time-varying parameters $\hat{P}(s)$ used for the **MRAC** problem in this study is given according to

$$\hat{P}(s) := \frac{\hat{y}_p(s)}{\hat{u}(s)} = \frac{\hat{\theta}_L(s)}{\hat{u}(s)}\bigg|_{\substack{DL(s)=0 \\ V_{CT}(s)=0}} = \frac{\dfrac{nk_a k_{e,u}}{R_a J_{eff}}}{s^2 + \left\{\dfrac{R_a B_{eff} + k_a(k_b + k_w)}{R_a J_{eff}}\right\}s + \dfrac{nk_a k_{e,yp}}{R_a J_{eff}}} \quad (1.1)$$

$J_{eff} = J_m + n^2 J_L$, $B_{eff} = B_m + n^2 B_L$

where (Jm, Bm, Ra, Ka, Kb) and (JL, BL) are the d.c. servomotor and the manipulator physical parameters, respectively. (kw, ke,u, ke,yp) are amplifier gains. VCT is the computed-torque compensation voltage. Jeff and Beff are the effective inertia and viscous friction, respectively, at the motor shaft. θL is the manipulator joint displacement. Gains ke,u and ke,yp are chosen large enough such that, in the absence of **MOEM** and computed-torque compensation, the effect on yp due to the presence of any step disturbance torque input DL can be minimized and output tracking can be achieved by setting the control input u to ym.

From Figure 1.2, the reference model transfer function $\hat{M}(s)$ for the **MRAC** problem is given by

$$\hat{M}(s) := \frac{\hat{y}_m(s)}{\hat{r}(s)} = \frac{\dfrac{nk_a k_{pr}}{R_a J_m}}{s^2 + \left\{\dfrac{R_a B_m + k_a k_b}{R_a J_m}\right\}s + \dfrac{nk_a k_{pr}}{R_a J_m}} \quad (1.2)$$

where ym is the reference model output representing the desired joint trajectory which yp is required to track, r is the reference command input, and kpr is a proportional gain applied to the error signal ym-r.

§ 2. ANALYSIS OF MRAC OF A ROBOTIC JOINT USING MOEM

The assumptions similar to those discussed in [2] are used. The controlled plant is a single-input single-output (SISO) linear time invariant (LTI) system described by a transfer function

$$\frac{\hat{y}p(s)}{\hat{u}(s)} = \hat{P}(s) = kp \frac{\hat{n}p(s)}{\hat{d}p(s)} \tag{A1}$$

where $\hat{u}(s)$ and $\hat{y}p(s)$ are the input and output of the plant, respectively. $\hat{n}p(s)$ and $\hat{d}p(s)$ are monic, coprime polynomials of degree m and n, respectively. m is unknown, but the plant is strictly proper (m≤n-1). $\hat{P}(s)$ is assumed to be minimum phase, but it is not assumed to be stable. A lower bound $|kp|_{min}$ on $|kp|$ is assumed to be known.

The reference model is described by

$$\frac{\hat{y}m(s)}{\hat{r}(s)} = \hat{M}(s) = km \frac{\hat{n}m(s)}{\hat{d}m(s)} \tag{A2}$$

where $\hat{n}m(s)$ and $\hat{d}m(s)$ are monic, coprime polynomials of degree m and n, respectively. The reference model is stable, minimum phase and km>0. The reference input r(•) is piecewise continuous and bounded in R.

The controller for **MRAC** of a robotic joint (controlled plant) using **MOEM** is shown in Figure 2.1. The control input u to the controlled plant is

$$u = c0\, r + \frac{\hat{c}(s)}{\hat{\lambda}(s)}(u) + \frac{\hat{d}(s)}{\hat{\lambda}(s)}(yp) \tag{2.1}$$

where c0 is a scalar, $\hat{c}(s)$, $\hat{d}(s)$, and $\hat{\lambda}(s)$ are polynomials of degrees n-2, n-1, and n-1, respectively. $\hat{\lambda}(s)$ is a monic polynomial such that $\hat{\lambda}(s) = \hat{\lambda}0(s)\,\hat{n}m(s)$ where $\hat{\lambda}0(s)$ is an arbitrary Hurwitz polynomial of degree n-m-1. Adaptive form of the controller in state space representation is shown in Figure 2.2. $\left(\Lambda \in R^{n-1 \times n-1},\ b_\lambda \in R^{n-1}\right)$ is chosen to be in controllable canonical form so that $(sI-\Lambda)^{-1} b_\lambda$ represents the transfer functions of stable adaptive observers with state equations [7]

$$\dot{w}_1 = \Lambda w_1 + b_\lambda u\, ,\ \dot{w}_2 = \Lambda w_2 + b_\lambda yp\, ;\ w_1, w_2 \in R^{n-1} \tag{2.2}$$

Initial conditions $w_1(0)$, $w_2(0)$ are arbitrary. Consider $c \in R^{n-1}$ be the vector of coefficients of the polynomial $\hat{c}(s)$ such that

$$\frac{\hat{c}(s)}{\hat{\lambda}(s)}(u) = \left[c^T (sI-\Lambda)^{-1} b_\lambda\right](u) = c^T w_1 \tag{2.3}$$

Similarly, there exists $d0 \in R$ and $d \in R^{n-1}$ such that

$$\frac{\hat{d}(s)}{\hat{\lambda}(s)}(yp) = \left[d0 + d^T(sI-\Lambda)^{-1}b_\lambda \right](yp) = d0\, yp + d^T w_2 \tag{2.4}$$

Thus, the adaptive control law for **MOEM** in the form of (2.1) is

$$u = c0\, r + c^T w_1 + d0\, yp + d^T w_2 := \theta^T w\;,\; \theta \in R^{2n},\; w \in R^{2n} \tag{2.5}$$

$$\theta^T := \left(c0, \bar{\theta}^T \right) := \left(c0, c^T, d0, d^T \right)\;,\; w^T := \left(r, \bar{w}^{-T} \right) := \left(r, w_1^T, yp, w_2^T \right) \tag{2.6}$$

where θ is a vector of controller parameters and w is a vector of signals that can be obtained without knowledge of the plant parameters.

Hence, from (2.6) the vector of nominal controller parameters such that a matching of the transfer function $r \to yp$ to $\hat{M}(s)$ is attained can be defined as

$$\theta^{*T} := \left(c0^*, \bar{\theta}^{*T} \right) := \left(c0^*, c^{*T}, d0^*, d^{*T} \right) \in R^{2n} \tag{2.7}$$

The corresponding parameter errors vector is

$$\phi := \theta - \theta^* \in R^{2n}\;;\; \bar{\phi} = \bar{\theta} - \bar{\theta}^* \in R^{2n-1} \tag{2.8}$$

It can be easily shown that matching of the transfer function of the overall closed-loop system $r \to yp$ to $\hat{M}(s)$ is achieved if and only if the following matching equality is satisfied [8]

$$1 - \frac{\hat{c}(s)^*}{\hat{\lambda}(s)} - \frac{\hat{d}(s)^*}{\hat{\lambda}(s)} \hat{P}(s) = \hat{M}(s)^{-1} c0^* \hat{P}(s) \tag{2.9}$$

The uniqueness of θ^* is guaranteed by the coprimeness of $\hat{np}(s)$ and $\hat{dp}(s)$ [6].

The output tracking error in **MOEM** is defined as

$$e := yp - ym \tag{2.10}$$

When matching is achieved ($e \equiv 0$), the joint angle yp can be expressed in the form [8]

$$yp = -\hat{M}\left(\tilde{\theta}^{*T} \tilde{w} \right),\; \tilde{\theta}(t)^{*T} := \left[\frac{1}{c0^*}\; \frac{\bar{\theta}^{*T}}{c0^*} \right] \in R^{2n},\; \tilde{w}(t)^T := \left[-u\; \bar{w}^{-T} \right] \in R^{2n} \tag{2.11}$$

where $\tilde{\theta}$ and $\tilde{w}$ are the nominal parameter and regressor vectors for **MOEM**.

Remark: Since both km, kp are non-zero, $c0^*$ is non-zero. Therefore θ^* is well-defined.

Solving for the reference command input r in (2.5) gives

$$r = \frac{1}{c0}\left\{ u - c^T w_1 - d0\, yp - d^T w_2 \right\} = \frac{1}{c0}\left\{ u - \bar{\theta}^{-T}\bar{w} \right\} \tag{2.12}$$

where $c0(t) \neq 0\; \forall\, t \geq 0$ is assumed. From (A2) and (2.12), ym can be expressed by

$$ym = \hat{M}\left\{ \frac{1}{c0}\left[u - \bar{\theta}^{-T}\bar{w} \right] \right\} = \hat{M}\left\{ -\tilde{\theta}^{T}\tilde{w} \right\},\; \tilde{\theta}(t)^T := \left[\frac{1}{c0}\; \frac{\bar{\theta}^{-T}}{c0} \right] \in R^{2n} \tag{2.13}$$

Subtraction of (2.13) from (2.11) yields the error equation for **MOEM**

$$e = \hat{M}\left[\left(\tilde{\theta}^T - \tilde{\theta}^{*T} \right)\tilde{w} \right] = \hat{M}\left(\tilde{\phi}^T \tilde{w} \right),\; \tilde{\phi} = \left(\tilde{\theta} - \tilde{\theta}^* \right) \in R^{2n} \tag{2.14}$$

where $\tilde{\phi}$ is the new parameter errors vector. $\tilde{\theta}^*$ and $\tilde{\theta}$ are given in (2.11) and (2.13), respectively. Next, an augmented error similar to that given in [3] is obtained in the form

$$\tilde{e} = \tilde{\phi}^T \hat{M}(\tilde{w}) = e + \tilde{\phi}^T \hat{M}(\tilde{w}) - \hat{M}(\tilde{\phi}^T \tilde{w}) := e + e_1 \qquad (2.15)$$

where e_1 is called the auxiliary error [4]. The augmented error signal $\tilde{e}$ is constructable because it is a function of measurable signals (in the input-output sense). The augmented error $\tilde{e}$ in (2.15) together with the normalized gradient algorithm with projection give an adaptation law for updating the elements of $\tilde{\theta}$ according to

$$\dot{c0} = g \frac{\tilde{e}\hat{M}(\tilde{w}_1)}{1 + \gamma \hat{M}(\tilde{w})^T \hat{M}(\tilde{w})} (c0)^2, \quad \frac{d}{dt}\left\{\frac{\tilde{\theta}_i}{c0}\right\} = -g \frac{\tilde{e}\hat{M}(\tilde{w}_{i+1})}{1 + \gamma \hat{M}(\tilde{w})^T \hat{M}(\tilde{w})} \qquad (2.16)$$

$$; g,\gamma > 0, \ 1 \leq i \leq 2n-1$$

and c0 is projected on the boundary $\partial\Theta$ defined by |c0|max = km/|kp|min when |c0| = |c0|max and the sign of $\dot{c0}$ favors c0 to exit $\partial\Theta$.

Mathematical proofs of global stability and parameter convergence for **MOEM** in the absence of DL and VCT can be proceeded as those outlined in [7]÷[8].

§ 3. NUMERICAL EXAMPLE

Numerical values of the physical parameters for the d.c. servomotor and gain kpr used in the reference model are

Jm = 30.0 x 10^{-6} kgm^2 Bm = 0.0 N/rad/s   (3.1)
ka = 17.0 x 10^{-3} Nm/amp   kb = 60.0 x 10^{-3} V/rad/s
Ra = 3.2 Ω kpr = 0.3253 V/rad

The robot manipulator is considered as a slender bar having a mass m=0.5 kg and a length l=0.5m. Hence, its moment of inertia JL about the joint axis is calculated as

$$JL = \frac{1}{3} ml^2 = 0.0417 \text{ kgm}^2 \qquad (3.2)$$

BL is assumed to be zero for simplicity. Gains kw, ke,u, and ke,yp are
kw = 156.9157 V/rad/s   ke,u = ke,yp = 968.9838 V/rad   (3.3)

From (1.3), (1.4), (A1), and (A2), $\hat{P}(s)$ and $\hat{M}(s)$ are determined

$$\hat{P}(s) = \frac{123.4565}{s^2 + 20.0s + 123.4565}, \quad \hat{M}(s) = \frac{57.6052}{s^2 + 10.6250s + 57.6052} \qquad (3.4)$$

The nominal values for the coefficients of the polynomials $\overset{*}{c}(s)$, $\overset{*}{d}(s)$, and gain $\overset{*}{c0}$ are determined using the procedure outlined in [2]. For this numerical example, they are

$$\overset{*}{c0} = 0.4666, \ \overset{*}{c1} = 9.3750, \ \overset{*}{d0} = -0.9094, \ \overset{*}{d1} = -7.9322 \qquad (3.5)$$

Therefore, the nominal parameter vector $\tilde{\theta}^*$ is

$$\tilde{\theta}^{*T} = \begin{bmatrix} 2.1431 & 20.0922 & -1.9490 & -17.0000 \end{bmatrix} \qquad (3.6)$$

In the presence of computed-torque compensation, the compensation voltage V_{CT} required to be added to the armature voltage is given by

$$V_{CT}(t) = \frac{R_a}{k_a} D_{L,est}(t) \qquad (3.7)$$

where $D_{L,est}$ is used for compensating a step D_L taken as the maximum gravity loading in the form

$$D_L(t) = 0.5 mgl, \quad D_{L,est}(t) = 0.5\, m_{,est}\, gl, \quad m_{,est} = \frac{3.0}{(nl)^2} \left\{ \frac{nJ_m c_0}{k_{pr}} - J_m \right\} \qquad (3.8)$$

The term $m_{,est}$ in (3.8) is the estimated mass of the manipulator linkage based on the estimated parameter c_0. For the case of a sinusoidal external disturbance, D_L and $D_{L,est}$ are given in the form

$$D_L(t) = 0.5 mgl\cos(y_p(t)), \quad D_{L,est}(t) = 0.5 m_{,est} gl\cos(y_p(t)) \qquad (3.9)$$

§ 4. SIMULATION RESULTS AND CONCLUSIONS

In Figure P4.1÷P4.2, the performance of **MOEM** in the absence of V_{CT} and D_L is shown. Figure P4.1 shows that the output tracking error e in (2.14) converges to zero asymptotically regardless to the persistent excitation of r. However, as shown in Figure P4.2, convergence of the controller parameter estimates (c_0, c_1, d_0, d_1) to their nominal values ($\overset{*}{c}_0, \overset{*}{c}_1, \overset{*}{d}_0, \overset{*}{d}_1$) can only be achieved when r is persistently excited.

Figure P4.3 illustrates that in the presence of a unit-step D_L, tracking of a unit-step desired joint trajectory is improved with the application of **MOEM** (without computed-torque compensation). It is observed that **MOEM** has resulted an overshoot in y_p but its output tracking error converges to zero asymptotically. In comparison, without **MOEM** there is no overshoot but a steady-state error of -0.1943 rad. is incurred.

Moreover, performance of **MOEM** with D_L and $D_{L,est}$ given by (3.8) is shown in Figure P4.4. No significant improvement in tracking performance is attained when computed-torque compensation is applied. Similar tracking performance is exhibited when D_L and $D_{L,est}$ are described by (3.9) and the same unit step r is used as illustrated in Figure P4.5. Effect of various values of $k_{e,u}$ and k_{e,y_p} on the response y_p in the absence of **MOEM** and computed-torque compensation is shown in Figure P4.6.

Lastly, in order to compensate for higher order D_L the gains $k_{e,u}$ and k_{e,y_p} must be replaced by higher order transfer functions so that undesired poles introduced by D_L can be cancelled reducing the effect of D_L in the adaptive control. This results in a higher order adaptive controller with additional complexity.

REFERENCES

[1]. **John J. Craig**, Adaptive Control of Mechanical Manipulator, Addison-Wesley Publishing Company, 1988.
[2]. **Shankar Sastry and Marc Bodson**, Adaptive Control: Stability, Convergence, and Robustness, Prentice-Hall Inc., 1989.
[3]. **Kumpati S. Narendra and Yuan-Hao Lin**, 'Design of Stable Model Reference Adaptive Controllers, 'Applications of Adaptive Control, Academic Press, 1980, pp. 69-130.
[4]. **Kumpati S. Narendra and Anuradha M. Annaswamy**, Stable Adaptive Systems, Prentice-Hall Inc., Englewood Cliffs, N.J. 07632, 1989.
[5]. **Kumpati S. Narendra, Yuan-Hao Lin, and Lena S. Valavan**, 'Stable Adaptive Controller Design, Part II: Proof of Stability, 'IEEE Transaction on Automatic

Control, Vol. AC-25, No.3, pp. 440-448.
[6]. **Marc Bodson and Shankar Sastry**, 'Input Error versus Output Error Model Reference Adaptive Control, ' Proceeding of the American Control Conference, Minneapolis, 1987, pp. 224-229.
[7]. **Yannis N. M. Papadakis and Stelios C. A. Thomopoulos**, 'A Modified Output Error Method for Model Reference Adaptive Control, Proceedings of 28th Annual Allerton Conference on Communication, Control, and Computing, Allerton House, Montecello, Illinois, October 3-5, 1990, pp. 1028-1037.
[8]. **Ricky Y. J. Tam**, 'Model Reference Adaptive Control of Robotic Manipulators using a Modified Output Error Method, 'Technical Report TR-SIU-EE-04-91, Southern Illinois University, Carbondale, IL 62901.

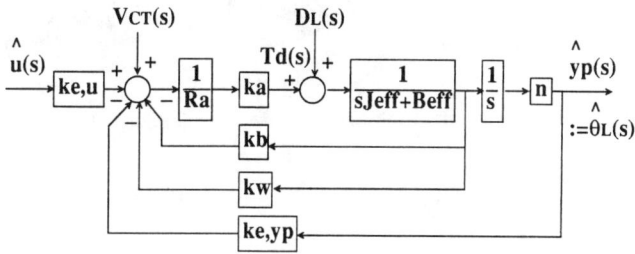

Figure 1.1: Block diagram of a d.c. servomotor driving a robotic joint

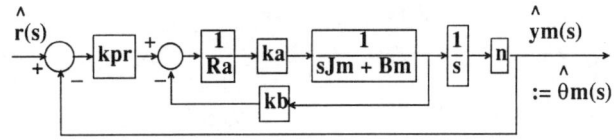

Figure 1.2: Block diagram of reference model

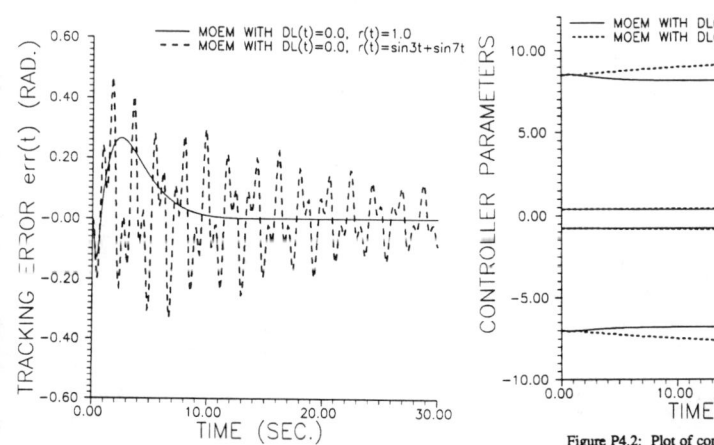

Figure P4.1: Plot of output tracking error err(t) in MRAC of manipulator using MOEM in the absence and presence of persistently excited r (DL=0.0, VCT=0.0).

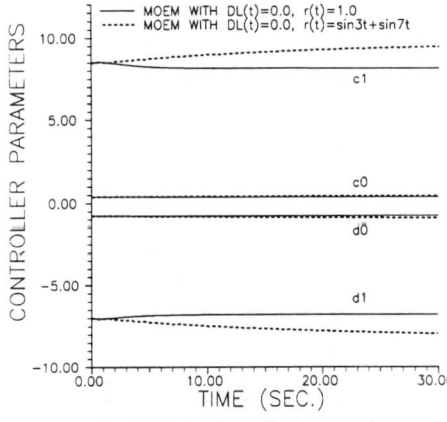

Figure P4.2: Plot of controller parameter estimates (c0, c1, d0, d1) with MOEM used in MRAC of manipulator in the absence and presence of persistently excited r (DL=0.0, VCT=0.0).

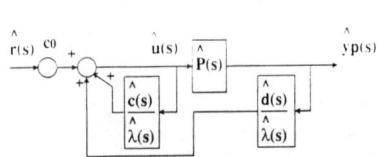

Figure 2.1: Basic Controller Structure

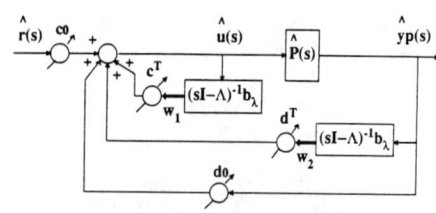

Figure 2.2: Adaptive Form of Controller Structure in State Space Representation

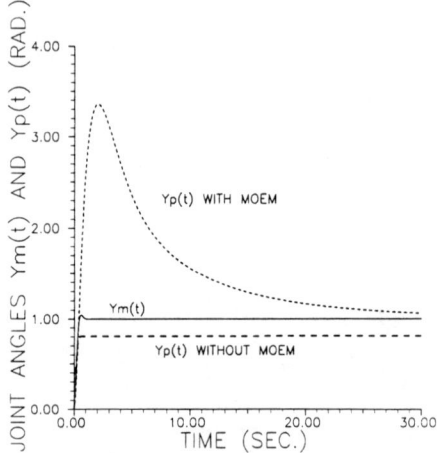

Figure P4.3: Plot of joint angles ym(t) and yp(t) in MRAC of manipulator in the presence and absence of MOEM with DL(t)=1.0 and r(t)=1.0.

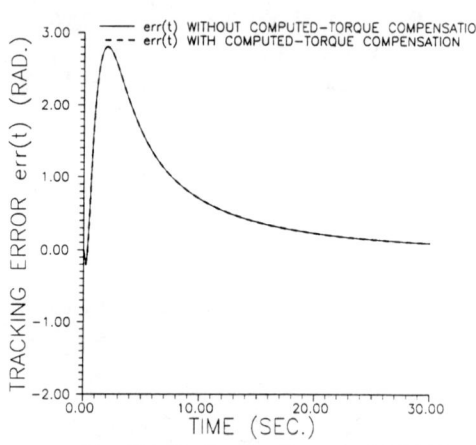

Figure P4.4: Plot of output tracking error err(t) in MRAC of manipulator using MOEM in the presence and absence of computed-torque compensation with DL(t)=0.5mgl and r(t)=1.0.

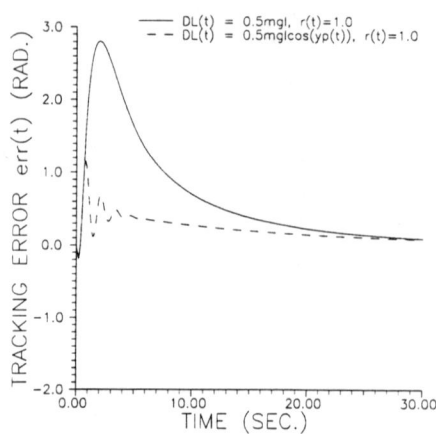

Figure P4.5: Plot of output tracking error err(t) in MRAC of manipulator using MOEM and computed- torque compensation with r(t)=1.0, DL(t)=0.5mgl and DL(t)=0.5mglcos(yp(t)).

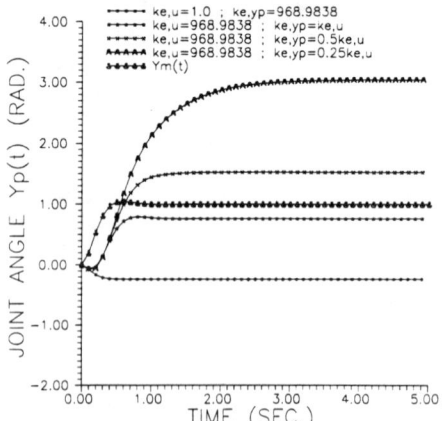

Figure P4.6: Plot of joint angle yp(t) due to various values of ke,u and ke,yp in the absence of MOEM and computed-torque compensation.

APPLICATION OF MODIFIED GENERALIZED PREDICTIVE CONTROLLERS TO TRAJECTORY CONTROL OF A ROBOT ARM

Sunay S. GENÇER, Yorgo İSTEFANOPULOS, Tamer ERGÜN
Boğaziçi University, Bebek, İstanbul, TURKEY

Abstract:
In this work, modified GPCs are used for the trajectory control of a three degree of freedom planar robotic arm. The algorithm proposed here is different from the standard GPC design in the following sense: firstly, in the minimization of the quadratic function a weighting sequence is used on the future predicted errors; secondly, a weighted average of several values of the control input predicted at different time instants, is applied as actual input; thirdly, dynamic weightings on predicted error is considered which appears to be an easier way of assigning weights on the future errors and assures a prespecified closed-loop response.

INTRODUCTION

In the concept of self-tuning control of discrete-time systems, many schemes are based on a Controlled Autoregressive Integrated Moving Average (CARIMA) model of the plant. A general purpose controller should be applicable to nonminimum-phase plants, open-loop unstable plants or plants with badly-damped poles, plants with unknown and/or varying dead-time and unknown order. Most of the self-tuning controllers such as minimum-variance, pole-placement or dead-beat techniques have problems in controlling systems with unstable open-loop zeroes, unknown model-order or varying dead-time.

However, Long Range Predictive Control (LRPC) has improved robustness properties over the other self-tuning methods. Dynamic Matrix Control by Cutler and Ramarker [1], Extended Prediction Self-Adaptive Control by De Keyser and Van Cauwenberghe [2], Model Algorithmic Control by Richalet et al. [3], Generalized Predictive Control by Clarke et al. [4] are all mainly based on the LRPC idea. A comparative study of these algorithms can be found in De Keyser et al. [5]. The strategy underlying LRPC methods is as follows

i) Given a mathematical model of the plant dynamics, predictions of the plant output over a time range called costing or prediction horizon are made.

ii) A sequence of future control signals over the control horizon are computed according to a specific control objective, which in general seeks to minimize some norm of future errors.

iii) The first element in the sequence is applied as control input to the plant. At the next sampling instant this procedure is repeated with new data set resulting in a receding horizon strategy.

GENERALIZED PREDICTIVE CONTROL

Among other LRPC methods, GPC is interesting since it provides a more general set of design criteria than others. GPC method is designed to overcome the problems arising from unstable open-loop zeroes, unknown model order etc. GPC algorithm is based on predicting the outputs of the plant at future sampling instants. Assuming that a set of future reference values are available, as would be the case in robotic applications, future errors can be predicted and the following quadratic cost function can be minimized.

$$J(N1,N2) = E\left\{\sum_{i=N1}^{N2}\left[w(k+i) - \hat{y}(k+i)\right]^2 + \sum_{j=1}^{NU}\lambda_j\left[\Delta u(k+j-1)\right]^2\right\} \qquad (1)$$

The design variables or tuning-parameters of GPC are the costing horizons (N1...N2), the control horizon (NU) and the control weighting λ.

The prediction involved in the GPC formulation requires a mathematical model of the plant dynamics. Usually a CARIMA model given below

$$(1-z^{-1})\, A(z^{-1})\, y(k) = z^{-l} B(z^{-1})\Delta u(k) + C(z^{-1})\, \zeta(k) \qquad (2)$$

is used to find optimal predictors. In the original approach [4], the $C(z^{-1})$ polynomial is assumed to be 1 and the following Diophantine identity is used to determine the i-step ahead output.

$$1 = A(z^{-1})\, F_i(z^{-1})\, (1-z^{-1}) + z^{-i}\, G_i(z^{-1})\, B(z^{-1}) \qquad (3)$$

Then the i-step ahead prediction with F_i of degree i-1, is

$$y(k+i) = F_i B \Delta u(k+i-1) + G_i\, y(k) + F_i e(k+i) \qquad (4)$$

Introducing the following equality

$$F_i B = R_i + z^{-i}\, R'_i \qquad (5)$$

for i= N1 to N2 the equation above can be written as

$$y = R\Delta u + y' + e \qquad (6)$$

where $y=[y(k+1),..., y(k+N2)]^T$, $\Delta u=[\Delta u(k),..., \Delta u(k+NU-1)]^T$, $e = [F_1 e(k+1),..., F_{N2}e(k+N2)]^T$, $y'_i = R_i \Delta u(k-1) + G_i y(k)$ and y' is the constrained prediction vector under the assumption that $\Delta u=0$, i.e. y' is a function of the past inputs and outputs only. As the predictions over the costing horizon (N1 to N2) is necessary, the number of Diophantine equations that must be solved is N2-N1. Solution of Diophantine equations can be simplified by using recursive relationship between each consecutive prediction polynomial F_i and G_i. An easier way of computing y' is noting that it is simply the so-called constraint prediction vector i.e. the response when $\Delta u[k+i]=0$ for i=1 to NU and disturbance is constant. So, N2-N1 one-step ahead predictions are necessary to compute y' [6]. Then, the control law minimizing J is obtained as

$$\Delta u = (R^T R + \lambda I)^{-1} R^T (w - R'\Delta u(k-1) - Gy(k) - Fe(k)) \qquad (7)$$

Notice that R is a N2xNU matrix and inversion involved in equation (7) reduces to a scalar operation when NU=1.

Many adaptations of GPC and guidelines for the selection of design variables of GPC can be found in Clarke et al. [7]. Clarke proposes constant observer polynomials as $C(z^{-1})$ to enhance robustness and shaping the disturbance response. Also, he proposes the use of auxiliary signals $P(z^{-1})y(k)$ and $P(z^{-1})u(k)$ to enhance the pole-placement capabilities of GPC. The latter approach is equal to placing a robustness filter in the internal model control loop of a GPC characterized in a 2-port structure [8]. Design variables of GPC can be effectively chosen according to the following rules for most of the cases

i) N1 should be equal to or larger than plant dead-time,
ii) N2 should be larger than degree of $B(z^{-1})$ and it may be as large as the rise-time of the plant.
iii) NU should be small as it reduces the computational complexity of the solution significantly. But if the system is complex such that it has unstable poles or poles near the unit circle then the choice of NU greater than the number of such poles gives stable control.
iv) λ could be zero or a very small number in order to increase numerical robustness involved in matrix inversion.

In this paper, modified GPC's are used for the trajectory control of a three degree of freedom planar robot arm. The modifications considered on GPC are

i) In the minimization of the quadratic function, a weighting sequence is used on the future predicted errors. Similar weighting functions are used in the Extended Predictive Self-Adaptive Control approach of De Keyser et al. An exponential weighting as $\alpha_i = \alpha_0^{N2-i}$ i = N1,..., N2 can be used leaving α_0 as design variable. Note that when $0 < \alpha_0 < 1$, there is more weighting on the remote future which would result in a smooth but sluggish control. As α_0 is increased a more jittery control with better output error performance can be obtained. The solution is given as

$$\Delta u = (R^T \mathcal{A} R + \lambda I)^{-1} R^T \mathcal{A} (w - y') \qquad (8)$$

where $\mathcal{A}$ is a diagonal matrix of the α_i weightings on prediction error.

ii) In the original GPC approach, when NU > 1 only the control input for the present instant is applied to the plant, but the other computed values of Δu vector are not used. Robustness of GPC can be increased by applying a weighted summation of the various input values computed at different instants to the plant as

$$u_{app}(t) = \frac{\sum_{i=1}^{NU} \gamma(i)\, u(k/k-i+1)}{\sum_{i=1}^{NU} \gamma(i)} \qquad (9)$$

The idea behind the weighted input GPC, is determining the actually applied control input as a weighted average of the available candidates for the control in order to reduce the effect of spurious on-off input signals.

iii) A third modification on GPC that will be introduced here is changing the original cost function to assure desired dynamics for future predicted errors, as follows

$$J(N1,N2) = E\left\{ \sum_{i=N1}^{N2} \left[P(z^{-1})\hat{e}(k+i) \right]^2 + \sum_{j=1}^{NU} \lambda_j \left[\Delta u(k+j-1) \right]^2 \right\} \qquad (10)$$

The advantage of this scheme is that a prespecified dynamics $P(z^{-1})$ response can be assigned for the resulting closed-loop error dynamics. This reduces the difficulty faced with in choosing α_i's in the first method.

CARIMA MODEL FORMULATION OF A ROBOTIC MANIPULATOR

The dynamics of an N joint manipulator is modelled by the well known nonlinear matrix differential equation.

$$\tau = M(q)\ddot{q} + V(q,\dot{q}) + G(q) \qquad (11)$$

where q, $\dot{q}$ and $\ddot{q}$ are N-dimensional vectors of joint positions, velocities, and accelerations respectively and $M(q)$ is the symmetric, positive-definite NxN inertia matrix, $V(q,\dot{q})$ is N dimensional vector representing Centrifugal and Coriolis forces and $G(q)$ is N dimensional vector of gravitational terms.

In order to partially compensate and decouple such a system, a compensator of the following form is proposed:

$$\tau = \widehat{M}(q)u + \widehat{V}(q,\dot{q}) + \widehat{G}(q) \qquad (12)$$

where the caret signifies the estimated values and u is the input vector to the compensated system. The joint dynamics will then be described by

$$\ddot{q} = u - [\widehat{M}(q)]^{-1}[(M(q)-\widehat{M}(q))\ddot{q} + V(q,\dot{q})-\widehat{V}(q,\dot{q}) + G(q)-\widehat{G}(q)] \qquad (13)$$

Let us now assume that the compensated system described by Eq. (13) can be modelled by N difference equations of the following form

$$A(z) y(k) = z^{-l} B(z) u(k) + \zeta \qquad (14)$$

where u(k) and y(k) are the system input and output at the sampling instant k, and l is the system time delay. $A(z^{-1})$ and $B(z^{-1})$ are polynomials of degree n and m respectively in the backward shift operator (z^{-1}). The $A(z^{-1})$ polynomial is taken to be monic. The parameters of these polynomials can be estimated using Recursive Least Squares (RLS).

The signal ζ is a disturbance term including errors due to linearization, discretization, decoupling as well as unmodelled dynamics and unmeasurable disturbance. It is assumed to be of the form

$$\zeta = C(z^{-1}) e(k) /\Delta + d \qquad (15)$$

where Δ is the differencing operator, i.e. $\Delta=1-z^{-1}$. Combining equations (14) and (15) results in a CARIMA model for each link of the manipulator.

SIMULATION STUDIES

In order to show the performance of the algorithms, and the effects of the modifications a number of simulation studies were carried out on a 3-DOF planar robotic manipulator. The manipulator has three rotational joints. Though it can be lifted along the z-axis to add a fourth degree of freedom, this movement is entirely decoupled. Although in practice we encounter manipulators of 6-DOF, this basic robot arm shows every characteristics of a general manipulator with non-linear and coupled terms.

In the simulations, desired trajectories in cartesian space are straight lines with sinusoidal time profiles to have continuous velocities and accelerations. The inverse kinematics algorithm is similar to [9] which minimizes the instantaneous accelerations resulting in the global minimization of the Euclidean norm of the joint velocities by making use of the redundancy in the manipulator. When this path planning algorithm is used with the decoupling scheme of the previous section the non-linear term $V(q,\dot{q})$ is also minimized. Very rough estimates are done in modelling the manipulator for controller design to show the effectiveness of the algorithms. Coriolis and centrifugal forces that depend on the joint velocities are neglected and torques are chosen as

$$\tau = \left[\widehat{M}(q)\right]^{-1} u \qquad (16)$$

so that independent GPCs are used for each joint assuming each joint as a second order system or double integrator. The double integrator plant for each joint can be represented by the following difference equation

$$y(k+1) = 2y(k) - y(k-1) + (T^2/2) [\, u(k) + u(k-1) \,] \qquad (17)$$

where y is the joint variable. With sampling period of T=0.01 s., the polynomials $A(z^{-1})$ and $B(z^{-1})$ are obtained as

$$A(z^{-1}) = 1 - 2 z^{-1} + z^{-2} \quad \text{and} \quad B(z^{-1}) = 0.00005 \,(1 + z^{-1})$$

The nonlinear terms in $V(q,\dot{q})$ are considered as disturbances to each SISO system. Although the controller does not make use of velocity feedback, and rough estimates of the masses are used in the evaluation of the inertia matrix, the simulations gave excellent results.

Simulation studies of GPC algorithm with different design parameters have been performed for trajectory control of a 3-link manipulator. In the results shown in the figures below, the joint angle error is given in radians. In these figures, solid lines (-) show the trajectory tracking error of link 1, broken lines (--) the error of link 2 and dotted lines (...) the error of link 3.

CONCLUSIONS

In this paper, different modified versions of Generalized Predictive Controllers are applied to the difficult problem of fast trajectory control of a manipulator, using minimal amount of knowledge of the arm.

All three algorithms increased the manipulability and robustness of original GPC by adding new design variables. But, the simulations also show that even crude assumptions on the system dynamics and uncareful choice of design variables do not affect performance of the algorithm in many cases. It is observed that N2 plays a significant role in the sensitivity of the GPC algorithm to design variables. Selecting N2 larger increases the robustness to the design parameters such as output error weights or pre-specified closed-loop dynamics.

ACKNOWLEDGEMENT

This research has been supported by the Research Fund of Boğaziçi University.

REFERENCES

[1] Cutler, C.R. and B.L. Ramarker (1980) Dynamic matrix control- a computer control algorithm, Proc. JACC, San Francisco, California.

[2] De Keyser, R.M.C., and A.R. Cauwenberghe (1979) A self-tuning multistep predictor application. Automatica, **17**, 167-174.

[3] Richalet, J. and A. Rault, J.L. Testud and J. Papan (1978) Model predictive heuristic control: applications to industrial processes, Automatica, B14T, 413-428.

[4] Clarke, D.W., C. Mohtadi and P.S. Tuffs (1987) Generalized predictive control. Parts 1. and 2., Automatica, **23**, 137-160.

[5] De Keyser, R.M.C. , Ph. G.A. Van de Velde, F.A.G. Dumortier (1988) A comparative study of self-adaptive long-range predictive control methods, Automatica, **24**, 149-163.

[6] Kaynak, O. and H. Hoyer (1988) Predictive control of a robotic manipulator, Robot Control: 1988 (SYROCO'88), Ed. U. Rembold, Pergamon Press, 213-218.

[7] Clarke D.W. and C. Mohtadi (1989) Properties of generalized predictive control , Automatica, **25**, 859-875.

[8] Gençer S.S., O. Kaynak, Y. İstefanopulos (1990) Internal model control of a robotic manipulator, BILCON'90, International Conference on Communication, Control and Signal Processing. Ed. E. Arıkan , Elsevier. Vol. 1, 756-765.

[9] Kazerounian, K. and Z. Wang (1988) Global versus local optimization in redundancy resolution of robotic manipulators., Int. J. of Robotics Research, **7**, No.5, 3-12.

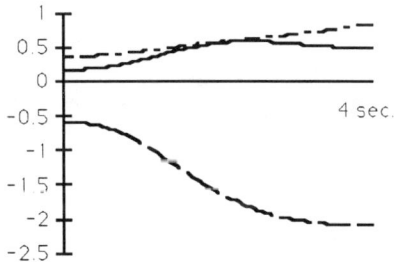

Fig 1. Reference Trajectories for joint angles designed to give a line on X-Y plane.

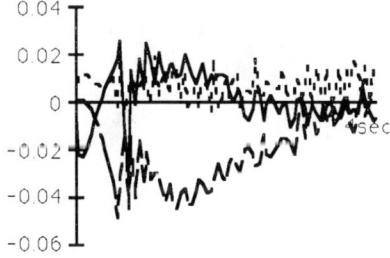

Fig 2. Joint angle errors for Standard GPC (N2=5 NU=1).

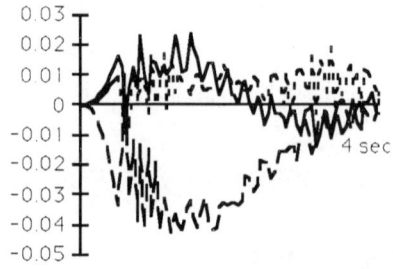

Fig.3 Errors for GPC with error dynamics (N2=5 NU=1) $P(z^{-1})=1-0.8z^{-1}+0.2z^{-1}$

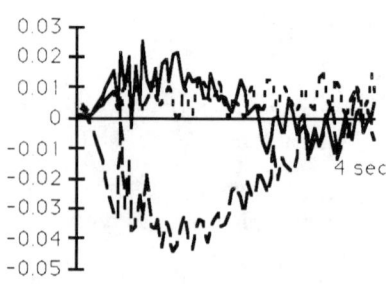

Fig.4 Errors for GPC with output error weighting (N2=5 NU=1 $\alpha=1\ldots5$)

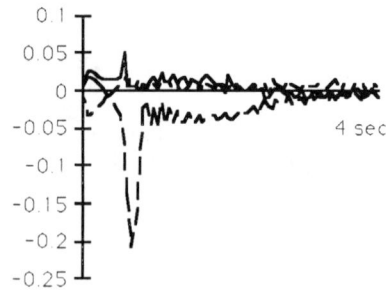

Fig.5 Errors for standard GPC (N2=5 NU=3)

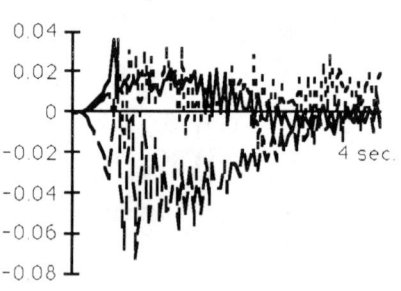

Fig.6 Errors for GPC with weighted control input (N2=5 NU=3 $\gamma=.5, .3, .2$)

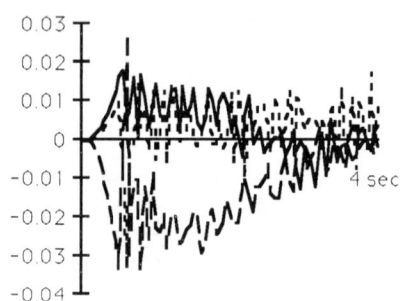

Fig.7 Errors for GPC with output error weighting (N2=3 NU=1 $\alpha=1,4,9$)

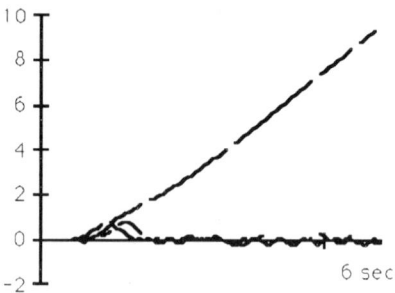

Fig.8 Errors for GPC with output error weighting (N2=3 NU=1 $\alpha=9,4,1$)

Design of Manipulator Trajectories with Minimum Motion Time and Specified Accuracy

Youlian SONG * and Robin M. C. DE KEYSER
Automatic Control Laboratory
University of Ghent, Belgium
* On leave from Shanghai Institute of Railway Technology, China

Abstract

In continuous path control of robotic manipulators, the trajectory can be planned in joint coordinates to avoid the big number of computations. The cubic spline function can easily be solved to interpolate joint key points, but deviations always exist between the actual hand path and the nominal path. By a proper selection of the time schedule, at which the key points are reached, the cubic spline function can be optimized. That results in a faster hand motion with smaller deviation. Increasing the number of key points can well control the deviation but prolongs the motion time. The new strategy is to combine the two methods together. Extra points are inserted at the spots where big errors appear, followed by the cubic spline interpolation and its optimization. This is done iteratively until the joint trajectories can ensure the hand path accuracy. Such a strategy yields a much faster motion with bounded deviation.

1. Introduction

In continuous-path conrol of robotic manipulations it is necessary for the hand to follow a specified trajectory in the working space with a defined accuracy. The task can be described on-line by a nontextual language known as 'teach-and-repeat' through moving the arm by 'lead-by-nose'. In this case the task is specified in joint coordinates. Industrial robots can also use high level textual language for programming off-line. In this case the individual positions and orientations of the hand are indicated in a world Cartesian coordinate system, in which an operator can easily visualize and measure. This is a more advanced approach. In this approach the required trajectory is specified pointwise. These points are known as key points which the hand should pass through. The selection of key points is mainly determined by technological requirements. The path between adjacent key points can be a straight line or low-order polynomials. In reality present robots use individual joint controllers to control joint actuator torques or forces and only understand joint coordinates. Therefore, with the trajectory planned in a Cartesian system, inverse Jacobian has to be done during every sampling period to compute the joint velocities [1]. Another method avoids to solve inverse Jacobian but samples on-line the Cartesian trajectory at fixed short time intervals and transforms the sampled points into joint coordinates as the input of joint servo mechanisms [2]. Both methods involve a large amount of real-time computation, which makes them unattractive in practical applications where fast motion is important.

Due to the heavy computational burden of conversion between joint and Cartesian coordinates, the second approach is adopted, i. e., to plan the trajectory at the joint level. In this approach the specified Cartesian key points are first transformed into sets of corresponding joint key points, each set for one joint. Then certain kind of curve fitting is done to link these sets of points to form joint trajectories. Piecewise straight lines can be used to form linear joint trajectories. This certainly is the easiest way of interpolation. However, during execution every arm pauses at the key points resulting in a discontinuous motion. In order to avoid abrupt changes of motion, a polynomial could be applied in the vicinity of joint key points to provide a continuous transition between two adjacent straight line segments. But the resultant hand trajectory then cannot reach the prespecified key points [3]. Polynomials may be introduced to fit joint key points to provide a smooth motion with continuous position, velocity and acceleration. The order of the polynomial depends on the given manipulator task [4,5].

Actually, the principle of spline functions can be introduced to simplify the computation. Among various kinds of spline functions the cubic spline interpolation requires the least computing time and has relatively small overshoot of angular displacement. When the number of key points is n, the solution of only n-2 equations is required to construct the cubic spline function for each joint [6]. If small path errors due to the fitting approximation are expected, one may give many points. In this case, a method that minimizes the sum of the square errors using (m-1) segment polynomials, where m < n, instead of interpolating all the n key points is proposed [7]. The disadvantage of the joint trajectory planning strategy lies in that it can only ensure the hand to pass through the specified key points, the actual path between the adjacent points is unpredictable. Errors always exist between the actual and the nominal path and sometimes the error is big. We can control the error within a given tolerance by sampling enough points on the nominal Cartesian path. Taylor presented a bounded deviation joint paths (BDJP) algorithm, in which the joint paths are piecewise straight lines with just enough midpoints and smooth transition between adjacent segments in order to keep the deviation less than a prespecified bound [8]. Choi and Kim modified Taylor's BDJP method by transforming the Cartesian velocities or path curve tangents and the positions at the selected points into joint coordinates. In this approach the cubic spline interpolation is not needed before estimating the maximum hand path deviation, thus the computational burden is considerably reduced. The modified algorithm converges more rapidly than Taylor's method but the continuity of acceleration is sacrificed [9]. The total motion time may also prolong as a result of the increasing number of points.

Indeed, the joint trajectories resulting from the cubic spline interpolation are determined by the times at which the hand passes the key points. Through optimizing the time schedule not only the deviation does decrease, but the total motion time can also be shortened [10]. However this algorithm cannot ensure a predefined accuracy. This paper presents a new strategy. It ensures a hand path within the deviation tolerance and gives the manipulator a much faster motion. The control programs are written in Turbo-Pascal, real-live experiments have been completed on a six DOF RTX SCARA-type robot. The nominal trajectory used in our examples is composed of straight line segments in the Cartesian domain. The principal advantages of straight line paths are the simplicity and predictability. In practical applications straight line segment paths frequently correspond to the desired motion of a manipulator.

2. Review of cubic spline function interpolation

The cubic spline proves to be an effective tool for interpolation. A cubic spline curve is a piece of cubic polynomial between two points to be interpolated. When a series of key points are specified, the piecewise cubic splines are smoothly joined together to form a continuous cubic spline function curve. These points become junction points, at which the cubic spline function has up to the second derivative.

In joint trajectory planning using cubic spline interpolation, the joint key points are computed from the corresponding Cartesian key points by inverse kinematic equations. For every joint there is a set of key points: q_{j1}, q_{j2},..., q_{jn}, j=1, 2,..., N, where N equals

DOF. For each joint the procedure for solving the cubic spline is the same, thus the subscript j can be omitted. Let $t_1 < t_2 < ... < t_n$ be an ordered sequence of time indices at which the hand passes the correspondent key points and let $Q_i(t)$ represent a piecewise cubic spline. If the velocities at the key points are not important, i.e. we only require the hand motion to be smooth and as fast as possible as in the material handling operation, the cubic spline can be solved from the linearity of the second derivative:

$$\ddot{Q}_i(t) = (t_{i+1}-t)/h_i \cdot \ddot{Q}_i(t_i) + (t-t_i)/h_i \cdot \ddot{Q}_i(t_{i+1}) \tag{1}$$

The cubic spline can be derived by integrating (1) twice and evaluating the constants of integration.

$$Q_i(t) = \ddot{Q}_i(t_i)/(6h_i)\ (t_{i+1}-t)^3 + \ddot{Q}_i(t_{i+1})/(6h_i)\ (t-t_i)^3 +$$

$$[q_i/h_i - h_i\ddot{Q}_i(t_i)/6]\ (t_{i+1}-t) +$$

$$[q_{i+1}/h_i - h_i\ddot{Q}_i(t_{i+1})/6]\ (t-t_i) \tag{2}$$

$$i = 1, 2,..., n-1 \qquad t_i < t < t_{i+1}$$

Accelerations $\ddot{Q}_i(t_i)$ are solved by the following set of linear equations

$$A(\ h\ )\ \ddot{Q} = B(\ q,\ h\ ) \tag{3}$$

Matrix A has a banded structure, this makes it easy to solve (3). In equation (3) **q** denotes the given joint key points and **h** denotes the motion time intervals. Obviously, **q** is fixed but **h** can be chosen subject to the maximum joint velocity constraints. Different **h** results in different $\ddot{Q}$ and hence a different cubic spline function.

3. Bounded deviation joint trajectories by increasing the number of key points

Let $\{ P_1, P_2,..., P_n \}$ represent specified key points in a Cartesian coordinate system and let $\{ t_1, t_2,..., t_n \}$ be the time sequence. These points are first transformed into sets of joint key points. Then cubic spline function interpolation is implemented for every set of joint points to obtain N joint trajectories $\{ Q_{J1}(t), Q_{J2}(t),..., Q_{JN}(t) \}$. When executing, each joint follows its own trajectory, the hand path is the combination of these joint motions. The actual hand path always deviates from the specified path, only key points can be exactly reached. Practical operations require the deviation to stay within certain limits. The traditional approach for solving the problem is increasing the number of key points. This approach can be summarized as follows.
(1). Transform all the specified key points $\{ P_1, P_2,..., P_n \}$ into manipulator joint space to get N sets of joint key points.
(2). Fit every set of joint points by cubic spline function interpolation. In this step N joint trajectories $\{ Q_{J1}(t), Q_{J2}(t),..., Q_{JN}(t) \}$ are formed.
(3). Evaluate the midpoint error of every path segment based on the assumption that for many common manipulator geometries the maximum deviation appears near the middle of the segment. For one segment it is done as follows:
 (a). compute $Q_{Jj}(t_i+(t_{i+1}-t_i)/2)$, j=1,2,...,N;
 (b). transform these joint mid-points to Cartesian coordinates;
 (c). evaluate the deviations at these points;
 (d). if the error is out of the specified limit then add the corresponding point as a supplementary key point.

(4). Repeat (2) and (3) until the actual hand path stays within the tolerance.
Theoretically, this approach can satisfy any demanded accuracy. But the fatal weakness of this method lies in its prolonged motion time, sometimes this trend is strong. Figures 1 and 2 show how an increasing number of points influences the path deviation and the motion time by three examples that correspond to three different Cartesian paths.

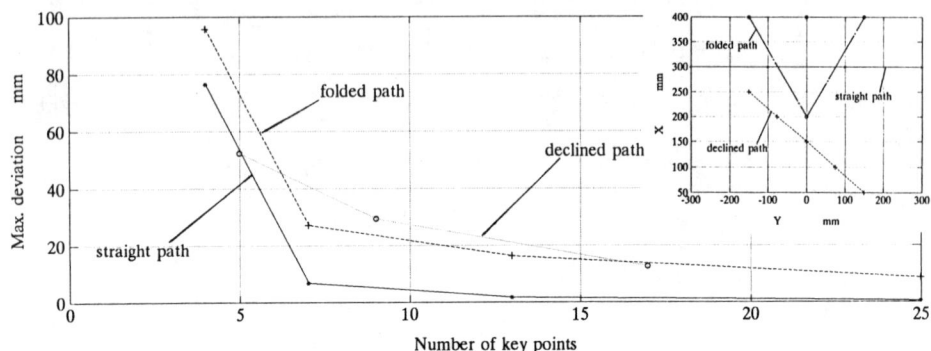

Fig. 1. The influence of adding points to the path deviation

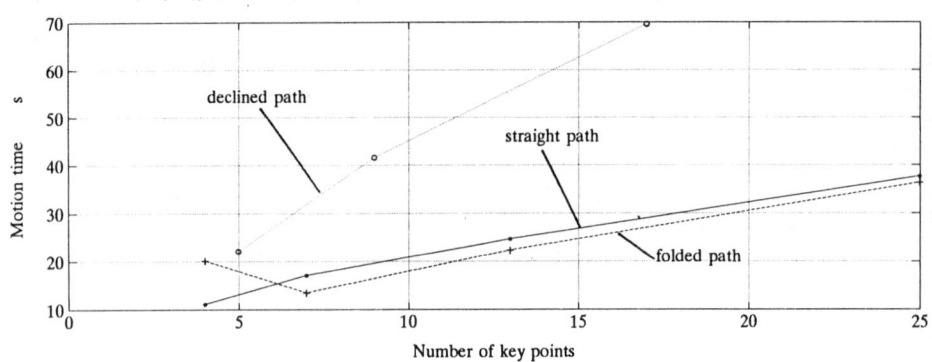

Fig. 2. The influence of adding points to the motion time

Let us explain the phenomenon. After computing the cubic spline function interpolation of joint trajectories, it must be inspected that the velocity, acceleration and jerk (the rate of change of acceleration) for each joint trajectory stays within the physical limits of the manipulator, i.e.,

$$|\max_{j} \dot{Q}_{Jj}(t)| \leq V_{cj} \qquad (4)$$

$$|\max_{j} \ddot{Q}_{Jj}(t)| \leq a_{cj} \qquad (5)$$

$$|\max_{j} \dddot{Q}_{Jj}(t)| \leq J_{cj} \qquad (6)$$

where $j=1, 2,..., N$

V_{cj} is the velocity constraint of joint j,
a_{cj} is the acceleration constraint of joint j,
J_{cj} is the jerk constraint of joint j.

When one of them exceeds its limit, we can multiply time t by a time scale S (S > 1) to slow down the motion.

$$S = \max (S_v, S_a^{1/2}, S_j^{1/3}, 1) \qquad (7)$$

where $S_v = \max_j | \dot{Q}_{Jj} / V_{cj} |$, $S_a = \max_j | \ddot{Q}_{Jj} / a_{cj} |$, $S_j = \max_j | \dddot{Q}_{Jj} / J_{cj} |$

This procedure is called " feasible solution converter " (FSC).

If the number and distribution of the points to be interpolated are changed, the resultant cubic spline function is certainly different. Then the maximum of $\dot{Q}_{Jj}$, $\ddot{Q}_{Jj}$, and $\dddot{Q}_{Jj}$ will also change. Practice proves that in most cases the trend for the maximum derivatives of $Q_{Jj}(t)$ to exceed the constraints becomes stronger when the number of key points increases. Thus the total motion time is prolonged to make the solution feasible. Figure 3 illustrates the influence of the number of points on the time scale S.
Due to this defect we developed a new planning algorithm [10].

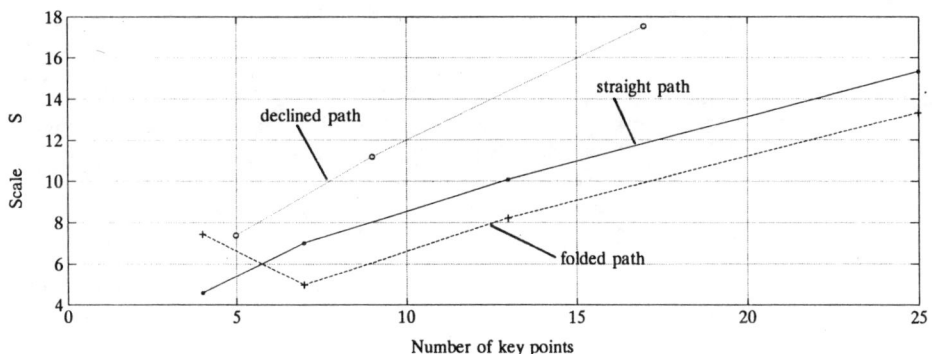

Fig. 3. The influence of adding points to the time scale

4. Optimization of cubic spline function interpolation

From equations (2) and (3) we know that for a specified series of points the cubic spline function can be changed by rearranging the time schedule $h = [h_1, h_2,..., h_{n-1}]$. The original h can be assumed according to the joint velocity constraints:

$$h_i = \max_j | (q_{j,i+1} - q_{j,i})/V_{cj} | \qquad (8)$$

$$j = 1, 2,..., N, \quad i = 1, 2,..., n-1$$

Normally this initial h is not quite interesting. The corresponding spline functions yield a poor hand path and a slow motion. An example is shown in figure 4. The nominal hand path is a straight line in the X-Y plane, the maximum deviation reaches 76.55 mm, the total motion time is 11.18 seconds.

In order to decrease the deviation and speed up the motion, the spline function interpolation has to be optimized. We define a loss function taking both deviation and motion time into consideration:

$$J = QD + RT \tag{9}$$

where   D (mm): absolute maximum hand path deviation
T (s): total motion time
Q, R : positive weighting factors

Q and R are chosen according to the requirements of a specific situation. The best ratio of Q to R can be evaluated through simulation.

The value of the loss function depends on the choice of **h**. By applying certain search method we can find the best time schedule **h** which minimizes the loss function J. If Q=1 and R=5, the above mentioned path deviation decreases from 76.55 mm to 13.14 mm, the total motion time is also shortened from 11.17 s to 9.74 s. The weakness of the method is that the improvement of the path accuracy is limited, the algorithm cannot ensure the actual hand path to stay within a prespecified error tolerance.

5. Joint trajectory design with fast motion and bounded deviation

The algorithms presented in the previous two sections are one-sided approaches. The add-point method only increases the number of key points but does not improve the time schedule **h**. This yields an accurate but slow trajectory. The optimization method only optimizes the spline interpolation itself but does not make use of the advantage of adding points of the known nominal path.

The new algorithm to be presented in this section combines the advantages of the above two methods and avoids their defects. The design procedure is as follows:
(1). give key points of the specified Cartesian path;
(2). transfer these points into joint space, obtain N sets of joint points;
(3). use cubic spline function to interpolate the N sets of joint points separately;
(4). optimize the cubic spline function by searching the time schedule **h** to minimize J;
(5). check the actual hand path segments between every two adjacent key points, find the spots where local maximum deviation is out of limit, then choose this spot as an extra key point;
(6). repeat steps (2) to (5) until the actual hand path is within the error tolerance.

For an experienced designer, step (4) may be omitted in the first one or two iterations. This can shorten the total computation time.

Obviously, the strategy presented above is an off-line trajectory planning strategy. All the data is computed only once, it is stored in the memory and retrieved when the robot is actuated. Since most manipulator applications involve very repetitive operations, the off-line planning is a valuable strategy. It also reduces the computational burden during operation a great deal because the total calculation is needed only once for a given application.

To evaluate the performance of this new trajectory design strategy, simulations and experiments were carried out on the RTX SCARA-type robot. The simulation and control program are written in Turbo-Pascal and run on the Tulip PC/AT 386/25. Here we introduce 2 examples: a straight line path and a folded line path. Both are in a X-Y plane.

Example 1: specified accuracy: 2 mm, original number of key points: 4, maximum deviation: 76.55 mm, motion time: 11.18 s, final number of points: 7, deviation: 1.40 mm, motion time: 6.17 s.

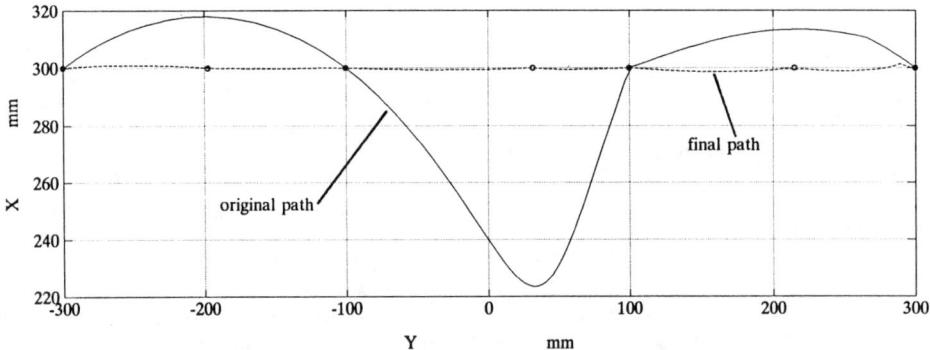

Fig. 4. Comparison of the straight line paths

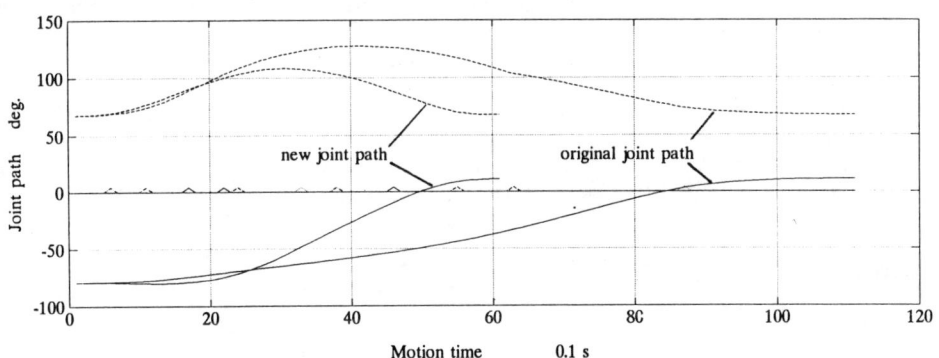

Fig. 5. Comparison of shoulder and elbow joint paths corresponding to fig. 4.

Example 2: specified accuracy: 5 mm, original number of key points: 4, maximum deviation: 95.74 mm, motion time: 20.17 s, final number of points: 9, deviation: 4.41 mm, motion time: 12.02 s.

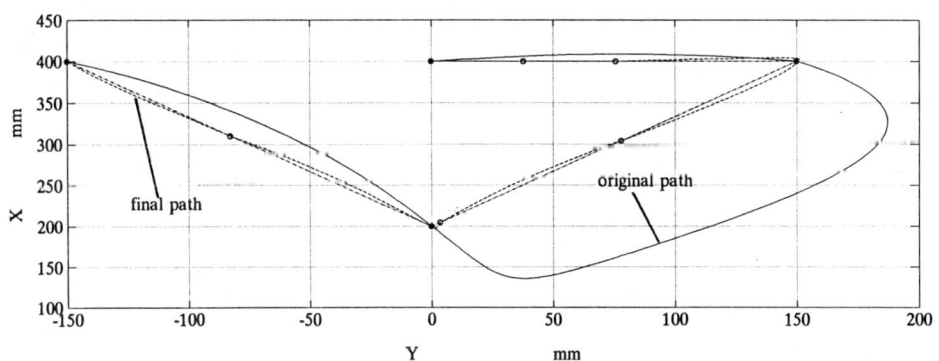

Fig. 6. Comparison of the folded line paths

6. Conclusions

(1). In continuous path control of robotic manipulators the control can be planned at the joint level to avoid the inverse kinematic transformation of the dense sampling points. Cubic spline interpolation is an effective off-line joint trajectory planning approach. For a repeated operation the calculation only needs to be carried out once before the robot starts to do the work. Hence the on-line computational burden is greatly reduced. Since most industrial manipulators work in a repeated manner, this is a practical approach.

(2). The actual hand path resulting from the cubic spline function joint trajectories can only ensure the hand to pass through the given points. Deviations always exist in the intermediate path between two adjacent points. Increasing the number of points may effectively improve the path accuracy but at the same time seriously prolong the total motion time.

(3). Cubic spline function changes when the time schedule is changed. Applying this feature the interpolation can be greatly improved by optimizing the time schedule based on an objective function which takes both motion time and path deviation into consideration. However there is a limit in the reduction of deviations. We cannot ensure a predefined path accuracy through this optimization.

(4). The new strategy combines these two approaches together so as to avoid their defects. After the improvement of the interpolation, points are added at the places where the local maximum deviation appears. This is done alternatively until the predefined accuracy is ensured. In doing this just enough intermediate points are added and the total motion time is greatly reduced.

References

1. D.E.Whitney, " Resolved motion rate control of manipulators and human prostheses," IEEE Trans. Man-Machine Systs., Vol. 10, No. 2, pp.47-53, June 1969.
2. R.Paul, " Manipulator Cartesian path control," IEEE Trans. System, Man, and Cybernetics., Vol. SMC-9, No. 11, pp. 702-711, Nov. 1979.
3. R.Paul, *Robot Manipulators: Mathematics, Programming and Control*, The MIT press, 1981.
4. A.J.Koivo, *Fundamentals for Control of Robotic Manipulators*, John Wiley & Sons, Inc. U.S.A., pp. 227-251, 1989.
5. P.G.Ranky and C.Y.Ho, *Robot Modeling - Control and Applications with Software*, IFS (Publications) Ltd., U.K. Springer - Verlag, pp. 95-105, 1985.
6. C.Lin, P.Chang, and J.Y.S.Luh " Formulation and optimization of cubic polynomial joint trajectories for industrial robots," IEEE Trans. Automatic Control, Vol. AC-28, No. 12, pp. 1066-1073, Dec. 1983.
7. C.Lin & P.Chang, " Joint trajectories of mechanical manipulators for Cartesian path approximation," IEEE Trans. Systems, Man, and Cybernetics, Vol. SMC-13, No. 6, pp. 1094-1102, Nov./Dec. 1983.
8. R.H.Taylor, " Planning and execution of straight line manipulator trajectories," IBM Journal of Research and Development, Vol. 23, No. 4, pp. 253-264, July 1979.
9. B.K.Choi & D.W.Kim, " Bounded deviation joint path algorithms for piecewise cubic polynomial trajectories," IEEE Trans. System, Man, and Cybernetics, Vol. 20, No. 3, pp. 725-733, May/June 1990.
10. Y.Song & R.M.C.De Keyser, " The optimization of cubic spline function interpolation of joint trajectories," Proc. of The International Symposium on Intelligent Robotics, pp. 544-555, Bangalore, India, Jan. 1991.

The Use Of Simulation Systems To Control Manually Operated Remote Manipulators, With Long Pure Time Delays.

Mike Griffin, Richard Mitchell
Department Of Cybernetics, University Of Reading,
Whiteknights, P O Box 225, Reading, Berks RG6 2AY, U.K.

Abstract

Various types of remote manipulating system have been made, where the actions of a distant human operator may be passed to a remote location and acted out as if by the operator himself. Cameras and other sensors may be used to convey the scene to the distant observer and to show him the results of his actions. If a large communications delay exists between the operator and the remote site, then a direct manipulation strategy may no longer be considered. To circumvent this problem, the concept of a "bridging simulator" is introduced, whereby the operator acts on a model, and the actions of the robot are defined from a remote copy of the same model. The design and feasibility of such a system is the subject of this paper.

Introduction

Current technology permits the projection of sensory information from a remote source to a human observer, using channels with which the observer is familiar. If controls are supplied to the observer such that a manipulation system at the remote location may be operated, the observer may interact with elements of this distant scene. Examples of such mechanisms exist in nuclear power stations or in tele-guided submersibles. Current technology will only permit direct robot interaction if the delay between command and observed response is small. If this interval is in the order of seconds, then a direct control strategy is not possible.

In these situations, a route planning approach is usually adopted, such that a CAD system is used to create a set of instructions for the robot, which is then acted out in a streamlike manner. This method works well for situations where the set of possible events is so highly constrained that the stream of instructions created may accurately perform the task without alteration. If this is not the case, however, a different approach has to be adopted. There are two possibilities, the first of these is to use an 'intelligent' robot which is able to achieve the required goal using its own intellectual resources, the other is to use a 'bridging simulation' approach.

In the proposed 'bridging simulation' system, the manipulator is jointly under the control of the user, and an automatic control schema. The human operator is presented with a scene which is partly real and partly synthetic in nature, eg the synthetic elements may be perceived to be a wireframe graphic overlaying the real image data. The wireframe scene is formed from a simulation running on a support system which is local to the user. This system is currently being updated from information gathered from the remote location. The operator manipulates elements within the simulation, and these are sent as instructions to a distant robot. The operator can then observe appropriate sensory feedback from his actions directly from the simulation, rather than waiting from the data stream to arrive from the remote unit. Finally, when this sensory information does arrive, the difference between the simulation and the actual interaction may be studied, and the discrepancies resolved.

A Proposed 'Ideal' Bridging Simulation System

The Bridging Simulation system is based around two models of the robot's environment, one of which is kept local to the user and the other local to the robot. Changes made to one model are sent down a communications link, and indirectly update the other model. This operation is bidirectional, and is a continuous attempt to keep the databases consistent with each other. The model contains geometrical and behavioural information sufficient to accurately simulate the behaviour of both the robot and its environment. The constraints necessary to prevent damage to either the robot or its surroundings are also an integral part of the simulation system.

The operator acts on this simulator to effect changes to his local model. The model contents are displayed to the operator, in a suitable sensory form, such that the instant a change is made to this simulation the sensory aspects of this change are presented. The model changes are then passed down the information channel to the remote model, and this is then updated. A process notes the changes in the model sent from the user, and from this a set of appropriate comands are issued to the manipulation system to actually bring this change about. Sensors within the domain of the robot examine the scene, and this information is then compared with the local robot model. Discrepancies between the real scene and the model are resolved, and the local model is updated. The robot model changes are then passed through the communications link, back to the operator's model. An overall block diagram of this schema is shown in fig1.

It can be seen that the key to the whole system is creation, maintainence and interaction of the various databases comprising the model element. A flexible modelling environment

Fig 1 - Block Diagram Of Ideal System

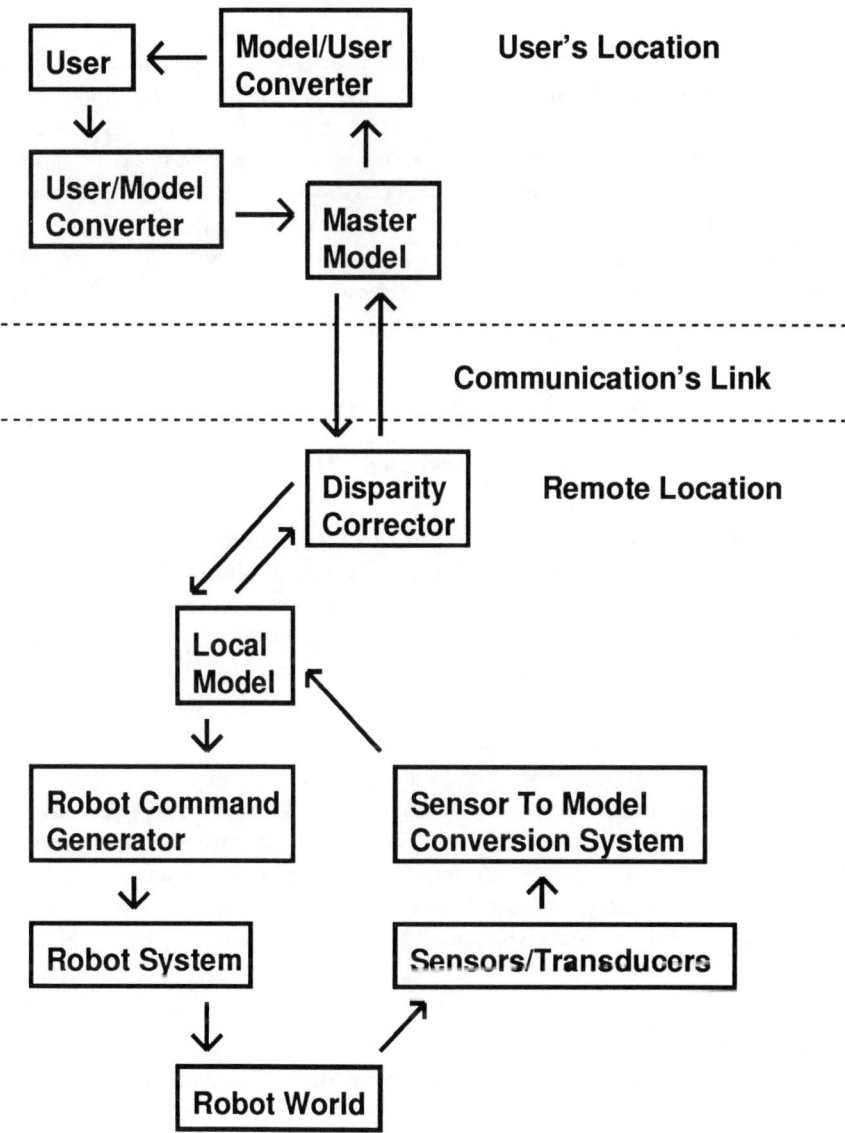

capable of capturing the structure, behaviour and meaning of a scene must be used, and must be readily manipulated. For this, it is suggested that an object-oriented modelling system is used. A programming environment known as LOKI has been developed for the Reading experimental system detailed in the next section. Due to potential complexities within an environment it may be noted that many situations may not be easily modelled.

It can be seen that, if the direct communications link between the two models is not used in this manner, and that the robot commands are generated local to the user and queued and verified before sending to the remote location, then this simplification would make the unit into a traditional path planning system. The above system can be used in a continuous mode of operation if the number of discrepencies between the robot model and the local model is not too large. If this is the case, then this approach may not function well since the alterations made by the user will have no bearing on the robot world, and thus the comands will be ignored. In this case, either a reduction in the number and rate of interactions made will be necessary, or the complete abandoning of this system in favour of a fully autonomous system.

Reading University Remote Manipulator System

It can be seen that the above system is a highly involved proposal, in technical terms. For development purposes, a reduced system has been proposed. This is partly completed, and is based around a Virtual Reality headset system, an object oriented system, and a telepresence head system built in the department. Block diagrams of this reduced proposal are shown in figs 2-3.

In this system, the remote robot is based around a Puma robot arm. The Puma is equipped with a head unit comprising of two cameras and two microphones. Their arrangement is supposedly similar to that of the observer's eyes and ears. The audio and video information produced by these sensors is passed down to a video and audio mixing system. Here, synthetically generated sound and video information from a modelling system is added in and the composite signals passed to a headset system. The headset system has two liquid crystal displays, and a pair of stereo headphones, such that the above data may be displayed to the observer.

An appropriate head tracking arrangement will be employed, so that as the operator's head moves the information may be passed to a computer system. The computer system contains the modelling system and the computer graphics generation arrangement. The user's interaction and movements are passed to this system, where the behaviour of the Puma can be modelled. Collision detection is performed within this environment, and movements which could damage the Puma are prevented and the resultant movements placed in a buffer

Fig 2 - System Block Diagram

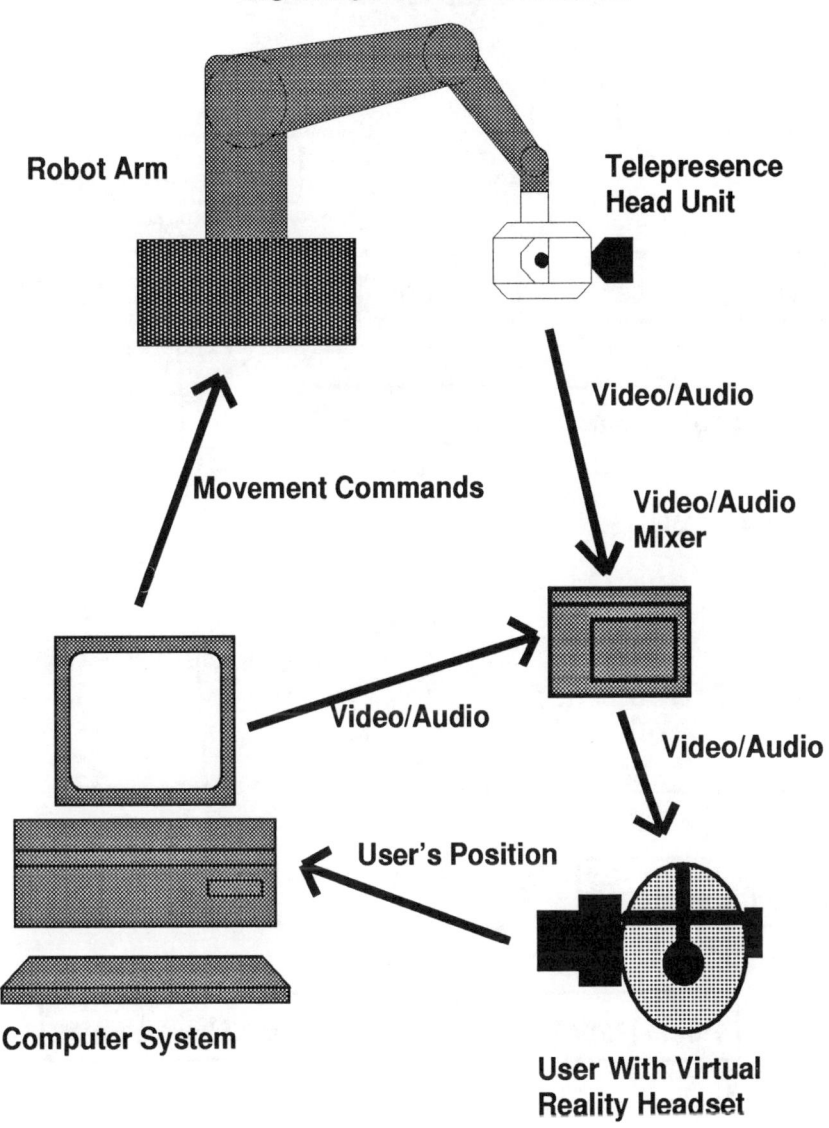

Fig 3 -Block Diagram Of Current Proposal

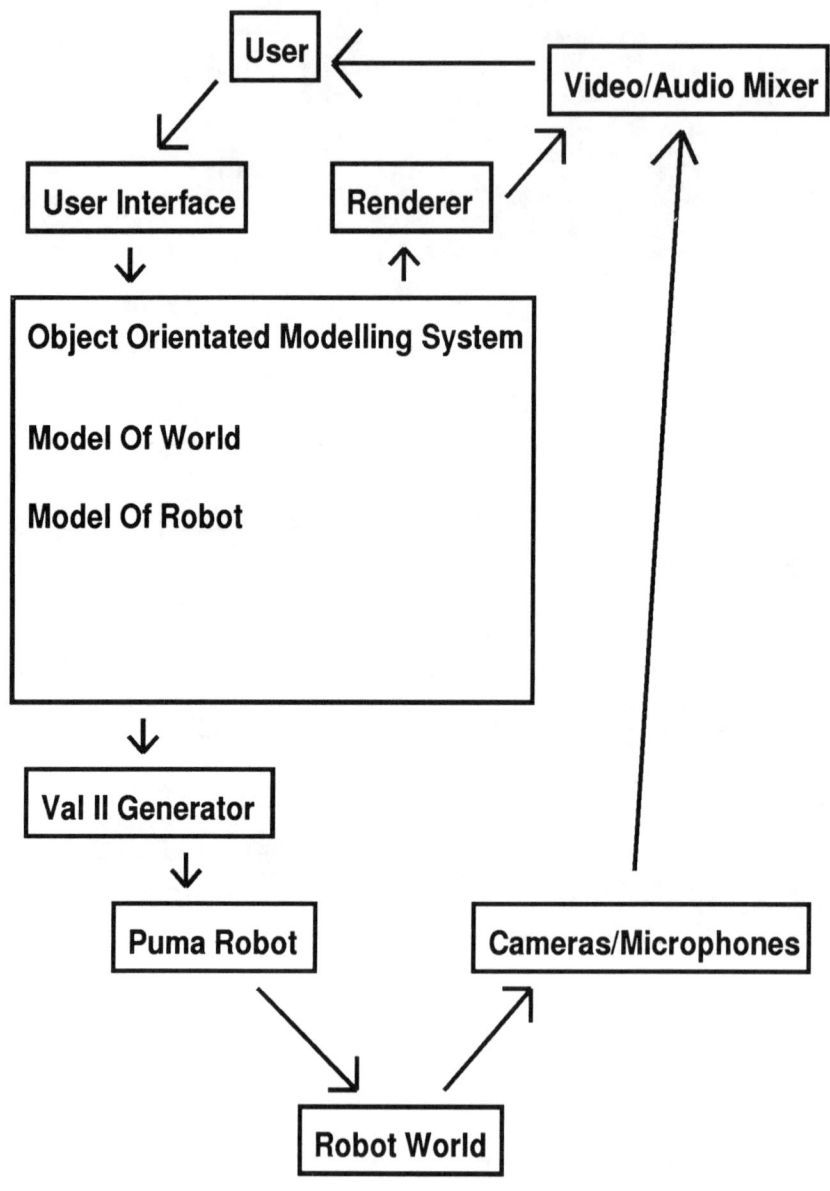

system. The buffer will be variable to allow for different pure time delays. From the buffer, the movement commands are given to a VAL II generator. This VAL generator creates the commands necessary for moving the robot. The command sequence is then passed out of the computer, along a serial line, to the Puma, where the onboard controller executes the commands.

This system is still under construction. The object oriented language 'LOKI' has been developed for general Virtual Reality work, but has been found suitable for this application. The head mounted display system and associated electronics have been made, except for the video mixing circuits, but the sound mixing is fully functional. The computer graphics interface and the robot telepresence head have been made, and are functioning. The design elements that still have to be made are the head tracking system, and the VAL II code generator.

Conclusions

It may be seen that for distant locations, direct manipulation of a robot by a human is impractical. Path planning arrangements may be used to alleviate this problem, but these are not suitable for situations where fast responses are critical. If a large time delay exists, or the complexity and changeability of the environment is high, then this system will also be inadequate. A bridging simulation system can act as a solution, if the level of the pure time delay is not too extreme, or the changeability of the environment too high. If this is the case, then the predominant operation would be the updating of the models, and the disparity level so high that a large command rejection rate would be created, and the robot effectively frozen.

The ideal bridging system was found to be too dificult to implement as a first stage project. It was decided to implement a simpler regime to test the validity of some of the suggestions. The Reading system was defined as a compromise, aiming towards the more complex solution. Implementation has been found to be much easier, and a working system not far off. The finer operational details and advantages and disadvantages to this arrangement will be found when construction and rigorous testing commences.

References

Alciatore, D.G., Hughes, P.J., Traver, A.E., and O'Connor, T.J., "Development And Simulation Of An Ergonomic Control System For A Large Construction Manipulator", Proceedings, 6th International Symposium On Automation And Robotics In Construction, June 1989.

Alciatore, D.G., Traver, A.E., and O'Connor, J.T., "Development Of A Heuristic Application-Specific Path Planner For Piping Construction", Proceedings, 6th International Symposium On Automation And Robotics In Construction, June 1990.

Barnes, J., Herbert, M., Kanade, T., Krotkov, E., Mitchell, T., Simmons, R., and Whittaker, W., "Ambler - An Autonomous Rover For Planetary Exploration", IEEE Computer, June 1989, pp 18-26.

Foley, J.D., "Interfaces For Advanced Computing", Scientific American, pp 83-90, October 1987.

Lozano-Perez, T., "Automatic Planning Of Manipulator Transfer Movements", IEEE Transactions On Systems, Man, And Cybernetics, Vol. SMC-11, No.10, October 1981, pp. 681-698.

Nitzan, D., "Development Of Intelligent Robots: Achievements And Issues", IEEE Journal Robotics And Automation, Vol RA-1, No.1, March 1985 pp. 3-13.

PATH COORDINATED CONTROL OF ROBOTIC ARMS

S.Arnaltes and J.Pérez Oria
Dpto Sist. Electrónicos y de Control. E.U.I.T.Telecomunicación
Universidad Politécnica de Madrid
Ctra de Valencia,Km.7. 28031 MADRID (SPAIN)

Abstract. A solution for solving the inverse kinematics problem of robotic arms is presented. The method gives a generalized solution for any class of robots, simple or redundant, having rotatory and/or slading joints. For solving the system of nonlinear equations that represents the manipulator end effector position and orientation in terms of the joint coordinates an iterative Newton-Raphson algorithm is used. The solution can be constrained in order to select the configuration of the manipulator.

1.INTRODUCTION.

An industrial robot can be modeled as an open loop articulated chain with several rigid links connected in series by revolute or prismatic joints, driven by actuators. The motion of the joints results in the motion of the links that positions and orientates the manipulator end effector.

Robot kinematics deals with the analytical description of the manipulator end effector with respect to a fixed reference coordinated system, to express the relations between the joint coordinates and the position and orientation of the end effector.

Two fundamental questions are of interest. First, given the joint coordinates of the arm what is the position and orientation of the end effector with respect to a reference coordinate system, which is usually referred to as the direct kinematics problem. Second, what are the necesary joint coordinates to reach a desired position and orientation of the end effector, known as the inverse kinematics problem. In most robotic applications, the inverse kinematics problem is used more frequently, since a task is usually stated in terms of the reference coordinates and the variables in a robotic arm are the joint coordinates.

In general, the inverse kinematics problem can be solved by several techniques. A geometric approach can be used for any six degree of freedom manipulator which has revolute or prismatic pairs for the first three joints and the joint axes of the last three joints intersect at a point [1]. However, this method suffers from the fact that it is not systematic. For the same class of simple manipulators a sytematic inverse transform technique using 4x4 homogeneous transformation matrices can be applied in solving the inverse kinematics problem [2]. There is a lack of indication on how to select an appropiate configuration of the manipulator from the several that place the end effector in the same position and orientation in this method. Although all commercial robots satisfy the above condition which makes the closed form arm solution possible, there are some special robots for which a solution is also necesary. Iterative methods provide such a solution. The iterative solution often requires more computation and it usually does not guarantee convergence to the solution. However, these methods are suitable for any class of robots, with any number of degree of freedom, including both simple or kinematically redundant robots [3].

We present an iterative Newton-Raphson method of obtaining a solution to the nonlinear kinematics equations based on the Denavit-Hatenberg matrices to formulate the kinematics relations, which allows to incorporate constraints on joint displacement systematically in the process of obtainig the solution.

2. THE BASIS OF THE PROBLEM.

Since the links of a robotic arm may rotate and/or translate with respect to a reference coordinate frame, the position and orientation of the end effector is a function of the angular rotations and/or linear translations of the links. Denavit and Hatenberg [4] proposed a systematic and generalized method to describe the geometric relations between the links of a robotic arm by attaching a coordinate frame to each link. This method uses 4x4 homogeneous transformation matrix to express the relative position and orientation of two adjacent frames defined as follows:

$$A_i = \begin{vmatrix} \cos\theta_i & -\cos\alpha_i \sin\theta_i & \sin\alpha_i \sin\theta_i & a_i \cos\alpha_i \\ \sin\theta_i & \cos\alpha_i \cos\theta_i & -\sin\alpha_i \cos\theta_i & a_i \sin\theta_i \\ 0 & \sin\alpha_i & \cos\alpha_i & d_i \\ 0 & 0 & 0 & 1 \end{vmatrix} \quad (1)$$

where (a_i, α_i), link parmeters, are the length and the twist angle of the link, which determine the structure of the link, and (d_i, θ_i), joint parameters, are the distance and the angle between the adjacent links and determine the relative position of neighboring links. Note that for a rotatory joint d_i is fixed and θ_i is a variable, and for a sliding joint θ_i is fixed and d_i is a variable.

The transformation T_n representing the position and orientation of the end effector with respect to a base frame is expressed as:

$$T_n = \prod_{i=1}^{n} A_i = \begin{vmatrix} n_x & o_x & a_x & p_x \\ n_y & o_y & a_y & p_y \\ n_z & o_z & a_z & p_z \\ 0 & 0 & 0 & 1 \end{vmatrix} = \begin{vmatrix} n & o & a & p \\ 0 & 0 & 0 & 1 \end{vmatrix} = \begin{vmatrix} R & p \\ 0 & 1 \end{vmatrix} \quad (2)$$

where n is the number of degrees of freedom, **n**, **o**, **a** are the orientation vectors given by their coordinates with respect to the base frame, **p** is the position vector, referred to the base frame, and **R** is a rotation matrix which represents the orientation of the end effector frame with respect to the base frame.

The above equation indicates that the arm matrix T_n is a function of the joint coordinates $q=(q_1, q_2, ..., q_n)^T$. Equating the elements of the matrix equation, we have twelve equations with n unknowns, joint coordinates, and these equations involve complex trigonometric functions. However, only six of these equations are independent. The three that indicate the end effector position and three of the nine that indicate the end effector orientation. For example, given n_x, n_y and o_x, the end effector orientation is fully determined. Thus, from $|\mathbf{n}|=1$, n_z is determined, from $|\mathbf{o}|=1$ and $\mathbf{n}\cdot\mathbf{o}=0$, o_y and o_z are determined, and **a** is obtained by the vectorial product **n**x**s**. Note that two coordinates of one orientation vector are always needed to completely determine the orientation.

In general, the desired position is given by the position vector $\mathbf{p}=(p_x, p_y, p_z)^T$, while the orientation is usually given by any suitable set of rotation angles with a predefined sequence of rotations. Since the 3x3 rotation matrix **R** can be expressed in terms of these rotations, we can obtain its element and select three of them for the system of equations. For example, for roll, pitch and yaw angles (ϕ, θ, Ψ), we have:

$$\mathbf{R} = \begin{vmatrix} C\phi C\theta & C\phi S\theta S\Psi - S\phi C\Psi & C\phi S\theta C\Psi + S\phi S\Psi \\ S\phi C\theta & S\phi S\theta S\Psi + C\phi C\Psi & S\phi S\theta C\Psi - C\phi S\Psi \\ -S\theta & C\theta S\Psi & C\theta C\Psi \end{vmatrix} \quad (3)$$

Equating the elements of the above matrix equation, we have:

$$\begin{aligned} n_x &= C\phi C\theta \\ n_y &= S\phi C\theta \\ n_z &= -S\theta \\ o_x &= C\phi S\theta S\Psi - S\phi C\Psi \\ o_y &= S\phi S\theta S\Psi + C\phi C\Psi \\ o_z &= C\theta S\Psi \\ a_x &= C\phi S\theta C\Psi + S\phi S\Psi \\ a_y &= S\phi S\theta C\Psi - C\phi S\Psi \\ a_z &= C\theta C\Psi \end{aligned} \quad (4)$$

We can now choose three of these equations to obtain the values for using in T_n. Note that the elements that require less computation in T_n must be chosen. For example, n_z, a_y and a_z for a PUMA 560 series robot.

For a six degrees of freedom robot there is, then, six equations with six unknowns, and the problem is determinated. For a kinematically redundant robot, n>6, the system of equations is undeterminated. For simplicity we assume that only 6-DOF robots are considered. The case of redundant robots is treated below.

3. THE APPROACH.

The inverse kinematics problem is the determination of a vector $\mathbf{q}^*$ which corresponds to a given target end effector position $\mathbf{p}^t$ and orientation $\mathbf{R}^t$. The approach we introduce is based on a residual vector definition $\_s$ which represents the residual position and orientation between the target and the current position and orientation, as indicated earlier only three of nine elements of **R** are used in defining the orientation:

$$\Delta \mathbf{s} = (\Delta p_x, \Delta p_y, \Delta p_z, \Delta n_x, \Delta n_y, \Delta o_x) \tag{5}$$

The elements of $\Delta \mathbf{s}$ representing the residual position are defined as:

$$\begin{aligned} \Delta p_x &= p_x^t - p_x^c \\ \Delta p_y &= p_y^t - p_y^c \\ \Delta p_z &= p_z^t - p_z^c \end{aligned} \tag{6}$$

and the elements of $\Delta \mathbf{s}$ representing the residual orientation:

$$\begin{aligned} \Delta n_x &= n_x^t - n_x^c \\ \Delta n_y &= n_y^t - n_y^c \\ \Delta o_x &= o_x^t - o_x^c \end{aligned} \tag{7}$$

The target joint coordinate vector $\mathbf{q}^*$ is obtained when the current end effector position and orientation coincides with the target position and orientation, i.e.:

$$\Delta \mathbf{s} = 0 \tag{8}$$

For solving these nonlinear equations, we use Newton-Raphson linearization technique to linearize (8) and then solve the system iteratively using the following expression:

$$\mathbf{q}^{k+1} = \mathbf{q}^k + \Delta \mathbf{q}^k \tag{9}$$

where $\Delta \mathbf{q}^k$ is the solution of the linear system:

$$\Delta \mathbf{s}^k = \mathbf{J}(\mathbf{q}^k) \Delta \mathbf{q}^k \tag{10}$$

where **J** is the 6x6 Jacobian matrix with J_{ij} defined as:

$$J_{ij}^k = [ds_i / dq_j]_{q=q^k} \tag{11}$$

The manipulator Jacobian matrix, which is required to solve (10), has to be determined in every iteration, and its inverse computed for the solution. The time required to compute the Jacobian matrix and its inverse can be considerable, so that the user may decide either to compute the Jacobian only after every m iterations, for example when the current and the target joint vectors are sufficient close you can consider the Jacobian matrix to be fixed, or aproximate it by some method, for example numerical interpolation.

3.1. Constraints.

The system of nonlinear equations (8) is always subject to constraints imposed on the solution such as:

$$\mathbf{q}^l \leq \mathbf{q} \leq \mathbf{q}^u \tag{12}$$

where $\mathbf{q}^l$ and $\mathbf{q}^u$ are the lower and upper limits on the joint displacements.

In general, for Newton-Raphson methods no control is exercised on the computed correction vector $\Delta\mathbf{q}^k$, at iteration step k. It is possible that $\mathbf{q}^{k+1}$ determinated from $\Delta\mathbf{q}^k$, in (9), does no satisfy (12) and the algorithm converges to a solution outside the permissible joint ranges of the robot. To prevent convergence to a nonfeasible solution $\mathbf{q}^{k+1}$ must be checked against the limits and the variables which values are out of range corrected to the nearest limit value.

This strategy can be used to select the appropiate configuration of the manipulator from the several that place the end effector in the same position and orientation. This is done by setting the lower and upper limits on the joint displacements to the appropiate values for the desired configuration of the manipulator.

3.2. Singularities.

The solution of (10) requires that matrix $\mathbf{J}$ has an inverse. It is possible that an arm can assume a configuration in which the Jacobian matrix has no inverse. Such singularities indicate that the current solution is a maximum or a minimum in the space of the joint coordinates. However, some singularities can be overcome if they appear in the process of obtaining the solution by controlling the step size of the Newton-Raphson iteration replacing (9) by the expression:

$$\mathbf{q}^{k+1} = \mathbf{q}^k + \lambda\Delta\mathbf{q}^k \qquad (13)$$

where λ is a positive number.

We can either reduce the step size ($\lambda<1$), if our solution is in the current region of the joint coordinate space, or augment the step size ($\lambda>1$), if the solution is in the next region with respect to the maximum or minimum.

4. THE ALGORITHM.

Newton-Raphson iteration method requieres a good initial estimation $\mathbf{q}^0$, sufficiently close to the solution of the system of nonlinear equations (8) for fast convergency. When such an estimation is available the algorithm does not require the Jacobian matrix to be updated in each iteration step. Although such an estimation is possible to obtain, we will consider that the initial estimate of $\mathbf{q}$ is the current robot configuration, that can be quite far from the solution. Then the algorithm consists of the following step by step procedure:

> Step 1. For the current robot configuration determinate the current end effector position and orientation, $\mathbf{s}^k=f(\mathbf{q}^k)$.
> Step 2. Determine the residual position and orientation between the target and the current position and orientation using (6) and (7).
> Step 3. Check for iteration termination, $\Delta\mathbf{s}^k<\lambda$. Where λ is a threshold value supplied by the user, usually the same order of magnitude as robot precision. If successful then terminate the process, the solution is $\mathbf{q}^*=\mathbf{q}^k$, else proceed to the next step.
> Step 4. Obtain the Jacobian matrix using (11) and compute its inverse.
> Step 5. If the Jacobian matrix is singular then: if it is the first iteration try a new initial estimate $\mathbf{q}^0$ of $\mathbf{q}$, else go to algorithm Step 7 in the former iteration and reduce or augment the step size using (13) as indicated before.

Step 6. Determine the correction for the current joint coordinate vector, $\Delta q^k = J^{-1}(q^k)\Delta s^k$.
Step 7. Update the joint coordinate vector using (9).
Step 8. Check for constraints, using (7) as indicated.
Step 9. Let q^{k+1} be the current joint coordinate vector and return to iteration Step 1.

In the algorithm Step 3 when the residual position and orientation is suffiently small we can proceed with the algorithm considering the Jacobian matrix to be fixed.

The following block diagram represents the procedure in a control-like manner.

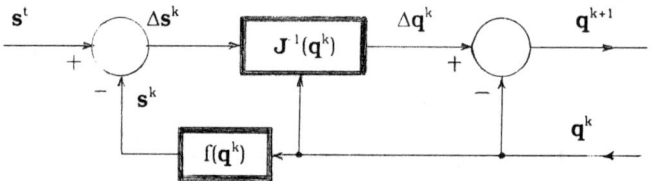

For kinematically redundant robots, with n>6, the system of nonlinear equations (8) is undeterminated. The 6xn Jacobian matrix is nonsquare, but its inverse can be obtained using pseudoinverses. Since n>6 in (10), the (6xn) Jacobian matrix can be partitioned into a (6x6) and a 6x(n-6) matrix using any objective function of the form minC=f(q) and determinating the optimum by a (n-6) dimensional optimization routine. This criterion will supply the extra equations needed to determinate the joint coordinates.

5. CONCLUSIONS.

A solution for solving the inverse kinematics problem of a robotic arm has been presented. The robot structure is analized by Denavit-Hatenberg transformation matrices. The method provides a generalized solution for any class of robots in an iterative manner, using a Newton-Raphson linearization technique. Certain constraints can be incorporated in the process of obtaining a solution that enable selection of one robot configuration from the several that place the end effector in the desired position and orientation. Other concepts such as singularities and computation reduction are treated. Due to its versality the method is a good candidate for off-line path planning and simulation applications. The above work may be used to provide coordinated control for path planning, since the same Jacobian matrix represents the relation between joint rates and position and orientation rates.

AKNOWLEDGEMENTS

This work is part of a research project supported by "Acciones Concertadas de Investigación de la Universidad Politécnica de Madrid" with project-id A9000200151.

REFERENCES

[1] D.L. Pieper, "The kinematics of manipulators under computer control," Stanford Artificial Intelligence Project, Stanford CA, Memo. AIM-72, 1968.

[2] R.P. Paul, B. Shimano and G.E. Mayer, "Kinematic control equations for simple manipulators," IEEE Trans. Syst. Man Cybern., vol. SMC-11, no. 6, pp. 456-460, 1981.

[3] A.A. Goldenberg, B. Benhabib and R.G. Fenton, "A complete generalized solution to the inverse kinematics of robot," IEEE J. Robot. Autom., vol. RA-1, no. 1, pp. 14-20, 1985.

[4] J. Denavit and R.S. Hatenberg, "A kinematic notation for lower-pair mechanisms based on matrices," ASME J. Appl. Mech., pp. 215-221, 1955.

COMMAND MATCHING FOR A CONSTRAINT ROBOT SYSTEM

P.N.Paraskevopoulos, F.N.Koumboulis and K.G.Tzierakis

NATIONAL TECHNICAL UNIVERSITY OF ATHENS
Division of Computer Science, Department of Electrical Engineering
15773 Zographou, Athens, Greece

ABSTRACT

In this paper, the problem of controlling a robot with end effector rigidly gripping an inertial load is studied using the command matching technique via proportional state feedback. The necessary and sufficient conditions for the problem to have a solution are established and the general analytical expressions of the controller matrices are derived. The stability properties of the resulting closed-loop system are also investigated.

I. INTRODUCTION

It was only recently that some first results have been reported for controlling robots strongly influenced by external contact forces [1]-[12]. The mathematical description of these robot systems involves dynamic, as well as algebraic equations, resulting in a generalized state space description [13], [14]. Such robot systems are for example, robots with end effector rigidly gripping an inertial load, robots with end effector in contact with a rigid surface, etc.

This paper is devoted to the problem of controlling a robot with end effector rigidly gripping an inertial load using the commnand matching technique via proportional state feedback. The command matching control strategy is one of the most desirable control strategies since it yields a closed-loop system the outputs of which are equal to the command signals (external inputs). This way, the performance of the system can be perfectly controlled.

The paper is organized as follows: In Section II, the overall equations of motion for the manipulator with end effector gripping an inertial load are linearized about its nominal state, resulting in a system of linear dynamic and algebraic equations, i.e. in a generalized state space description. In Section III, the problem of command matching via proportional state feedback for the manipulator gripping an inertial load is studied wherein the necessary and sufficient conditions for the problem to have a solution are established and the general analytical expresssions of the controller matrices are derived. The resulting closed-loop system is proven to be asymptotically stable after appropriately choosing the degrees of freedom of the controller matrices. Finally, in Section IV, the above results are applied to the UMS-2 manipulator. It is noted that all above results are the first in the field.

The problem of command matching may be viewed as a special case of either the decoupling problem or the exact model matching problem. These problems are solved for generalized state space systems via proportional state feedback in [15] and [16], respectively. In this paper, the command matching of robots with end effector gripping an inertial load is treated using a modified version of the approach reported in [15].

II. EQUATIONS OF MOTION OF A ROBOT WITH END EFFECTOR GRIPPING AN INERTIAL LOAD

Based on the well known Lagrangian formulation, the equations of motion of a robot with n links, taking into acount the contact force, are given by

$$\Phi(q)\ddot{q} + \hat{G}(q,\dot{q}) + G(q) = T + J^T(q)f \qquad (2.1)$$

where $q \in \mathbb{R}^n$ is the vector of the joint angles, $\Phi(q) \in \mathbb{R}^{n \times n}$ is the inertial matrix function, $\hat{G}(q,\dot{q}) \in \mathbb{R}^n$ is the vector of Coriolis and centrifugal forces, $G(q) \in \mathbb{R}^n$ is the vector of the gravity forces, $J(q) = \frac{\partial L(q)}{\partial q} \in \mathbb{R}^{3 \times n}$ is the manipulator Jacobian, $T \in \mathbb{R}^n$ is the vector of the generalized forces applied at each joint, and $f \in \mathbb{R}^3$ is the vector of the contact force. The equations of motion of the pure inertial load, taking into acount the contact and gravitational forces, are

$$m\ddot{p} = -f - mg \qquad (2.2)$$

where m denotes the mass of the load, $g \in \mathbb{R}^3$ denotes the gravity vector and $p \in \mathbb{R}^3$ denotes the position vector of the load in a fixed workspace coordinate system. Assuming that the load is rigidly gripped by the robot end effector, the additional constraint

$$p = H(q) \qquad (2.3)$$

must also be satisfied, where $H(q)$ is the vector function representing the direct kinematic relations of the robot. This last relation represents an equation of constraint rather than a definition of kinematic relations. With regard to the matrix function $J(q)$ it is mentioned that it can also be given by the following relation [6]

$$J(q) = \frac{\partial H(q)}{\partial q} \qquad (2.4)$$

Equations (2.1)-(2.3) may be expressed more compactly as follows

$$\begin{bmatrix} \Phi(q) & 0 & 0 \\ 0 & mI & 0 \\ 0 & 0 & 0 \end{bmatrix} \begin{bmatrix} \ddot{q} \\ \ddot{p} \\ f \end{bmatrix} = \begin{bmatrix} T - \hat{G}(q,\dot{q}) + J^T(q)f \\ -f - mg \\ H(q) - p \end{bmatrix} \qquad (2.5)$$

We assume that the force applied to the end effector of the robot, due to the inertial load, is such that the amplitude of variations in joint position, velocity and acceleration about some nominal value remains small. This permits the manipulator dynamics to be represented by a perturbation model about the nominal state given by

$$[q_0^T, \dot{q}_0^T, p_0^T, \dot{p}_0^T, f_0^T] \qquad (2.6)$$

where q_0 is the nominal manipulator joint configuration, $\dot{q}_0$ is the nominal joint velocity, p_0 is the nominal position vector of the inertial load, $\dot{p}_0$ is the nominal inertial load velocity, and f_0 is the nominal contact force vector between the robot and the load.

To linearize the nonlinear dynamic equations given by (2.5) about the nominal state (2.6) we work as follows: At the nominal operating point, the manipulator is at rest, hence $\dot{q}_0=0$, $\ddot{q}_0=0$, $\dot{p}_0=0$, $\ddot{p}_0=0$ and $\dot{f}_0=0$, resulting in simplification of the linearized equations of motion. Writing (2.1) in the form

$$T = T(q, \dot{q}, \ddot{q}, f^T) \tag{2.7}$$

and expanding the function T in a multivariable Taylor series about the nominal state (2.6), and upon dropping all higher order terms, yields

$$\delta T = \Phi(q_0)\delta\ddot{q} + \frac{\partial}{\partial q}(G-J^Tf)\bigg|_{q_0} \delta q - J^T(q_0)\delta f \tag{2.8}$$

where

$$\delta q \equiv q-q_0,\ \delta\dot{q} \equiv \dot{q}-\dot{q}_0,\ \delta\ddot{q} \equiv \ddot{q}-\ddot{q}_0,\ \delta p \equiv p-p_0,\ \delta\dot{p} \equiv \dot{p}-\dot{p}_0,\ \delta\ddot{p} \equiv \ddot{p}-\ddot{p}_0$$
$$\delta T \equiv T-T_0,\ \delta f \equiv f-f_0,\ \delta\dot{f} \equiv \dot{f}-\dot{f}_0$$

and where use was made of the fact that $\hat{G}(q,\dot{q})$ is homogeneous in $\dot{q}$, as reported in [5] and [9]. Linearization of the constraint equation (2.3) yields

$$\delta p = \frac{\partial}{\partial q}H(q)\bigg|_{q_0} \delta q = J(q_0)\delta q \tag{2.9}$$

where use was made of (2.4). Using (2.2), (2.8) and (2.9), the linearized equations of motion of a robot with end effector gripping an inertial load have the following dynamic description:

$$E\dot{x}(t) = Ax(t) + Bu(t) \tag{2.10}$$

where $x(t) \in \mathbb{R}^{2n+9}$, $u(t) \in \mathbb{R}^n$, $E, A \in \mathbb{R}^{(2n+9)\times(2n+9)}$, $B \in \mathbb{R}^{(2n+9)\times n}$ and where

$$x(t) = [\ \delta q^T\ \delta p^T\ \delta\dot{q}^T\ \delta\dot{p}^T\ \delta f^T\ ]^T, \qquad u(t) = \delta T(t) \tag{2.11a}$$

$$E = \begin{bmatrix} I & 0 & 0 & 0 & 0 \\ 0 & I & 0 & 0 & 0 \\ 0 & 0 & 0 & mI & 0 \\ 0 & 0 & \Phi(q_0) & 0 & 0 \\ 0 & 0 & 0 & 0 & 0 \end{bmatrix},\ A = \begin{bmatrix} 0 & & 0 & I & 0 & 0 \\ 0 & & & 0 & 0 & I & 0 \\ 0 & & & 0 & 0 & 0 & I \\ -\frac{\partial(G-J^Tf)}{\partial q}\bigg|_{q_0} & 0 & 0 & 0 & J^T(q_0) \\ -J(q_0) & & I & 0 & 0 & 0 \end{bmatrix},\ B = \begin{bmatrix} 0 \\ 0 \\ 0 \\ -I \\ 0 \end{bmatrix} \tag{2.11b}$$

System (2.10) is a generalized state space (g.s.s.) system. However, since $\det E = 0$, system (2.10) is a singular system [13], [14]. In order to control the position vector p of the load rigidly gripped by the robot end effector and the vector of the robot joint angles q, we propose the following output (performance) matrix

$$C(q_0) = \begin{bmatrix} \hat{J}(q_0) & 0_{(n-3)\times 3} & 0_{(n-3)\times n} & 0_{(n-3)\times 3} & 0_{(n-3)\times 3} \\ 0_{3\times n} & I_3 & 0_{3\times n} & 0_{3\times 3} & 0_{3\times 3} \end{bmatrix} \quad (2.12)$$

where $\hat{J}(q_0)$ is an $(n-3)\times n$ selection matrix such that

$$\text{rank} \begin{bmatrix} \hat{J}(q_0) \\ J(q_0) \end{bmatrix} = n \quad (2.13)$$

Hence, the n-dimensional performance vector **y** is given by

$$\mathbf{y} = C(q_0)\mathbf{x} = \begin{bmatrix} \hat{J}(q_0)\delta q \\ \delta p \end{bmatrix} \quad (2.14)$$

The performance vector **y** involves the coordinates of δp (which are the most important variables to be controlled), as well as the variations of n-3 joint angles. Using (2.13) and (2.9), the variations of the rest three joint angles are determined by a proportional (not dynamic) matrix relation from δp and the variations of the joint angles involved in **y**. Thus, the variations of these three joint angles can be indirectly controlled via **y**. Note that in order to control independently all coordinates of the position of the center of gravity of the inertial load, it is necessary that

$$\text{rank} J(q_0) = 3 \quad (2.15)$$

Relation (2.15) results from the necessity of invertibility of the open-loop system. Clearly, according to (2.13) and (2.15), the matrix $\hat{J}(q_0)$ can readily be determined.

In order to control the vector **y**, the control strategy of command matching will be used. The use of this strategy is presented in the next section.

III. COMMAND MATCHING FOR A ROBOT WITH END EFFECTOR GRIPPING AN INERTIAL LOAD

1. Formulation of the Problem

In the frequency domain, the g.s.s. system (2.10), (2.14) takes on the form

$$sEX(s) = AX(s) + BU(s) \quad , \quad Y(s) = C(q_0)X(s) \quad (3.1)$$

System (3.1) is assumed to be solvable, i.e. it is assumed that $\det[sE-A]\neq 0$. Using (2.11b), the solvability assumption takes on the form

$$\det\left[s^2[mJ^T(q_0)J(q_0) + \Phi(q_0)] + \frac{\partial}{\partial q}(G-J^Tf)\bigg|_{q_0}\right] \neq 0 \quad (3.2)$$

To system (3.1) apply the proportional state feedback law

$$U(s) = FX(s) + N\Omega(s) \quad ; \quad F \in \mathbb{R}^{n\times(2n+9)}, \; N \in \mathbb{R}^{n\times n} \quad (3.3)$$

where $\Omega(s)$ is the Laplace transform of the external input vector $\omega(t)\in\mathbb{R}^n$. The problem of command matching consists in finding controller matrices **F** and **N** such that the transfer function matrix of the closed-loop system (3.1), (3.3) is the unity matrix **I**, i.e. such that

$$C(sE-A-BF)^{-1}BN = I \quad (3.4)$$

Relation (3.4) may be written as follows [15]
$$C(sE-A)^{-1}B[I-F(sE-A)^{-1}B]^{-1}N = I \qquad (3.5)$$
Clearly, the closed-loop system (3.1), (3.3) is solvable, i.e. $\det(sE-A-BF) \neq 0$, iff $\det[I-F(sE-A)^{-1}B] \neq 0$. From relation (3.5) we readily conclude that it is **necessary** that the two matrices $C(sE-A)^{-1}B$ and N are invertible. Hence, from (3.5) we get
$$C(sE-A)^{-1}B = [\Gamma-\Phi(sE-A)^{-1}B] \quad ; \quad \Gamma=N^{-1}, \quad \Phi=N^{-1}F \qquad (3.6)$$
Equation (3.6) is equivalent to (3.5) iff the following condititons hold
$$\det[C(sE-A)^{-1}B] \neq 0 \qquad (3.7a)$$
$$\det\Gamma \neq 0 \qquad (3.7b)$$
From (3.7) and premultiplying (3.6) by N, we readily conclude that $\det[I-F(sE-A)^{-1}B] \neq 0$, i.e. that the closed-loop system (3.1), (3.3), is solvable for every F and for every N satisfying (3.6) and (3.7b). Equation (3.6) may be expressed equivalently as follows
$$\Psi(sE-A)^{-1}B = I \quad ; \quad \Psi = \Gamma^{-1}[C-\Phi] = NC-F \qquad (3.8)$$
Thus far, the problem has been reduced to that of solving (3.8) for Ψ, subject to the condition (3.7a). Clearly, if (3.8) can be solved for Ψ, there always exists an invertible matrix N and a matrix F satisfying both relation $\Psi=NC-F$ and condition (3.7b).

2. Necessary and Sufficient Conditions

Substituting (2.11) to $(sE-A)^{-1}B$, yields

$$(sE-A)^{-1}B = \begin{bmatrix} I \\ J(q_0) \\ sI \\ sJ(q_0) \\ -s^2 mJ(q_0) \end{bmatrix} \left\{ s^2[mJ^T(q_0)J(q_0) + \Phi(q_0)] + \left.\frac{\partial(G-J^Tf)}{\partial q}\right|_{q_0} \right\}^{-1} \qquad (3.9)$$

Note that the invertibility of the matrix $s^2(mJ^T(q_0)J(q_0)+\Phi(q_0)) + \left.\frac{\partial}{\partial q}(G-J^Tf)\right|_{q_0}$ is guaranteed from the solvability of the open-loop system (relation (3.2)). From (2.15) and (2.12) we readily conclude that the open-loop system is invertible, i.e. condition (3.7a) is satisfied. According to (3.9), equation (3.8) may be rewritten as follows

$$\Psi_1 + \Psi_2 J(q_0) + s[\Psi_3 + \Psi_4 J(q_0)] - s^2 \Psi_5 mJ(q_0) = \left.\frac{\partial(G\ J^Tf)}{\partial q}\right|_{q_0} + s^2[mJ^T(q_0)J(q_0)+\Phi(q_0)]$$
$$(3.10a)$$

where the matrix Ψ has been partitioned as
$$\Psi = [\Psi_1 \,\vdots\, \Psi_2 \,\vdots\, \Psi_3 \,\vdots\, \Psi_4 \,\vdots\, \Psi_5] \quad ; \quad \Psi_1,\Psi_3 \in \mathbb{R}^{n\times n}, \quad \Psi_2,\Psi_4,\Psi_5 \in \mathbb{R}^{n\times 3} \qquad (3.10b)$$
Equating like powers of s in both sides of (3.10) we derive the following set of equations
$$\Psi_1 = \left.\frac{\partial}{\partial q}(G-J^Tf)\right|_{q_0} - \Psi_2 J(q_0) \qquad (3.11a)$$

$$\Psi_3 = -\Psi_4 J(q_0) \tag{3.11b}$$

$$m\Psi_5 J(q_0) = -mJ^T(q_0)J(q_0) - \Phi(q_0) \tag{3.11c}$$

Equations (3.11a) and (3.11b) are always solvable for Ψ_1 and Ψ_3 while

$$\Psi_2, \Psi_4 : \text{arbitrary} \tag{3.12}$$

Equation (3.11c) may be rewritten as follows

$$ZJ(q_0) = \Phi(q_0) \quad ; \quad Z = -m[\Psi_5 + J^T(q_0)] \tag{3.13}$$

The above equation can be solved with regard to Z iff

$$\text{rank}\begin{bmatrix} J(q_0) \\ \Phi(q_0) \end{bmatrix} = \text{rank}[J(q_0)] = 3 \tag{3.14}$$

where use has been made of (2.15).

We have thus established the following theorem.

Theorem 3.1. The necessary and sufficient conditions for the solution of the problem of command matching via proportional state feedback for a robot gripping an inertial load are

$$\text{rank}\begin{bmatrix} J(q_0) \\ \Phi(q_0) \end{bmatrix} = \text{rank}[J(q_0)] = 3 \tag{3.15}$$

3. General Solution of the Controller Matrices

Assuming that (3.15) holds, then the general solution of (3.13) is

$$Z = \Phi(q_0)J^T(q_0)[J(q_0)J^T(q_0)]^{-1} \tag{3.16}$$

Substitution of (3.16) in (3.13) yields

$$\Psi_5 = -[m^{-1}\Phi(q_0) + J^T(q_0)J(q_0)]J^T(q_0)[J(q_0)J^T(q_0)]^{-1} \tag{3.17}$$

From (3.11a), (3.11b), (3.12) and (3.17) we readily conclude that the general solution of Ψ is given by

$$\Psi = [\Psi_1 \vdots \Psi_2 \vdots \Psi_3 \vdots \Psi_4 \vdots \Psi_5] = D_0(q_0) + \Lambda D_1(q_0) \tag{3.18a}$$

where

$$\Lambda = [\Psi_2 \vdots \Psi_4] \tag{3.18b}$$

$$D_0(q_0) = \begin{bmatrix} \frac{\partial}{\partial q}(G - J^T f)\big|_{q_0} & 0_{n\times 3} & 0_{n\times n} & 0_{n\times 3} & -[m^{-1}\Phi + J^T J]J^T[JJ^T]^{-1} \end{bmatrix} \tag{3.18c}$$

$$D_1(q_0) = \begin{bmatrix} -J(q_0) & I_3 & 0 & 0 & 0_{3\times 3} \\ 0 & 0 & -J(q_0) & I_3 & 0_{3\times 3} \end{bmatrix} \tag{3.18d}$$

Substitution of (3.18) in (3.8) yields

$$F = -D_0(q_0) + NC(q_0) - \Lambda D_1(q_0) \tag{3.19}$$

where N is an arbitrary invertible matrix. We have thus established the following theorem.

Theorem 3.2. Assume that the conditions of Theorem 3.1 are satisfied. Then, the general analytical expressions of the controller matrices N and F appearing in the control law (3.3) which result in command matching for a robot with end effector gripping an inertial load are

$N(q_0)$: arbitrary invertible 3×3 matrix (3.20a)

$$F(q_0) = -\left[\frac{\partial}{\partial q}(G \cdot J^T f)\bigg|_{q_0} \quad 0_{n\times n} \quad 0_{n\times 3} \quad 0_{n\times n} \quad -[m^{-1}\Phi + J^T J]J^T[JJ^T]^{-1}\right]$$

$$+ N(q_0)\begin{bmatrix} \hat{J}(q_0) & 0_{(n-3)\times 3} & 0_{(n-3)\times n} & 0_{(n-3)\times 3} & 0_{(n-3)\times 3} \\ 0_{3\times n} & I_3 & 0_{3\times n} & 0_{3\times 3} & 0_{3\times 3} \end{bmatrix}$$

$$- \Lambda \begin{bmatrix} -J(q_0) & I_3 & 0 & 0 & 0_{3\times 3} \\ 0 & 0 & -J(q_0) & I_3 & 0_{3\times 3} \end{bmatrix} \quad (3.20b)$$

The free parameters in (3.20) are the elements of the arbitrary invertible matrix $N \in \mathbb{R}^{n\times n}$ and the completely arbitrary matricx $\Lambda \in \mathbb{R}^{n\times 6}$.

Remark 3.1. Expressions (3.20a) and (3.20b) involve only arbitrary elements as well as the dynamic and kinematic parameters of the robot with end effector gripping an inertial load. This property is very useful for on-line control. Also, note that if we choose $\Lambda = 0$, then no feedback of $\delta \dot{q}$ and $\delta \dot{p}$ is involved, a fact which greatly facilitates the implementation of the feedback law.

4. Command Matching and Stabilizability

In this subsection we will investigate if asymptotic stability can be achieved together with command matching using the degrees of freedom of the matrix F given in (3.20b). As it is known from [13] and [17], the closed-loop system (3.1), (3.3) is asymptotically stable iff

$$\text{rank}[sE - A - BF] = 2n + 9, \quad \forall s : \text{Re}\{s\} \geq 0 \quad (3.21)$$

Substitution of (3.20b) and (2.12) in (3.21), yields

$$\text{rank}\left\{s^2 \Phi(q_0)\left\{I + J^T(q_0)[J(q_0)J^T(q_0)]^{-1}J(q_0)\right\} + N\begin{bmatrix}\hat{J}(q_0) \\ J(q_0)\end{bmatrix}\right\} = n, \quad \forall s : \text{Re}\{s\} \geq 0 \quad (3.22)$$

Condition (3.22) may be written equivalently as follows

$$\text{rank}\left\{w\Phi(q_0)\left\{I + J^T(q_0)[J(q_0)J^T(q_0)]^{-1}J(q_0)\right\} + N\begin{bmatrix}\hat{J}(q_0) \\ J(q_0)\end{bmatrix}\right\} = n, \quad \forall w \in \mathbb{C} \quad (3.23)$$

In order to satisfy (3.23), using the arbitrary matrix N, it is necessary and sufficient that the following condition holds [13]

$$\text{rank}\begin{bmatrix} w\Phi(q_0) \\ \hat{J}(q_0) \\ J(q_0) \end{bmatrix} = n, \quad \forall w \in \mathbb{C} \quad (3.24)$$

According to (2.13) we readily observe that (3.24) is always satisfied. Thus, we have established the following corollary.

Corollary 3.1. Assume that the conditions of Theorem 3.1 hold. Then, the closed-loop system (3.1), (3.3), with **F** and **N** given in (3.20), can always become asymptotically stable by appropriate choice of the degrees of freedom in **F**.

IV. APPLICATION TO THE UMS-2 ROBOT

In this section all aspects studied in the previous sections will be applied to the UMS-2 robot [9]. This robot is a manipulator with three joints. The dynamic equations of motion of this manipulator in unconstraint form are given by

$$T_i = \Phi_i \ddot{q}_i + H_i \qquad (4.1)$$

where

$$\Phi_1 = J_{z1} + J_{z2} + J_{z3} + m_3(q_3+l_3)^2 , \quad \Phi_2 = m_2+m_3 , \quad \Phi_3 = m_3 \qquad (4.2a)$$

$$H_1 = 2m_3(q_3+l_3)\dot{q}_1\dot{q}_3 , \quad H = (m_2+m_3)g , \quad H = -m_3(q_3+l_3)\dot{q}_1^2 \qquad (4.2b)$$

The position vector **p** is given by

$$\mathbf{p} = \begin{bmatrix} (q_3+l_3)\cos q_1 \\ (q_3+l_3)\sin q_1 \\ q_2 \end{bmatrix} \qquad (4.3)$$

The command matching strategy for this case has the following direct interpretation: To control independently the three coordinates (in the fixed workspace coordinate system) of the position vector of the center of mass of the object gripped by the end effector of the UMS-2 manipulator. Since the UMS-2 manipulator has three joints, i.e. n=3, we have that

$$\mathbf{C} = [0_{3\times3} \ \mathbf{I}_3 \ 0_{3\times9}] \qquad (4.4)$$

The transfer function matrix of the open-loop system, according to (4.4) and (3.9), is given by

$$\mathbf{C}(s\mathbf{E}-\mathbf{A})^{-1}\mathbf{B} = \mathbf{J}(q_0)\left\{s^2[m\mathbf{J}^T(q_0)\mathbf{J}(q_0) + \Phi(q_0)] + \frac{\partial}{\partial q}(\mathbf{G}\cdot\mathbf{J}^T\mathbf{f})\Big|_{q_0}\right\}^{-1}$$

For the present case n=3. Hence, according to (2.14), the matrix $\mathbf{J}(q_0)$ must be a 3×3 inverible matrix and thus the conditions of Theorem 3.1 are satisfied. Therefore, the problem of command matching for the UMS-2 robot has always a solution.

Using (4.4) and the invertibility of $\mathbf{J}(q_0)$, the general analytical expressions for the controller matrices reduce as follows:

$\mathbf{N}(q_0)$: arbitrary invertible 3×3 matrix (4.5a)

$$\mathbf{F}(q_0) = -\left[\frac{\partial}{\partial q}(\mathbf{G}\cdot\mathbf{J}^T\mathbf{f})\Big|_{q_0} \ 0_{3\times9} \ -m^{-1}\Phi\mathbf{J}^{-1}\cdot\mathbf{J}^T\right] + \mathbf{N}(q_0)[0_{3\times3} \ \mathbf{I}_3 \ 0_{3\times9}]$$

$$- \Lambda \begin{bmatrix} -\mathbf{J}(q_0) \ \mathbf{I}_3 & 0 & 0 & 0_{3\times3} \\ 0 & 0 & -\mathbf{J}(q_0) \ \mathbf{I}_3 & 0_{3\times3} \end{bmatrix} \qquad (4.5b)$$

Note that, according to Corollary 3.1, the closed-loop system, resulting after the application of the control law (3.3) (with **F** and **N** given by (4.5)) to system (3.1), can always become asymptotically stable.

V. CONCLUSIONS

In this paper, the following aspects regarding the problem of command matching of a robot with end effector gripping an inertial load have been resolved: The necessary and sufficient conditions for the problem to have a solution via proportional state feedback (Theorem 3.1), the general analytical expressions of the proportional state feedback controller matrices (Theorem 3.2), and the stability properties of the closed-loop system (Corollary 3.1). Other related problems, such as the control of a robot with end effector in contact with a rigid surface, are currently under investigation.

REFERENCES

[1] H.Hemami and B.F.Wyman, "Modeling and control of constraint dynamic systems with applications to biped locomotion in the frontal plane", *IEEE Trans.Automat. Contr.*, vol.24, pp.526-535, 1979.

[2] H.West and H.Asada, "A method for the control of robot arms constraint by contact with the environment", *Proc. American Control Conf.*, Boston MA, 1985, pp.383-386.

[3] N.H.McClamrock and H.P.Hung, "Dynamics of a closed chain manipulator", *Proc. American Control Conf.*, Boston, 1985, pp.50-54.

[4] H.Kazerooni, P.K.Houpt and T.B.Sheridan, "Robust compliant motion of manipulators, Part I: The fundamental concepts of compliant motion", *IEEE J. Robotics and Automation*, vol.2, pp.83-92, 1986.

[5] H.Kazerooni, P.K.Houpt and T.B.Sheridan, "Robust compliant motion of manipulators, Part II: Design method", *IEEE J. Robotics and Automation*, vol.2, pp.93-105, 1986.

[6] N.H.McClamroch, "Singular systems of differential equations as dynamic models for constraint robot systems", *Proc. IEEE Int. Conf. on Robotics and Automation*, San Francisco, CA, Apr.8-10, 1986, pp.21-28.

[7] H.Hemami, C.Wongchaisuwat and J.L.Brinker, "A heurisitic study of relegation of control in constraint robotic systems", *ASME J. Dyn. Syst. Meas. Contr.*, vol.109, pp.224-231, 1987.

[8] D.E.Whitney, "Historical perspective and state of the art of robot force control", *Int. J. Robotics Res.*, vol.6, pp.3-14, 1987.

[9] J.K.Mills and A.A.Goldenberg, "Force and position control of manipulators during constraint motion tasks", *IEEE Trans. Robotics and Automat.*, vol.5, pp.30-46, 1989.

[10] H.P.Huang, "The unified formulation of constrained robot systems", *Proc. IEEE Int. Conf. on Robotics and Automation*, Philadelphia, 1988, vol.3, pp.1590-1592.

[11] N.H.McClamrock and D.Wang, "Feedback stabilization and tracking of constraint robots", *IEEE Trans. Automat. Contr.*, vol.33, pp.419-426, 1988.

[12] H.P.Huang and W.L.Tseng, "Asymptotic observer design for constraint robot systems", *IEE Proc.-D*, vol.138, pp.211-216, 1991.

[13] L.Dai, **Singular Control Systems**, Springer Verlag Berlin, 1989.

[14] G.C.Verghese, B.C.Levy and T.Keilath, "A generalized state-space for singular systems, *IEEE Trans. Automat. Contr.*, vol.29, pp.1076-1082, 1984.

[15] P.N.Paraskevopoulos and F.N.Koumboulis, "The decoupling of generalized state space systems", *IEEE Trans. Automat. Contr.*, to appear in July 1991.

[16] P.N.Paraskevopoulos, F.N.Koumboulis and D.F.Anastasakis, "Exact model matching of generalized state space systems", *Journal of Optimiz. Theory and Applications*, accepted for publication.

[17] F.N.Koumboulis and P.N.Paraskevopoulos, "On the stability of generalized state space systems", submitted.

Control of a Flexible Arm for Prescribed Frequency Domain Tolerances: Modelling, Controller Design, Laboratory Experiments

Matei Kelemen, Jacob Starrenburg* and Arunabha Bagchi

Department of Applied Mathematics and
Mechatronics Research Centre Twente and
* Department of Electrical Engineering
University of Twente
P.O.Box 217, 7500 AE Enschede
the Netherlands

Abstract

In this article we solve the problem of achieving quantitative specifications for the tip of a rotating flexible beam with uncertainty in some of its (physical and geometric) parameters. The frequency domain design method used here is also helpful in clarifying some limitations on the feedback loop capabilities due to the distributed nature of the problem. Finally, time domain simulations and laboratory experiments confirm the validity of the design method suggested in this work.

1 Introduction

In recent years a significant research effort has been devoted to the study of flexible beams. Apart from the theoretical interest, this attention is motivated by important applications in robotics and flexible structures.

There are two (not decoupled) directions of research, with results both in state space and frequency domain modelling and control of flexible beams. Most of the control problems for flexible beams solved so far dealt with optimization, (robust) stabilization, vibration suppression, and related problems.

Our goal in this article is to go a step further, namely to achieve (whenever possible) desired frequency domain performance for the closed loop, despite large uncertainty in the plant parameters. The importance of this problem cannot be overemphasized in view of the increasing need for accurate, but at the same time for fast and less energy consuming industrial robots.

The desired performance is given by *quantitative specifications*, and the actual design should take into consideration both the extent of uncertainty and the narrowness of performance tolerances to reduce the cost of feedback (i.e. the overdesign). An effective design method for approaching this type of problems is the *quantitative feedback theory* (QFT). QFT was proposed in the early 1970's for lumped systems but its structure allows for various extensions, including that to distributed systems.

Specifically, we consider a slender flexible arm, i.e. a beam which is rotated by a motor in a horizontal plane, about an axis through the fixed end of the arm. The arm may carry a payload on its tip, and there are frequency-domain specifications on the movement of the tip of the arm. Our problem will be to design a feedback loop for controlling the motor torque in such a way that the specifications should be met (whenever possible) in spite of the uncertainty about some of the parameters of the plant, or in the presence of unknown disturbances.

To solve this problem we first develop a mathematical model of the cantilevered beam and the motor. We then express our task as a standard control problem. Finally, the feedback loop is found by extending QFT to our distributed case. In the process we point out some limitations in the feedback loop capabilities appearing in the control of flexible beams. We work with the *non-reduced* model of the plant, which proves to be useful in the robust stabilization of the system.

At the same time we want to obtain maximum benefit from a feedback loop based on a single and relatively simple measurement; namely, the angle of the motor shaft. While it is known that more measurements can improve the performance of the controller, our approach is justified by the fact that in many practical situations there are quite severe constraints imposed on measurements. At the same time an advantage of our frequency domain approach is that the design method allows for much more flexibility in the location of sensors than in a state space approach.

NOTE: A more detailed presentation of the ideas contained in this article and a more complete list of references is to be found in (Kelemen *et al*, 1991).

2 Model of Open Loop Elements

2.1 The Beam

Most authors obtain the model of the flexible beam in rotating cartesian coordinates. However, for our control purpose it is more appropriate to have it in fixed, polar coordinates.

Thus we have a beam of length R which is rotated by a motor, horizontally, about an axis through its fixed end. We would like a model of the beam in which we could visualize the beam rotating. This is why we shall consider the plane in which the beam moves, relative to *fixed* axes (ox, oy), but in *polar* coordinates (r, θ), i.e. $x = r\cos\theta$, $y = r\sin\theta$,

Now if we consider the arc $\phi = r\theta$ as the dependent variable we obtain the following (linearized) dynamical behavior of the beam

$$\Sigma(\phi(t,r)) := \rho\,\ddot{\phi} + 2\epsilon(EI)\frac{\partial^5 \phi}{\partial t \partial r^4} + (EI)\frac{\partial^4 \phi}{\partial r^4} = 0. \tag{2.1}$$

Here EI is the flexural rigidity, ρ is the mass per unit length of the beam, and ϵ is the internal (viscous) damping coefficient, all of them being uniform over the length R of the beam.

The boundary conditions for this linear partial differential equation with constant coefficients are

$$\phi(t,0) = 0, \quad \frac{\partial \phi}{\partial r}(t,0) = \theta_0(t), \quad \frac{\partial^2 \phi}{\partial r^2}(t,R) = 0, \quad \frac{\partial^3 \phi}{\partial r^3}(t,R) = 0, \tag{2.2}$$

where $\theta_0(t)$ is the angle of the motor shaft. These conditions are a direct consequence of the fact that the beam is a cantilever one, i.e. at one end it is clamped on the motor shaft, and the other end is free. We shall assume that the initial conditions are zero.

REMARK 2.1: From (2.2) one may view the angle of the motor shaft, θ_0, as a boundary *input* to the beam, which is for our purpose a plant to be controlled, while arc ϕ is the *output*.

2.2 The Motor

The motor we shall consider in this work is a *dc* one, having an amplifier of speed-feedback type.

Its simplified (but still realistic) model is given by the equation

$$\dot{\theta}_0 = \frac{1}{k_f}(V_{ref}(t) + \frac{1}{k}\beta(t)). \tag{2.3}$$

Here $\beta(t) = (EI)\partial^2\phi/\partial r^2(t,0)$ is the bending moment of the beam which works as a reaction on the motor shaft, while V_{ref} is the input-speed reference voltage to the amplifier, and k, k_f are gain constants.

Thus the motor is represented by an integrator with *output* the angle of the motor shaft θ_0, and with two *inputs*: a reference input V_{ref}, and the reaction input β.

3 The Closed Loop

We want to position the tip of the beam by an appropriate positioning of the motor shaft. Due to uncertainties in the parameters of the plant (i.e in the payload and in the flexural rigidity of the beam) and due to the accuracy required in positioning the tip of the beam, we have to use a feedback loop.

The representation of the closed loop, with its two interlaced feedback loops, is given in figure 1. Here B is the plant (beam) described by (2.1) and (2.2), and M is the actuator (motor) described by (2.3). The elements of the loop which should be designed in order to achieve the desired behavior of the closed loop are G, the compensator (which generates the input reference voltage to the motor) and F, the prefilter; $c(t)$ is the command input (e.g. a reference).

We shall proceed by deducing the closed loop transfer function from $c(t)$ to $\phi(t,r)$ for an arbitrary $r \in [0,R]$, in the presence of a new input: the inertial force of a symmetric payload of mass m, located at the point r_ℓ on the beam. Then, by invoking (2.1), we can write

$$\Sigma(\phi(t,r)) = m\frac{d^2\phi_{t,r_\ell}}{dt^2}\delta_{r_\ell}(r),$$

where δ is the Dirac delta function. Applying the Laplace transform in the time variable we obtain

$$\Phi(s,r) = \Theta_0(s)P_{0r}(s) + H\Phi(s,r_\ell)P_{r_\ell r}(s).$$

As usual, we denoted the transform variable by s and used capital letters for the transformed functions. Also, $H = -ms^2$.

We have considered the boundary condition $\theta_0(t)$ as an independent input. Therefore, by superposition of inputs, we have that $P_{r_1 r_2}$ is the transfer function from an input located at r_1 to an output evaluated at r_2; when $r_1 = 0$ the transfer function is from the boundary condition $\Theta_0(s)$, seen as an input, to the output at the point r_2. $P_{r_1 r_2}$ can be computed by using standard methods in the theory of ordinary differential equations (with s considered a parameter). For our design purpose it is enough to have $P_{r_1 r_2}$ in numerical form.

The feedback loops are described by the following relation

$$\Theta_0 = ((CF - \Theta_0)G + (EI/k)\Phi^{(2)}(0))M,$$

see also figure 1. Here we have dropped the s argument, as we shall do from now on. A straightforward computation shows that $\Phi(r) = T_C C$, where

$$T_C = F\frac{L}{1+L}D_1. \tag{3.1}$$

Here $L = GP_{eqv}$,

$$P_{eqv} = \frac{M}{1 - C_1}, \tag{3.2}$$

and

$$D_1 = P_{0r} + \frac{HP_{r_\ell r}P_{0r_\ell}}{1 - HP_{r_\ell r_\ell}},$$

$$C_1 = (EI/k)M(P_{00}^{(2)} + P_{r_\ell 0}^{(2)}\frac{HP_{0r_\ell}}{1 - HP_{r_\ell r_\ell}}).$$

Thus, in relation (3.1) T_C is the closed loop transfer function from the command input $c(t)$ (applied at the root of the beam) to the arc $\phi(t, r)$, and L is the loop transmission corresponding to an equivalent subplant P_{eqv}. A brief computation shows that the "complete" plant is given by $P_O = P_{eqv}D_1$.

4 The Design Problem

The design problem we shall consider in this article concerns the achievement of a desired response to the command input of the flexible structure described previously.

We now formulate our design problem which calls for both stability and quantitative performance specifications to be satisfied at the point $r = r_\ell$ (where the payload is located) on the controlled beam.

PROBLEM : Given the range of uncertainty for the payload, $0\,[kg] \leq m \leq 0.5\,[kg]$, and for the flexural elasticity of the beam, $20\,[Nm^2] \leq EI \leq 25\,[Nm^2]$, find (if possible) a real rational, strictly proper *fixed* compensator G and prefilter F, such that the family indexed by m and EI of closed loop transfer functions T_C with $r = r_\ell$

(i) is stable
(ii) satisfies the stability margin condition given by $20ln|\frac{L}{1+L}| \leq 2.3\ dB$
(iii) satisfies the frequency domain quantitative specifications

$$20ln|\frac{0.98}{1+s}| \leq 20ln|T_C(s)| \leq 20ln|\frac{1.02}{1+s/1.4}|, \quad s = j\omega. \tag{4.1}$$

The considered ranges of the uncertain parameters are physically realistic. Note that the allowed variation of EI is quite small since, usually, it can be measured with reasonable accuracy. As for other data relating to the beam, we use values identified in the flexible arm used in (Kruise et al, 1989). The "fixed" data of the problem are: the length of the beam $R = 1.9\ [m]$, the mass density per unit length of the beam $\rho = 0.648\ [kg/m]$, the damping factor $\epsilon = 0.00138\ [s]$, and the motor constants (section 2.2) $\frac{1}{k_f} = 2.58\ [rad/s/V]$, and $\frac{1}{k} = 0.025\ [V/N/m^2]$. Also, since the payload is not placed exactly on the tip of the beam, we have assumed that the payload is located at $r_\ell = 0.99R$

By stability we mean that all the poles should be in the left half of the complex plane, bounded away from the imaginary axis (i.e. exponential stability). This can be checked by the Nyquist criterion. Due to formula (3.1) the closed loop is stable if we design the feedback loop in a stable way, and if the factor D_1 (unchanged by feedback) is stable. This explains why, in (ii), the stability margin is imposed on $\frac{L}{1+L}$ and not on T_C. We recall that the peak value of $|\frac{L}{1+L}|$ is related to the damping factor of the closed loop system, that is to the attenuation of vibrations in which we are interested.

5 The Design Procedure

The design methodology is an extension of the quantitative feedback theory (QFT), as it is presented in (Horowitz and Sidi, 1972), to our distributed case. The method makes use of the logarithmic complex plane, i.e. the Nichols Chart, for translating the specifications on the closed loop into bounds on the (open) loop transmission. This is accomplished by taking advantage of the fact that the Nichols Chart has two systems of coordinates imprinted on it: a rectangular one for the magnitude, in dB, and the phase, in degrees, of a complex number H, and a curved one for the magnitude and phase of the complex number $\frac{H}{1+H}$.

The compensator G and the prefilter F are designed separately. By G we assure that the *variation* in $|T_C|$ (when the parameters go over their uncertainty range) is within the maximum variation allowed in (iii), see section 4. By F we assure that the *actual spread* of (the family) $|T_C|$ is between the bounds given in (iii).

5.1 Design of G

STEP 1 : Build the (uncertainty) templates of the plant.
That is, one has to find the locus in the Nichols Chart, for each ω, of $P_{eqv}(j\omega, m, EI)$, when m and EI vary over their range of uncertainty, i.e. $0 \leq m \leq 0.5$

and $20 \leq EI \leq 25$. Actually, only a few templates are enough for performing the design.

STEP 2: Find bounds on a nominal loop transmission L.

As nominal point we have chosen ($m = 0.5$, $EI = 20$). However, the design does not depend on this choice.

Let us consider T_C (see (3.1)) as a map from the set $\mathcal{P}$ of plants denoted by P_{eqv} and indexed by m and EI, to the complex plane,

$$T_C(P_{eqv}) = F\frac{L}{1+L}D_1.$$

Then, if Δ is the difference operator,

$$\Delta ln T_C = \Delta ln(\frac{L}{1+L}D_1). \tag{5.1}$$

Satisfying (iii) (from section 4) over $\mathcal{P}$ will result, via (5.1), in bounds on the nominal $L(j\omega)$ for a certain range of frequencies. Actually, the bounds are obtained by manipulating the templates in the Nichols Chart. Typical bounds are shown in figure 2. The admissible regions are the shaded ones.

Thus by (5.1), (iii) is satisfied at a given frequency if

$$\Delta|\frac{L}{1+L}|_{dB} \leq T_{max} - T_{min} - (D_{max} - D_{min}) \tag{5.2}$$

over *all* the parameter uncertainty range. Here T_{max}, T_{min} are the functions appearing at the right hand side and left hand side of the inequalities (4.1) respectively, and D_{max}, D_{min} are the maximum and minimum value of $|D_1|_{dB}$ over the parameter uncertainty range. Also, the right hand side of (5.2) represents the actual specifications on $\frac{L}{1+L}$.

From here we can see that a necessary condition to satisfy (iii) is that

$$D_{max} - D_{min} \leq T_{max} - T_{min}. \tag{5.3}$$

Numerical results summarized in (Kelemen et al, 1991) show that for $\omega = 1.5$ the inequality (5.3) is still satisfied, however for $\omega = 2$ it is not. But even for $\omega = 1.5$ the margin is so small (i.e. $\frac{L}{1+L}$ has to be so insensitive to parameter uncertainty) that it requires an extremely high gain feedback. (We recall that this is not practical since it greatly amplifies the sensor noise at the input of the plant.) That is why we chose to satisfy (iii) on the smaller frequency range, $0 \leq \omega \leq 1$.

REMARK 5.1: We stress that the limitation in the feedback loop capabilities given by (5.3) does not appear in the lumped case where $D_1 = 1$. They appear in our case, i.e. specifications *cannot* be satisfied for $\omega \geq 2$, because we want to control the behavior of a plant distributed in space, by using a point (i.e. not distributed in space) feedback loop.

The remaining requirements of the design problem, (i) and (ii), refer to stability. The Nyquist stability criterion can be traced back on the Nichols Chart (the critical point being now $0\ dB$ and -180 degrees). To apply this criterion we have to know the number of unstable poles of the plant; by frequency domain simulations we concluded that the family of plants P_{eqv} is stable.

In our case, we have not only to stabilize, but also to achieve a given stability margin (ii), in a robust way with respect to (m, EI). Therefore we shall again use (uncertainty) templates, even for frequencies larger than $\omega = 1$, which was the limit for achieving performance specifications (iii). This time we use them for achieving the robust stabilization as given by the stability specification (ii), thus obtaining more bounds to be satisfied by the nominal loop transmission. In figure 2 a bound can be seen corresponding to $\omega = 7$.

STEP 3 : Shape the loop transmission, by means of the compensator G, to satisfy both the performance and stability bounds obtained at step 2. This is done by a usual rational function compensation procedure.

The QFT methodology enables a suboptimal design, in the sense of reducing the *cost of feedback* (roughly speaking the effect of the sensor noise at the input of the plant, or the bandwidth of the system). However, the closer the suboptimal design is to an optimal one, the more complex structure is needed for the compensator. Since in our case the frequency range for achieving performance specifications is very small it does not seem necessary to try to satisfy it with a compensator of complicated structure.

Therefore, our first approximation for a compensator was a second order system. To satisfy the stability bounds as well, we had to consider also a first order numerator for the compensator. By shaping the loop transmission we found

$$G = \frac{1.66(s + 1.33)}{(s + 4)(1 + s/20)}$$

REMARK 5.2: As can be seen from figure 2, for $\omega = 3$ and $\omega = 4$ our design is not far from the corresponding stability bounds. We could afford this closeness since we worked with the exact, i.e. the *non-reduced* model of the plant, as the one provided in section 2. This shows the usefulness of the exact model of the plant to the robust stabilization of distributed systems.

5.2 Design of F

F should be designed in such a way that (iii) of section 4 will be satisfied. Therefore,

$$T_{min} \leq |F|_{dB} + |\frac{L}{1 + L}|_{dB} + |D_1|_{dB} \leq T_{max}$$

over all the uncertainty range.

Since G has already been designed, L can be readily computed. With numerical results from (Kelemen et al, 1991) and another rational function compensation procedure, we obtained the prefilter

$$F = \frac{0.54}{1 + s/1.57},$$

which completes the design of the feedback loop.

The designed feedback loop solves the problem of section 4; specification (iii) is satisfied for $0 \leq \omega \leq 1$.

REMARK 5.3: The gain of the compensator we designed is relatively small. This was necessary to preserve the stability of the closed loop system, endangered by the *non-minimum phase* property of P_{eqv} (graphically illustrated in figure 5). The price to be paid is a relatively slow time respose.

6 Implementation of the controller

6.1 Experimental Setup

To implement and test the design, the experimental setup available at the Control Laboratory of the department of Electrical engineering, University of Twente, has been used. A sketch of this setup, including the parameters of interest, is shown in figure 3.

One end of a flexible strip is clamped on a vertical shaft, which is driven by a DC motor. Therefore, the link rotates in a horizontal plane. The other end of the strip is free. The angular position of the load axis is measured with a resolver, and converted by a Resolver-to-Digital Converter (RDC) into a 12 bit representation, with an accuracy of about $0.1°$. The voltage of the motor is controlled via a 12 bit Digital to Analog Converter (DAC) and a power amplifier.

An IBM-compatible personal computer with an INTEL-80286 microprocessor and a 80287 mathematic coprocessor, both running with a clock frequency of 12 MHz has been used to implement the control algorithms; the programming language is TurboPascal.

6.2 Sampling Frequency

We now want to estimate the maximum sampling period h [s], required for implementing our design. When implementing a continuous controller by means of a discrete-time computer algorithm, the sources of approximation errors are:

- Computation time delay.
- Sampling continuous time signals into, and reconstruction of continuous signals from, discrete time series of values.
- Discretization of the values of the signals.

While executing the control algorithm, the following sequence of activities are performed during each sampling period (see also figure 1):

- Compute the prefilter (F) output.
- Measure the angle of the motor shaft θ_0 at sampled time-points (the sampling frequency is $1/h$ [Hz]).
- Compute the compensator (G) output V_{ref}.
- Send this V_{ref} value to the DAC.

- Update values of all variables.
- Wait for the beginning of the next sampling period and start this procedure again.

The computation time delay in the above algorithm between the measurement of θ_0 and the update of the output voltage V_{ref} is found to be smaller than half a sampling period $h/2$.

The error due to sampling $\theta_0(t)$ to a discrete-time value sequence and the physical reconstruction of $V_{ref}(t)$ out of a computed discrete time series, can be expressed as an additional time delay of half a sampling period $h/2$ (Åström and Wittenmark, 1984).

The last effect, value discretizing, can be neglected in our case. This is due to the 12 bits representation used for the measured angle and the motor voltage, which as mentioned before, results in a very good accuracy.

Thus the overall discretization error can be expressed as a time delay which does not exceed h.

Comparing the discretized controller to the designed continuous one, this time delay will generate in the frequency domain (ω) a *phase lag*

$$\psi(\omega) = \frac{h\,\omega}{2\,\pi}360 \;[\text{degrees}] \tag{6.1}$$

In the Nichols chart, this corresponds to a (frequency dependent) shift to the left of the discretized $L_{discr.} = G_{discr.}P_{eqv}$ compared to the continuous $L = GP_{eqv}$.

The required sampling frequency can now be estimated as follows. Discretizing G should not cause too much phase shift when L approaches the crossover frequency, because it could endanger the (robust) stability of the closed loop. From figure 2 we see that this frequency is about $\omega = 3$, and the maximum phase shift allowed is 15°. Then from (6.1) we obtain $h = 0.087$ [s], that is, the sampling frequency should be 12 Hz or more. (Actually, we have worked with 20 Hz.)

This is a rather low frequency which allows the laboratory experiments to be performed with a much cheaper equipment than what has actually been used in the laboratory.

6.3 Resulting Performance

For implementing the feedback control loop, F and G are discretized by using the Tustin approximation, see (Åström and Wittenmark, 1984).

In figure 4 and 5 we have collected measured tip position step responses for the extreme cases: the most favorable ($m = 0$ [kg]), and the most unfavorable one ($m = 0.5$ [kg]). The measured value of the flexural rigidity is $EI = 22.3$ [Nm^2].

The experiments show a very satisfactory performance of the designed feedback loop, taking into account the fact that it is based only on a single measurement.

REMARK 6.1: The laboratory experiments introduce an additional limitation to the theoretical feedback loop capability due to the motor friction. This clearly has a nonnegligible adverse influence on both the steady state error and the speed of the closed loop response.

Therefore, we have added a small extra voltage to the computed V_{ref}. At the same time we have assured a smooth 0-crossing of the applied motor voltage, to avoid undesired oscillations, see (Kelemen et al, 1991).

7 Conclusions

In this article we have solved the following problem: achieving (to the extent possible) quantitative specifications for the movement of the flexible robot arm carrying a payload. This has been done by applying a quantitative feedback methodology for designing a feedback loop to control the flexible arm, and compensating for the uncertainty in the payload and flexural rigidity of the beam. The same design procedure has clarified some limitations in the feedback loop capabilities.

The feedback loop is based on a single measurement, the angle of the motor shaft. Since this type of constraint is encountered in many practical situations, our technique shows how to achieve benefit of feedback under these conditions. In the design we have used the *non-reduced* model of the plant, which proved to be useful in the robust stabilization of the closed loop. Time domain simulations (not reported here) and the laboratory experiments confirm the qualities of the design procedure presented in this work.

8 References

1. M. KELEMEN, J. STARRENBURG AND A. BAGCHI (1991). "Modelling and feedback control of a flexible arm of a robot for prescribed frequency-domain tolerances", TW Memorandum, *Dept. of applied mathematics, University of Twente*, Enschede, the Netherlands.

2. I.M. HOROWITZ AND M. SIDI (1972). "Synthesis of feedback systems with large plant ignorance for prescribed time-domain tolerances", *Int. J. Contr.*, vol. 16, pp. 287-309.

3. L. KRUISE, J. VAN AMERONGEN AND P. LÖHNBERG (1989). "Modeling and control of a flexible robot link", in *Dynamics of Controlled Mechanical Systems*, G. Schweitzer and M.Mansour (eds.), Springer, Berlin.

4. K.J. ÅSTRÖM AND B. WITTENMARK (1984). *Computer controlled Systems*, Prentice Hall, Englewood Cliffs.

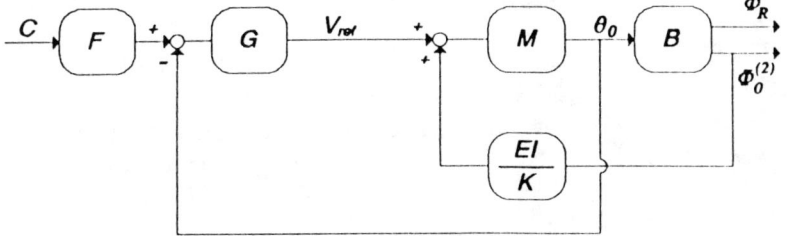

Figure 1: The closed loop

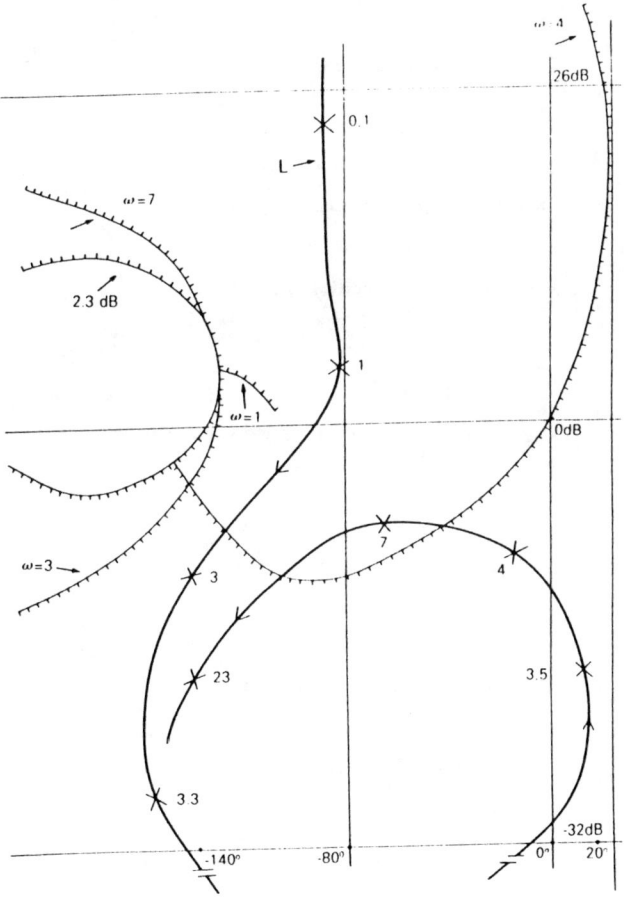

Figure 2: Nichols Chart design of G

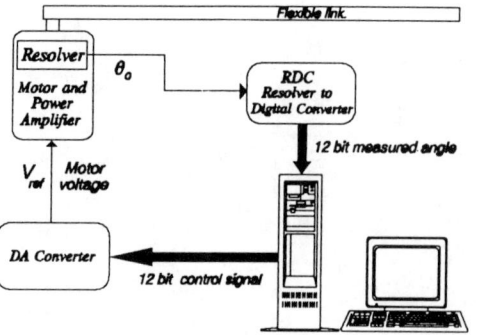

Flexible link constants:
Length L=1.9 [m]
Mass per unit length ρ=0.684 [kg/m]
Damping ε=0.00138 [s]
Flexural rigidity EI=22.3 [Nm^2]

Figure 3: Sketch of experimental setup

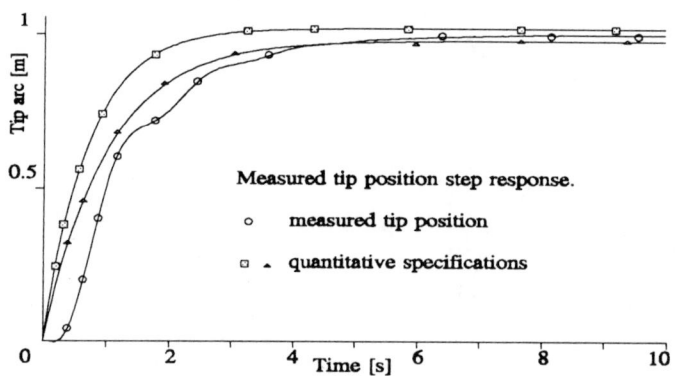

Figure 4: Results for payload m=0 [kg] with friction compensation

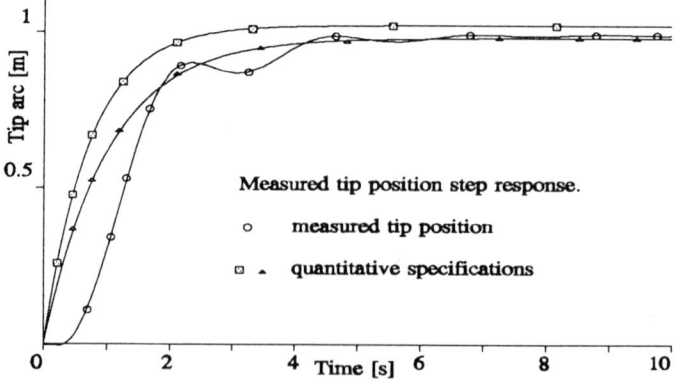

Figure 5: Results for payload m=0.5 [kg] with friction compensation

PART 3
ROBOT PATH PLANNING AND TRACKING

High-Speed Robot Path Planning in Time-Varying Environment Employing a Diffusion Equation Strategy

G. Schmidt and W. Neubauer
Chair and Laboratory for Control Engineering (LSR)
Technische Universität München
POB 20 24 20, D - 8000 München 2, Germany

Abstract

This paper discusses a method for combined map and sensor-based planning of smooth collision-free robot motion paths. Application of an unsteady diffusion equation model to path planning in a time-varying environment distinguishes this approach from prior work in the field. Collision-free robot paths between a starting point and a goal point are generated very rapidly by on-line simulation of the diffusion process and by repetitively evaluating the gradient of the computed concentration distribution functions. Related numerical algorithms can be easily implemented on a massively parallel distributed processor architecture. Applications of this approach to high-speed dynamic path planning for a robot vehicle operating in various types of time-varying 2-D environment are presented. Furthermore, the usefulness of the approach for robot arm path planning in configuration space is demonstrated.

1 Introduction

Path planning is a well-known problem in robotics. In its simplest form, robot path planning is concerned with finding a collision-free motion path between a given starting point S to a fixed goal position G in a known obstacle congested environment or space. Research on path planning can be classified into two major categories. In the first category, a polyhedral object, such as a robot vehicle or a robot arm, is to be moved in an environment filled with fixed polyhedral obstacles. Obstacle sizes, positions and orientations etc. are assumed to be a priori known. A collision-free path is generated through search in a graph formed out of connectivity relations of obstacles or patches of free motion space. Numerous approaches to this problem are reported in the literature; see for example [1], [2]. They prove to be computationally intensive and most suitable for off-line application with robots operating in an unchanging, time-invariant world.

In the second category various types of physical analogies are employed for path planning purposes. Representatives of this class are the family of potential field methods, the wave propagation and the diffusion equation strategies. An early example of the potential field methods is the charge distribution model proposed by Khatib [3]. Some of the deficiencies of this method [4] are avoided by more recently reported potential field models satisfying Laplace's equation together with Dirichlet [5] or Neumann boundary conditions [6]. These methods are ideally suited to implementation on massively parallel distributed processor systems, either digital [5] or analogue [6]. A wave propagation strategy to optimal path planning in metric configuration space is treated in [7]. An unsteady diffusion equation strategy, as first described by Ritter [8], shows close relationship to the before mentioned Laplace equation methods, particularly in case of path planning in time-invariant environment. This is an assumption made by most path planning approaches reported in the literature to date.

Although not explicitely mentioned in [7], the unsteady diffusion equation strategy provides an unexploited potential for application to high-speed robot path planning in case of moving and/or changing obstacles as well as for vehicle evasion or pursuit applications. Furthermore, map and sensor-based planning techniques can be easily incorporated in one approach, leading to what we denote as "dynamic path planning".

This article presents the theoretical background and the algorithms of the dynamic path planning approach. It discusses the possibility of implementing the algorithms on a massively parallel distributed processor facility. Application of dynamic path planning to an interesting class of robot vehicle problems and to robot manipulators are reported.

2 Path Planning Strategy

2.1 Physical analogy

Assuming that a robot can be considered a point moving in a congested environment, the following physical analogy can be employed for path planning purposes. Goal point G of the robot's collision-free motion path is considered the location of a virtual source which emits some gaseous substance, e.g. a scent or a perfume. While concentration at point G is kept constant, substance diffuses steadily into the surrounding space. Complete and instantaneous absorption of any substance reaching obstacle points as well as natural and artificial space boundaries is assumed, resulting in concentration values equal to zero at the corresponding point sets. As a result of the diffusion process, a concentration distribution develops over time and space. In its equilibrium state, the distribution shows a monotonously decreasing concentration along a path between the goal point G, i.e. the point of maximum scent concentration, and an arbitrary robot starting point S within the substance filled space. The unique path between S and G can be easily found by following a steepest ascent strategy. The physical concepts can be formalized as follows.

2.2 Diffusion process

The diffusion process is modelled by an unsteady or dynamic diffusion equation of the form

$$\frac{\partial u}{\partial t} = a^2 \cdot \nabla^2 u - g \cdot u \tag{1}$$

with boundary functions
$$u(t; \underline{x}_G) = 1, \quad \underline{x}_G \in \Omega \subset R^n \tag{2}$$
$$u(t; \underline{x}_O) = 0, \quad \underline{x}_O \in \delta\Omega \subset R^n \tag{3}$$

and an initial function $u(0; \underline{x}) = 0$ for $\underline{x} \neq \underline{x}_G$.

$u(t; \underline{x})$ denotes the concentration distribution function over time and Euclidean space with $\underline{x} = [x, y, ...]^T \in \Omega \cup \delta\Omega \subset R^n$. Ω defines a closed connected region, the workspace, including goal point $\underline{x}_G$. The boundary set $\delta\Omega$ is formed by obstacle and free workspace boundaries. $\nabla^2 = \partial^2/\partial x^2 + \partial^2/\partial y^2 + ...$ represents the Laplace operator; a^2 and $g \geq 0$ are the diffusion constant and the substance disintegration rate, respectively. X and $\delta\Omega$ are fixed sets for a time-invariant workspace, while a changing environment implies $\Omega(t)$ and/or $\delta\Omega(t)$.

2.3 Simulation of diffusion process

Numerical methods are very well suited to simulation of the diffusion equation model (1) to (3). For simplicity, the following discussion will treat 2-D diffusion, although the method can be easily extended to 3-D problems.

Let $u(t_k; x_i, y_j)$ be a discrete-time regular sampling of $u(t; x, y)$ on a uniform rectangular grid. By application of standard finite difference methods, the following time and space discretized analogon of (1) can be obtained for a grid point or node $r = (i, j) \in \Omega' \cup \delta\Omega'$.

$$(u_{k+1;r} + u_{k;r})/\tau = a^2/h^2 \cdot \sum_{\substack{m=1 \\ m \in N(r)}}^{M} (u_{k;m} - u_{k;r}) - g \cdot u_{k;r} \quad (4)$$

τ denotes the time step size and h grid width. $N(r)$ and M represent the set and number of neighboring nodes of r considered for an approximation of the Laplace operator, Fig. 1.

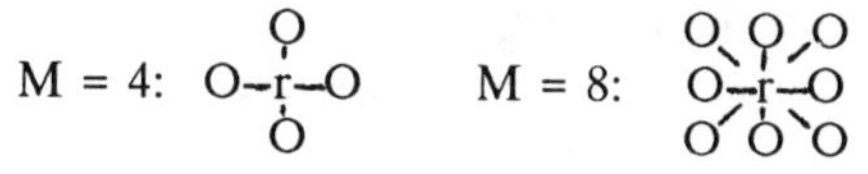

$$M = 4: \quad \begin{matrix} & \circ & \\ \circ & -r- & \circ \\ & \circ & \end{matrix} \qquad M = 8: \quad \begin{matrix} \circ & \circ & \circ \\ \circ & -r- & \circ \\ \circ & \circ & \circ \end{matrix}$$

Fig. 1: Neighbourhoods $N(r)$ for approximating the Laplace operator

For $g = 0$ fastest convergence of the difference scheme is achieved by selecting parameter $a^2 = h^2/\tau \cdot M$, resulting in the simple recursion-relation form of (4)

$$u_{k+1;r} = 1/M \cdot \sum_{m=1}^{M} u_{k;m} - \tau \cdot g \cdot u_{k;r} . \quad (5)$$

Scheme (5) shows oscillatory convergency, which should be avoided when applied to path planning in case of a time-varying workspace and/or goal point. Reducing a^2 to $a^2 = h^2/(1+M)\tau$ leads to the non-oscillatory recursion-relation

$$u_{k+1;r} = 1/(1+M) \cdot (\sum_{m=1}^{M} u_{k;m} + u_{k;r}) - \tau \cdot g \cdot u_{k;r} . \quad (6)$$

Based on (6) and the boundary conditions defined by (2) and (3) the following simulation algorithm can be formulated for the unsteady diffusion process:

$$u_{k+1;r} = \begin{cases} \text{see eq. (6)}, & \text{for } r \in \Omega' \\ 0, & \text{for } r \in \delta\Omega' \\ 1, & \text{for } r = r_G \end{cases} \quad (7).$$

$u_{0;r} = 0$, for $r \neq r_G$; $k = 0, 1, 2, ...$

This algorithm updates for each time step k the distribution function $u_{k;r}$. For k large and Ω' and $\delta\Omega'$ fixed, the function converges to its equilibrium state $u_{\infty;r}$. $u_{\infty;r}$ shows a single peak at r_G with a monotonous slope leading to any other point r in the workspace. For $g = 0$, $u_{\infty;r}$ can also be interpreted as an electrical potential function with the goal point on constant potential $u_{k;rG} = 1$ and boundary or obstacle points on ground, i.e. $u_{k;r0} = 0$.

2.4 Steepest ascent technique

Based on the above-mentioned properties of the distribution function, a steepest ascent path, following the gradient of $u_{\infty 0;r}$ from an arbitrary starting point r_S to the goal point r_G, can be computed.

A simple *ascent algorithm 1*, using information immediately available from the diffusion simulation, proceeds as follows:

- pick the starting point r_S;
- in the set of M=8 immediate neighbors of r_S select grid point r_{P1} with maximum concentration $u_{\infty;r}$;
- continue with the same procedure starting from r_{P1} until r_G is reached.

The shortest path possible is given by the sequence of grid points r_S, r_{P1}, r_{P2}, ..., r_{Pi}, ..., r_{Pn} = r_G. This path proves to be comparatively rough with a tendency to hit obstacle corners, as shown in the occupancy grid representation of a laboratory environment, Fig. 2(a). Black strokes at free space grid points indicate computed gradient directions.

Smoother paths can be generated by application of a gradient interpolation method during steepest ascent search and by permitting the selection of appropriate path points r_{Pi} outside the set of discrete grid points r. *Ascent algorithm 2* proceeds as follows:

- pick the starting point r_S;
- evaluate gradient direction at r_S based on second order polynomial interpolation of $u_{\infty;r}$ over r_S and its M=8 immediate neighbors;
- select $r_{P1} = r_S + \eta \cdot$ grad $u_{\infty;rS}$;
- evaluate gradient at grid point next to r_{P1} as mentioned above;
- select $r_{P2} = r_{P1} + \eta \cdot$ grad $u_{\infty;rP1}$;
- continue with the same procedure starting from r_{P2} until r_G is reached.

For step size η in the range 0.3 to 0.5·h the sequence of path points r_S, r_{P1}, r_{P2} ....,r_G generated by algorithm 2, leads to quite satisfactory results, as demonstrated in Fig. 2(b). Algorithm 2 is used in the applications reported next.

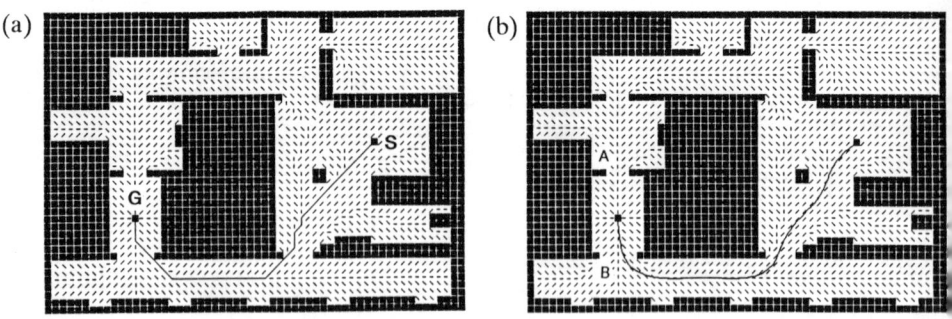

Fig. 2: Path planning in static 2-D environment, occupancy grid representation
(a) steepest ascent algorithm 1; (b) steepest ascent algorithm 2
L=39; h=0.5 m; η=0.4·h.

3 Applications

3.1 Path planning for a robot vehicle

Static environment

Fig. 2(b) shows a basic result of path planning for a fixed and a priori known laboratory environment. The plane is represented by an occupancy grid, consisting of 40 x 58 nodes with grid width h = 0.5 m. Squares and dots denote obstacles and free space, respectively. Obstacles are actually "grown" to cope with the size of the real robot, which is approximated by a circle with radius 0.5 m. The path generated by means of the interpolated gradient scheme indicates various favourable properties, such as:

- smoothness;
- obstacle corners are passed at a safe distance;
- path center line is followed in corridors or regions between obstacles;
- in case of two or more posible path candidates, the path with wider passages or gateways is automatically generated.

Because of the properties of the diffusion process there exists a minimum delay time before path computation from S to G can be actually started. When diffusion simulation is initiated at k = 0 with initial function u(0; $\underline{x}$) = 0, a minimum of L time steps are required before diffusion reaches r_S, i.e. $u_{L;r_S} \neq 0$. L denotes the number of grid nodes passed by the shortest path possible from S to G. L = 39 in the examples shown in Fig. 2. For a time-invariant environment with fixed start and goal point, the following rough estimate for L can be given

$$L = c \cdot \sqrt[n]{n_g}, \quad c = \text{constant} \tag{8}$$

with n and n_g denoting dimension of space R^n and total number of grid nodes, respectively.

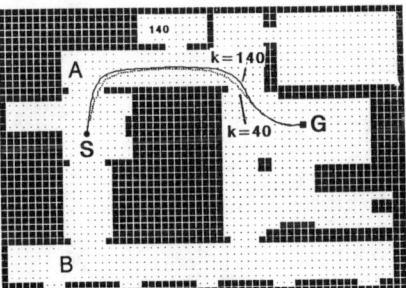

Fig. 3: First path estimate and final path

As soon as diffusion has reached the starting point, a first path estimate can be computed L = 35. Fig. 3 shows the dotted path computed at k = 40 and the full line quasi-equilibrium path computed at k = 140. A comparison of both paths indicates that for k > L only minor path improvements are achieved. In Fig. 3 local path error e_p < 0.1 · h = 0.05 m. The experiment demonstrates that a reasonable first path can already be generated before an equilibrium state has been reached. This feature is one major reason for the high-speed operation of the proposed path planning strategy. It has also a major impact on planning in case of environmental changes.

Changing environment

Let us assume next, that the planning process, i.e. diffusion simulation and ascent computation, is continuously active and that diffusion is more or less in its equilibrium state. At some point of time, passage A in Fig. 2(b) or 3 is partly or completely blocked; the environmental change is made known to the diffusion process through updating of its boundary conditions. Since simulation will continue from its current state before blockage, the blockage at A has only a small effect on the basic shape of the distribution function. An alternative path from S to G passing through B is already generated during the transient phase of the diffusion process, and it is steadily improved during further time steps. No restart of diffusion simulation from zero initial conditions is required, as considered in [5]. This feature is another reason for the high-speed operation of our planning strategy.

Dynamic path planning

Next we will assume that in Fig. 2(b) or 3 the mobile robot at S starts moving with velocity v_R according to the computed path. Its current position in the plane is made known to the diffusion process by inputing from time to time a new start point S' or $r_{S'}$. The change from S to S' will have no effect on the state of the active diffusion process. The planning strategy adapts to the robot's motion by immediately generating an updated path between S' and G. For low-speed robot motions, S(t) and G will always remain connected by an updated path. This will also be the case when G is moving, i.e. G(t), or when both S and G are simultaneously moving within the workspace. An application is shown in Fig. 4, where sample steps of on-line dynamic path planning for a vehicle pursuit problem are presented. A mobile robot R is supposed to follow a vehicle V. If we identify $R \equiv S(t)$ and $V \equiv G(t)$ and if the current positions of both are made known to the diffusion strategy, then a path between R und V will be generated and continuously updated when motion is started. Based on this path information, R pursues V and eventually docks at V, provided that the speed difference $v_R - v_V > 0$. This experiment demonstrates that robot path planning and

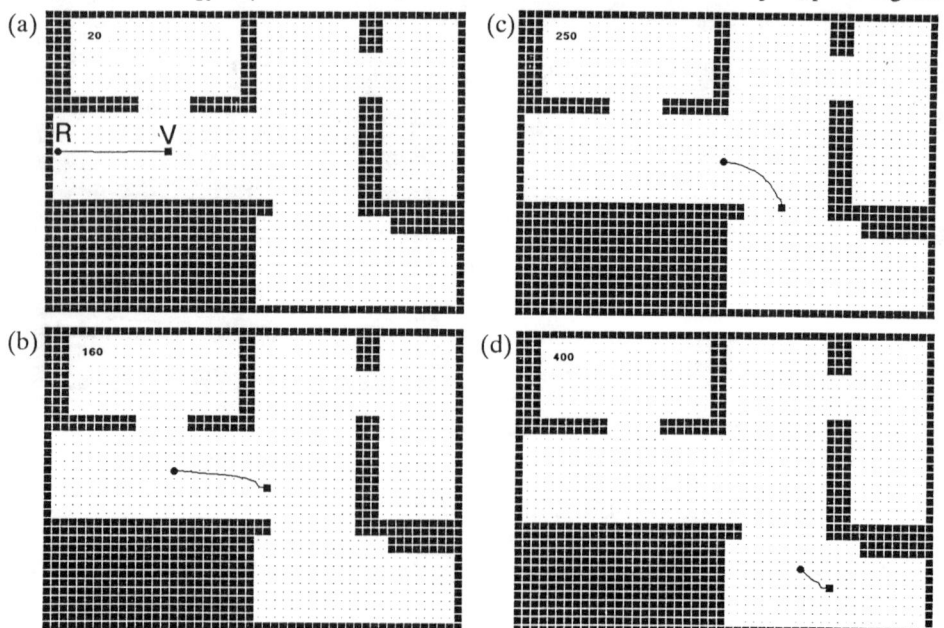

Fig. 4: Path planning for vehicle pursuit problem
R: mobile robot (follower); V: vehicle (leader)

motion become closely interrelated activities, which led us to the notion "dynamic path planning".

Results of an evasion experiment are shown in Fig. 5, where a randomly moving obstacle M, e.g. another extended vehicle, is crossing the path of a mobile robot R heading to a fixed goal point G. Information about current obstacle position and shape is again made known to the planning strategy, either by appropriate sensors onboard R or fixed in the environment. As long as M is approaching the repetitively generated path from R to G, the path will only be slightly modified in order to avoid collision, Fig. 5(a). When M is getting close, Fig. 5(b), a reconfiguration of the path is initiated with the path moving from the lower side of M (Fig. 5(a)) to its upper side (Fig. 5(c)), i.e. M is able to pass R without collision. Passing causes a major disturbance in the distribution function, which even gets into a transient state. As a consequence a temporary "planning uncertainty" may occur. The cause of this uncertainty are the formation of additional local peaks in the distribution function as caused by rapidly changing boundary conditions in the diffusion process. Consequently, the steepest ascent algorithm will temporarily track the closest local peak, instead of the global peak at r_G, and it will produce an incomplete path, as shown in Fig. 5(b). However, after a short settling time, e.g. from k=260 to 270 in Figs. 5 (b) and (c), the uncertainty disappears and a full collision-free new path is generated. Path uncertainty does not cause major problems as long as speed of motion of R and M remains small, compared to the updating rate of the planning strategy; see Section 4 for more details.

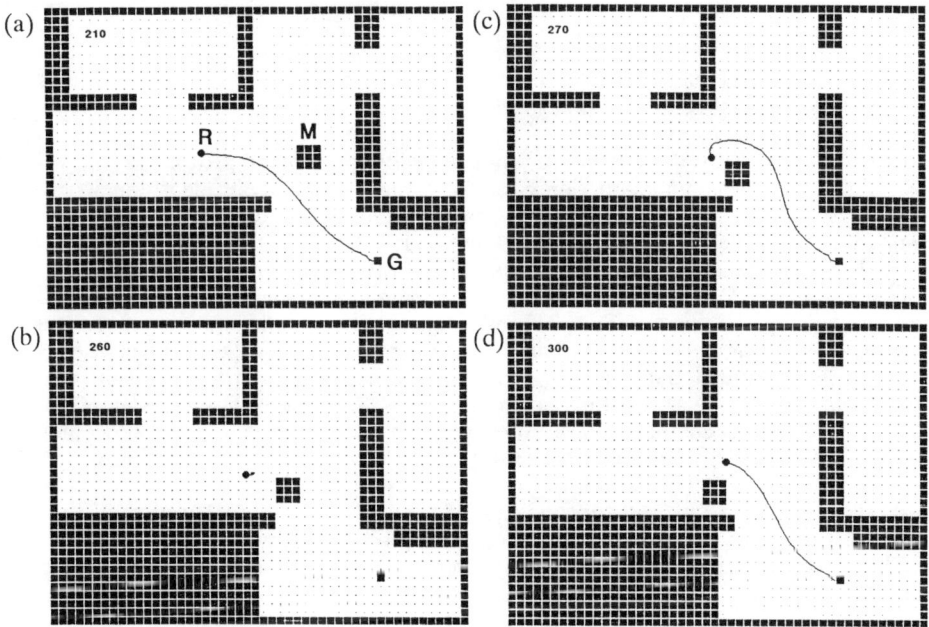

Fig. 5: Moving obstacle M crossing path of R travelling to fixed goal point G

3.2 Path planning for a robot arm

The same path planning strategy can be used for a robot manipulator, provided that the obstacle congested Cartesian task space is transformed first into configuration space. Fig.

6(a) shows as an example a two-degree-of-freedom SCARA type robot within a field of four obstacles. The configuration space together with the planned motion path are presented in Fig. 6(b), while 6(a) shows in addition traces of the resulting collision-free arm motion from S to G in task space.

4. Implementation Issues and Performance

The algorithms described in Section 2.3 and 2.4 can be directly implemented on a parallel processor system. For example, each grid node may be represented by a processor communicating with its immediate M neighbors only. In case of a completely parallel implementation computation time t_c for one time step k of diffusion simulation becomes independent of the total number of grid nodes n_g.

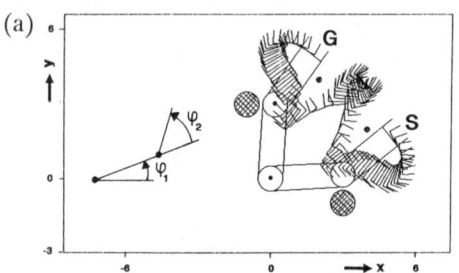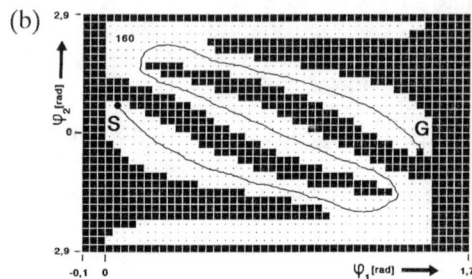

Fig. 6: Path planning for 2-link robot arm
(a) task space; (b) configuration space

However, even with the algorithm implemented on a mono-processor, such as a PC-AT-486, small computation times can be achieved. With the algorithm written in Turbo-Pascal language, including an assembler routine for diffusion simulation, the computation time per time step and per 1,000 grid nodes proves to be in the order of 3.5 msec. This means that for the 2,320 nodes in Fig. 2(a) a total computation time T_c = 8.12 msec per time step k is required. With L = 39 the minimum time delay during initialization of path planning is $T_L = L \cdot T_c$ = 317 msec. In case of dynamic path planning situations, typically K = 10 time steps are required to get a meaningful (without effects caused by planning uncertainty) path update from a continuously running diffusion process simulation. This means for the given examples with sensor-based updating of the occupancy grids that computation of a new path update should take place at a minimum updating interval of $T_u = K \cdot T_c$ = 81.2 msec.

In the given examples velocities of moving objects amounted up to 1 grid width h per 1 updating interval T_u. This data implies a maximum object velocity of v = 0.5 m / 81.2 msec = 6.15 m/sec, which seems to be rather high for the typical environments considered in the examples.

The given numbers demonstrate however the excellent performance and real-time capabilitiy of the proposed planning strategy, even if a mono-processor implementation is chosen.

For the numerical representation and processing of the distribution function $u_{k;r}$ at least 32- or even 64-bit floating point arithmetic needs to be used, since distribution function values decrease very rapidly with distance between r and source location r_G. This disadvantage can be compensated by reformulating the diffusion equation model with a set of Neumann boundary conditions; see also [6].

Lastly, let us discuss how to use parameter g in the diffusion equation model (1). A small positive value given to the substance disintegration constant may accelerate the settling of transients in the diffusion process during dynamic path planning situations. The choice of g needs however to be carefully balanced against possible numerical problems caused by the distance dependent reduction of u values, as mentioned before.

5. Conclusions

The application of an unsteady diffusion equation strategy offers a high-speed method for robot path planning in fixed as well as in time-varying environments. The method allows us to combine map and sensor-based path planning. Furthermore, the artificial separation of path planning and path execution can be eliminated by the proposed dynamic planning approach. It can be applied to various types of mobile robot as well as robot arm problems. Even with the implementation of the related algorithms on a mono-processor, the planning method proves to be highly efficient. A tremendous speed-up of path planning can be achieved through implementation of the simple algorithms on a massively parallel processor architecture. Further enhancements of the path planning strategy are under investigation.

Acknowledgement

The work reported in this paper was partially supported by the Deutsche Forschungsgemeinschaft, as part of an interdisciplinary research project on "Information Processing in Autonomous Mobile Robots (SFB 331)".

References

[1] Lozano-Pérez, T.: "Spatial Planning: A Configuration Space Approach". IEEE Transaction on Computers, 32 (1983) 2.
[2] Kampmann, P.; Schmidt, G.: "Multilevel Motion Planning for Mobile Robots Based on a Topologically Structured World Model". IAS-2, An Int. Conf., Amsterdam (1989), 241-252.
[3] Khatib, O.: "Real-Time Obstacle Avoidance for Manipulators and Mobile Robots". Int. Journal of Robotics Research. 5(1986) 1, 90-98.
[4] Koren, Y.; Borenstein, J.: "Potential Field Methods and Their Inherent Limitations for Mobile Robot Navigation". Proc. IEEE Int. Conf. on Robotics and Automation, (1991), 1398-1404.
[5] Connolly, C.I.; Burns, J.B.; Weiss, R.: "Path Planning Using Laplace's Equation". Proc. IEEE Int. Conf. on Robotics and Automation, (1990), 2102-2106.
[6] Tarassenko, L.; Blake, A.: "Analogue Computation of Collision-Free Paths". Proc. IEEE Int. Conf. on Robotics and Automation, (1991), 540-545.
[7] Dorst, L.; Trovato, K.: "Optimal Path Planning by Cost Wave Propagation in Metric Configuration Space". Proc. of SPIE - The International Society for Optical Engineering. Mobile Robots III, Vol. 1007 (1988), 186-197.
[8] Ritter, H.: "Selbstorganisierende neuronale Karten (Self-Organizing Neural Maps)". Doctoral Dissertation, TU München, 1988.

Path Planning of Transfer Motions for Industrial Robots by Heuristically Controlled Decomposition of the Configuration Space

G. Duelen, C. Willnow
Fraunhofer Institut für Produktionsanlagen und
Konstruktionstechnik (IPK)
Pascalstraße 8-9, 1000 Berlin 10

Abstract: A fundamental part of a fully automated task-level programming system for industrial robots is a path planner for collision free transfer motions. The most common approach is to perform the planning task in the configuration space (C-space) of the robot. The exponential growth of the C-space with the number of joints of the robot makes it practically impossible to investigate the C-space completely for robots with more than three joints. The approach proposed in this paper is based on a hierarchical decomposition of the C-space that allows to concentrate the investigation on the parts necessary to find a solution. The procedure starts with a rough decomposition and consists in constructing hypothetical paths based on estimates about collisions in sub-C-spaces. Non-collision-free parts of the hypotheses are further refined until a solution is found. Tests have demonstrated that the number of collision tests necessary to find a solution for a robot with 5 joints is reduced by the procedure by a factor of more than 10^4.

1 Introduction

Fully automated task-level programming systems for industrial robots have to plan two major kinds of robot motions: on the one hand technology dependent motions like welding paths or assembly motions, and on the other hand transfer motions which move the robot's effector from one position of action to another. This article concerns transfer motions, a subject which has received increasing attention during the last couple of years /1 - 9/.

Most published approaches solve the planning tasks in the joint or configuration space (C-space) of the robot instead of using the real world 3D-space. The C-space has the advantages of being more general by abstracting concrete robot kinematics and of being a representation ready for the use of standard planning methods. But it has also a severe disadvantage:

Let $Q_i = [q_{i\,min}, q_{i\,max}] \subset \Re$ be the joint range of the i-th joint between the joint limits $q_{i\,min}, q_{i\,max}$ of a robot with N joints. Then

$$Q = \underset{i=1}{\overset{N}{\times}} Q_i$$

is the set of all possible robot poses, the so-called C-space.

Unfortunately, there is no general procedure known to transform the obstacles in the robot's environment from the real world 3D-representation to the C-space. For this reason, procedures based on the C-space discretize it, transform each granulum to real world 3D-space, and do the collision test there. But the complexity of this method is

$$O(|Q|) = O\left(|Q_i|^N\right)$$

The number of collision tests that have to be performed grows polynomial with the number of granuli per joint $|Q_i|$ and with the number of joints N even exponentially.

Approaches to reduce the number of collision tests by smarter representation of the C-space have been made by /3/ and /8/. Their approaches summarize regions of the C-space enabling them to achieve better resolutions, but both approaches still investigate the C-space completely and for that reason they are not able to pass the limit of 3 joints. /5/ pass this limit. The main part of the path is still planned for 3 joints, but in difficult situations (start and goal) they plan for 5 joints in a limited C-space. /2/ propose a method designed for a robot with 7 joints. They achieve impressing results by strongly focusing the attention (calculation capacity). But in difficult situations the procedure loses the perspective of the whole situation.

The approach proposed here can be placed between the mentioned approaches. It summarizes regions of the C-space, it intensifies investigations in difficult situations and concentrates on regions of the C-space promising a solution. The method starts with a rough description of the C-space and consists in the construction of hypothetical paths based on estimates about collisions in sub-C-spaces. Non-collision free parts of the hypotheses are further refined until a solution is found. The requirements of local concentration and global perspektive are controlled by heuristics.

2 Hierarchical decomposition of the C-space

To simplify the subsequent discussion we will assume the joint limits of the joints of the robot to be normalized to intervals between 0 and 1:

$$Q = [0,1]^N, Q_i = [0,1], i = 1,...,N.$$

The C-space and the sub-C-spaces will be represented by the position of their centre p and their breadth b. Then the complete C-space s_C on the highest level of decomposition is denoted as

$$s_c = (p_c, b_c) = ((0.5, 0.5,..., 0.5), 1.0).$$

The chosen method of decomposition divides a C-space one time in each coordinate, creating 2^N new sub-C-spaces. These sub-C-spaces are called children of the parent C-space and are defined for a C-space $s = ((p_1,..., p_N), b)$ as

$$\text{children}((p_1,...,p_N), b) = \{((q_1,..., q_N), b/2) | q_i \in \{p_i \pm b/4\}\}.$$

Fig. 1a,b shows decompositions for N=2 and N=3. R is a parent C-space and

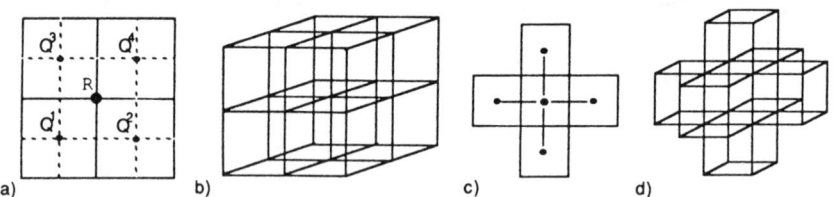

Fig.1: Decomposition of C-spaces with N = 2 (a) and N = 3 (b) and the even level neighborhood N1 for N = 2(c) and N = 3(d)

$Q_1,..., Q_4$ are child C-spaces. The definitions determine the set of all C-spaces S as

$$S = \{s | s = s_c \vee s \in \text{children}(r), r \in S\}.$$

The process of decomposition starts with the complete C-space s_c and consists of the decomposition of s_c and of C-spaces which have been created by preceding decompositions. Each newly created C-space will then be tested for collisions and is available for the process of generating hypotheses.

In this way the process of decomposition successively constructs the set of all investigated C-spaces $S^I \subset S$. The process of generating hypotheses will take place on the most elementary, not further decomposed C-spaces $S^E \subset S^I \subset S$. The relation between S^E, S^I and S is shown in Fig. 2a for a 1-dimensional C-space.

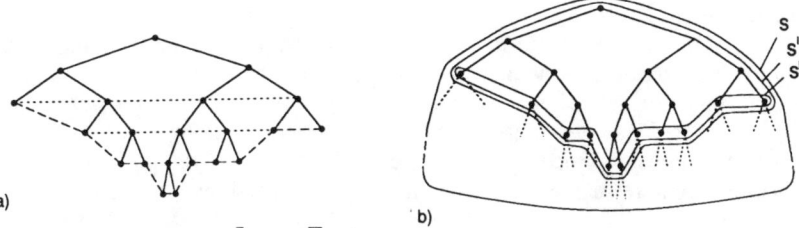

Fig. 2: a) The relation of S, S^I and S^E in a decomposed C-space for N = 1 b) The relation between tree structure of C-spaces and neighborhood N1(...) and N2(---)

The neighborhood between C-spaces is implicitly given by their tree structure, but has to be made explicit for efficient processing. This is done by first constructing the even level neighborhood N1 which defines neighbors on single levels of decomposition. Based on it the level changing neighborhood N2 will be constructed which will be used for the generation of hypotheses. The chosen neighborhood N1 is shown in Fig. 1c,d for N=2 and N=3. Its relation to the tree structure of spaces is shown in Fig. 2b.

It is obviously not very meaningful to investigate a region of the C-space in much detail while regions close to it are only roughly investigated. But this can happen up to now as shown in Fig. 3a. To avoid such discontinuities of attention, it should be continuously distributed. Because of the discrete nature of the levels of decomposition it can only be called quasi-continuous. A suitable rule to maintain that is: a space is decomposed with quasi-continuous distribution of attention if each decomposed space has all N1-neighbors. A decompostion corresponding to the decomposition in Fig. 3a which meets the demand of quasi-continous distribution of attention is shown in Fig. 3b. The quasi-continuous distribution

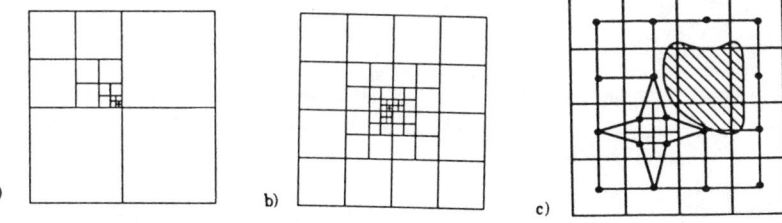

Fig 3.: Decomposition of a C-space with discontinuous (a) and quasi-continuous (b) distribution of attention c) Mapping of a decomposed C-space to a graph

has on one hand heuristic aspects because it maintains the requirement of global perspective and influences in this way the generation of hypotheses. On the other hand it serves as an invariant that simplifies the level changing neighborhood N2. Because now we know for each investigated elementary C-space $s \in S^E$ that its N2-neighbor is
- on the same level of decomposition or

- on the next lower level or
- on the next higher level.

The relation of the N2-neighborhood to the N1-neighborhood and the tree structure of C-spaces is shown in Fig. 2b.

3 Collision Testing

It is the task of the collision tester to support the path planner with propositions about the passability of C-spaces. But these propositions may not only be of a 2-level range ("collision" or "no collision") such as the ones given by a collision tester for single configurations. The collision tester has to be able to make propositions about complete sub-C-spaces which requires intermediate values like "maybe passable" or "probably passable". Sub-C-spaces that require such estimates about passability are shown in Fig. 4.

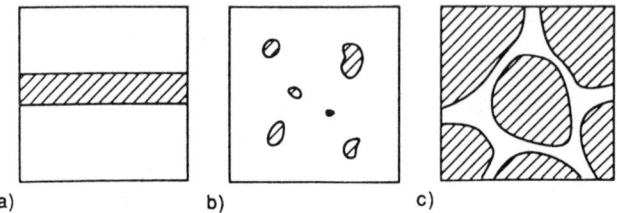

Fig. 4: a) C-spaces requiring multi-valued collision testing functions

The path planner described here does not require a spezial procedure for collision testing, but defines an interface that permits to adapt suitable collision testers. The kernel of this interface are the predicates

full: S → BOOL
free: S → BOOL
passable: S → BOOL,

with the properties that for $s \in S$

full (s)   → ¬passable(s)
free (s)   → passable(s)

are true.

Given these predicates, we can define the set of passable C-spaces S^P where the generation of hypotheses will take place, and the set of free C-spaces S^F which has to be a superset of the sought collision free path

$$S^P = \{S \in S^E | passable(s)\} \qquad S^F = \{s \in S^E | free(s)\}$$

To supply the path planner with more detailed information about the passablility of C-spaces, we add a further function to the interface

passablility: $S \rightarrow [0,1]$.

This function has to make estimates about the passability of C-spaces to give the planner the ability to choose the more passable C-spaces for a hypothesis when the occasion arises. For this function

full (s)   → passability (s) = 0
free (s)   → passability (s) = 1

for a given C-space $s \in S$ have to be true.

4 Generating Hypotheses

The chosen procedure for generating (hypothetical) paths is based on graph theoretical minimum-path methods. To do this, the set of elementary passable C-spaces and their N2-neighborhood are mapped to the nodes N and arrows A of graph $G = (N, A)$. Furthermore, a cost function c is defined.

The set of nodes N is given by the set of passable C-spaces S^P
$N = S^P$.

The neighbors N_n of a node n are given by the N2-neighborhood of C-spaces
$$N_n = N2(n) \cap S^P$$
where the function $N2: S \to Q(S)$ gives the N2-neighbors of a C-space.

The set of arcs of the graph is now defined as
$A = \{(n,m) | n \in N, m \in N_n\}$.

Fig.3c shows an example for the mapping of a decomposed C-space to a graph. The primarily relevant influences on the cost function $c: S \times S \to [0,1]$ are the breadth of the C-spaces and their passability. A simple cost function for an arrow $a = (v,w)$ can then be defined as

$$c(v,w) = c((p_v, b_v), (p_w, b_w)) = (b_v + b_w + \text{passability}(v) + \text{passability}(w))/4$$

In some cases the influence of the passability of C-spaces in the neighborhood led to an improvement in computation time. Regarding the level of decomposition allows to control the local concentration of the procedure.

Based on these definitions it is now possible to use one of the standard methods for minimal paths (eg the method of Dijkstra) to construct hypothetical paths. In general they consist of assigning each node a minimal distance to the goal node and moving then in the obtained distance field from the start to the goal node. Unfortunately, the standard methods lack in the ability to work on a changing graph such as the graph of C-spaces does in the case of decomposition. The procedure for path planning described here was significantly improved by developing a method to update the distance field locally instead of creating it newly after each step of decomposition.

5. Path Planning

Now we can define the process of path planning consisting of 9 steps:

1. Decompose the C-spaces containing start and goal configuration until they are collision-free or reach a predefined finest level. Assign each elementary C-space a passability value.

This step creates an initial decomposition of the C-space for generating hypotheses.

2. If the start or the goal C-space is not collision-free, then STOP.

In this case the procedure can not guarantee that a path is collision-free in the start or goal position. For that reason the procedure has to stop.

3. Generate the initial distance field.

If the initial distance field assigns the start C-space an infinite value, the start and goal C-spaces are unconnected parts of an incoherent graph. Both parts are divided by FULL C-spaces, and there exists no collision-free path.

4. If the distance value of the start C-space is infinite, then STOP.

Now we can build a first hypothetical path from start to goal.

5. Build a hypothesis.

The hypothesis consists of a sequence of connected C-spaces. If they are all collision-free, then the sought collision-free path is found.

6. If all C-spaces of the hypothesis are collision-free, then STOP. Report "path found".

If not all C-spaces of the hypothesis are collision-free, the not collision-free sections have to be further investigated.

7. Decompose all not collision-free C-spaces of the hypothesis.

This step leads to a changed graph with a probably changed distance field. For this reason, the distance field has to be updated.

8. Update the distance field.

The new distance field will now lead to new knowledge about the structure of the C-space. The cycle of generating a hypothesis and decomposition of the C-space can begin again.

9. Goto 4.

The further behaviour of the path planner depends on the decisions made by the generation of hypotheses. It can decide to refine the current hypothesis or to give it up and to look for alternatives.

Fig. 5 shows the solution steps for two example tasks for $N=2$. The first task in Fig. 5 consists of finding a path round an obstacle without colliding with a second one. Fig. 5a shows the task, Fig 5b shows the initial decompostion, and Fig. 5c shows the first hypothesis. Figs. 5d-f show the intermediate decompositions and hypotheses. Fig. 5g shows the final decompostitions and the sought collision free path. Note how the procedure keeps the granularity as rough as possible and prefers free space.

The second task is more difficult, because the path planner has to find the narrow chanels that lead round a big central obstacle. Fig. 5h shows an intermediate decompostions and the relating hypothesis. Fig. 5i shows the final decomposition with the resulting path.

The solution of the first task took about 2sec CPU-time on a micro-VAX 3200. The finest C-space is on the 4th level of decompostion, resulting in a granularity of 16 per joint and a problem size of 256. But the path planner did only investigate 49 C-spaces (about 19%).

The solution of the second task took about 32sec CPU-time. The finest C-space is on the 7th level of decomposition, resulting in a granularity of 128 per joint and a problem size of $1.6 * 10^4$. But the planner did only investigate 389 C-spaces (2.43%).

As we can see, the economization of investigation and computation increases with the size of the problem space. This effect will continue for the application given in the following section.

Fig. 5: Solution steps of example tasks for N =2.

6 Application

The Robot Technology Experiment ROTEX of the D2-Mission of the German Space Lab which will deliver first experiences in the operation of industrial robots for automation in space will be a practical application for the path planner. The task of the path planner will be to transfer the robot in case of program abortion to a defined position for restart.

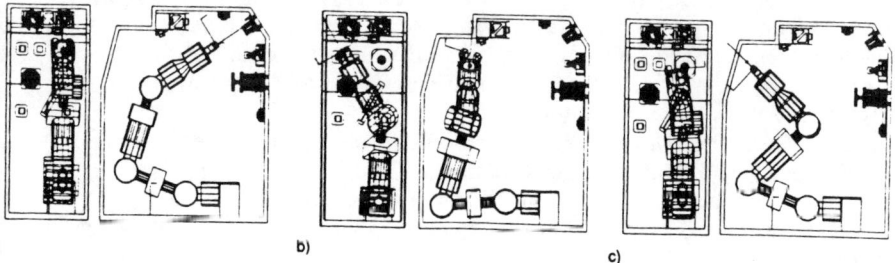

Fig. 6: The ROTEX-robot in its cell in start-(a), intermediate-(b), and goal position (c) of a planned task.

The main difficulty of this task lies in the relatively small size of the box containing the robot (see Fig.6) which hardly allows the robot to move, and the high amount of collision tests that have to be performed to investigate a single sub-C-space. The robot was modelled by one sphere and seven cylinders and its environment was modelled by 37 objects.

Calculation times for planning tasks in this environment have been measured between 17min and 1h 54min. Fig.11 shows the task to turn joint 4 of the robot from a negative value (Fig. 6a) to a positive value (Fig.6c). Fig. 6b shows the difficult intermediate position.

The calculation time to solve the task was 1h 54min on a micro-VAX 3200. This was about the same time it took a human programmer to solve the task on first trial. The discretion of the C-space was up to 64 granuli resulting in a problem size of about 10^9. But the procedure did only investigate 7521 C-spaces in trying 11 hypotheses which lead to an economization of more than 10^4.

7 Conclusion

The described procedure for path planning for industrial robots is based on a hierarchical decomposition of the C-spaces that is controlled by heuristically generated hypothetical paths.

It is able to reduce the amount of collision tests that have to be performed during the search by a factor of more than 10^4 for a robot with 5 joints. By this it is possible to pass the limit of 3 joints and to maintain acceptable resolution and calculation time.

Further improvements can be achieved by stronger focusing decomposition methods. Current research concentrates on the ability of the system to save the already made experience with a robot cell and to reuse this knowledge in subsequent planning tasks.

8 References

/1/ P. Adolphus, D. Nafziger: Schnelle kollisionsvermeidende Bahnplanung im Konfigurationsraum mit entfernungsfeldern; Robotersyteme 6, 1990.

/2/ Pierre E. Dupont, Stephen Derby: An Algorithm for CAD-Based Generation of Collision-Free Robot Paths, Proceedings of the NATO Advanced Research Workshop on CAD-Based Programming for Sensory Robots; Il Chiocco, Italy, July 4 - 6 1988.

/3/ B. Faverjon: Obstacle Avoidance using an Octree in the Configuration Space of a Manipulator; IEEE Proceedings International Conference on Robotic Atlanta March '84.

/4/ W. Gerke: Collision-Free Control of Industrial Robots using Dynamic Programming; International Conference Control '85 IEEE Conference Publication No 252 July '85.

/5/ T. Hasegowa, H. Teresaki: Collision Avoidance: Divide-and-Conquer Approach by Space Characterisation and Intermediate Goals; Transaction on System, Man and Cybernetics, Vol. 18. No 3, May/June 1988.

/6/ M. Herman: Fast, Three-Dimensional Collison-Free Motion Planning, Proceedings of the IEEE International Conference on Robotics and Automation, San Francisco 1986.

/7/ K. Hoermann: A Cartesian Approach to Findpath for Industrial Robots, NATO Advanced Research Workshop on Languages for Sensor Based Control in Robotics, Castelveccio, Italy 1 - 5 Sept. '86.

/8/ T. Lozano-Perez: A Simple Motion Planning Algorithm for General Robot Manipulation; IEEE Journal of Robotics and Automation Vol. RA-3, No 3, June '87.

/9/ P.W. Verbeck, L. Dorst, B.J. H. Verweer, F. C. A. Groen: Collision Avoidance and Path Finding Through Constrained Distance Transformation in Robot State Space; Intelligent Autonomous System, International Conference, Amsterdam '86.

Planning and optimization of geometrical trajectories inside collision-free subspaces with the aid of high order Hermite splines

T.A. Rieswijk, P. Schalkwijk and G. Honderd

Delft University of Technology,
Faculty of Electrical Engineering, Control Laboratory.
Mekelweg 4, P.O. Box 5031, 2600 GA Delft. The Netherlands.

Abstract

The work, described in this paper, is performed as part of a collision-free motion-planning system for two cooperating robots in an assembly cell. The robots are of different types, an anthropomorphic type and a scara type. This is further described in [1].

This collision-free motion-planning is part of an umbrella project: the Delft Intelligent Assembly Cell (DIAC) -project. The aim of the DIAC project is to obtain within four years a reliable working demonstration of a flexible assembly cell.

In this paper an algorithm is introduced for the planning of optimally shaped trajectories in joint-space through complicated collision-free subspaces. For this purpose a new approach on the area of robot-trajectory planning is introduced using Hermite splines.

Problem description

The collision-free movements are planned in a high dimensional joint-space with one dimension for each axis of the robots. The motion-planning is performed in two phases:

Phase 1: search for a collision free path for a generalized n-dimensional robot (for a 4D scara + a 6D anthropomorphic robot n=10). The search is performed by using a swept volume approach in workspace combined with a binary search in joint space, as described in [2]. The result of this is a sequence of n-dimensional boxes in joint space as shown in figure 1 with ϕ_i for joint i. The boundaries of each box represent transition states for both robots that have to be passed synchronously.

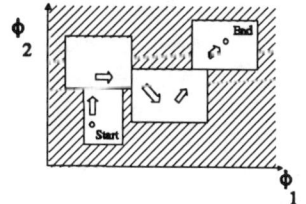

Figure 1, 2D-example of collision-free area

Phase 2: planning of a collision-free trajectory in the combined joint space, describing the movements of all joints as a function of time. This trajectory must pass through the resultant sequence of collision-free segments of phase 1.

The total trajectory-planning of phase 2 is split into two parts:

Part 1: planning of the collision-free trajectory in the combined joint space as a function of one common variable: λ. This part of the total movement planning is described in this paper.

Part 2: optimization of the relation of the common variable λ with time as described in [3].

The planned trajectory, which will be the result from part 1, must fulfil the following demands:
- restriction of the total trajectory inside the complex collision-free area.
- the movements of all axes must be described as functions of one variable: $\underline{\phi}(\lambda)$. In this way the movements of all axes are synchronized.
- continuity in position, first and second derivative to λ : respectively $\underline{\phi}(\lambda), \delta\underline{\phi}(\lambda)/\delta\lambda$ and $\delta^2\underline{\phi}(\lambda)/\delta\lambda 2$. This is important for the planning of a time-dependent trajectory with continuous: $\underline{\phi}(t), \delta\underline{\phi}(t)/\delta t$ and $\delta^2\underline{\phi}(t)/\delta t^2$.
- possibility of manipulation and optimization of the shape in jointspace.

To fulfil these demands the trajectory-planning is separated in

1. placement of a series of N fixed through or via-points: $\underline{p}_j$ (j=1..N), which will be called control-points.

2. planning of parametric curves along or through these control-points. Through usage of these parametric curves, which will be called spline-functions, the amount of parameters, describing the movement, can be limited. Each degree of freedom is described as a separate spline-function. The notation $s(\lambda)$ will be used for a general spline-function.

1. Introduction splines

As described in [4] spline-functions can be divided in 2 separate groups: splines with infinite and finite support. The most frequently used representation for splines with infinite support is the truncated power representation as given in (1).

In (1) q_K is a polynomial of degree K (<=m). The function $s(\lambda)$ passes along or through all knot-values y_i for $:\lambda = \lambda_i$ (i=1..N). The function $s(\lambda)$ is continuous through its (m-1)-th derivative with respect to λ for $\lambda_1 \leq \lambda \leq \lambda_N$. All values c_i (i=1..N) and all parameters of $q(\lambda)$ are influenced, when one values of y_i is changed. This property is called infinite support. Due to this infinite support long series of knot-values can easily lead to very ill-conditioned linear systems causing numerical instability.

$$s(\lambda) = q_K(\lambda) + \sum_{i=1}^{N} c_i (\lambda - \lambda_i)_+^m \qquad (1)$$

$$\lambda_+^m = \begin{cases} \lambda^m & \lambda > 0 \\ 0 & \lambda \leq 0 \end{cases}$$

$$s(\lambda) = F(\{\lambda_{i-k1}, y_{i-k1}\}, \{\lambda_{i-k1+1}, y_{i-k1+1}\}, ..., \{\lambda_{i+k2}, y_{i+k2}\}) \quad \text{for } \lambda_i \leq \lambda \leq \lambda_{i+1} \qquad (2)$$

The second type of splines has parameters which are calculated on basis of a limited subset (from i-k1 to i+k2 in (2)) and therefore has a finite support. Due to this limited support this type of splines doesn't suffer from the disadvantage of possible numerical instability.

On the area of spline-methods with a finite support different approaches can be distinguished. To fulfil the demand of restriction inside a collision-free subspace, some types are especially suitable: B-splines and Hermite splines.

B-splines are approximating splines with minimal support. As described in [5] a B-spline $s(\lambda)$ of m-th order is continuous through its (m-1)-th derivative to λ and is locally dependent on m+1 supporting control-points. A Convex Hull around each subset of m+1 control-points limits the B-spline. Two different types of B-splines can be distinguished: rational and non-rational B-splines.
- Non-rational B-splines have a very smooth shape, which makes this type of splines suitable for robot-trajectory-planning as shown in [6]. A disadvantage however is that they can only be influenced by the places of the control-points which in our case are strongly restricted by the demand to keep their Convex-Hull within the prescribed collision-free area.
- Rational B-splines offer an extra attraction-factor in each control-point to influence the spline, but suffer from a more complex parametrization containing $1/q(\lambda)$-type of functions and from the fact that the influence of these attraction-factors is limited and sometimes quite unpredictable.

Hermite splines don't suffer from the mentioned disadvantages, which occur when B-splines are used, and therefore seem more suitable for our purpose.

2. Trajectory-planning with Hermite splines

$$f_j(x) = \sum_{k=0}^{5} a_{k,j} x^k \quad (j=1..6)$$

$$\phi(\lambda) = \sum_{j=1}^{6} f_j(\lambda - \lambda_i) \underline{w}_{j,i} \quad \text{for } \lambda_i \le \lambda \le \lambda_{i+1} \quad (3)$$

$$\phi(\lambda_i) = \underline{p}_i \quad \phi(\lambda_{i+1}) = \underline{p}_{i+1}$$

Hermite splines are interpolating splines with a finite support, which are calculated from a weighted summation of a small set of standard-functions (f_j in (3)). Between every pair of control-points the same set of functions $f_j(x)$ is used within the same fixed interval for x. In our case the fixed interval: $0 \le x \le 1$ is chosen with $x = \lambda - \lambda_i$ for $\lambda_i \le \lambda \le \lambda_{i+1}$.

Hermite splines have a relatively high degree (2*m-1 for continuity through the (m-1)-th derivative to λ) but offer with the weight-factors $\underline{w}_{j,i}$ the possibility of a more direct influence on the geometrical shape than B-splines. For continuity up to $\delta\phi^2/\delta\lambda^2$ a summation of six Hermite functions of degree 5 has to be used as shown in formula (3) and suitable choices for the weight-factors have to be made.

To simplify the continuity-demands to $\underline{w}_{j,i}$ and the shape-manipulation suitable choices of the boundary-values of standard functions $f_j(x)$ (j=1..6) are made.

	j=1	j=2	j=3	j=4	j=5	j=6
$f_j(0)$	1	0	0	0	0	0
$f_j(1)$	0	1	0	0	0	0
$\delta f_j(0)/\delta\lambda$	0	0	1	0	0	0
$\delta f_j(1)/\delta\lambda$	0	0	0	1	0	0
$\delta^2 f_j(0)/\delta^2\lambda$	0	0	0	0	1	0
$\delta^2 f_j(1)/\delta^2\lambda$	0	0	0	0	0	1

Table 1, Boundary-values for Hermite-functions

In table 1 the suitably chosen boundary-values of the six standard functions are given, which will result in the following set of standard functions:

$$f_1(x) = -6x^5 + 15x^4 - 10x^3 + 1 \qquad f_4(x) = -3x^5 + 7x^4 - 4x^3$$
$$f_2(x) = 6x^5 - 15x^4 + 10x^3 \qquad f_5(x) = -\frac{1}{2}x^5 + \frac{3}{2}x^4 - \frac{3}{2}x^3 + \frac{1}{2}x^2 \quad (4)$$
$$f_3(x) = -3x^5 + 8x^4 - 6x^3 + x \qquad f_6(x) = \frac{1}{2}x^5 - x^4 + \frac{1}{2}x^3$$

2.1 Calculation of the weight-vectors

From (3) and (4) the demands to the weight-vectors $\underline{w}_{j,i}$ for continuity of the spline through its second derivative to λ can now be formulated in (5).

For derivation of the weight-vectors the objective influence on the robot-trajectory must be formulated. In [7] length and curvature are mentioned as important criteria to find minimum-time geometrical robot-trajectories. For a proper trajectory-optimization we must find the weight-vectors, which directly influence these criteria in a predictable way.

$$\phi(\lambda_i) = \underline{w}_{2,i-1} = \underline{w}_{1,i}$$
$$\frac{\delta\phi(\lambda_i)}{\delta\lambda} = \underline{w}_{4,i-1} = \underline{w}_{3,i} \quad (5)$$
$$\frac{\delta^2\phi(\lambda_i)}{\delta\lambda} = \underline{w}_{6,i-1} = \underline{w}_{5,i}$$

Calculation of $\underline{w}_{1,i}$ and $\underline{w}_{2,i}$

In order to guarantee interpolation the expressions for $\underline{w}_{1,i}$ and $\underline{w}_{2,i}$ directly follow from (3) and (4) and are described in (6). The weight-vectors $\underline{w}_{j,i}$ (j=3..6) can be used to manipulate the shape of the resulting curve.

$$\underline{w}_{1,i} = \phi(\lambda_i) = \underline{p}_i$$
$$\underline{w}_{2,i} = \phi(\lambda_{i+1}) = \underline{p}_{i+1} \quad (6)$$

In [8] three types of shape-manipulating parameters are introduced for Hermite splines of degree 3: tension, continuity and bias. For our purpose of optimization of robot-trajectories through usage of parameters, which directly influence length and curvature, the tension-parameters are most suitable because of their most direct influence. Continuity and bias parameters will not be used. Usage of the continuity parameters will disturb the continuity of $\delta\phi(\lambda)/\delta\lambda$ and $\delta^2\phi(\lambda)/\delta^2\lambda$. The bias-parameters are not usable, because they only influence the places were overshoot appears but not the amount of overshoot. Six important properties of the method, as described in [8], are changed or added:

1. Usage of Hermite functions of degree 5 instead of functions of degree 3.
2. Extension from a one-dimensional tension-parameter to a vector with one tension value for each dimension in joint space.
3. Introduction of extra smoothing of the trajectory by proper choices of $\underline{w}_{5,i}$ and $\underline{w}_{6,i}$.
4. Analytic calculation of the set of tension-parameters with approximately the least amount of overshoot.
5. Iterative approximation of the upper and lower limits for all tension-parameters, which still give a collision-free trajectory.
6. Avoidance of undesired loops and cusps in the trajectory.

Calculation of $\underline{w}_{3,i}$ and $\underline{w}_{4,i}$

$$\underline{w}_{3,i} = \frac{1}{2}(I-H)\{(\underline{p}_i-\underline{p}_{i-1})+(\underline{p}_{i+1}-\underline{p}_i)\}$$

$$\underline{w}_{4,i} = \frac{1}{2}(I-H)\{(\underline{p}_{i+1}-\underline{p}_i)+(\underline{p}_{i+2}-\underline{p}_{i+1})\}$$

$$H = \begin{pmatrix} h_1 & 0 & \cdots \\ \cdots & \cdots & \cdots \\ \cdots & 0 & h_n \end{pmatrix} \quad (7)$$

Calculation of the weight-vectors $\underline{w}_{3,i}$ and $\underline{w}_{4,i}$ is performed on basis of [8] with the extension from one tension-parameter to a vector with tension-parameters: $\underline{h}$ on the diagonal of n x n matrix H in (7).

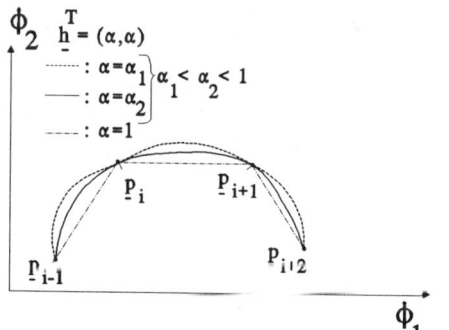

Figure 2, 2D Example of influence of $f_3(x)$ and $f_4(x)$ on $\underline{\phi}(\lambda)$ with $\underline{w}_{5,i}=\underline{w}_{6,i}=\underline{0}$

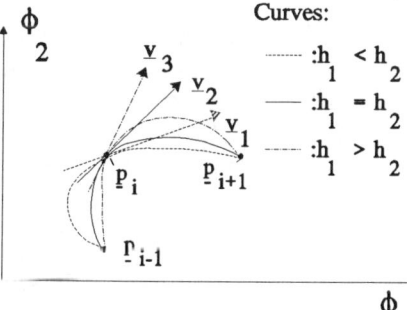

Figure 3, 2D-example of the influence of mutually different h_i-elements on $\underline{\phi}(\lambda)$

When a tension-vector $\underline{h}$ has equal elements $\underline{h}^T=(\alpha,\alpha,..,\alpha)$, the contribution of $(\underline{p}_{i+1}-\underline{p}_{i-1})$ to the local tangent of $\underline{\phi}(\lambda)$ around $\underline{p}_i$ (i=1..N) can be increased through decrease of α. In figure 2 the effect of decrease of α on $\underline{\phi}(\lambda)$ is shown, when $\underline{w}_{5,i} = \underline{w}_{6,i} = \underline{0}$.

When the tension-parameters are mutually different, the tangent in each control-point $\underline{p}_i$ differs from $(\underline{p}_{i+1} - \underline{p}_{i-1})$. In figure 3 this effect is shown. When $h_1 = h_2$, the tangent-vector of the curve in $\underline{p}_i$ ($\underline{v}_2$ in figure 3) is parallel to $(\underline{p}_{i+1} - \underline{p}_{i-1})$. Curves with $h_1 < h_2$ and $h_1 > h_2$ have respectively $\underline{v}_1$ and $\underline{v}_3$ as their tangents.

Calculation of $\underline{w}_{5,i}$ and $\underline{w}_{6,i}$

$$\max\left|\frac{\delta^2\phi_k(\lambda)}{\delta^2\lambda}\right| - \left|\frac{\delta^2\phi_k(\lambda^*)}{\delta^2\lambda}\right| \quad \text{with} \quad \begin{cases} \lambda^* = {}^L\lambda_i & ({}^L\lambda_i < {}^L\lambda_i < \lambda_i + \tfrac{1}{2}) \\ \lambda^* = {}^R\lambda_i & (\lambda_i + \tfrac{1}{2} < {}^R\lambda_i < \lambda_{i+1}) \end{cases} \tag{8}$$

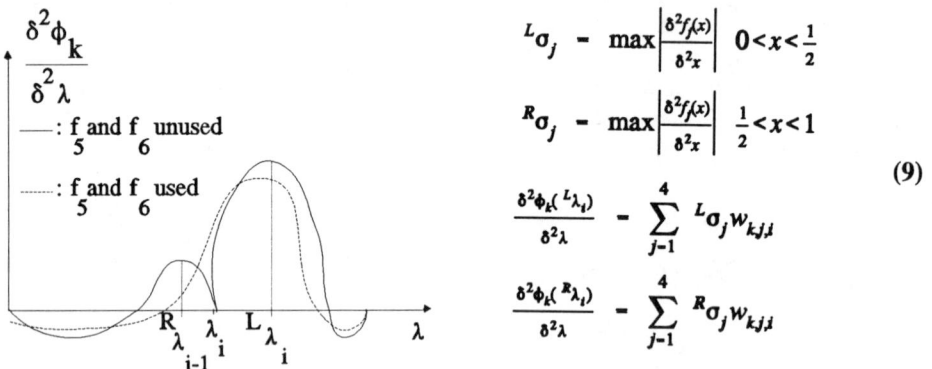

$$\begin{aligned} {}^L\sigma_j &= \max\left|\frac{\delta^2 f_j(x)}{\delta^2 x}\right| \quad 0 < x < \tfrac{1}{2} \\ {}^R\sigma_j &= \max\left|\frac{\delta^2 f_j(x)}{\delta^2 x}\right| \quad \tfrac{1}{2} < x < 1 \\ \frac{\delta^2\phi_k({}^L\lambda_i)}{\delta^2\lambda} &= \sum_{j=1}^{4} {}^L\sigma_j w_{k,j,i} \\ \frac{\delta^2\phi_k({}^R\lambda_i)}{\delta^2\lambda} &= \sum_{j=1}^{4} {}^R\sigma_j w_{k,j,i} \end{aligned} \tag{9}$$

Figure 4. Improvement of $\delta^2\phi_k/\delta^2\lambda$ through usage of f_5 and f_6

For each joint the function $\delta^2\phi_k/\delta^2\lambda$ can now be derived, which is calculated using $w_{k,1,i}$, $w_{k,2,i}$, $w_{k,3,i}$ and $w_{k,4,i}$ as described in (6) and (7) and $w_{k,5,i} = w_{k,6,i} = 0$ (k=index of element in $\underline{w}_{j,i}$). The resulting function $\delta^2\phi_k/\delta^2\lambda$ has 2 extremes for each interval $(\lambda_i, \lambda_{i+1})$ as given in (8) and shown in figure 4. This is a function with much oscillatory effects, which will disturb the smoothness of the trajectory-shape. Through choosing $w_{k,5,i} = 1/2 * \{\delta^2\phi_k({}^R\lambda_{i-1})/\delta^2\lambda + \delta^2\phi_k({}^L\lambda_i)/\delta^2\lambda\}$ and $w_{k,6,i} = w_{k,5,i+1}$ the Hermite functions $f_5(x)$ and $f_6(x)$ can be used to suppress these undesired oscillatory effects in $\delta^2\phi_k/\delta^2\lambda$. The resulting improved function is shown as the striped line in figure 4.

The assumption, that $\delta^2 f_j(x)/\delta^2$ with (j=1..4) have their extremes for approximately the same x-values, leads to the formulation in (9).

$$\underline{w}_{5,i} = \frac{0.47*(I-H)(2\underline{p}_i - \underline{p}_{-2} - \underline{p}_{+2})}{-2.89*(2\underline{p}_i - \underline{p}_{-1} - \underline{p}_{+1})} \tag{10}$$

$$\underline{w}_{6,i} = \underline{w}_{5,i+1}$$

Combining the expressions in (9) in with: $w_{k,5,i} = 1/2 * \{\delta^2\phi_k({}^R\lambda_{i-1})/\delta^2\lambda + \delta^2\phi_k({}^L\lambda_i)/\delta^2\lambda\}$ and $w_{k,6,i} = w_{k,5,i+1}$ leads to the new expressions for $\underline{w}_{5,i}$ and $\underline{w}_{6,i}$ in (10).

2.2 Calculation of the tension-limits

Two different type of limits for the tension-parameters are distinguished: 1. hard limits $\underline{h}_{Hmin}$ and $\underline{h}_{Hmax}$ and 2. soft limits $\underline{h}_{Smin}$ and $\underline{h}_{Smax}$. Hard limits describe the intervals for $\underline{h}$, which restrict the trajectory within the collision-free area. The soft limit-intervals are within the hard limits : $h_{k,Hmin} < h_{k,Smin} < h_k < h_{k,Smax} < h_{k,Hmax}$ and describe the intervals for $\underline{h}$ where unnecessary loops and cusps in the trajectory are avoided.

$$\eta_i(\lambda) = \min\{\|\phi(\lambda-\lambda_i)-(\underline{p}_i+\gamma(\underline{p}_{i+1}-\underline{p}_i))\|; 0<\gamma<1\} \tag{11}$$

The soft maxima are the highest possible tension-parameters with the minimum amount of extremes for $\eta_i(\lambda)$, as described in formula (11). The soft maxima are analytically calculated on basis of the following approximations: $f_3(x) - 10.f_5(x) \approx 0$ and $f_4(x) + 10.f_6(x) \approx 0$. This gives the following values : $h_{k,Smax} = 0.423$ (k=1..n). A trajectory with $\underline{h} = \underline{h}_{Smax}$ closely approximates straight lines between control points.

When, starting from $\underline{h} = \underline{h}_{Smax}$, the tension-parameters are decreased, at some moment the amount of extremes of $\underline{\phi}(\lambda)$ will increase causing loops and cusps in the trajectory. The soft minima are defined as the lowest possible tension-parameters which give the same amount of extremes for $\underline{\phi}(\lambda)$ as the soft maxima. Starting from the soft maxima these soft minima are iteratively calculated through stepwise decrease of the tension-parameter of each dimension until the amount of extremes increases.

The hard maxima and minima are iteratively approximated through respectively a stepwise increase and decrease of the tension-parameters, starting from the soft maxima, until the trajectory exceeds the collision-free area.

3. Optimization of the trajectory

The optimization is performed in 2 steps: 1. optimization of the placement of the control-points and 2. optimization of the tension-parameters. For both optimization-steps suitable criteria have to be found.

In (12) a combination of minimum length and minimum curvature is given, which was mentioned in [7] as suitable criterium to approximate true time-optimal geometrical movement.

$$\min\left\{\int_{\lambda=0}^{\lambda_N} \frac{\rho(\lambda)}{\left\|\frac{\delta\phi(\lambda)}{\delta t}\right\|_{max}} d\lambda\right\} \quad \rho(\lambda) = \frac{\delta\|\phi(\lambda)\|}{\delta\lambda} \tag{12}$$

The control-points are placed on the (n-1)-dimensional subspaces between two neighbouring collision-free segments. This point-placement has the property that straight lines between those points and trajectories with $\underline{h} = \underline{h}_{Smax}$, which approximate these straight lines, are always kept within the collision-free area.

$$\text{Min. length}: Min\left\{\sum_{i=1}^{N-1} \|\underline{p}_{i+1}-\underline{p}_i\|\right\} \quad \text{Min. bending}: Min\left\{\sum_{i=1}^{N-2} \frac{(\underline{p}_{i+2}-\underline{p}_{i+1})\cdot(\underline{p}_i-\underline{p}_{i+1})}{\|\underline{p}_{i+2}-\underline{p}_{i+1}\|\cdot\|\underline{p}_i-\underline{p}_{i+1}\|}\right\} \tag{13}$$

Optimization of the point-placement is performed using the dynamic programming approach. As suitable criteria 2 different possibilities are suggested: minimum euclidean length and minimum bending of the series of straight lines between the control-points as given in (13).

The minimum bending point-placement gives the smoothest interconnection of straight lines between the control-points. Minimum length point-placement gives in most cases smaller h_{Hmin}-values than minimum bending, because minimum-length placement offers more space for overshoot of the trajectory in the direction of the tangent in each control-point. The last property makes this placement most suitable.

$$\min\left\{\int_{\lambda=0}^{\lambda_N} \rho(\lambda) \cdot \sqrt{\left\|\frac{\delta^2\phi}{\delta^2\lambda}\right\|^2 - \left|\frac{\frac{\delta\phi}{\delta\lambda} \cdot \frac{\delta^2\phi}{\delta^2\lambda}}{\left\|\frac{\delta\phi}{\delta\lambda}\right\|}\right|^2} \, d\lambda\right\} = \min\left\{\int_{\lambda=0}^{\lambda_N} \frac{\delta length(\lambda)}{\delta\lambda} \cdot \frac{\delta curvature(\lambda)}{\delta\lambda}\right\} \quad (14)$$

Because in a final situation trajectories in a n-dimensional (n=10) joint-space must be optimized, the amount of simultaneously optimized parameters, the search-space for these parameters and the calculation-time for calculation of the criteria must be kept small. This is partly already done through separate optimization of both point-placement and tension-parameters and restriction of the possible places of the control-points to the (n-1)-dimensional subspaces between the collision-free segments. Using the criterium in (12) for tension-optimization the edge of the forbidden area for $\delta\phi/\delta t$ must be calculated which will cost much extra calculation-time. As a first attempt to combine both length and curvature in a criterium for tension-optimization without this extra calculation of the edge of the forbidden area, the purely geometrical criterium in (14) is suggested with $\rho(\lambda)$ as given in (12). Due to its purely geometrical nature usage of the criterium in (14) will in most cases not result in the finding of the exact time-optimal trajectory but, through combining the important properties of trajectory-length and curvature, in a reasonably shaped trajectory, while the calculation-time is suppressed.

In figure 5 a 2D example shows the curves with respectively $\underline{h}=\underline{h}_{Hmax}$, $\underline{h}=\underline{h}_{Hmin}$, $\underline{h}=\underline{h}_{Smax}$, $\underline{h}=\underline{h}_{Smin}$ and $\underline{h}=\underline{h}_o$ ($\underline{h}_o$: optimal tension using (14)) for a minimum bending point-placement. From figure 5 can be concluded that setting the tension-parameters directly affects length and curvature of the trajectory.

The sharp bendings in the trajectories with $\underline{h}=\underline{h}_{Hmax}$, $\underline{h}=\underline{h}_{Smax}$ and $\underline{h}=\underline{h}_{Hmin}$ as shown in figure 5 will locally suppress the maximum possible velocity. The trajectories with

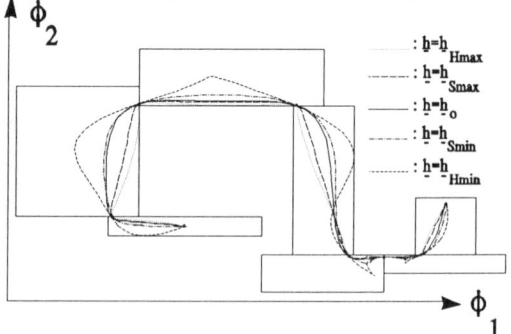

Figure 5, 2D-example of calculation of the tension-parameters

$\underline{h}=\underline{h}_{Hmax}$ and $\underline{h}=\underline{h}_{Hmin}$ contain cusps and small loops which lead to longer trajectories with more curvature. Trajectories with $\underline{h}=\underline{h}_{Smax}$ and $\underline{h}=\underline{h}_{Smin}$ are free of loops and cusps. From both trajectories with $\underline{h}=\underline{h}_{Smax}$ and $\underline{h}=\underline{h}_{Smin}$ the first one has the smallest length and the second one the smallest bending in all control-points. The optimal trajectory is a compromise between the shortest and least 'sharp' trajectory and is found through optimization within the soft tension-limits: $\underline{h}_{S_{min}}$ and $\underline{h}_{S_{max}}$ instead of the hard tension-limits. This keeps the search-space for the tension-parameters relatively small which is important for search in a 10-dimensional joint-space.

4 Conclusions and suggestions

The introduced interpolation technique offers strong possibilities of manipulation of the length and curvature of a trajectory using tension-parameters. The trajectory can be limited inside complex collision-free areas and unnecessary loops and cusps can be avoided. Through separate optimization of placement of the control-points and the tension-parameters, usage of purely geometrical criteria and restriction of the search-spaces for both optimization-steps an attempt is made to suppress the necessary calculation-time. Other criteria and suitable optimization techniques are further investigated.

References

[1] Meijer, B.R. and Jonker, P.P.,*"The architecture and philosophy of the DIAC (Delft Intelligent Assembly Cell)"*, proceedings of the 1991 IEEE International conference on Robotics and Automation, April 9-11, Sacremento CA.,pp. 2218-2223, 1991

[2] Verwer, B.J.H.,*"A multiresolution work space, multiresolution configuration space approach to solve the path planning problem"*,1990 IEEE Int. Conf. on Robotics and Automation, May 13-18, 1990, Cincinatti, USA, pp.2107-2112.

[3] Rieswijk, T.A., Sirks, M., Honderd, G. and Jongkind, W.,*"A fast and efficient algorithm for the computation of path constrained time-optimal motions."*,Euriscon "91,Greece ,1991

[4] Greville, T.N.E.,*"Theory and applications of spline-functions"*,Academic Press, New York, 1969.

[5] Mortenson, M.E.,*"Computer Graphics: An introduction to the Mathematics and Geometry"*, Industrial Press Inc.,New York, 1989.

[6] Shiller, Z. and Dubowsky, S.,*"Robot path planning with obstacles, actuator, gripper and payload constraints"*,The International Journal of Robotics Research, Vol. 8, No. 6, 1989, pp. 3-18.

[7] Shin, K.G. and McKay, N.D.,*"Selection of Near-Minimum Time Geometric Paths for Robotic Manipulators"*,IEEE Transactions on Automatic Control, Vol. AC-31, No. 6, 1986, pp. 501-510.

[8] Kochanek, D.H.U. and Bartels, R.H.,*"Interpolating splines with local tension, continuity and bias"*, Computer graphics, Vol. 18, No. 3, pp. 33-41, 1984.

A fast and efficient algorithm for the computation of path constrained time-optimal motions

T.A. Rieswijk, M. Sirks, G. Honderd and W. Jongkind

Delft University of Technology,
Faculty of Electrical Engineering, Control Laboratory.
Mekelweg 4, P.O. Box 5031, 2600 GA Delft. The Netherlands.

Abstract

The work, described in this paper, is performed as part of a collision-free motion-planning system for two cooperating robots in an assembly cell. The robots are of different types, an anthropomorphic type and a scara type.

This collision-free motion-planning is part of the Delft Intelligent Assembly Cell (DIAC) -project. The aim of the DIAC project is to obtain within four years a reliable working demonstration of a flexible assembly cell. This is further described in [1].

In this paper a modified approach of earlier algorithms is presented to calculate the time-optimal motion of a manipulator along a precalculated trajectory in joint space. The necessary motor-torques can be limited inside given boundaries through usage of a priori knowledge of robot-dynamics. To reduce the necessary calculation-time a new approach of simultaneous integration is introduced.

Problem-description

For two cooperating robots in an assembly-cell simultaneous movements must be planned in partly shared working areas. To allow simultaneous motion of these robots in their partly shared working areas, the movements of both robots are planned in a high dimensional joint-space with one dimension for each degree of freedom of the robots. The motion-planning is performed in two phases:

Phase 1: search of a collision free path for a generalized n-dimensional robot (for a 4D scara + a 6D anthropomorphic robot n=10). The search is performed by using a swept volume approach in workspace combined with a binary search in joint space ([2]). The result of this is a sequence of collision-free boxes in an n-dimensional joint-space. The boundaries of each box represent transition states for both robots that have to be passed synchronously.

Phase 2: planning of a collision-free trajectory through the resultant sequence of collision free segments in this combined joint space, describing the movements of all joints as a function of time.

For the planning of the time-dependency, knowledge of the robot-dynamics is applied.

$$\underline{\tau}_m = M(\underline{\phi})\underline{\ddot{\phi}} + \underline{\dot{\phi}} C(\underline{\phi})\underline{\dot{\phi}} + \underline{G}(\underline{\phi}) \quad \text{with: } \underline{\tau}_{min} \leq \underline{\tau}_m \leq \underline{\tau}_{max} \tag{1}$$

In formula (1) a frequently used description of robot-dynamics is given. $M(\underline{\phi})$ is the n x n inertia matrix, $C(\underline{\phi})$ is the n x n x n array of the centrifugal and Coriolis forces, $\underline{G}(\underline{\phi})$ is the n x 1 vector of gravity forces and $\underline{\tau}_m$ are the motor-torques. For these motor-torques limits are also given in formula (1).

The total trajectory-planning of phase 2 is split into 2 parts:

Part 1: planning of the collision-free trajectory in the combined joint space as a function of one common variable λ as described in [3].

Part 2: optimization of the relation of λ with time.

In part 1 a smoothly shaped contour in joint space is planned with emphasis on the property of restriction inside the collision free subspace. The resultant geometrical trajectory is bounded inside the collision-free area and describes the motion of all axes as functions of one variable: $\underline{\phi}(\lambda)$, which is continuous in $\underline{\phi}(\lambda), \delta\underline{\phi}(\lambda)/\delta\lambda$ and $\delta^2\underline{\phi}(\lambda)/\lambda^2$ for $\lambda_0 \leq \lambda \leq \lambda_f$.

In part 2 the algorithm, as described in this paper, is applied.

1. Introduction phase-plane search

Common property of most algorithms, which calculate the time-optimal motion along a prescribed trajectory in joint space or cartesian space, is the fact that the time-optimal movements of all joints are calculated through optimizing the time-dependency of the common parameter λ. This reduces the search-space for the optimization.

The time-optimization of λ takes place in a λ,μ-plane ($\mu=\delta\lambda/\delta t$). Each coordinate in this plane defines the complete joint-position and velocity in joint-space as described in formula (2). A transition between 2 coordinates in the (λ,μ)-plane can be related to the necessary functions for joint-acceleration and the motor-torques.

$$\phi_i = \phi_i(\lambda)$$
$$\dot{\phi}_i = \frac{\delta\phi_i}{\delta\lambda} \cdot \frac{\delta\lambda}{\delta t} = \frac{\delta\phi_i}{\delta\lambda}\mu \quad \text{with } (i = 1..n) \; \wedge \; (\lambda_0 \leq \lambda \leq \lambda_f) \tag{2}$$

$$\tau_{i,min} \leq {}^\lambda M_i \frac{\delta\mu}{\delta t} + {}^\lambda Q_i \mu^2 + {}^\lambda S_i \leq \tau_{i,max} \tag{3}$$

$${}^\lambda M_i = \sum_{j=1}^n M_{ij} \frac{\delta\phi_j}{\delta\lambda} \quad {}^\lambda Q_i = \sum_{j=1}^n \sum_{k=1}^n C_{i,j,k} \frac{\delta\phi_j}{\delta\lambda} \frac{\delta\phi_k}{\delta\lambda} + \sum_{j=1}^n M_{ij} \frac{\delta^2\phi_j}{\delta^2\lambda} \quad {}^\lambda S_i = G_i$$

Combining (1) and (2) the demands for the motor-torques can be derived as functions of λ,μ and $\delta\mu/\delta t$ in (3). In (3) each set of torque-limits $\tau_{i,min}$ and $\tau_{i,max}$ yields lower and upper bounds for δμ/δt respectively LB_i and UB_i with $LB_i \leq δμ/δt \leq UB_i$.

$$LB_i = \frac{\tau_{i,\min}(if\ ^\lambda M_i>0) + \tau_{i,\max}(if\ ^\lambda M_i<0) - (^\lambda Q_i \mu^2 + {}^\lambda S_i)}{^\lambda M_i}$$

$$UB_i = \frac{\tau_{i,\max}(if\ ^\lambda M_i>0) + \tau_{i,\min}(if\ ^\lambda M_i<0) - (^\lambda Q_i \mu^2 + {}^\lambda S_i)}{^\lambda M_i}$$
(4)

Furthermore the limits LB_i and UB_i as given in (4) must hold for all n joints. This gives one pair of final upper and lower bounds for δμ/δt: $GLB(\lambda,\mu)$ and $LUB(\lambda,\mu)$ as formulated in (5).

$$\left.\begin{array}{l}GLB(\lambda,\mu) = \max_i\{LB_i, i=1..n\}\\ LUB(\lambda,\mu) = \min_i\{UB_i, i=1..n\}\end{array}\right\} GLB(\lambda,\mu) \leq \frac{\delta\mu}{\delta t} \leq LUB(\lambda,\mu) \quad (5)$$

In order to find a suitable algorithm for time-optimization in the (λ,μ) phase-plane a comparison is made between different approaches with respect to computation-load and usage of memory.

For the time-optimization in the (λ,μ)-plane 3 different types of approaches can be distinguished:
1. Grid-search: Search in a fully discretized (λ,μ)-grid. In [4] the dynamic programming approach is used to obtain the optimal connections between the columns with discrete μ-values belonging to pairs of neighbouring discrete λ-values in the (λ,μ)-grid. Connections in the (λ,μ)-grid which cause excess of certain limits, like the motor-torques, can be excluded. This technique is numerically very robust but suffers from a relatively high calculation- and memory-load, because all possible connections between 2 neighbouring columns with μ-values must be evaluated.
2. Binary-search: Search through stepwise increase of the individual μ-values, which belong to fixed discretized λ-values. In [5] a binary-search method is applied for optimization in a (λ,μ)-plane with a discretized λ-axis. When suitable initial step-sizes for μ are chosen, this technique will, compared to the grid-search technique in [4], consume less calculation-time and memory. Main disadvantage of this technique is that suitable initial step-sizes for μ cannot be given beforehand and a wrong choice will introduce unnecessary calculation-work.
3. Optimal switching: Search through integration from fixed points at the border of a forbidden area for μ in the (λ,μ)-plane with maximum acceleration or deceleration. Compared to grid-search and binary-search techniques the optimal switching techniques, as described in [6],[7] and [8], are numerically less stable but more efficient in their usage of calculation-time and memory. Due to the importance of the last advantage, the new proposed method belongs to this type of methods.

2 Optimal switching techniques

Optimal switching methods explicitly use information about a forbidden area in the (λ,μ)-plane. This forbidden area can be defined as the region, where $LUB(\lambda,\mu) < GLB(\lambda,\mu)$ and

no solution for δμ/δt can be found satisfying the condition in (5).
All optimal switching methods consist of 2 basic parts:
- the search of suitable starting points for the time-optimal curves on the border between forbidden and allowed area in the (λ,μ)-plane. These starting points will be referred to as contact-points.
- the calculation of the time-optimal curves through numerical integration from these starting points using extreme values of δμ/δt.

2.1 Search for contact-points

Basic property of each contact-point (λ_j^t,μ_j^t) (with $j=1..N_t$, $\lambda_j^t > \lambda_{j-1}^t$) is that it marks the place on the border between forbidden and allowed area, where optimal curves can touch. These points can be found through comparison of all values of $\delta\mu(\lambda,\mu)/\delta\lambda$ on the border with the tangent along the border until they are equal. The value of $\delta\mu(\lambda,\mu)/\delta\lambda$ can be derived using (6).

$$\frac{\delta\mu(\lambda,\mu)}{\delta\lambda} = \frac{\delta\mu(\lambda,\mu)}{\delta t} \cdot \frac{1}{\mu} \qquad (6)$$

Dependent on the tolerance for δμ/δt in a contact-point 3 different types of contact-points can be distinguished:
1 Tangency-points with $\delta\mu(\lambda_j^t,\mu_j^t)/\delta t$ fixed to only 1 value: $\delta\mu(\lambda_j^t,\mu_j^t)/\delta\lambda = GLB(\lambda_j^t,\mu_j^t)/\mu_j^t = LUB(\lambda_j^t,\mu_j^t)/\mu_j^t$.
2 Critical points with a finite interval for $\delta\mu(\lambda_j^t,\mu_j^t)/\delta t$: $GLB(\lambda_j^t,\mu_j^t)/\mu_j^t \leq \delta\mu(\lambda_j^t,\mu_j^t)/\delta\lambda \leq LUB(\lambda_j^t,\mu_j^t)$.
3. Zero-inertia points with an infinite interval for $\delta\mu(\lambda_j^t,\mu_j^t)/\delta t$.
In [7] special attention is paid to the detection of these 3 types of points as contact-points.

2.2 Calculation of the time-optimal curves

The time-optimal curves are calculated through numerical integration in the (λ,μ)-plane

$$\mu(\lambda^*) = \mu_j^t + \int_{\lambda-\lambda_j^t}^{\lambda^*} \frac{\delta\mu(\mu,\lambda)}{\delta\lambda}d\lambda \qquad (7)$$

from all contact-points as described in (7). In (7) j is the index of the used contact-point. The integration takes place in forward and backward direction for lambda with $\delta\mu(\lambda,\mu)/\delta\lambda = LUB(\lambda,\mu)/\mu$ in forward and $\delta\mu(\lambda,\mu)/\delta\lambda = GLB(\lambda,\mu)/\mu$ in backward direction.

All earlier introduced strategies integrate one total curve at a time. The integration of an optimal curve is stopped when:
- a curve intersects with the forbidden region, or
- a curve intersects with another, earlier calculated, curve, or
- the value of λ leaves the interval $[\lambda_0,\lambda_f]$.
The intersection-points of forward and backward integrated curves in the (λ,μ)-plane will be called switching-points.

In figure 1 the forward and backward integrated curve from respectively the starting-point (λ_0,μ_0) and end-point (λ_f,μ_f) in the (λ,μ)-plane intersect in switching-point (λ_s,μ_s). The part of the forward integrated curve (solid line in figure 2) from $\lambda = \lambda_0$ to $\lambda = \lambda_s$ and the part of the backward integrated curve (dotted line in figure 2) from $\lambda = \lambda_f$ to $\lambda = \lambda_s$ form together the optimal curve. The remaining part of the forward integrated curve from $\lambda = \lambda_s$ until the intersection with the forbidden region is not further used.

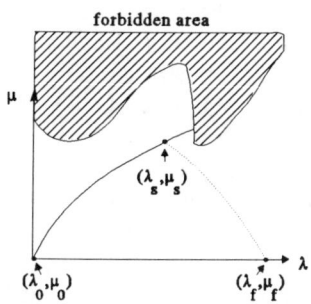

Figure 1, Switch-point-calculation through intersection

When initial and final curve don't intersect, new optimal curves have to be calculated from the contact-points. These contact-points can either be calculated before or during the computation of the optimal curves.

The algorithms, which calculate new contact-points during the computation of the optimal curves, use the following mechanism ([6], [7] and [8]):

Mechanism I

step 1: Integrate from (λ_0,μ_0) the initial curve in forward λ-direction and stop, when the curve intersects with either the forbidden area, or another already calculated curve, or when $\lambda > \lambda_f$.

step 2: Integrate from (λ_f,μ_f) the final curve in backward λ-direction and stop, when this curve intersects with either the forbidden area, or the initial curve, or when $\lambda < \lambda_0$.

step 3: If initial and final curve intersect in (λ_s,μ_s), then an optimal curve can be constructed from the initial curve with $\lambda_0 \leq \lambda \leq \lambda_s$ and the final curve with $\lambda_s \leq \lambda \leq \lambda_f$. The algorithm is finished.

step 4: If the 2 curves under consideration don't intersect and the forward integrated curve intersects with the forbidden region, then start searching from this intersection-point in forward λ-direction for a new contact-point (λ_j^i,μ_j^i).

step 5: Start integrating a curve backward from the new contact-point (λ_j^i,μ_j^i) until it intersects with the earlier calculated forward integrated curve. The part of the intersecting forward integrated curve before this intersection-point and the part of the new backward integrated curve from the intersection-point up to the contact-point will be stored as a part of the optimal curve.

step 6: Start integrating a curve forward from the new contact-point (λ_j^i,μ_j^i) until it either intersects with the earlier calculated backward integrated curve, or with the forbidden area, or when $\lambda > \lambda_f$.

step 7: If the curve in step 6 intersects with the backward integrated curve, then the part of the new forward integrated curve before from (λ_j^i,μ_j^i) up to the intersection-point and the part of the intersecting backward integrated curve from its starting point up to the intersection-point will be stored as a part of the optimal curve.

step 8: If the curve in step 6 intersects with the forbidden area, then go to step 4.

Algorithms, which are based on mechanism I, are only suitable, when viscous friction is left out of the robot-model which is used for trajectory-planning. In those situations the true time-optimal trajectory in the (λ,μ)-plane will be found.

The inclusion of viscous friction in the used robot-model can introduce forbidden isles in the allowed area where no solution for (5) exists. These isles introduce the possibility of multiple solutions to connect (λ_0,μ_0) with (λ_f,μ_f). The solution with the highest µ-values will then be time-optimal.

When isles appear in the (λ,μ)-plane through the introduction of viscous friction in the robot-model, the search of neighbouring contact-points <u>during</u> the computation of the optimal curves, as described in mechanism I, becomes complex. To guarantee that, in the presence of isles, at least a sub-optimal solution will be found, the proposed method will perform a search of all contact-points <u>before</u> the computation of the optimal curves. To search for the true time-optimal curve in a (λ,μ)-plane with isles and multiple curves between (λ_0,μ_0) and (λ_f,μ_f) special algorithms must be used as described in [6]. The proposed method can only find the true time-optimal solution in situations without forbidden isles in the (λ,μ)-plane. On basis of the same principle as the proposed one and with some modifications the method, which is introduced in this article, can also be made suitable for the search of the true time-optimal curve in the presence of forbidden isles.

For all earlier proposed optimal switching methods, as described above, the essential common property is that only one total curve is integrated at a time. As shown in the example of figure 1 this can lead to the removal of important parts of earlier integrated curves for which a lot of integration-steps had to be performed. For each integration-step of a curve a complete evaluation must be performed to calculate the next (λ,μ) values as described (7) using (2)-(6) with the descriptions of the robot-dynamics and the geometrical movement. Reduction of these unnecessary integration-steps will therefore give an important reduction of the amount of calculation-time.

3. Proposed algorithm for optimal switching

To suppress unnecessary integration-steps the proposed algorithm will integrate more curves simultaneously, in contradiction with all other optimal-switching methods which only integrate one total curve at a time.

Due to this simultaneous integration the information, belonging to each integrated curve and parts of the optimal curve, must be stored in temporary lists during optimization. The proposed algorithm uses the following lists:

- a list with elements, describing integration-direction and start and end-values of all integrated curves given as $^a\psi^i$ in (8). This list will be called the list of active curves. The list-element $^a\psi^i$ contains information about the integrated curve with index i (i=1..N_a) and $^a\psi^{i+1}_i > {}^a\psi^i_i$. Each list-element $^a\psi^i$ in (8) contains: the λ-value for the

$$^a\psi^i = \begin{pmatrix} \lambda^t_j \\ \lambda^e_i \\ dir_i \\ fl_i \\ po^a_i - \{\mu^a_i(\lambda)\} \end{pmatrix} \quad ^o\psi^k = \begin{pmatrix} \lambda^t_{j1} \\ \lambda^t_{j2} \\ po^o_k - \{\mu^o_k(\lambda)\} \end{pmatrix} \quad (8)$$

contact-point λ_j^t, from where the integration of the curve started, the temporary last λ-value during integration λ_i^e and a pointer po_i^a, which points to an array with all μ-values of curve i as a function of λ: $\mu_i^a(\lambda)$. For each active curve also extra other information must be stored: the direction dir_i for the direction of the integration ($dir_i=-1$ for backward and $dir_i=+1$ for forward integration) and the flag fl_i to indicate if the active curve is normally integrated ($fl_i=0$) or temporary inhibited ($fl_i=1$).
- a list with elements, describing the temporary parts of the optimal curve. The list-element $^o\underline{\psi}^k$ contains information about the temporary part of the optimal curve with index k (k=1..N_o) and $^o\psi_1^{k+1} > ^o\psi_1^k$. Each list-element $^o\underline{\psi}^k$ in (8) contains the λ-values belonging to the contact-points from where the part of the optimal curve starts: λ_{j1}^t and ends: λ_{j2}^t and a pointer po_k^o to an array with all μ-values of optimal curve-part k as a function of λ : $\mu_k^o(\lambda)$.
- a list with all contact-points (λ_j^t,μ_j^t) with index j (j=1..N_t) and $\lambda_{j+1}^t > \lambda_j^t$, including starting point $(\lambda_1^t,\mu_1^t) = (\lambda_0,\mu_0)$ and final point $(\lambda_{Nt}^t,\mu_{Nt}^t) = (\lambda_f,\mu_f)$.

Proposed new calculation-mechanism:

step 1 Calculate all contact-points and store them in a list (λ_j^t,μ_j^t) including the initial and final points in the (λ,μ)-plane : $(\lambda_1^t,\mu_1^t) = (\lambda_0,\mu_0)$ and $(\lambda_{Nt}^t,\mu_{Nt}^t) = (\lambda_f,\mu_f)$.

step 2 Put the initial contact-point (λ_1^t,μ_1^t) on the list of active curves : $^a\psi_1^1 = \lambda_1^t$, $^a\psi_3^1 = 1$, $^a\psi_4^1 = 0$ and create memory for the μ-values with pointer $^a\psi_5^1$.

step 3 Put the final contact-point $(\lambda_{Nt}^t,\mu_{Nt}^t)$ on the list of active curves : $^a\psi_1^2 = \lambda_{Nt}^t$, $^a\psi_3^2 = -1$, $^a\psi_4^2 = 0$ and create memory for the μ-values with pointer $^a\psi_5^2$.

step 4 Perform one integration-step for the active curve which has the lowest value of $\mu_i^a(\lambda)$ of all active curves which are not inhibited ($^a\psi_4^i = 0$), in direction for λ as indicated by $^a\psi_3^i$.

step 5 If the value of $\mu_i^a(^a\psi_2^i)$ becomes larger than μ_j^t of a contact-point, which has not been used as a start-point of an active curve, then insert two new active curves: $^a\underline{\psi}^i$ and $^a\underline{\psi}^{i+1}$ in the list of active curves with the contact-point (λ_j^t,μ_j^t) as start-point and $^a\psi_1^{i-1} < \lambda_j^t < ^a\psi_1^{i+2}$. Curves $^a\underline{\psi}^i$ and $^a\underline{\psi}^{i+1}$ will be integrated in respectively backward and forward direction: $^a\psi_3^i = -1$ and $^a\psi_3^{i+1} = 1$. Both curves are activated through: $^a\psi_4^i = ^a\psi_4^{i+1} = 0$ and the pointers $^a\psi_5^i$ and $^a\psi_5^{i+1}$ will point to the memory space for respectively $\mu_i^a(\lambda)$ and $\mu_{i+1}^a(\lambda)$.

step 6 If integrated curve $^a\underline{\psi}^i$ passes under neighbouring active curve $^a\underline{\psi}^{i-1}$ { if $^a\psi_2^i < ^a\psi_1^{i-1}$ and $\mu_i^a(^a\psi_2^i) < \mu_{i-1}^a(^a\psi_1^{i-1})$ }, then delete curve $^a\underline{\psi}^{i-1}$ from the list of active curves and if $^a\underline{\psi}^{i-2}$ has the same starting point as $^a\underline{\psi}^{i-1}$, then delete also $^a\underline{\psi}^{i-2}$.

step 7 If integrated curve $^a\underline{\psi}^i$ passes under $^a\underline{\psi}^{i+1}$ { if $^a\psi_2^i > ^a\psi_1^{i+1}$ and $\mu_i^a(^a\psi_2^i) < \mu_{i+1}^a(^a\psi_1^{i+1})$ }, then delete $^a\underline{\psi}^{i+1}$ from the list of active curves and if $^a\underline{\psi}^{i+2}$ has the same starting point as $^a\underline{\psi}^{i+1}$, then delete also $^a\underline{\psi}^{i+2}$.

step 8 If integrated curve $^a\underline{\psi}^i$ passes under optimal curve $^o\underline{\psi}^k$ { if $\mu_i^a(\lambda^*) < \mu_k^o(\lambda^*)$ for $^o\psi_1^k \leq \lambda^* \leq ^o\psi_2^k$ }, then delete $^o\underline{\psi}^k$ from the list of optimal curves.

step 9 If the integrated active curve doesn't intersect with: the forbidden area, another active curve or temporary parts of the optimal curve and if $\lambda_0 \leq ^a\psi_2^i \leq \lambda_f$, then go to step 4.

step 10 If forward integrated curve $^a\underline{\psi}^{i1}$ intersects with backward integrated curve $^a\underline{\psi}^{i2}$ (i1=i or i2 =i), then:

- if the intersection-point lies under optimal curve k { if $\mu_{intersect} < \mu_k^o(\lambda_{intersect})$ }, then first delete $^o\underline{\psi}^k$ from the list of optimal curves.
- delete $^a\underline{\psi}^{i1}$ and $^a\underline{\psi}^{i2}$ from the list of active curves, insert the new optimal curve-element k in the list of optimal curves with $^o\psi_2^{k-1} < {}^a\psi_1^{i1} < {}^o\psi_2^{k+1}$, put the lowest μ-values of both curves in a new array with $^o\psi_3^k$ as its pointer and fill the new element $^o\underline{\psi}^k$: $^o\psi_1^k = {}^a\psi_1^{i1}$ and $^o\psi_2^k = {}^a\psi_1^{i2}$.

step 11 If integrated curve $^a\underline{\psi}^i$ intersects with the forbidden area or its $^a\psi_2^i$-value leaves the interval $[\lambda_0, \lambda_f]$, then inhibit the curve through setting $^a\psi_4^i = 1$.

step 12 If the active curve list is empty or contains only inhibited curves and the optimal curve is not connected over some λ-intervals ($^o\psi_2^{k-1}, {}^o\psi_1^k$), then:
- if unused contact-points (λ_j^t, μ_j^t) with $^o\psi_2^{k-1} < \lambda_j^t < {}^o\psi_1^k$ are available, then insert the unused contact-point (λ_j^t, μ_j^t) with $^o\psi_2^{k-1} < \lambda_j^t < {}^o\psi_1^k$ and the lowest μ_j^t-value as start-point for 2 new active curves $^a\underline{\psi}^i$ and $^a\underline{\psi}^{i+1}$ with $^a\psi_1^{i-1} < {}^a\psi_1^i < {}^a\psi_1^{i+2}$. Initiate them with: $^a\psi_3^i = -1$, $^a\psi_3^{i+1} = 1$ and $^a\psi_4^i = {}^a\psi_4^{i+1} = 0$. The pointers $^a\psi_5^i$ and $^a\psi_5^{i+1}$ point to new created memory-space for the μ-values.
- if unused contact-points (λ_j^t, μ_j^t) with $^o\psi_2^{k-1} < \lambda_j^t < {}^o\psi_1^k$ are <u>not</u> available, then the algorithm finishes without a proper solution.

step 13 If the active curve list is empty or contains only inhibited curves and the optimal curve is totally connected from start to end, then the algorithm finishes with a proper solution.

step 14 If the algorithm is not finished, then go to step 4.

An example of the working of the proposed algorithm is given in figure 2. In figure 2 the solid curves represent the temporary situation for 4 active curves $^a\underline{\psi}^i$ (i=1..4), which are simultaneously integrated. The last calculated μ-values of all curves at some moment during the integration of the curves are marked by the horizontal striped line with $\mu = \mu_i^a$. Curves $^a\underline{\psi}^2$ and $^a\underline{\psi}^3$ start from the contact-point on the forbidden area: (λ_2^t, μ_2^t). The dotted parts of all curves show the final integration-steps until the optimal curve is completed.

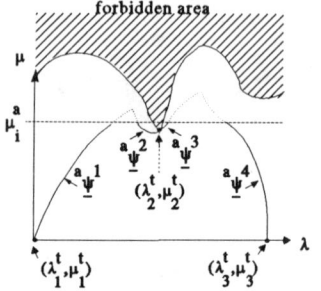

Figure 2. Proposed parallel integration

4. Conclusions + suggestions

Through the introduction of <u>simultaneous curve-calculation</u> in combination with the already very efficient principle of <u>optimal switching</u>, the proposed algorithm is expected to further reduce the calculation-load for time-optimization in the (λ,μ)-plane.

The quasi parallel character of the proposed algorithm can be made really parallel, when parallel computing devices are available.

For inclusion of viscous friction in the robot-model, which is used for the optimization, some of the steps in the algorithm must be changed, but the principle idea of simultaneous integration can still be used.

References

[1]   Meijer, B.R. and Jonker, P.P,*"The architecture and philosophy of the DIAC (Delft Intelligent Assembly Cell)"*, proceedings of the 1991 IEEE International conference on Robotics and Automation, April 9-11, Sacremento CA.,pp. 2218-2223, 1991

[2]   Verwer, B.J.H.,*"A multiresolution work space, multiresolution configuration space approach to solve the path planning problem"*,1990 IEEE Int. Conf. on Robotics and Automation, May 13-18, 1990, Cincinatti, USA, pp.2107-2112.

[3]   Rieswijk, T.A., Schalkwijk, P. and Honderd, G.,*"Planning and optimization of geometrical trajectories inside collision-free subspaces with the aid of Hermite splines."*,Euriscon "91,Greece ,1991

[4]   Shin, K.G. and McKay, N.D.,*"A dynamic programming approach to trajectory planning of robotic manipulators."*,IEEE transactions on Automatic Control, vol. AC-31, No. 6, pp. 491-500, 1986.

[5]   Shin, K.G. and McKay, N.D.,*"Robust trajectory planning for robotic manipulators under payload uncertainties."*,IEEE transactions on Automatic Control, vol. AC-32, No. 12, pp. 1044-1054, 1987.

[6]   Shin, K.G. and McKay, N.D.,"Minimum-Time Control of Robotic Manipulators with Geometric Path Constraints",IEEE transactions on Automatic Control, Vol. AC-30, No. 6, pp. 531-541, 1985.

[7]   Pfeiffer, F. and Johanni, R.,*"A concept for manipulator trajectory planning."*,IEEE journal of robotics and automation, vol. RA-3, No. 3, pp. 115-123,1987.

[8]   Bobrow, E., Dubowsky, S. and Gibson, S.,*"Time-optimal control of robotic manipulators along specified paths"*,The int. journal of robotics research, Vol. 4, No. 3, pp. 3-17, 1985.

Three-Dimension Abstraction of Convex Space Path Planning

P.K. Sinha and Pi-Luen Ho

Department of Engineering,
University of Reading,
P.O. Box 225,
Reading, RG6 2AY,
United Kingdom

KEYWORD: path planning.

ABSTRACT: In this paper, a novel framework, called path abstraction, is proposed to describe the trajectories of robot joints in joint-interpolated space. This is based on partitioning the overall joint paths into separate convex or concave sub-paths and using second-order polynomials to fit the partitioned paths. The algorithms to determine the parameters of these polynomials are outlined and described to demonstrate the effectiveness of the proposed technique.

1. INTRODUCTION

For path control of manipulators, a continuous-time path information is needed for the trajectory controller. There are two major approaches which have been proposed[1]: (a) the joint-interpolated approach; and (b) the cartesian space approach. The joint-interpolated approach plans polynomial sequences that yield smooth joint trajectory. Since the trajectory is planned directly in terms of the joint variables (joint angular position, velocity, and acceleration), the trajectory planning can be implemented in real-time. For faster computation and less extraneous motion, lower-degree polynomial sequences are preferred. For cartesian space approach, translation and rotation are needed to accomplish the desired motion of the manipulator joints. Sequences of cartesian points may be specified directly by the robot programming languages. Paul[2] and Taylor[3] developed techniques for straight-line path in cartesian space. Lin et al.[4] used cubic joint polynomials to spline n-interpolation points selected by user on the desired straight-line path in joint-interpolated space. All of them used polynomials for each section of moving path of manipulator hand or joints to approximate the exact path of a manipulator.

In this paper, a different concept, called path abstraction, is proposed to describe the trajectory of each joint in joint-interpolated space. It is based on a second-order polynomial which three real-number parameters. By determining these parameters in a specific way, the desired path of a single joint may be obtained with a reasonable degree of accuracy.

2. PATH ABSTRACTION

For a path of the single robot joint, by partitioning its waveform into a group of convex or concave sub-waveforms, the overall waveform may be represented by the following form:

$$p = p_1|_{t=T_1} + p_2|_{t=T_2} + \ldots$$
$$= \sum_{i=1}^{\infty} p_i|_{t=T_i} \tag{1}$$

where each p_i is either a convex or a concave waveform.

If each sub-waveform p_i may be abstracted into a second-order polynomial as:

$$p_i|_{t=T_i} = (k_1 + k_2 t + k_3 t^2)|_{t=T_i \text{ to } T_{i+1}} \tag{2}$$

where k_{i1}, k_{i2}, and k_{i3} are the parameters corresponding to the sub-path p_i, then the overall path may be constructed by using these parameters: k_{i1}, k_{i2}, and k_{i3}, ($i=1,\ldots,\infty$). Since all sub-paths are either convex or concave, the second derivative of the sub-path function in equation 2 is either negative or positive. The second-order function in equation 2 is useful in describing such a curve. An adequate determination of k_{i1}, k_{i2}, and k_{i3} may generate a waveform similar to the overall path. Derivation of these parameters is illustrated below.

For sub-path p_i where time t varies from T_i to T_{i+1}, parameter k_{i1} may be derived as:
$$k_{i1} = p_i(T_i) \tag{3}$$

From equation 2, it gives:
$$p_i(T_{i+1}) = k_{i1} + k_{i2} \cdot T + k_{i3} \cdot T^2 \tag{4}$$

where T is equal to $(T_{i+1}-T_i)$. To obtain the best fitted curve, using the least-square data fitting method[5,6] gives:

$$k_{i2} = \frac{7k_{i1}+3p_i(T_{i+1})}{-2T} + \frac{30}{T^3}\int_0^T p_i(t)(t-\frac{t^2}{T})dt \tag{5}$$

and

$$k_{i3} = \frac{p_i(T_{i+1})-k_{i1}-k_{i2}T}{T^2} \tag{6}$$

where an error function is defined as:

$$E = \int (p_i(t)-p(t))^2 dt$$
$$= \int (k_{i1}+k_{i2}t+k_{i3}t^2-p(t))^2 dt \qquad (7)$$

The values of parameters k_{i1}, k_{i2}, and k_{i3} may be computed by an on-line algorithm. It is also possible to use hardware to obtain the parameter values. Figure 1 shows the functional block diagram of such an abstraction mechanism, where path is represented by x variable; and H denotes the zero-order sample-and-hold device.

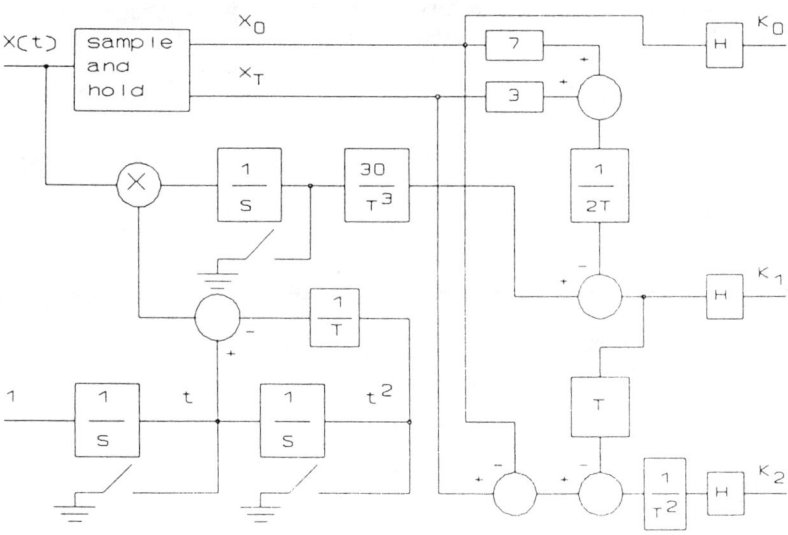

Figure 1. The functional block diagram of the abstraction mechanism.

Back-tracking of the path may be used in some specific applications using the information of the derived parameters (k_{i1}, k_{i2}, and k_{i3}). For example, by defining the following 3×3 matrix:

$$B = \begin{bmatrix} 1 & T & T^2 \\ 0 & -1 & -2T \\ 0 & 0 & 1 \end{bmatrix} \qquad (8)$$

the parameters of the time function of back-tracking path p_{bi} may be derived by:

$$[k_{bi1},k_{bi2},k_{bi3}]' = B \cdot [k_{i1},k_{i2},k_{i3}]' \qquad (9)$$

where ' denote the matrix transpose. The back-tracking matrix B has following properties:

$$B \cdot B = I \; ; \; B = B' \qquad (10)$$

I being an identity matrix.

3. APPLICATIONS AND CONCLUSIONS

As an example, a 1.0 Hz sinusoidal path is examined. Figure 2 shows the waveform of such path, where time t varies from zero to 1.0 second. If this path is partitioned into six equivalent sub-paths (the first sub-path is from t=0 to 1/6 sec., the second sub-path is from t=1/6 to 2/6 sec. etc), then each sub-path may be approximated as a convex or concave waveform. By using the abstraction mechanism in Figure 1 (or using an algorithm to implement equations 3, 5, and 6), the best fitted curve of each sub-path may be obtained. The corresponding parameters k_{i1}, k_{i2}, k_{i3} are given below:

sub-path	1st	2nd	3rd	4th	5th	6th
k	0.0	0.8660	0.8660	0.0	-0.8660	-0.8660
k	6.8448	3.1555	-3.6893	-6.8448	-3.1555	3.6893
k	-10.0821	-18.7462	-8.6641	10.0821	18.7462	8.6641

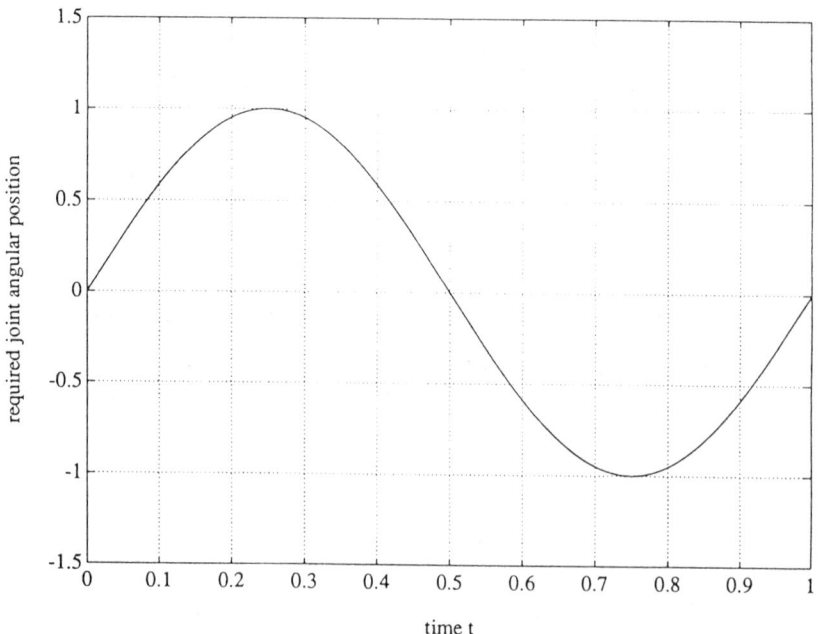

Figure 2. The original 1.0 Hz sinusoidal path waveform.

Applying these information into equation 2, the approximated path may be obtained. Figure 3 shows the abstracted path; and Figure 4 is the plot of the abstraction error.

The back-tracking path of the sinusoidal path may be obtained by using the derived parameters k_{i1}, k_{i2} and k_{i3} and the back-tracking matrix B (equation 8). Figure 5 shows the results of such a back-tracking path.

With an angular waveform as the joint path, Figures 6, 7, and 8 show the original path, the abstracted path, and the abstraction error, where six partitions are designed to form the abstraction.

In Figure 4, the abstraction error is very small, compared with the peak-to-peak value of the original signal. Therefore, this approach allows designers to approximate the desired path by using very small amount of information.

In Figure 8, the abstraction error is larger than the error shown in Figure 4. This larger error is basically caused by the shape of the original path, which is a triangular waveform with some discontinuous points. These discontinuous points may also be seen in a square-like waveform. However, most robot trajectory planning of industrial applications may not use such discontinuous path, since the main robot joint actuators are based on servo motors. The triangular waveform is a worst case for practical applications. Further work on the implementation of this abstraction method for industrial robots is under progress.

4. REFERENCES

1. Fu, K.S., Gonzalez, R.C., and Lee, C.S.G, Robotics - control, sensing, vision, and intelligence, McGraw-Hill, Inc., 1988.
2. Paul, R.P. "Manipulator cartesian path control", IEEE, trans., systems, man, cybern., SMC-9, pp.702-711, 1979.
3. Taylor, R.H. "Planning and execution of straight line manipulator trajectories", IBM J., 23, pp.424-436, 1979.
4. Lin, C. S., Chang, P. R., and Luh, J. Y. S., "Formulation and Optimization of Cubic Polynomial Joint Trajectories for industrial Robots", IEEE Trans. Automatic Control, AC-28 pp. 1066-1073, 1983.
5. Ralph H. Pennington "Introductory Computer Methods and Numerical Analysis", The MACMILLAN Co. New York, 1965.
6. Roland Glowinski "Numerical Methods for Nonlinear Variational Problems", SPRINGER-VERLAG Co. New York, 1984

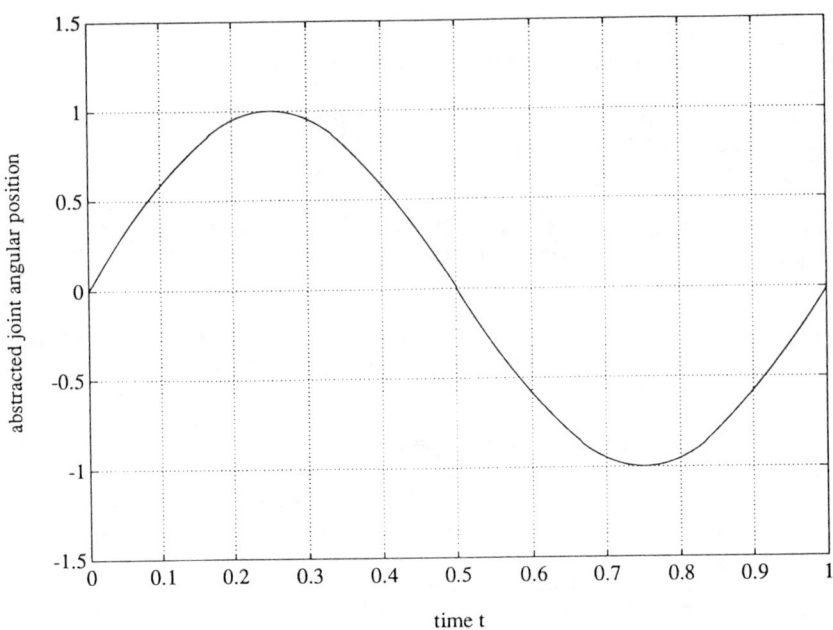

Figure 3. The abstracted path of the 1.0 Hz sinusoidal path waveform.

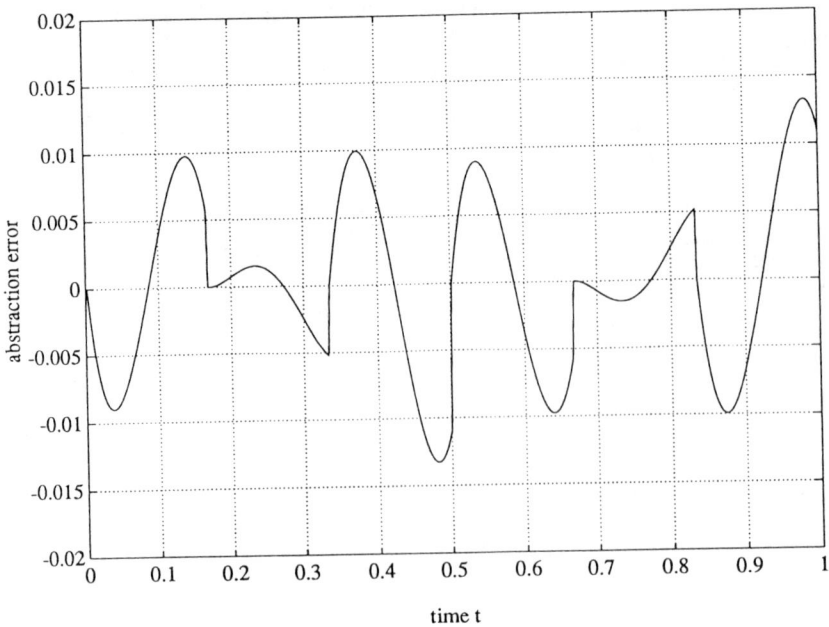

Figure 4. The abstraction error of the 1.0 Hz sinusoidal path waveform.

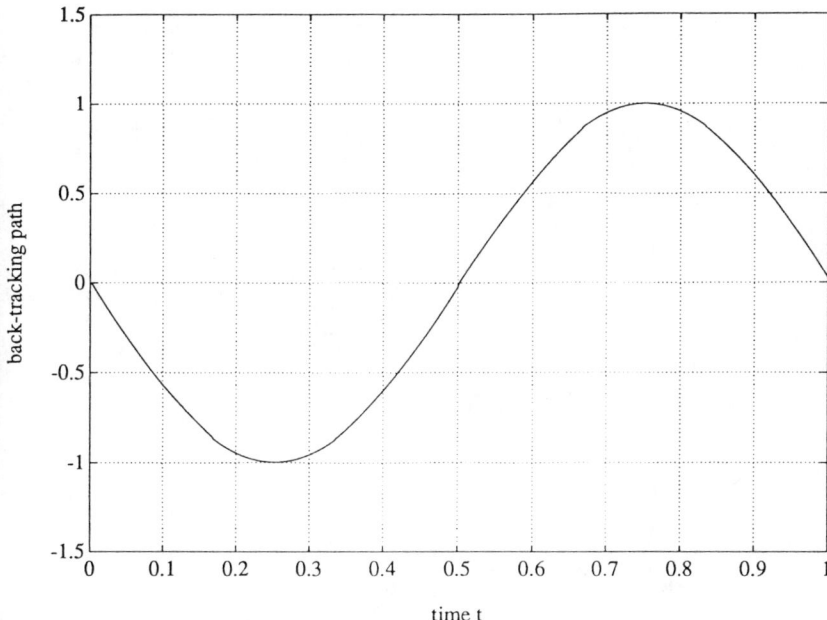

Figure 5. The back-tracking of the 1.0 Hz sinusoidal path.

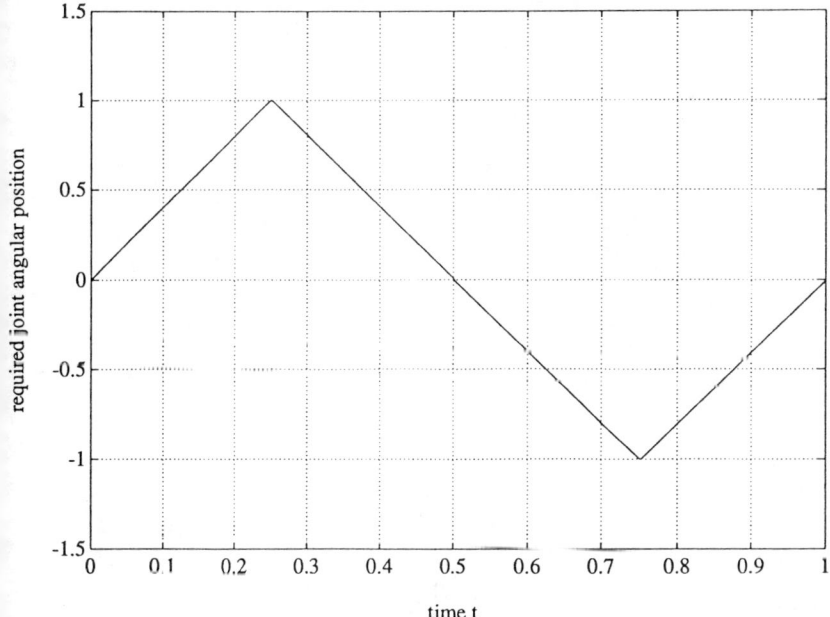

Figure 6. The 1.0 Hz triangular path waveform.

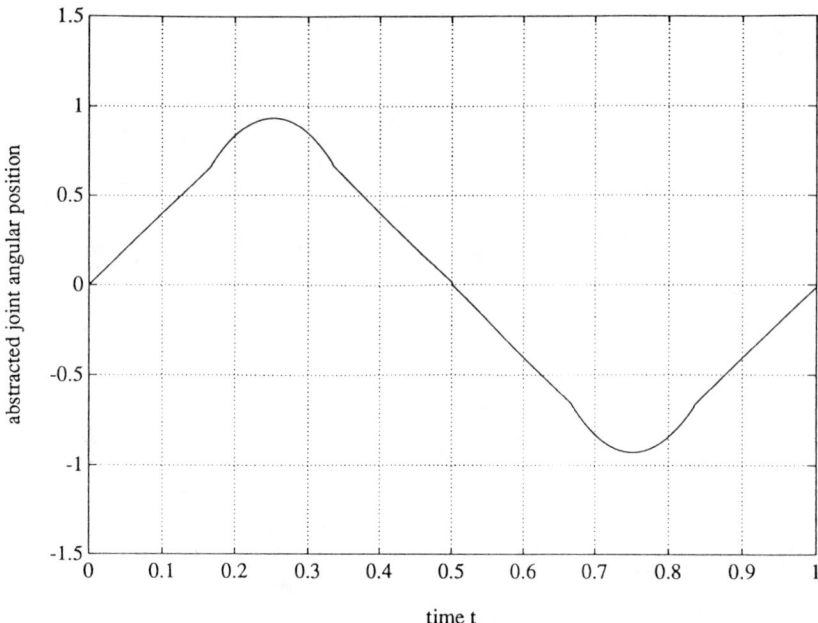

Figure 7. The abstracted path.

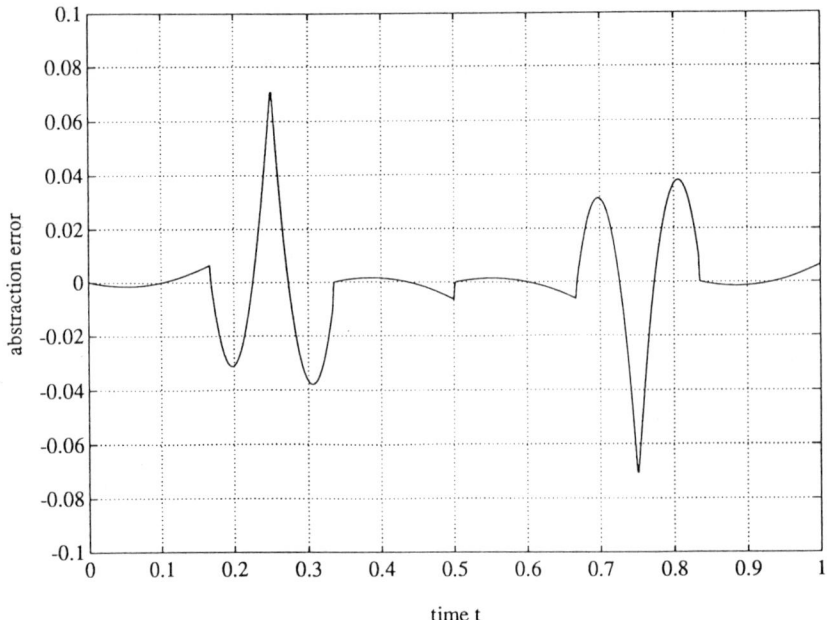

Figure 8. The abstraction error.

An Approach to Real-Time Flexible Path Planning

Alex C. Meng* Simeon Ntafos† Markos Tsoukalas‡

Abstract: In this paper we present an approach for finding real-time solutions to the path replanning problem when the specified target is changed at mid-flight. We refer to this as the changing target problem. The digital map data are converted into a geometric representation of the terrain. To achieve real-time performance, we precompute shortest paths from selected points (jump points) in the terrain using shortest path maps and shortest path trees. When the target is changed, an appropriate jump point is selected, the shortest path from the jump point to the new target is recovered and adjusted to obtain a new path from the current position of the craft to the new target. The set of jump points is chosen using a combination of geometric optimization and heuristics. Jump point selection is flexible, ranging from pre-assigning jump points to searching for a good jump pooint (depending on the available time).

1 Introduction

Automated mission planning is a central problem in Robotics research and has been studied extensively. The most commonly used techniques rely on dynamic programming [2, 9]. However the massive amount of data in digital maps for typical applications limit the applicability of this approach to pre-mission planning. For in-flight dynamic planning problems, most approaches have resorted to heuristic methods or some local optimization technique. The quality of the paths generated varies from scenario to scenario. Real-time in-flight path planning that obtains optimal or near-optimal solutions remains a difficult problem.

In this paper we discuss an approach to real-time dynamic flight planning by considering the following problem:

The changing target problem: A pre-mission path is given or planned for given start and target positions in a terrain described by a digital map. As the craft

---

*Computer Science Center, Texas Instruments, Dallas, TX 75265.
†Computer Science Program, The University of Texas at Dallas, Richardson, TX 75083-0688. Supported in part by NSF Grant IRI-9000470 and a grant from Texas Instruments.
‡Computer Science Program, The University of Texas at Dallas, Richardson, TX 75083-0688.

flies along the path, the target may change to a new position. The problem is how to plan a new path from the current position of the craft (which can be anywhere along the pre-planned path) to the new target in real-time.

In the next section we describe the general approach we use for solving the changing target problem. In sections 3, 4 we present the algorithms we use in preprocessing and real-time respectively.

2   General Approach

The objective of our solution to the changing target problem is to do robust flight path planning with real-time performance. To achieve this objective, we basically trade memory for time. Our approach uses a regional classification scheme called the **shortest path map** from Computational Geometry [4, 11]. A shortest path map allows us to reduce the infinite number of possible target positions to a small number of equivalence classes. Shortest path maps are defined with respect to a given start point. Then, we still have to deal with the infinite number of possible positions along the preplanned path which the craft could occupy at the time the target is changed. We do this by selecting important **jump positions**. The selected jump positions are used as possible start points from which shortest path maps (and hence shortest paths) are computed. Jump positions can be selected with respect to the known flight path or, in a more general framework, to handle a sequence of changes to the target position.

We use geometric shapes to describe the terrain and then apply algorithms from computational geometry. We assume that the terrain elevation data from the Defense Map Agency is available for the targeted area. Research has shown that doing path planning directly on the 3-dimension model is difficult [1]. Instead, we propose to approximate the 3-dimensional terrain with a sequence of 2-dimensional contour maps. For a fixed altitude, we can generate a 2-dimensional **contour map** that represents the horizontal slice of the terrain at that altitude. We may need to expand the terrain regions in the contour map for horizontal clearance, and to avoid possible errors in the terrain data. The storage capacity and mission criteria will determine the number of slices or contour maps for the mission. The major work in map processing is to construct the geometric representation of the contour map as a set of polygons. We can use the algorithm described in [12]. One advantage of this algorithm is that the degree of approximation is settable. This is important not only with respect to processing time but also with respect to the size (number of corners) in the resulting polygons.

Given the polygonal representation of the terrain and a fixed start position, we can compute the shortest path map which essentially represents all shortest paths from the start position to any target position in the map. To plan a path from the current position to any position in the map, we just need to plan a local path to

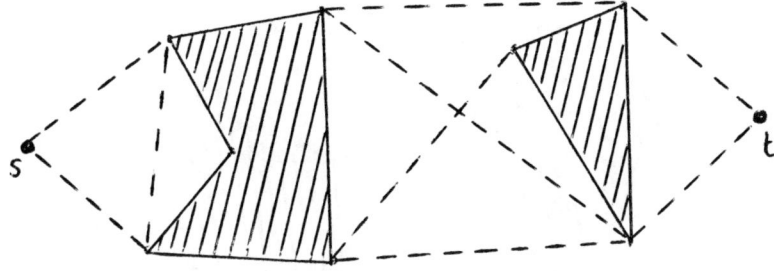

Figure 1: Visibility graph

an appropriate jump position and recover the precomputed shortest path from the selected jump position to the new target. In the following we define some concepts that we use in later sections.

- **visibility graph**
  For a given set of polygons, the corners of the polygons form the vertex set of the visibility graph. Consider the edges that connect any two vertices. The visibility graph contains those edges that do not intersect the interior of any polygon as illustrated in Figure 1. The start and target points may be included (as degenerate polygons).

- **the shortest path**
  For the given start and target positions S and T, if we use the distance of travel as the measure of cost, the shortest path is a continuous piecewise linear path from S to T with the least cost. It is a well-known property of visibility graphs that *the shortest path will travel on edges of the visibility graph* as shown in Figure 1.

- **the shortest path map (SPM)**
  Given a set of polygons in a 2-dimensional area and a start position S, the shortest paths from S to points in free space navigate the space amidst the obstacles in at most $O(n)$ ways [4, 11] (these represent distinct ways for a shortest path to get around the obstacles in its way). For a fixed start position S, the shortest path map is a partition of free space into regions such that for all points in the same region, the shortest paths from S to these points navigate through the obstacles the same way, i.e., the paths are the same except for the

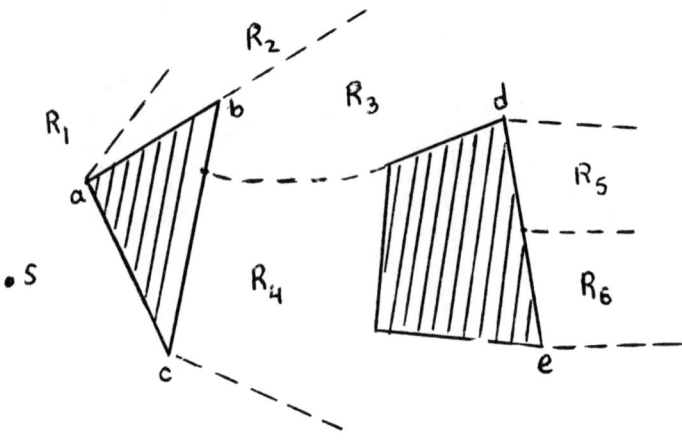

Figure 2: The shortest path map from S

last leg. Figure 2 shows one example with dashed lines indicating the last legs of shortest paths from S.

- **the shortest path tree (SPT)**
  From the shortest path map for a fixed start position S, we can find the shortest path from S to any position in the area. To facilitate the search, we keep the shortest path information in the SPM in the form of a shortest path tree. A shortest path tree is a data structure such that the shortest path from S to each region (and hence each point) can be quickly retrieved. The leaves of the shortest path tree represent the regions in the SPM. The internal nodes of the shortest path tree are corners of obstacles and the root is the start position S. Each node x (besides the root) contains a pointer to its parent and the parent is the vertex where the shortest path from S to x (or points in x if x is a region) makes its last turn. A path from a leaf to the root consists of the vertices along the shortest path from S to points in the region corresponding to the leaf (in reverse order). Figure 3 shows the shortest path tree for the shortest path map in Figure 2.

3 Pre-mission Processing

An essential structure for our approach is the visibility graph for the given polygons. We construct a reduced visibility graph by including only edges that are locally tangent at both endpoints. This can be easily done in $O(n^3)$. Faster, but more complex, algorithms for constructing the visibility graph are available [7, 13].

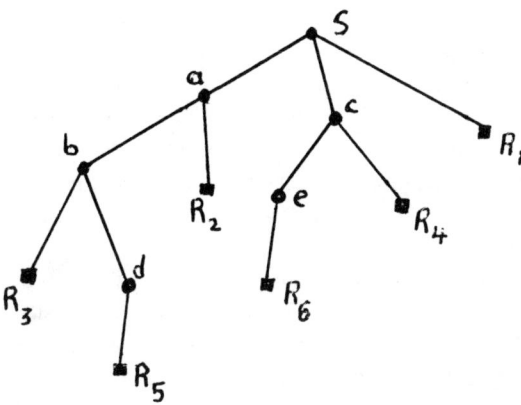

Figure 3: The shortest path tree

The shortest path map, defined in the preceding section, may contain regions that have quadratic curves in their boundaries (as shown in Figure 2). A description of the regions is needed in order to run point location algorithms when the target is changed, i.e. we find the region in an appropriate SPM that contains the new target and then obtain the shortest path to it from the corresponding shortest path tree. Most existing point location algorithms work on planar subdivisions with straight line boundaries. Adapting them to handle quadratic curves is straight forward. An alternative is to avoid dealing with quadratic curves altogether by constructing a modified shortest path map that disregards curved boundaries but keeps sufficient information about the resulting regions so that point location can still be performed correctly. We use the following two steps:

- find the shortest paths from S to every vertex in the visibility graph
- derive a modified shortest path map by extending the edges of the shortest paths

To find all the shortest paths from a fixed start position S, we use Dijkstra's shortest path algorithm and construct the shortest path tree at the same time as described in [3]. The complexity of the shortest path algorithm is $O(n^2)$. More complex implementations that use Fibonacci heaps improve this to $O(n \log n + E)$ [5]. By extending the edges of the shortest path tree, we can then use the shortest path tree to construct the modified shortest path map.

Jump points may be selected using optimization criteria related to the changing target problem. Given a shortest path, we can partition it into segments that have similar visibility properties, e.g., we can consider the variation of SPMs as the start point moves along the path. Then we can assign a jump point to each segment so that minimal adjustment will be needed to obtain an overall path to the new target. However, the number of jump points is likely to be small because of memory limitations. Finding the best placement for a specified number of jump points is an interesting open problem. If the target will be changed repeatedly, jump points are not as dependent on the original shortest path and different criteria can be used. For example, selecting a subset of the nodes of the Voronoi diagram of the scene should result in good solutions.

In practice, other considerations may determine jump point location. Likely candidates for consideration include navigation check points, positions with distinct landmark features, positions that mask the craft from threats and positions that maintain communication lines. Since we are going to compute the SPM for each jump point, we need to trade off among three parameters: the number of jump points, the quality of the replanned path and the run time performance. With more jump points, we can find better paths with less time. However, it will increase the storage capacity required. Conversely, less jump points means more search time for a quality path at run time.

Once we have determined the jump position for a given altitude, we will construct the SPM for each jump position. When the target position is changed at mid-flight, we select a subset of the jump points to replan a path. We need to determine the region that the new target belongs to in each of the corresponding shortest path maps at run time. To reduce the search time, we preprocess the SPMs using the so-called *slab method* [10].

4 Real-time Processing

After the preprocessing, for any given current position of the craft and the new target position, we can reduce the run-time path replanning to the following:

- select a jump point X and locate the region that contains the new target in SPM(X).

- retrieve the precomputed shortest path from the jump point to the target and adjust it to obtain a path from the current position to the new target.

Depending on the amount of time available, we can search for a jump point in several different ways:

- pre-assign jump points to segments of the shortest path.
- find the nearest jump point and fly the shortest path to it.
- search the jump points locally within a small range and select the best one.

- search the jump points globally and find the best one.

This step is flexible relative to the time available for searching. In local or global searches, the new target is located in each SPM, the shortest paths to it are recovered and the jump point is selected based on the quality of the overall path from the current position to the new target through the jump point.

For the new target position, we can use the slab regions to quickly determine the region where it belongs in the shortest path map. We use the Y-coord of the target as the key in a binary search that determines the horizontal slab containing the target. Then we use the X-coord of the target in a binary search of the slab from the previous step to identify the region containing the target. Once we locate the jump point, we can retrieve the shortest path from the jump point to the new target position by looking up the region containing the target in the shortest path tree for the SPM of the jump point. Then we plan a local path to the jump point and adjust to obtain an overall path to the new target.

5 Conclusion

We presented an approach for real-time dynamic path planning for the changing target problem. Due to the dynamic nature of the problem, the target and threats can change repeatedly and our algorithm can also be used repeatedly. By compiling shortest paths in shortest path maps for a set of jump points, we reduce the run-time computation to a minimum.

An interesting application for the changing target problem is in dealing with unexpected threats. Upon detection of a new threat, if possible, the craft should plan a threat avoidance path to increase its survivability. Note that the threat avoidance problem can be viewed as a sequence of changing target problems. That is, once the new threat is detected, we can select one (or a sequence of) new target off to the side of the threat and have the craft move to it (them). When the craft bypasses the threat, the target is changed again to the original one and the craft reaches its destination.

A problem with the above approach is that the threat is modeled rather crudely. A much better model for areas with varying degrees of risk and for threat penetration situations is provided by using weighted regions [8]. Finding least cost paths through n weighted regions takes $O(n^7)$ time which may well be impractical [8]. We have obtained $O(n^2)$ algorithms for the spacial case in which the weights are zero (to indicate risk-free space), one (to indicate areas of risk) and infinity (to indicate obstacles or very high risk areas) [6]. Furthermore, this algorithm can handle linear features that have arbitrary weights assigned to them. This provides a way to model threats with an appropriate set of linear features (placed so that the craft's path is forced to cross them) so that we still have a reasonably good model of the threat but we can use the faster algorithm to do path planning. While the complexity is significantly reduced, real-time response is unlikely.

References

[1] Canny, J., "Complexity of Robot Motion Planning," PhD Dissertation, MIT, 1987.

[2] Chapoton, C., "AI Applications to the ETMP Program," Texas Instruments Engineering Journal, Vol. 3, 1986, pp. 24-33.

[3] Cormen, T., C. Leiserson and R. Rivest, "Introduction to Algorithms," Mc-Graw Hill, 1990.

[4] Franklin, W., V. Akman and C. Verrilli, "Voronoi Diagrams with Barriers and on Polyhedra for Minimal Path Planning," *The Visual Computer*, pp. 133-150, 1985.

[5] Fredman, M. and R. Tarjan, "Fibonacci Heaps and Their Uses in Improved Network Optimization Algorithms," *Journal of ACM*, Vol. 34, No. 3, pp. 596-615, July 1987.

[6] Gewali, L., A. Meng, J. Mitchell and S. Ntafos, "Path Planning in 0/1 Weighted Regions with Applications," Proc. 4th ACM Symp. on Computational Geometry, June 1988.

[7] Ghosh, S. and D. Mount, "An Output Sensitive Algorithm for Computing Visibility Graphs," Proc. of 28th FOCS, pp. 11-19, 1987.

[8] Mitchell, J. and C. Papadimitriou, "The Weighted Region Problem" Proc. 3rd ACM Conf. on Computational Geometry, 1987.

[9] Nordmeyer, R., "Enhanced Terrain Masked Penetration," Final Report, Texas Instruments, Inc. 1986.

[10] Preparata, F.P., M. I. Shamos, *Computational Geometry, An Introduction*, Spring-Verlag, 1985.

[11] Reif, J., and J. A. Storer, "Shortest Paths in Euclidean Space with Polyhedral Obstacles," Technical Report CS-85-121, CS Department, Brandeis University, 1985.

[12] Sklansky, J., V. Gonzales, "Fast Polygonal Approximation of Digitized Curves," *Pattern Recognition*, Vol 12, pp 327-331, 1979.

[13] Welzl, E., "Constructing the Visibility Graph for n Line Segments in $O(n^2)$ Time," *Information Processing Letters*, Vol. 20, 1985, pp. 167-171.

A PATH PLANNING METHOD FOR MOBILE ROBOTS IN A STRUCTURED ENVIRONMENT

Fivos V. Hatzivasiliou(*) and Spyros G. Tzafestas
Intelligent Robotics and Control Unit (IRCU),
Computer Science Division,
National Technical University of Athens,
Zografou 15773, Athens, Greece
(*) Also with : Hellenic Air Force Research Center,
Terpsithea, A.Glyfada, Athens 16501 , Greece .

ABSTRACT

This paper addresses the collision-free path planning problem for mobile robots, that operate in a structured environment. The collision avoidance algorithm proposed is based on the geometric description of all the objects (as polyhedra), and the geometric and kinematic description of the mobile robot. The algorithm transforms the obstacles so that they represent the locus of forbitten positions for a reference point that stands for the mobile robot. The generation of the locus of forbitten positions is based on an efficient "prunning" method developed here, and paths here are found by searching a network between known start and goal points.

The above algorithm implemented in the Intelligent Robotics and Control Unit., and evaluated by simulating the generation of obstacles on a graphics monitor. Therefore comparative results were obtained , by means of time and optimal length measures.

1. INTRODUCTION

The path planning problem is of primary interest in mobile robot applications and is quite different from path planning problem for manipulators. A mobile robot is a mobile machine capable of autonomous operation in a structured and/or unstructured environment.

It is known that current mobile robots up to now have limited capabilities to perform tasks recognizable as 'intelligent thinking'. However, modern intelligent mobile robots are able to employ the 2-D or 3-D world of a vision system to assist various tasks (such as visual navigation, obstacle avoidance etc.).

Path planning in the presence of obstacles can be classified in one of two major categories:
(i) map based (high level control approache), where assuming that a precompiled map of the terrain is available, high level path planners provide paths which guarantee a collision - free motion through an arbitrary ensemble of known obstacles,and
(ii) sensor-based (low level control approache) where the interaction with the real world is the concern of the low-level part of the control system, whose task is the execution of elementary operations that guide the robot along the precisely specified path. The main

desire is to shift parts of the path planning task to a low level control system, that makes better use of low-level control capabilities in performing real time operations.
In the first case the problem is simple: Given an object with initial location (and orientation), a goal position and orientation, and a set of obstacles located in space, find a continuous path for the object from the initial position to the goal position, avoiding collisions with obstacles along the way.

The algorithms for obstacle avoidance by mobile robots can be grouped into the following classes:
(1) **Hypothesize and test,**
(2) **Penalty Function ,and**
(3) **Explicit free space.**

The earliest algorithm is the "hypothesize and test" method. A simple path from the start to the goal point, usually a straight line, is hypothesized, and then the path is tested for potentional collisions. If there are collisions, then a new path is proposed. This is repeated (usually using information about the previous collision) along the path. In summary the steps for this algorithm are [1]:

(1) Calculate the volume swept out by the moving object along the proposed path.
(2) Determine the overlap between the swept volume and the obstacles, and
(3) Propose a new path.

The main advantage of the "hypothesize & test" method is its simplicity. The technique's basic computational operations are the detection of potentional colissions and the modification of proposed paths to avoid collisions.

The second category of algorithms is based on the introduction of a penalty function that encodes the presence of obstacles. In general, the penalty is infinite for configurations that causes collisions, and drops off sharply with distance from obstacles. The total penalty function is computed by adding the penalties from individual obstacles, and, possibly , adding a penalty term for deviations from the shortest path.

The third category of obstacle avoidance methods is based on explicit representations of subsets of robot configurations that are free of collisions, the so called free space. Obstacle avoidance is then the problem of finding a path within these subsets that connects the initial and final configurations.The algorithms differ primarily in the particular subsets of free space they represent and in the representation of these subsets. The advantage of free space methods is that they use an explicit characterization of free space which allows them to define search methods that are guaranteed to find paths, if one exists, within the known subset of free space. Moreover it is feasible to search for short paths, rather than simply finding the first path that is safe. The disadvantage is that the computation of the free space may be expense, since the motion of a mobile robot constrained by the presence of obstacles needs geometric descriptions of objects. There are also additional constraints on motion imposed by the kinematic structure of the robot itself.

Regarding the geometric description there are different ways for the manipulation of the available data.For the second and third class of algorithms discussed above , there are two general approaches. The first is to search the free space directly . Explicit representations of free space using overlapping generalized cones ,called freeways, are used in [3],[4]. Translations are made along freeways with rotations performed at the intersections of freeways. One useful feature of this approach is that it typically generates paths for the mobile robot that stay well away from the obstacles, although in some instances this leads to paths which are considerably longer than the shortest path. When the workspace is moderately populated with obstacles, this method is fast and quite effective.

A more refined representation of free space can be obtained using the Generalized Voronoi Diagrams(GVD). A GVD is the locus of points which are equidistant from two or more obstacle boundaries including the workspace boundary. Now if a collision-free path exists (from source to goal), then a collision-free path exists which traverses the GVD except for its initial and final phases which correspond to entering and exiting the GVD.

A second general approach for solving the path planning problem is the motion of an object through a space of obstacles transformation of the into an equivalent but simpler problem of planning the motion of a point through a space of enlarged configuration-space obstacles. The configuration space provides an effective framework for investigating a variety of robot-motion planning problems. This technique is very effective when applied to a purely translational motion in a plane where the optimal path can be found. If the obstacles are modeled as collections of convex polyhedra, the position constraints can be stated in terms of the position of the vertices of the moving object relative to the planes of the obstacle surfaces, and the above technique (configuration-space obstacles) becomes extremely powerful.

The algorithm presented in this paper uses an accurate growing operation which, in conjuction with a prunning technique, minimizes the blind areas in 2D space. Use of a graph searching technique for path finding which produces optimum 2D paths when only translations are involved. This kind of technique can easily be generalized to deal with 3D obstacles[1]. For concreteness, the discussions and examples of this paper assume that all objects are modeled as sets of (possibly overlapping) convex polyhedra.

2. 2-D COLLISION AVOIDANCE

Consider the problem of moving a point object A from position Start (S) to position Goal (G) avoiding the obstacles (Fig.1). The main property of this path is that it is composed of straight lines connecting the S and G points via a sequence (possibly empty) of vertices of obstacles. In the case of planar motion with arbitrary polygonal objects, the shortest collision-free path that connects any two accessible points always has this property[1].

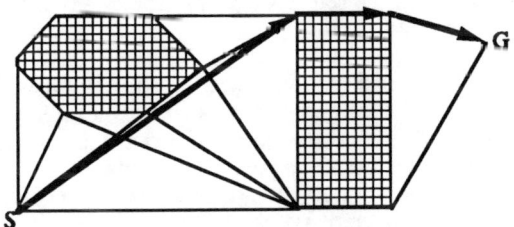

Fig.1

The undirected graph is denoded by VG(N,L). The node set N is V U {S,G} where V is the set of all vertices of obstacles, and the Link set L is the set of all links (n_i,n_j) such that a straight line connecting the ith element of N to the jth does not overlap with any

obstacle. The graph VG(N,L) is called the visibility graph (**VGRAPH**) of N (since connected vertices in the graph can see each other, see Fig.1). The shortest collision-free path from S to G on the plane is the shortest path in the VGRAPH from the node corresponding to S to that corresponding to G when the Euclidean metric is based on the links.
This method of finding collision free paths for a point by finding the shortest path in a visibility graph is called the VGRAPH algorithm.The simplicity of the VGRAPH algorithm stems from the fact that the moving object A is a point. This is a good approximation when the moving objects are small in relation to the obstacles, but causes problems otherwise.In this paper we will show how a more general form of the collision avoidance problem can be accurately reduced to the VGRAPH problem.

A simple generalization of the problem is to make the moving object A a circle with nonnegligible radius r_A. The VGRAPH algorithm has been adapted to this situation by moving the vertices away from the obstacles so that they are at least r_A away from all the sides (Fig.2). Moving A so that its center point moves through the new displaced vertices will still produce a minimum distance, collision-free path. However the path found is different from that in Fig.1 .

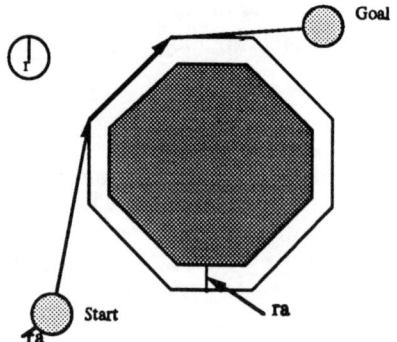

Fig.2

In the VGRAPH algorithm, if the moving object is not a point, a new set of obstacles must be computed which are the forbidden regions of some reference point on the moving object. These new obstacles must describe the locus of positions of this reference point which would cause a collision with any of the original obstacles.The procedure for computing a new obstacle O' from an original obstacle O and a moving object A is called "growing O by A". This name reflects that the obstacles are being enlarged so that the moving object can be shrunk to the reference point. The result of this operation has been indicated as **GOS(A)**.

A very common case is shown in Fig.3 where the moving point A is a rectangular solid. This figure demonstrates how the process of growing obstacles allows the representation of A as a point. The "growing" operation has been defined as the computation of the locus of positions of the moving object's reference point that would cause a collision with a given obstacle. The position of the moving object is represented by its (x,y) position. This is an arbitrary but natural choice. Different types of moving objects would call for different choices.The use of polyhedra as the basic unit of shape description influences our choice of obstacle representation . Polyhedra (polygons on the plane) have boundaries which are linear equations in the coordinate variables. This property makes them computationally attractive. In this section the objects were represented as polygons in a planar cartesian coordinate system.

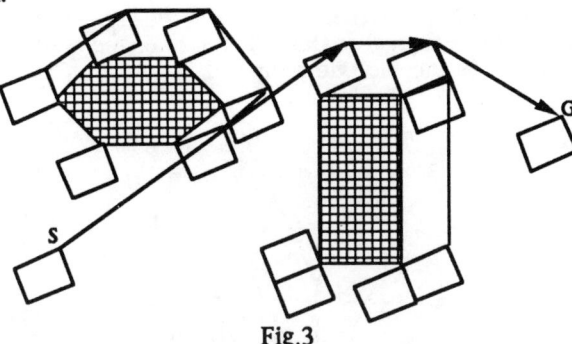

Fig.3

3. OBSTACLE PRUNNING

In this section we expand the VGRAPH algorithm and compute in more detail the shape of the forbidden regions for the moving object's reference point. The situation where the moving object is 2D and nearly circular will be examined (This can be easily expanded in 3D for cylindrical objects). The problem for rectangular or polygonal moving objects has been thoroughly studied in [1], [3], [6].

In the 2D world the degrees of freedom (for growing and prunning), are the x and y positions of the moving object. Consider the case of growing a polygonal obstacle by a circular solid as in Fig.4 .

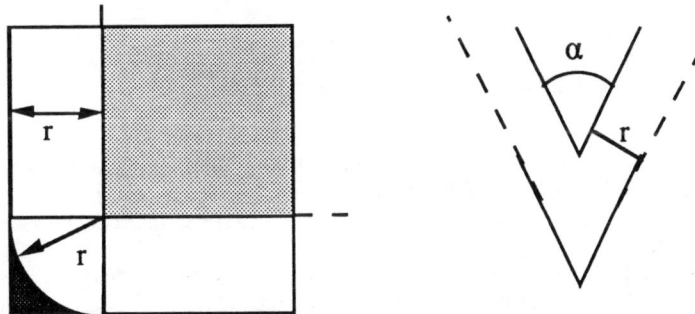

Fig.4   Fig.5

The growing algorithm moves each of the sides of the original obstacle by a constant amount r_A and then intersects the lines to obtain the vertices of the enlarged polygon. The main drawback of this algorithm is that it generates wasted space near pointed corners, as seen by the dark shaded regions in Fig.4.

The growing operation for a polygon vertex is depicted in Fig.5. With known r and θ one can calculate the wasted space by:

$$r^2\left(\frac{1}{\tan\theta}-\frac{\pi-2\theta}{2}\right) \quad \text{(Eq.1)},$$
$$\theta=\alpha/2 \text{ (rad)}$$

To reduce this wasted space for a feasible solution, the following prunning technique is introduced.
The line from the old to the new vertex is drawn (after the "growing" operation). This line is also the bisector for the two equal angles. On that line, and at distance r, we draw the vertical line (Fig.6) and define the two points of intersection as the new vertices in VGRAPH.

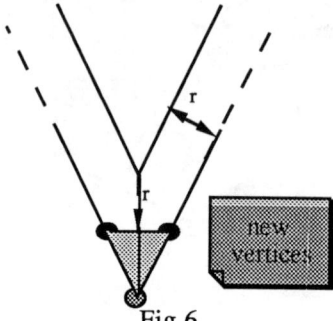

Fig.6

Following the above technique the additional available new free space is equal to:

$$r^2 \frac{(\sin\theta-1)^2}{\sin\theta\cos\theta} \quad \text{(Eq.2)}$$

(Dashed space in Fig.6)

Now the area, that is not included is equal to:

$$r^2 \left(\frac{1}{\tan\theta} - \frac{\pi-2\theta}{2} - \frac{(\sin\theta-1)^2}{\sin\theta\cos\theta} \right) \quad \text{(Eq. 3)}$$

Since the size of moving object in 2D is proportional to r^2 the main criterion for evaluating the method and defining a kind of heuristic is the angle θ. The main drawback of this method is that it creates more complexity for the VGRAPH search. This will be discussed in the next section.
Certainly the above technique can be repeated many times to minimize the wasted space, but this will generate a combinatorial explosion in the VGRAPH search computational requirements. The first derivative of the quantity (2) without the r^2 (for the above mentioned reason) is:

$$\frac{\partial}{\partial \theta} \left(\frac{(\sin\theta-1)^2}{\sin\theta\cos\theta} \right) = \frac{2\cos\theta(\sin\theta-1)}{\sin\theta\cos\theta} - \frac{(\sin\theta-1)^2(\cos^2\theta-\sin^2\theta)}{\sin^2\theta\cos^2\theta}$$

(Eq.4)
for angle $\theta=[0,\pi/2]$ or $\alpha=[0,\pi]$
The plots for eqns (2) and (4) are:

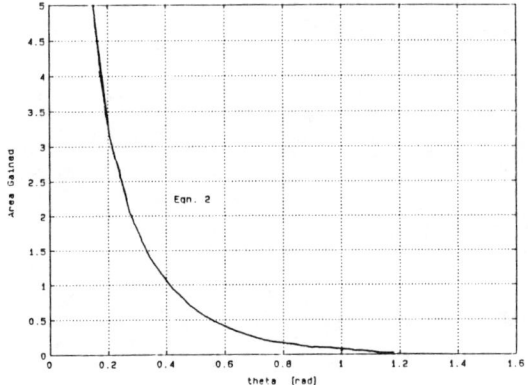

Fig.7

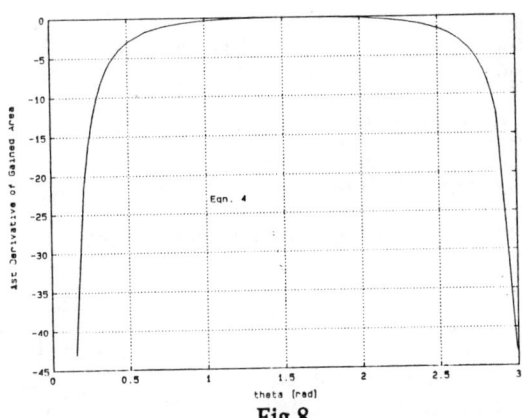

Fig.8

Both become zero for $\theta=\pi/2$ (ie $\alpha=\pi$), and this is reasonable since in this case there is neither angle nor vertex, so there is no wasted space at all. The ratio between the area defined in (2) and the total wasted area (1) is:

$$\frac{\text{Equation (2)}}{\text{Equation (1)}} = \frac{\left(\frac{(\sin\theta-1)^2}{\sin\theta\cos\theta}\right)}{\frac{1}{\tan\theta}-\frac{\pi-2\theta}{2}}$$

(Eq. 5)

and the plot has the form:

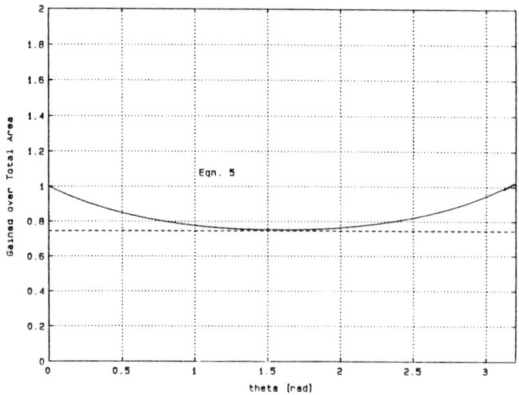

Fig.9

The plot shows that this ratio is between 100% for $\theta=0$ and 75% for θ close to $\pi/2$. This ratio gives a good measure for the efficiency of the above prunning technique. The result of prunning a vertex X (of a grown obstacle) will be indicated as **PG(X)** *and is the two new vertices that replace X in the Graph.*

4. COMPLEXITY ISSUES

The generalized visibility graph VG(N,L) has been defined as:
i) a node set N and,
ii) a link set L_{ij} of links between node pairs (n_i,n_j), such that the visibility function $l(n_i,n_j)$ is true.

A node is a representation of a region in an n-dimensional parameter space. The path finding problem is defined as : Given a node set N with an associated parameter vector set, a start node n_s, and a goal node n_g, find a sequence of nodes, from n_s to n_g, by means of an ordered set of intermediate nodes n1,..., nk (which may be null), such that the visibility function for each node pair in sequence is true and a cost function is minimized.

A direct approach for finding an optimum path is to enumerate all possible paths and choose the one for which the cost is minimum. The algorithm of Moore and Dijkstra [7] can be used to compute the shortest path from the start vertex (numbered 1). The above algorithm computes the shortest path from the start vertex to all the others. The algorithm works in $O(M) + O(N^2)$ time where M is the number of links. In a dense graph (where M is of order N^2) the term N^2 is acceptable. In a sparse graph the above time becomes $O(MlogN)$. This is found using priority queues, since it is possible to represent a priority queue in the computer such that each of the basic operations (insert an element, delete an element (by name), delete the element with the smallest numerical value) needs $O(log(K))$ elementary operations, where K is the number of elements in the structure. In the more special case , where all the lengths are equal, the shortest path algorithm works in $O(M)$ time. Thus in the worst case ($O(N^2)$) the prunning algorithm doubles the vertices (nodes), and so the complexity becomes $O(4N^2)$ (order $O(N^2)$).

Certainly we can use a simple heuristic approach (as is evident from Fig.9) and apply the above algorithm to angles greater than a predetermined value.
The same prunning algorithm can also be used in conjuction with other graph-searching techniques, since for node sets whose cardinality is of practical interest (>50) the compunational load of the direct method is prohibitive, and a more efficient heuristic based search may be used.
The A* algorithm allows the use of efficient heuristic information. For each node, an estimate **hhat** is determined of the cost h to travel from it to the goal node. Initially, n_s, is placed on a list of candidate nodes for examination. At each step of the algorithm, the node with the minimum total path cost estimate is moved onto another list, and its minimum cost estimate visible successor nodes are placed on the first list. Finally the cost function may be tailored to suit the requirements of a particular problem environment, e.g. "distance to be traveled in a subspace of the parameter space","junctions of distances in parameter space"etc.. Finally the form of the visibility function depends on the semantics of the parameters.

5. CONCLUSIONS

In this paper a variation of the configuration space method was presented, which adds flexibility to the VGRAPH search methods (either direct or heuristic approaches).
The above variation was implemented using Pascal, and its efficiency was evaluated by simulating randomly generated obstacles on a graphic monitor. Since the prunning algorithm was applied only to circular moving objects, we didn't examine the case of rectangular moving objects that has been thoroughly examined in [1],[3],[6].
Ofcourse the real world is 3-dimensional and not just a 2D surface map If the robot is to be allowed to continue co-habitation with humans it must be able to model the 3-dimensional world. The worlds where mobile robots will do useful work are not constructed from exact simple polyhedra. However since polyhedra are useful models of a realistic world, it is useful to build a special world for behavioural experiments.

REFERENCES

[1] *Tomas L. Perez & Michael A. Wesley*,An Algorithm for Planning Collision - Free Paths Among Polyhedral Obstacles, Communications of the Acm,Oct. 1979,Volume 22,Num.10, pp.560-570.

[2] *Fu,Gonzales,Lee*, Robotics, Mc Graw Hill

[3] *R.A. Brooks,* Solving the find path problem by good representation of free space, IEEE Trans. System, Man & Cyber., vol SMC-13, no 3 , pp.190-197, Mar./Apr. 1983.

[4] *S.K. Kambhampari & L.S. Davis* , Multiresolution path planning for mobile robots, IEEE J. Robotics Automat., vol. RA-2, no 3, pp. 135-145,Sept. 1986.

[5] *S.M. Udupa*, Collision Detection & avoidance in computer controlled manipulators, Proc. 5th Intern. Conf. on Artificial Intelligence,MIT, Cambridge, MA, Aug. 1977, pp.737-748.

[6] *R.A. Brooks & Lor. Perez*, A subdivision Algorithm in Configuration Space for Findpath with Rotation, IEEE Trans. System, Man & Cyber., vol SMC-15, No.2, March/April 1985,pp.224-233.

[7] *M.Gondran, M. Minoux & S.Vajda*, Graphs & Algorithms, New York, NY:Wiley,1984.

[8] *O. Takahashi & R. Schilling*, Motion Planning in a Plane Using Generalized Voronoi Diagrams ,IEEE Trans. Robotics Automat.,Vol.5, No2, April 1989, pp.143-150.

[9] *L.Gewali, S. Ntafos & I. G. Tollis*,Path Planning in the Preesence of Vertical Obstacles, IEEE Trans. Robotics Automat.,Vol.6. No.3, June 1990,pp.331-341.

[10] *N. Ayache & O. Faugeras*, Maintaning Representations of the environment of a Mobile Robot,IEEE Trans. Robotics Automat.,Vol.5, No 6, Dec.1989, pp.804-819.

MINIMUM–TIME MOTION PLANNER FOR MOBILE ROBOTS ON UNEVEN TERRAINS

Alain LIEGEOIS and Christophe MOIGNARD
Laboratoire d'Automatique et de Microélectronique de Montpellier
Université de Montpellier II, FRANCE

ABSTRACT

This paper presents a method for generating the minimum–time path and motions of a vehicle, taking into account, the relief, the obstacles, the ground surface characteristics, and the vehicle dynamics, including engine and gear.
The terrain is modelled by a triangulation, obtained from the level curves and characteristic points. Monodimensional elements like rivers and roads can be added as polygonal lines, and regions of various surface characteristics are sets of such mono– or bi–dimensional elements, the obstacles being affected by a vehicle mobility equal to zero. The other edges of the model are candidates for generating parts of the optimal path.
A A*–like search is used for that. Taking the travel time as the cost function, the velocity of the vehicle is computed from its dynamic model : the power/rpm engine characteristic is approximated by a third degree polynomial from which the driving force is computed. The model of the terrain supplies the required resistant forces due to slope and friction. In climbing motions the best gear is selected automatically by an appropriate algorithm while the downhill parts can be passed either at maximum velocity or kept the same as when climbing. Realistic examples are given which demonstrate the ability of the method for generating the orders required by the pilot : heading angle, gear, and speed.

1. INTRODUCTION

A great amount of work has been devoted in the past to the problem of path planning for autonomous vehicles in 2D–worlds with obstacles. In these cases the speed is low and can be assumed constant on a flat floor, so that the vehicle dynamics is ignored and the problem becomes a geometrical one : i.e. finding a trajectory, the shortest path or not, of a vehicle from a starting configuration (position and heading angle), to a goal configuration, while avoiding collisions with obstacles.
More realistic situations for outdoors vehicles are considered in the so–called "weighted–region problem" [MIT87, MIT88, ROW90] which, despite its two–dimensional character, leads to an enormous computational complexity for finding approximate numerical solutions.
On the contrary, very little attention has been paid to vehicles on uneven surfaces in spite of the potential applications of Robotics in surveillance, fire–fighting, agriculture, planets exploration and so on. This may be due to the increased theoretical complexity of the problem as compared to the two–dimensional one :
a) The ground is uneven and non–homogeneous, and its characteristics may vary according to weather variations,

b) The vehicles may be driven by other actuators than electric motors, for example by combustion engines,
c) Their velocities may be high,
d) The performance index is no longer the path length, it may be the time, the fuel consumption, with additional constraints such as risk, stability, etc.

For these reasons, new problem formulations and solutions are needed. In the open literature, the minimum–time motion has retained the attention of the researchers. Gaw and Meystel [GAW86] consider a world representation based on polygonized "isolines" (or level–curves). A similar approach, but using a different vehicle model and including a mobility coefficient, has been treated by Manaoui and Liégeois [MAN88,LIE90], while Shiller and Chen [SHI90] consider a bicubic surface representation, on which the time–optimal motion under constraints is computed along a geodesic curve.

Having in mind the search for simplified and fast computations, the method presented in this paper is based upon a discretization of the environment and a graph search technique. It starts from a triangulation (Section 2) preserving the edges of the discretized level–curves, that ensures a reasonable number of successors to each node. Then the cost function is computed from a realistic dynamic model of the vehicle and of its engine (section 3) as in [MAN88] and pure or modified A* search is finally used to compute the optimal path, gear ratio and speed (section 4). The cost function is shown to be not too sensitive to reasonably small deviations from the obtained path and to variations of the mobility factors, which justifies the search on the triangles edges and the weighting of the heuristic cost, all things which accelerate the search.

2. TERRAIN REPRESENTATION

Terrain modelling is the key of motion planning when the vehicle dynamics is to be considered. At every point of a trajectory, the model must be able to provide the slope and the friction coefficient, which are necessary to compute the speed and thus the duration of motion. The model must also be consistent with additional constraints –for example the presence of rocks, ice, smog– which may modify the speed. Since a graph search technique will be used in order to be sure to find, if exists, the optimal motion between two points on the surface, some discretization of the terrain is to be called for. Regular grids lead to a very large number of points so that a triangulation between the vertices of polygonized level curves has been preferred.

Three–dimensional Delaunay triangulation has been successfully used elsewhere for modelling solid objects [BOI 88]. It is based on strong mathematical properties [PRE85] but is computationnaly heavy when the shape complexity requires a great number of points to describe the surface. If one considers the relatively smooth characteristics of most reliefs on Earth seen with a resolution of, say, tens of meters, and if overhangs are disregarded, we may search for a 2–D type Delaunay triangulation able to work much faster than a 3–D one.

Polygonation

The discretization of the level curves is preliminary to obtaining further a "good" triangulation preserving the initial main information.

Our method consists in limiting the maximum length of any edge of the polygonation as a function of the difference of altitude between two consecutive level curves and of an estimated maximum slope of the terrain, in such a way that the length of an edge is always smaller than the length of the projection of any vector joining vertices on level curves immediately above and under the considered edge.

Another method involves constraining the triangulation to level curves by inserting the middle points of the edges of the level polygonized level curves which are omitted in the "first" triangulation (below).

The first triangulation
The method used in this step is similar to Watson's algorithm [SLO84].The algorithm is shown experimentally to work in time $O(n^{3/2})$.

Insertion of additional points
In practice, such "characteristic" points are for example saddle–points, bridges, fuel stations, points ensuring a radio–link communication, where the robot is likely to pass through. As the motion planner will give a path following some edges of the triangulation, the characteristic points must be added as new vertices. Theoretical works have demonstrated that though the addition of a single point may break all the original triangulation[FIO85], it is not the case in practice and this operation can be done in constant time, by applying for example Guibas and Stolfi's algorithm [GUI85]. Our first approach described below was still simpler, since an exact Delaunay triangulation is not necessarily required. It is based of the fact that the incircle test is equivalent to constructing triangles with are as equilateral as possible. Consider the ideal case of a regular triangular grid, a part of which is show in figure 1.

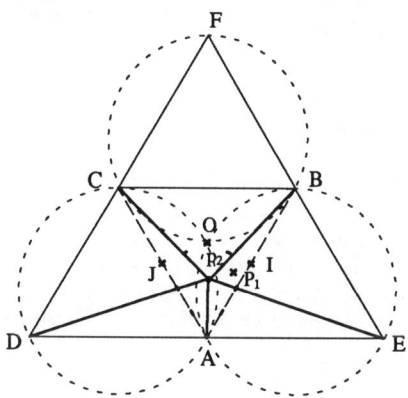

Figure 1 : simple insertion.

Without loss of generality (by symmetry), we assume that the point P to be inserted lies inside the triangle ABC and, moreover, inside the quadrilateral AIOJ, which means that A is the nearest–neighbour of the new point. It is known [GUI85] that the three edges issued from the new point and ending respectively at A, B and C are edges of the Delaunay triangulation after insertion, but one can also claim that at least one of the two edges AB and CA must be deleted : the new point is either interior to the circumcircle of triangle AEB (point P_1) resp. ACD, or to both circles (point P_2). The result is that new edges are P_1E or P_2D and P_2E. The actual topology of the "initial triangulation" being the same as the regular one, the following algorithm can be applied :

step 1. locate the new point in an existing triangle and find the nearest vertex, say A,
step 2. add the edges PA, PB and PC,
step 3. consider the adjacent triangles, say AEB and ACD, and apply the incircle test in order to add EP (and delete BA) and/or PD (and delete AC).

The method does not ensure a Delaunay result but its implementation is very simple. Since the complexity of *steps 2* and *3* are O(1), the overall computational complexity is that of *step 1*, which is known to be O(log n) [KIR83]. Actually, we have used a practical "bucketing" technique since preprocessing is done offline. The set of n points are located into rectangular buckets, each containing approximately m points. Locating a new query point thus begins with a search in a quad–tree [SAM82] and then finds the nearest–neighbour among the subset of m points.

Figures 2 and 3 show the triangulation resulting from the level–curves and additional characteristic points.

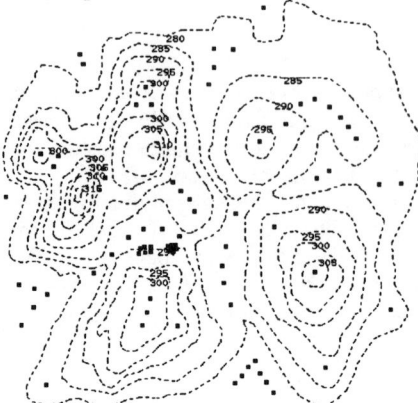

Figure 2 : level curves and additional characteristic points.

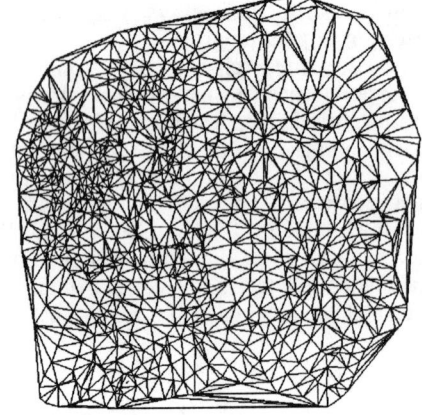

Figure 3 : triangulation.

3. VEHICLE DYNAMIC MODEL AND OPTIMAL SPEED COMPUTATION

The minimum–time motion computation requires the knowledge of the optimal gear–ratio and speed on each segment of a trajectory approximated by a polygonal line formed by consecutive edges of the triangulation. More precisely, the vehicle is assumed to follow line segments parallel to the corresponding edges, located at an infinitely small distance from them on the triangles where the vehicle pitch axis is the most horizontal, in order to prevent slippage or tip–over [SHI90].

The dynamics can be expressed as :

$$m \, dV/dt = F(V) - f(V) - m \, g \, (\sin \theta + K_f \cos \theta) \tag{1}$$

where m is the vehicle mass, g the acceleration of gravity, K_f the friction coefficient on the path segment, V the vehicle speed, t the time and θ denotes the angle of the slope with the horizontal plane. F(V) is the active force and f(V) includes other types of friction, for example the aerodynamic drag, which can be modelled as polynomials : $f(V) = aV^2 + bV + c$. By neglecting the transients, dV/dt is set to zero so that V is the positive maximal solution, if exists, of a polynomial equation as long as F(V) is polynomial. It is For this purpose that the torque and power versus engine speed characteristics are considered. For most of the atmospheric engines, it is found that the torque characteristic $C(\omega)$ is fairly well approximated by a parabola, which corresponds in turn to a third–order polynomial for the power $P(\omega)$:

$$C(\omega) = -\alpha \, \omega^2 + \beta \, \omega + \gamma \tag{2}$$
$$P(\omega) = \omega \, C(\omega). \tag{3}$$

If one denotes by $K_i = \omega/V$ the transmission ratio of gear number i and by r_i the mechanical efficiency ($FV = r_i P$), the above equations yield :

$$F(V) = -A_i \, V^2 + B_i \, V + C_i \tag{4}$$
where $A_i = \alpha \, r_i \, K_i^3$, $B_i = \beta \, r_i \, K_i^2$ and $C_i = \gamma \, r_i \, K_i$.

At this stage, it can be noted that the coefficients α, β and γ can be computed either by identification of experimental curves or by the knowledge of the engine maximum torque and power, and at least of one of the corresponding speeds [MOI89].
Then, the friction coefficients are subtracted :
$A_i = A_i - a$, $B_i = B_i - b$, $C_i = C_i - c$.
For computing the maximum speed along a climbing path, the algorithm works as follows:
1. select the high gear : $i = i_{max}$,
2. (loop) compute the best solution of $-A_iV^2+B_iV+C_i-mg(\sin(\theta)+K_f\cos(\theta)) = 0$.
      If it is real then return i and the best speed consistent with the maximum allowed RPM else $i = i-1$.
      If i=0 then return "infinite cost", else continue.
For descending a slope, the same algorithm can be used (replacing $\sin(\theta)$ by $|\sin(\theta)|$) else one can decide to select the maximum allowable speed, which is more risky. In the former case, the computed speed is :

$$V_{comp} = \frac{B_i + \sqrt{B_i^2 + 4A_i(C_i - mg(|\sin\theta| + K_f\cos\theta))}}{2A_i} \tag{5}$$

and the optimal speed is

$$V = \begin{cases} V_{max} & \text{if } V_{comp} \geq V_{max} \\ \text{else} & V_{comp} \end{cases} \tag{6}$$

$$\text{where : } V_{max} = \frac{\omega_{max}}{K_i} \quad \text{with : } \omega_{max} = \frac{\beta + \sqrt{\beta^2 + 3\alpha\gamma}}{3\alpha}.$$

Finally, the time is computed by $T = D/V$ where D is the segment length.

4. MOTION PLANNER IMPLEMENTATION AND EXAMPLES

The motion planner is responsible for computing the time-optimal path from a starting point to a goal point (both added to the triangulated terrain model) in following some edges of the triangulation. It provides also the corresponding gear ratios and speeds (figure 4). For that purpose, a A* algorithm is called for, which uses classically $f = g + h$ as the cost function, where g is the computed time from the starting point to the current node (vertex) and h is some evaluation of the minimum time required for reaching the goal. In the general case of the application considered in this paper, the problem may not be symmetric, i.e. two different costs can be associated with the same part of a path, depending upon the direction of motion, since different strategies for descending slopes can be used. At each step of the algorithm which builds the search graph, the cost functions g and h of all the successors of the current node are computed. The "quasi-Delaunay" triangulation used for modelling the terrain ensures that this is done in constant time, since it is known [PRE85] that the mean number of edges originating from a vertex is equal to six. This result, associated with the fast computation of adding eventually on line new vertices, makes the proposed method more able to cope with uncertainties (reflexive planning) than the others [GAW86, MAN88] where the number of successors ("visible nodes") and the computing time cannot be easily estimated.
Another feature lies upon the choice of the heuristic evaluation function h. It has been proved [NIL84] that if $h \leq h^*$ minimum cost to the goal, the A* algorithm is admissible, i.e. it finds the optimum, if exists. In the case considered here, setting $h = h_0$ provides admissibility, where $h_0 = D_0/V$, D_0 being the straight-line distance between the current point and he goal, and V chosen by one of the following methods :

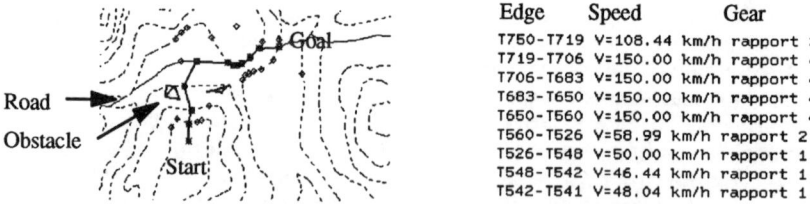

Road
Obstacle

Edge	Speed	Gear
T750-T719	V=108.44 km/h	rapport 3
T719-T706	V=150.00 km/h	rapport 4
T706-T683	V=150.00 km/h	rapport 4
T683-T650	V=150.00 km/h	rapport 4
T650-T560	V=150.00 km/h	rapport 4
T560-T526	V=58.99 km/h	rapport 2
T526-T548	V=50.00 km/h	rapport 1
T548-T542	V=46.44 km/h	rapport 1
T542-T541	V=48.04 km/h	rapport 1

Figure 4 : a trajectory with gear–ratios and speeds

1) $V = V_{max}$,
2) V is computed by equation (5), assuming an hypothetical straight–line motion in the three–dimensional space and taking the minimum value of K_f.

The time given by 1) is the inferior limit, that provides a h_0 much more less than h^* and leads A* to construct a leafy search graph. For that reason, solution 2) is preferred. Moreover, since getting more rapidly a path is more important in practice than ensuring exact optimality (taking into account the previous approximations concerning the terrain discretization and the neglected dynamic transients), the modified cost function evaluation $f = g + k\,h_0$ has been used, where $k \geq 1$.

As it is known, the corresponding algorithm works as pure A* when $k = 1$ and follows the gradient of h_0 when k is large. Increasing k thus reduce the number of nodes developed during the search, and consequently, the computation time. A large number of experiments [MOI 89] confirmed this behavior and a "good" choice of k was looked for. Figure 5 gives an example of the CPU time as a function of k. It has been found that $k = 1.2$ almost provides the optimal result and that $k=1.5$ induces only a 3% error, which is tolerable while the computing time is reduced by a factor of ten.

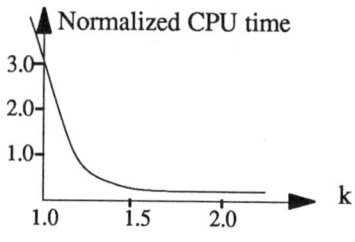

Figure 5 : influence of coefficient k.

The algorithms presented in this paper have been implemented on a graphics VAXstation, in COMMON LISP language, as an interactive tool for simulations. The level curves come from a digitizing tablet, but any digital elevation model could have been used as well, after suitable processing for retaining only the most informative points. After the triangulation procedure, regions and lines are defined (including "obstacles" in the large), the cost of following their edges being altered ("weighted") by attributing corresponding values of the friction coefficient K_f. Different vehicle and engine characteristics can be chosen.

Figures 6 and 7 illustrate examples of simulations, where it is shown that the A*–like algorithm provides not only the optimal path and motion along it, but also, about it, the explored nodes which constitute good candidates for changing dynamically the motion in the case of a unexpected local disturbance. Furthermore, if it is presumed that in a given open field, no maze situation is likely to be encountered, the time necessary for planning is decreased by letting the planner "forget" parts of its previous trials, which is done by limiting the size of the ordered "OPEN" list of nodes in the search.

5. CONCLUSION
The methods presented in this paper have leaded to a motion planner which can cope with realistic situations, due to a compromise between simplicity and good approximations of the geometric and dynamic models. Experiments have shown that if the terrain is not crowded by obstacles and mazes, the A*–like computational cost is not much larger than that of a gradient method.
However, further improvements could be added, which concern for example the path–smoothing, or the vehicle dynamic stability [SHI90], the cost of dynamic transients, and more sophisticated engine torque models.Finally the search algorithms must be extended by further research if reactive planning, in the case of moving obstacles and goals for example, is required.

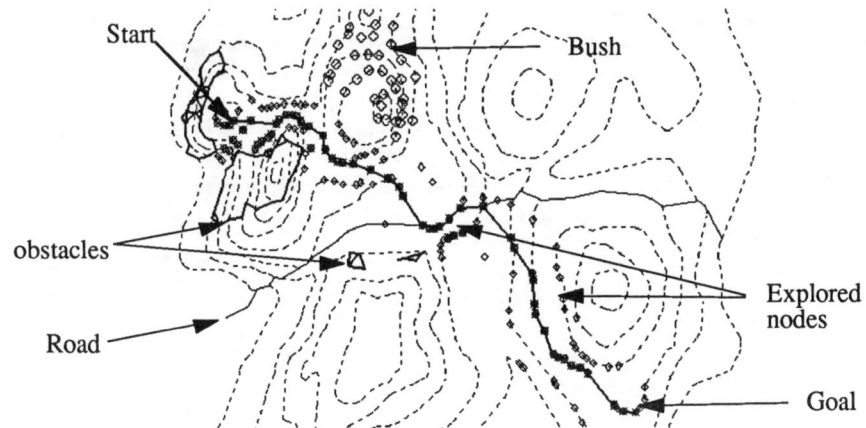
Figure 6 : an optimal trajectory in uneven terrain with obstacles, difficulties and a road.

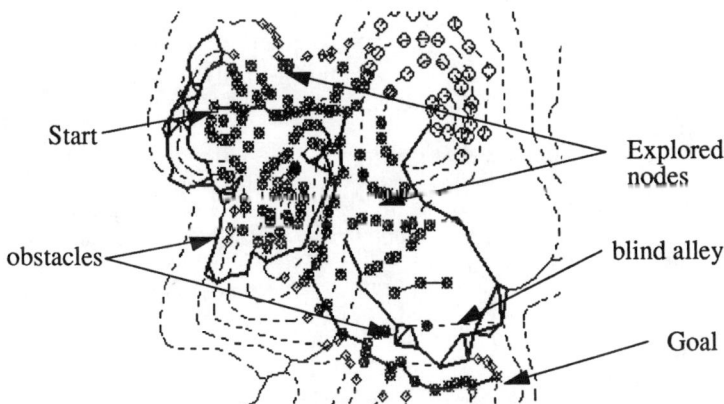
Figure 7 : a complex path throught a maze.

ACKNOWLEDGEMENTS.
Parts of this work have been supported by D.R.E.T.–D.G.A. under contracts No 86/117 and 89/445.

REFERENCES

[BOI88] Boissonnat, J.D., Shape reconstruction from planar cross sections, *Computer Vision, Graphics and Image Processing*, 44, 1988, pp. 1–29.

[FIO85] De Fioriani, L., Falcidieno, B., and Pienovi, C., Delaunay–based Representation of surfaces Defined over Arbitrarily Shaped Domains, *Computer Vision, Graphics and Image Processing*, 32, 1985, pp. 127–140.

[GAW86] Gaw, D., and Meystel, A., Minimum–Time Navigation of an Unmanned Mobile Robot in a 2–1/2D World with Obstacles, *Proc. 1986 IEEE Int. Conf. on Robotics and Automation*, vol. 3, pp. 1670–1677.

[GUI85] Guibas, L., and Stolfi, J., Primitives for the Manipulation of General Subdivisions and the Computation of Voronoï Diagrams, *ACM Trans. on Graphics*, vol. 4, No 2, April 1985, pp. 74–123.

[KIR83] Kirkpatrick, D., Optimal search in Planar Subdivisions, *SIAM J. Comput.*, vol. 12, No 1, February 1983, pp. 28–35.

[LIE89] Liégeois, A., Emulation d'algorithmes optimaux de mouvements de mécanismes complexes en espace encombré, *Rapport de synthèse finale*, Convention DRET–LAMM No 86/117, June 1989.

[MAN88] Manaoui, F.O., Etude et simulation d'algorithmes de navigation pour robots mobiles autonomes sur terrain inégal, *Thèse de doctorat*, Université de Montpellier II, France, December 1988.

[MIT87] Mitchell, J.S.B., and Papadimitriou, C.H., The Weighted Region Problem, (extended abstract), *Proc. 1987 ACM Conf. on Computational Geometry*, pp. 30–38.

[MIT88] Mitchell, J.S.B., An algorithmic approach to some problems in terrain navigation, *Artificial Intelligence*, 37, 1988, pp. 171–201.

[MOI89] Moignard, C., Génération de trajectoire pour un robot mobile autonome en terrain inégal, *Rapport de DEA*, Université de Montpellier II, France, July 1989.

[NIL84] Nilsson, N.J., Principles of Artificial Intelligence (Springer–Verlag, 1984).

[PET87] Petrie, G. and Kennie, T.J.M., Terrain modelling in Surveying and Civil Engineering, *Computer Aided Design*, vol. 19, No 4, May 1987, pp. 171–187.

[PRE85] Preparata, F.P., and Shamos, M.I., Computational Geometry–An Introduction (Springer–Verlag, 1985).

[ROW90] Rowe, N.C., and Richbourg, R.F., An Efficient Snell's Law Method for Optimal–Path Planning across Multiple Two–Dimensional Irregular, Homogeneous–Cost Regions, *The Int. J. of Robotics Research*, vol. 9, No 6, December 1990, pp. 48–66.

[SAM82] Samet, H., Neighbour finding techniques for images represented by quad–trees, *Computer Graphics and Image Processing*, No 18, 1982, pp. 37–42.

[SHI90] Shiller, Z., and Chen, J.C., Optimal Motion Planning of Autonomous Vehicles in Three Dimensional Terrains, *Proc. 1990 IEEE Int. Conf. on Robotics and Automation*, vol. 1, pp. 198–205.

[SLO84] Sloan, S.W., and Houlsby, G.T., An implementation of Watson's algorithm for computing 2–dimensional Delaunay triangulations, *Adv. Eng. Software*, 1984, vol. 6, No 4, pp. 192–197.

MOBILE ROBOT TRAJECTORY PLANNING

P. K. Sinha and A. Benmounah

Department of Engineering, University of Reading, P.O.BOX 225, Reading RG6 2AY, U.K.

Keyword : Mobile robot, Transputer, Path planning, Parallel processing.

ABSTRACT
This paper describes a methodology for the navigation/steering of an autonomous sensor-guided vehicle capable of (a) moving in a shop-floor (X-Y plane) along a planned trajectory avoiding known, and (b) reacting to unexpected obstacles. The vehicle, equipped with obstacle detection sensors, is capable of planning its own trajectory between its home (current) position and the user-defined destination. An experimental vehicle (Reading Mobile Platform, RMP), with on-board sensors, has been built to demonstrate the feasibility of using transputers in implementing such a trajectory planning technique.

1. INTRODUCTION
Path planning is one of the most vital issues in the design of autonomous mobile robots, which can be viewed as finding a safe and short (optimum) path through an environment from some starting point to some destination without colliding with any obstacles in the process [1]. In this paper, a novel approach of path planning is presented; where the path with known obstacles is pre-planned while the robot is stationary (before action). It also consists of local avoidance of unknown obstacles while following the planned trajectory [2, 3]. The actual path followed is a global trajectory, where superimposed on the motion control, is a requirement for local obstacles to be avoided which were not present at the initial starting time. A detailed description of one such trajectory planning scheme and its implementation using a transputer and the occam language are presented.

The primary objective in the design of this experimental AGV like the (Reading Mobile Platform, RMP, Fig 1), was to study a number of navigational/routine schemes for application in different environments.

2. MULTI-GOAL PATH PLANNING
When the initial position, final goal location, and the obstacle map in the prescribed workspace are given, the robot must find out the appropriate path from start to goal point such that it can travel smoothly through the area without colliding with obstacles which may be blocking its way [4]. Initially the robot is placed at (X_0, Y_0) origin where the middle of the rear wheels put at T/2 is coinciding with the origin of X and Y axis Fig 2. Since it is designated to follow the reference of the robots' path, this centre point of the rear wheels is called "GUIDEPOINT". Although arbitrarily chosen, the guidepoint has some effects on path planning and path guidance algorithms. These effects are generally

caused by the space occupied by the robot, its dynamics and kinematics [5]. To send the robot from one position to another in X-Y plan, two cases are considered [6]:(a) when there is obstacle between START-GOAL points, and (b) when the obstacle is far enough from the straight path between START-GOAL points, or when there is no obstacle at all.

(a) When X_n and Y_n (goal coordinates), the obstacle dimensions and its location in X-Y plan (A1,A2,A3,A4) are given, the algorithm must find a short collision-free path with minimum number of segments that lead the robot to the goal while keeping far from obstacle (Fig 3). This collision-free path is calculated in the following manner:First the angle of the obstacle substended by the line A1 and A2 is calculated to identify the "forbidden" area occupied by the obstacle (double hashed area shown of Fig 3), this is defined by

$$\alpha_{obstacle} = \alpha_{A1} - \alpha_{A2} \qquad (1)$$

where

$$\alpha_{A1} = \arctan\left(\frac{Y_1}{X_1}\right), \quad \alpha_{A2} = \arctan\left(\frac{Y_2}{X_2}\right) \qquad (2)$$

Second the angle between the START-GOAL path, and Y axis is calculated in order to obtain more information for the algorithm. This will enable the robot to take the side of the shortest path (left side or right side) for optimum avoidance using the following formula.

$$\alpha_{path} = \arctan\left(\frac{X_n}{Y_n}\right) \qquad (3)$$

If the shortest planned path is on the obstacle's left-side then

$$\alpha_{A1} > \alpha_{path} > \left(\alpha_{A2} + \left(\frac{\alpha_{obstacle}}{2}\right)\right) \qquad (4)$$

but if it is on the right side, then

$$\alpha_{A2} < \alpha_{path} < \left(\alpha_{A2} + \left(\frac{\alpha_{obstacle}}{2}\right)\right) \qquad (5)$$

If equation 4 is valid, the robot first steers left by the angle (Fig 3)

$$\alpha_{avoid} = \alpha_{A1} + \alpha_{SECUR} \qquad (6)$$

where α_{A1} and α_{SECUR} are, respectively, the angle between X axis and the line drawn from the X-Y plan origin to the point A1 (first point of the obstacle), and the angle constant value (15°) that should be added to α_{A1} to ensure that the robot is far enough from colliding with the obstacle. The robot then moves forward by a segment of length L_2, steers back by α_{avoid} angle, and then moves along the segment L_{23} (Fig 3) in order to reach the goal point (as formulated by the following three equations):

$$L_2 = \frac{Y_n}{\sin\alpha_{avoid}} \qquad (7)$$

$$L_{21} = L_2 * \cos\alpha_{avoid} \qquad (8)$$

$$L_{23} = X_n - L_{21} \qquad (9)$$

The same sequence is followed when the shortest path is on the right, bearing in mind the value of the angle α_{A2}.

(b) When the straight path between START-GOAL points is far enough from the obstacle on its right (Fig 5c), this is checked by

$$\alpha_{path} > (\alpha_{A1} + \alpha_{SECUR}) \qquad (10)$$

and by

$$\alpha_{path} < (\alpha_{A2} - \alpha_{SECUR}) \qquad (11)$$

if the straight path is far enough from the obstacle on its left (Fig 5d).

If equation (10) is valid, the robot must steer by

$$\alpha = \arctan\left(\frac{X_n}{Y_n}\right) \qquad (12)$$

and move by a distance

$$L_1 = \sqrt{(X_n^2 + Y_n^2)} \qquad (13)$$

Same steps are followed when there is no obstacle in the workspace.

3. LOCAL OBSTACLE AVOIDANCE

After the shortest path has been calculated and planned, and while the robot begins to move, the algorithm must also be able to cope with any sudden change in the environment, which might be the case of unexpected obstacle blocking the pre-determined (anticipated) collision-free path. In the case of a presence of unexpected obstacle blocking the planned path, similar method of obstacle avoidance in [7] is adopted for local avoidance. The difference here, is that in [7] when the presence of an obstacle is detected, the propulsion drive comes to a stop and a mechanism for detecting the relative distance of the two edges of the obstacle is triggered. After edge distances comparison, the robot must be able to take the (shortest) side to avoid the obstacle, and contour it. But in this paper the robot must detect the unknown obstacle, avoid it, and then carry on along the planned trajectory. For example, if the shortest path is taken on the left side of the known obstacle, the avoidance of the unknown obstacle should be taken on the left side. Otherwise, a collision of the robot with the known obstacle is almost inevitable, particularly when this planned path is closer to the known obstacle. This may be due to the size of the robot and the space between the known and unknown obstacles which might be smaller than the size of the robot. In this paper, only the case of Figure 5b, where there is an obstacle between START-GOAL point is considered. All the steps of this case are illustrated by the flow-chart of Fig 6. The schematic configuration of the hardware for the RMP is shown in Fig 7.

4. EXPERIMENTAL RESULTS AND CONCLUDING COMMENTS

The robot is supposed to follow the (calculated) planned path in order to reach the GOAL point. During the process, and because of some factors such as uneven surface of the workspace, robot design, and mechanical deformation, some error (deviation) in following the planned path may occur. This may lead the robot far from the GOAL point. In order to check the difference between the planned path and the path followed by the robot while it is moving, the actual path was marked on the floor and then compared to the ideal one as shown in Fig 8. In this Figure, it is seen that the difference between the two plots is very small and the reached GOAL point is closer to the prescribed one. When an unknown obstacle blocking the path is detected and avoided by the robot, the error is slightly higher. The error between the calculated path and the followed path is smaller when there is no unknown obstacle.

ACKNOWLEDGEMENT
The work here forms part of a project supported by the UK SERC and Inmos.

REFERENCES
[1] Lin, C-H., and Fu, L-C., "A centre-line oriented robot navigation", IEEE International Conference on Robotics and Automation, 1988.
[2] Growley, J.L., "Navigation for an intelligent mobile robot", IEEE Journal of Robotics and Automation, Vol. RA-1, NO.1, March 1985.
[3] Parodi, A.M., "Multi-Goal real-time global path planning for an autonomous land vehicle using a high-speed graph search processor", IEEE International Conference on Robotics and Automation, Published by IEEE Computer Society Press, 25-28 March 1985.
[4] Takahashi, O., and Schilling, R.J., "Motion planning in a plane using generalized veroni diagrams", IEEE Transactions on robotics and automation, Vol. 5, NO.2, April 1989.
[5] Kambhpati, S., and Davis, L.S., "Multi-resolution path planning for mobile robots", IEEE Journal of robotics and automation, Vol.RA-2, NO.3, Sept 1986.
[6] Benmounah, A., "Transputer control of an AGV:design, construction, and testing of a mobile platform", Ph.D Thesis, University of Reading, (UK), June 1991.
[7] Sinha, P.K., and Benmounah, A., "AGV control using parallel processing techniques", 21st International Symposium on Industrial Robots (ISIR), IFS Publications UK, Copenhagen 23rd-25th Oct 1990.

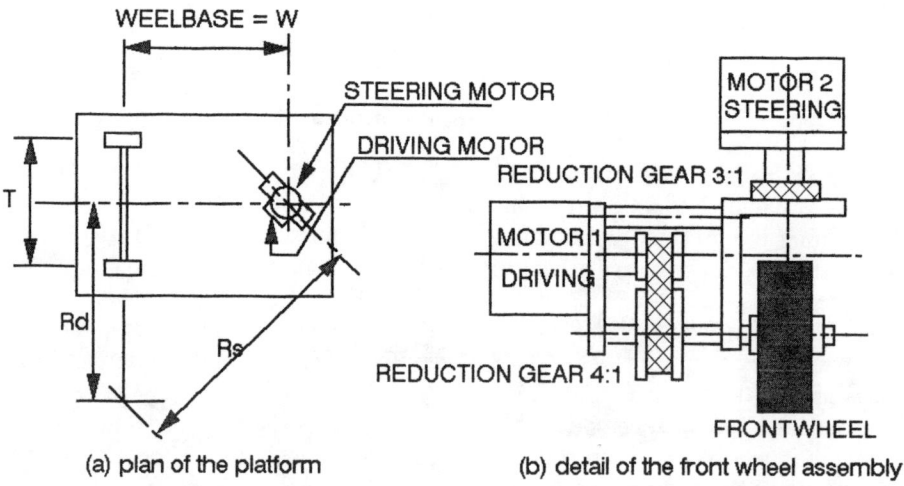

(a) plan of the platform

(b) detail of the front wheel assembly

(c) general view of the AGV

Fig 1 Architecture of the AGV designed

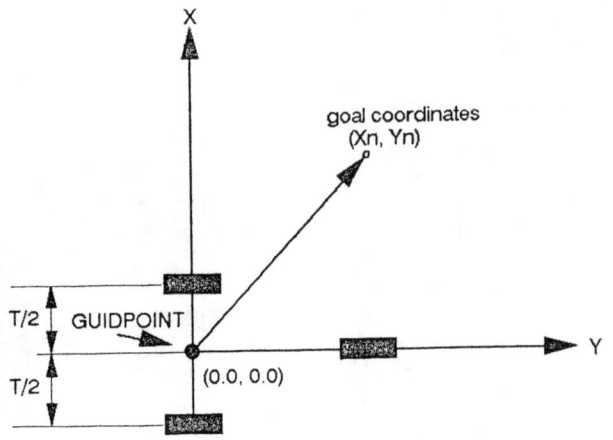

Fig 2 Robot position initialisation

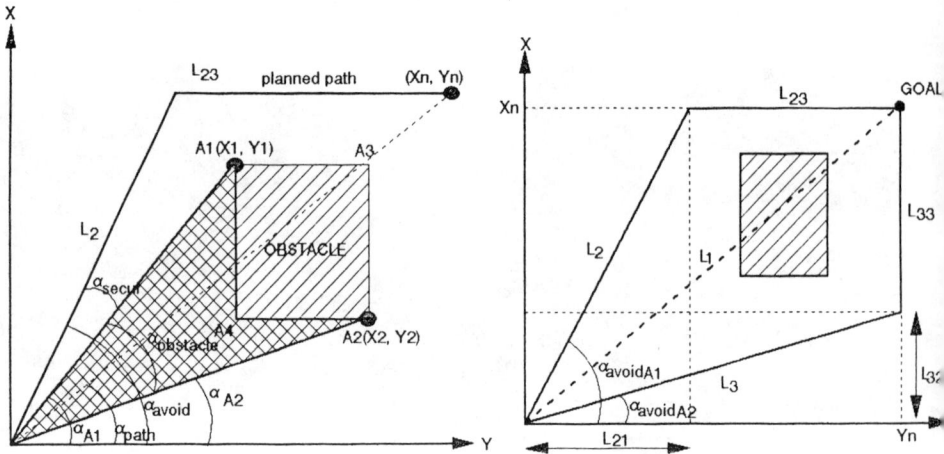

Fig 3 Collision-free path

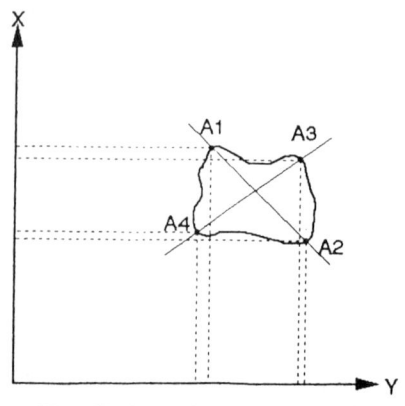

Fig 4 Obstacle approximate shape

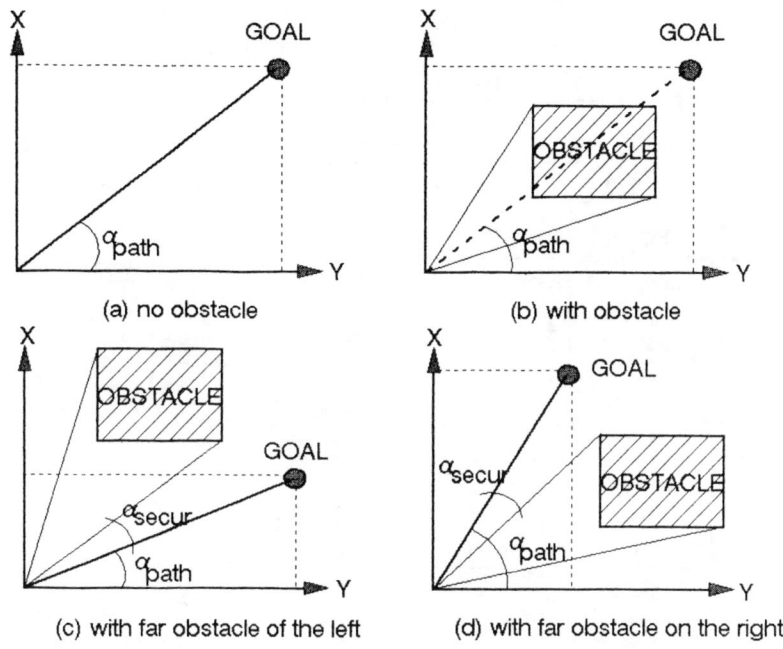

Fig 5 Different obstacle position vis-avis the goal visibility

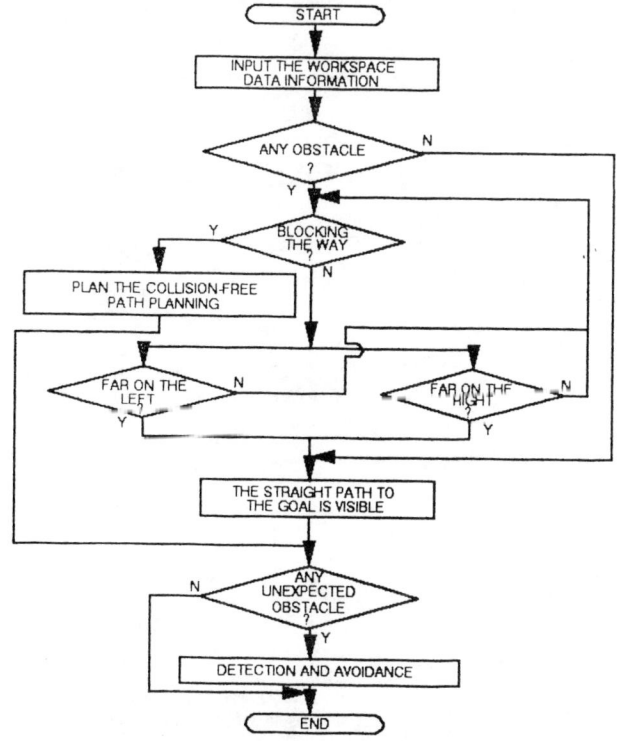

Fig 6 Optimal path planning flow-chart

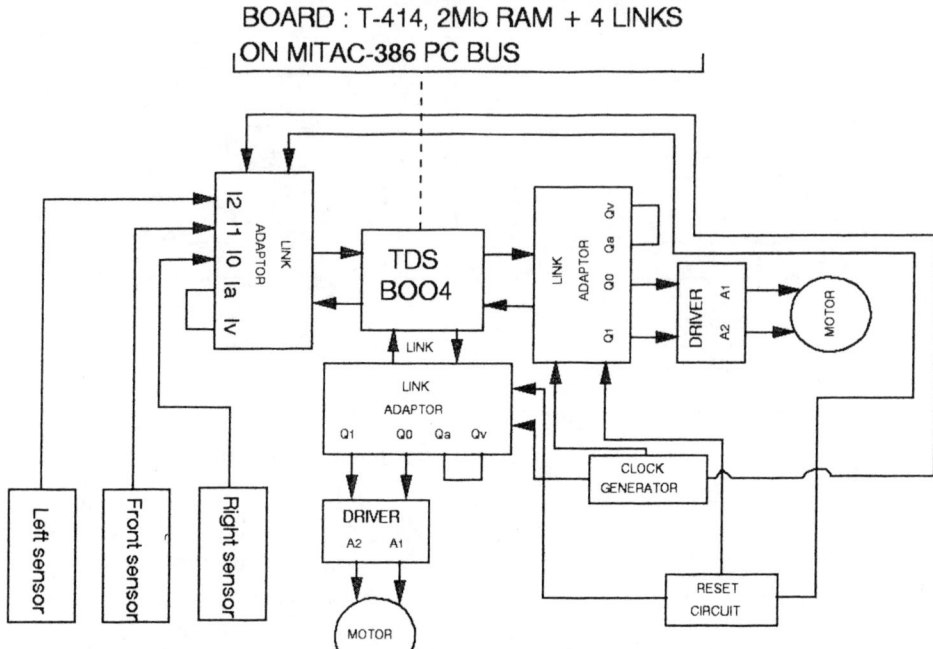

Fig 7 AGV control block diagram

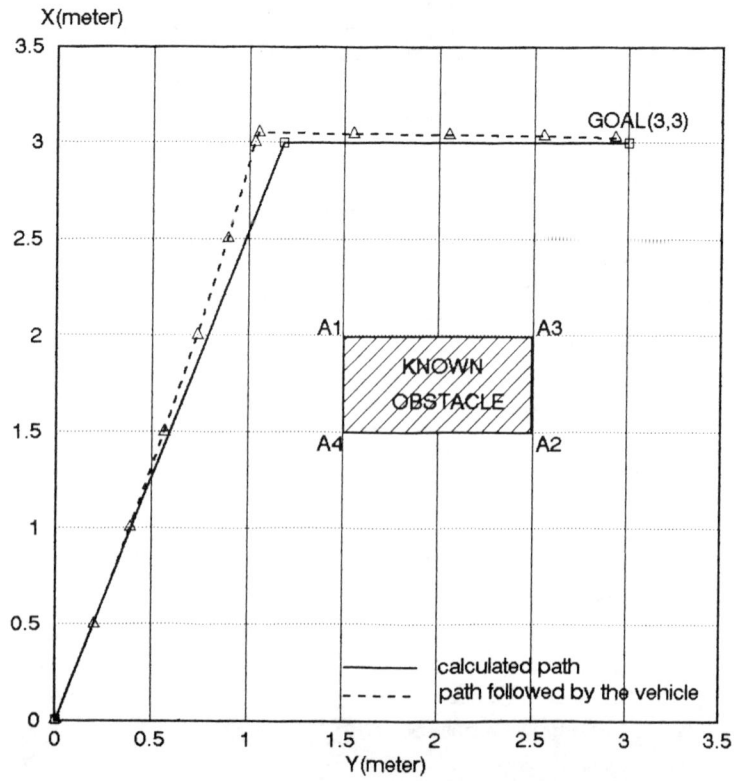

Fig 8 Experimental results of the planned path

TASK MODELING FOR PLANNING AN AUTONOMOUS MOBILE ROBOT

Valérie SCHAEFFER & Anne MAUBOUSSIN

Laboratoire d'Informatique Fondamentale, Université Pierre-et-Marie-Curie
B.166 - 4, place Jussieu - 75252 PARIS CEDEX 05 - FRANCE

1. INTRODUCTION

Considering the state of art in mobile robot navigation relying on environment perception and understanding, it is now necessary to develop mobile robot programming at a sufficiently high level so that it can describe a complex and abstract mission that will further be realized in an autonomous way by the robot. Such a programming tool must be coupled to a mechanism that finally expresses this abstract mission as a sequence of elementary commands to the robot.

The aim of our work is to realize a robotic application modeling tool coupled to an off-line planner in order to organize the global activity of the robot in its environment. Our case study is the autonomous mobile robot ROMEO[1] (equipped with a camera and ultrasonic sensors as perception means) moving about in a structured and partially known factory-like environment. The robot is in charge of achieving given top-level goals depending on various *activities*. We have considered the four following types of activity that cover a wide range of possible applications: transportation, distribution, collecting and control.

Our modeling tool covers the modeling of: the environment, the robot, the activities and the tasks. It is used at first by an administrator to describe a specific robotic application. Then in standard use, it enables a robot programmer to create and enrich a library of complex missions combining various pre-defined top-level goals. The library is used for giving the current mission to the robot. This mission is decomposed into a sequence of elementary actions by the planner: The first step consists in computing a global plan that optimizes the global traject of the robot and satisfies succession constraints between tasks as well as constraints on shareable resources; The second step consists in providing a sufficient detailed plan at a local level so that it can be executed by the robot.

In this paper we choose to focus on the higher levels of robotized task modeling and planning (labelled with * in Fig. 1). In Section 2, we present the robot environment modeling. We explain our conceptual environment representation and its possible structuration in clusters that can be used in mission specifications. Section 3 is about robotized task modeling used in planning and the three levels organization of our system. In Section 4, we describe the mission specification language. Section 5 is a short

---

[1] ROMEO is a project of the Laboratoire de Robotique de Paris; the Laboratoire d'Informatique Fondamentale collaborates on the aspects of this project concerned with Artificial Intelligence.

S. G. Tzafestas (ed.), *Robotic Systems*, 287–295.
© 1992 *Kluwer Academic Publishers. Printed in the Netherlands.*

description of the global planning algorithm that takes as input a complex mission defined by the user and sequences the various sub-tasks involved in it.

THE APPLICATION FRAME	➡	THE PLANNING TOOLS

* Environment Modeling
  Robot Modeling
  Activities Modeling
* Tasks Modeling
	USED IN	* Mission Specification
		* Global Planning
		  Local Planning

Figure 1. Modeling for Planning.

2. ROBOT ENVIRONMENT MODELING

The real world in which the robot moves and acts cannot be represented as a whole since the data size is prohibitive. Thus we decided to represent only an *a priori* known part of the environment [3, 12]. The choices we have made for this modeling have been guided by our will to be as near as possible to the representation of the world that a human is able to describe, and to abstract ourselves all the more of a purely geometrical and metrical representation of the environment.

Thus we represent the free space where the robot moves by *main roads* between *viewpoints* located by a layout of *beacons*. These information are contained in a *navigational graph* coupled to an *inter-reachability graph* that joins each beacon to every other one the robot is able to reach from it. The navigational graph is used for rapidly moving in the environment; The inter-reachability graph is used for reaching a particular place where the robot must achieve a given task or for recovering the initial trajectory in case of failure.

We regard the robot as an agent in charge of realizing various activities in a man-made environment. Managing and organizing the robot activity at this high level of abstraction requires an environment knowledge that is structured according to the activities to realize. So a third graph called *global graph* is added to the two others; It expresses the connections between *world units* (that can be assimilated to rooms) via *transitions* between these world units. Each world unit is linked to main-roads, viewpoints and beacons it contains; It also "knows" the inside objects that can be involved in robotized tasks. Conversely the objects are linked to their world unit and to their components relevant for the robotized tasks.

Furthermore, the environment can be structured following logical criteria related to the robot activities. The operator can define *clusters* that are associations of selected environment objects of the same type. For example (cf. Fig. 2), a cluster can be: a sector (a cluster of world units), a load (a cluster of transportable objects) or a machine set. These global entities are used as parameters in missions: "control *sector_1*", "transport *load_1* from *R11* to *R4* on *bench*", "fill up *machine_set_1* with *fuel*" (distribution mission).

3. TASK MODELING FOR PLANNING

In the domain of robot task planning, a controversy exists between strategic planning and tactical or reactive planning [14].
- *Strategic* planning is an off-line process based on a predictable knowledge of the robot environment. It is implemented using traditional planning techniques, such as linear

planning [5], hierarchical planning [11] or case-based planning [13]. These techniques are obviously limited for planning in a real world environment; A difference between the perceived environment and the world model, on which the current plan is computed, leads to a failure in the plan execution.
- New techniques have been recently developed for dealing with uncertain environment [4]. In *reactive* on-line planning, also called situated activity, plan formation and execution interleave.

Until recently, research in those two domains has been conducted separately; Yet the robotic systems would now take advantage of realizing the synthesis of these ideas.

In case of a rather predictable environment such as a man-made one, which is precisely our concern, it is useful and even necessary to plan the robot activity at a sufficiently high level of abstraction, so that it does not become trapped in local minima as it acts. The job of the planner is then to think about the future enough to have appropriate activities ready to go as current activities finish. Planning should allocate the needed resources for doing execution, and this allocating should optimize the use of the robotic resources; For example, several tasks can be partially executed at the same time sharing common paths.

Another step is necessary to refine such a global plan in order to obtain a sequence of commands. At this level it seems convenient to take into account the perceived environment before locally completing and executing parts of the global plan. Delaying the interaction with the environment does not assume that the global plan will never be reconsidered; Some of the perceived environment states could invalidate parts of the global plan. Nevertheless the flexible local planning and the high predictability of the man-made environment lead us to assume that global replanning will seldom be necessary.

Our planning system is structured in three expert modules dealing with planning at different levels of abstraction of the robotized tasks: *missions*, *objectives* and *actions*. A multi-experts control architecture [6] realizes the link between these modules. The highest level module enables the user to define the robot current mission, the second level module produces a global plan realizing this mission in the environment and the lowest level module locally refines the global plan, interleaving the plan formation and execution.

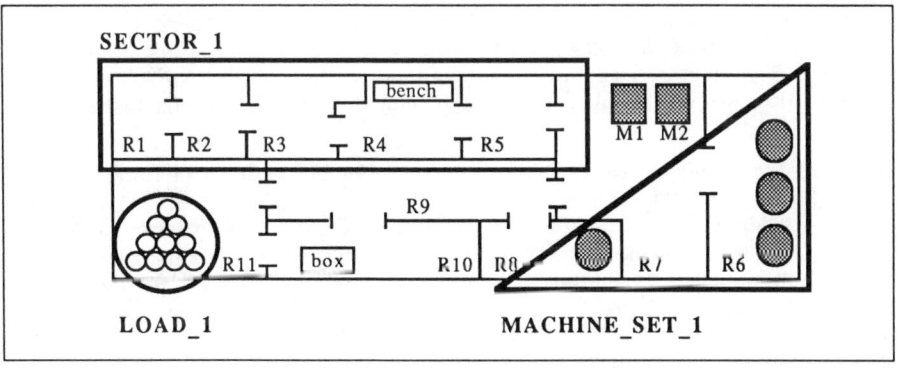

Figure 2. An Example of Clusters.

3.1. Task modeling

The action level deals with action means of the robot (e.g. trans-palette, control sensors, brushes, pump, tank) and with navigation functions. An *action* is a low-level routine of commands that realizes either the activation of a robot action means on an environment object or a moving in the environment. According to our environment modeling, a robot moving is either a moving between viewpoints or an object docking in order to act on this object. In the following "action" will only denote the first meaning we introduced above, i.e. the activation of a robot action means.

The objective level deals with the robot functionalities in its environment such as transportation, distribution, collecting or control. Each functionality corresponds to the activation of one action means. An *objective* is linked to: one action means, the world units where this action means will be activated and the objects on which it acts (physical entities such as boxes, machines, etc.). An objective is then defined by the sequence of actions and movings that realizes the function. Objectives enable the operator to describe the robot tasks at a functional level rather than in detailed actions (e.g. "transport an object from one place to another"). Thus an objective is a high level robotized task very similar to a finalized task as defined in [1].

The mission level deals with global description of tasks. A *mission* is a complex task recursively decomposed into collections of sub-missions. A *sub-mission* is either an already created mission or an *objective* (a lowest level sub-mission). A *collection of sub-missions* can be a sequence of ordered sub-missions, a set of sub-missions or a repeated sub-mission. The mission specification language, bound to a user-friendly graphical interface, enables the human operator to create a *library of reusable mission descriptions* and to manage the robot activity at a higher level of abstraction than with most of the task-level languages [7, 10]. This library also contains the clusters of environment objects defined by the operator. The progressive enrichment of the library enables the operator to process experimentaly in the construction of global command tools accurately adapted to the activities and the environment of the robot.

3.2. Task planning

The library and the associated mission specification language make the operator able to define the *current mission* of the robot. The current mission is decomposed by the system into a set of objectives and a set of associated elementary succession constraints. These two sets are used by the objective level module in computing the global plan for the current mission.

The objective-level module uses the global graph to search a path between the world units involved in the objectives of the current mission. This search takes into account the succession constraints as well as the resources constraints associated with every objective. The ability to share some of the robot resources is used for minimizing the path between world units (it favours the sharing of identical sub-paths in respect of path reliability criteria). The search is realized with a simulated annealing algorithm [8], and produces a global plan that gives the sequence of visited world units and for each world unit the set of actions that must be realized in it.

At last, the local planning sequences the various actions in each world unit and computes the local movings from the navigational graph. At this level the formation of the plan concerning one world unit interleaves with its execution in the real world unit.

4. MISSION SPECIFICATION LANGUAGE

For a robot in charge of realizing numerous and various objectives in its environment, it is useful to have a mission description tool that organizes complex mission objectives following structuring and/or precedence criteria. The mission specification language provides various operators for combining objectives.

4.1. Objectives and missions

Viewing an objective as a basic element in the mission specification language, it is defined by its name and a set of typed parameters. The actual parameters values are objects of the environment (world units or physical entities). Environment objects are typed and classified according to their role in robot activities. A type can be decomposed into a hierarchy of sub-types; For example, the type "transportable object" is split into "fragile transportable object" and "tough transportable object". An objective parameter can be instantiated with any object of the same type or of a sub-type of the parameter type. According to the type of the argument, the planned actions realizing the objective can be different; For example, the transportation of fragile objects is not necessarily refined in the same way as the transportation of rough objects. Thus objectives are *generic objectives*.

Some dependency relations can exist between the parameters of an objective, as shown by the arrows in Fig. 3. They reflect the relations, described in the environment modeling, between the objects, their components and the world units. Thus, some of the actual parameters may be omitted, they are then evaluated using the various cross-references, e.g. the initial world unit of a transportable object, the world unit or the filling of a machine can be omitted and retreived from the object description or from the machine description.

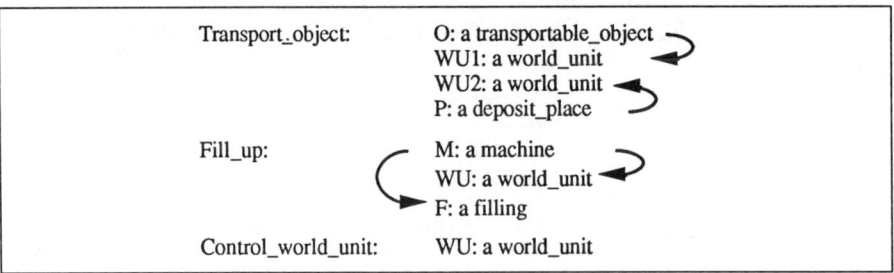

Figure 3. Examples of Generic Objectives.

Missions can be: *generic missions*, i.e. mission schemes that can be applied to various sets of environment objects, or *instantiated missions* defined for a given set of environment objects.

At the lowest level, an elementary mission is either a collection of generic objectives (with various possible organization) or a totally or partially instantiated generic objective. Each already defined mission can be used in the definition of other missions: It then becomes a *sub-mission*. Therefore a mission is either a collection of sub-missions and generic objectives, or a totally or partially instantiated sub-mission. In the following the term *sub-mission* will also refer to generic objectives.

4.2. Mission building

The mission specification language enables the user to define a mission either as a collection of sub-missions organized in set, sequence or loop, or as a totally or partially instantiated sub-mission.

Composition

The user can choose one of the three following operators to create a mission by composing sub-missions.

The *set-operator* defines a mission as a set of sub-missions. In this case, no precedence between sub-missions is specified; It means that the sub-missions can be realized in any order. They even can have common sub-parts.

The *sequence-operator* defines a mission as a sequence of sub-missions that must be realized in the given order. Because of its execution duration, a sub-mission can be considered as a time interval. Thus, various orders based on Allen's relations on time intervals [2] can be used for sequencing sub-missions. At the present time, the precedence relation between two sub-missions is strict, i.e. a sub-mission ends before the next one begins. Providing other precedence relations will enable the user to express more precise constraints on the beginning or/and the end of the sub-missions.

The *loop-operator* defines a mission as the repetition of a particular sub-mission given a set of values for one or more loop-parameters. These sets of values can be clusters as presented in Section 2. According to the type of the clusters and to the type of the sub-mission, the repetition generates either a set or a sequence of instantiated sub-missions.

For each of these three composition operations, the parameters of a mission are imported from its sub-missions, together with the dependency relations between the parameters.

Instantiation

A mission can be specified in another way: the instantiation, total or partial, of the parameters of a generic sub-mission. A total instantiation produces an *instantiated mission*; A partial instantiation creates a new generic mission.The parameters number of this new mission is less than the instantiated sub-mission one. A parameter is instantiated with an environment object, or a set of objects or a cluster (in case of a loop-operator), of the same type as the parameter type.

The human operator can specify a mission, at his convenience, from already existing sub-missions, as well as from further defined sub-missions.

4.3. Mission library

The various missions created by the user, generic ones as well as instantiated ones, can be stored in the mission library, which is initialized with generic objectives. The user manages the library by creating, specifying, deleting, renaming, saving a mission; He can also display a complex mission as a hierarchical tree of sub-missions, and assign the current mission to the robot.

4.4. Current mission

The ultimate goal of the mission construction tool is to assign a current mission to the robot. This mission is either chosen in the mission library among the instantiated missions, or created as an instantiated mission. It then can be saved in the library. Fig. 4 gives an example of a current mission defined by the human operator of the robot, on the basis of the environment described in Fig. 2 and the generic objectives of the Fig. 3.

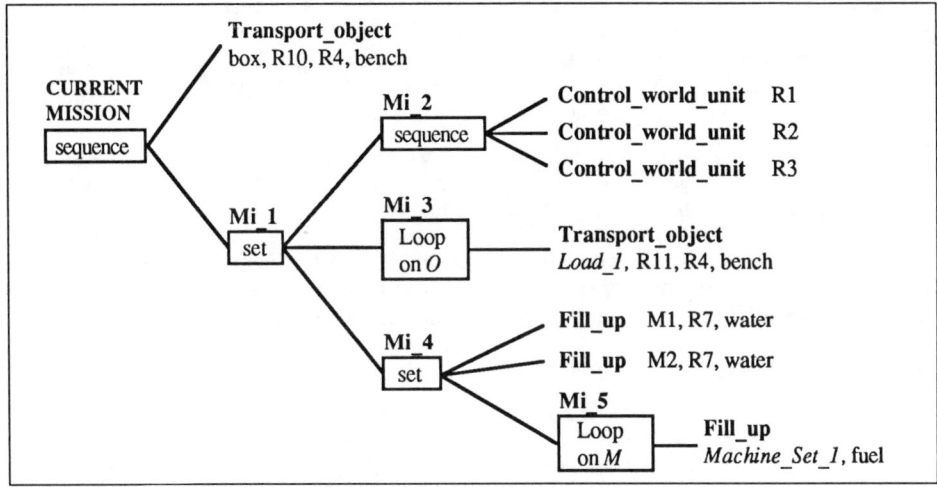

Figure 4. Example of a Current Mission.

The current mission description let the user specify priorities in the execution of sub-missions, thanks to the sequence-operator. The next section shows how the planning process uses these priorities to produce a global plan.

5. GLOBAL PLANNING

The aim of global planning is to produce an abstract plan realizing the current mission at a world unit level. The environment description used for computing the global plan is the global graph that describes links between world units. The planned solution gives the sequence of world units which constitutes the robot route and, for each world unit, the set of actions that must be realized in this world unit.

The global planning arranges the objectives described in the current mission, in order to minimize the cost of the solution. For the moment, the cost is evaluated in term of route reliability, but it could integrate other criteria such as estimated execution time of the objectives. The method that builds and evaluates the plan leads to share robotic resources: displacement (common sub-routes between objectives) as well as robot action means (e.g. transportation of many objects at once), provided that the resultant reliability is not reduced. This method verifies succession constraints between objectives that come from priorities described in the current mission organizing; It also takes into account the succession constraints linked to the decomposition of an objective into actions, and the robotic resources constraints that are involved in the current mission objectives.

5.1. Decomposing the current mission

The abstract description of the current mission given to the robot is recursively decomposed into sub-missions, finally resulting in the set of all the objectives involved in the mission. Concurrently, priorities between sub-missions are translated into succession constraints between objective couples.

This translation relies on two sets linked to each sub-mission, the *First-Objectives* set (FO-set) and the *Last-Objectives* set (LO-set). The first one contains the objectives of the mission that can be executed first (at the time of the mission execution) according to the priorities between sub-missions at any level of the mission. Similarly the second one contains the objectives that can end the mission. For every new specified mission, the two sets are built incrementally starting from the objectives.

The two sets associated to a mission are used to compute the succession constraints on current mission objectives. A succession constraint is a couple of objectives that expresses their precedence order in the plan. For computing succession constraints, the only relevant case is when a mission is a *sequence* of sub-missions. For each sequence of sub-missions of the current mission (at any decomposition level), and for each couple of consecutive sub-missions in the sequence, a succession constraint exists between every element of the first sub-mission LO-set and every element of the second sub-mission FO-set.

5.2. Decomposing the objectives

Some objectives are realized in various world units, e.g. "Transport O from WU1 to WU2 on P". These objectives are translated by the system into a sequence of actions that take place in one world unit. For example, the objective of transportation is translated into the sequence of two actions: "take O" (in WU1) before "Put O on P" (in WU2). Some succession constraints, which we call internal succession constraints, are attached to the actions. The succession constraints previously attached to the objective that has been decomposed into actions are now attached to the action starting or ending the sequence.

5.3. Representing and computing the global plan

The global plan is computed on the basis of the couples: action, world unit in which it is realized. Each couple comprises static knowledge (the succession constraints it must satisfy) and dynamic knowledge (the path to the world unit of the next couple in the plan, the cost of this path and the state of the robotic resource the action uses). Dynamic knowledge is updated during the building of the plan. The planning computes the minimal global path that satisfies succession and resources constraints.

The method followed to build a global plan is the simulated annealing algorithm. Three points are essential:

- Computing an initial solution: We build the initial solution by simply juxtaposing the objectives decomposed in sequences of couples action—world_unit (no robotic resource is shared, each objective is realized separately). Yet the objectives are arranged so that the succession constraints between them are satisfied. So each type of constraint, succession constraint as well as internal succession constraint or resources constraint, is satisfied.

- Finding a new solution in the neighbourhood of the current solution: A new solution is created by random permutting two couples action—world_unit in the current solution. The succession constraints and the resources constraints are then checked in order to validate the solution.

- Choosing adequate values for the algorithm parameters such as initial temperature, stop temperature, temperature decrement at each step and number of iterations at a given temperature.

6. CONCLUSION

The modeling tool is under experimentation; It is implemented using an hybrid language (frame-object) and is coupled to the simulated annealing algorithm written in C. The tool depends on the type of the robotic application (mobile robot, factory-like environment, nature of the activities). Nevertheless the modeling frame enables to represent various domains of application. After the characteristics of a specific application have been described, the operator is able to define a robot mission at a high level of abstraction without worrying about the robot elementary working. With regard to traditional task level programming languages, our mission specification language provides an additional level of abstraction for combining robotized tasks.

The planner of the system produces a sequence of actions and movings that optimizes the route in respect of reliability criteria. Reliability criteria rely on the ability of the robot vision means to recognize the perceived environment and the ability of the pilotage to follow a route. These capabilities have been studied in [9] and validate our environment modeling. A next step is to realize the cooperation between the planning module and the pilot module, by a control manager, in order to execute the plan in the real environment and to possibly replan it. Our model will also be enriched with temporal features in the environment description as well as in the task and mission description.

REFERENCES

[1] R. Alami, R. Chatila, M. Ghallab, C. Laugier, J.-P. Laumond, "Robotique et intelligence artificielle", Actes des 3èmes journées nationales PRC-GDR IA, France, Hermès 1990.

[2] J. F. Allen, "Towards a General Theory of Action and Time", Artificial Intelligence 23, pp123-154, 1984.

[3] C. Bauer, V. Schaeffer, P. Novikoff, J.-M. Allée, "Modélisation conceptuelle de l'environnement d'un robot mobile autonome", Actes de la Convention IA'90, Paris, Hermès 1990.

[4] R.A. Brooks, "A Layered Intelligent Control System for a Mobile Robot", Proc. of the 3rd International Symposium of Robotic Research, Gouvieux, France, October 1985.

[5] R. E. Fikes, N. J. Nilsson, "STRIPS: A New Approach to the Application of Theorem Proving to Problem Solving", Artificial Intelligence 2, pp189-208, 1971.

[6] M.-P. Gleizes, P. Glize, "Les systèmes multi-experts", Hermès 1990.

[7] A. Haurat, J.-L. Perrard, "Les langages de programmation des robots industriels", Hermès 1989.

[8] S. Kirkpatrick, C.D. Gelatt Jr, M.P. Vecchi, "Optimization by simulated annealing", Science Vol. 220, n° 4598, 1983.

[9] P. Novikoff, "De l'Interaction Perception-Navigation en Robotique Mobile", Thèse de Doctorat de l'Université Pierre-et-Marie-Curie (Paris 6), Juillet 1991.

[10] M. Parent & C. Laurgeau, "Les Robots, Tome 5, Langages et Méthodes de Programmation", Hermes 1983.

[11] E.D. Sacerdoti, "A structure for plans and behaviour", Elsevier-North Holland 1977.

[12] V. Schaeffer, C. Bauer, P. Novikoff, J.-M. Allée, and J.-P. Bénéjam, "A vision guided multi-modules control system for an autonomous mobile robot", Proc. of the ISRAM'90, Burnaby British Columbia, July 18-20, 1990.

[13] G.J. Sussman, "A Computer Model of Skill Acquisition", New-York Elsevier 1975.

[14] W. Swartout, "DARPA Santa Cruz Workshop on Planning", W. Swartout Editor, AI Magazine, Summer 1988.

AN OPTIMAL SOLUTION TO THE ROBOT NAVIGATION PLANNING PROBLEM BASED ON AN ELECTROMAGNETIC ANALOGUE.

V.Petridis and T.D.Tsiboukis
Department of Electrical Engineering
Faculty of Engineering
University of Thessaloniki
Thessaloniki 540 06
GREECE

Abstract

The robot navigation planning problem in the case of either structured or unstructured environment is solved by means of an electromagnetic analogue. The robot's environment is represented by a two-dimensional working area within which there exist certain obstacles of arbitrary shape. The robot and the target are represented by perfect conductors while the working area is represented by a sheet of finite conductivity. The obstacles are represented by regions within the working area of zero conductivity. If a voltage is applied between the robot and the target the problem of navigation planning is transformed to the one of finding the flux lines of the current flow established. The shortest flux line represents the optimal path. The working area and the obstacles can be of any shape. The technique employed is sufficiently fast to warrant its application to actual problems.

1. Introduction

A considerable amount of work has been published on the problem of robot navigation planning in the case of a structured and known environment [1]. In this case the optimal path to a certain target has to be computed taking into account the size of the robot. In the case of unstructured or dynamic environment the robot must employ its sensors to explore the environment in order to construct a model of it (world modelling) and then compute a path to the desired target (planning) [2-4]. In either case there exists the problem of navigation planning, i.e. that of computing an optimal (or suboptimal) path to the desired target.

There are a number of approaches to the path planning problem. Polygonal boundaries that approximate the obstacles [4] have been employed. In [5] the terrain is represented by a cubic B-spline patch and the path is represented by a B-spline curve. Two dimensional certainty grids have been employed in [3]. In this paper the problem of navigation planning is solved by employing an electromagnetic analogue. The technique presented can be employed to either structured or unstructured environment. A similar work has been reported in [6].

The main advantages of the technique presented are: a) the optimal path can be calculated, b) the working area can be of any shape c) the obstacles can be of any shape, d) it can be extended to 3-D environment.

2. The electromagnetic analoque

The robot' enviroment can be simulated by a two-dimensional working area within which there exist a certain number of obstacles. The robot and the target, at any instant, are represented by perfect conductors (Fig.1) (i.e. infinite conductivity) while the working area is represented by a conducting sheet of finite conductivity . The obstacles are represented by regions within the working area of zero conductivity (perfect insulators) as shown in Fig.1.

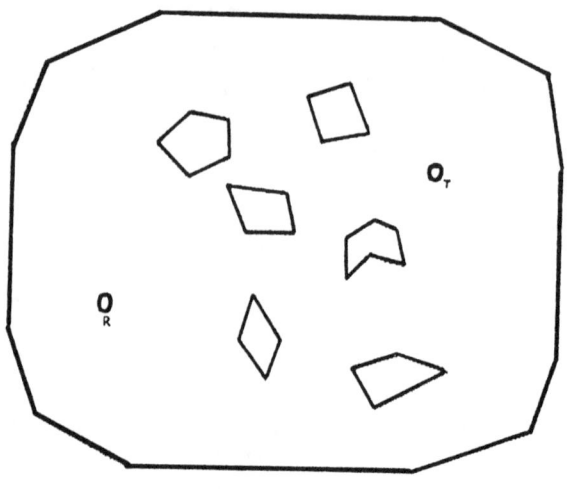

Fig. 1

When a constant voltage $U=\Phi_1-\Phi_2$ is applied between the robot and the target a current flow is established within the conducting sheet, governed by Ohm's law

$$J=\sigma E \tag{1}$$

and the continuity equation

$$\nabla J=0 \tag{2}$$

where J and E are the electric current density and the electric field intensity respectively.

The shortest path between the robot and the target, that avoids the obstacles is specified by the shortest flux line of the electric field established. Therefore, according to the electromagnetic analogue, the calculation of the optimal solution of the robot navigation planning problem has been transformed to the calculation of the shortest flux line of the electric field.

3. The field equations

The problem can be expressed in terms of the scalar potential Φ by replacing the electric field E with the negative gradient of the potential function Φ. Thus, the potential function Φ is governed by the scalar equation

$$\nabla \cdot (\sigma \nabla \Phi)=0 \tag{3}$$

subject to the following two boundary conditions:
(a) Φ is specified on the robot's and target's boundary (Dirichlet condition), and (b) the normal component J_n of the current density vanishes at the interfaces between the obstacles and the conducting sheet, as well as at the external boundary of the working area. This requirement is equivalent to the vanishing of the normal derivative (homogeneous Neumann condition).

4. Finite element solution

Due to the complicated geometry of the system the problem can not be solved analytically. For a numerical treatment, a finity difference (FD) or a finite element (FE) procedure can be adopted. The most appropriate method for the problem under consideration is, undoubtedly, the FE method since the FD method lacks the geometrical flexibility in fitting the irregular boundary shapes. In the finite element procedure the conducting region is subdivided into a finite set of subregions (usually triangular elements (Fig. 2), within which the unknown field quantity is approximated by suitable interpolation functions that contain as unknowns the values at vertices (nodes) of the element.

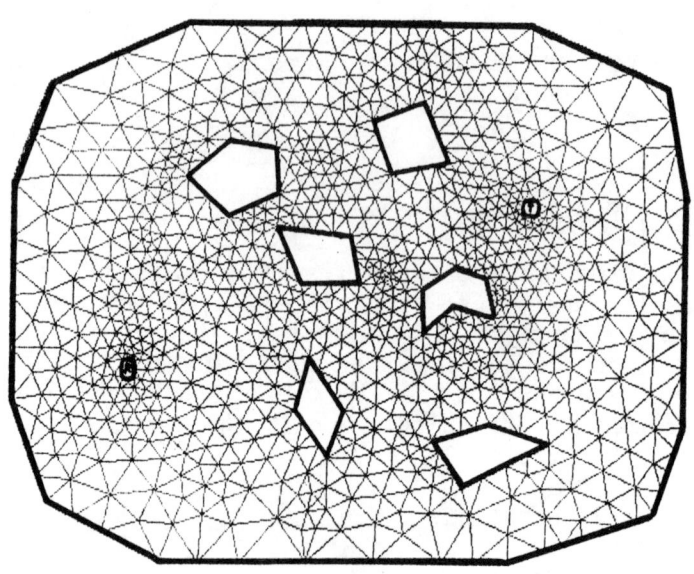

Fig. 2

The application of the Galerkin method when using the interpolation functions leads to an algebraic system of equations of the form

$$[S] [\Phi] = [b] \qquad (4)$$

where [S] is a very sparse symmetric-stiffness-matrix, [Φ] is a column vector containing the unknown potential values at the nodal points and [b] is a column vector of known terms. The solution of the matrix equation (4) can be obtained by either direct or iterative methods. An efficient iterative method, applied in this paper, is the ICCG (Incomplete Choleski Conjugete Gradient) [7].

The FE procedure consists of three phases, namely the preprocessing (the preparation of the input data) the main processing, and the postprocessing (the determination of the shortest flux line).
In order that the computational cost be reduced adaptive finite element analysis based on mesh modification (Delauney criterion) and error estimation algorithms (Bt-criterion) have been used [8,9].

The investigation of the flux lines is, usually, carried out by drawing appropriate arrows (Fig.3) in a discrete set of points giving simultaueously the direction and the modulus

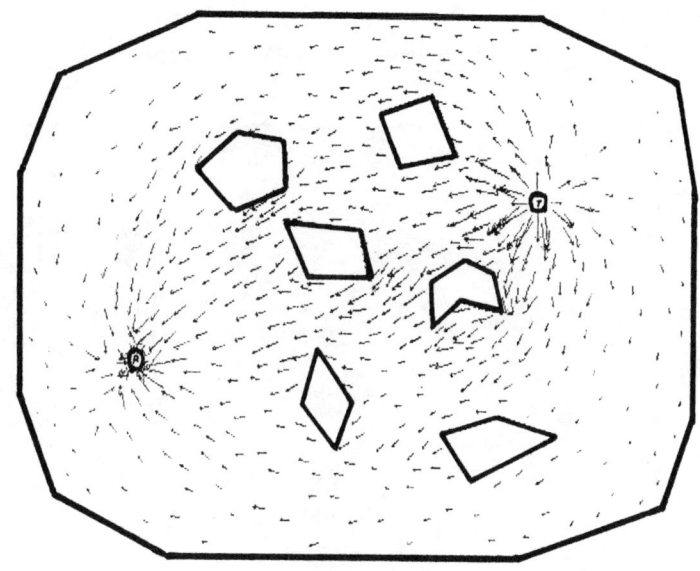

Fig. 3

of the field vector at each point. This approach, although very popular in finite element analysis, is not appropriate in this case. Instead, a better field representation can be obtained by drawing [10], in a continuous way, the current tubes which have their ends at the robot's and target's position. Then the determination of the shortest flux tube is a very easy task.

5. An example

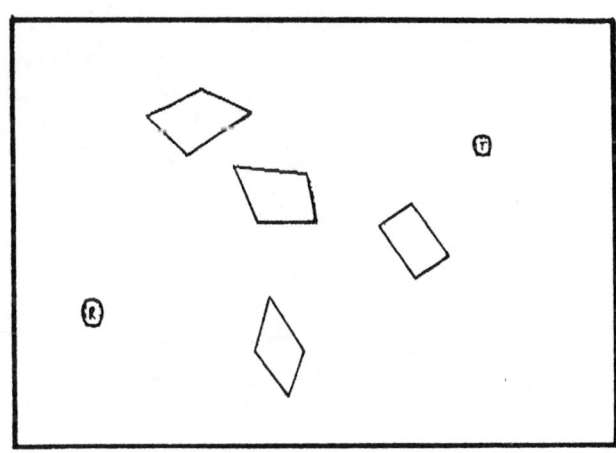

Fig. 4

Let us consider the working area shown in Fig.4 containing four typical obstacles positioned randomly.

The first order mesh shown in Fig. 5 has been generated automatically employing adaptive finite element analysis based on Delauney criterion (1268 elements and 678 nodes).

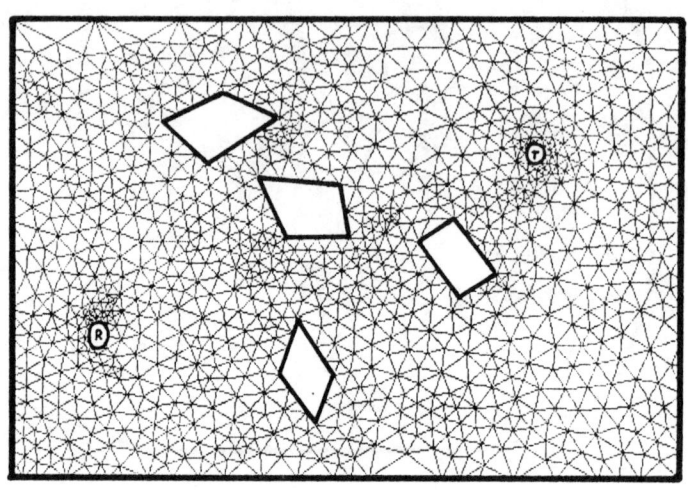

Fig.5

The preprocessing time has been 15sec and the main processing time has been 40sec. The determination of the shortest flux line(Fig.6) took 10 sec. The computer employed was a 386SX with 387 cooprocessor.

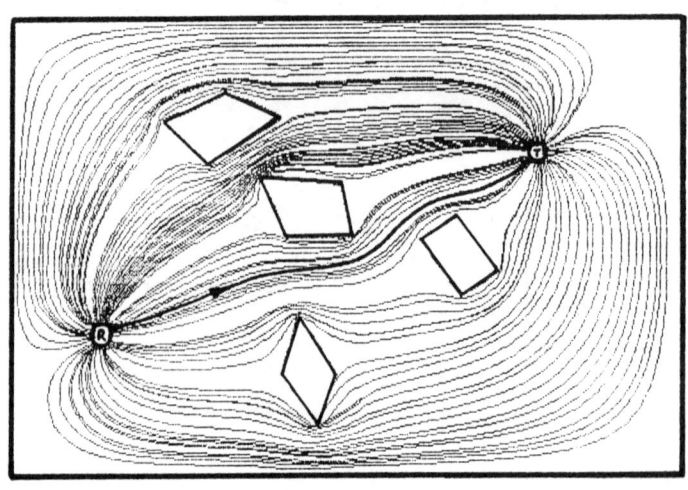

Fig. 6

6. Conclusions

The technique presented in this paper computes the optimal path with an accuracy determined by the density of the finite element grid used. The working area and the obstacles can be of any shape. The presented technique can be extended to the case that the terrain is any 2-D surface in the 3-D space. It can be employed to automatically generate the optimal path in real-world situations since the time required for computation is quite reasonable.

References

1. S.H.Whitesides, "Computational Geometry and Motion Planning in Computational Geometry" Elsevier Sc. Publ. G.T. Toussaint editor, N.York, 1985.
2. "Intelligent Systems and their Application" feature article, IEEE Expert, vol.2, Winter 1987.
3. A.Elfes "Sonar-Based Real-World Mapping and Navigation" IEEE Trans. on Rob. and Aut., vol.3, No.3, June 1987 pp. 249-265.
4. S.V.N.Rao et al,"Robot Navigation in an Unexplored Terrain" Journal of Robotic Systems, vol.3, 1986, pp. 389-407.
5. Z.Shiller and Y.R.Gwo, "Dynamic Motion Planning of Autonomous Vehicles" IEEE Tr. on Rob. and Aut., vol.7, April 1991, pp.241-249.
6. L.Tarassenko, M.J.Brownlow, A.F.Murray", VLSI neural networks for autonomous robot navigation" Proc. of INNC-90 Paris, July 1990, pp.213-216.
7. D.S.Kersaw, "The Incomplete Cholesky-Conjugate Gradient Method for the Iterative Solution of Systems of Linear Equations", J.Comp. Phys.26, 1978, pp.43-65.
8. Z.J.Cendes, D.N.Shenton and H.Shahnasser, "Magnetic Field Computation using Delauney triangulation and complementary finite-element methods" IEEE Trans.Magn.,MAG-19,1983 pp.2551-2554.
9. J.Penman and M.D.Grieve, "Self-adaptive mesh generation technique for the finite element method", IEE Proc.,Pt A, vol. 134, 1987, pp.634-650.
10. D.Savalle, G.Meunier, J.C.Sabonnadiere, J.C.Verite, "Tridimensional modelling of current and magnetic flux lines" IEEE Trans. on Magn., vol 24, No.1, 1988, pp.378-380.

REAL-TIME LQG ROBOTIC VISUAL TRACKING *

N. P. Papanikolopoulos and P. K. Khosla

Department of Electrical and Computer Engineering
The Robotics Institute
Carnegie Mellon University
Pittsburgh, Pennsylvania 15213

Abstract

In this paper, a modified LQG technique is proposed for the solution of the robotic visual tracking problem (eye-in-hand configuration). The problem of robotic visual tracking is formulated as a problem of combining control with computer vision. A cross-correlation method provides the object's motion measurements which are used to update the system's measurement vector. These measurements are fed to a discrete steady state Kalman filter that calculates the estimated values of the system's states and of the exogenous disturbances. Then, a discrete LQG controller computes the desired motion of the robotic system. Experimental results are presented to show the effectiveness of the approach.

1. Introduction

One of the most desirable characteristics of a robotic manipulator is its flexibility. Flexible robots can quickly adapt to the evolving requirements of an unknown task, and can properly react to sudden changes in the environment. Flexibility and adaptability can be achieved by incorporating vision and generally, sensory information in the feedback loop. This sensory information enhances the robot's capability by continuously updating the robot's view (or model) of the world and the task. The completeness and accuracy of this view depends on the existence of a framework for the integration of sensory information with the other components of a robotic system.

Research in computer vision has traditionally emphasized the paradigm of image understanding which focuses on the static analysis of one image. Recently, more emphasis has been given to the dynamic analysis of a sequence of images [1]. This sequence of images is produced either by a moving camera that captures views of a static environment, or by a static camera that captures views of a moving object. A characteristic example of the second category is the motion analysis area where vision information is used for tracking moving objects [2, 3, 4]. Little research [5, 6] has been conducted in using vision information in the dynamic feedback loop (moving camera and target). Particularly, in the motion analysis research, Roach and Aggarwal

---

*This research was supported by the Defense Advanced Research Projects Agency, through ARPA Order Number DAAA-21-89C-0001 and by Innovision Inc.

[7] have presented a scheme for tracking rigid convex polyhedra. Their scheme was based on image segmentation which is time-consuming. A stereo system for tracking known 3-D targets was presented by Gennery [4]. Wallace and Mitchell [3] have used complex Fourier series for obtaining a solution to the same problem. Hunt and Sanderson [8] have presented algorithms for visual tracking based on mathematical prediction of the position of the object's centroid. Lee and Wohn [9] have used image differencing techniques to track a moving object.

This paper addresses some of the issues associated with the use of vision information in the dynamic feedback loop. In particular, we deal with the problem of robotic visual tracking of a moving target (eye-in-hand configuration). To achieve our tracking objective, we combine computer vision techniques, for detection of motion, with a simple LQG control strategy. A cross-correlation technique (SSD optical flow) is used for computing the vector of discrete displacements in real-time. The paper is organized as follows: In Section 2, we review the definition of the optical flow and present methods for computation of the vector of discrete displacements. The mathematical formulation of the visual tracking problem is described in Section 3. The LQG controller in conjunction with the cartesian robotic control schemes are discussed in Section 4. The hardware configuration of our experimental testbed (DD Arm II) and some experimental results are presented in Section 5. Finally, in Section 6, the paper is summarized.

2. Optical Flow

This Section will present an outline of our vision techniques in order to illustrate their characteristics (noise, computational complexity, quantization errors). We assume a pinhole camera model with a frame R_s attached to it. In addition, we assume a perspective projection and the focal length to be unity. A point **P** with coordinates (X_s, Y_s, Z_s) in R_s projects onto a point **p** in the image plane with image coordinates (x, y) given by:

$$x = X_s/Z_s \quad \text{and} \quad y = Y_s/Z_s. \tag{1}$$

Let us assume that the camera moves in a static environment with a translational velocity $\mathbf{T} = (T_x, T_y, T_z)^T$ and with an angular velocity $\mathbf{R} = (R_x, R_y, R_z)^T$ with respect to the camera frame R_s. The optical flow equations are given by:

$$u = [x\frac{T_z}{Z_s} - \frac{T_x}{Z_s}] + [xyR_x - (1+x^2)R_y + yR_z] \tag{2}$$

$$v = [y\frac{T_z}{Z_s} - \frac{T_y}{Z_s}] + [(1+y^2)R_x - xyR_y - xR_z] \tag{3}$$

where $u = \dot{x}$ and $v = \dot{y}$. Values u and v are also known as the optical flow measurements. Instead of a static object and a moving camera, if we were to assume a static camera and a moving object then we would obtain the same result as in equations (2) and (3) except for a sign reversal. The computation of u and v has been the focus of much research and many algorithms have been proposed [10, 11]. For accuracy reasons, we use a modified version of the matching-based technique [11] also known as the Sum-of-Squared Differences (SSD) optical flow. For every point $\mathbf{p_A} = (x_A, y_A)$ in image A, we want to find the point $\mathbf{p_B} = (x_A + u, y_A + v)$ to which the point $\mathbf{p_A}$ moves in image B. It is assumed that the intensity values in the neighborhood L of $\mathbf{p_A}$ remain almost constant over time, that the point $\mathbf{p_B}$ is within an area S of $\mathbf{p_A}$, and that velocities are normalized by time T to get the displacements. Thus, for the point $\mathbf{p_A}$ the SSD algorithm selects the displacement $\mathbf{d} = (u, v)$ that minimizes the SSD measure:

$$e(\mathbf{p_A}, \mathbf{d}) = \sum_{m,n \in N} [I_A(x_A+m, y_A+n) - I_B(x_A+m+u, y_A+n+v)]^2 \tag{4}$$

where $u, v \in S$, N is an area around the pixel we are interested in, and I_A, I_B are the intensity functions in images A and B, respectively. The different values of the SSD measure create a surface called the *SSD surface*. The accuracy of the measurements of the displacement vector can be improved by using multiple windows. The selection of the best measurements is based on the confidence measure of each window. Efficient confidence measures for the selection of the most accurate measurements are described in [12]. The next step of our algorithm involves the use of these measurements in the visual tracking process. These measurements should be transformed into control commands to the robotic system. Thus, a mathematical model for this transformation must be developed. In the next Section, we present a mathematical model for the visual tracking problem.

3. Mathematical Model For The 2-D Visual Tracking Of An Object

Consider a target that moves in a plane with a feature, located at a point **P**, that we want to track. The projection of this point on the image plane is the point **p**. Consider also a neighborhood S_w of **p** in the image plane. The problem of 2-D visual tracking of a single feature point can be defined as: "find the camera translation (T_x, T_y) with respect to the camera frame that keeps S_w stationary in an area S_o around the origin of the image frame". It is assumed that at initialization of the tracking process, the area S_w is brought to the origin of the image frame, and that the plane of motion is vertical to the optical axis of the camera. The problem of visual tracking of a single feature point can also be defined as [2]: "find the camera rotation (R_x, R_y) with respect to the camera frame that keeps S_w stationary in an area S_o around the origin of the image frame". Assume that the optical flow of the point **p** at the instant of time kT is $(u(kT), v(kT))$ where T is the time between two consecutive frames. It can be shown that at time $(k+1)T$, the optical flow is:

$$u((k+1)T) \approx u(kT) + u_c(kT), \quad v((k+1)T) \approx v(kT) + v_c(kT) \tag{5}$$

where $u_c(kT), v_c(kT)$ are the components of the optical flow induced by the tracking motion of the camera. Equations (5) are based on the assumption that the optical flow induced by motion of the feature does not change in the time interval T. Therefore, T should be as small as possible. To keep the notation simple and without any loss of generality, equations (5) will be used with k and $(k+1)$ instead of kT and $(k+1)T$ respectively. If the camera tracks the feature point with translation $T_x(k)$ and $T_y(k)$ with respect to the camera frame, then the optical flow that is generated by the motion of the camera with $T_x(k)$ and $T_y(k)$ is:

$$u_c(k) = -\frac{T_x(k)}{Z_s}, \quad v_c(k) = -\frac{T_y(k)}{Z_s}. \tag{6}$$

We assume that for 2-D visual tracking the depth Z_s remains constant. The same model can be used for keeping the feature point stationary in an area S_r different from the origin.

Consider a target that moves in a plane which is vertical to the optical axis of the camera. The projection of the target on the image plane is the area S_w in the image plane. The problem of 2-D visual tracking of a single object can be defined as: "find the camera translation (T_x, T_y) and rotation (R_z) with respect to the camera frame that keeps S_w stationary". It is assumed that the target rotates around an axis Z which at $k=0$ coincides with the optical axis of the camera. The mathematical model of this problem in state-space form is (a formal derivation is given in [13]):

$$x(k+1) = Ax(k) + Bu_c(k) + Ed(k) + Hv(k) \tag{7}$$

where $A = H = I_3^{**}$, $B = E = T I_3$, $x(k) \in R^3$, $u_c(k) \in R^3$, $d(k) \in R^3$, and $v(k) \in R^3$. The vector

---

**The symbol I_n denotes the identity matrix of order n.

$\mathbf{x}(k) = (x(k), y(k), \theta(k))^T$ is the state vector, $\mathbf{u}_c(k) = (u_c(k), v_c(k), R_z(k))^T$ is the control input vector, $\mathbf{d}(k) = (u(k), v(k), \omega(k))^T$ is the exogenous disturbances vector, and $\mathbf{v}(k) = (v_1(k), v_2(k), v_3(k))^T$ is the white noise vector. $x(k), y(k), \theta(k)$ are now the X, Y and roll component of the tracking error, respectively. The measurement vector $\mathbf{y}(k) = (y_1(k), y_2(k), y_3(k))^T$ is given by:

$$\mathbf{y}(k) = \mathbf{C}\mathbf{x}(k) + \mathbf{w}(k) \tag{8}$$

where $\mathbf{w}(k) = (w_1(k), w_2(k), w_3(k))^T$ is a white noise vector ($\mathbf{w}(k) \sim N(0, \mathbf{W})$) and $\mathbf{C} = \mathbf{I}_3$. The measurement vector is obtained by using the SSD algorithm. First, the tracking error of the projections of the two different feature points on the image plane is computed. Then, an algebraic system of four equations (two tracking error equations per point) is formulated. The solution of the system is the X, Y and roll component of the tracking error. If the projections of the two feature points on the image plane are not the same, it is guaranteed that the system of equations has a solution. It is assumed that each one of these features at time $t = 0$ is located at its desired position. The LQG control strategy that keeps the target stationary is discussed in detail in the next Section.

4. LQG Controller

A useful control technique for this type of problem is the LQG (Linear Quadratic Gaussian) control scheme. Neglecting for the time being the white noise terms of our system, we will consider the more general problem of determining the matrices $\mathbf{G}$ and $\mathbf{G}_d$ in the linear control law:

$$\mathbf{u}_c(k) = -\mathbf{G}\mathbf{x}(k) - \mathbf{G}_d \mathbf{d}(k). \tag{9}$$

The reason it is necessary to separate the exogenous variables from the process state vector $\mathbf{x}(k)$, rather than deal directly with the metastate vector $\mathbf{x}_M^T(k) = (\mathbf{x}^T(k), \mathbf{d}^T(k))$, is that in developing the theory for the design of the gain matrix we assume that the underlying process is controllable. If we try to create a new system with metastate vector $\mathbf{x}_M^T(k) = (\mathbf{x}^T(k), \mathbf{d}^T(k))$, we can show, by using the controllability matrix, that the new system is uncontrollable. Thus, we should work with the form of the system (Eq. (7) and (8)) that was developed in Section 3. One can observe that the matrices $\mathbf{E}$ and $\mathbf{B}$ are equal, so equations (7) and (8) can be rewritten as:

$$\mathbf{x}(k+1) = \mathbf{A}\mathbf{x}(k) + \mathbf{B}\mathbf{u}_o(k) + \mathbf{H}\mathbf{v}(k) \tag{10}$$
$$\mathbf{y}(k) = \mathbf{C}\mathbf{x}(k) + \mathbf{w}(k) \tag{11}$$

where $\mathbf{u}_o(k) = \mathbf{u}_c(k) + \mathbf{d}(k)$. A performance criterion that can be minimized for the selection of the optimum gain matrix $\mathbf{G}$ is:

$$J = \sum_{k=0}^{\infty} [\mathbf{x}^T(k) \mathbf{Q} \mathbf{x}(k) + \mathbf{u}_o^T(k) \mathbf{R} \mathbf{u}_o(k)] \tag{12}$$

where $\mathbf{Q} = \mathbf{Q}^T \geq 0$ where $\mathbf{R} = \mathbf{R}^T > 0$. The performance criterion contains a quadratic form in the state vector $\mathbf{x}(k)$ plus a second quadratic form in the vector $\mathbf{u}_o(k)$. Physically, the first quadratic form represents a penalty for the tracking error and the second corresponds to a modified cost of control. The performance criterion is minimized by selecting an appropriate gain matrix $\mathbf{G}$. Taking into consideration the white noise terms of our system, the controller becomes an LQG controller and the control law is:

$$\mathbf{u}_o(k) = -\mathbf{G}\hat{\mathbf{x}}(k) \tag{13}$$

where $\hat{\mathbf{x}}(k)$ is the estimated value of the state vector $\mathbf{x}(k)$. Thus, $\mathbf{u}_c(k)$ is given by:

$$\mathbf{u}_c(k) = -\mathbf{G}\hat{\mathbf{x}}(k) - \hat{\mathbf{d}}(k) \tag{14}$$

where $\hat{\mathbf{d}}(k)$ is the estimated value of the disturbance vector $\mathbf{d}(k)$. The performance criterion now is the expected value of J. The optimal control gain matrix G is $G=(B^TPB+R)^{-1}B^TPA$ with P being the unique symmetric positive definite solution of the matrix algebraic Ricatti equation:

$$A^T[P-PB(B^TPB+R)^{-1}B^TP]A+Q=P. \tag{15}$$

The design parameters are the elements of the matrices Q, R. By selecting these, one can obtain the desired gain matrix G. There is no standard procedure for the selection of the elements of these matrices. One technique [14] is the optimization approach. The next step in our algorithm is the computation of the vectors $\hat{x}(k)$ and $\hat{d}(k)$. We design an observer for the estimation of the metastate vector $x_M(k)$. The state-space model of equations (7)-(8) can be rewritten as:

$$x_M(k+1) = A_M x_M(k) + B_M u_c(k) + H_M v_M(k) \tag{16}$$

$$y(k) = C_M x_M(k) + w(k) \tag{17}$$

where

$$A_M = \begin{bmatrix} 1 & 0 & 0 & T & 0 & 0 \\ 0 & 1 & 0 & 0 & T & 0 \\ 0 & 0 & 1 & 0 & 0 & T \\ 0 & 0 & 0 & 1 & 0 & 0 \\ 0 & 0 & 0 & 0 & 1 & 0 \\ 0 & 0 & 0 & 0 & 0 & 1 \end{bmatrix}, \quad B_M = \begin{bmatrix} T & 0 & 0 \\ 0 & T & 0 \\ 0 & 0 & T \\ 0 & 0 & 0 \\ 0 & 0 & 0 \\ 0 & 0 & 0 \end{bmatrix},$$

and

$$C_M = \begin{bmatrix} 1 & 0 & 0 & 0 & 0 & 0 \\ 0 & 1 & 0 & 0 & 0 & 0 \\ 0 & 0 & 1 & 0 & 0 & 0 \end{bmatrix}, \quad H_M^T = \begin{bmatrix} 0 & 0 & 0 & T & 0 & 0 \\ 0 & 0 & 0 & 0 & T & 0 \\ 0 & 0 & 0 & 0 & 0 & T \end{bmatrix}.$$

As it was mentioned before, the measurement vector consists of the measured translational components of the tracking error $x(k)$, $y(k)$ and of the roll component of it, $\theta(k)$. A steady state Kalman filter [15] can be designed for the estimation of the metastate vector $x_M(k)$. The assumptions are that $Q_e = E[v_M(k) v_M^T(k)]$, $R_e = E[w(k) w^T(k)]$, $x_{Mo} = E[x_M(0)]$ and $E[v_M(k) w^T(j)] = 0$ for all k, j. The state update equation is:

$$\hat{x}_M(k+1) = A_M \hat{x}_M(k) + B_M u_c(k) + K_e(y(k) - C_M \hat{x}(k)) \tag{18}$$

where

$$K_e = A_M P_e C_M^T [C_M P_e C_M^T + R_e]^{-1} \tag{19}$$

and P_e satisfies the matrix algebraic Ricatti equation

$$A_M[I - P_e C_M^T (C_M P_e C_M^T + R_e)^{-1} C_M] P_e A_M^T + H_M Q_e H_M^T = P_e. \tag{20}$$

The time-invariant steady state Kalman filter can be implemented easily and does not require a large number of calculations. In addition to the steady state Kalman filter, we use the time-varying discrete Kalman filter which constantly updates the Kalman gain matrix K_e. This improves the performance of our observer but it is computationally more expensive than the time-invariant Kalman filter. The state update equation of the new state observer is the same as (18) while the other equations are:

$$K_e(k) = A_M P_e(k) C_M^T [C_M P_e(k) C_M^T + R_e]^{-1} \tag{21}$$

$$P_e(k+1) = [A_M - K_e(k) C_M] P_e(k) A_M^T + H_M Q_e H_M^T \tag{22}$$

where $\hat{x}_{Mo} = E[x_M(0)]$ and $P_e(0) = E[(x_M(0) - \hat{x}_{Mo})(x_M(0) - \hat{x}_{Mo})^T]$. The performance of the observer depends on the selection of the Q_e and R_e matrices. We should mention that the white noise model is only an approximation to the actual noise model of the camera. Thus, the selection of the Q_e and R_e matrices is done empirically and a search for the best set of noise variances is conducted. The initialization of the vectors $\hat{x}(k)$ and $\hat{d}(k)$ is given by:

$$\hat{x}(1) = y(1) \, , \, \hat{d}(1) = T^{-1}(y(1) - 0). \tag{23}$$

The next step of our algorithm is the calculation of the triple $(T_x(k), T_y(k), R_z(k))$. The calculation of the $T_x(k)$ and $T_y(k)$ is done by using equations (6) which require the knowledge of the depth Z_s. $R_z(k)$ is given directly as the computed control signal. The knowledge of the depth Z_s can be acquired in two ways. The first way is direct computation by a range sensor or by stereo techniques [1]. The use of stereo for the recovery of the depth is a difficult procedure because it requires the solution of the correspondence problem. A more effective strategy that requires the use of only one visual sensor is to use adaptive control techniques. The control law is based on the estimated on-line values of the model's parameters that depend on the depth. More details about our adaptive control schemes can be found in [16]. After the computation of the translational $T(k)$ and rotational $R(k)$ velocity vectors with respect to the camera frame R_s, we transform them to the end-effector frame R_e with the use of the transformation ^eT_s. The transformed signals are fed to the robot controller. We experimented with a cartesian PD robot control scheme with gravity compensation. The mathematical model of the robot's dynamics is:

$$D(q)\ddot{q} + c(q,\dot{q}) + g(q) = \tau \tag{24}$$

where q is the vector of the joint variables of the robotic arm, D is the inertial acceleration related matrix, c is the nonlinear Coriolis and centrifugal torque vector, g is the gravitational torque vector and τ is the generalized torque vector. The model is nonlinear and coupled. The PD control scheme assumes that all velocities in the dynamics equations are zero. This implies that $\dot{q} = J = c(q,\dot{q}) = 0$ ($J(q)$ is the manipulator's Jacobian). Thus, the actuators' torque vector τ is given by:

$$\tau = J^T(q) F + g(q) \tag{25}$$

$$F = K_p \Delta x + K_v [\dot{x}_d - J(q)\dot{q}] + \ddot{x}_d \tag{26}$$

where F is the generalized force vector, $\Delta x^T = (\Delta x_p^T, \Delta x_o^T)$ is the position and orientation error vector, and K_p and K_v are diagonal gain matrices. The subscript d denotes the desired quantities. In our experiments, $\ddot{x}_d$ is selected to be zero.

5. Hardware And Some Experimental Results

A number of experiments were performed on the CMU DD Arm II robotic system. This robotic system consists of: a) a Sun 3/260 host system on a VME bus, b) Multiple Ironics M68020 boards, c) a Mercury 32000 Floating Point Unit, d) an IDAS/150 image processing system, d) a Panasonic industrial CCD color camera, Model GP-CD1H, e) six Texas Instrument TMS320 DSP processors, each controlling one joint of the CMU DD Arm II system, f) sensors such as a tactile sensor and a force sensor, and g) a six degrees of freedom joystick. The IDAS/150 contains a Heurikon 68030 board as the controller of the vision module and two floating point boards, each one with computational power of 20 Mflops. The software is organized around 3 processes: a) **Vision process** which does all the image processing calculations and has a period of 150 ms, b) **Interpolation process** which reads the data from the vision system, interpolates the data and sends the reference signals to the robot cartesian controller, and c) **Robot controller process** which drives the robot and has a period of 3.33 ms. During the experiments, the camera is mounted on the end-effector and has a focal length of 7.5mm. The objects (books, toys, pencils) are moving on a plane (average depth $Z_s=680$mm). The user, by moving the mouse around, proposes to the system some of the object's features that he is interested in. Then, the system evaluates on-line the quality of the measurements, based on the confidence measures described in [13]. Currently, four features are used and the size of the attached windows is 10x10. The experimental results are plotted in Fig. 1 and 2 where the dotdashed trajectories correspond to the trajectories of the center of mass of the moving objects. The vector Mez_P represents the position of the end-effector in the world frame. The simple PD produces oscillations around the desired trajectory. In this example, along with the translational motion, the object performs a rotational motion around an axis that passes through the center of mass of the object. Even with noisy measurements, the LQG seems to perform well. This becomes obvious, when one reduces the number of the windows which are used (increased noise in the measurements), and the LQG controller continues to keep the target at the desired position.

6. Conclusions

In this paper, we considered a LQG approach to the robotic visual tracking problem (eye-in-hand configuration). We formulated the problem as a control and vision problem and discussed the issues related with this formulation. A cross-correlation technique (SSD Optical Flow) was used to provide accurate measurements of the object's motion parameters. An LQG regulator in conjunction with cartesian robotic controllers were studied as possible solutions to the robotic visual tracking problem. The vision and control techniques were tested on a real robotic environment, the CMU DD Arm II. Experimental results show that the methods are quite accurate, robust and promising. Their most important characteristic is that they can be implemented in real-time. Future research efforts should be focused on the extension of the techniques to the 3-D robotic visual tracking problem, the explicit use of the target model in the whole mathematical formulation, the solution of the problem of vanishing features, and finally, the direct integration of the robot dynamics in the feedback loop.

References

1. L. Matthies, R. Szeliski, and T. Kanade, "Kalman filter-based algorithms for estimating depth from image sequences", Tech. report 88-1, Carnegie Mellon University, The Robotics Institute, 1988.

2. D. Tsakiris, "Visual tracking strategies", Master's thesis, Department of Electrical Engineering, University of Maryland, 1988.

3. T.P. Wallace and O.R. Mitchell, "Analysis of three-dimensional movement using Fourier descriptors", IEEE Trans. PAMI, Vol. 2, No. 6, 1980, pp. 583-588.

4. D.B. Gennery, "Tracking known three-dimensional objects", Proc. AAAI 2nd Natl. Conf. on AI, 1982, pp. 13-17.

5. J.T. Feddema, C.S.G. Lee, and O.R. Mitchell, "Automatic selection of image features for visual servoing of a robot manipulator", Proc. of the IEEE Intern. Conf. on Robotics and Automation, May 1989, pp. 832-837.

6. L.E. Weiss, A.C. Sanderson, and C.P. Neuman, "Dynamic sensor-based control of robots with visual feedback", IEEE Journal of Robotics and Automation, Vol. RA-3, No. 5, October 1987, pp 404 417.

7 J.W. Roach and J.K. Aggarwal, "Computer tracking of objects moving in space", IEEE Trans. PAMI, Vol. 1, No. 2, 1979, pp. 127-135.

8. A.E. Hunt and A.C. Sanderson, "Vision-based predictive tracking of a moving target", Tech. report CMU-RI-TR-82-15, Carnegie Mellon University, The Robotics Institute, January 1982.

9. S.W. Lee and K. Wohn, "Tracking moving objects by a mobile camera", Tech. report MS-CIS-88-97, Department of Computer and Information Science, University of Pennsylvania, November 1988.

10. B.K.P Horn and B.G. Schunck, "Determining optical flow", Artificial Intelligence, Vol. 17, 1981, pp. 185-204.

11. P. Anandan, "Measuring visual motion from image sequences", Tech. report COINS-TR-87-21, COINS Department, University of Massachusetts, 1987.
12. N. Papanikolopoulos, P. K. Khosla, and T. Kanade, "Vision and control techniques for robotic visual tracking", *Proc. of the IEEE Int. Conf. on Robotics and Automation*, 1991, pp. 857-864.
13. N. Papanikolopoulos, P. Khosla, and T. Kanade, "Robotic visual tracking: Theory and experiments", Tech. report, Carnegie Mellon University, The Robotics Institute, 1990.
14. F.L. Lewis, *Optimal control*, John Wiley & Sons, New York, 1986.
15. A. Gelb, *Applied optimal estimation*, MIT Press, Cambridge, 1974.
16. N. Papanikolopoulos, P. Khosla, and T. Kanade, *Adaptive robotic visual tracking*, Accepted to the American Control Conference, 1991.

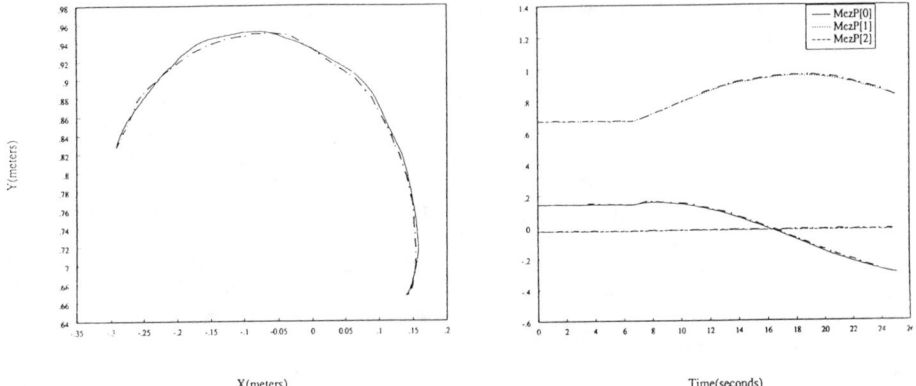

Figure 1: LQG controller in conjunction with a cartesian PD robotic controller with gravity compensation.

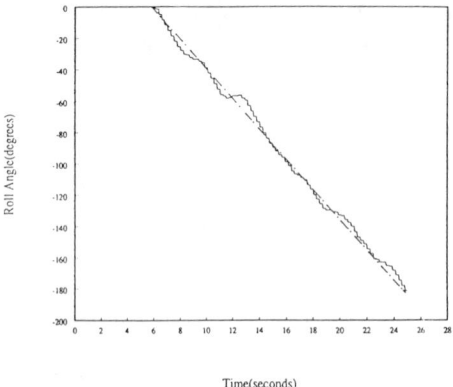

Figure 2: Roll trajectory of the robot end-effector and the object in the previous example.

Trajectory Tracking for Mobile Robot

S. Delaplace, P. Blazevic, J. G. Fontaine, N. Pons, J. Rabit
Laboratoire de Robotique de Paris, URA 1305, UPMC, Tour 66, 2ème étage,
4, Place Jussieu, 75252 Paris Cedex 05, B164, FRANCE

1. Introduction

During the ROMEO project (Experimental Mobile RObot number 0) development at the L.R.P, we have conducted a study on trajectory tracking for mobile robot in a constrained environment.

The path planning is only used when low level control loops are present to ensure good tracking and local recovery of deviation provoked by unmodeled perturbations. This automatic recovery feature is essential. It permits not to reconsider the control objectives permanently and to lighten the task of higher planning levels [BARR 89][CANN 88]

The feedback control of an autonomous robot vehicle can sometimes present subtle and surprising problems. In particular, it is important to realize that a vehicle's position or pose is represented by three parameters (X, Y, θ) two for translation and one for orientation, when we assume only planar motions. However, for common tricycle and differential drive configurations, there are two degrees of freedom for control, that are steering angle and velocity or the independant velocity of the two wheels of a differential drive vehicle. Such systems are referred to as nonholonomic [KANE 61], [LAUM 86], [LOBA 80], [LOBA 81].

The experimentation support, ROMEO, is an autonomous vehicle with two rear driving guide wheels and two idle wheels in front. Eight ultrasonic sensors are fixed on the periphery of the robot and one is situated on a turret with two degrees of freedom which also supports a CCD camera. Two optical encoders mounted on the rear wheels give the robot's position and orientation by direct odometry and control DC motors through specialized circuits. Computer control of ROMEO is assumed by separated modules physically characterized by several specialized cards. This structure allows to use a compatible IBM PC AT on board.

In this paper, we first study the ROMEO model using a schematic diagram of the robot following a wall. The robot model is controlled by constraining a point located in front of the two rear wheels. Then, results of a feedback control simulation are verified using experimental results.

We have used a similar model up to follow a trajectory analytically represented in the (X, Y) plan. Finally, simulation and experimentation are developed for straight line tracking, circle tracking and right angle trajectories.

2. ROME0 model

2.1 Hypothesis
This robot model is controlled so as to constrain a point located in front of the rear wheels, to be at a constant distance from a wall and to maintain the robot parallel to this surface.

The following hypothesis are used to build this model :

a/ The angle between the robot and the wall is small ($\alpha < 15°$)

b/ The system input is the desired distance between the reference and the robot

c/ The two wheels control is carried out using the velocity control :

$$\text{Right wheel}: V_r = V + \Delta V$$
$$\text{Left wheel}: V_l = V - \Delta V$$

in which V is the fixed M point velocity (Fig. 1.1) and $V = \dfrac{V_r + V_l}{2}$.

To obtain the distance between the robot and the wall, we use an ultrasonic sensor situated in front of the rear wheels at a distance d (Fig. 1).

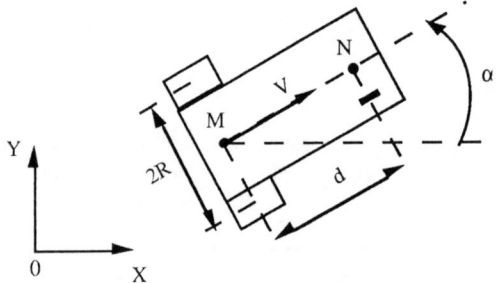

Fig. 1 : General description

2.2 Useful relations
The scalar velocity components of the point M, for small values of α, are given by :

$$\begin{cases} \dfrac{dX_m}{dt} = V \\ \dfrac{dY_m}{dt} = V\alpha \\ \dfrac{d\alpha}{dt} = \dfrac{\Delta V}{R} \end{cases}$$

and (X_n, Y_n) coordinates of the point N :
$$X_n = X_m + d$$
$$Y_n = Y_m + d\alpha$$

In Laplace form the relation is written :

$$Y_n(s) = \dfrac{\Delta V}{R}(\dfrac{V + ds}{s^2})$$

where $Y_n(s)$ is the model output representing the distance from N to the vertical face (Fig. 2). Then, the whole system can be described by the feedback control diagram presented in (Fig. 3).

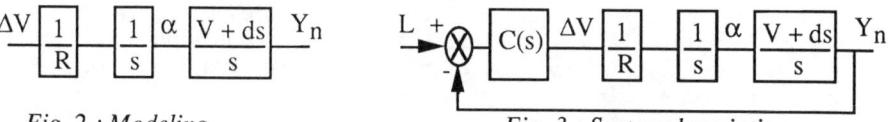

Fig. 2 : Modeling   Fig. 3 : System description

Remarks :
- L is the system input
- $l = Y_n(s)$ is the distance from the robot to the wall, measured by the ultrasonic sensor
- $C(s)$ represents the correction system for optimizing the feedback control. In our application, $C(s)$ is a proportional correction $(d>0)$.[BLAZ 91]

3. Simulation and experimental results

3.1 Simulation
The system behavior is simulated with an input step.
According to ROME0 geometrical values, the corrector $C(s)$ is optimized by a classical method and $C(s) = 100$.

3.2 Experimental
The robot control law is :

$$\Delta V = (L - l).C(s)$$

The recorded value is the difference between the input and output.
The results are presented for an input $L = 100\ mm$ (Fig.4).

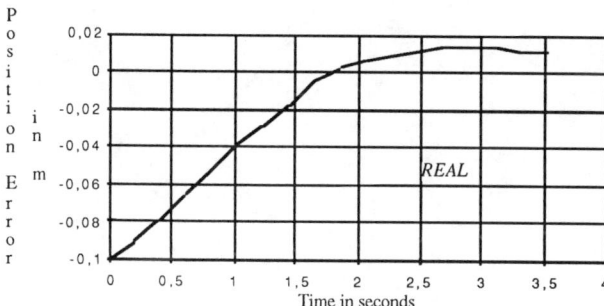

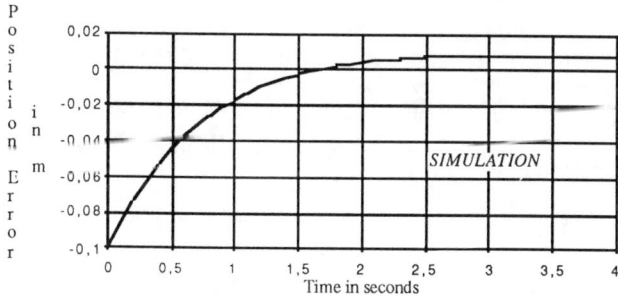

Fig. 4 : Simulation and experimental results for a step input

The response delay is less than one second. The robot reaches the desired position very quickly. The position error is zero after 20 seconds The simulation and experimental results are very close. So, this low level control can be considered like a reflex one used when the robot moves into a corridor for instance.

4. Trajectory tracking

Although there has been a significant amount of work on the control of robot manipulators, there has been little theory on the control of autonomous mobile vehicles presumably because of the newness of the field [HONG 87].

Given such a reference trajectory, it is necessary to select the guide point on the vehicle that is required to track this trajectory. For common steered-wheeled and differential-drive vehicles, it is usual to select the midpoint of the (rear) axle, since the direction of the vehicle is always tangent to the trajectory of this point. For position and orientation control this choice is not very good because there does not exist a stabilizing feedback control. Indeed, in this case the system is not controlable. The lack of controlability results simply because the cart cannot move along the wheels' axle instantaneously. So, it is more easy to control a cart's point located at a distance d from wheels' axle [SAMS 90].

So, we want seek to control a virtual point N of the robot on the trajectory defined by the path planning unit. The robot model minimizes the distance l between this point and tangent (T) to the trajectory in a point B just as (NB) perpendicular to (T) (Fig. 5).

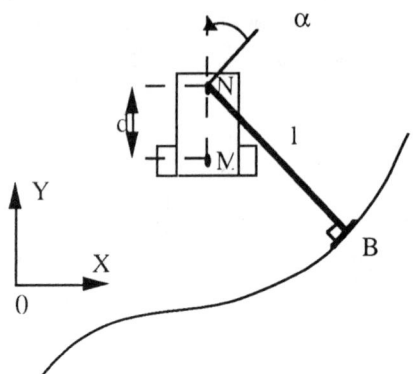

Fig. 5 : Tracking along trajectory

Then we have the same system of kinematic equations for the robot as the one we have when an ultrasonic sensor is placed at a distance d of the middle point M of the rear wheels.

To obtain the robot position and orientation we have implemented on the robot an odometric task. The (X_m, Y_m, θ) parameters, in the absolute reference, are measured by the optical encoders [TSUM 78] and :

$$X_m(k) = X_m(k-1) + \Delta U_k \cos(\frac{\theta(k) + \theta(k-1)}{2})$$

$$Y_m(k) = Y_m(k-1) + \Delta U_k \sin(\frac{\theta(k) + \theta(k-1)}{2})$$

with
$$\theta(k) = \theta(k-1) + \Delta\theta_k$$

$$\Delta\theta_k = \frac{\pi r}{NR}(\Delta N_l - \Delta N_r)$$

$$\Delta U_k = \frac{\pi r}{N}(\Delta N_l - \Delta N_r)$$

- ΔN_l and ΔN_r are the pulse number provided by the encoders for the left and right wheel between t_{k-1} and t_k.
- r is the radius of the two rear wheels
- R is the half distance between the two rear wheels.

Now, it is necessary to calculate the distance l and the robot attitude α, in relation to the curve, as a function of the odometric variables.

4.1 Determination of l and α

To calculate this, we suppose that the analytic description of trajectory in the plan (X, Y) is known.

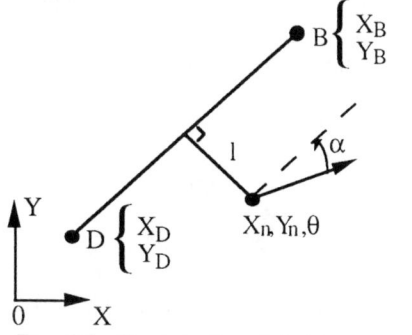
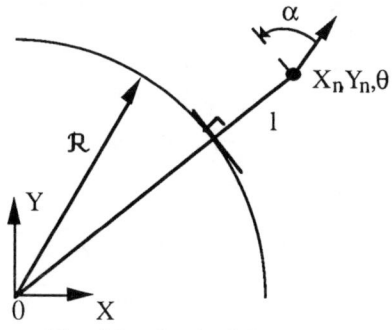

Fig. 6.1 : Straight line definition   Fig. 6.2 : Circle definition

For a straight line *(DB)* (Fig. 6.1) defined by :
$$a(X - X_D) + b(Y - Y_D) = 0 \text{ where } \begin{array}{l} a = Y_D - Y_B \\ b = X_B - X_D \end{array}$$

The distance l is given by :
$$l = (X_n - X_D)\sin\varphi - (Y_n - Y_D)\cos\varphi$$

and the angle α :
$$\alpha = \varphi - \theta$$
with $\varphi = \arctan(\frac{Y_B - Y_D}{X_B - X_D})$

In the circle case, l is given by :
$$l = \sqrt{(X_n^2 + Y_n^2)} - \mathcal{R}$$

and the angle α :

$$\alpha = arctg(\frac{Y_n}{X_n}) + \frac{\pi}{2} - \theta$$

In these relations, X_n and Y_n are calculated with the odometric results and :

$$X_n = X_m + d\cos\theta$$
$$Y_n = Y_m + d\sin\theta$$

4.2 Mobile robot control on the trajectory

We have seen in paragraph 2.1 the equation's system which controls the robot. Without angular approximation, this system can be written :

$$\begin{cases} \frac{dX_n}{dt} = V\cos\theta - d\frac{d\theta}{dt}\sin\theta \\ \frac{dY_n}{dt} = V\sin\theta + d\frac{d\theta}{dt}\cos\theta \\ \frac{d\theta}{dt} = \frac{\Delta V}{R} \end{cases}$$

This system must assume the robot convergence to the trajectory governed by α and l parameters.

So, we have chosen a control law that permits to constrain robot orientation and position :

$$\frac{d\theta}{dt} = K_2(K_1 l + \alpha)$$

This command contains a orientation loop and a position loop. Then, the velocity control is given by :

$$\Delta V = RK_2(K_1 l + \alpha)$$

4.3 Simulation and experimental results

We simulate and experiment the control law for a linear trajectory and circular trajectory.

To realize these trajectories, we determine the gain values K_1 and K_2 to obtain a ΔV command compatible with the robot. Nevertheless, we introduce a ΔV saturation in order to protect the electronic components used to command the DC motors. We choose a robot linear velocity V very slow (V = 0,04 m/s). The distance d, representing the distance (MN), is arbitrarly fixed and equal to 0,350m.

The linear trajectory tracking is obtained for a reference straight line which has an angle of slope of $\frac{\pi}{4}$. The results of simulation and experimentation noticed on the figure 7, are very similar. We can see that there is no position error and that the robot needs only two meters to reach the reference trajectory.

For a circular tracking, we choose a cercle radius of one meter. In this case, the robot describes the circle with a position error (Fig. 8). An integrator is inserted in the direct chain to correct this error.

An example of the right angle tracking is done as an application of the previous straight line following (Fig. 9). We treated this example using two values of the distance d (0,350m and 0,0m). Thus two trajectory envelops are obtained for each distance d. Consequently it is obvious that joining two straight lines does not require to use circle pieces.

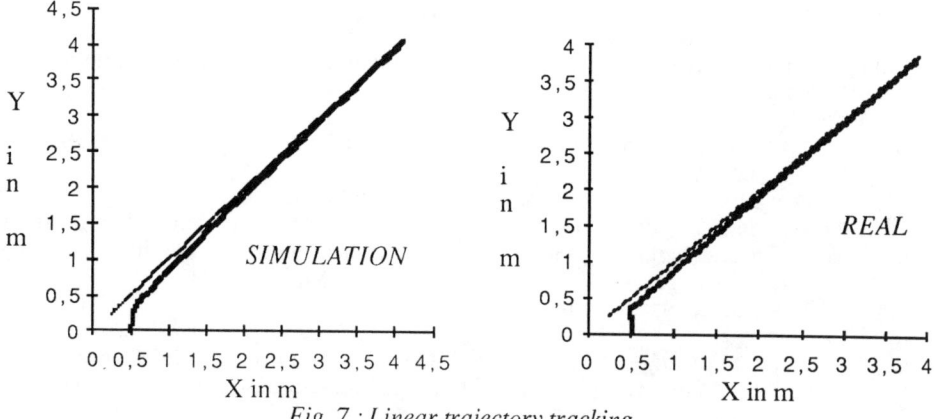

Fig. 7 : Linear trajectory tracking

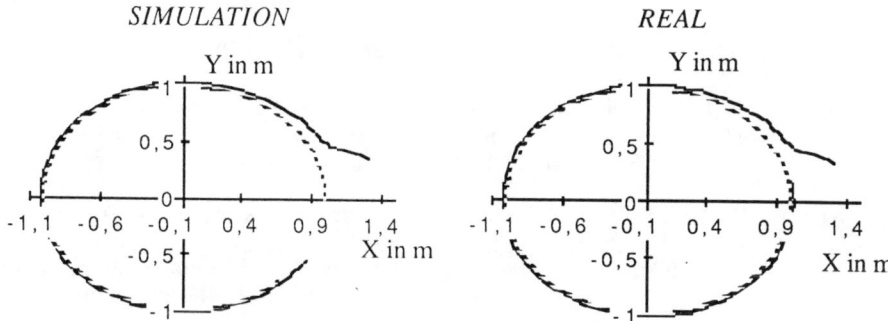

Fig. 8 : Circular trajectory tracking

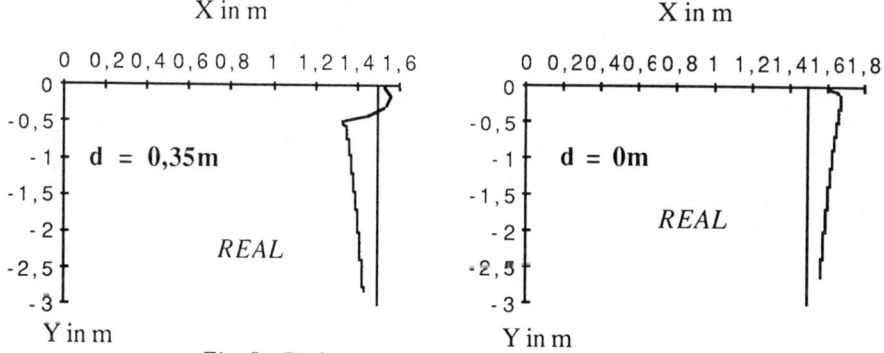

Fig. 9 : Right angle trajectory tracking

5. Conclusion

In this paper we have presented three main applications we have implemented in our mobile robot Romeo.
Namely :
- wall following using one ultrasonic sensor
- trajectory tracking
- experimental validation of the robot control model

The experimental results obtained are good and validate our theoretical approach. In the future we plan to use the previous tools to provide to the robot a set of abilities such as :
- reflex navigation
- any trajectory tracking

[BARR 89]: J. Barraquand, J. C. Latombe, " On Nonholonomic Mobile Robots and Optimal Maneuvring ", Revue d'Intelligence Artificielle, Vol. 3 - n°2, pp 77-103, 1989.
[BLAZ 91]: P. BLAZEVIC, S. DELAPLACE, J. G. FONTAINE and J. RABIT, " Mobile Robot Using Ultrasonic Sensors. Study of a Degraded Mode. ", Robotica 1991, To be published.
[CANN 88]: J. F. Canny, " The Complexity of Robot Motion Planning ", MIT Press 1988.
[HONG 87]: T. HONGO, H. ARAKAWA, G. SUGIMOTO, K. TANGE and Y. YAMAMOTO, " An Automatic Guidance System of a Self-Controlled Vehicle ", IEEE Transactions on Industrial Electronics, Vol. IE-34, n°1, pp 5-10, 1987.
[KANE 61]: R. KANE, " Dynamics of Nonholonomics Systems ", J. Application Mechanics, Vol 28, n°4, pp 574-578, 1961.
[LAUM 86]: J. P. LAUMOND, " Feasible Trajectories for Mobile Robots with Kinematic and Environment Constraints ", Int. Conf. on Itelligent Autonomous Systems, Amsterdam, pp 346-354, 1986.
[LOBA 80]: L. G. LOBAS, " Nonholonomic System which Models Rolling of a Tricycle Along a Plane ", Soviet Appl. Mechanics, Vol 16, pp 346-352, 1980.; 1
[LOBA 81]: L. G. LOBAS, " Trajectories of a Two-Stage Mechanical System with Rolling ", Soviet Appl. Mechanics, Vol 16, pp 1084-1089, 1981.
[SAMS 90]: C. SAMSON and K. AIT-ABDERRAHIM, " Mobile Robot Control. Part 1. Feedback Control of a Nonholonomic Wheeled Cart in Cartesian Space. ", Rapport de Recherche 1288, Oct 1990.
[TSUM 78]: T. TSUMURA, N. FUJIWARA, "An Experimental System for Processing Movement Information of Vehicle.", Proc. of the 28th IEEE Vehicular Technology, Mars 1978, 163-168, Denver.

A Robust Tracking System for Mobile Robot Guidance

Juan Frau, Albert Larré, Eduard Montseny & Gabriel Oliver

Departament ESAII, Universitat Politècnica de Catalunya
c/Pau Gargallo, 5. 08028-Barcelona, Spain.

ABSTRACT

This paper describes a robust tracking vision system which provides an estimate of the target position on the image plane with high speed performance. The system can be divided in two main parts: the recognition module and the motion estimation/tracker module, which supplies the data to the control subsystem in order to perform the feedback and thus the mobile robot guidance. The recognition module faces the problem by means of correlation matching and vertex detection. These techniques provide two confidence levels that allow the recognition of the target as well as the position estimation on the image plane. The application of an optimized polynomial regression algorithm allows an oriented search inside a constrained area of the whole image as a result of the motion estimation level.

INTRODUCTION

Visual feedback from the environment is a very important technique to enhance the performance of any robotic system. Common tasks in manufacturing plants can be intelligently performed when introducing image processing in addition to other sensory information. Material handling, inspection and manipulation, or automatic assembly are some examples of this kind of applications [1]. In some cases, visual information can become an essential part of a manipulator controller that needs to grasp moving objects or manipulate parts of unknown shapes [2][3]. Automatic mobile robot guidance can also be done from path identification or target tracking [4][5]. Motion estimation from a moving object can be approached by using optimal filtering or designing heuristic algorithms. In the first case the most usual technique is the well-known Kalman-Bucy filter [6][7]. Heuristic approaches try to solve the correspondence problem by applying an optical flow method or a matching method [8]. The first one determines a 2-D field of instantaneous velocities from the information contained in gray level images. Optical flow allows calculating the relative velocity of the object with respect to the camera. On the other hand, matching methods deal with the relationships between some features of the moving object that finally provide a description of the target motion from its model parameter computation.

This paper describes a predictive target tracking system that solves the correspondence problem from a heuristic point of view and deals with the prediction of the moving object by means of an optimized filter algorithm. The latter is carried out by means of regression analysis. Data obtained from the recognition subsystem permit building up a three-level model (geometrical, contour, motion). At the same time, the predicted motion vector becomes the feedback information for a mobile robot controller. A preprocessing level based on contour extraction and thinning as well as vertex detection -in a constrained window- allows correlating possible candidates with the target model. Finally, the matching step is reached and the motion vector is updated. A time delay inherent in the calculated position and orientation of the target is the main cause of difficulties in the implementation of the system. This delay is circumvented by predicting the location of the moving object. In the general case, the dynamics of the target is unknown. Moreover, it is supposed that noisy images may be used. That is why we put forward a regression algorithm as a filter step and a sequence of significance tests, which dynamically self-manage the optimal order from residual analysis.

The paper is organized as follows. An accurate contour-thinning extraction algorithm, which supplies the primitive representation of the objects from multilevel images, is given in Section II. The vertex detection scheme, as well as a brief review of the contour-matching paradigm, is developed in Section III. The motion estimation and the prediction algorithm is explained in Section IV. Finally, we discuss some results obtained from the experiments in Section V. Section VI contains the conclusions.

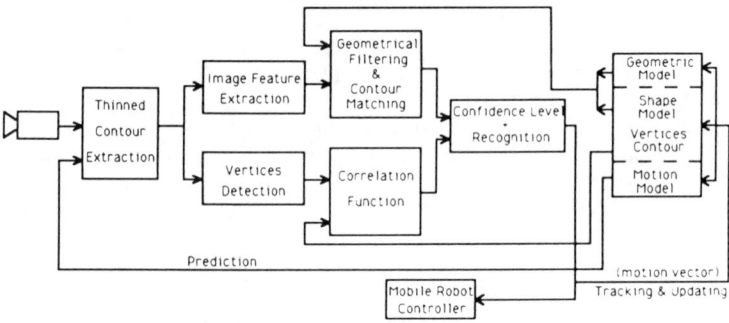

Figure 1. Basic scheme for the target tracking system.

CONTOUR EXTRACTION AND THINNING SCHEME

The contour extraction subsystem takes as input a multilevel image the dimensions of which are 256x256 pixels with 256 grey levels for pixel. The output of that subsystem can be an image of the same dimensions of the input image, or four times this resolution (512x512). As the algorithm works with subpixel precision, it offers the option of more resolution in the output image and so, it tells us more accurately where the contour is. In both options all the contours of the output image are one pixel wide. In this case we use as output a 512x512 resolution.

The algorithm can be divided into three parts as shown in figure 2. The first block carries out the gradient extraction of the image with windows of different dimensions. In the second part, a selection of pixels candidate for being contour is obtained. Finally, we get the definitive contour from the pixels candidate.

In the first step of the algorithm, the derivative operators are calculated by means of two pairs of windows of different dimensions. The results of every pair of windows gives the (x, y) components of the gradient, from which the polar coordinates (i.e. module and argument) are computed. The gradient vectors obtained with even dimensional windows are placed in the intersection of every four pixels in the input image. This means that there are gradient vectors in the four corners of every pixel. On the other hand, if odd dimensional windows are used, the gradient vectors computed are applied in the center of every pixel in the input image. Thus, if we take the results of an even window and the results of an odd one, we will obtain a map of vectors the density of which is greater than the pixels in the original image. The discusion above is represented in figures 3 and 4. In the second step of the algorithm the input data is the gradient vector map (every vector is a module and an argument), from which the pixels candidate to be contour are selected. The set of pixels that will finally perform the contour is contained in the set of candidate pixels extracted in this step. The criterion of selection of candidates depends on the arguments of the gradient vectors only, and it is specified as follows: For every point with a vector obtained from an even or an odd window, all the vectors contained in an ellyptical region around it, are analyzed. The pixel is said to be a candidate if all the arguments of the gradient vectors contained in the ellyptical region have a coherence among them.

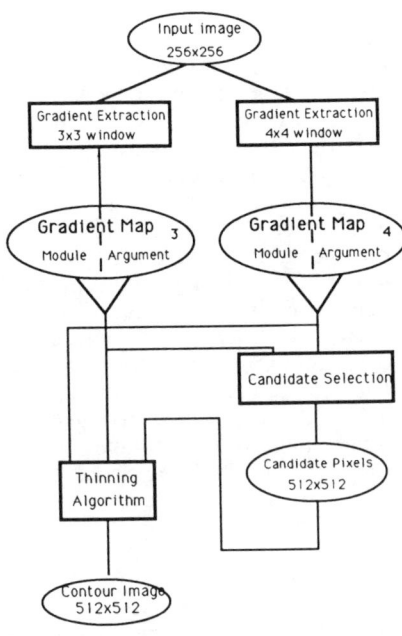

Fig. 2. Block diagram of thinned contour extraction algorithm.

-1		+1
-2		+2
-1		+1

+1	+2	+1
-1	-2	-1

-1	-1	+1	+1
-1	-2	+2	+1
-1	-2	+2	+1
-1	-1	+1	+1

+1	+1	+1	+1
+1	+2	+2	+1
-1	-2	-2	-1
-1	-1	-1	-1

Fig. 3.-Derivative extractor operators.

Once it is determined the pixels that are inside the ellipse, the coherence analysis consists in computing the difference between all the arguments and that of the central one. If this difference are lower than a fixed threshold, the pixel is said to be a candidate for contour. The dimensions of the axis of the ellipse are fixed experimentally. The major axis, which is

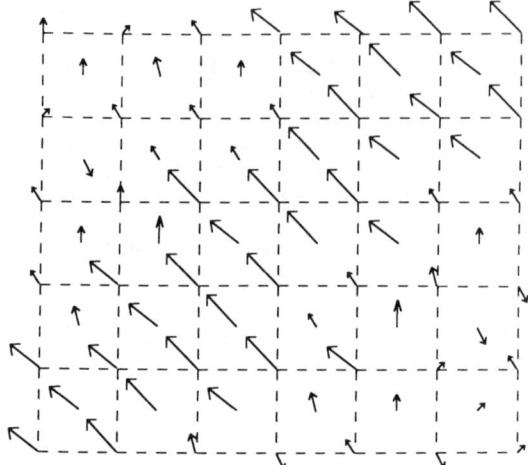

Fig. 4. Situation of gradient vectors computed with respect to the input image.

taken perpendicular to the argument of the central gradient vector under study, determines the minimal radius of the contour segment that can be detected. If a small axis is choosen, the algorithm will detect contour segments with a smal radius of courvature.

The last block of the algorithm carries out the thinning of the contour. The inputs of this block are two candidates' maps. The first is the map of candidate points which are in the intersection of every four pixels of the original image, and the second one, the points which are on the center of every pixel. This input image is an unthinned contours image, which can be three or four pixels in width. For every candidate pixel, the direction of the gradient vector is analyzed. Finally, a candidate is choosen as contour pixel if the module of its gradient vector reaches a maximum among all the vectors which are over the line fixed by its argument.

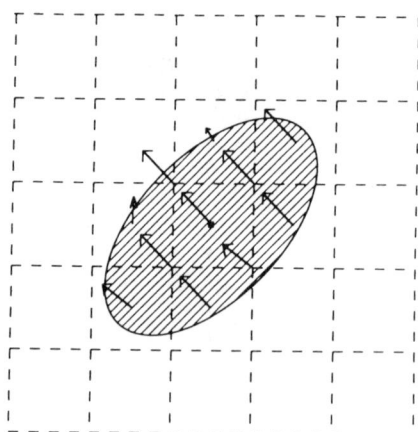

Fig. 5. The elliptical region analyzed in the candidate for contour determination.

VERTEX DETECTION / CONTOUR MATCHING ALGORITHMS

Our system of polygonal contour description is divided into two stages. In the first stage, a certain number of pixels are marked out as candidates for vertex making use of local information. In the second stage, those candidates go through a process of analysis where their significance to carry out a polygonal representation based on that set of vertices is determined. If its degree of significance is not high enough, the candidate for vertex is crossed out from the initial list, which may affect the degree of representativity of some candidates. This process lasts as long as the crossing out of candidates is at work, until a stable representation is finally reached.

In figure 1, the contour-thinning extraction module can be seen to give the contour coordinates to the vertex detection module. Thus, the latter works on the hypothesis that there exists a continuity between contour pixels.

The determination of candidates for vertex is based on the analysis of the information contained in a 7x7 window centered on the pixel under study (see figure 6). In this process we analyse the 3x3, 5x5 and 7x7 crowns detailed in the figure separately. With 3x3 crown, we obtain some information on the degree of curvature of the contour at a very local level, whereas subsequent 5x5 and 7x7 crowns provide a much more global information. Taking the contour as an ordered set of pixels, for each P_i belonging to that contour, a numerical value $W3(\ )$ is associated to preceding P_{i-1} and subsequent P_{i+1} in 3x3 crown. The difference between these values:

$$C3(P_i) = |\ W3(P_{i-1}) - W3(P_{i+1})\ | \qquad (1)$$

indicates the degree of curvature of the contour in the proximity of P_i. In the same way, we can obtain values $C5(P_i)$ and $C7(P_i)$ for the 5x5 and 7x7 crowns. We determine that a pixel is a candidate for vertex if each of the values C3, C5 and C7 verify:

$$C_k \geq L_k \qquad (k = 3, 5, 7)$$

L_k being a threshold established independently for each crown.

The coordinates of the pixels initially selected as candidates for vertex go through the second module, where this information will be processed so as to eliminate some unnecessary or uncertain vertices, leaving only the most significant ones, which will determine the representation of the object. This vertex selection is carried out by an iterative process described below.

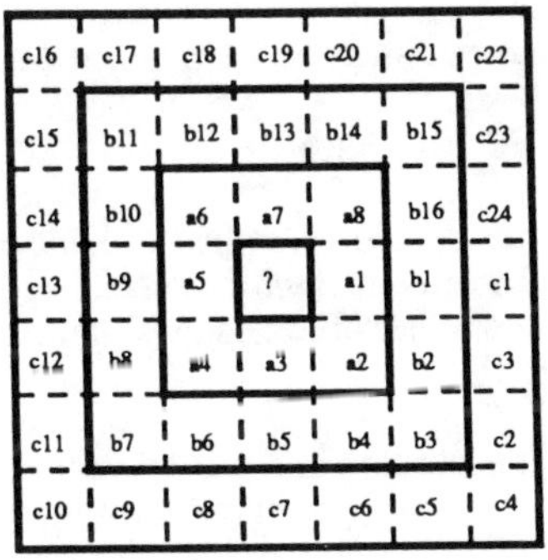

Fig. 6.- 7x7 window detailing the three crowns used in the determination of the candidates for vertex.

For each pixel marked as candidate V_i, the distance with the previous candidate and that with the following candidate are calculated, as well as that between the previous candidate and the following one:

$$a_i = d(V_{i-1}, V_i) \qquad b_i = d(V_i, V_{i+1}) \qquad c_i = d(V_{i-1}, V_{i+1}) \qquad (2)$$

from these data we calculate the angle associated to V_i: θ_i. This allows us to calculate for each V_i three values μ_a, μ_b and μ_θ (the functions of which are expressed in figure 7) that we will associate to the degree of stability of the vertex, which may be determined according to the following criteria:

1) The bigger the difference between a candidate's associated θ_i value and π radians, the more stable it is [9].

2) The bigger the distances a_i and b_i are, the more stable the candidate is.

As may be seen, the stability criteria refer to the possibility of a contour pixel considered as a candidate really corresponding to a vertex in the figure and, as a consequence, of its appearing when the object is recorded and digitalized at different times.

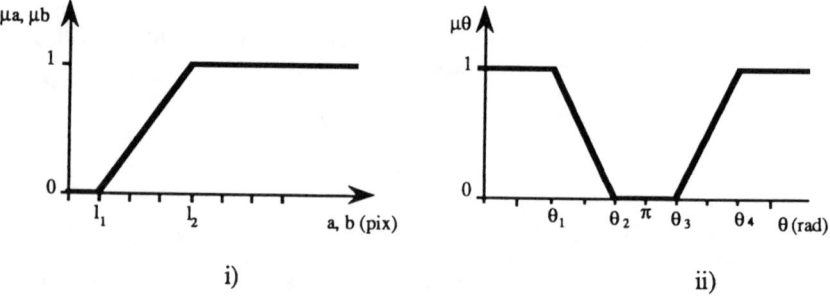

Fig. 7.- Functions used to express the stability criteria.

This information can be summarized in a unique value μ_i for each candidate V_i that one can interpret as the 'vertex-like' characteristic. We have defined that value as:

$$\mu_i = \mu_a \cdot \mu_b \cdot \mu_\theta \qquad (3)$$

The main advantage of this vertex detection method is that the first stage, in which the list of candidates is obtained, can be implemented with a specialized processor that operates in real time and reduces drastically the number of pixels to be treated in subsequent stages, where a bigger amount of calculus is needed.

Geometrical filtering is realized according to a feature vector associated to each object in the searching window (area, center of gravity, axis of least inertia with respect to the absolute reference,..). Contour matching is obtained as a result of a set of discrimination functions built up from the possible rotation invariant contour description associated to each object contained in the searching window. Invariant descriptors are based on polar codings of every candidate, taking as a relative reference the main axis of inertia and its perpendicular straight line through the c.g. [10].

MOTION ESTIMATION / PREDICTION SCHEME

We have constrained the basic parameters which define the motion vector z to its linear and angular position on the image plane. The recognition module supplies the vector $z=[Xc,Yc,\emptyset]^t$, where (Xc,Yc) are the coordinates of the center of gravity (c.g.) of the target projection, and $\emptyset$ is the angle formed by the axis of least inertia and the axis OX of the image plane.

Therefore, target motion is separated into c.g. translation and rotation around itself. These motions have been approximated by time-depending polynomial models, as follows [4]:

$$Xc(t) = a_0 + a_1.t + ... + a_p.t^p + w_x$$

$$Yc(t) = b_0 + b_1.t + ... + b_q.t^q + w_Y \qquad (4)$$

$$\emptyset(t) = c_0 + c_1.t + ... + c_r.t^r + w_\emptyset$$

Generally, it can be assumed that these coordinates are uncoupled, and also that the optimal polynomial orders are different. Therefore, it is suitable to calculate the corresponding coefficients in a parallel way. Least squares parameter estimation criterion leads to solving a system of equations for each one of the motion vector components. An optimized procedure, which takes advantadge of the uniformly spaced samples, was implemented to recursively calculate vectors involved in the parameter estimation algorithm, and to avoid matrix inversions from sample to sample [11].

In this latter procedure the polynomial orders p, q and r were assumed to be known. In order to estimate the optimal order of each model -associated to Xc, Yc and $\emptyset$- the system must know the highest order to start the significance tests. Assuming that the maximum order considered in the analysis is m, we might think that the bigger m is, the better fitting becomes.

Let us assume that Xc, Yc and $\emptyset$ correspond to continuous functions; the lowest bound of all possible values of the approximation errors, are denoted by:

$$E_m(Xc) = |Xc-{^\wedge}Xc|_\infty \qquad E_m(Yc) = |Yc-{^\wedge}Yc|_\infty \qquad E_m(\emptyset) = |\emptyset-{^\wedge}\emptyset|_\infty \qquad (5)$$

The most immediately available methods for constructing polynomial approximations often provide approximations the maximal errors of which are significantly larger than E_m, and one can not be sure that error goes to zero as m $\to \infty$, even if Xc, Yc and $\emptyset$ are smooth or many times differentiable.

Moreover, one could guess that, for example, in the case of a constant acceleration of the target, the best order must be the second one; therefore, it should be quite inefficient to attempt fitting with higher orders. This kind of models would easily show an "overparameterization", that is, the coefficients of the polynomial associated to higher degrees than the optimal would be statistically insignificant (buried in the noise).

As a result, we propose an algorithm that has been called Self-Management Order Procedure (SMOP). The problem consists in choosing which of the following polynomials is the best one, in the case of Xc:

$$\hat{X}_m = a_0' + a_1' t + \ldots + a_m' t^m + w'_x$$

$$\hat{X}_{m-1} = a_0'' + a_1'' t + \ldots + a_{m-1}'' t^{m-1} + w''_x \quad (6)$$

$$\hat{X}_1 = a_0 + a_1 t + w_x$$

Many systems try to solve the approximation problem by expanding the model in terms of orthogonal polynomials in the form

$$m) \ \hat{X}_m = \Gamma_0 \Psi_0(t) + \Gamma_1 \Psi_1(t) + \ldots + \Gamma_m \Psi_m(t) + w'_x$$

$$m-1) \ \hat{X}_{m-1} = \Gamma_0 \Psi_0(t) + \Gamma_1 \Psi_1(t) + \ldots + \Gamma_{m-1} \Psi_{m-1}(t) + w''_x \quad (7)$$

$$\ldots$$

$$1) \ \hat{X}_1 = \Gamma_0 \Psi_0(t) + \Gamma_1 \Psi_1(t) + w_x$$

where

$$\Psi_0(t) = A_0$$

$$\Psi_1(t) = B_0 + B_1 t \quad (8)$$

$$\ldots$$

$$\Psi_m(t) = N_0 + N_1 t + \ldots + N_m t^m$$

For each orthogonal polynomial, from 1 to m, it could be possible to calculate the sum of square errors

$$SQ_R = \sum_{i=1}^{N} (X_{ci} - \hat{X}_i)^2$$, where N is the number of data considered in the analysis. Assuming that data noise is normally distributed, SQ_R's have a chi-square distribution with $N-(p+1)$ degrees of freedom, where $p = m, m-1, \ldots, 1$. Consequently, m significance tests could be established in a parallel way:

$$m) \ H_{om}: \Gamma_m = 0 \quad m-1) \ H_{o(m-1)}: \Gamma_{m-1} = 0 \ \ldots \quad 1) \ H_{o1}: \Gamma_1 = 0 \quad (9)$$

For every polynomial the ratio F_k can be calculated as follows:

$$F_k = \frac{(SQ_{R,H0} - SQ_R)/q}{SQ_R/(N-p')} \quad (k=1, \cdots m) \quad (10)$$

where: SQ_{R,H_0} is the sum of square errors, taking into account the corresponding null hypothesis, q is the number of constraints for each hypothesis (one in all cases) and p' is the number of parameters for every model. (i.e. if order=p then p'= p+1).

In this case, F-statistics are distributed as an F of Snedecor having $v_1=1$ and $v_2=N-p'$ degrees of freedom:

$$F_k \sim F_{v1,v2}$$

The algorithm we have implemented has avoided parallel computation of significance tests without losing high speed performance, allowing an important simplification of the final processor. At the same time, the procedure deals with checking hypotheses without calculating the coefficients of any orthogonal polynomial. Taking into account that SQ_R's are the same if they are calculated from time-depending polynomials as (4), F-statistics defined as (10) can also be obtained without computing the family of orthogonal polynomials.

Assuming a significance level $\alpha=5\%$, the SMOP scheme can be described as follows

m) $H_{om}: \Gamma_m = 0$?

    if $F_m > F_{ma}$ then

        H_{om} is rejected; Γ_m is not null
        optimal order = m
        stop checking hypotheses

        otherwise

        H_{om} is accepted; $\Gamma_m = 0$
        optimal order < m
        check test m-1; $H_{0(m-1)}: \Gamma_{m-1} = 0$?
        ...

    endif

To start the SMOP scheme, it is necessary to estimate the polynomials associated to m) and m-1) orders. This way, the coefficients of time-depending polynomials are calculated only if needed.

RESULTS

The final implementation of the described tracking system has led to the results we discuss below. The target to be tracked is moving around a 3D environment according to the commands received from a remote control unit. The sudden changes in its trajectory provide sharp alterations in its shape and contour. Figures 9a and 9b show the results achieved for Xc/Yc coordinates of the z vector when tracking the target from the SMOP scheme, with N=9 observations. Maximal order considered in the system is m=4. At the bottom each graph shows the evolution of residuals, for every coordinate.

Figure 10 shows the same results on the image plane. It seems clear that predictive-tracking presents a good level of accuracy, even though there are some difficult points in the trajectory. Figures.11a and 11b show the time evolution of the estimated optimal-orders for Xc and Yc, in the lorry sequence tested. It is possible to appreciate that most of the time the system associates second and third orders to the target motion.

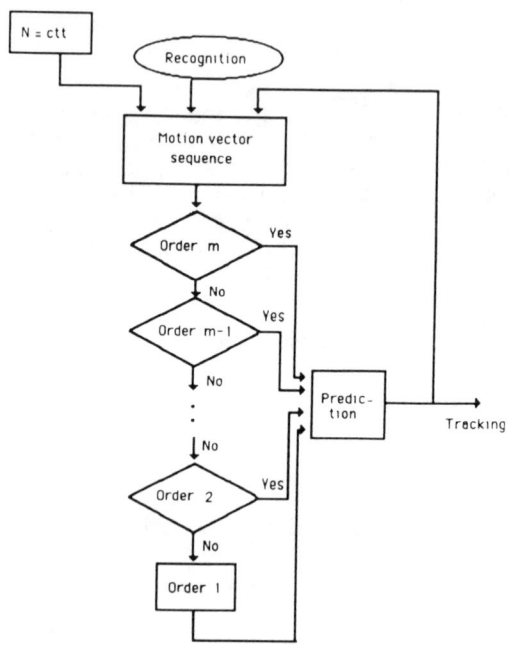

Fig.8.- Basic bloc diagram of the Self-Management Order Procedure (SMOP).

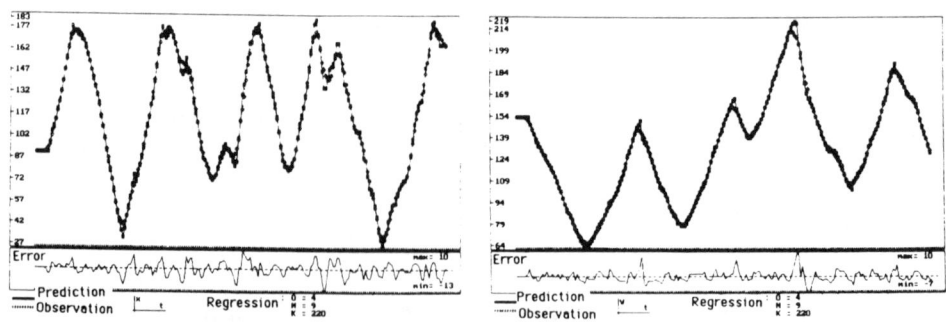

Fig. 9a and 9b.- Predictive-tracking results of the SMOP scheme, when N=9.

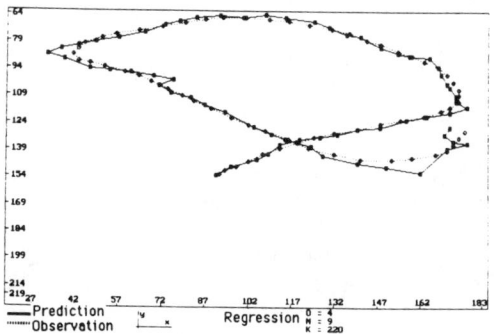

Fig. 10.- Tracking results on the image plane.

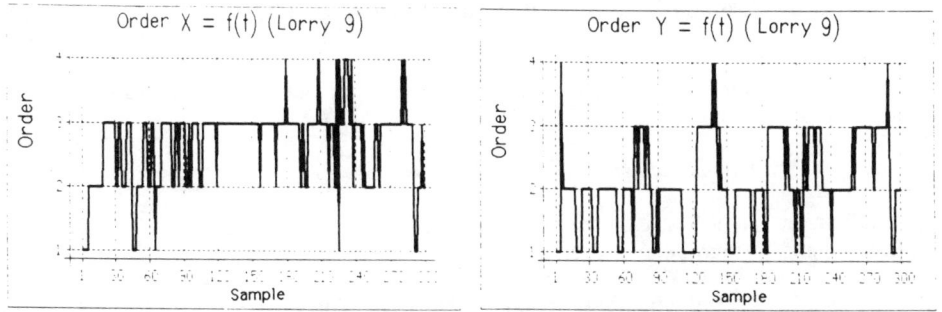

Fig. 11a and 11b. Time evolution of the optimal-order for Xc and Yc.

As a matter of fact, the whole algorithm could be even more optimized if the target described a trajectory with a special order in each coordinate. In that case, the SMOP scheme could be temporarily avoided, speeding up the overall system performance, but increasing the likelihood of target loss.

CONCLUSIONS

In this paper we have presented a tracking system which deals with a segmentation algorithm based on contour extraction and thinning algorithm over a gray level window of every image, followed by a vertex detection and contour-matching subsystem. As a result, a sequence of data corresponding to the target motion become noise input observations of the motion estimation / prediction subsystem. The latter associates an optimal order of the motion model, for each coordinate, by means of the "Self-Management Order Procedure" (SMOP), allowing the prediction of the vector z. The whole system has shown its competence as a feedback vision module in mobile robot control tasks as well as navigation. The trade-off between system robustness and high-speed performance has allowed dealing with the adaptive tracking of targets with a near-predictive behaviour. The main topics we are going to focus from on now on are, on the one hand the exploitation of the intrinsic parallelism of the contour extraction / thinning algorithm and the SMOP scheme, and, on the other hand, the extension of the approach to track maneuvering targets.

ACKNOWLEDGMENTS

We acknowledge partial financial support from research project CICYT- ROB90-0730.

REFERENCES

[1] Trivedi,M.M.,et. al., "A Vision System for Robotic Inspection and Manipulation," *IEEE Trans. on Computers*, **C-22**, pp 91-98, (1989).

[2] Weiss,L.E. et.al., "Dynamic Sensor-based Control of Robots with Visual Feedback," *IEEE Journal of Robotics and Automation*, vol 3, n 5, pp 404-417, (1987).

[3] Koivo,A.J. and Houshangi,N., "Real-time Vision Feedback for Servoing Robotic Manipulator with Self-tunning Controller," *IEEE Trans. Syst., Man Cyber.*, **SCM-21**, 1, (1991).

[4] Frau,J. et. al., "Motion Estimation from Target Tracking," NATO ASI series on Expert Systems and Robotics, vol F71, pp 445-458, Corfú, (1990).

[5] Bhanu,B. and Burger,W.,"Qualitative Motion Detection and Tracking of Targets from a Mobile Platform," Image Understanding Workshop, pp 289-318, (1988).

[6] Maybeck,P.S. *Stochastic Models, Estimation, and Control*, Academic Press, (1988).

[7] Goodwin,G.C. and Sin,K.S. *Adaptive Filtering Prediction and Control*, Prentice-Hall Information and System Sciences series, (1984).

[8] Aggarwall,J.K. and Nandhakummar,N. "On the Computation of Motion from a Sequence of Images-A Review," *Proceedings IEEE*, vol 76, pp 917-935, (1988).

[9] Huntsberger,T.L. et. al. "Representation of Uncertainty in Computer Vision Using Fuzzy Sets," *IEEE Trans. on Computers*, **C-35**, 2, pp 145-156, (1986).

[10] Frau,J. and Llario,V. "3D-Tracking and Adaptive Motion Prediction of a Target from a Mobile Robot," *IEEE Int. Conference on Intelligent Motion Control*, pp 433-437, Istambul, (1990).

[11] Frau,J. et. al. "Polynomial Regression Analysis for Estimating Motion from Image Sequences," SPIE Symposium on Advances in Intelligent Systems, pp 329-340, Boston, (1990).

A Real-Time Multiple Lane Tracker for an Autonomous Road Vehicle

Klaus Peter Wershofen and Volker Graefe

Institut für Meßtechnik
Universität der Bundeswehr München
8014 Neubiberg, Germany

Abstract

An algorithm is being developed which allows a vision guided autonomous road vehicle running on a multi-lane highway to determine continuously whether additional lanes exist to the left and to the right of the presently used lane.

The present state of this algorithm is described. Having a hierarchical structure, it first detects and tracks the lane currently used by the vehicle. The location of this lane in the image, and its width, determine those image sections in which additional lanes are then looked for. A white lane marker in such a search space indicates the presence of an additional lane and is then tracked, too.

The algorithm was implemented on one single parallel processor of the robot vision system BVV 3. It was tested with Autobahn scenes recorded on a video tape and played back in real time. When conditions were favorable the own lane, and one or two adjacent lanes, whenever they were present, were tracked reliably with a cycle time of 20 ms. The tracker functions reliably, even under certain unfavorable conditions, including sunshine with shadows from trees, windshield wipers moving in front of the camera, and passing vehicles temporarily concealing the marker of an adjacent lane. The tracker is able to detect its own failure. It then re-initializes itself and resumes tracking within 160 ms. This ability greatly improves its robustness.

Introduction

A vision guided autonomous road vehicle, driving with high speed on a motorway, is a topic of research <Graefe, Kuhnert 88>. One basic behavior required by such a vehicle is changing from one lane to another one in order to avoid an obstacle <Graefe 90a> or to pass another vehicle. The autonomous vehicle then needs to decide whether a desired lane change is practicable. As a first step towards developing such a competence a multiple lane tracker working in real time is developed. It is based on the principles of dynamic vision <Dickmanns, Graefe 88a, b>. In its final version it will be a basis for determining the number of lanes to the left and to the right of the vehicle and the nature of the lane markers in the image. It will also detect changes in the number of lanes as they occur, for example, at entries to, or exits from a motorway.

Initialization and subsequent tracking are the two modes of operation of the algorithm. The initialization may be performed even while the vehicle is running. Moreover, requiring only 160 ms on one parallel processor of the real-time vision system BVV 3 <Graefe 90b>, it is fast enough to be repeated regularly during the operation of the vehicle, providing an additional dimension of robustness.

Types of Roads

With regard to the problem of lane detection and tracking, real roads may be classified into three basic types: roads with lines marking the boundaries of each lane, partially marked roads, and unmarked roads. Different approaches must be taken to recognize the lanes on these specific roads.

Lane detection on an **unmarked road** is a difficult task. The lane boundaries may not even be visible in the image. Often only the two borders of the road can be detected. The location of the lanes is then determined by dividing the image of the road geometrically into lanes. On **partially marked roads** the recognition of lanes may be based primarily on detection of the marked lane borders, and additionally on detection of unmarked lane borders, the latter often being the transitions between the road and the soil next to the road. On a **fully marked road**, such as a typical freeway, each lane border is characterized by a line painted on the surface of the road. The recognition of the lane may then be reduced to the recognition of the well defined lane markers.

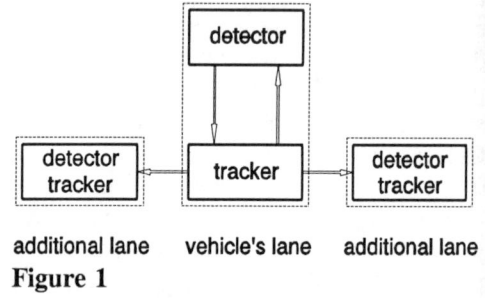

Figure 1
Structure of the multiple lane tracker

At this time the multiple lane tracker is designed to detect and track lanes on well marked roads.

The Multiple Lane Tracker

Figure 1 gives an overview of the multiple lane tracker. It consists of four basic parts: the **detector** for the vehicle's lane, the **tracker** for the vehicle's lane, and two **detectors and trackers** for one additional lane on each side of the vehicle's own lane.

The vehicle's lane will always be present, therefore, its white borderlines are used as reference lines for the search of additional lanes on both sides of this lane. If an additional lane is detected its markers are tracked. Otherwise the search for additional lanes is repeated in each subsequent image. This makes it possible to recognize a change in the number of lanes of the road.

The Detector for the Vehicle's Lane

The task of the detector for the vehicle's lane is to find the two white lines limiting the vehicle's lane during the initialization phase of the system and also during later re-initialization phases.

When driving on a highway, usually the appearance of the own lane in the image is rather well predictable, even if the camera parameters and the vehicle's location on the road are not exactly known. In a typical driving situation it may be assumed that the lane has a constant width and that it is either straight or moderately curved. The optical axis of the camera will be nearly parallel to the axis of the lane. Figure 2 shows the appearance of the road in such a typical situation. The lane markers appear as two (nearly) straight lines whose approximate locations and slopes in the image are more or less predictable.

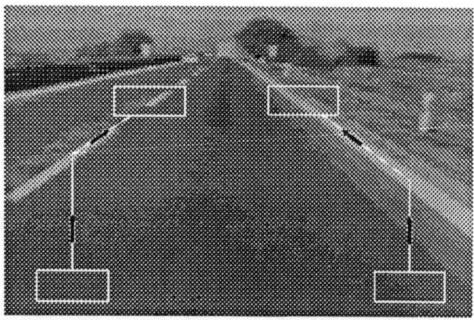

Figure 2
Search strategy for finding the vehicle's lane during initialization. The lower boxes mark the initial location of the search spaces, the upper ones their final locations as used during tracking.

This expectation concerning the visual appearance of the lane markers is the basis for finding them in the image. Two line finders are placed at the lower part of the image. From there they start the search for the white lines which mark the vehicle's lane by following a vertical search path. One of them searches the left line, the other one the right line. As soon as a line has been detected it is followed upwards in the image. In this way the slope of the line may be determined in addition to its location, but this has not been implemented yet. Such information may be helpful for guiding the tracker of the vehicle's lane in selecting the correct features.

In the subsequent tracking phase the lane should be tracked in a certain predetermined look-ahead distance. To do this, it is assumed that a known relationship exists between the apparent width of the lane in the image and the look-ahead distance. This assumption is justified since the parameters governing the projection of the lane into the image (e.g. focal length, pitch angle of the camera, width of the lane) are more or less known. The line finders, therefore, follow the lines up to those points in the image where the horizontal distance between the lines has a certain value corresponding approximately to the predetermined look-ahead distance. The image coordinates of the lane markers at those points are then passed to the lane tracker.

The entire initialization is performed on one single image out of the sequence. Subsequent images are ignored by the multiple lane tracker until the initialization has been completed. Therefore, the time required for the initialization has to be short enough to ensure that the location of the lane markers in the image does not change too much before it has been completed, otherwise the locations determined by the detector would no longer be valid when the tracker is started.

The Tracker for the Vehicle's Lane

The basic function of the tracker for the vehicle's lane is to repeatedly locate the lines marking this lane. The search for the lines in each new image is based on the expectation that they will appear in well-defined regions of the image, called search spaces.

In the beginning the search spaces are centered at those coordinates which the detector for the vehicle's lane has reported (Figure 2). The vertical coordinate of the search spaces is determined by the desired look-ahead distance. It will not be changed during tracking. The horizontal coordinates of the search spaces, however, are continuously adjusted in order to keep the lane markers centered in the search spaces.

To do this, an expectation as to where the lines will be found in the next image is generated after processing each image. The expectation is calculated by low-pass filtering the coordinates reported by the line finders in previous images. Normally, it would suffice to expect the line to be found exactly where it was found in the last image. This strategy, however, fails when interrupted lines are to be tracked. In this case it may happen that the line finders search the line exactly in the gap between visible sections. If some unrelated feature resembling the line happens to be present near the expected location of the lane marker, it may instead be found, causing the tracker to lock on to the false feature.

It is always possible for any tracker to loose its target, therefore, the lane tracker continuously monitors its own operation.

- If one line finder has failed to find its line for more than a certain length of time, the search space of this line finder is repositioned. The new location of the search space is calculated from the location of the lane marker found by the opposite line finder, and the nominal width of the vehicle's lane.
- If both line finders have been unable to deliver a line coordinate for more than a certain length of time, the tracking of the vehicle's lane is discontinued temporarily.
- The distance between the lines tracked by the lane tracker is continuously compared with the nominal width of the vehicle's lane. If the difference exceeds a certain limit, it is assumed that the vehicle's lane is no longer being tracked correctly, and the tracking is discontinued temporarily.

Whenever the tracking has been discontinued, the lane detector is automatically triggered to re-initialize the tracker. Correct tracking of the lane will then normally resume within a fraction of a second. This ability of quickly recovering from a tracking failure is of great importance for the robustness of the system, enabling it, for instance, to handle situations like entering or leaving a tunnel where, due to the rapid change of the illumination level, the camera is temporarily unable to provide useful images.

The Detector and Tracker for One Additional Lane

The search for additional lanes is initiated simultaneously with the start of tracking the vehicle's lane. One lane is searched to the left of the presently used lane, and one to the right.

The search and tracking of additional lane markers is guided by the assumption that the additional lanes have the same width as the vehicle's own lane. One line finder is placed on the left side and one on the right side of the own lane. In contrast to the tracker for the vehicle's lane, their search spaces are not centered around the lane markers detected, nor is the width of the detected lane in the image checked.

A Line Finder as the Basic Component of the Tracker

The detector for the vehicle's lane, the tracker for the vehicle's lane, and the detectors and trackers for additional lanes are based on a line finder which finds sections of lane markers in the image.

The line finder operates in two steps. First, it searches individual edges in the image that could correspond to the left or right edge of a lane marker. Then it attempts to group two adjacent edges of opposite signs into a line corresponding to a lane marker. The method of controlled correlation <Kuhnert 86, 88> is used to detect individual edges in the image. It is a generalization of the well-known correlation method, finding features in an image by correlating a small area of the image with a prototype of the searched feature, a so-called mask. The correlation is performed only along a one-dimensional search path. The result of the controlled correlation is a discrete function, the so-called correlation function. An extremum of the correlation function indicates the location of the searched feature.

For efficiency, the line finder uses ternary masks (having elements with values of +1, 0, and -1 only). Figure 3 shows a typical ternary mask as implemented in the line finder. It is selective for an edge with an orientation of 120 degrees. Other masks, also fitting into a square of 16·16 pixels, are used for other orientations. The search path is a horizontal line.

A lane marker may be described by an edge at its left side with a change in the grey level from dark to bright, and an edge on its right side with a change in the grey level from bright to dark. Those changes in the grey level correspond to two extrema of the correlation function having opposite signs and being located close to each other. If exactly one such pair of correlation extrema exists within the search space, the line finder reports a white line and returns the coordinates of the detected edges. In any other case, the detector states a failure.

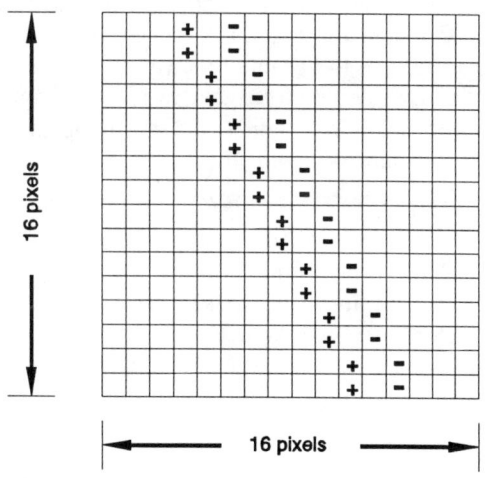

Figure 3
Ternary mask for detecting an edge with an orientation of 120 degrees. A "+" stands for +1, a "-" for -1.

Implementation, Experiments, and Results

The performance of the multiple lane tracker was evaluated in real-time experiments using video tapes recorded on an Autobahn while driving with speeds between 70 km/h and 85 km/h (Figure 4). The videos were taken on a highway with normal traffic, most vehicles driving faster than the vehicle carrying the camera. The tests included Autobahn sections with two lanes, three lanes, and more than three lanes, for example at the crossing of two highways, or at exits. Many lane changes were performed while driving. The focal length of the camera was changed several times between 8 mm and 16 mm, also the pitch angle of the camera was varied to determine the robustness of the system.

The multiple lane tracker was tested with these scenes played back in real time. The vehicle's own lane could be detected and tracked during most of the time, regardless of which lane the vehicle was driving in. Also, one additional lane marker on each side of the own lane was tracked most of the time when the corresponding additional lane was actually present.

All phases of operation of the multiple lane tracker, including the initialization, are fully automatic, requiring no intervention or assistance by a human operator.

The time required by the detector for the initialization phase is about 160 ms. This time is short enough to allow a re-initialization while the vehicle is running (at a speed of 100 km/h, a time of 160 ms corresponds to a distance of about 5 m).

During the tests with the video scenes re-initializations were automatically executed many times, for example, whenever the vehicle performed a lane change causing an adjacent lane to become its own lane. A re-initialization of the tracker is also triggered if the pitch angle of the camera is changed or when insufficient image data is available, for example, in a tunnel. After leaving the tunnel, the vehicle's lane is automatically detected and tracked again.

Whenever the tracking of the own lane failed, an automatic re-initialization was performed. This corrected the problem.

The tracking of the lanes was done at a look-ahead distance of about 25 to 50 meters. For tracking four lane markers less than 20 ms are necessary. The search regions and the detected lines may be visualized on a TV monitor. Figure 4 shows a typical situation while tracking. The vehicle is using the middle lane of the Autobahn. The white parallelograms are marking the search spaces, the white rectangles the detected lines.

Problems

Although the multiple lane tracker in its present form is working satisfactorily in a certain variety of situations, a number of problems still need to be solved in order to make it truly robust.

Some of the problems are related to dashed lines. During the initial search for the lane being used it may happen that the vertical search path of the line finder (see Figure 2) passes between two sections of a dashed line. This may cause the borderline of another lane to be followed, or it may result in some other features to be detected if they have a similar appearance as sections of lane markers. Such features are then tracked as long as the distance between them is nearly equal to the nominal width of the lane.

The detection of the own lane is based on the assumption that the surface of the road is homogeneous. In parts of the recorded scenes, however, the road had been repaired in such a way that in the middle of the lane the surface was brighter than at the lane's borders. This bright region sometimes looks rather similar to a lane marker and may lead to a false detection.

Figure 4
Typical situation while tracking. The white parallelograms mark the search spaces, the white rectangles mark the detected lines.

When driving on the outer left or on the outer right lane of the road the search region in which an additional lane marker is to be detected is placed beside the road. Objects looking similar to lane markers, such as guardrails, and sometimes even white posts along the road, can cause the line finder to report a line. The tracker, however, is presently not able to distinguish them from a lane marker.

Tracking of wrong features may also sometimes be caused by other vehicles passing or driving in front of the vehicle. This may happen if those vehicles have bright marks on their surface.

One possible way to reduce the problems mentioned above is to have the tracker process larger sections of the image, thus utilizing the fact that the lane markers extend over large parts of the image. Such a strategy, however, needs a more sophisticated internal model of the lane to be detected.

In any case, it is worth noting how many problems may be discovered when testing a basically robust system in a larger variety of real-world scenes. Such experiments are tedious and expensive, but there is no substitute for them if the goal is to understand how to make robot vision truly robust and flexible.

Conclusions

A multiple lane tracker was presented which is able to track three lanes of an Autobahn in real time by tracking the corresponding lane markers. The cycle time of the tracker is 20 ms for three lanes on a single parallel processor of the real-time vision system BVV 3.

The performance of the tracker was evaluated in real-time experiments with video tapes recorded on an Autobahn while driving with speeds between 70 km/h and 85 km/h.

When conditions were favorable the own lane and one or two adjacent lanes were tracked consistently whenever they were present.

The tracker functions reliably even under certain types of unfavorable conditions, including sunshine with shadows from trees, windshield wipers moving in front of the camera, and passing vehicles temporarily concealing the marker of an adjacent lane. If the tracker does fail, it normally detects its own failure. It then re-initializes itself and resumes tracking within 160 ms. This ability greatly improves its robustness.

Certain problems that remain to be solved were discovered during the experiments. Some of them relate to interrupted lines, others to the tracking of wrong features, such as guardrails instead of lane markers.

References

Dickmanns, E.D.; Graefe, V. (88a): Dynamic Monocular Machine Vision. Machine Vision and Applications 1 (1988), pp 223-240.

Dickmanns, E.D.; Graefe, V. (88b): Applications of Dynamic Monocular Machine Vision. Machine Vision and Applications 1 (1988), pp 241-261.

Graefe, V. (90a): An Approach to Obstacle Recognition for Autonomous Mobile Robots. IEEE/RSJ International Workshop on Intelligent Robots and Systems (IROS'90). Tsuchiura, pp 151-158.

Graefe, V. (90b): The BVV-Family of Robot Vision Systems. In O. Kaynak (ed.): Proceedings of the IEEE Workshop on Intelligent Motion Control. Istanbul, pp IP55-IP65.

Graefe, V.; Kuhnert, K.-D. (88): Towards a Vision Based Robot with a Driver's License. Proceedings, IEEE International Workshop on Intelligent Robots and Systems, IROS'88. Tokyo, pp 627-632.

Kuhnert, K.-D. (86): A Model Driven Image Analysis System for Vehicle Guidance in Real Time. Proceedings of the Second International Electronic Image Week. CESTA, Nice, pp 216-221.

Kuhnert, K.-D. (88): Zur Echtzeit-Bildfolgenanalyse mit Vorwissen. Dissertation, Fakultät für Luft- und Raumfahrttechnik der Universität der Bundeswehr München.

Acknowledgement

Part of the work reported has been performed with support from the Ministry of Research and Technology (BMFT) and from the German automobile industry within the project PROMETHEUS.

PART 4
MOBILE ROBOTS: ARCHITECTURES, PERCEPTION, NAVIGATION AND CONTROL

An Architecture for Intelligent Mobile Robots *

João Sequeira   João Sentieiro
CAPS/LRPI, IST - Technical University of Lisbon
Av. Rovisco Pais 1, 1096 Lisboa Codex, Portugal
Tel. +(351) 1 3524309, FAX +(351) 1 3523014
email: M785@BETA.IST.RCCN.PT.

Abstract

This paper proposes a new approach for the design of intelligent mobile robot architectures exhibiting behaviour emergence.

A robot is described by a set of tasks whose accomplishment may be necessary to satisfy the robot objectives. The inovation consists on using variable structure stochastic automata agregates, as decision mechanisms.

A mobile robot reaching for a target in an environment which may contain several obstacles, hiding or not the target, is simulated.

Introduction

The design of Intelligent Mobile Robot (IMR) architectures is currently based on a hierarchical approach. Depending on how the hierarchic structure is organized, two main classes of architectures can be identified in the literature: a) each level performs a given function, such as path planning or modelling (see for instance [3] or [4]) b) each level emulates a particular behaviour, such as avoiding collisions or looking for the power supply (the main example is [2]). Figure 1 shows the main differences between these classes.

The architecture proposed by Brooks has the interesting property of being incremental. For a reduced number of behaviours the resulting robot behaves, macroscopicaly, like an insect (robotic insect) [2]. Brooks argues that intelligence can be achieved as a side effect by means of coalitions in a robotic insects population.

This approach provides the motivation for the study of the low complexity mobile robot presented in this paper.

---

*This work was partially financed by JNICT grant 409/88.

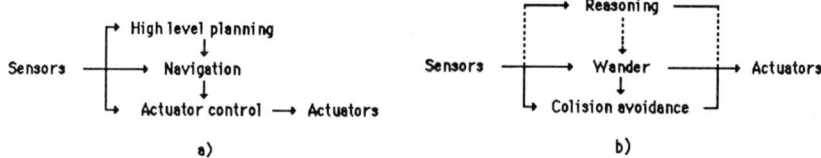

Figure 1: Architecture classes.

Learning Automata in a Random Environment

A variable structure stochastic automaton (or learning automaton) is a finite state machine in which the transitions between states are assumed to be stochastic and time varying. The automaton is assumed to be placed in an environment with unknown characteristics. At some states, an action, of a given set, is performed over the environment. The answer returned to the automaton provides a qualitative measure of the quality of the chosen action.

Formally a variable structure stochastic automaton can be defined by the 6-tuple:

$$\{B, A, \Phi, p(n), T, G\} \qquad (1)$$

Where

- $B = \{0, 1\}$, is a set of possible environment answers where 0 stands for *Reward the action execution* and 1 stands for *Penalize the action execution* [1].

- $A = \{a_0, \ldots, a_r\}$, is the set of possible actions.

- $\Phi = \{\phi_1, \ldots, \phi_m\}$ is a set of states available to the automaton.

- $p(n) = [p_1(n), \ldots, p_m(n)]$ is a probability vector, where $p_i(n)$ represents the probability of the automaton state to be ϕ_i at the instant n.

- $T : \Phi \times B \to \Phi$ is the state updating algorithm.

- $G : \Phi \to A$ is the output function.

In this work $m = r$ and G is a deterministic function that maps each state into an unique action.

Modelling a real environment can be a difficult task, so the answers returned to the automaton may be erroneous. To model this uncertainty the automaton

---
[1]This environment model is known in the literature as *P-Model* and the answers are also known as *Success* and *Failure*.

is assumed to be placed in a random environment. Each a_i has a probability $c_i = \Pr[\beta(n) = 1 \mid a(n) = a_i]$ of receiving a penalty.

At present a great variety of updating algorithms with learning properties is known. Most of them have been studied for stationary environments (the c_i being constant). Among these, four algorithms can be identified as being described by the following equations:

if $a(n) = a_i$

$$\begin{array}{ll} p_i(n+1) = p_i(n) + a[1 - p_i(n)] & \text{if } \beta(n) = 0 \\ p_j(n+1) = (1-a)p_j(n) & j \neq i \\ p_i(n+1) = (1-b)p_i(n) & \text{if } \beta(n) = 1 \\ p_j(n+1) = \frac{b}{r-1} + (1-b)p_j(n) & j \neq i \end{array}$$

For $b = 0$ the algorithm obtained is called *Linear Reward-Inaction* or L_{R-I}. For $a = b$, $b \ll a$ and $a \ll b$, the algorithms are called, respectively, *Linear Reward-Penalty* (L_{R-P}), *Linear Reward-ϵPenalty* ($L_{R-\epsilon P}$) and *Linear ϵ-Reward-Penalty* ($L_{\epsilon R-P}$).

All the algorithms described above have the property of achieving a lower mean penalty than that obtained when each of the actions is chosen with equal probability. This property, usually called *expediency*, indicates the learning capability of these automata. The L_{R-I} algorithm has a set of absorbing states so it is not appropriate for non stationary environments [6].

None of the other three algorithms has absorbing states and their action probabilities converge in distribution to a random variable p^* at all the points where the distribution function of p^* is continuous [2] [6]. The $L_{\epsilon R-P}$ has a poorer performance than the L_{R-P} or the $L_{\epsilon R-P}$ algorithms [5]. Nevertheless, it will be seen ahead that it is useful for the robot implementation.

Automata hierarchies (see figure 2) allow an intuitive problem structuring [1] and have computational advantages over a single automaton with equal number of simple actions [6]. It can also be shown that they have ergodic behaviour when the component automata are ergodic (case of the L_{R-P}, $L_{\epsilon R-P}$ and $L_{R-\epsilon P}$ algorithms).

Proposed Architecture

The approach proposed in this paper is greatly inspired in the Maslow hierarchical model used in Psychology Motivation Theory. Briefly, the Maslow model divides a human in a hierarchy of necessities and a set of behaviours. Every time a necessity of a given level emerges the adequate behaviours are activated and

---

[2]This is a weaker form of convergence relatively to the L_{R-I} scheme.

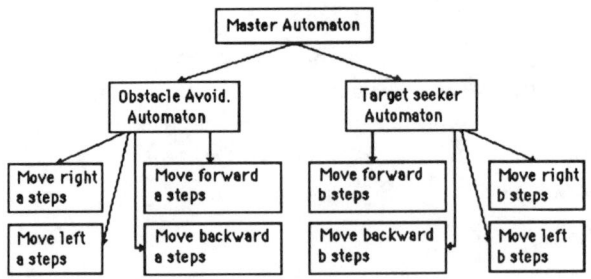

Figure 2: Automata hierarchy for the robot architecture.

the upper levels of the hierarchy are inhibited of activating their associated behaviours. Note that, similarly to the Brooks architecture, a given behaviour can be activated by distant levels in the hierarchy.

The robot to study moves in a 2-dimensional area (for the sake of simplicity only static circular obstacles are allowed) and is defined by a set of simple behaviours, or actions, and by a set of decision mechanisms which determine, at every instant, the action to be executed.

The following simple behaviours are considered:

1. Move s steps to the right.

2. Move s steps to the left.

3. Move s steps forwards.

4. Move s steps backwards.

The robot is supposed to perform two complex behaviours: It must learn how to avoid obstacles and how to reach a fixed target. The architecture proposed can be seen in figure 2.

The *Master* automaton (A_0) is responsible for deciding which behaviour is activated at a given instant. The actions leading to collision avoidance (automaton A_1) and target following (automaton A_2) have step lengths of $s = 3$ e $s = 4$, respectively.

A set of sensors is assumed to provide the necessary information to define the following classification indices:

Automaton A_0 - *Success* if the robot has not collided with an obstacle.

The robot is discouraged from moving into an area defined around each obstacle. If the robot is in this area it is said to be in a collision danger situation:

Automaton A_1 - *Success* if there is no collision danger.

Automaton A_2 - *Success* if target is at sight and closer to the robot.

Consider an environment without obstacles. If A_0 chooses to avoid collisions both automata, A_0 and A_1, are rewarded regarding the action chosen by A_1, so the probability of choosing the target following decreases. By choosing the A_0 automaton of $L_{\epsilon R-P}$ type, the small increase in the probability of choosing A_1, for each reward, will stop it from forgetting A_2.

Results

Automaton A_0 has learning parameters $a = 0.001$ and $b = 0.1$. Automata A_1 and A_2 are both of L_{R-P} type and have identical parameters, $a = b = 0.1$.

Figure 3 shows a set of simulations in different environments. The obstacles are represented by the ovals (scaled circles with radius 100). The "critical area" around an obstacle is a circular crown with dimension 4 in environments 2, 5, 6, 7 and 8, and dimension 2 in the others. Starting and target points are, respectively, (500, 0) and (100, 600) in environments 1 to 8, and (100, 600) and (500, 0) in environments 9 and 10.

Figure 4 shows the action probabilities for the simulation with environment 2. For the *Master* automaton actions 0 and 1 are, respectively, avoid collisions and target following. For the collision avoidance and target seeking automata, actions 0, 1, 2 and 3 correspond to the right, left, forward and backward displacements, respectively.

In environment 2 the target is at sight since the beginning of the simulation. Until collisions start to happen the *Master* automaton does not show any preference between A_1 and A_2. After the obstacle is by-passed both behaviours choose actions that lead the robot to reach for the target.

A_1 learns that in order to avoid obstacles it should not step backwards. A_2 shows preference to step left.

After the target is reached A_1 and A_2 choose actions with opposing effects.

As for environment 2 it can also be shown that in environments 1, 3 and 5 both behaviours cooperate (in the sense that both choose actions that made the robot reach the target), while in environments 9 and 10 there is no cooperation [7].

In environments 3, 4 and 5 it is possible to find explicitly a behaviour hierarchy. Collision avoidance emerges as the leading behaviour while target seeker shows no preference between its actions.

In environment 6 the target is never at sight and in environment 7 time is not enough for the robot to learn how to follow the target.

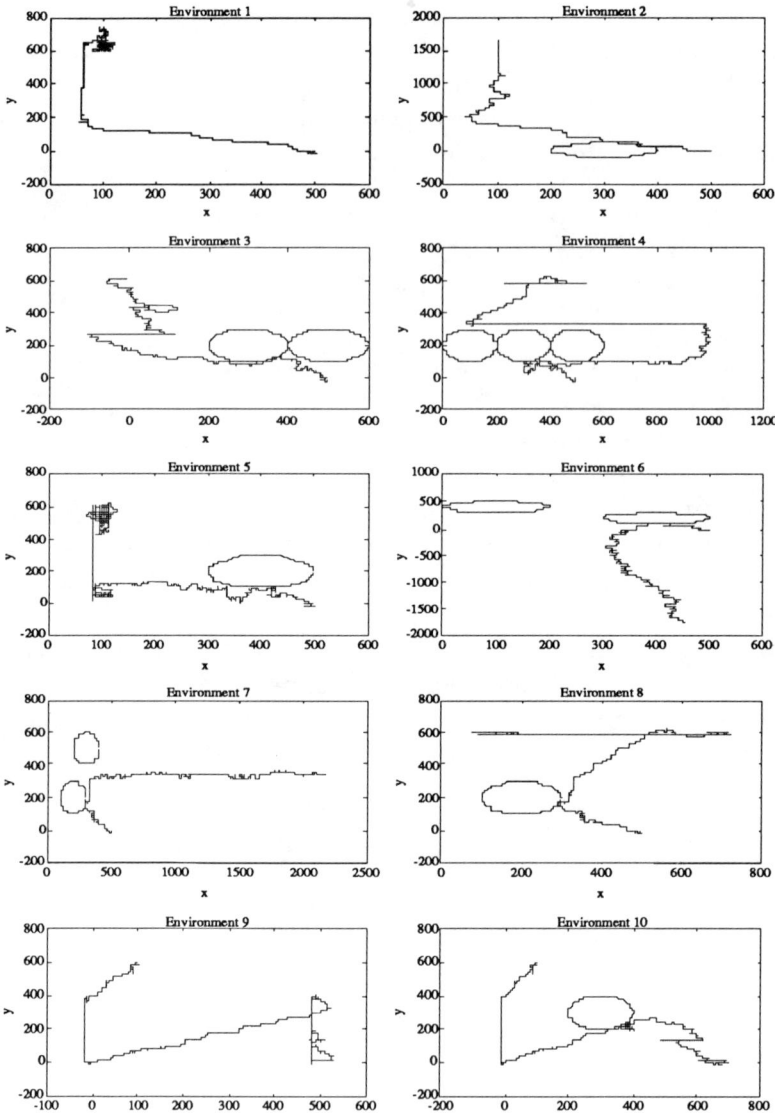

Figure 3: Simulations.

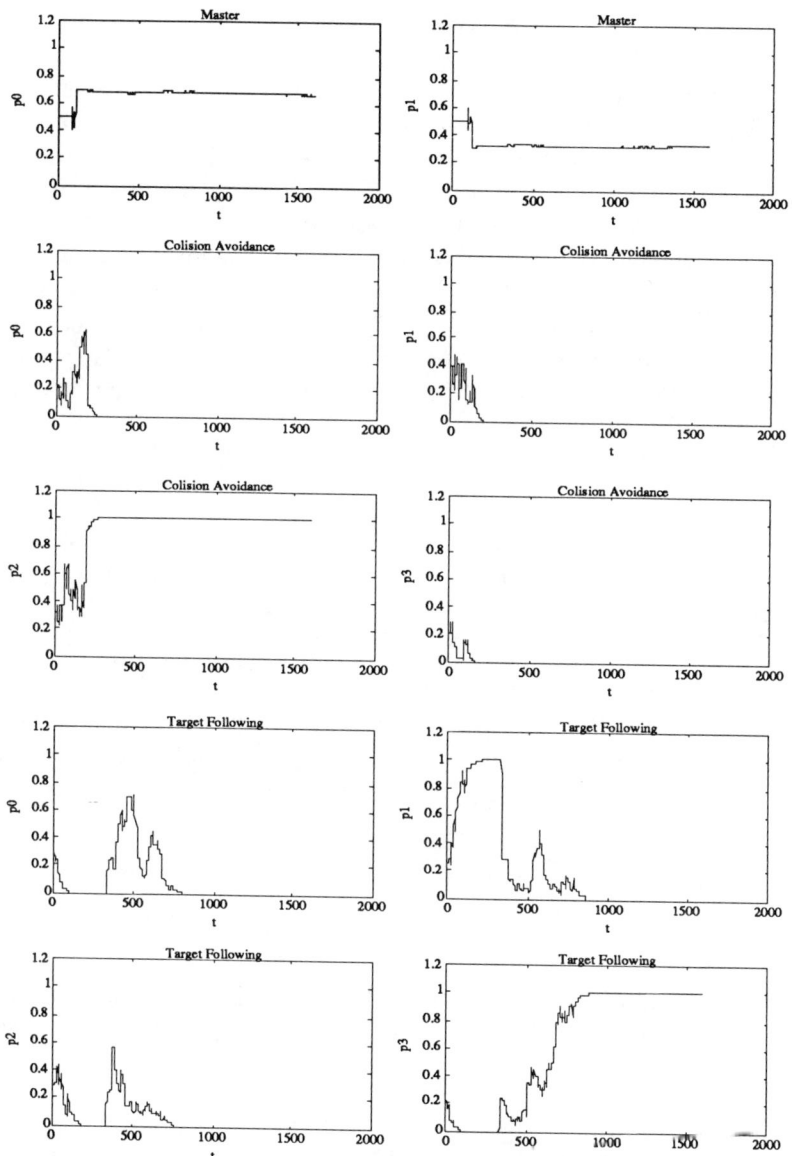

Figure 4: Action probabilities in environment 2.

Conclusions

In this paper a mobile robot architecture was presented. At every instant, stochastic decision mechanisms choose one action to be executed. The key idea when using variable structure learning automata as decision mechanisms is to take advantages of the learning properties of these (very simple) algorithms in order to avoid complex modelling of the environments.

Although the low complexity of the architecture, the robot has shown to be able of learning how to avoid static obstacles and how to seek for a fixed target.

The results presented show that a single low complexity robot can exhibit behaviour emergence. Since the architecture is easily expanded to a larger number of behaviours better results can be expected in future developments.

References

[1] Baba, N. "New Topics in Learning Automata Theory and Applications," *Lecture Notes in Control and Information Sciences*, Springer-Verlag, 1984.

[2] R. A. Brooks. "Achieving Artificial Intelligence Through Building Robots," *A.I. Memo*, M.I.T. Artificial Intelligence Laboratory, May 1986.

[3] C. Isik & A. M. Meystel. "Pilot Level of a Hierarchical Controller for an Unmanned Mobile Robot, *IEEE Journal of Robotics and Automation*, Vol.4, No.3, June 1988.

[4] G. Giralt & R. Chatila & M. Vaisset. "An Integrated Navigation and Motion Control System for Autonomous Multisensory Mobile Robots, *The First International Symposium on Robotics Research*, M.I.T.. Edited by Michael Brady and Richard Paul.

[5] Narendra, K. S. & Thathachar, M. A. L. "Learning Automata - A Survey," *IEEE Transactions on Systems, Man and Cybernetics*, Vol.SMC-4, No.4, 1974.

[6] Narendra, K. S. & Thathachar, M. A. L. "Learning Automata - An Introduction," Prentice-Hall, 1989.

[7] Sequeira, J. "Autómatos com Capacidade de Aprendizagem - Aplicações em Produção e Robótica," MSC thesis, IST - Technical University of Lisbon, Portugal, 1991.

Software Architecture for an Autonomous Manipulator

J.B. Thevenon   E.J. Gaussens   P. Le Page   F. Arlabosse

June 12, 1991

Abstract

The work presented here [1] is an attempt to make the robot software architecture to evolve from "model based" organisation to "perception based" framework.

In our understanding, "perception based" framework enhances autonomy (and adaptability) of the robot (or manipulator). It uses logical sensors to instantiate the situation perception, in order to allow, to tune, choose, or adapt the strategies fitting best with these situations.

1 Introduction

Talking about architecture makes sometimes difficult to see the concerned level of abstraction. Let us recall the different terms used here :

A Function is an abstraction that describes the activity of the robot (plan a path or grasp an object and so on). By defining the set of functions as well as their relationship we get the so-called *Functional architecture*.

An Application is a set of real world problems. It contains the definition of various goals and requirements (explore a planet, co-operation between robots and so on...). An application specifies a subset of the functions needed.

A Mode of Operation is the circumstancial manner in which a particular function is to be performed. The same robot can potentially work on the same application in alternative operation modes, performing the same functions. The *Operational Architecture* indicates the functions allocations by specifying where it is done

---

[1] The work reported here is supported the Esprit II project "MARIE" (EP2043).

in the architecture, therefore giving more details on the functions interaction.

An Implementation is the actual technological solution in selection, allocation and configuration of the computer hardware. The *Implementation Architecture* specifies how it is done in the real-time kernel and also the hardware architecture.

Our work is focused on all these aspects of an architecture for autonomous robot. The application is to make automatic decision about the state of a pipeline by setting a transducer on different part of it. Each measurement point or target is defined on-line based on the results of previous measurements and hypothesis of default in the pipe. The target location is unknown but the expert system making the diagnosis specifies in which area on the pipe the robot must find one to deliver new measurements. The initial knowledge about the world is the pipe location. Some unknown obstacles may be found during the motion. The perception system is made of an Ultrasonic sensor belt located at the end effector (for scanning in the x,y plane), a Ultrasonic sensor scanning in the z axis and below the robot (it is located near the end-effector) and a camera also looking in the z direction and located near the end-effector. The robot is fixed on a linear table moving along the pipe. The two controls (arm and table) are independents. The chosen arm is a kind of scara robot but with the z translation located as the first joint instead of being located at the wrist.

For such an application we must ask to ourself what autonomy means ? A first answer is the definition proposed in [7] :

Autonomous controller have the power and ability for self-governance in the **performance** *of control* **functions**. *It must be able to perform a number of* **functions** *in addition to conventional control function. One of them is to tolerate failure.*

This definition recover two main ideas :

- a self-governance is needed in the performance of function. The self-governance is to be considered with regard to the environment. Therefore, a robot moving in an unpredictable world should be able to perform the same function but in various way because each way to perform a function is dependent on the acquired / predefined knowledge about the nature of the environment. Moreover, each operation mode has its own limits or boundary conditions which must be very well known to decide which mode is the best in front of a unpredictable situation. We claim that by knowing *a priori* a set of various and generic unpredictable situations we will be able to select the good algorithms performing the same function. The consequence is that a lot of effort must be done to identify these sets of unpredictable situations called here Exceptional Situations since they are only unpredictable regarding their occurence but not regarding their generic description.

- some new functions have to be added to the conventionnal control function. The failure tolerance can be ensured by the fact that the Exceptional Situation sets are identified so that we can **predefine** some recovery strategies to be used in the case where the current situation falls into one of these sets. For us, the new function corresponding to failure tolerance is in fact an **Exception Handling function**. Moreover, there is a strong need to make periodic situation assessment either on the world or on the robot and its conventional control functions. In another terms we need a new function which gives the capability to *reason about the system computations from both the perception and control sides*.

To be more complete, and since our application is such that the robot evolves always in the same environment, therefore another new function is *Learning*. Each new external information has to be stored in such a way that our belief about the environment increases. At last, the robot should be able to ensure its own safety by the capability of *Reflex Functions*. This paper is focused on the first two functions. For the other functions see [1].

2 The Units : a way for representing robot "competencies"

In this section we present what we have called Units which are a representation of functions containing the foreseen mode of operations and each including a subfunction for reasonning about the computation. This subfunction is called During Motion Checking (DMC). Then, the way the Units are connected give the functional architecture. We have done this conceptual work because we wanted to bring explicitely the constraints due to the environment at the functional level. Let us see on an example what there is behind this term.

In our application we will need to follow the pipe in an automatic way. Then we have set here a particular robot competency. Various algorithms may be used to perform this function :

- a segmentation in the camera image can produce the pipe spine and gives via points in the camera frame. Once these viapoints are transformed in the cartesian space they feed a cartesian interpolation function. But, if an obstacle is discovered on the pipe, the segmentation need also an interpretation of the image which can be much more complex and time consuming. Let us assume that this algorithm (called Traditional Feedback Control or TFC) is given for the parts of the environment above the pipe, which are known.

- the camera image can also be used in a Task Function Approach (TFA) of robot control [6]. The derived sensor-based control may authorise some uncertainty in the knowledge of certain parameters. It can be used for partially known environment,

- the fuzzy control theory [4] can be used for unknown environment i.e. with unknown obstacles on the pipe.

This sharing between algorithms and environment conditions can appear unjustified in the sense that fuzzy control could also be used for well known environment. We made it for two purposes, all meant to increase adaptivity and robustness :

- the selection of the best suited algorithm regarding robot behaviour versus environment constraints,

- the introduction of a kind of redundancy in the Units.

Because we have matched algorithms with environment characteristics we can define with a good precision what are the possible problems during the computations

- under a TFC if an obstacle appears, even if we have selected this algorithm because the current belief was that there was no obstacles, a change of algorithms to perform the function must be done. The relevant information is coming from the perception side and the choice of the new algorithm is made based on a fast situation assessment "inside" the Unit (aim of the DMC),

- with the TFA some positivity constraints on the task jacobian have to be checked. If such a constraint is violated, a decision is made by the DMC. Here the problem is coming from bad computation in the control algorithm itself.

- the camera itself and the associated algorithms may fall into trouble and produce either strange data or no data. In this case, the function itself is failling and the recovering can not be decided "inside" the Unit, this decision is then passed to a higher level (the Exception Handling one).

In this general example we have seen that we consider the Unit as a sense-act cycle with the addition of a DMC subfunction for situation assessment based on known relevant parameters and decision making based on prior analysis of what is

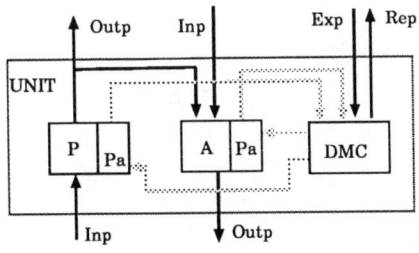

P : Perception Side i.e. a rigid hierarchy of perception algorithms

A : Action Side i.e. a set of algorithms having the same functionnality regarding the traditionnal control hierarchy

Pa: Parameter of the algorithms

DMC : During Motion Checking for diagnosis

Inp : Input;   Outp : Output ;   Exp : Expectation ;   Rep : Response

⋯⋯⋯⋯ a read of intermediate results or variables values used for situation assessment or a write of a new parameter value or a switch of algorithm

⟶ Data flows for perception-action cycle expectations and warning responses or for communication with other Units

Figure 1: The concept of Unit

to be done in these cases. In our study case the Units are conceptually described in figure 1.

The strong idea behind the DMC subfunction is that if we have a good understanding of the control methods it is possible to "focus" the DMC algorithm on a restricted set of relevant parameters and to determine simple strategies to solve the problem without having to report to the Exception Handling mechanism. Then, the purpose of the DMC is to : first, try to solve the problem at the concerned level and second, to report when it is unable to tackle the problem. These two ways of actions are found in the implementation : in the real-time kernel each control algorithm is a real-time task and inside each of these task a call to a subtask is made to evaluate the relevant parameters. This subtask can either to return new parameter value to the calling task or to generates an alarm which specifies the end of the current task which in turn reports to the Execution Controller with details on the failure. Then, a particular real-time task called DMC decides for a change of algorithms and returns the new task to be called with the arguments to the Execution Controller or directs an alarm to the Execution Controller and transmitted to the Exception Handling mechanism.

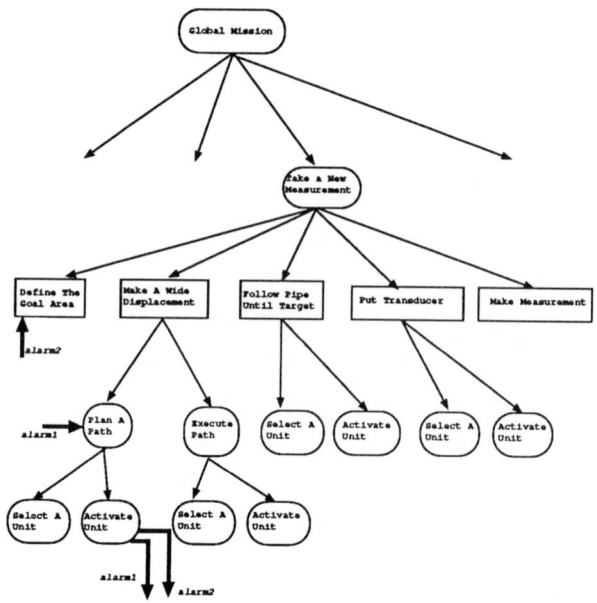

Figure 2: Task Decomposition and Backtracking for Exception Handling

3 Exception Handling Mechanism

With the presence of a DMC subfunction we can expect to reduce the number of possible failures existing in the classical algorithms. However, we should be able to make selective backtrack based on the alarm given by the concerned DMC. For example, the Unit running path planning algorithms may fail but for many reasons : the expectation given to the Unit indicates too coarse a granularity in the free space decomposition (*alarm1*) or there is no path whatever the granularity is (*alarm2*).

This means two different backtracking points :

1. *alarm1* means to rerun the Units with different expectations, it implies a backtrack to the task "Plan A Path" which computes these expectations and then calls the subtasks again,

2. *alalrm2* means to redefine the goal area, which in turn will determine if it is possible or not. If not, the "Take a New Measurement" fails and need a backtrack to higher level.

This behaviour of the task planning and Exception Handling is figured on picture 2.

Since we have adopted a deterministic approach for the problem of exception recovery we must predefine all the recovery strategies i.e. to which point in the task tree decomposition the system has to backtrack. This means in particular, that at the begin of a mission the system "knows" all the possible ways it has to tackle a set of possible problems. If a problem appears and if it has not been foresee then the system will not be able to recover from it. Therefore, the main effort to ensure robot autonomy is due to the designer. We don't think that an approach fully based on symbolic reasonning with a rule based system for decision making, allows us to reach our objectives especially regarding the time needed to achieve the mission. On the other hand, it seems impossible to predefine all the exceptional situations without having an unrealistic complexity, therefore the use of qualitative classes of such situations evaluated by simple rules or heuristics shortcuts this complexity and the design using the Unit concept allows to decompose the complexity of the system without introducing abstraction levels.

So, even if an exception is generally defined as an unpredicted situation, the fact to identify them beforehand will as the system's design goes on, allow to incorporate them into well circumscribed reactions of the robots, somehow then turning exceptions into well predicted situations ! While defining the task tree, and its associated control loops, we focus on the boundary conditions linked with each task/sub-task, an **exception is then defined as an "out of boundary" situations, at a specific stage of our design**.

We may summarize the major articulations of our approach :

1. Developing within the task analysis, the algorithms/symbolic systems to perform the task,

2. Incorporating exception analysis in the tasks identification,

3. Defining Exception Handling Models [2],

4. Using a planning system which embeds diagnostic/recovery techniques in the case of exceptions,

5. Designing the software architecture to provide specialised mechanisms for integrated exception diagnostic/recovery .

4 Conclusion

This paper presents our view on how to ensure autonomy for a robot equiped with various sensors and achieving a complex mission. We have developped the concept of Exception Centered Architecture which is, in our understanding, at the core of the futur robotic system development. Our belief is that Artificial Intelligence should not be a particular technique "above" the real system but a technique integrating various algorithms and know-how in perception and robot control. This view allows us to well define the notion of exception in a complex system and application and to deliver an Exception Handling mechanism as weel as a way to handle locally disfunctions by introducing a DMC subfunction.

Up to now we have defined the task tree, the different exception classes (we have found 13 of them) and the code of our task planner and Exception Handler containing this expertise. We have already introduced some minor modifications in the task tree due to some low level problems . This shows that it is not possible to have a good system by considering the Artificial Intelligence techniques as decoupled from control problem arising on the hardware and the real-time software.

References

[1] MARIE Project Consortium. *Public Deliverables D1 and D2*. Technical Report, CEE, 1990.

[2] Meijer G.R. *Autonomous Shopfloor Systems; A study into exception handling for robot control*. PhD thesis, University of Amsterdam, 1991.

[3] Barr A. H. Superquadrics and angle-preserving transformations. *IEEE Computer Graphics and Applications*, 1981.

[4] Sugeno M. An introductory survey of fuzzy control. *Information Sciences*, 1985.

[5] Khatib O. Real-time obstacle avoidance for manipulator and mobile robots. In *IEEE International Conference on Robotics and Automation*, 1985.

[6] Espiau B. Samson C., le Borgne M. *ROBOT CONTROL : The Task Function Approach*. Clarendon Press - Oxford, 1990.

[7] Antsaklis P.J. - Passino K.M. - Wang S.J. Towards intelligent autonomous control systems : architecture and fundamental issues. *Journal of Intelligent and Robotic Systems*, 1989.

[8] Lozano-Perez T. Spatial planning : a configuration space approach. *IEEE Transactions on Computers*, 1983.

[9] Dombre E. - Khalil W. *Modelisation et commande des robots*. Hermes, 1988.

The VAHM Project (Véhicule Autonome pour Handicapés Moteurs)
(Autonomous Vehicle for the Disabled)
Automatic Control of Mobility
MOUMEN K., PRUSKI A.

Laboratoire d'Automatique et d'Electronique Industrielles

METZ University

Fax : 87 30 24 44   Tel : 87 30 58 40

Ile du Saulcy, 57045 METZ CEDEX 1, FRANCE.

SUMMARY

The VAHM project (Autonomous Vehicle for the Disabled) is a federating theme of research in the field of robotics. Several aspects are dealt with:
* the design and control of an on-board manipulating arm;
* the analysis of man/machine communication;
* safety: developing sensors, feasibility analysis;
* mobility.

This article describes the means used to implement mobility. Three problems are analysed:
* environment modelling;
* path planning;
* trajectory generation.

INTRODUCTION

The vehicle which we are developing aims at assisting the physically handicapped in their everyday movements in order to increase their autonomy. The mobility aspect of the project consists in studying the basics of the controls bringing an assistance during motion. Two levels of control are available:
* basic controls: following a wall, going through a door, ... ;
* macrocontrols naming the goal to reach.

The map of the environment of motion is introduced by the handicapped person, by a third party or automatically builds itself up during a training period. Three methods are analysed:
* interactive graphics;
* scanning architect's blueprints;
* training.

The way the environment is represented in the computer will determine the amount of memory necessary as well as speed a path is found. The various methods of environment modelling to be exploited can, be summed up in four categories:
* description of the points travelled trough [L0Z 79];
* areas with particular characteristics: convex polygons [CHA 81], generalised cones [BRO 83], Voronoï's diagrams, Delaunnay's triangulation , ...
* grids [THO 84], [PRU 89] ;
* codification of a set of cells in a grid [PRU 90], [SAM 80] .

Path seeking consists of using the advocated model to deduce a set of points travelled through, by knowing the coordinates of both the starting and finishing points. These points make up the basic elements needed to establish the vehicle's control characteristics before the physical generation of trajectory. This article will describe the methods of modelling, path planning and trajectory generation.

PRINCIPLE OF MODELLING METHOD

The technique used for modelling the environment described in [PRU 90] constitutes a method of representation of free areas by a set of rectangles (2D) or parallepipeds (3D) with particular characteristics. Each of them is represented by a code, defining either a multiple value number or a set of values. The difference between these two concepts constitutes the possibility of applying either arithmetic or set operations to them. These numbers are called multivalue numbers (MVN) or multivalue codes (MVC).

The modelling method consists in laying a grid on the environment, every cell of which has an x and a y coordinate (in the case of 2D). A set of adjoining cells (set to "1" for a free space and "0" for an occupied space) is represented by a CMV.

<u>Definition 1</u> : A multivalue code is a numerical representation of a number in which each element may take three states: 0, 1 and X. State X represents states 0 and 1 simultaneously. Let's build the set of multiple numbers:

$$T = a_n a_{n-1} \ldots a_1 a_0$$

with a_i one of the states defined above.
In base 10, we have:

$$(N)_{10} = a_n 2^n + a_{n-1} 2^{n-1} + \ldots a_0 2^0$$

Example:
$T = 1XX0$ $(N)_{10} = 2^3 \times 1 + 2^2 \times X + 2^1 \times X + 2^0 \times 0$
$(N)_{10} = (8, 10, 12, 14)$

In modelling the environment we only consider a sub-set of the CMV, the codes with continuous values. This sub-space enables us to represent only one homogenous area. This type of code results in a loss of space in the memory, but also a significant time gain in the various operations.

<u>Definition 2</u> : A multivalue number with continuous values belongs to a sub-set of NMV for which:

$a_0 = X$ and $a_j = X$ only if $a_i = X$ for $j = i + 1$

<u>Definition 3</u> : A multidimensional NMV is a set of NMVs. In a computer a NMV is represented by two words:
* the CODE
* the VALIDATION

The code represents the NMV for which the state X is replaced by any state (0 or 1). The word VALIDATION associated to the code contains a bit for each CODE bit whose state is "1" if the associated bit in CODE is explicit (0 or 1), or "0" if associated bit in CODE is not explicit (X).

Example:
$T = 101XXX$
$NMV = \{CODE, VALIDATION\} = \{101000, 111000\}$

BASIC PROPERTIES

* The CODE of which the X states are forced to "0" represents the lowest value of the set of values.
* A logical operaration OR between the CODE and VALIDATION's "1" complement defines the greatest value of the set.
* The VALIDATION's 2 complement represents the cardinal of the set.
* The smallest of the values of the MNV is always a multiple of the cardinal of the code.

Let's take a binary grid which represents the environment. The cells at "1" represent a free area, the cells at "0" represent an occupied area. Let's consider the monodimensional case. The grid is made up of an n bit word WORD, and the coding is achieved according to the following algorithm:

1. $i = 1 ; j = 0 ; q = 1 ;$
2. DO
   Calculation of the maximum mx of bits to "1" possible from the position
   $mx = 2^k$; (with the first bit's position to "1" in j from the left).
   DO
   with the mask M(j,i) containing m bits, of witch i bits
   consecutive from the jth position are set to "1"
   $R = WORD \wedge M(j,i)$
   If $R = M(j,i)$ then {there is not code i = mx;}
   Else {code(q) = j ; valid(q) = i/2 - 1 ; j = j + i/2 ; q = q + 1 ; i = mx}
   WHILE (i < mx)
   WHILE(j < n)

This algorithm's complexity is in O(n). In two dimensions, the gird is defined by a list of variable words (bits). The coding first consists in performing logical AND ($\wedge$) between the word 2k and 2k+1 (k=0,...,m/2) and then in coding the result of logical AND as it was written in the monodimensional case. The list of the resulting codes is minimal and optimal.

Although the list structure constitutes a concise method for area representation, the processing of the codes in the list requires a lengthly searching time. The NMV structure is very interesting and can be represented by a tree structure. The multivalue codes tree (MCT) is based on the following concept: let's consider three types of nodes, one for each possible state of the elements of an MVN.

* A is a parent of B and C   and represents an X state
* B is a son of A   and represents a 0 state
* C is a son of A   and represents a 1 state

The MCT is built starting with the source node, to which a type B or C nodule is associated, depending on the state of elements of the MVN taken one by one from left to right. According to definition 2, when an X element is met, all the elements to its right are X. Thus the first X represents the leaf node of the code. In the multidimensional case, the first dimension leaf node becomes root to a second tree.

Example of environment modelling

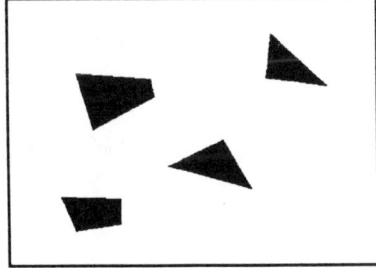
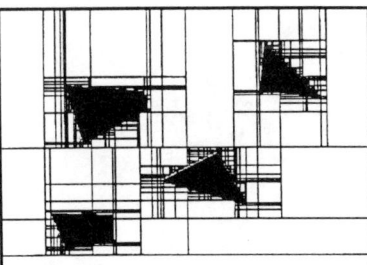

PATH PLANNING

Path seeking consists in finding, in the MCT model, the set of adjoining codes from a source code, which includes the goal point. To seek the codes adjoining an i code the intersection must be checked between the code on the boundary of code i and certain codes from the tree. The boundary code is deduced by a mere incrementation or decrementation, as the case may be, of code i. The smaller the boundary code cardinal, the faster this is. The complexity, in the general case, is O(mlogn) with m the cardinal of the code and n the number of cells in line or column.

The choice of the boundary code (four possibilites in 2D) is defined according to the minimal distance to the goal code. This is worked out immediately on the CMVs.

$$Disq(Aq,Bq) = Min\{|\overline{VALIDATION\ Aq} + (Code\ Aq - Code\ Bq)|,$$
$$|-VALIDATION\ Bq + (Code\ Aq - Code\ Bq)|\} \quad \text{with} \quad q = x\ OR\ y$$

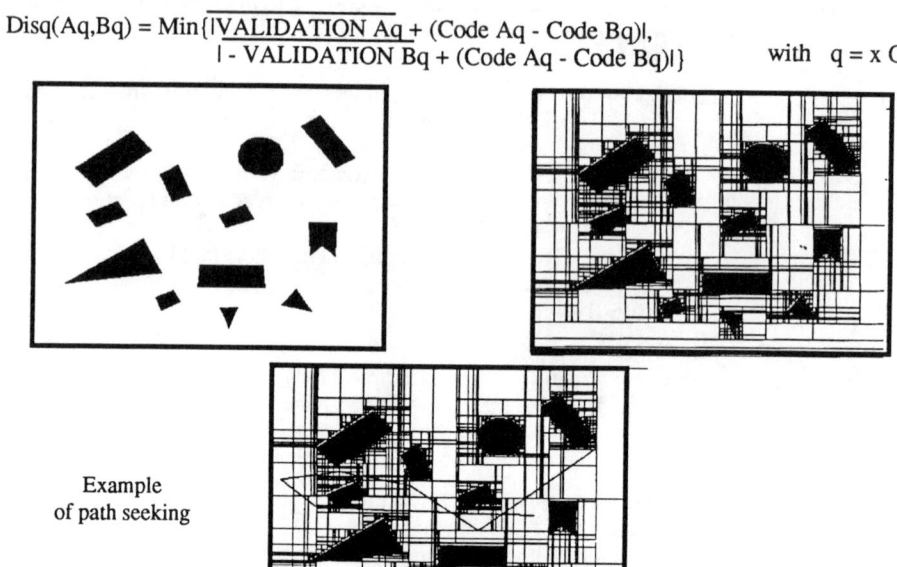

Example of path seeking

THE VEHICLE'S ARCHITECTURE

The vehicle we are studying has four wheels: two driving wheels mounted on the same axis, and two free wheels. Each driving wheel is moved by an independently controlled motor. This mechanical structure allows a rather simple kinematic control. The vehicle's position is given, at any time, by two Cartesian coordinates (x,y) and a direction angle θ.

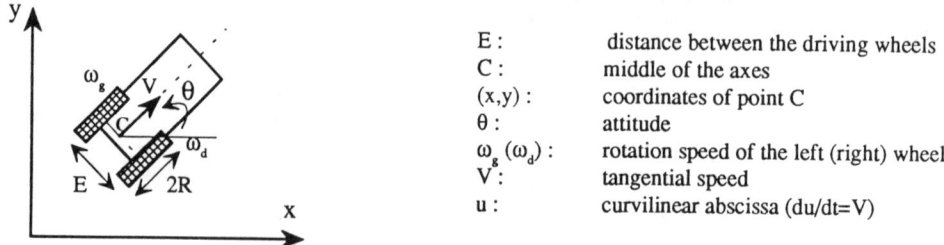

E :	distance between the driving wheels
C :	middle of the axes
(x,y) :	coordinates of point C
θ :	attitude
ω_g (ω_d) :	rotation speed of the left (right) wheel
V :	tangential speed
u :	curvilinear abscissa (du/dt=V)

The expressions giving the direction θ, the tangential speed V, the variations of the Cartesian coordinates dx and dy and the curvature radius ρ are:

$$\theta(t) = \theta_0 + R/E \int_0^t (\omega_d - \omega_g)d\tau \qquad (1)$$

$$V = du/dt = (\omega_d + \omega_g)R/2 \qquad (2)$$

$$\begin{cases} dx = du.\cos\theta \\ dy = du.\sin\theta \end{cases} \qquad (3)$$

$$\rho = |(\omega_d + \omega_g)/(\omega_d - \omega_g)|.E/2 \qquad (4)$$

ACTUATORS

The activators used are direct power and permanent magnet motors. These motors work according to the following equations:

$$\begin{cases} U = k\omega + RI + L.(dI/dt) \\ J.(d\omega/dt) = \Gamma_{em} - \Gamma_{rés} = kI - \Gamma_{rés} \end{cases} \quad (5)$$

where

R and L : represent the resistance and the inductance of the armature, respectively
k : the constant of torque and of the counter-electromotive force
Γ_{em} : the electromagnetic torque
J and $\Gamma_{rés}$: respectively, the inertia moment and the resistant torque brought back to motor's shaft
ω : the rotating speed ($\omega = N\omega_{exterior}$; N : reduction ratio of speed reductor)
V and I : respectively, the tension and the power of the armature

The variation of power tension U is achieved with chopper.

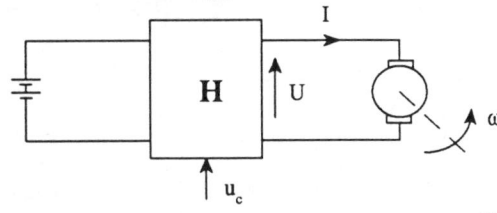

Because of the way it works, the chopper is discontinuous system. Introducing, downstream, a filter and a freewheel diode, and taking into account the inductive aspect of the charge, the chopper can be assimiled to perfect amplifier.

So we have:
$$U = k_h u_c \quad (6)$$

The sign of U is obtained with a polarity shunting circuit. On an established rate of flow, the angular speed of the corresponding motor is given by:

$$\omega = k^{-1}U - Rk^{-2}\Gamma_{rés} \quad (7)$$

The speed of each driving wheel is obtained from the speed of the corresponding motor by:

$$\omega_d \text{ (or } \omega_g\text{)} = \omega/N$$

TYPE OF SYSTEM TO BE CONTROLLED

The equations (1) ,(2) , (5) and (6) ruling the vehicle's movement are linear, the constants of electrical (L/R) and mechanical (propotional to J) time of the second order model (speed/tension) expressed by equation (5) are not of the same order of importance. So the system may be reduced to a first order model keeping the mechanical time constant. These considerations enable us to contemplate a certain number of strategies for easily usable numerical controls.

STRATEGIES OF CONTROL (of mobility)

The real trajectory is a curve extrapolated by a broken line. Defined as upstream on one level, this broken line takes into account both the shortest path and the bulk of the vehicle.

Several control strategies may be considered [KAN 88], [MAK 89] . The principles of these controls differ according to the mechanical architecture of the moving object. In the case of our vehicle's type of architecture, the essential problem lies in smoothing out the trajectory while imposing orders of tangential speed which remains constant or varies by stages. From the easiest to the most complicated, the following strategies can be quoted:

1) a succession of pure translations (along the segments)
        and of pure rotations (at the apexes of the broken line)
2) a translation on the majority of the segments

and circular arcs to join these segments
3) a translation on the majority of the segments
and a variation of curvature radii near the apexes
4) a translation on the majority of the segments
and a smoothing out at the apexes while imposing a minimal "cost" on the variations of the speeds and of controls.

- The first strategy results in $\omega_d = \omega_g$ for the translations and $\omega_d = -\omega_g$ for the rotations; therefore, the same chopper controls for $\omega_d = \omega_g$ and opposed controls for $\omega_d = -\omega_g$
- The second strategy boils down to $\omega_d = \omega_g$ for the translations and $\omega_d - \omega_g$ = constant for the rotations, so the same chopper controls for $\omega_d = \omega_g$ and difference of controls $u_{cd} - u_{cg}$ = constant for the rotations.
- The third strategy imposes the same controls for the translations and variations of the controls while keeping the sum $u_{cd} + u_{cg}$ constant when attitude of θ varies.
- The fourth strategy corresponds to $u_{cd} = u_{cg}$ for the translations and imposes, for the variations of attitude θ, controls which satisfy an optimisation criterion.

CHOICE OF THE STRATEGY OF CONTROL

The vehicle's relatively slow motion (1 to 2m/s) is compatible with the maintaining of the constant tangential speed, even for small curvature radiuses.

So, the most convenient strategy of control for our application is the one which corresponds to curvature radii which are constant by stages (infinite for the translation). Depending on the limitations (width of the free space and bulk of the vehicle) this curvature radius may be zero in certain cases. Then we switch from strategy (2) to strategy (1).

It should be noted that for each change of direction, the first strategy imposes the total stoppage of both wheels followed by their driving in opposite directions. Such an operation results in a considerable slowing down of the global movement. This strategy will therefore only be resorted to in case of necessity.

DISCUSSION OF THE STRATEGY OF CONTROL

We propose to maintain a constant curvilinear speed (du/dt = V) for the whole distance. This condition, corresponding to $\omega_d + \omega_g$ = constant, can be performed without any practical difficulty.

On the other hand, switching from translation to rotation with a constant curvature radius ρ, means, according to expression (4) ($\rho = |(\omega_d+\omega_g)/(\omega_d-\omega_g)|.E/2$) that we must instantly switch from $\omega_d - \omega_g = 0$ to $|\omega_d - \omega_g| = (E.V)/(R.\rho)$, which is theoretically impossible.

However, the model in speed (system 5), as written above, can, with a very good approximation, be assimilated to a first order system. The variations of ω_d and ω_g take the following form, for controls on complementary levels.

Tending towards the strategy to be adopted boils down to shortening the duration of transitions as much as possible, when switching from the «translations» to the «rotation with ρ constant» mode and vice-versa. This condition is absolutely accessible if a regulation which sufficiently speeds up this type of transitory is introduced. These accelerations of the mode changes are relatively small considering the speed of the motion. So they are compatible with our application. The non-

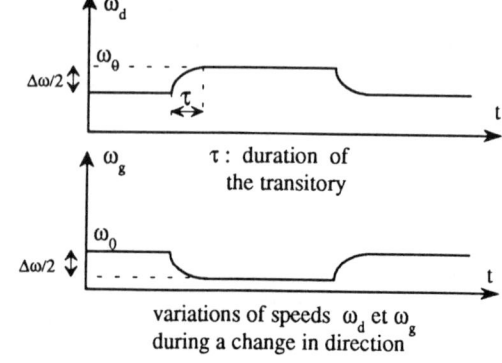

τ : duration of the transitory

variations of speeds ω_d et ω_g during a change in direction variation in level of the speeds ω_d et ω_g has no effect on kinematic precision. Indeed, switching from one mode to the other is performed with $\omega_d + \omega_g$ constant. The curvilinear speed du/dt given by expression (2) remains unchanged. As for attitude θ (expression (1)), the delay it shows when leaving the translation mode (integral of the transitory) is compensated when it returns to this mode (same integral).

WORKING OUT THE PARAMETERS OF THE CHOSEN STRATEGY

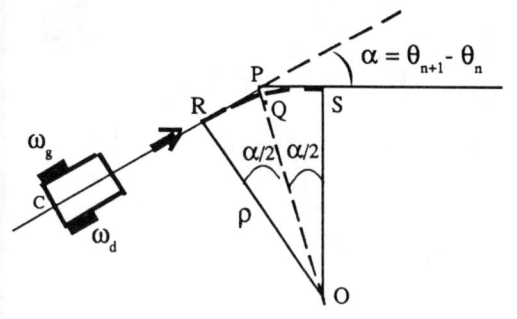

R: priming point of the smoothing out mode

S: end of that mode

The broken line and gaps $\varepsilon_i = P_i Q_i$ are defined as upstream.

Considering the diagram above, we get:

$RP/\rho = tg(\alpha/2)$ and $OP/RP = (\rho + QP)/RP = \sin(\alpha/2)$

We can deduce from these two relations:

$$\rho = \varepsilon \cos(\alpha/2)/|1-\cos(\alpha/2)| \tag{8}$$

As the tangential speed is constant ($V = R\omega_o$) we get:

$$\begin{cases} \omega_d + \omega_g = 2\omega_0 \\ \text{and} \\ |\omega_d - \omega_g| = \Delta\omega \end{cases} \tag{9}$$

Relation (4) becomes:

$$\rho = E.\omega o/\Delta\omega$$

which gives us:

$$\begin{cases} \Delta\omega = E.\omega_0/\rho \\ \text{and} \\ \omega_d = \omega_0 + \text{sgn } \Delta\omega/2 \\ \omega_g = \omega_0 - \text{sgn } \Delta\omega/2 \end{cases} \tag{10}$$

(in expression (10) sgn equals +1 for a left turn and -1 for a right turn.)

ALGORITHM OF CONTROL

On most translations, the verification is performed by speed regulation. When the smoothing out mode is about to be energized, a verification of position is carried out, for precision.

In smoothing out mode, while checking speed regulation and difference of speeds, the variation of attitude θ is counted. The turning process is stopped and we switch to translation mode when $|N_d - N_g|$ becomes equal to $N_{2\pi}.|\alpha|/2\Pi$, with N_d and N_g the number of impulsions delivered by two incremental captors respectively mounted on the right and the left driving wheels, and $N_{2\pi}$ the number of impulsions per revolution of the driving wheels.

The following diagrams illustrate some examples of environment modelling, path seeking and motion simulation obtained with the strategy presented in this article.

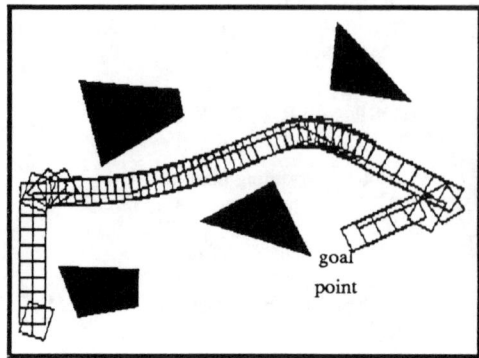
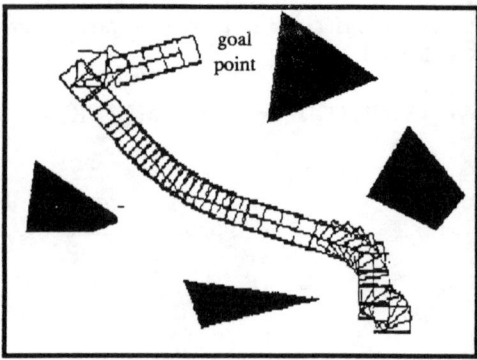

Examples of trajectory generating

CONCLUSION

In the case of our application (VAHM Project), mobility control requiries fast path seeking and a trajectory generation which meets the demands of aesthetics and comfort. Path seeking is performed from a modelling of the environment which takes up little space in the memory. While meeting the above-mentioned demands, the control strategy takes into account the limitations of bulk, which it translates into motion parameters in a simple way. At any time, thanks to multiple regulations, the vehicle's position is accurately determined. If an unexpected obstacle was to be detected, this would allow for the resumption of the initial trajectory after the obstacle had been bypassed. The results of simulation tests are very interesting and, during a second stage, they will allow the mounting of this automatic mobility control onto the real system.

REFERENCES

[CHA 81]   CHATILLA R. "système de navigation pour robot autonome: modélisation et processus de décision",Thèse de Doctorat, Toulouse 81.

[THO 84]   THORPE "Path relaxation: path planning for a mobil robot", Pittsburg, Rapport Technique CMV-RI-TR-84-5.

[LOZ 79]   LOZANO PEREZ T., WESLEY M."An algorithm for planning collision-free path among polyedral obstacles",Conm. ACM, Vol 22,1979 p. 560-570.

[SAM 80]   SAMET M. "Region representation quatrees from an binary array", Comp. Graphics and Image Processing 13, 1980 pp 88-93.

[PRU 89]   PRUSKI A., BOSCHIAN V. "Guid modelisation for autonoms robot", IFAC-INCOM 189, Madrid, Ap. 89.

[PRU 90]   PRUSKI A."Multivalue coding: application to the autonomous robots", Factories of the Future, Dec. 90, Nor Folk, USA.

[KAN 88]   KANAYAMA, YUTA "Vehicle path specification by a sequence of strait lines", IEEE Journal of robotics and automation, Vol 4 (3), Juin 1988.

[MAK 89]   MAKINO H., SUDA H. "Continuous path control using clothoïdal interpolation", 20ème ISIR, Tokyo, Oct. 1989.

A MODULAR APPROACH TO MOBILE ROBOT DESIGN

G.Barrall and K.Warwick
Department of Cybernetics,
University of Reading, U.K.

ABSTRACT

A modular approach to mobile robot design is described in some detail. The modules are joined together by a new type of bus specially designed for the purpose of mobile robot control (although it may easily be extended to other robot control tasks). Also described in the paper is a new 'very' high level programming language, designed for the control of such a modular bus system which includes the necessary flexibility to allow for research advances in the field of mobile robotics as they occur.
The approach of the paper is to show that by creating a standard modular design we allow for a greater degree of cooperation and coordination between research teams in the field.

1.INTRODUCTION

Many bodies and institutions around the world are currently designing either whole or parts of mobile robots[1-8,10-12]. Currently these designs show expertise in one area (e.g. vision[4], path planning[7][11], manipulation [13], etc.) and it is often the case that a body will design a first class vision system but have to mount it on a poor (or even nonexistent) chassis for mobility. The reverse is, of course, also often true.
A possible solution to this problem would be if such units were available "off the shelf". If it was possible to buy a navigator from some source and link it to a prototype vision system this would save time and would enable researchers to focus their efforts on their own personal speciality[9].
However the problem remains of compatibility. If you were to buy a navigator it would probably require some complex interface to allow it to be connected to the chassis and vision system. Many robots are designed around standard bus systems (e.g. VME [12], LAN [2], etc.) which require standard interfaces but still pose the problem of different data codes being used on the bus to represent the same things, therefore once again, requiring a complex interface to be designed.
However, if a standard interface could be agreed upon, then any body that wishes to find a potential market for their product just needs to add a standard interface such that a purchaser knows exactly how to pass data to it.
In fact, taking the idea a step further, if a standard bus system could be agreed upon, then a robot designer could build their robotic structure around it and plug commercial units designed by others straight onto the robot.
This paper describes an example for a new standard which is currently under research at Reading University (England) which hopefully could lead the way to greater robotics

cooperation and unity over Europe.

II. THE PROPOSED STANDARD

Early in the development of our system it was realized that we would need both hardware and software unity if we were to succeed in making a standard that was flexible enough to survive. Therefore not only has a hardware design been proposed but also a new language is currently under development to control it.

This paper tackles each topic as separately as is possible....

IIa. THE HARDWARE

The hardware consists of a standard BUS which links all of the robot's modules together (A MODULE being defined as a unit to carry out a task e.g. Navigator, Vision System, Manipulator Controller, Reflex generator, etc.). Integral to the bus is a microprocessor which receives commands in the standard control language (See IIb) and sends the relevant control codes to each of the buses connected modules(See fig. 1).

This connected processor is termed as the BUS MASTER.

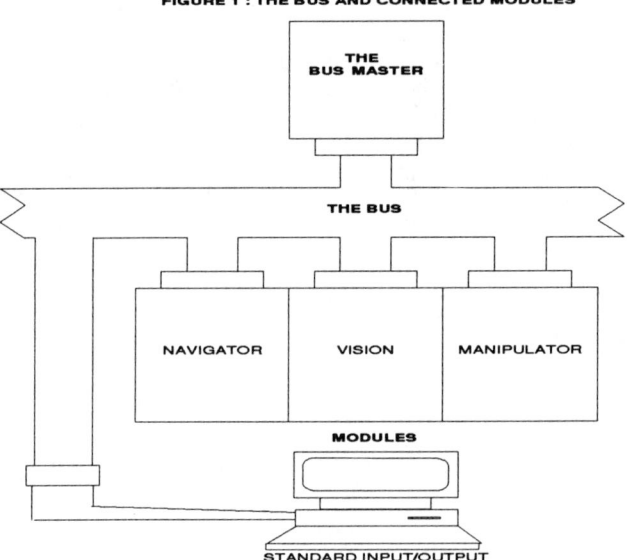

FIGURE 1 : THE BUS AND CONNECTED MODULES

The bus itself is similar to any standard system, although it accepts either eight or sixteen bit codes from the modules inserted into it. Most but not all data transfers between modules are passed via the bus master, however it is still possible for one module to talk directly to another by taking direct control of the bus.

Each module consists of ;
a) The necessary interface to the bus.
b) The necessary logic to calculate its function (i.e. Path planning, vision decoding, etc.)
c) The hardware for controlling all input output devices the module needs(e.g. Motor controllers, Camera servos, etc.).
d) All the input output devices needed by the module(e.g. Wheels, Cameras, Sonic radar, etc.).

An example navigator module is shown in figure 2.

In addition to these standard modules we have also defined the "Reflex Module". The reflex module borrows greatly from Brooks 'subsumption' architecture [5] which has recently become so popular in robot design. The reflex module is able to interrupt the bus at any time (obviously the interrupt will not occur until after the end of the current cycle!) and inform the bus master of an impending disaster (e.g. a contact sensor hits something, or dangerous levels of heat are detected, etc.). After receiving this warning the bus master will call an emergency stop and inform the path planner to replot a course. If many reflex emergencies occur in a short time the action of the bus master will be to sound any alarm modules it has available and then wait for help!

Reflex modules can also be designed that monitor the robots internals and warn of system breakdown (e.g. If both wheels go to full speed despite no path planner activity the module can shut the robot down). Also along these lines is a low power warning module which sends a GO POWERPOINT command to the path planner. There are, of course, many more possibilities.

Presented here is the power up sequence of the robot bus from power on.
1) The bus master first activates all its systems and runs its operating system.
2) Once all other modules on the bus are initialized the bus master requests standard identity codes from each module so that it knows what is connected where!
3) Now the bus master has located a Standard I/O terminal on the bus and requests a program from this. If no Standard I/O is found the bus master will load a program from any internal store that it may have.
4) The bus master now executes the loaded program step by step, sending any relevant codes off to the appropriate module(e.g. GO commands sent to the path planner, etc.).
5) Whilst executing the program the bus master watches for background conditions (See next section for details...).

FIGURE 2 : EXAMPLE NAVIGATOR MODULE

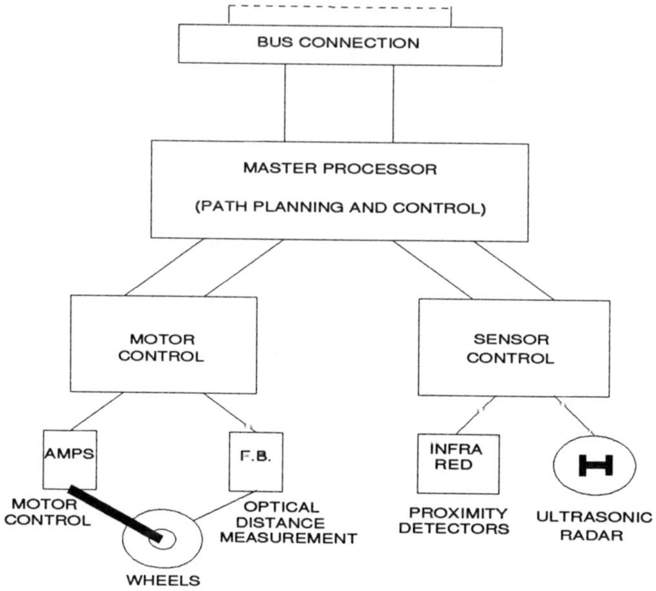

IIb. THE SOFTWARE LANGUAGE (RPL)

As previously stated the bus master follows a set program loaded from a Standard. I/O. module on the robot. This module may be an actual user terminal, preset instructions or may be even an A.I. block which generates its own RPL program for the robot as the robot's situations change.

The programming language itself is similar to many other high level languages in the way it handles Mathematics and Logic but is even higher when you consider the commands provided for direct control of the robot.
For example;
  i) GO KITCHEN : Instructs the path planner to take the robot to the kitchen.
  ii) LOCATE BOX : Directs the vision system to return the x,y,z coordinates of a box before it.
  iii) TAKE BOX : Directs the manipulator module to pick up a box located by the vision system.
  iv) etc.

The actual commands in the basic RPL language just cover the standard logic, iteration, and mathematics. The more complex "higher level" commands are imported in from "C-type" include files. It is suggested that one of these files exists for each module so that a new include file is easily substituted for the previous one when a module is upgraded. Also when a module is purchased an include file should be supplied with it to provide all the commands it understands and the codes the commands translate into. This feature of the language provides upwards compatibility and does not restrict the commands understood by future modules.

In fact import modules which include procedures made from existing (non-included) commands seems a likely feature to be added in the near future.

Given here is an example RPL program along with explanatory comments.

```
: Example RPL Code
: G.S.Barrall 1991
: First include module commands
        Include GSB_Nav;
        Include GSB_Vision;
        Include GSB_Manipulate;
: Second set up background commands
        if SEE box display "Box!";
: Send map to path planner
        LOADMAP "CYBBUILD.MAP"
: Off we go...
        GO OFFICE;
        LOCATE PENCIL;
        TAKE PENCIL;
        GO LAB;
: Save map with obstacles added!
        SAVEMAP "CYBBUILD.MAP"
: End of program.
```

IIc.A DEMONSTRATION (of the operation of the previous program)

Figures 4.1-4.6 show a simulation of a robot following the above program. The specification of the robot is shown in figure 3 and is based upon our in-house robot at the Reading Cybernetics Department. It has five fixed ultrasound radars and a sixth ultrasound radar mounted on a rotating head. The navigator uses constant feedback from the fixed radars to keep the robot from bumping into walls and uses information from the rotating radar periodically to head the robot on a more direct course towards its next waypoint (if that is at all possible).

The simulation assumes that the robot is fitted with an advanced vision processing system (using both depth ultrasound scans [13] and stereo camera vision) which can recognize a few pretaught objects(using a neural net or similar system.) as well as a fairly accurate but light weight gripper.

FIGURE 3 : MOBILE ROBOT DESIGN

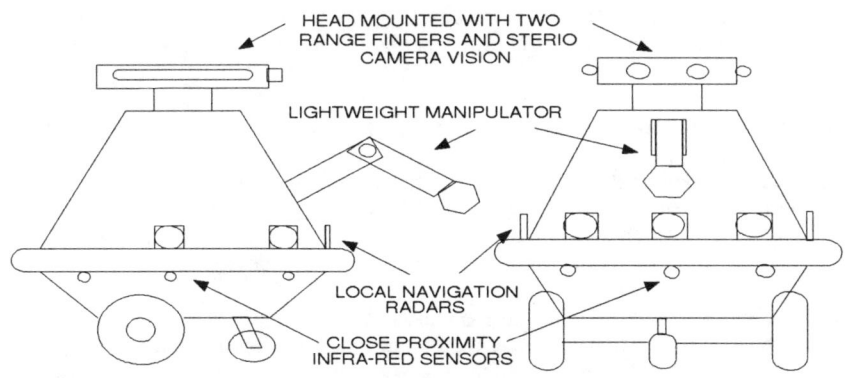

FIGURE 4 : Simulation results

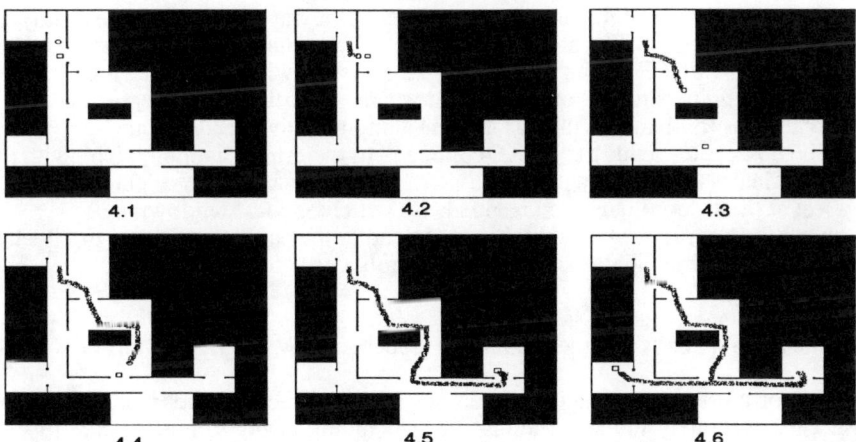

Please note that the circle represents the robot, the square the next waypoint and the grey trail represents the path taken so far....

A layout of the modules connected to the bus inside the robot is shown in figure 1. The actions of each module in figures 4.1-4.6 are as follows;

4.1) The robot's modules reset and initialize.
The bus master identifies each module on the bus.
The bus master downloads the program from Standard.I/O.
It then downloads the include files.
Next it notes that any "box" found (value token of "box" is already defined by GSB_Vision include file) by the vision system will require a message to be sent to the Standard.I/O (Std I/O).
Finally it downloads the .MAP file from Std..I/O. and sends it to the navigator.
The bus master sends the navigator the code for GO and the token for kitchen as defined in GSB_Nav.
The navigator's path planner calculates the best path using calculated waypoints and informs its motor controller to start moving towards the first waypoint.

4.2) The robot arrives at the first waypoint and moves onto the next. The bus master waits for conformation of destination reached (Monitoring the bus to see if a box is discovered!).

4.3) The robot moves past the fourth waypoint on its path and encounters a large(!) object blocking its path. The navigator stops the robot and requests the bus. Upon being granted the bus it asks the vision system to identify the object.
The navigator then asks its path planner to find a way round the obstacle. The vision system meanwhile takes a snapshot of the object and tries to identify it. The path planner suggests a new course and the robot moves off.

4.4) After the robot is past the obstacle the vision system identifies the object as a box. It takes control of the bus and informs the navigator of the obstacles description. Being a very good navigator it tags the map file with a new obstacle at the location of the box. The bus master notices that a box was found by the vision system and sends the characters 'B o x !' to the Std.I/O. for display as commanded earlier.

4.5) The robot arrives at the office (hopefully the last waypoint is near the desk!) and the navigator switches into standby mode. Noting this the bus master sends the LOCATE PENCIL command to the vision system which scans the area for the most likely 'pencil like' object. It returns these coordinates in a register accessible by the bus (probably in the bus master) along with the PENCIL tag.
The bus master sends the TAKE command to the gripper module along with the coordinates connected with the PENCIL tag. The intelligent gripper takes the object at these coordinates and responds by switching into standby mode.
Now the bus master sends the GO command to the navigator along with the label for LAB. The navigator again moves the robot as before.

4.6) The robot arrives at the lab.
The program ends and the bus master requests a new one from Std.I/O.

The simulation plots in figures 4.1 - 4.6 demonstrate the GO commands from the program and use very accurately modelled ultrasonics (including beam width and side lobes as well as taking into account incidence angles beyond which reflection does not take place). This simulation was written as a base to test both the RPL language and also the navigator itself.

COMPARISONS

Finally in this paper the authors would like to show how some of the worlds most popular robots could use (or add to) a new bus system. It is not at all intended to cause any offence or to show any design flaws with these robots current control systems but instead to show how a greater degree of cooperation could be obtained overall.

i) I.M.A.S. as presented in reference [3] is an example of a navigation system where the navigator actually includes vision processing rather than just radar as in figure 2. An excellent navigator and path planner is (or was) being developed here but can not be expanded upon by other developers. A standard bus would allow for the connection of a standard manipulator to load the vehicle or a weapons load, etc.

ii) The G.S.R. project [2] also had an excellent navigation system which was going to be (in 1986) connected to a colour vision system and a planning subsystem via a L.A.N.. If a standard had been around maybe they could have purchased these items and saved time by not having to design them in-house.

iii) To simulate a mobile robot such as "Mister Ed" [6] a navigator would have to be designed that would have an inbuilt joystick or the joystick could be part of a module that sent the relevant commands to other modules on the bus.

iv) An excellent navigation system has been designed by the Robotics Research Group at Oxford[7] and could be used and expanded by many other research groups if a standard interface was fitted and the product marketed.

v) The Stanford Cart[4] included a vision system that could label points of interest in a video image. By fitting a standard interface that returned these coordinates to a bus master, a robot could guide itself towards a specific object or use the coordinates to control a gripper to take a specific object. This would be very useful to a lot of current research projects where people are duplicating this work.

vi) The CMU rover[4] was constructed from a camera, sonar and infrared proximity sensor system. This system is very similar to the one given in this paper to demonstrate the working of the bus. If these simple units and the bus had been available at the time the CMU rover could have been constructed very quickly from common standard modules.

CONCLUSION

This paper has described the new robot bus system which is under development in the Cybernetics Department at Reading University.

It has been shown how the bus may be used to design just about any robot from AGVs (of all sizes!) to walking legs, vehicles or even stationary manipulator systems.

Also the authors hope it has been shown that introducing a standard bus will allow robotic systems designers a possible market place to buy and sell standard modules and therefore add a large incentive for robot funding in industry and universities.

It is hoped that this paper may be a small first step in bringing together a more concerted effort across Europe in the field of robot design. The authors would welcome all feedback (a good Cybernetics concept!) from readers and would hope that the many bodies involved in robot design could come together now and agree upon a standard robot bus for a unified future.

REFERENCES

[1] Ren C.Luo and M.G.Kay, "Multisensor Intergration and Fusion in Intelligent Systems", in IEEE Trans. on Man and Cybernetics, Vol.19, No.5,Sept/Oct 1989,pp 901-931.

[2] S.Y.Harmon, "The Ground Surveillance Robot (GSR)", in IEEE J. of Robotics and Automat, Vol.RA-3, No.3, June 1986, 266-279.

[3] Can Isik, "Pilot Level of a Hierarchical Controller for an Unmanned Mobile Robot", in IEEE J. of Robotics and Automat., Vol.4, No.3, June 1988, pp.241-255.

[4] Hans P.Moravec, "The Standford Cart and the CMU Rover", in Proc. IEEE, Vol.71, No.7, July 1983, pp.872-884.

[5] Rodney A.Brooks, "A Robust Layered Control System For a Mobile Robot", IEEE J. of Robotics and Automat., Vol RA.-2, No.1, March 1986, pp.14-23.

[6] J.Cornnell and P.Viola, "Cooperative Control of a Semi-Autonomous Mobile Robot", in IEEE International Conference on Robotics and Automation 1990, Vol.2, pp. 1118-1121.

[7] B.Steer and T.Atherton, "Design for Navigation", in IEEE International Conference on Robotics and Automation 1990, Vol2, pp.942-947.

[8] Hans P.Moravec, "Three Degrees for a Mobile Robot", Internal Paper, The Robotics Institute Carnegie-Mellon University, Pittsburgh, 6th May 1984.

[9] D.F.Sherry and D.L. Schacter, "The Evolution of Multiple Memory Systems", in Psychological Review, Vol.94, No.4, pp.439-454, 1987.

[10] J.Bornstein and Y.Koren, "Obstacle Avoidance with Ultrasonic Sensors", in IEEE J. of Robotics and Automat., Vol.4, No.2, pp.213-218, April 1988.

[11] P.Hoppen, T.Knieriemen, E.von Puttkamer, "Laser-Radar based Mapping and Navigation", in IEEE International Conference on Robotics and Automation, Vol.2, pp.948-953, 1990.

[12] G.McKee, "Notes on Rudyard", Internal papers, Department of Computer Science , University of Reading, Reading, England, 1990.

[13] A.J.Vayda and A.C.Kak, "Geometric Reasoning for Pose and Size Estimation of Generic Shaped Objects", in IEEE Conference on Robotics and Automation 1990, Vol.3, pp.782-789

Environment Representation by a Mobile Robot using Quadtree Encoding of Range Data

Leon Piotrowski
EDF/DER, 6 quai Watier, 78400 Chatou, France

INTRODUCTION

A method is presented for building and updating the spatial representation of a robot's environment using quadtree/octree[1] encoding of range finder measurements. The intended applications are environment sensing and modelling, robot navigation (trajectory planning) and self-localisation for both autonomous and fixed robots that need to move about within their workspace. This work is preliminary in that it only considers the two-dimensional (2-D) obstacle space of a mobile robot (quadtrees). The extension to 3-D (octrees) follows naturally.

The aim of this study is to evaluate the ease and efficiency with which quadtree representations of range measurements can handle the following two fundamental problems encountered by mobile robots :

1. real structurally-complex environments where the presence of obstacles prohibits the sensing of the whole environment from any given location;

2. the creation of a global representation of the environment from a sequence of local acquisitions taken at different (unknown) positions within the environment. (In reality, one should be able to estimate the robot's approximate position at any time. It is erroneous, however, to state that its precise location within the environment is known without confirmation, especially after several consecutive displacements.)

It logically follows that self-localisation of the robot can be achieved by "matching" the presently perceived environment with the known (global) representation obtained from all the previous acquisitions.

This study is conducted under the following conditions :

- a real structurally-complex environment, (Figure 1);

- no assumptions are made on the type of objects present;

- arbitrary and unknown rotations and/or translations of the robot within its environment when moving from one location to another;

- only a part of the whole environment can be perceived from any given position due to the presence of obstacles;

- no preprocessing of the range finder measurements.

Quadtree-type data structures are known to have several very powerful advantages for representing spatial relationships :

- a compact representation of occupied, free and unknown space with fast access to this information;

- large parts of the environment can be ignored for analysis;

- the possibility of having a multi-resolution representation available at all times.

Of equal importance, one can cite a number of well-known inconveniences of quadtrees :

- a rigid decomposition of space with an implicit origin;

- the imposed grid structure is not well-suited for the rotation of nodes and often leads to a loss in spatial precision;

- the difficulty of merging local quadtrees into a global tree when the relative displacement of the local origins is unknown.

A simple algorithm is presented showing that the advantages outweight the inconveniences for creating a global spatial representation of the robot's workspace and for determining the robot's location within its global environment.

This work possesses a number of advantages over other related studies. It is less intensive in computer time than the image-processing approach of Hong and Shneier[2]. In their work, the robot's (octree) environment is reconstructed from a sequence of images taken after each displacement of the robot. Unfortunately, their method insists that the camera position must be known in advance. Chen and Tsai [3] performed preliminary investigations on robot location using curved object surface patches. These patches were needed to be known and, in addition, the CCD camera had to be carefully calibrated. Another approach [4] uses the quadtree and pyramid structures to analyse range images for surface curvatures with the aim of obtaining viewing-direction-independent methods. However, the extraction of the surface curvatures depends greatly on the noise present in the range measurements.

The method presented here is independent of these constraints.

Finally, the quadtree approach used here should be compared to the hypergraph representation[5].

MATERIAL AND ENVIRONMENT

Figure 1(a) shows a photograph of the environment used in this study. The physical dimensions of the room are approximately 15m by 7m. The quadtree representation of a typical ranger finder scan of the room (in a horizontal plane which is 1 m above and parallel to the floor) is shown in Figure 1(b). A 360 degree scan is presented with distance measurements taken in 1 degree intervals. The distances were obtained by a RIEGL LD 90 directional range finder under PC control. This range finder is capable of measuring distances from 1 m to 50 m using a time-of-flight pulsed IR laser. A measurement time of 1 second was pre-programmed into the range finder (implying a measuring precision of $\pm$ 1 cm). The angular rotation of the range finder was obtained by placing it on a computer-controlled rotating base (MICROCONTROL). The complete system (range finder plus rotating base) was mounted on a wheeled table which was physically pushed to different locations within the room. Each displacement involved an arbitrary translation and an arbitrary reorientation of the whole system.

All quadtrees have 9 levels of spatial resolution implying a maximum precision of 256 x 256 elements. Each distance measure is stored in the appropriate leaf at acquisition. The algorithm presented in the next section was programmed in the C language.

METHOD

This study uses the following three types of quadtrees :
1) WORLD : a quadtree representing the robot's complete environment. The origin is arbitrarily chosen at creation and remains fixed. The aim is to update this world model with information obtained at each successive displacement of the robot using the Acquisition quadtree.
2) ACQUISITION : a quadtree representation of what the robot is able to perceive in its environment from any given location. This representation is purely local with the measured distances indicating the position of objects relative to the robot's position (always taken to be at the quadtree origin; leaf (128, 128) in Cartesian coordinates). No information is available concerning the robot's location within the World quadtree.
3) SCATTERGRAM : essentially a multi-resolution (2-D) histogram covering the World space. Each node at each tree level defines a finite spatial region of the World and contains an associated counter preset to zero.

An exhaustive matching of the Acquisition and World quadtrees at the leaf level is prohibitive on account of the vast number of tests that would be necessary. Furthermore, the Acquisition and World trees are most likely to contain some objects that are not common to both (because the robot has moved). One has no way of knowing, a priori, which individual measurements of the Acquisition quadtree correspond to which in the World tree. In addition, individual measurements may contain errors due to the different viewing angles and the different nature of the object surfaces.

The algorithm presented is an iterative multi-resolution approach which uses the spatial relationships among the nodes existing within each quadtree level. In this approach, multi-resolution is somewhat synonymous to iteration because each pass of the method descends one tree level and is therefore able to match the two quadtrees more precisely. (Presented another way, one can equally say that the unlikely matches are eliminated, within the precision of any given tree level : "tree pruning"). Our method will show that in order to overcome some of the inconveniences of the quadtree structure (rigid boundaries and loss of precision for arbitrary rotation of nodes) it is necessary to consider more than one possible path for descending the tree from levels k to (k-1). If the number of paths is small then computer execution will be very fast. This situation depends on the structural similarity of the World and Acquisition quadtrees.

Let the World and Acquisition quadtrees have n+1 levels of spatial resolution where level n corresponds to the root and level 0 indicates the leaves. In our study n=8. Furthermore, at any given tree level, k, let N_{wk} and N_{ak} signify the number of existing gray or black nodes in the World and Acquisition trees, respectively. The population of world nodes can therefore be written as w_{ik}, $i=1,..,N_{wk}$. Similarily, a_{ik}, $i=1,...,N_{ak}$, represents the population of gray or black nodes at level k within the Acquisition quadtree.

For any pair of nodes w_{ik} and w_{jk} in the World quadtree and any other pair a_{ik} and a_{jk}, i<>j, in the Acquisition quadtree an attempt is made to calculate what we will call the "mapping parameters" that will map the node a_{ik} to the node w_{ik} as well as mapping a_{jk} onto w_{jk}. The first requires a simple spatial translation; mapping a_{jk} onto w_{jk} will usually require both a rotation of a_{jk} about a_{ik} followed by the same translation. Of course, the calculation of the mapping parameters might be impossible owing to the particular choice of w_{ik}, w_{jk}, a_{ik} and a_{jk}. Should it fail, then another choice of nodes is made. The mapping of a k-level node is taken to be the mapping of its center. The node onto which it is mapped is also taken to be a k-level node and corresponds to that which contains the mapped center within its boundaries.

Let us consider the case where we have chosen four such nodes at quadtree level k and that the mapping parameters have been computed successfully. There is no garantee that using these mapping parameters to project the whole Acquisition quadtree into the World quadtree space will result in a reasonable matching of the two trees. This is because only two pairs of nodes have been examined and the overall spatial structure within each quadtree has not yet been considered. The best we can say is that it is possible that nodes a_{ik} and a_{jk} correspond to nodes w_{ik} and w_{jk}, within the precision allowed by k-level quadtrees.

The next action to take is to assume that these mapping parameters are correct and to map the robot's position in the Acquisition quadtree (leaf $(2^{n-1}, 2^{n-1})$) into a k-level Scattergram node. The corresponding counter of this node is then incremented by one to indicate that this region of the world space has just been estimated for the robot's world position.

To appreciate the computational simplicity, it must be emphasized that only a single leaf of the Acquisition tree, representing the robot's local position, is mapped into a k-level Scattergram node. The whole Acquisition tree is not mapped.

When the above procedure has been repeated for all possible mapping parameters (i.e. for all possible combinations of w_{ik}, w_{jk}, a_{ik} and a_{jk}) the k-level Scattergram resembles a 2-D spatial histogram of estimations of the robot's position within the World quadtree. It should be peaked to reflect the common structures in both the World and Acquisition quadtrees. The degree of spread of the Scattergram is directly related to the structural similarity of the two quadtrees (at level k) and the computational error introduced by the mapping operations.

The information contained within the Scattergram permits us to refine our analysis and pass from tree level k to tree level (k-1). What it tells us is that there are some k-level size regions of the world space that contain a relatively high number of computed estimates for the robot's position. These estimates can be made more accurate by repeating the whole procedure with (k-1)-level sons. Deciding which sons to use is simple : one only considers the sons of w_{ik}, w_{jk}, a_{ik} and a_{jk} that gave high Scattergram values at the k level. This will lead to a new Scattergram at the (k-1) resolution level which will also be peaked. The iteration is continued until the leaf level is reached.

Some criterion obviously needs to be programmed for selecting those Scattergram k-level nodes that indicate the robot's (most likely) approximate position in the world. Choosing the node that contains the highest counter value was found to not always give the best (or even correct) results. Trial and error showed that it was necessary to use a threshold value to be sure that future iterations would always consider the sons that permitted the method to converge to the correct matching of the World and Acquisition quadtrees. This is the only way of overcoming the inconveniences of the quadtree structure mentioned earlier, without the additional complexity of some back-tracking algorithm between tree levels. In this investigation, one ignored those Scattergram nodes having counter values not among the M greatest values. Typically, M varied from 1 to 12. For those situations when the World and Acquisition quadtrees were structurally similar, M tended to be around 12 (implying that differences could only be resolved at the higher resolution levels, k=2,1,0).

The last iteration, at the leaf level, gives 1 or possibly more estimates of the robot's precise location within the world. These same mapping parameters are then used to map the whole Acquisition quadtree into the World quadtree space, thus enabling an update of the world model.

RESULTS

The results of three different (real world) applications are presented in Figures 2 to 4. In each case, the image on the left shows what the range finder is capable of sensing at its current position superimposed on what is assumed to be the known world prior to moving to this location. The symbol "+" indicates the last known position of the range finder in the world. Note that there are some environmental structures which are common to both sets of measurements and that there also exist objects which are not present in both acquisitions. The image on the right shows how the present acquisition is matched to the world representation. The symbol "X" indicates the (new) position of the range finder within the updated world model. No a priori information about the range finder displacement was used.

Execution times for a Compaq 386 PC running under DOS varied from 15 seconds to 5 minutes and depended on the number of measurements in the two quadtrees as well as their structural similarity.

CONCLUSION

This investigation shows that it is not only possible but also quite efficient to use quadtree encoding of range measurements for monitoring the spatial layout of the robot's workspace and for determining the robot's position within its workspace. Certain inconveniences and limitations of the quadtree/octree data structure (such as a fixed origin, rigid spatial boundaries and loss of precision for the arbitrary rotation of nodes) are overcome by considering more than one possible path for descending the tree from levels k to (k-1).

The multi-pass multi-resolution approach used by this method avoids the combinatorial explosion that frequently occurs in matching 2-D patterns. By first starting at a low resolution level (e.g. k=n-1 or n-2) the number of possible $(w_{ik}, w_{jk}, a_{ik}, a_{jk})$ combinations is small. The "pruning" of the Scattergram quadtree at level k also guarantees that the number of possible combinations of the sons will also be relatively small.

Future investigations will concentrate on optimizing the decision of which k-level Scattergram nodes should be retained so as to determine the appropriate sons for the level (k-1) pass. Alternatively, an algorithm that incorporates a back-tracking capability between adjacent tree levels should also be examined to see if it provides a simpler and more robust solution.

REFERENCES

[1] H. Samet, "The Design and Analysis of Spatial Data Structures", Addison-Wesley, 1990.
[2] T.H. Hong and M.O. Shneier, "Describing a Robot's Workspace Using a Sequence of Views from a Moving Camera", IEEE Trans. PAMI, vol.7, no.6, pp.721-6, Nov. 1985.
[3] S.Y. Chen and W.H. Tsai, "Robot Location using Surface Patches of Curved Objects", Int. J. Rob. Aut., vol.4, no.3, pp.123-33, 1989.

[4] H.S. Yang, "Range image analysis via quadtree and pyramid structure based on surface curvature", SPIE, vol.1002, Int. Robots & Comp. Vis., pp.597-608, 1988.
[5] K.D. Rueb and A.K.C. Wong, "Structuring Free Space as a Hypergraph for Roving Robot Path Planning and Navigation", IEEE Trans. PAMI, vol.9, no.2, pp.263-73, Mar. 1987.

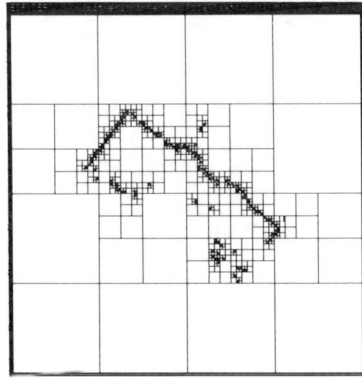

Figure 1. The environment (above) and a quadtree representation of a typical scan of it (left). Note the structural complexity of the environment and the variety of object surfaces that the range finder laser must deal with.

Figures 2-4 (next page). Three examples (top, middle, bottom) showing the World and Acquisition (left) and the updated World (right). "+": last known robot position; "X": robot's present position.

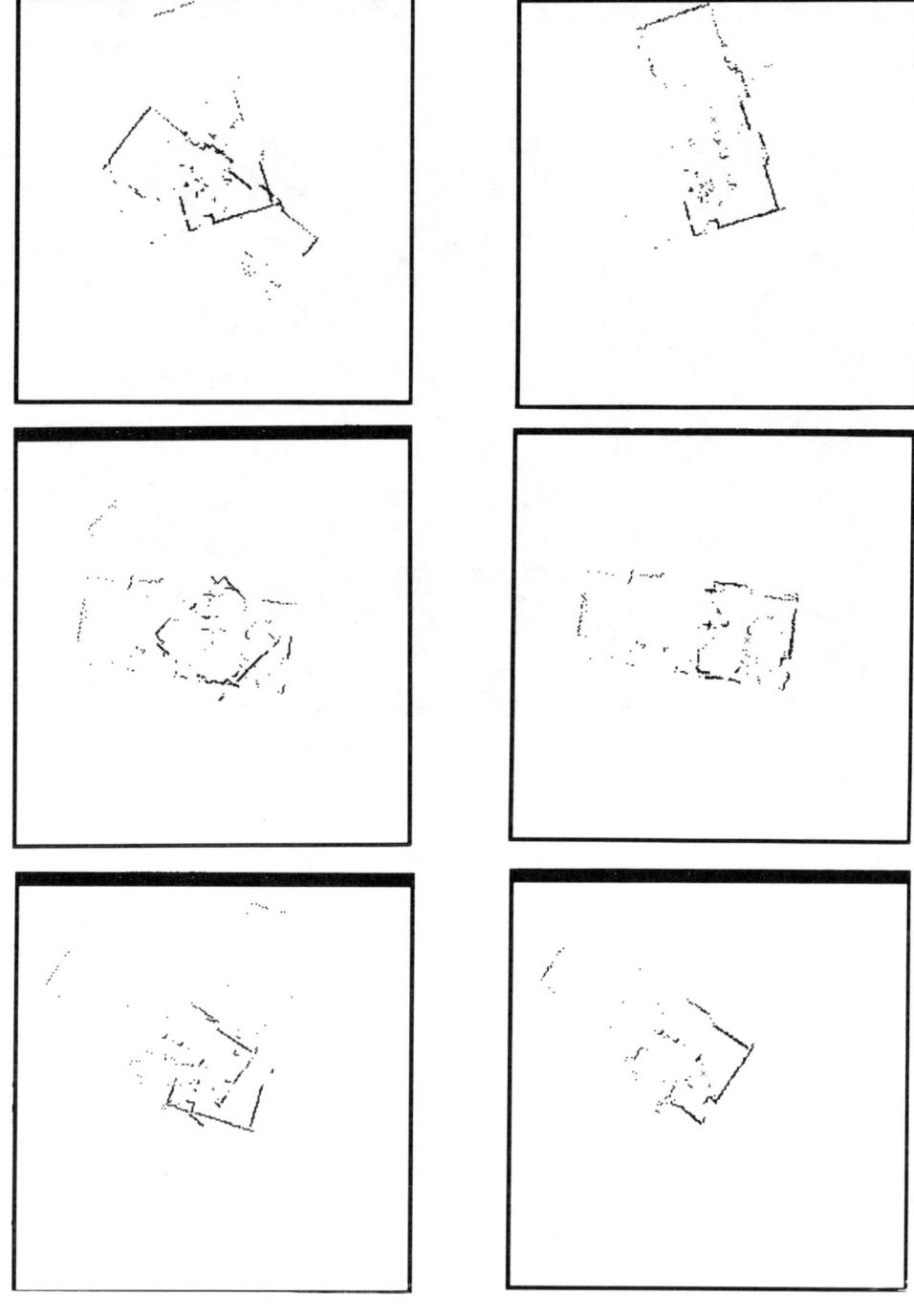

PERCEPTION PLANNING IN PANORAMA

António Miguel de Campos
Maria Margarida Matos
Pedro Fogaça

LNETI - DEE
Estr. Paço do Lumiar 22
1699 Lisboa Codex Portugal

ABSTRACT

In this paper, the general approach to perception planning in the frame of the PANORAMA ESPRIT project is presented. The main goal of this project is the development of a perception and navigation system for mobile robots moving autonomously in partially structured, and partially unknown outdoor environments.

Perception planning consists of two phases: an initial phase, where an analysis of perception opportunities along a planned path is done, to enable the definition of an high level perception plan; and an execution phase where the plan is detailed and monitored. Model-driven perception is used, and attention focusing data structures are associated to each individual perception action command. The high level perception plan is viewed as a list of intentions, that may be changed dynamically as the plan is actually executed.

1. INTRODUCTION

The main objective of the PANORAMA (Perception and Navigation System for Autonomous Mobile Applications) project is the development of a perception and navigation system for autonomous vehicles, which will allow the automatic determination of trajectories, and safe continuous motion in partially structured, and partially unknown outdoor environments.

The performance of the system shall be demonstrated in two real application domains: a forest vehicle used to carry wood logs from an harvesting site to a work station in a road, and a rock drilling vehicle for surface drilling in large open quarries. These applications involve autonomous navigation in outdoor environments for which knowledge about the working environment, and road network, is available and must be used for correct navigation. This information is stored in a global map data base, and includes terrain characteristics and the precise location of certain visual landmarks.

Perception exists to allow obstacle-avoiding navigation and as an aid to the correction of inertial system localization errors. Its main role is to detect the presence of obstacles in front of the vehicle, using a laser range finder, and to recognize the presence of certain visual landmarks, at criteriously chosen times, using a video camera.

It is desirable that perception actions are focussed and model-driven so that the system is not overloaded with unrelevant sensory data and processing. To achieve this objective, attention focusing data structures, which are used to define the recommended conditions to successfully execute the command, and to give clues to help the recognition tasks, are associated to each individual perception action command.

obstacle detection

The focus of attention for the laser scanner must be defined so that it is assured that it can detect potential obstacles sufficiently in advance to allow a safe reaction to its presence (e.g., break the vehicle before reaching it). Notice that, to guarantee this, some constraints must be imposed to the speed of the vehicle (e.g., the vehicle speed must decrease in curves).

landmark recognition

Landmark recognition tasks must be planned to assure that an efficient correction of position and attitude errors, accumulated mainly due to the drift in the inertial-system-based localization system. Landmark recognition is the method used to manage keeping the robot's positional and angular uncertainty, relative to the a priori model of the world, within reasonable limits. When a landmark is recognized, its position in the image restricts the possible orientations, and possibly the spatial position of the robot relative to it. These restrictions can, in certain circumstances, imply a significant decrease in the robot's positional and (or) angular uncertainty.

A landmark data base is used to store convenient model representations for the reference objects. These can be either objects that were already present in the environment (natural or man-made) or special objects that were laid down at precise locations to partially structure the environment. The models are used to predict the geometrical appearance of the objects in screen space, and to ease landmark recognition by the image processing routines, by giving clues for matching.

As the robot's positional and angular uncertainty, relative to the a priori map of the environment, is continuously estimated, it is possible to define the image boundaries within which a landmark should be searched for, so that the processing time is kept to a minimum. The availability of these estimates also enable the prediction of the most efficient sensor orientation and the best robot location to successfully recognize a landmark.

2. PERCEPTION ANALYSIS

When the local navigation system receives information that the vehicle must traverse a specific driving section of an overall path found by a global route planner, an initial coarse path is determined by the local path planner for the whole section. This path will then be used to generate subgoal points for fine path planning and to determine perception opportunities.

obstacle detection

An initial perception analysis step is used to select the areas to be scanned by the laser beam, in order to detect and be able to react to the presence of obstacles. The areas to scan must be such that an obstacle can be detected an the amount of time before reaching it, so that there is time to "see it" (detection time) and to react to it (reaction time), by stopping the vehicle, if necessary.

The distance to the area to scan is computed based on the planned vehicle speed and on estimates of the detection and reaction times. The minimum and maximum pan angles necessary to scan an area limited by the left and right border of the planned corridor are also determined.

The analysis also includes the detection of path locations where the laser beam may touch known objects or where the vehicle is close to a known object.

landmark recognition

To prepare the traversing of a new driving section, the perception analysis step must select observation areas along the planned path from where landmarks can be advantageously perceived by the laser beam or the video camera. The driving section information contains pointers to these landmarks which can be useful during its traversing.

During this step, geometrical reasoning is done, based in sensor models and object and obstacle information, to find out locations from where an object is advantageously perceivable, using the appropriate sensors, while the vehicle moves inside the initial coarse path corridor, computed by the path planner. From these locations the object is not occluded by another object in the map, its camera image has a convenient height, and the expected angle of vision of the object is contained within an convenient observation angle, centered in respect to the robot heading.

The estimated errors in the localization of objects, in respect to the vehicle, due to the accumulated robot localization errors, are taken into account. As each future path position is considered, the position of each object becomes more uncertain. This uncertainty increases by an amount which, in a first approach, is considered to be proportional to the travelled distance.

The search for good visibility areas is done taking into consideration that the robot may be anywhere inside a corridor area centered in, possibly incorrect, trajectory points. So, when a certain perception area is selected, we have strong confidence that the object image will be found.

For each object present in the map, a visibility structure is created which stores information about visibility conditions along the path and will be used to determine the observation areas along the path which correspond to good observation opportunities.

The size of the object image, for a given focal distance, is estimated, and tested to see if its value is within the accepted limits, and if so, the visibility structures are scanned to search for objects whose angle of vision overlaps the one of the current object. Such objects may occlude the current object, or may be occluded by the current object.

The good visibility points in the path are finally aggregated into continuous good perception areas. This enables the determination of useful perception opportunities, at some corridor areas, which will be used to determine the general desired behaviour for the focus of attention for landmark recognition by the video camera.

3. PERCEPTION PLANNING

The local perception planner must compose a high level perception plan, that will be refined at execution time. In this plan, each action actually corresponds to the low level perception goals to try to reach, while navigating through the next section of the environment. It represents a list of intentions, that will probably be changed as the plan is actually executed.

Perception planning must take into consideration the estimated drift in the inertial system, the opportunity constraints produced by the analysis, and the perception strategies and priorities defined for the different tasks and situations. Relevant objects that can be used in the near future to correct localization errors must be identified. Each one of these objects is a potential candidate for observation, which should be regarded as a high priority goal if the localization error estimates exceed a certain threshold.

In the composition of the plan, the results of the perception analysis are checked to select adequate observation locations for each of the perceivable objects. The possible visualization of known objects, or obstacles, by the two sensors or the visualization of two objects in the same video image must be taken into consideration.

The general desirable behaviour for the focus of attention (tilt and pan angles) of the laser scanner is determined using the obstacle detection analysis results. Laser beam actions must be planned to assure that close objects will be properly perceived so that the perception map has enough information to guarantee safe movement. A sequence of laser commands is computed so that new commands are sent only when there is a significant change in the sensor parameters.

4. PERCEPTION EXECUTION AND MONITORING

The execution of the plan is then initiated and monitored. Whenever a convenient location area is approached, each action of the perception plan is detailed. The execution of each perception plan action will, in general, imply the actual execution of a convenient sequence of focused and model-driven perception commands by the sensor management module.

The actual location where the commands are executed, and the degree of success in the perception action are fed back to the planner by the sensor management module. During execution, if new constraints appear, or any of the assumptions about the local environment is proved to be invalid, the perception planner may be asked to redo the planning, partially or completely, to determine a more adequate sequence of low level perception goals.

This is usually the case when an obstacle is detected. It is advisable to consider it as a candidate for observation (e.g. using the laser range finder) if its precise location is relevant to replan the robot's path.

Notice that, if the successful execution of a perception action triggers a significant decrease in vehicle localization errors, this will probably imply a decrease in the priority of the next landmark recognition tasks. The perception planner will possibly then re-write its list of intended high level goals, i.e. the plan, and, if there are no time limitations, a new perception analysis phase may even be initiated to produce new constraints.

focus of attention

Each command is associated to a focus of attention structure, containing information about the recommended sensor orientation and the location to successfully execute the command. In the case of reference object detection tasks, during execution, a data structure, called the predicted model, which gives clues for matching, is associated to each action, at the sensor management level. This structure contains the description of the predicted image features of an object, and the region of sensor space in which data processing routines should be applied for searching and localizing it.

object modelling

Object models, which are stored in a object model data base, are basically implemented as geometric models to which additional features can be attached (e.g. surface properties, and recommended procedures for localization and recognition).

Different model types will be used for different object types (landmark, road, road crossing, beacon). The general object representation uses a polygon mesh, modelling rigid solids by information about its boundaries.

For conical, or cylindric objects, a simple representation based in non-planar polygonal faces is used. These faces correspond to a half model of the objects, which is rotated so that the apparent view of the object is correct. This representation was introduced to enable the fast prediction of the appearance of objects of circular section, which have the same appearance for the observer independently of a rotation over their main axis. Without their introduction, more sophisticated models would have to be used, and these type of objects would have to be treated in a more elaborated way.

feature predicting

Each object present in the a priori map is described by its dimensions, and position coordinates and attitude in relation to a world frame. The values of these parameters are used to produce the matrix transformation to apply to the matrix of points of the model to produce a world model of the object.

After a backface removal algorithm is used to select the visible faces, and a clipping algorithm is applied to reject segments which are behind the image plane, the perspective transformation is applied.

The sensor plane coordinates are determined for each vertex of the clipped face. Then, each line is clipped to the image area. The area of each clipped face, and the coordinates of its centroid, are computed, in pixel coordinates, as the lines are clipped.

region of interest

The predicted model includes the region of sensor space to which the data processing routines should be applied for searching and localizing the predicted features of an object. This region, which is defined taking into account the spatial and angular position uncertainty of the vehicle, in respect to the object under observation, restricts the perceptual processing associated to each landmark to reasonable limits.

5. SIMULATION RESULTS

A Sunview-based simulator was developed to test our approach. A map of the location for the first real demonstration was used. The topview of the environment used can be seen in Figure 1. The path considered consists of a partial tour around a building. The objects in the map include 7 conic beacons , 2 road gates, and a garage building.

In figure 1, the results of the laser scanner analysis are shown. The areas in black correspond to situations where it was detected that the laser beam may touch known objects or where the vehicle may be over or close to a known object. It may be viewed as a recommendation to use the scanner for object analysis or for fine detection. In the popup window, the recommended pan and tilt angles for the laser scanner are shown.

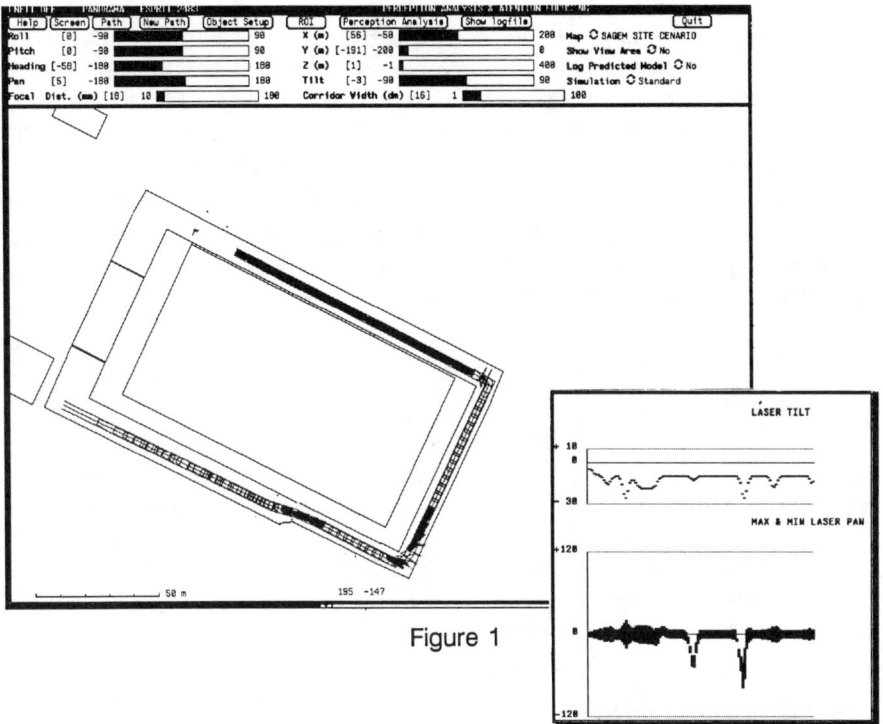

Figure 1

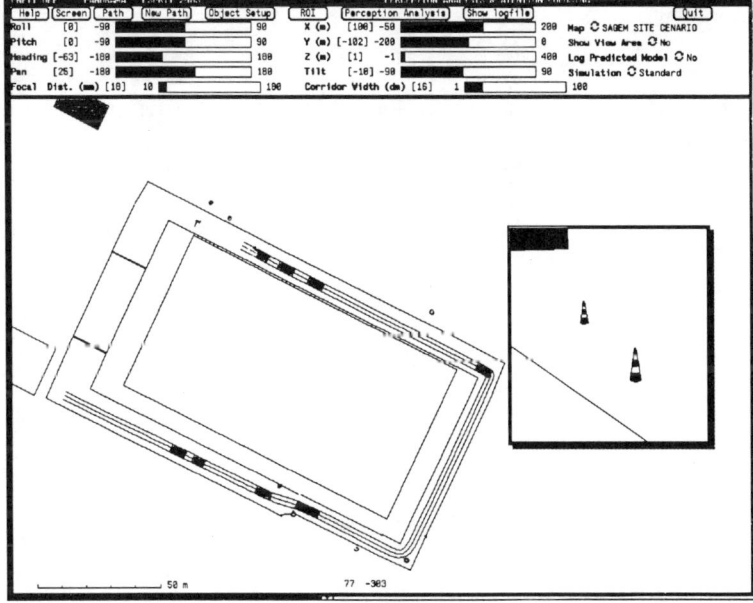

Figure 2

In figure 2, the selected perception areas for the observation of different known objects can be seen. These results correspond to a simulation run in which the estimated errors in the localization of objects, in respect to the vehicle, were not taken into account. In the popup window, the screen image predicted for the focus of attention of the last perception area is shown. It corresponds to a situation where the same image can be used to find two different conic beacons.

If the estimated errors were taken into account, the last three perception opportunities would not exist. In fact, the accumulated errors in the position of the robot after the second curve are too big to guarantee any good observation, if no relocalization is achieved before reaching it.

6. CONCLUSIONS

The general approach proposed to solve the perception planning problem in the frame of the PANORAMA ESPRIT project was presented. It was tested satisfactorily in a SUN workstation, using a Sunview-based simulator, and a first integration of the complete vision cycle was demonstrated at the SAGEM plant, in Osny, Paris, in last April, aboard the REMI vehicle, which is the main testbed for the development of the perception and navigation system for PANORAMA.

In the future, we intend to develop perception strategies to handle obstacle analysis, and landmark and obstacle detection, using a rule based approach. A final implementation solution using an integrated pattern-directed reasoning system for planning is under study.

ACKNOWLEDGEMENTS

The work presented was done with the partial financial support of the EEC, under ESPRIT contract n. 2483.

NAVIGATION AND PERCEPTION APPROACH OF PANORAMA PROJECT

G. Frappier, P. Lemarquand, T. Van den Bogaert : SAGEM (France)

ABSTRACT

This paper presents the general approach and philosophy retained within the PANORAMA project for Navigation, Localization and Perception applied in autonomous mobile robotics. In addition an overview of the project objectives and first achievements will be provided. PANORAMA Navigation relies on a combination of proprioceptive sensors and exteroceptive sensors. This is a balanced approach between inertial dead-reckoning navigation and external means such as GPS-NAVSTAR (Global Positioning System) or visual information as road edges, landmarks or beacons detection. Thus the perception system is mainly working with a prediction / verification strategy. The prediction is possible thanks to the combination of inertial navigation data and "map" data. The PANORAMA general philosophy is to combine and merge in a optimal way the exteroceptive and proprioceptive data in addition of the map data.

1. PANORAMA PROJECT OVERVIEW

PANORAMA is an ESPRIT Project, funded by the European Commission and launched in March 1989, which is intended to prove the feasibility of an autonomous transporting system replacing a man controlled vehicle evolving outdoors in a partially structured environment such as in agriculture, construction sites, forest fields and open mines.

It is planned to construct two demonstrators available in 1993, one of these being an experimental platform derived from a forest forwarder machine supplied by RAUMA REPOLA, the other is dedicated to open mines experimentation and supplied by TAMROCK. These machines should evolve in a real environment as Finnish forest and open mines at the begining of 1994 and should demonstrate autonomous operation.

The Panorama project aims to develop an advanced perception and navigation system for automated industrial vehicles dealing with partially structured and partially known <u>outdoor</u> environments.

The main features of the project are the following :

- a <u>hierarchical</u> decomposition of system control from task planning level to action level,

- a <u>combination</u> of proprioceptive sensing (inertial navigation system, odometers) and exterioceptive sensing (vision, laser range finder) for navigation and guidance purposes,

- an <u>incremental</u> approach of development and integration : a test - bed (4 wheels drive Mercedes) is being set up and should allow a continuous assessment of robot capabilities from simple and restricted situations to more and more complex scenarios.

Panorama is a collaborative project which gathers 7 major european industrial companies associated to 7 famous universities and research institutes, belonging to 6 countries of the European Community and to Finland as an EFTA country : SAGEM (F), BRITISH AEROSPACE PLC (UK), CEA-LETI (F), SEPA SPA (I), EASAMS (UK), TAMPELLA LTD TAMROCK (SF), RAUMA REPOLA (SF), EID (P), LNETI (P), CRIF / WTCM (B), UNIVERSIDAD POLITECHNICA DE MADRID (E), TECHNICAL RESEARCH CENTRE OF FINLAND (SF), HELSINKI UNIVERSITY OF TECHNOLOGY (SF), UNIVERSITY OF SOUTHAMPTON (UK).

2. PROBLEMS AND DIFFICULTIES

The applications concerned by the PANORAMA project are automatic material transportation in forest, agriculture, open quarries, docks, ... and special purpose machines such as harvesters, rock drilling machines, etc. The generic mission has then the following characteristics :

- outdoor environment (partially known),
- mobility on various surfaces such as tared road (with or without white lines), lanes, tracks or open field,
- 3 D terrain (not only flat terrain),

- total travelled distance from 100 m up to several kilometers,
- difficult and various weather and brightness conditions.

In this context, the main purpose of an autonomous mobile robot is to move from one point to another and avoiding obstacles in a cluttered environment. Depending on situations, this environment may be completely known, partially structured or unstructured. This emphasizes the complexity of geometric reasoning which has to manage balancing between a priori knowledge and perceived data, for its geometric representation part, and balancing between optimal control and real time constraints for its motion planning part. In any case, an autonomous mobile robot will have to face this complexity, because it will be impossible for it to have all the a priori knowledge to perform optimal motion planning for, at least, two reasons : first, it is impossible to represent and to store in memory all the enough accurate data at the right level of resolution and, second, the mobile robot has to cope with a dynamic world that continuously modifies the constraints of the problem.

The navigation problem could be split into two categories :

- off-road navigation,
- on road navigation.

The word "road" is taken here as a general meaning for good tared road with white or yellow line marks, tared road without lines, lanes or tracks, that could usually be identify by human eyes and possibly by image processing.

It is clear that in the case of off-road navigation, which is a poorly structured environment, modelling is quite impossible and, as such dead reckoning navigation is the only way to achieve navigation task with the complementary aid of additional perception of landmarks or passive beacons for localization updating. In this case the remaining problem is mainly to detect and to avoid obstacles.

For road navigation, the natural solution is to use an antropomorphic approach only based on artificial vision. But in this case, it is very difficult to built a robust navigation which relies only on road edges detection and works in all weather and brightness conditions.

3. NAVIGATION / PERCEPTION APPROACH

Considering partially known environments, the general approach of PANORAMA navigation is based on a optimal combination of **proprioceptive information** (such as inertial attitude reference, odometer, ...), **exteroceptive information** (such as road/lane edges tracking, landmark or beacon visual detection, Lidar, GPS,...) and **a priori knowledge data base** (mainly containing the map information with roads, beacons data,...). Briefly said, this hybrid and robust solution is to monitor the vehicle trajectory by using the vehicle localization and attitude in order to pick-up in the map data base the relevant information to help and to increase the performance of the perception sub-system. As a feedback localization updatings are then possible. In this way, the perception module (passive and active vision) is considered as a local help working with a prediction / verification strategy. This prediction is possible thanks to the combination of accurate localization / attitude sub-system (using inertial sensors) and the map information. This approach is illustrated on the balance figure below.

On the other side, the INS have one major technical drawback : their accuracy is altered with time. Here again the balance is helpfull since updatings of navigation drifts are possible, after detection of particularly well known beacons or landmarks.

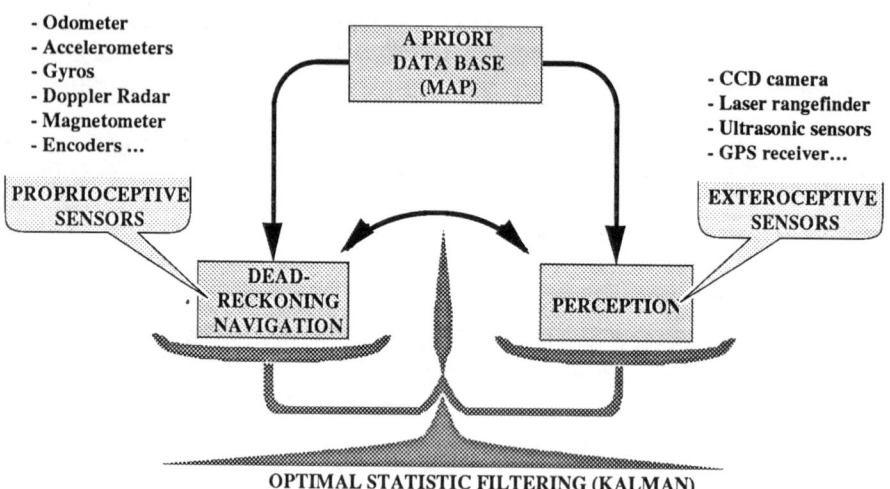

Fig. General balanced approach

The adjustement of the balance is a difficult and sensitive task which closely depends on the data base contents and accuracy. The combination of this three components is practically realized and implemented by using Kalman filtering technics to obtain the best estimation and prediction in the sense of a quadratic criteria.

The PANORAMA approach was retained mainly for two reasons :

- the strong need for accurate enough localization and especially heading orientation and attitude data (roll and pitch) which are needed for outdoors off-road applications but also to help the vision sub-system. In fact, every body wants inertial data and their advantages but at low prices !
- the industrial needs in real applications : robustness and reliability.

4. INERTIAL HYBRID DEAD-RECKONING NAVIGATION

To obtain land vehicle position in a geographical coordinate system, there are two general solutions :

- <u>absolute localization</u> : the principle is typically to use certain known ground points (such as landmarks or beacons) or satellites (GPS, LORAN, ...) to compute the vehicle position with triangulation - like technics. The system gives directly the absolute position, but the accuracy highly depends on the known reference points and on the distances or the angles measurements. The main drawback is the need of a specific and often heavy infrastucture.

- <u>dead - reckoning navigation</u> : that is a method of calculating position with respect to an initial known position by integrating measurements of velocity (module and direction) to obtain position. The inertial navigation is the upper class of dead - reckoning navigation. The major drawback of this navigation process is that the uncertainty grows with the time or/and the travelled distance. Thus the navigation needs to be regularly updated by other means.

In the PANORAMA project, the localization approach is to use dead - reckoning navigation based on gyroscopique heading and attitude (roll and pitch) and distance measurement (like odometers) associated with global positioning system (GPS - NAVSTAR) in differential mode (in open sites) and/or visual landmarks (in forest).

5. PERCEPTION

Perception (composed of laser range finder, vision, ultrasonic and infrared proximeters) is devoted to help the Navigation System based on proprioceptive sensing (inertial system, odometers, doppler radar). The vision tasks are activated upon request of the Planner, thanks to an A Priori Data Base modelling the environment : "Goal - Driven" process using a prediction - verification strategy.

Herefore, the three main Vision tasks are :

- Beacon and Landmark detection (Navigation updating) :

In order to update the navigation drifts of the Inertial Unit, a recognition and localization of visual beacons and landmarks is performed when needed.

- Road Tracking (Local path planning, Guidance) :

Navigating in free fields is a suitable task for the dead - reckoning process, but the location drifts are not allowed when driving on a road or a lane. The road tracking process intends to perform a visual servoing of the Robot on the road edges with the help of the previous stored road data (length, curvature,...) and the vehicle location.

- Obstacle Detection (Safety purpose) :

The indispensable task to be associate to navigation is the obstacle avoidance. The obstacle detection is performed by a continuous scanning mode of the Laser range finder in addition of the classical proximity sensors. Vision in this case will be an help, trying to identify the detected object which is not already mentioned in the A Priori Data Base.

6. GEOGRAPHICAL DATA BASE : MAP

This "A Priori Data Base" is constituted of the road network (in a vectorized presentation including crossroads, road width,...), the crossable surfaces, the landmarks and beacons information (position, orientation, model, size,...).

In order to update efficiently the navigation, it is clear that information with better accuracy than the navigation itself is required. As order of magnitude, a typical INS that can be used for PANORAMA will have a drift of 10 meters after a 1 km mission : then the precision of the landmarks/beacons position must be better than 5 meters (typically one meter) to achieve efficient updatings. Such an accuracy is not so obvious to obtain because the main existing geographical data bases and maps do not usually guarantee such a precision. However some information can be digitized by classical means such as the one used by surveyors or more recently by GPS in phase differential mode. Another approach is to use a learning mode consisting in driving on the roads with an inertial surveying system.

Concerning the road network map, the accuracy needed for road navigation could be not so good than for localization updating. That is to say that 10 meters or more accuracy could be sufficient, depending of course of the level of difficulty in "road" edges detection (good tared road "easy" to detect or bad lane).

7. REALIZATION AND FIRST ACHIEVEMENTS : REMI TEST-BED

The first concretization of this developments is the robotization of a commercial 2 tons, 4 wheels drive vehicle (REMI) which is now fully equipped with actuators, perception and navigation sensors and computing hardware. This vehicle is used as a project test-bed allowing continuous integration and tests of system functions, and it is expected the autonomy capabilities of the vehicle to upgrade continuously as software modules are integrated and refined.

Up to date the first version of the following functions have been successfully tested on REMI : Visual beacons detection (SEPA and CEA), Attention focusing (LNETI), environment modelling (CRIF), Local path planning (CEA), localization (SAGEM), piloting (VTT, CEA, SAGEM), perception sensors management and simple obstacle laser detection (SAGEM). The automated land vehicle features controlled actuation of steering, throttle and brakes and is able to navigate autonomously in a blind mode, outdoors in an industrial site.

The more advanced integration of the on-board part of the Panorama system in the test-bed vehicle is planned by the end of this year (1991) and it is foreseen the vehicle to perform autonomy with environment interaction ability (i.e., obstacle avoidance) in a structured environment by beginning of 1992. In a second stage more and more complicated scenarios will be considered and final system capabilities will be demonstrated within forest and open mine environments by the end of 1993.

Another important issue of the project is an original configuration of the computing hardware architecture which allies advantages of well-known

VME based systems (for interfaces with sensors and actuators) and the high computing power potentialities of Transputer based architectures (parallelisation of high level processing for modelling, planning and control / supervision processes), enabling an easy upgrading of system performances and large flexibility in development.

8. CONCLUSION

The application-oriented approach of the project made us to consider new solutions for navigation and localization of industrial vehicles. This led SAGEM to the definition and prototype realization of a navigation system based on the hybridation of an inertial attitude measurement unit, GPS and processing of cartographic data. As a main component of this navigation system, the issue of a prototype of an inertial strap-down vertical and azimuth reference unit "CARO" (Centrale d'Attitude Robotique), especially designed for robotics, is a relevant spin-off of Panorama and prepare future industrial developments/applications in vehicles automation.

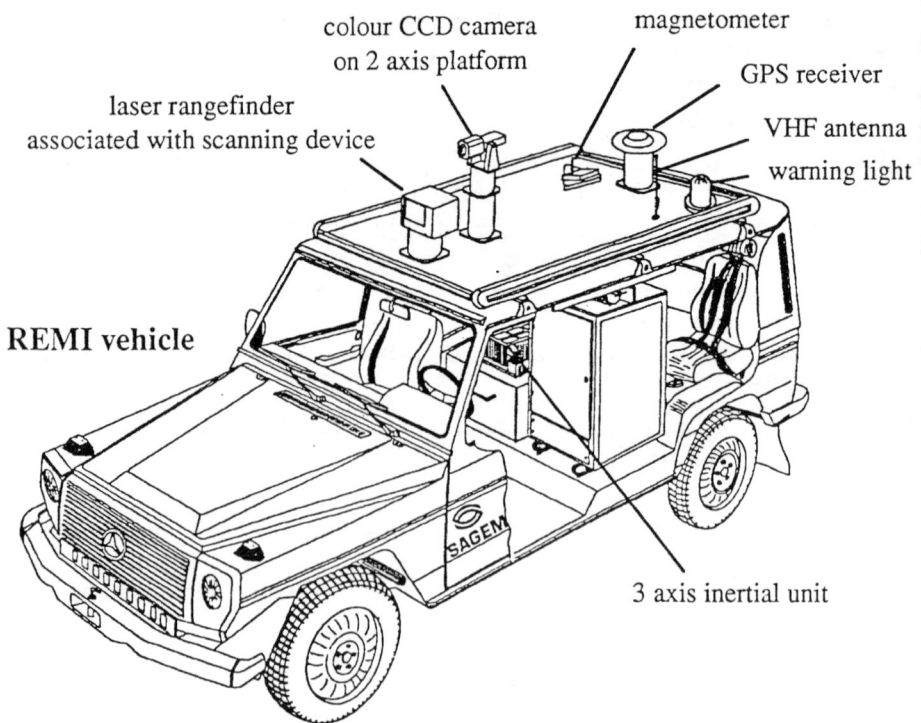

Trajectory generation for mobile robots with clothoids

G.M. van der Molen
Industrial Control Unit
University of Strathclyde
Glasgow, United Kingdom

Abstract

Clothoids are spirals which have a curvature that is continuous with the arc length. This continuity makes them very suitable for use as trajectory segments in mobile robot path planning. The main difficulty in their application lies in the fact that there is no closed expression for clothoids between two given points in a Cartesian space. Some algorithms which provide an iterative, numerical solution to this problem are presented in this paper.

1. Introduction

New advanced free-ranging robots are being developed to replace existing Automated Guided Vehicles (AGVs) which have become common transport systems in (semi-)automated factories. Commercial AGVs often depend on electrical wire guidance or other fixed beacon systems for position determination and data transfer. Modern Flexible Manufacturing techniques demand more flexibility from the transport system in terms of replanning and rerouting facilities than current AGV systems can deliver. The need to replace wire guidance networks or other beacons can be an elaborate and expensive process. It is therefore important that reliable means of positioning are developed for free-ranging robots to compensate for the loss of a (continuous) beacon system. One essential factor is a reliable set of sensors, another is to restrict the allowed command space to commands which the robot is capable of fulfilling. A part of the latter is the selection of the exact trajectory the robot is to follow.

For practical purposes, such as path planning between obstacles, most trajectories consist of a series of consecutive sections. Straight line and circle segments are popular "building blocks" but have the disadvantage that the curvatures in their connecting points are not continuous. Other types of curves which have been investigated as alternatives or coupling sections between lines and circles include clothoids (Kanayama and Miyake [2] and Sadowski [5]) and polynomials of various orders (Kanayama and Hartman [1] and Nelson [3], [4]).

A clothoid (also called Euler's spiral or Cornu's spiral) is a curve with the special property that the curvature is proportional with the arc length. The curvature is defined as the derivative of the orientation (the angle of the tangent of a trajectory with a fixed reference axis) to the arc length, therefore a trajectory with a continuous curvature will result in continuity for the steering angle, smoothing the motions of the vehicle.

In this paper an iterative algorithm is developed for symmetric clothoid segments, with a connecting circle segment, between two postures. It offers the opportunity to plan paths with a higher average curvature than a pure clothoid pair, when keeping the same maximum curvature. This gives better manoeuvrability in cluttered environments, because the required turning space is smaller.

2. Basic definitions, equations and restrictions

The basic mathematical equations upon which the results in this paper are based, are presented in this section. The defining property of a clothoid is that the curvature (the derivative of the orientation α with regard to the distance s) is a linear function of the distance :

$$\frac{d\alpha}{ds} = c(s) = c_0 + ks \qquad [2.1]$$

It follows that the orientation α is a quadratic function of s :

$$\alpha(s) = \alpha_0 + c_0 * s + \frac{k}{2}s^2 \qquad [2.2]$$

The basic relations between the distance (or arc length) s and the Cartesian coordinates x and y are the differential equations [2.3] and [2.4] as illustrated in Figure 1 :

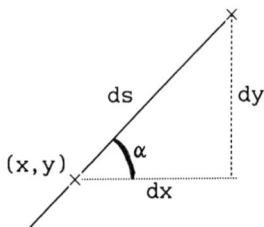

Figure 1

$$dx = \cos \alpha \, ds \Leftrightarrow \frac{dx}{ds} = \cos \alpha \qquad [2.3]$$

and

$$dy = \sin \alpha \, ds \Leftrightarrow \frac{dy}{ds} = \sin \alpha \qquad [2.4]$$

or in integral notation :

$$x(s) = x_0 + \int_0^s \cos(\alpha_0 + c_0*w + \frac{k}{2}*w^2) \, dw \qquad [2.5]$$

and

$$y(s) = y_0 + \int_0^s \sin(\alpha_0 + c_0*w + \frac{k}{2}*w^2) \, dw \qquad [2.6]$$

Other useful equations are derived from these basic equations; the most important are:

$$[2.1], [2.2] \Rightarrow c_{e1} = \frac{2(\alpha_1 - \alpha_0)}{s_1} - c_0 \qquad [2.7]$$

$$[2.1], [2.2] \Rightarrow k = \frac{2(\alpha_1 - \alpha_0 - c_0*s_1)}{s_1^2} \qquad [2.8]$$

For clothoids with a zero starting curvature ($c_0 = 0$) and fixed end orientation α_1 the Cartesian end coordinates x_1 and y_1 are linear in the arc length s_1 (and therefore in each other), or:

$$x(s) = x_0 + a_c*s \qquad [2.9]$$

and

$$y(s) = y_0 + a_c*b_c*s \qquad [2.10]$$

where a_c and b_c are dependent on α_1.

Some restrictions are placed on the start and end postures of the trajectory sections to make the problem more tractable; the start and end curvatures c_0 and c_e are taken to be zero:

$$c_0 = c_e = 0 \qquad [2.11]$$

and it is assumed that:

$$-\pi < \alpha_e - \alpha_0 < \pi \qquad [2.12]$$

There are two special cases. If $\alpha_e = \alpha_0$ the clothoid found is a straight line, or two clothoid pairs are necessary. If $\alpha_e - \alpha_0 = \pm \pi/2$, $\tan(\alpha_e - \alpha_0)$ does not exist. However, with some slight changes to the equations involved this problem is easily overcome.

3. Algorithm for symmetric clothoid pairs

The simplest case in planning clothoids between two postures is a combination of two clothoids and one or two straight trajectory segments. In this case a symmetric clothoid pair connecting two points on the tangents through the postures can be planned. The main variables in the algorithm are illustrated in Figure 2:

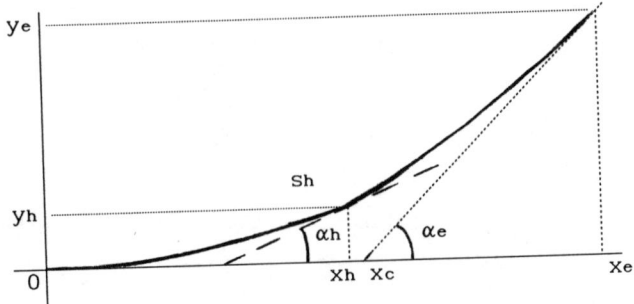

Figure 2

The start of the algorithm is to locate the intersection point (x_c, y_c) of the tangents through the start and end postures. The next step is to choose the start and end points (x_0, y_0) and (x_e, y_e) of the clothoid pair on the above tangents, symmetrically with regard to the intersection point. A coordinate transformation is then performed to simplify the notation in the remainder of the algorithm; in the new coordinates $(x_0, y_0, \alpha_0) = (0, 0, 0)$. The result is that:

$$(x_c, y_c) = (x_{ee} - y_{ee}/\tan \alpha_e, 0)$$

$$= (\frac{1 + \tan^2 \alpha_h}{2} x_e, 0) \qquad [3.1]$$

In the following step the arc length s_h of both clothoids is determined, together with the sharpness k and the coordinates (x_h, y_h) of its end point (thus the starting point of the second). In the first place point (x_h, y_h) must lie on the line through (x_c, y_c), with an angle of $\pi/2 + \alpha_h = \pi/2 + \alpha_e/2$ with the x-axis because of the symmetry of the clothoid pair, see Figure 2. As equation:

$$x_c = x_h - y_h/\tan(\pi/2 + \alpha_h) = x_h + y_h * \tan \alpha_h \qquad [3.2]$$

The other relation between x_h and y_h arises from the property that x and y are linear functions of s when clothoids with a fixed end orientation are considered, as in [2.9] and [2.10]. In the new coordinate system equations [2.5] and [2.6] simplify to:

$$x(s) = \int_0^s \cos\left(\alpha_1 \frac{w^2}{s^2}\right) dw \qquad [3.3]$$

and

$$y(s) = \int_0^s \sin\left(\alpha_1 \frac{w^2}{s^2}\right) dw \qquad [3.4]$$

With $\alpha_1 = \alpha_h$ substituted, these equations can be solved for one specific value of s, for example s = 1, which then yields the coefficients a_c and b_c. From [2.9] and [2.10] it follows that :

$$y_h = b_c * x_h \qquad [3.5]$$

which combines with [3.2] to :

$$x_h = \frac{1}{1 + b_c * \tan \alpha_h} x_c \qquad [3.6]$$

and together with [2.9] and [3.1] :

$$s_h = \frac{1 + \tan^2 \alpha_h}{2 * a_c * (1 + b_c * \tan \alpha_h)} x_e$$

$$= \frac{1}{a_c * (1 + \cos \alpha_e + b_c * \sin \alpha_e)} x_e \qquad [3.7]$$

The sharpness k can now be found with equation [2.8] :

$$k = \alpha_e / s_h^2 \qquad [3.8]$$

and the maximum achieved curvature is c_{e1}, as in [2.7].

4. Algorithm for symmetric clothoid pairs with a circle segment

A trajectory section consisting of a symmetric clothoid pair with a circle segment in between may be preferable over a pure clothoid pair because the total path length is smaller. This can be advantageous either for reasons of efficiency (fuel saving) or to limit the required manoeuvring space. The curvature of the circle segment is assumed to be given by a fixed value c_c.

The principle of the algorithm for the determination of the clothoids is that a clothoid arc length s_1 is to be found, such that the end posture of the combination clothoid pair - circle segment is the desired end posture. The boundary condition is that the end curvature c_{e1} of the first clothoid is the circle curvature c_c. The algorithm starts by choosing c_c and clothoid arc lengths $s_{min} = s_1$ and $s_{max} = s_2$, with $x_e(s_1) < x_e$ and $x_e(s_2) > x_e$, for example $s_1 = 0$ (on-the-spot change of curvature) and $s_2 = \alpha_e/c_c$. These two values of s are required to assure convergence of the regula falsi iteration.

The next step is to compute the end posture (x_1, y_1, α_1) of the first clothoid segment with length s_1 ; subsequently the end coordinate $x_e(s_1)$ of the trajectory section is computed by :

$$x_e(s_1) = 2\cos \alpha_h * (x_1 * \cos \alpha_h + y_1 * \sin \alpha_h + \frac{1}{c_c}\sin(\alpha_h - \alpha_1)) \qquad [4.1]$$

When $x_e(s_1)$ is known, the regula falsi iteration interval $[s_{min}, s_{max}]$ can also be updated. The derivation of equation [4.1] can be illustrated by Figure 3 :

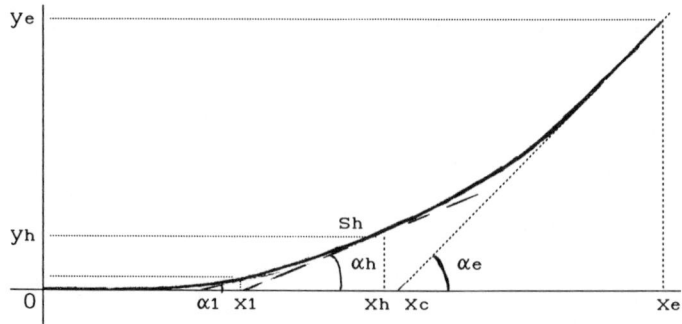

Figure 3

and follows from straightforward geometric considerations.

The objective now is to find an s_i such that the end coordinate $x_e(s_i)$ of the clothoid pair is equal to the desired end coordinate x_e :

$$f(s_i) \equiv x_e(s_i) - x_e \qquad [4.2]$$

$$= 0 \qquad [4.3]$$

This also assures that $y_e(s_i) = y_e$, because of the symmetry of the desired end coordinates and the clothoid pair with regard to (x_c, y_c). As there is no direct method of computing the appropriate s_i, an iterative algorithm is used. To have both guaranteed and fast convergence a combination of the regula falsi and Newton's methods is implemented. If the new "Newton" iterate falls within the current "regula falsi" interval $[s_{min}, s_{max}]$, this value is taken as the new $s_{(i+1)}$ and the interval is updated, otherwise a regula falsi iteration step is taken. For Newton's method the derivative of $f(s)$ to s must be known. It follows from [4.1] :

$$f'(s_i) = \left.\frac{df}{ds}\right|_{s_i} = \frac{1}{s_i} \cos \alpha_h \, (x_1 * \cos \alpha_h + y_1 * \sin \alpha_h) \qquad [4.4]$$

Next, the new Newton iterate is calculated :

$$s_{(i+1)} = s_i - \frac{f(s_i)}{f'(s_i)} \qquad [4.5]$$

If the new iterate lies outside the current regula falsi interval, an alternative new iterate is calculated with the regula falsi method :

$$s_{(i+1)} = s_{max} - \frac{s_{max} - s_{min}}{f(s_{max}) - f(s_{min})} f(s_{max}) \qquad [4.6]$$

The algorithm then continues with the next iteration, until convergence is sufficient.

5. Boundaries for trajectories with clothoid pairs

For obstacle avoidance in the trajectory generation stage of path planning the envelope formed by the robot around the trajectory should be free of objects. It is assumed that the environment and the robot are represented in a polygonal world model. In such a model it is convenient if a reasonable approximation of the area occupied by the vehicle driving a clothoid can be given by straight lines and circle segments.

The case best suited to finding these "enveloping" boundaries is that where the trajectory consists of a symmetric clothoid pair with a connecting circular arc of specified curvature. For simplicity it is assumed that the robot motion can be completed by driving forward only.

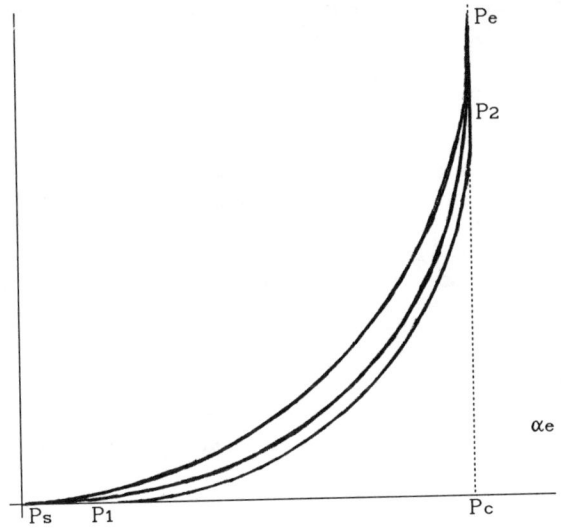

Figure 4

Also, the total turning angle α_e is restricted : $-\pi < \alpha_e < \pi$, and $\alpha_e \neq 0$. The start posture is $(0,0,0)$ and the start and end postures P_s and P_e are taken symmetrical with respect to the intersection point of their tangents $P_c = (x_c, 0)$, because any connecting straight lines would be retained in the approximation.

Let c_{min} be the curvature of the circle connecting P_s and P_e, with the circle tangents in these points parallel to the posture orientations as in Figure 4. Then the circle segment curvature c_c is larger than c_{min}, and two new points P_1 on the line section $P_s P_c$ and P_2 on $P_c P_e$ can be found so that a circular arc with curvature c_c contains P_1 and P_2, and $P_s P_c$ and $P_c P_e$ are its tangents in those points. The clothoid pair with circle segment connecting P_s and P_e is then bounded by the lines $P_s P_1$ and $P_2 P_c$ and the circular arcs $P_1 P_2$ with curvature c_c and $P_e P_s$ with curvature c_{min}. These two lines and circular arcs thus provide an area which, if it is free of obstacles, guarantees that the clothoids-with-circle trajectory is collision-free, without having to compute the trajectory exactly in advance.

6. Conclusions

Under certain conditions regarding the positions of the start and end points of a trajectory section, a connecting symmetric clothoid pair with an additional straight line segment can be calculated in a one-step algorithm. A more complex, iterative algorithm has been developed for the case of a symmetric pair with a circle segment in between. This provides shorter paths with a smaller maximum curvature. All algorithms assume that the curvatures at the start and end of the trajectory sections are zero. This condition is easily satisfied when the trajectory is planned off-line for a known environment, but not when on-line replanning is done in an environment with unknown and possibly moving obstacles.

7. Acknowledgements

The author would like to acknowledge the support of the ESPRIT programme of the European Communities, as the work for this paper was done as part of ESPRIT-II project 2043 "MARIE" (Mobile Autonomous Robot in an Industrial Environment), and the fruitful cooperation with the other partners in the project. Thanks especially to M.A. Johnson for the suggestions for improvements of the paper.

8. References

[1] Y. Kanayama and B.I. Hartman, "Smooth local path planning for autonomous vehicles", Proc. IEEE Conf. on Robotics and Automation, p.1265-1270, 1989.

[2] Y. Kanayama and N. Miyake, "Trajectory generation for mobile robots", Robotics Research : The Third International Symposium, ed. O. Faugeras and G. Giralt, Cambridge, Mass., p.333-340, 1986.

[3] W.L. Nelson, "Continuous steering-function control of robot carts", IEEE Tr. on Industrial Electronics, vol.36, no.3, p.330-337, August 1989.

[4] W.L. Nelson, "Continuous-curvature paths for autonomous vehicles", Proc. IEEE Conf. on Robotics and Automation, p.1260-1264, 1989.

[5] R. Sadowski, "Aspects of the traditional approach on parallel parking", Internal MARIE document WP4-BOS-WOP-900605, 1990.

Digital models for autonomous vehicle terrain - following

N. Christou (*), K. Parthenis (*), B. Dimitriadis (*), N. Gouvianakis (*)

Abstract

Terrain elevation digital data are potentially useful for a variety of applications. Research is already under way to anticipate the ways in which digital elevation data can be made most useful for particular applications.

Invariably, the availability of terrain digital data has shown that this information can be effectively used to support Earth resource analysis through simulation modelling techniques. Furthermore, digital terrain elevation and feature data displays are particularly useful for mission planning analysis due to the realism, efficiency, and flexibility possible with advanced processing and display technology.

Perspective visual simulation is perhaps one of the most effective display methods since it creates a natural appearing image, allowing the user to closely associate geographic data with the real world.

Currently, we are developing techniques and software modules that allow us to extract digital terrain models from conventional contour maps which combined with satellite imargery data can create realistic perspective 3-dimensional terrain scenes. These 3-dimensional terrain scenes are examined as models for an autonomous vehicle terrain following system.

Specific radar or electro-optical sensors may be digitally modelled to produce sensor simulations using digital elevation and feature data. Sensor simulations are valuable to mission planning and may serve as enroute navigation check points. Algorithms for trajectory planning of a terrain autonomous vehicle are also being developed.

1. Introduction

This article attempts to examine some of the aspects involved in the analysis and development of what could be a completely passive method of navigation, which derives motion information from digital land and/or sea floor terrain elevations.

(*)   HITEC S.A. 18 Poseidonos Str., GR 176 74 Kallithea, Athens

The availability of terrain digital data has shown that this information can be effectively used to support earth resource analysis and evaluation. Furthermore, digital terrain elevation (DTE) and feature data displays are particularly useful for mission planning, efficiency, and flexibility, possible with advanced processing and display technology.

Specific radar or electro-optical sensors may be successfully integrated with a - priori knowledge of digital elevation and feature data which can be valuable elements to mission planning and may serve for enroute navigation check points. In general, the underlying principles in such planning operations are based on recursive real-time registration of sequential images of the changing surface scene beneath the moving vehicle. This position - velocity estimate combined with other conventional on-board sensors (e.g., altimeters, compasses) and sparce storage of digital terrain elevations (DTEs) can constitute a passive autonomous terrain - following navigation system.

Currently, we are developing techniques and software modules that allow us to extract digital terrain models from conventional 1:50,000 contour maps which we combine with imagery data to create realistic perspective 3-dimensional terrain scenes, which later can be examined as models for autonomous vehicle terrain - following systems. Two general areas of application of such systems we deem important are land and sea-floor relief selective, mapping, and collision avoidance and rescue salvage systems.

2. Digital Terrain models

A digital terrain model (DTM) is a numerical representation of terrain surface and can be used to depict the topography of the land or sea-bottom in the computer. The most basic form of a DTM is a collection of discrete data points with known position and height or depth which can be distributed randomly or arranged regularly over a rectangular grid. Hence, there are two major types of DTMs, random DTM and regular DTM (over a rectangular or square grid).

It is useful to have a mathematical surface defined over a DTM. Such a surface allows the generation of 3-D perspective views, the calculation of volume integrals, automatic contouring, slope and aspect transformations, relief shading, solid modelling and other analysis of the terrain surface.

A regular DTM has the following advantages:

- the data structure of a regular DTM is simple. The most commonly used method to represent it is a matrix structure (image-like structure), i.e., it does not require the storage of every x,y coordinate of each grid node, but only the x,y of one corner point and grid step size.

- it is extremely easy to obtain profiles parallel to the coordinate axes (image rows or columns) or profiles making 45° angle with either one of the implied axes.

- it is easy to generalize by consolidating grid cells (image resampling)

- it allow fast processing operation to generate derivable products (low, medium and high level image processing image statistics, etc.)

However, a regular DTM has the following disadvantages:

- it is not adaptive to the distribution of data points. Unlike a random DTM for which the spacing of the points can be adjusted to acccomodate the nature and complexity of the terrain surface, the spacing of the data points in a regular DTM is fixed. An obvious solution in this case is to use progressively finer grids at areas of steep gradients, very mulch like the recursive subdivision schemes used in quad-tree structures.

- most of the times the original terrain data points are not regularly spaced and hence one or another form of interpolation is required. As a result, the original data points are rarely included in the formation of a regular DTM, which can be considered as containing error. This situation is invariably appearing when a regular DTM is obtained from digitizing (or scanning) contour maps.

- as a result of the previously stated limitations, a regular DTM does not handle terrain - specific lines (such as breaklines) as easily as a random DTM.

From the above discussion, it is obvious that the regular DTMs are easier to store, manipulate and graphically depict; they render themselves for faster visualization techniques and are "second-hand" surfaces suffering from lower fidelity in terrain representation.

The generation of DTMs is useful not only for contour generation and depiction of terrain relief on maps, but also for their direct usefulness and applicability in problems such as coverage maps for radar operations, intervisibility of transmitting equipment in telecommunications, etc.

Currently, we are generating DTMs from 1:50,000 topographic relief maps. The procedure we employ involves the following basic steps:

STEP 1: Use of a CCD camera to capture the contour map excerpt in an image frame, which is subsequently enhanced (using an enhance - 9 type digital filter) and stored as a 256 levels gray-tone image in a computer compatible file (CCF).

STEP 2: An Image edge - enhancement procedure follows in order to accentuate the individual contours (in conjuction with procedures that reduce the noise level in the image).

STEP 3: A line-thinning procedure is followed by a line - following one and finally a line connectivity procedure allows us to generate a vector file containing ifnormation on the individual contours (x,y and h values).

STEP 4: The map-contour image is coregistered with an image of the road network extended over the same area using a 6-parameter transformation, a set of control points and a resampling (rectification) algorithm for the superposition of the second image (road network) on the contour one. The road network image generation has also followed the same steps as STEP 1, 2 and STEP 3 above (except for the height information).

The role of the road-network file at this point is to facilitate the simulation of a predetermined navigation trajectory for route planning.

STEP 5: From the vector contour file a regular DTM file is generated using as grid size the original pixel-map size and two dimensional price-wise polynomial surface interpolation techniques. The generated DTM is used to obtain 3-D perspective views and relief shading of the terrain surface. The last stage (i.e., the 3-D perspective is not yet fully implemented in our library system).

STEP 6: Although not yet implemented, this step involves the generation of intervisibility maps from the digital elevation models. The plan is to generate a suite of maps for low-, medium - and high - intervisibility maps.

3. Navigation & Guidance concepts:

Two of the most commonly and most frequently asked questions are: "Where am I?" (NAVIGATION) and "Which way can I reach my destination?" (GUIDANCE). In todays era, man has ventured forth on ships, land vehicles, aircraft, and spacecraft. Each one of these classes of vehicles has evolved a distinctive complement of Navigation and Guidance equipment and algorithms.

Navigation and Guidance are two very closely related functions (operations). The notion of Navigation used in the present article is confined to the measurement of position (absolute or relative, on-board or external, real-time or post-mission). The notion of

Guidance is used in the context of determining the required directional information to steer without necessarily calculating absolute position. What is worth noting right from the beginning is that, usually, guidance systems are much less expensive than navigation systems.

Some of the most critical parameters involved in navigation and guidance are: cost, accuracy, autonomy, spatial coverage, time coverage and sampling, and human interfaces.

COST: The total cost of a navigation sytem apart from users receiving equipment includes the construction and maintenance of transmitters, the preparation and dissemination of charts and correction tables and the construction and maintenance of monitor/control stations.

ACCURACY: It is the most important attribute of any system. Users' needs and application areas dictate the required or desired accuracy levels of a system. "Differential techniques" (in which a stationary surveyed station broadcasts its measured offsets that are applied by nearby receivers to correct their own measurements or location fixes) improve accuracy, generally, by an order of magnitude.

AUTONOMY: Autonomy may conveniently be subdivided into five classes:

    a. Passive self-contained systems (users neither receive nor transmit signals, e.g., inertial navigators)

    b. Active self-contained systems (users emit but do not receive externally generated signals, (e.g., sonar, radar devices);

    c. Active radio navaid systems (users exchange signals with navigation stations, e.g., distance measuring equipment, tactical information distribution systems, position location reporting systems, collision - avoidance systems);

    d. Passive radio navigation systems (users receive signals from ground - based and/or space-based transmitters but do not emit signals, e.g., LORAN, OMEGA, VOR, TRANSIT, NAVSTAR / GPS).

    e. Natural navigation aids (e.g. magnetic compass, stars, landmarks).

SPATIAL COVERAGE:
> According to their geographic extent of coverage, navigation systems may be classified as line-of-sight (LOS) systems, regional and global coverage systems (beyond LOS).

TIME SAMPLING AND COVERAGE:
> Some navigation systems give continuous indications of position and velocity; others are intermittent. They might have a time-delay build-in by nature (scanning of a radar beam) or adjustable measurement sampling interval, or post-fitted position solutions (from minutes to many hours depending on the application).

HUMANINTERFACE:
> Navigation has become automated as on-board computing facilities increased. Hence, instead of requiring many human skills on each vehicle, a smaller number of highly specialized personnel is needed. However, still in most manned vehicles, humans select the way points that define a route and humans monitor equipment performance and failures.

Measured terrain-altitude profile match to a stored profile, is one of the most promising techniques for terrain - following. The measured profile can be obtained by subtracting radar altitude from barometric altitude.

Today, for low-altitude aircraft flights used to avoid detection by surface - based radars, a manual terrain - following system displays pitch cues on a head-up display (HUD) based on airborne radar scanning of peak terrain ahead of the aircraft.

Terrain - following technology has advanced enough to permit an airborne radar to measure terrain profiles at several azimuths and issue pitch an azimuth commands. Such a system is the "Terrain Following - Terrain Avoidances" system (TFTA).

TFTA can choose the best path and best vertical profile for maximum concealment, subject to a pilot selected risk of ground contact.

4. Route planning principles

In order to support planning operations where knowledge and information about the real world is imprecise, and/or uncertain, or partially/totally lacking, one must combine evidence provided by multiple, diverse knowledge sources. Digital land or sea-floor terrain models (DTMs), landmarks and imagery availability can be a possible solution for developing effective plans for navigation.

Drawing conclusions from consideration of appropriate information acquired by sensor data and prestored maps of varying degrees of reliability, precision and completeness requires evidential reasoning. A DTM, for example, can be effectively used for constructive an intervisibility map of a geographic area, depicting maximum height above ground level or minimum heigh (all the way down to the ground). An illustrative example is the calculation of the potential risk associated with flying over a rapidly changing mountainous terrain under very bad visibility conditions using a perstored DTM.

Route planning principles are currently examined and analysis will soon follow.

Navigation of a Mobile Robot

P. van Turennout & G. Honderd
Delft University of Technology
Department of Electrical Engineering
Control Laboratory
P.O. Box 5031, 2600 GA Delft, the Netherlands

Abstract
Navigation of a mobile robot consists of three elements: sensor data processing, planning and trajectory control. The sensors provide information about the internal state of the robot and they perceive the environment with respect to the robot. With this information a path is planned or replanned and finally this path is executed by the control. This process as a whole is complex and, as a consequence, slow. To avoid this, a fast low-level navigator is implemented consisting of position estimation and trajectory control. Most known mobile robots use dead-reckoning algorithms for the calculation of their position. These methods rely on relative data, being the distance travelled and change of orientation of the robot between two samples. From these data the position coordinates can be obtained by use of a numeric integration algorithm. Many variants of these algorithms exist and some of them have been tested in simulation upon accuracy versus computational complexity. The trapezoid rule proved to be well suited here. Still, notorious error sources such as unknown wheel radii, misalignment and slip of the driving wheels contribute to erroneous navigation. To eliminate the dependence upon relative data, the use of absolute position measurements seems appropiate. This is a task of the high-level part of the navigation, since it is more complex and asks for time consuming sensor data processing. Also, reference points (beacons) or a detailed map of the environment should be available. Matching the sensor data with the environment is a slow process compared to the driving speed of the robot. Therefore, the robot must rely on its dead-reckoning for some time.

1 Introduction

To accomplish an autonomous behaviour of a mobile robot the cooperation is needed between 1) a navigator, which does all the planning and control functions, 2) the mechanical structure of the robot itself, including the actuation and 3) the sensor system available to the robot. These three elements are linked together in a closed loop, as depicted in figure 1. The robot structure and the sensor system are briefly

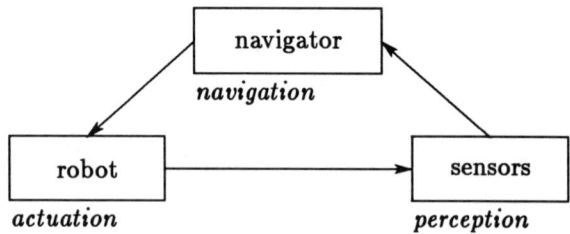

Figure 1: *The three cooperating elements in an autonomous mobile robot.*

considered here, the navigator is explained in the next section. The type of mobile robot that is considered here, is a wheeled mobile robot. The wheel configuration consists of driven and/or sensed wheels and wheels for support, either on fixed or steered axes. The actuation system is thought of as a servo system with speed setpoints for the driven wheels as inputs. The sensors can be divided into two types: those measuring the internal robot states (wheel speeds by tachometers, wheel angles and steering angles by encoders, for example) and those perceiving the environment with respect to the robot. The latter is done by vision systems, distance measuring devices (laser range finders, ultrasonic or infrared sensors) and even collision detectors. Some basic sensor data processing to eliminate noise or image distortion is supposed to be included.

2 The navigator

This is the decision-making level. This implies that all information needed to make the decisions is concentrated here. The navigator has two main functions. First, a planning and control function is needed to calculate an output to the actuation system must be provided, corresponding to a specified task by an operator and based on sensor data inputs. Second, for a robot to really behave autonomously, or even intelligently, it must learn from its actions. Map building is a good example of this. The main attention goes to the planning and control function. Also, a division of the navigator in hierarchical levels is considered.

2.1 Functions of the navigator

For the planning and control function, the navigator receives tasks or commands from an operator (not necessarily human). It also receives information from the sensor system. The final output consists of control outputs to the actuation system. Between the input and output three processing stages can be distinguished (see figure 2): sensor data processing, path planning and trajectory control. Also shown in this figure is a large data base containing all relevant information. This

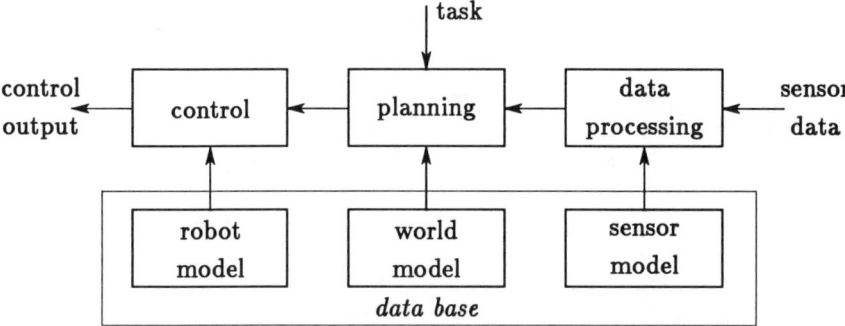

Figure 2: *Three processing stages in the planning and control function.*

information is accessible for all three processing stages; the connections indicated in the figure are not exclusive for each stage.

The sensor information must be processed to obtain information relevant to the planning. This implies the knowledge of sensor characteristics and also the placement of the sensors on the robot. This kind of knowledge is gathered in a sensor model. With this knowledge can be compensated for sensor specific properties (e.g., the beamwidth of ultrasonic sensors) and it tells *how* relevant information can be taken from the data (e.g., feature extraction in images). It is also important to indicate a measure of uncertainty or tolerances in the data: how reliable is it for further processing and does this reliability increase or decrease with time.

The processed sensor data enters the planning stage. With the estimated robot position, the local map and with use of the map stored in the world model (either an a priori given map or one built from local maps) a path is planned (or replanned) with respect to the given task. Once the path, or at least a part of it, has been established, it has to be transformed into a trajectory. Here, boundary values of the robot parameters have to taken into account. Knowledge about the present position, maximum accelerations, speeds and steering angles is given by the robot model. The trajectory may be specified as a set of velocity profiles and desired positions. The final part is to control the robot with respect to this trajectory.

The second function of the navigator is to maintain and, if possible, update the data base. Three models have been mentioned: the sensor model, the robot model and the world model. When an element in one of these models is incorrect, the navigator should invoke a learning mode. In this mode the omission in the model should be corrected by use of the other models and new data. For example, when a perfect robot position estimate can be made and there is little or no uncertainty in the local map from the sensor data, this local map can be added to the global world model. As another example, a local map matched with the global world model provides an update to the position estimate (see for example Durrant-Whyte, 1989).

However, in practice the uncertainty is spread over all three the models: neither

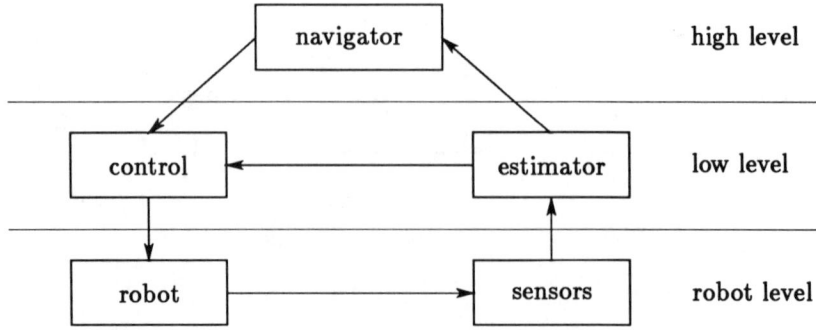

Figure 3: *Division of navigator into two hierarchical levels.*

the robot parameters are known to calculate the position, nor the sensor data is perfectly reliable, nor the world model is known exactly. Only if the uncertainty of one of them is significantly larger than the others, it can be reduced by means of this kind of adaptation. Often this is indicated by the term learning. It does not necessarily imply the use of artificial intelligence, but it can also be implemented as Kalman-filters, for example.

2.2 Hierarchical levels

The scheme of figure 2 has one large disadvantage: it is too slow. Every sample a control output has to be caculated from the sensor data via the planning stage. Of course, all available information is taken into account to arrive at an optimal (re)planned trajectory and a corresponding control output, but this need not be necessary at every sample. Furthermore, it would require a huge processing capacity to do it, or the robot speed must be reduced such that a larger sample time becomes acceptable. For this reason the navigator is splitted into a fast low-level part and a slower high-level part (figure 3). Basically, the low-level part consists of the trajectory controller and a position estimator. This level acts as a fast inner loop. The high-level part operates parallel to the low-level part. It provides the controller with the latest desired trajectory and the estimator with position updates. This may be done at a slower rate. The remaining part of this paper deals with the trajectory controller and the position estimator.

3 Trajectory control

The trajectory of a mobile robot can be specified by velocity profiles for the translation speed $\underline{v}$ and the rotation speed ω. The translation speed is characterized by a vector in the Cartesian space. For an omnidirectional robot the speed vector can

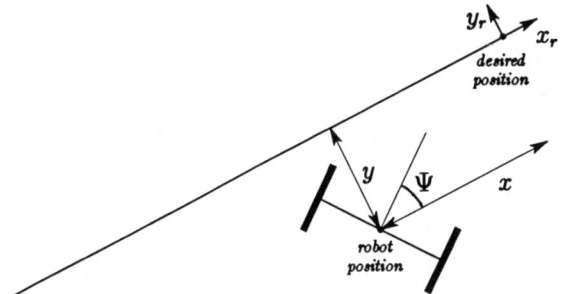

Figure 4: *Definition of robot coordinates.*

point in any direction, whereas for other, more conventional, robots this vector is fixed to the robot orientation.

Feeding these profiles to the robot as input to the servo system of the motors (after a transformation to desired angular velocities of the driven wheels) results in the desired trajectory, but only in the ideal case. Disturbances acting on the robot, and in particular on the contact between the driven wheels and the floor, cause deviations in the robot path. This can be corrected for when also the desired positions are specified. The robot position $\underline{x}$ is given by two coordinates x and y in the plane and its orientation Ψ: $\underline{x} = [x\ y\ \Psi]^T$. The difference $\underline{\varepsilon}$ between the desired and measured robot position is fed back by a gain matrix K. The total control vector $\underline{u}$ becomes

$$\underline{u} = \underline{v}_{\text{desired}} + K\underline{\varepsilon} \qquad (1)$$

where $\underline{v}_{\text{desired}}$ is a vector containing the desired velocities. Similar approaches are used by Song et. al. (1989) and Cox (1991).

By placing the origin of the coordinate system in the desired robot position, the actual robot location can directly be interpreted as the position error with respect to the trajectory. This is illustrated in figure 4 for the case of a straight line as the trajectory. The translation speed vector can be decomposed into a component v_L aligned with the robot orientation and a component v_P perpendicular to the robot. The latter is usually neglected for a conventional (i.e. non-omnidirectional) robot. The time derivatives of the robot coordinates can be expressed as

$$\dot{x} = v_L \cos \Psi - v_P \sin \Psi \qquad (2)$$
$$\dot{y} = v_L \sin \Psi + v_P \cos \Psi \qquad (3)$$
$$\dot{\Psi} = \omega \qquad (4)$$

or, written in a compressed form,

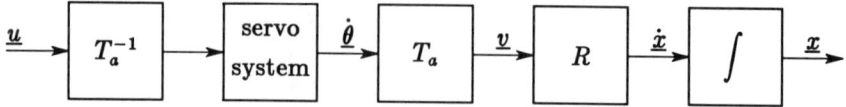

Figure 5: *Robot model for the actual position.*

$$\dot{x} = R\,v, \quad \text{with } R = \begin{pmatrix} \cos\Psi & -\sin\Psi & 0 \\ \sin\Psi & \cos\Psi & 0 \\ 0 & 0 & 1 \end{pmatrix}, \quad v = \begin{pmatrix} v_L \\ v_P \\ \omega \end{pmatrix} \quad (5)$$

Equations (2) and (3) express the so-called nonholonomic constraints of a mobile robot. These constraints are non-integrable. To calculate the control u according to equation (1), the position error $\varepsilon = x$ is needed. This error can not be measured directly, but must be reconstructed from sensor data. A simple way to get an estimate of the robot position is by means of dead-reckoning.

4 Dead-reckoning

Consider a mobile robot with actuators and sensors on its wheels. The actuators and sensors need not be collocated, but encoders and motors may be on the same axes. Kinematic equations describe how the angular velocities of the wheels relate to the robot translation and rotation speeds. The actuation is given by

$$v_a = T_a\,\dot{\theta} \quad (6)$$

with $\dot{\theta}$ the vector of angular velocities of the wheels and T_a is the map representing the actuator configuration. The robot speed is denoted as v_a to indicate the *actuated* (or actual) speeds. In the same way, the speeds as they are *sensed*, are given by

$$v_s = T_s\,\dot{\theta} \quad (7)$$

where T_s is the map representing the sensor configuration. The speeds as they are sensed, should preferably equal the actual speeds, i.e. $v_s = v_a$. Muir and Neuman (1987) have given a methodology to obtain the maps T_a and T_s. In the case of collocation, T_a and T_s are equal.

Figure 5 shows how the actual position of the robot results from the control input u. The calculated control u is transformed by the inverse of the map T_a to get the control signals to the servo systems of the actuated wheels. The robot speed $\dot{x}$ is obtained from the angular velocities $\dot{\theta}$ by the map T_a and the rotation matrix R from equation (5). The position x is the integrated value of these speeds.

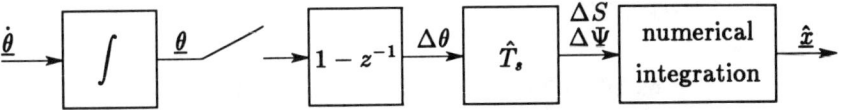

Figure 6: *Calculation of the estimated robot position.*

The goal is now to obtain a position estimate $\hat{x}$ from the sensor data. Dead-reckoning, or odometry, is a well-known method for this. The wheel angles $\underline{\theta}$ are measured by encoders. The placement of the encoders is given by the map T_s. In figure 6 is illustrated how the encoder data is processed to get a position estimate. The difference between two samples of the encoder data is taken to obtain the change in wheel angles $\Delta \underline{\theta}$. In the figure, z^{-1} is the delay operator over one sample. The displacement ΔS and the change of orientation $\Delta \Psi$ are calculated from the mapping $\hat{T}_s$. This is in fact the discrete equivalent of equation (7). The implementation of this mapping uses an estimate $\hat{T}_s$, since it contains parameters like wheel radii and distances between wheels which are only approximately known. These parameters must be obtained by calibration. The mapping $\hat{T}_s$ can even become nonlinear due to effects like slip and misalignment of wheels. The latter, for example, causes the robot to behave differently for clockwise and counterclockwise changes of orientation. It also causes a drift speed perpendicular to the robot orientation (the speed component v_P). At the moment, this component and the resulting displacement are neglected, leaving the displacement ΔS only in the longitudinal direction.

The final step in the calculation of the estimate $\hat{x}$ is the discrete equivalent of the integration of equation (5). It has already been stated that these equations are non-integrable for the x- and y-coordinates. Therefore, this has to be approximated by a numerical integration algorithm. Many different algorithms exist, ranging from fairly simple (first-order Euler method) to highly complex (high-order variable-step methods). Since the dead-reckoning has been introduced in the fast low-level part of the navigator, complex (hence slow) implementations should be avoided. Also, the total accuracy is determined by both the mapping $\hat{T}_s$ and the numerical integration approximation. The algorithm should balance accuracy versus complexity. In a comparitative study (Van der Oord, 1990) the trapezoid rule proved to be well suited. From equations (5) this results in

$$\hat{\Psi}(k+1) = \hat{\Psi}(k) + \Delta \Psi \tag{8}$$
$$\hat{x}(k+1) = \hat{x}(k) + \tfrac{1}{2}\Delta S (\cos \hat{\Psi}(k+1) + \cos \hat{\Psi}(k)) \tag{9}$$
$$\hat{y}(k+1) = \hat{y}(k) + \tfrac{1}{2}\Delta S (\sin \hat{\Psi}(k+1) + \sin \hat{\Psi}(k)) \tag{10}$$

Each new sample, only one sine and one cosine need to be calculated together with some multiplications and summations. The sine and cosine can be digitally calculated very fast by a CORDIC-algorithm (Volder, 1959). This special algorithm can even be implemented in hardware.

5 Final remarks

The navigation of a mobile robot incorporates a variety of tasks: sensor data processing, map matching, path planning, world modelling, trajectory generation and control. The trajectory controller and position estimator are the basic tasks for the robot being able to drive accurately.

Dead-reckoning is an open-loop position estimator and therefore suffers from cumulative errors. Also, an error in the initial position can not be corrected. By a good calibration of the mapping $\hat{T}_s$ and a careful implementation of the numerical integration, large errors can be avoided. In an implementation for the mobile robot at our laboratory, an accuracy is obtained of 1 cm in longitudinal direction and 2 to 3 cm in perpendicular direction after a 30 m straight drive. The robot has two independently driven wheels with collocated encoders.

Still, an update of the position by means of an absolute position measurement remains necessary. A first attempt at this will be made. With the availability of straight walls in the environment it becomes possible to incorporate position updates by use of ultrasonic sensors. While the robot is driving along a wall, these sensors continuously measure the distance to this wall. In fact, the y-coordinate of the robot is measured directly. From a series of measurements also the orientation relative to the wall can be estimated.

References

[1] Cox, I.J., Blanche — An Experiment in Guidance and Navigation of an Autonomous Robot Vehicle, *IEEE J. of Robotics and Automation*, Vol. 7, No. 2, pp.193–204, april 1991.

[2] Durrant-Whyte, H.F. and J.J Leonard, Navigation by Correlating Geometric Sensor Data, *Proc. IEEE/RSJ Int. Workshop on Intelligent Robots and Systems*, pp. 440–447, 1989.

[3] Muir, P.F. and C.P. Neuman, Kinematic Modelling of Wheeled Mobile Robots, *J. of Robotic Systems*, Vol. 4, No. 2, pp281–340, 1987.

[4] Oord, P.J.C van der, Navigation algorithms for the mobile robot, Delft University of Technology, Control Laboratory, report T90.071, 1990 (in Dutch).

[5] Volder, J.E., The CORDIC Trigoniometric Computing Technique, *IRE Trans. on Electronic Computers*, Vol. EC-8, pp. 330–334, september 1959.

[6] Song, K.T., J. De Schutter and H. Van Brussel, Design and Implementation of a Path-Following Controller for an Autonomous Mobile Robot, *Proc. Int. Conf. on Intelligent Autonomous Systems*, pp.253–263, 1989.

ROBOT NAVIGATION AND EXPLORATION IN AN UNKNOWN ENVIRONMENT

Raashid Malik and Samuel Prasad
Dept. of Electrical Engineering and Computer Science
Stevens Institute of Technology
Hoboken, NJ 07030, USA

Abstract - Path planning algorithms are concerned with providing sequences of robot motion commands which enable a robot to efficiently travel from its current position to a specified destination without colliding with environment obstacles. Two distinct categories of such robot path (or motion) planning strategies exist. In the first category complete knowledge of the environment is assumed including the position and shape of all obstacles. In the second category little or no knowledge of the environment is required. This paper is concerned with issues that arise in the latter category.

The problem of traversing from one point to another in an unknown environment is actually part of a larger problem. Emphasis in the literature has often been placed on discovering algorithms that always complete the task (if indeed the task can be completed). Scant attention is paid to the fact that in typical robot environments, the task (or variations of it) often has to be repeated. Our contention is therefore that not only should path planning algorithms converge but simultaneously learn as much as possible about the environment. Hence the initial traversal may be inefficient but all subsequent traversals will not; because the problem will be transferred into the first category where optimal shortest path strategies have been developed.

We refer to the environment learning task as mapping. In this paper we describe a mapping algorithm based on touch or proximity-range sensors. The algorithm has been implemented in a simulated environment and the results of these simulations are presented.

I. INTRODUCTION

A significant difference between flora and fauna is the ability of animals to wander about and encounter new experiences. The evolution of superior sensory capabilities in animals, in comparison to plants, is probably a consequence of this "wander-ability". Superficially at least, autonomous mobile robots resemble creatures from the fauna whereas robot manipulators, which are rooted to factory floors bear similarities to species in the flora. The mobile robot, because of its "wander-ability", must have the capability of sensing the environment in its vicinity. In addition, this robot needs the faculty to make "reasonable" decisions based on its sensory input. Some researchers [1] have argued and demonstrated that extremely limited intelligence, associated with elementary sensor decisions, can make mobile robots emulate the seemingly complex behaviour observed in insect preambulations. Insects, like other animals, are presumably driven by instincts and stimuli (e.g. danger or hunger). Likewise most mobile robots are driven, or programmed, to achieve specific objectives or accomplish pre-defined tasks. One may thus question how much "intelligence" is required to carry out some of the mundane tasks required of mobile robots. Probably the most fundamental operation required of a mobile robot is the traversal from its current position to some given destination. Lumelsky [2] has shown that a point automaton, armed with nothing more than a primitive touch sensor and with no prior knowledge of its environment is capable of reaching a given destination (if this destination is in fact, reachable). The robot motion (or navigation) algorithms used in [2] are selected with a single criterion - guaranteed convergence, i.e. the algorithms, regardless of the shape or number of obstacles in the robot's path, move the robot to its destination (whenever possible). It has been further shown [3] that including more sophisticated (vision or range) sensors on the robot does not guarantee a more "efficient" path to the destination, when compared to a robot with a simple touch sensor. What appears to be lacking in the motion behaviour of these automatons is "wisdom". Wisdom is often associated with knowledge and memory whereas intelligence is more often linked to perspicacity that is characterized by the skillful use of any available knowledge to efficiently acheive one's goals. Assume one has partial knowledge of the environment; how does one incorporate this knowledge to make the robot motion more efficient in terms of time taken or distance travelled to reach the destination? Reaching the destination may be the current and immediate objective; however there may be other goals which could be more efficiently achieved by learning and remembering the obstacles in the robot's path. We use the term "wisdom" to refer to the process of acquiring and utilizing any partial knowledge about the environment. An objective of this paper is to attempt to make robot navigation in unknown environments a bit wiser (but hopefully not offensive).

II. BACKGROUND

Robot navigation in known environments has been extensively studied *[4], [5]*. The robot path to a destination can be pre-planned and optimized in a completely known environment, whereas such planning and optimization is not possible in unknown environments. *Navigation* in this context is mostly concerned with algorithms that specify motion directives which enable the robot to reach its destination. These motion directives do not always result in an optimal path because the environment is unknown and only information about the immediate vicinity of the robot is available. No attempt is made to learn the environment. Another approach is to first learn the environment completely and then apply the navigation methods of known environments. This learning process is referred to as *terrain-acquistion* in *[6], [7]* or *mapping* in *[8], [9]*. The goal in *navigation* is to reach a **given destination**, whereas in *mapping*, the objective is to learn the shape and location of all objects, surfaces or walls in a **given region**. The paramount objective in the *navigation-driven* approach to robot motion planning is reaching the destination; and environment sensing becomes an incidental activity *[10]*. The *mapping-driven* approach on the other hand proceeds to first completely learn the environment before attempting to chart a path to the destination. The second method is the subject of this paper. Since the *mapping-driven* method also allows flexibility in defining the region of interest, it becomes equivalent to the *navigation-driven* method when the region of interest is confined to a nominal path to the destination, such as the M-line described in *[2]*.

III. MODEL

A. Region Envelopes: If the robot is confined to a closed region such as an apartment, the natural envelope of the region is the boundary of the closed region (Fig. 1a). There may however be restrictions on the operational region of the robot, which would necessitate defining a virtual boundary or an envelope (Fig. 1b).

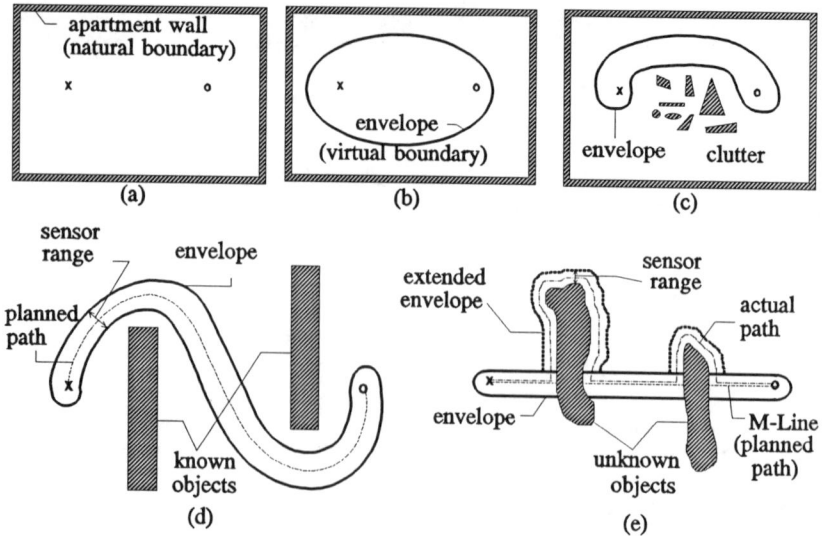

Fig. 1. Mapping Regions with (a) natural boundaries, (b) restricted range, (c) clutter, (d) partially known region (navigation-driven) and (e) unknown region (navigation-driven) (x - initial robot location, o - possible destination.

The concept of an envelope may also be used to avoid traversing (or mapping) through cluttered or inconsequential regions (Fig. 1c). Similarly by limiting the breadth of the envelope to the visual range of the robot sensors, the robot motion can be transformed from *mapping-driven* to *navigation-driven* (Fig. 1d). This concept can be used in unknown environments as well, by defining the envelope to stradle the M-line and allowing envelope extensions around any obstacle intersecting the M-line (Fig. 1e).

Mapping is not possible without precise knowledge of the robot's location. A sensed surface can be correctly charted on the map only if the sensed point on the boundary is accurately known relative to some coordinate system. Robot position knowledge allows the envelopes to function as virtual boundaries restricting the motion of the robot, whereas natural boundaries (walls and objects) are avoided by using the robot sensors.

B. Mapping Regions: The mapping strategy we use is based on systematically expanding the rim of the explored region. Our region is assumed to be two-dimensional consisting of either free or occupied space. Free space is that portion of the environment in which the robot is allowed to maneuver, and consists of explored and unexplored regions. Occupied space also consists of two types: object space and forbidden space. The relationships between these space types and their associated boundary types are shown in Fig. 2a.

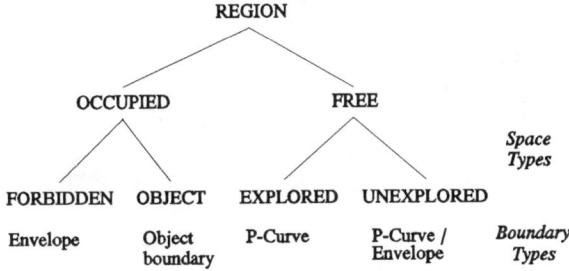

Fig. 2a. The environment map is defined in terms of the space types. These space types are enclosed within the boundary types indicated.

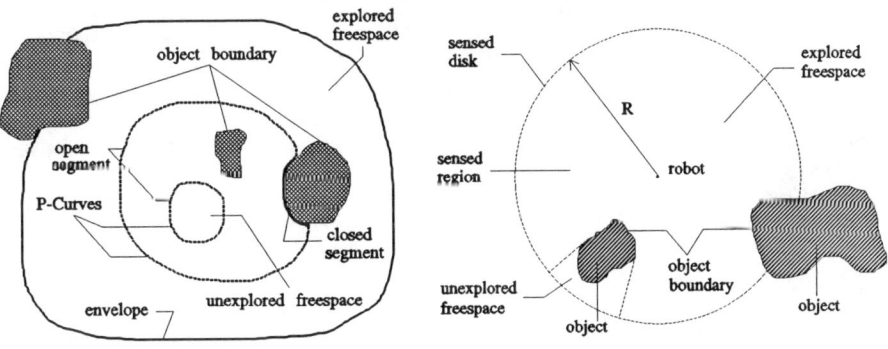

Fig. 2b. Map illustration. *Fig. 2c. Sensed Region.*

Forbidden space is the area on the far side of the envelope that is beyond the operational range of the robot. Explored space is that part of the environment which has been completely sensed by the robot. The boundary of this connected area is called the peripheral curve (P-Curve). The explored space may contain pockets of unexplored space, resulting in a number of distinct P-Curves as shown in Fig. 2b. A

P-Curve consists of a connected sequence of *open segments* and *closed segments*. The open segment of a P-Curve separates explored and unexplored space, whereas a closed segment separates explored space and occupied space - object or forbidden.

C. Robot and Sensor Models: The robot is assumed to be a point automaton, moving about in free space and capable of sensing its surroundings within a disk of radius R, which defines the extent of the sensed region accurately conveyed to the robot. In conforming with constraints on vision and range sensors we assume that information concerning free or occupied (object) space is obtained only radially from the sensor (Fig. 2c).

IV. MAPPING PROCESS

The mapping process proceeds by extending the outer P-Curve. The robot moves along an open segment of this curve. Its movement allows it to sweep over and sense unexplored space, while itself remaining inside explored space. A new P-Curve is constructed by extending (where possible) the *current P-Curve* R units outward into freespace. The *new P-Curve* is marked *open* whenever the sensed disk intersects unexplored freespace, and marked *closed* whenever occupied space is sensed (or determined from the position/envelope system) as shown in Fig. 3a. This method allows the explored freespace to grow and fill a defined envelope. Exceptions to the process are discussed below.

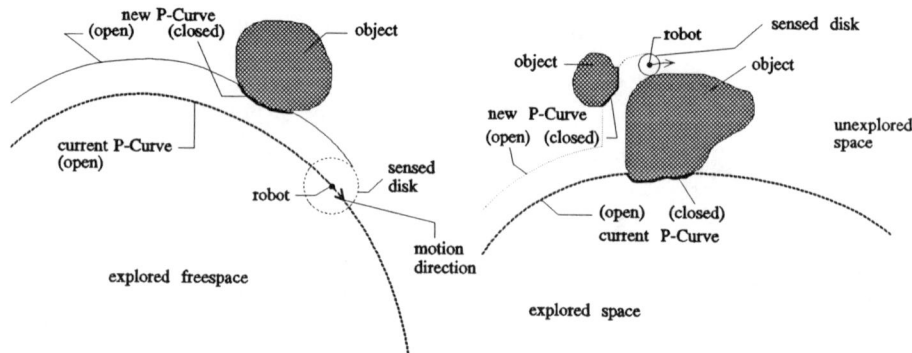

Fig. 3a. Creation of a new P-Curve on an open segment of the current P-Curve.

Fig. 3b. The new P-Curve created due to a closed segment of the current P-Curve.

When the robot approaches a closed segment of the current P-Curve (indicating the presence of an object or envelope) the robot resorts to a *boundary-hugging* mode. In this mode, while following the boundary of the object, the new P-Curve is extended to encompass the entire object. The process is illustrated in Fig. 3b.

Creation of the new P-Curve is complete when the robot returns to the original starting point. If the new P-Curve consists entirely of a single closed segment the mapping process is essentially complete unless other P-Curves with open segments exist.

A. Exceptions: The normal extension of the explored region is modified whenever the new P-Curve intersects itself. Some of these situations are illustrated in Fig. 4a, where the new P-Curve is shortened and/or an auxiliary P-Curve is also created. Note that auxiliary P-Curves enclose regions of unexplored freespace. These regions are explored by the robot after the main P-Curve is completely closed.

A closed segment of the P-Curve may represent the boundaries of multiple objects or be part of the boundary of an object with many closed segments on the current P-Curve. When there are multiple objects then the situation is handled by circumnavigating each object in turn. Note that the presence of multiple objects in a closed segment is indicated by discontinuities in the curve representing that segment. Construction of the P-Curve is illustrated in Fig. 4b.

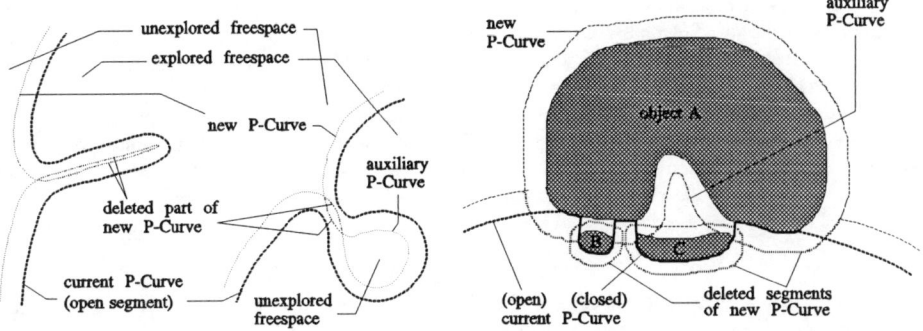

Fig. 4a. *The looped portions of the new P-Curve are deleted to extend outwardly the explored region; Auxiliary P-Curves are created when pockets of unexplored regions exist.*

Fig. 4b. *New P-Curve construction due to multiple objects.*

B. *Algorithm:* The robot motion algorithm (**MAPPER**) is shown in Appendix A1. The process periodically activates and deactivates a concurrent process referred to as **CHARTER**. **MAPPER** is concerned only with robot motion, i.e. traversing the current P-Curve on open segments and hugging object boundaries in closed segments. **CHARTER** is concerned with generating the new P-Curve and the construction of the environment map. The main algorithm, for clarity has been written without invoking the envelope concept, hence will only terminate in a fully enclosed region. Appendix A2 shows the added statements required in **MAPPER** to incorporate the envelope concept. When multiple objects occur in a closed segment then the procedure in Appendix A3 is employed to circumnavigate these objects in sequence.

V. SIMULATION

A preliminary simulation of the mapping process was conducted. The robot environment was represented as a binary matrix, whose element values were either *zero* (freespace) or *one* (object-occupied space). The robot's sensor was modelled as an elementary proximity sensor capable of detecting occupied or freespace in the immediate vicinity of the robot, defined by the matrix elements adjacent to the element representing the position occupied by the robot. The sensed information was incorporated into an evolving partial map represented by a matrix whose elements could take on any one of the four values: zero (explored freespace), one (explored object-occupied space), E (envelope or forbidden space) and ? (unexplored space).

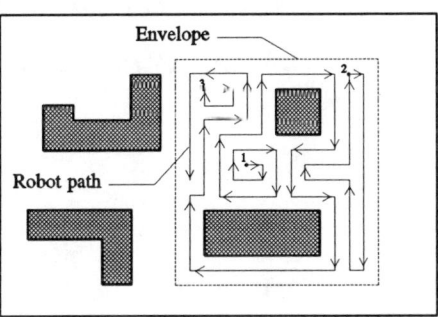

Fig. 5a. *Mapping-Driven Approach.*

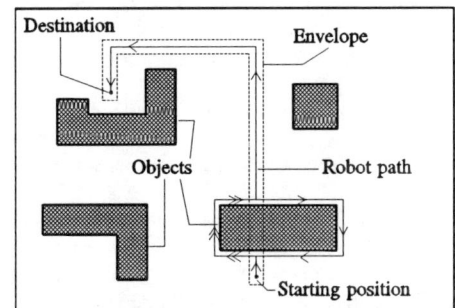

Fig. 5b. *Navigation-Driven Approach.*

A number of arbitrary environments were constructed (in matrix form) and the robot's initial location in freespace was randomly chosen. In all cases **MAPPER** accomplished the mapping task successfully. Results of simulations in which the robot was made to operate in *mapping-driven* and *navigation-driven* modes are shown in Figs. 5a and 5b respectively.

In Fig. 5a, the robot path generated by **MAPPER** in a *mapping-driven* mode is shown. This is quite different from the *navigation-driven* approach (Fig. 5b) where mapping is not the primary objective, rather reaching the destination from an initial location is. This is achieved by confining robot motion to a narrow area (defined by an envelope) containing the initial location and destination.

VI. CONCLUSION

We have described a method for specifying robot movement in an unknown environment for the primary purpose of learning the shape and location of all objects in a region. The method allows for restricting the mapped region to areas of interest using imaginary envelope boundaries to limit the area explored by the robot. The mapping process was simulated in a computer model to demonstrate the utility and effectiveness of the procedure.

We are currently working on proof methods to ensure that the mapping algorithm accomplishes its task in any arbitrary 2-D region. We are developing deatiled boundary-following motion algorithms for implementation on a mobile platform.

APPENDIX A1

The following data structutres are used by **MAPPER**, the robot motion algorithm described in this paper.

a) P-Curve - A list of points (xy-coordinates) defining the *open and closed segments* make up a P-Curve. A P-Curve may contain open and/or closed segments. Let the jth $(0 \leq j \leq m_j)$ closed segment be represented by (H^j, L^j), where H^j and L^j are the starting and ending points of the closed segment respectively. A closed segment may or may not contain discontinuities (indicating the presence of multiple objects). Let n_j $(0 \leq i \leq n_j)$ be the number of discontinuities. When $n_j = 0$ there are no discontinuities and there is only one object between (H^j, L^j).
b) P-List - A list of pointers to the main and auxiliary P-Curves.

MAPPER:
1. Initialize - Set P-list to initial robot location. The P-list now contains one P-curve made up of a single-point *open segment*. Go to Step 2.
2. Examine P-List.
   2.1. If P-List is empty terminate procedure. **Mapping is complete.**
   2.2. If P-List is not empty set the first P-curve on the list to be the current P-curve. Go to Step 3.
3. Examine current P-Curve for open segments.
   3.1. If there is at least one open segment, move robot to any point on the open segment. Mark this point as *s*. Go to Step 4.
   3.2. If there are no open segments, delete current P-Curve from the P-list. Go to Step 2.
4. Begin *charting*. Go to Step 5. /* activate the **CHARTER** concurrent procedure */
5. Move clockwise along the open segment of the P-Curve until one of the following occurs.
   5.1. The robot reaches *s*. Terminate *charting*. /* deactivate **CHARTER** */
      Delete current P-Curve from the P-list. Add new P-Curve(s) created as a result of the *charting* process to the P-list. (*¹). Go to Step 2.
   5.2. The robot reaches the jth closed segment (H^j, L^j) containing $0 \leq i \leq n_j$ discontinuities.
      If i = 0 (single object)
         Circumnavigate the object if the object was not previously visited. Mark object as visted.
      Else

PROCESS-CLOSED-SEGMENT.
EndIf.
Terminate *charting*. /* deactivate **CHARTER** */
Move robot to L^j . /* end of the closed segment */
Add new P-Curve(s) created as a result of the charting process to the P-list. (*²).
Go to Step 4.

APPENDIX A2

The following modifications to **MAPPER** are required to support the envelope concept, including these additional data structures.

a) E-Point - A point (xy-coordinates) representing the intersection between the envelope and a closed segment.
b) E-List - A list of E-Points.

Modifications and additions to **MAPPER**:

i) Replace Step 2.1 by: If P-List is empty go to Step 6.
ii) Insert at *¹: Add any E-Points to E-List.
iii) Insert at *²: Add any E-Points to E-List.
iv) Add Steps 6, 7 and 8:
    6. Examine E-list.
       6.1. If E-List is empty terminate procedure. **Mapping is complete**.
       6.2. If E-List is not empty mark the first E-point on the list as e_s . Go to Step 7.
    7. Circumnavigate object until robot reaches e_s . Go to Step 8.
During circumnavigation, if robot is within envelope perform *charting*.
    8. Robot is now at e_s .
Add any new P-Curve(s) created as a result of the charting process to the P-list.
Add any E-points to the E-list. Go to Step 2.

APPENDIX A3

PROCESS-CLOSED-SEGMENT:
For $k = 0$ to n_j do
    Move the robot to H_k^j (if it is not already there).
    Check if H_k^j is on an object that was previously not visited.
    If yes, then circumnavigate the object on which H_j^k is located. Mark object as visited.
EndFor

REFERENCES

[1] Rodney A. Brooks, *"A Hardware Retargetable Distributed Layered Architecture for Mobile Robots"*, Proc. IEEE Robotics and Automation, April 1987.
[2] V. J. Lumelsky, *"A Comparative Study on the Path Length Performance of Maze-Searching and Robot Motion Planning Algorithms"*, IEEE Trans. on Robotics & Automation, Feb 1991.
[3] V. J. Lumelsky, Tim Skewis, *"Incorporating Range Sensing in the Robot Navigation Function"*, IEEE Transactions on Man, and Cybernetics, October 1990.
[4] T. Lozana-Perez, *"Spatial Planning: A Configuration Space Approach"*, IEEE Trans. on Computers, Feb 1983.
[5] Phillip John Mckerrow, *"Introduction to Robotics"*, Addisson Wesley, 1990.
[6] V. J. Lumelsky, S. Mukhopadhyay, K. Sun, *"Dynamic Path Planning in Sensor-Based Terrain Acquisition"*, IEEE Trans. on Robotics & Automation, Aug 1990.
[7] Nageswara S. V. Rao, S. S. Iyengar, B. John Oommen, R. L. Kashyap, *"On Terrain Model Acquisition by a Point Robot Amidst Polyhedral Objects"*, IEEE Journal Of Robotics and Automation, Aug 1988.

[8] Raashid Malik, Samuel Prasad, *"Map Generation from Range Sensor Data"*, Stevens Inst. of Tech, Technical Report #9011, Dec 1990.

[9] Raashid Malik, Samuel Prasad, *"Robot Mapping with Range Data"*, Proc. Int. Conf. on Advanced Robotics, June 1991.

[10] M. P. Turchen, A. K. C. Wong, *"Low Level Learning for a Mobile Robot: Environment Model Acquisition"*. Proc. of 2nd Conf. on AI Applications, Dec 1985.

THE OPTIMAL NEXT EXPLORATION: UNCERTAINTY MINIMIZATION IN MOBILE ROBOT SELF-LOCATION *

Vincenzo Caglioti
Dipartimento di Elettronica, Politecnico di Milano, Italy

1. Introduction

The accurate self-location of a mobile robot is often required, when moving from a navigation phase to a manipulation phase. In many cases, the self-location can take advantage from the knowledge of the geometric models of some of the objects present in the environment of the mobile robot (e.g. architectonic elements). The sensor detections aimed at locating the robot with respect to the environment are noise sensitive, and hence yield an uncertain position. Therefore, in order to reduce the position uncertainty, redundant sensor detections must be utilized [1]. In this paper the use of range measurements is considered. In order to expedite the robot location, the number of the required sensor detections must be restricted: therefore detections must be selected, that maximally reduce the residual positional uncertainty. An intuitive approach to the selection of a detection, that minimizes the a posteriori variance of the robot position estimate, would involve the examination of the whole set of the possible exploration directions. In this paper a criterion is adopted, based on the variance of the translational parameters of the position. In the case the uncertainty affecting the single range measurements is supposed to be constant (as assumed in [2]), some useful geometric properties can be shown that allow, once an arbitrary exploration direction has been chosen, to *i)* evaluate its distance from the optimum; and *ii)* remove from further consideration all the exploration directions, having criterion not better than that of the chosen exploration direction.

2. Problem formulation

A two dimensional curve L is given, which describes the environment within which an object has to be localized. For instance, L may represent the internal contour of a room, within which the position and orientation of a mobile robot has to be determined. We formulate the localization problem as if the curve L, whose shape is given, had to be localized relative to the object. Therefore we can represent the equation of L, when in its *nominal* position, as $g_0(x, y) = 0$, where x and

---

* Work partially supported with grants of CNR, PFR 3.2 "TISANA"

y are the coordinates relative to an object reference (let O be its origin). Since the position of the curve L is uncertain, then its equation can be represented by $g_0(x', y') = 0$, where (x', y') indicate the coordinates relative to a *displaced* reference. The displacement is described by the parameters $(\delta x, \delta y, \delta\theta)$, which in the sequel will be denoted by the *curve* parameters, as follows:

$$\begin{bmatrix} x' \\ y' \end{bmatrix} = \begin{bmatrix} \cos\delta\theta & \sin\delta\theta \\ -\sin\delta\theta & \cos\delta\theta \end{bmatrix} \cdot \begin{bmatrix} x - \delta x \\ y - \delta y \end{bmatrix} \tag{1}$$

We suppose that an initial estimate of the curve parameters is known (from, e.g., odometry), together with their covariance matrix Λ:

$$\Lambda = \begin{bmatrix} \sigma_{XX} & \sigma_{XY} & \sigma_{X\vartheta} \\ \sigma_{YX} & \sigma_{YY} & \sigma_{Y\vartheta} \\ \sigma_{\vartheta X} & \sigma_{\vartheta Y} & \sigma_{\vartheta\vartheta} \end{bmatrix}.$$

In addition, the position of a given object point $C : (x_C, y_C)$ has to be estimated with the greatest possible accuracy. Symmetrically, the variance of the translational curve parameters, when referred to C (in stead of O), must be minimum. This problem arises, for instance, when a mobile robot moves from a navigation phase to a manipulation phase: in this case the absolute position of the robot end-effector has to be accurately estimated. This problem also arises, when the object must move in a cluttered environment, and the position of a point of the external contour of the object must be accurately estimated in order to avoid collisions.

We suppose that an orientable range sensor is available, such as a laser range finder or a proximity sensor mounted on the robot gripper, which can measure the distance between the sensor and a point on the curve L with a certain accuracy. The measurement accuracy is supposed to be constant and, without loss of generality, it is set equal to one: $\sigma^2 = 1$.

Indicating by $(\delta x_C, \delta y_C, \delta\theta_C)$ the curve parameters relative to the object point C, the accuracy $\sigma^C_{\rho\rho}$ of the absolute position of C is represented by the trace of the covariance matrix Λ^C_{xy} of the translational parameters $(\delta x_C, \delta y_C)$. This quantity coincides with the mean value of the square of the distance ρ between the true position of C and its estimated position.

$$\sigma^C_{\rho\rho} = E[\delta x_C^2 + \delta y_C^2] = \sigma^C_{xx} + \sigma^C_{yy}$$

The problem formulation is then the following: given a current estimate of the curve parameters and their covariance matrix, which is the point of the curve, whose distance from the range sensor has to be measured, in order to minimize the a posteriori criterion $\sigma_{\rho\rho}$?

When translating the reference from O to C, the covariance matrix of the environment position and orientation parameters is modified as follows:

$$\Lambda_C = \begin{bmatrix} 1 & 0 & -y_C \\ 0 & 1 & x_C \\ 0 & 0 & 1 \end{bmatrix} \cdot \Lambda \cdot \begin{bmatrix} 1 & 0 & -y_C \\ 0 & 1 & x_C \\ 0 & 0 & 1 \end{bmatrix}^T$$

The variances of the parameters δx_C and δy_C are then given by the two first diagonal elements of Λ_C. These are given by the following relations:

$$\sigma_{xx}^C = \sigma_{xx} - 2y_C \sigma_{x\theta} + y_C^2 \sigma_{\theta\theta}$$
$$\sigma_{yy}^C = \sigma_{yy} + 2x_C \sigma_{y\theta} + x_C^2 \sigma_{\theta\theta} \qquad (2)$$

Notice that there is a value $(-\sigma_{y\theta}/\sigma_{\theta\theta}, \sigma_{x\theta}/\sigma_{\theta\theta})$ of (x_C, y_C) for which both σ_{xx}^C and σ_{yy}^C are minimum. Their minimum values are easily derived from the last relations. In this case, also, $\sigma_{x\theta}^C$ and $\sigma_{y\theta}^C$ are null. The point, with respect to which the variance of the translational parameters is minimum, is called the *information center*, and it is indicated by I.

To complete the diagonalization of the covariance matrix, a rotation of an angle α around the information center I is needed, given by:

$$\tan 2\alpha = \frac{2\sigma_{xy}^C}{\sigma_{xx}^C - \sigma_{yy}^C}$$

The available information can then be represented synthetically by the variance of the rotational parameter θ and by an *information ellipse*: the information ellipse is defined by the ellipse, centered on I, whose major and minor axes are rotated by an angle α relative to the coordinate axes; the length H of the "horizontal" axis of the ellipse is given by the inverse of the second diagonal element of the diagonalized covariance matrix Λ_I, while the length V of the "vertical" axis of the ellipse is given by the inverse of the first diagonal element of Λ_I.

By relation (2):

$$\sigma_{\rho\rho}^C = \sigma_{xx}^C + \sigma_{yy}^C = \sigma_{\rho\rho}^I + \Delta^2 \cdot \sigma_{\theta\theta}, \qquad (3)$$

where Δ is the distance between I and C. In the next Section the expression of the terms in relation (3) will be derived as related to the curve L.

3. Relationship between criterion and curve: basic properties

The equation of the curve L is first translated into the following form which indicates the distance from the curve at least in a sufficiently strict proximity of L:

$$f_0(x, y) = \frac{g_0(x, y)}{|\nabla g_0(x, y)|} = 0$$

The curve parameters describe the displaced reference in (1), so that the equation of the displaced curve is as follows:

$$f_{\delta x, \delta y, \delta \theta}(x, y) = f_0(x', y') = 0$$

In the sequel it is supposed that the position uncertainty is small enough, so that this last expression can be well approximated by linearization:

$$f_{\delta x, \delta y, \delta \theta}(x,y) \approx f_0(x,y) + \frac{\partial f_0}{\partial \delta \theta}\delta\theta + \frac{\partial f_0}{\partial \delta x}\delta x + \frac{\partial f_0}{\partial \delta y}\delta y,$$

where the partial derivatives are calculated for the nominal curve position. Therefore the derivatives in the last relation assume the following form:

$$\frac{\partial f_0}{\partial \delta \theta} = f_x(x,y) \cdot y - f_y(x,y) \cdot x = -M$$

$$\frac{\partial f_0}{\partial \delta x} = -f_x(x,y)$$

$$\frac{\partial f_0}{\partial \delta y} = -f_y(x,y)$$

In the last equations, M indicates the moment of the vector $\nabla f_0(x,y)$ (which is of unit modulus) with respect to the point O, while f_x and f_y are used to synthetically indicate the partial derivatives with respect to δx and to δy, respectively, calculated in the null position error situation.

We now give the expression of the conditional probability distribution of the position and orientation parameters, where the conditioning event is the measurement of a set $\{P_1 \ldots P_N\}$ of points. If we suppose that the range measurement errors are independent zero mean gaussian distributions with $\sigma^2 = 1$ variance, and that the a priori distribution of the parameters is uniform within a large domain in the parameter space (i.e., no a priori information is available), then the a posteriori joint probability distribution of the parameters, from Bayes theorem, is of the following form:

$$p(\delta x, \delta y, \delta\theta) = K \cdot \exp(-\frac{1}{2}\sum_{i=1}^{N}(f_0(x_i,y_i) - M_i\delta\theta - f_x(x_i,y_i)\delta x - f_y(x_i,y_i)\delta y)^2), \quad (4)$$

where K is a normalization factor, and (x_i, y_i) indicates the coordinates of the i-th measured point. Since the a posteriori distribution is normal, it can be put in the form:

$$p(\delta x, \delta y, \delta\theta) = K' \cdot \exp(-\frac{1}{2}[\delta x - \overline{\delta x} \quad \delta y - \overline{\delta y} \quad \delta\theta - \overline{\delta\theta}] \cdot \Lambda^{-1} \cdot \begin{bmatrix} \delta x - \overline{\delta x} \\ \delta y - \overline{\delta y} \\ \delta\theta - \overline{\delta\theta} \end{bmatrix})$$

If the mean value $(\overline{\delta x} \quad \overline{\delta y} \quad \overline{\delta\theta})^T$ of the parameters is the null vector, then the coefficients of the first degree terms in the quadratic expression in (4) are null. This condition, which is useful in order to update the estimate of the curve parameters, is not discussed here. If the information center I_N (after the measurement of the N-th point) coincides with the origin O, then the coefficients of the terms $\delta x \cdot \delta \theta$ and $\delta y \cdot \delta \theta$ are null in the quadratic expression in (4). In fact, if all the non diagonal elements of the, say, third column are null in the symmetric matrix of the coefficients, then the elements in the same position of the covariance matrix (which is the inverse of the coefficient matrix) are null too. This condition on the information center results in the following equation:

$$\sum_{i=1}^{N} M_i \cdot \begin{bmatrix} f_x(x_i, y_i) \\ f_y(x_i, y_i) \end{bmatrix} = 0,$$

where $M_i = x_i f_x(x_i, y_i) - y_i f_y(x_i, y_i)$ is the lever arm of the normal to the curve L through P_i with respect to the origin O; therefore M_i coincides with the distance between the origin and the normal to L through P_i. The above condition states that the sum of the vectors $\vec{b}_i$ connecting the information center O to the nearest point on the normal to L through the measured points P_i is null:

$$\vec{B}_O = \sum_{i=1}^{N} \vec{b}_i = 0 \qquad (5)$$

The covariance matrix $\Lambda_{yx}^{I_N}$ of the translational parameters of the curve, as referred to the information center I_N, is then given by:

$$\Lambda_{yx}^{I_N} = \begin{bmatrix} \sum_i f_y(x_i, y_i)^2 & \sum_i f_x(x_i, y_i) f_y(x_i, y_i) \\ \sum_i f_x(x_i, y_i) f_y(x_i, y_i) & \sum_i f_x(x_i, y_i)^2 \end{bmatrix}^{-1},$$

while

$$\sigma_{\theta\theta} = \frac{1}{\sum_i M_i^2}.$$

If we define α_N as the angle of the reference rotation needed to diagonalize the covariance matrix (α_N can be calculated as shown in the last Section), then the information ellipse is characterized by I_N, α_N, and by the axis lengths H_N and V_N:

$$H_N = \sum_{i=1}^{N} \cos^2 \vartheta_i \qquad V_N = \sum_{i=1}^{N} \sin^2 \vartheta_i$$

where ϑ_i indicates the orientation angle of the vector $\vec{b}_i$ (i.e. the orientation angle of the tangent to the curve near the point P_i), with respect to the rotated reference, i.e., $\vartheta_i = \theta_i - \alpha_N$. Notice that neither the trace of the covariance matrix of the translational parameters nor its determinant is modified by this reference rotation. It can be shown that:

$$\text{tr}(\Lambda_{xy}^{I_N}) = \text{tr}(\Lambda_{yx}^{I_N}) = \frac{H_N + V_N}{H_N \cdot V_N} = \frac{H_N + V_N}{\sum_{1 \le i < j \le N} \sin^2(\vartheta_i - \vartheta_j)}$$

The current state of the information available on the curve parameters is completely represented by the current information ellipse, and by the current variance $\sigma_{\theta\theta}$ of the orientation angle.

The evolution of the information ellipse is now derived in correspondence of the measurement of a new point P. Let the current information ellipse be represented by the information center I, by the axis lengths H and V, while its orientation is taken into account simply by referring the coordinates to the rotated reference centered on I. The new measured point is represented by the lever arm $D = x'f_{y'} - y'f_{x'}$, with respect to I, of the normal to the curve L through P, and by the orientation angle ϑ of the tangent to L in the nearest point to P on L. The inverse of the new covariance matrix of the translational parameters, relative to the *current* information center, assumes the following form:

$$\Lambda''_{yx} = \begin{bmatrix} H_x & V_x \\ H_y & V_y \end{bmatrix}^{-1} = \begin{bmatrix} H + \cos^2 \vartheta & \sin \vartheta \cos \vartheta \\ \sin \vartheta \cos \vartheta & V + \sin^2 \vartheta \end{bmatrix}^{-1} \tag{6}$$

The trace of the new covariance matrix Λ'_{xy} relative to the *new* information center I' is given by:

$$\sigma'_{\rho\rho} = \mathrm{tr}(\Lambda'_{xy}) = \frac{H + V + 1}{HV + H\sin^2\vartheta + V\cos^2\vartheta} \tag{7}$$

The position of the new information center I' is now determined. A coordinate reference is adopted, centered on the current information center I and whose axes are parallel to the axes of the information ellipse. As we move from the current reference (which is centered on I) to a new reference (x, y), the vector $\vec{B}_{xy}$ becomes non zero. In particular:

$$\vec{B}_{xy} = \begin{bmatrix} x \cdot H \\ y \cdot V \end{bmatrix}$$

This non zero contribution must be balanced by the contribution $\vec{b}$ due to the new measured point P, if (x, y) are the coordinates of the new information center. The orientation angle of $\vec{b}$ is ϑ, and its module is D. The coordinates of the new information center I' are shown to be given by:

$$\begin{bmatrix} x_{I'} \\ y_{I'} \end{bmatrix} = \frac{D}{HV + H\sin^2\vartheta + V\cos^2\vartheta} \begin{bmatrix} V\cos\vartheta \\ H\sin\vartheta \end{bmatrix}$$

and from this relation the distance Δ' between the point C and I' can be calculated. The variance of the orientation angle θ after the new measurement is given by:

$$\sigma'_{\theta\theta} = \left(\frac{1}{\sigma_{\theta\theta}} + \frac{HVD^2}{HV + H\sin^2\theta + V\cos^2\theta}\right)^{-1}. \tag{8}$$

Let us now denote by (x_C, y_C) the coordinates of the object point C with respect to the reference attached to the information ellipse. The criterion to be minimized, which is given by (3), is then expressed as a function of D and ϑ as follows:

$$J(D,\vartheta) = \frac{H+V+1}{HV + H\sin^2\vartheta + V\cos^2\vartheta} + \frac{\Delta'^2}{\frac{1}{\sigma_{\theta\theta}} + \frac{HVD^2}{HV + H\sin^2\vartheta + V\cos^2\vartheta}}$$

where
$$\Delta'^2 = d_\vartheta^2 + (h_\vartheta \cdot D - P_{C\vartheta})^2,$$

and
$$d_\vartheta^2 = \frac{(x_C H \sin\vartheta - y_C V \cos\vartheta)^2}{V^2 \cos^2\vartheta + H^2 \sin^2\vartheta},$$

$$h_\vartheta = \frac{\sqrt{V^2 \cos^2\vartheta + H^2 \sin^2\vartheta}}{HV + H\sin^2\vartheta + V\cos^2\vartheta},$$

$$P_{C\vartheta} = \frac{x_C V \cos\vartheta + y_C H \sin\vartheta}{\sqrt{V^2 \cos^2\vartheta + H^2 \sin^2\vartheta}}.$$

It can be observed that, as D goes to the infinity, the criterion value is independent of ϑ, and it coincides with $(H+V)/HV$. Notice that J has a discontinuity in I (i.e., for $D=0$), since there it depends on ϑ.

4. Uncertainty minimization

In addition to L, two more curves are to be considered. The first curve is the envelope of the normals to L; this curve, which is indicated by E, can be constructed a priori since the shape of the curve L is known. E also coincides with the locus of the curvature centers of L. The second curve is called the *detection* curve, and it is indicated by N. N is the locus of the points Q with polar coordinates (D,ϑ), relative to the reference attached to the current information ellipse, such that there exists a normal to L through Q which is orthogonal to ϑ. Notice that the detection curve can only be defined once the current state of the information is specified: this curve will not be constructed explicitly, otherwise the scanning of the whole curve L would be necessary. Notice that our problem is actually the minimization of the criterion $J(D,\vartheta)$ within the detection curve N. The method proposed in this Section involves an off-line pre-processing, and an on-line phase which is executed once L is known.

During the pre-processing, the plane (x,y) is subdivided into sufficiently small square cells by means of a square grid. The envelope E of L is constructed. To each cell the following data are associated: *i)* the range of the normals to L that intersect the cells; *ii)* the range of the normals to E that intersect the cell. Each local extremum of the curvature in L corresponds to a cusp in E.

An intersection between E and a normal to E through I generally corresponds to a local extremum of D along N. An intersection between E and a tangent to E (i.e., a normal to L) corresponds to a point on N, for which $D=0$ (i.e., to a local minimum for D). A bounded domain Π containing the detection curve N can therefore be defined, and it consists of the set union of a number of circular sectors centered on I. Notice that the explicit construction of the domain Π can be avoided, since the condition for a point in the plane to belong to Π can directly be verified without requiring the enumeration of the cells contained in Π.

One of the above extrema of D along the detection curve N is taken as an initial candidate detection point $(\bar{D}, \bar{\vartheta})$ (e.g., the extremum characterized by the minimum value of the criterion $J(D, \vartheta)$). Let be $K = J(\bar{D}, \bar{\vartheta})$: the level curve K_J defined by $J(D, \vartheta) = K$ is then considered. The gradient to the curve is calculated locally, in order to ascertain whether lower criterion values are in the interior or in the exterior of K_J. Suppose that the interior of the curve K_J is characterized by lower values of J. Then the contour of the intersection between the region Π and the interior of K_J is scanned cell by cell. During the scanning process, for each cell S the center $\bar{S}$ is considered, and the intersection S_K between the normal to the segment $\bar{S}I$ through $\bar{S}$ and the level curve K_J is determined and S is associated to the cell containing S_K. Then S contains the intersection between a circle with diameter $S_K I$ and the curve K_J. Therefore, the orientation of the segment $S_K \bar{S}$ is an extremum of one range of the possible orientations γ of the normals to L through $\bar{S}$, characterized by criterion values lower than K. These ranges on the orientation γ of the normals to L through the cell center are associated to each cell.

Finally the level curve K_J is scanned again, and for each cell only those normals to L through the cell are considered, whose orientation γ is within the specified ranges. In this way only the detections with a criterion value lower than that of the first candidate detections are considered, and a minimization of the criterion can easily be obtained.

5. Conclusions

A method has been presented for the determination of the range measurement which minimizes the variance of the translation parameters of a point within a known environment. The examination of the whole set of the possible sensor observations can be avoided, if the variance of the observation error may be considered constant. The presented method is currently being tested on a mobile robot equipped by a laser range finder.

References

[1] H.Durrant-White, – "On Uncertain Geometry in Robotics", IEEE Journal on Robotics and Automation, Vol. RA-4, (1988)
[2] P.Whaite, F.P.Ferrie, – "From Uncertainty to Visual Exploration" – to be published

A Parallel Blackboard Model for Mobile Robotics control

Michel OCCELLO, Claudine CHAOUIYA, Marie-Claude THOMAS
I3S - Equipe APARSA - Robotique, Université de Nice, CNRS URA 1376
Bât.4, Av. A. Einstein, SOPHIA ANTIPOLIS 06560 Valbonne, FRANCE

1 Introduction

Current techniques for autonomous mobile robots control fall into two main classes: the hierarchical approach [2] and the blackboard approach.[8]. *In the first, a vertical decomposition of the robot programming problem is carried out.* Each layer of control gives the robot new abilities relying upon abilities it already possesses [2]. *The second is based on a horizontal decomposition which is more functional than the previous approach.* A Blackboard structure allows cooperation between all the subsystems, considered as knowledge sources performing complete specialized tasks [8, 7]. The classical blackboard architecture is presented as a shared database, knowledge sources (KS), and a control unit. KS are specialised reasoning agents which read and write on the diverse abstraction levels defined in the blackboard structure. The control unit consists of a monitor observing the blackboard state to notice changes and a scheduler triggering sources as dictated by the circumstances.

A. Elfes and S.L. Talukdar [4] suggested a robotics software architecture split into a *real world modelling module, a planning and simulation module and a distributed control module.* For Telerobotics, a rather similar decomposition into several panels of blackboards has been proposed [1]. In this paper, *we define an architecture which is based on three blackboards* and aims *to give gradual modes of sophistication between autonomy and teleoperation. The hard real time constraints due to the robotics context, motivate the specification of a parallelised control module.*

Section 1 describes the whole architecture, illustrating its behaviour and the interaction between the three modules involved by means of an example. The next sections focus on the distributed control module. Section 2 reviews parallelisation approaches for blackboard architectures. We conclude from this that there has not been sufficient study in this area with regard to software layers in robotics, and so this justifies original architecture studies. Our proposal for a parallel blackboard is then presented. Section 3 completely describes this architecture using Petri nets formalism. Finally, we summarise the avantages of our approach and give suggestions for further study.

2 The Control Software Architecture through an Operational example

The proposed architecture supports an advanced real time control system, allowing both supervisory control and autonomous modes. It consists of three main parts:

*the user control module (**UCBB**), the mission dynamical execution module (**EBB**) and the distributed control module (**CPBB**). Each module is based on an adapted*

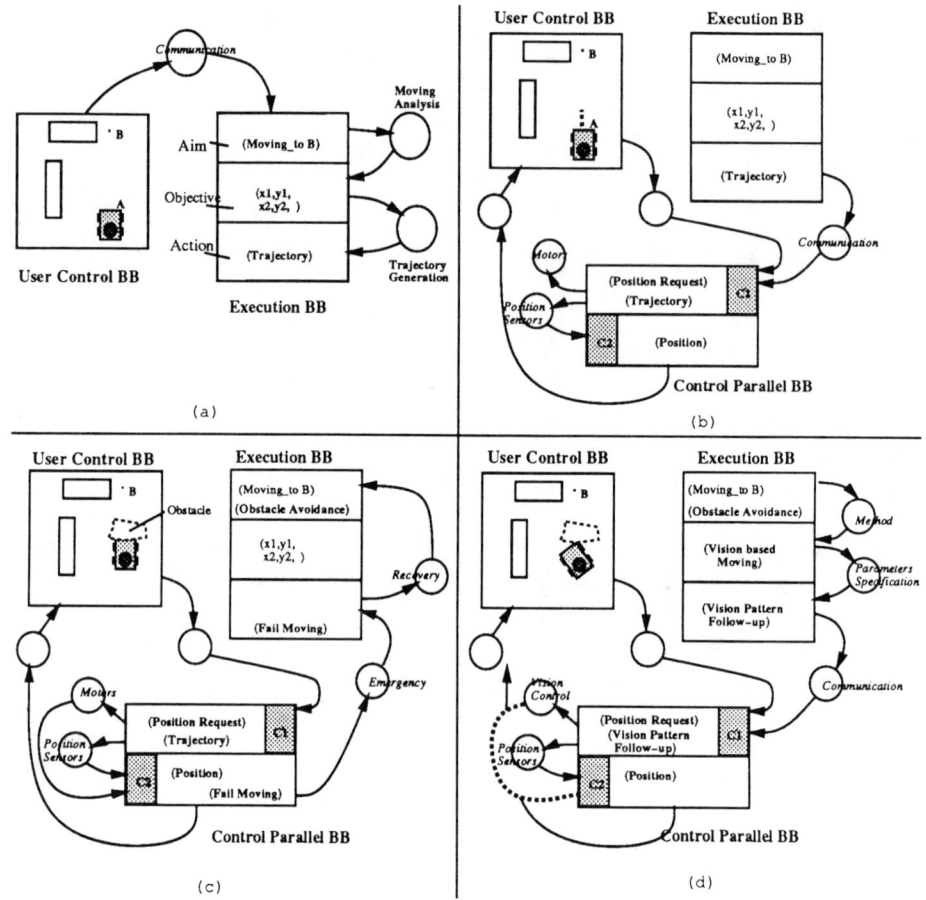

Figure 1: The four phases of the operational example

blackboard architecture. The first has a graphical aspect, the second may appear to perform reactive reasoning, while the third is concerned with real time execution. We name such an architecture a Multi-Shape Blackboard Architecture.

This section develops a very simplified operational example for this software architecture for mobile robotics, emphasising the cooperation between the three blackboard-based modules. Figure 1 illustrates four phases which are described below.

(a) To perform an action on the robot, the operator specifies the order on the **UCBB** through choices on menus and a pointing device; in this example, the operator chooses a move from point A to point B.

A communication specialist attached to the UCBB signifies an event "Move-to B" to the **Aim Level of the EBB**.

The income of this event triggers a *Moving Analyse Specialist* which writes in the **Objective Level** the coordinates of the routing postures and the directions.

The Controller of the **EBB** then chooses one of the trajectory generation methods, depending on the context and user preferences. The selected KS then places the resulting trajectory on the **Action Level**.

(b) The trajectory is now sent to the **Interpretation Level of CPBB**. The level controller activates the appropriate routine to execute the move. While the motion is going on, the **UCBB** periodically requests the position value in order to update the robot representation.

(c) It is the uncertainty in the real which makes blackboard incremental event processing of interest. Let us assume that an unidentified obstacle makes the intended action fail. *Motors relative KS* writes an event "Fail moving" on the **Diagnosis Level of the CPBB**. The *Emergency KS* ensures the arrival of a event "Fail moving" on the lower level of the **EBB** and the trajectory is then deleted. The new event is processed by a *Recovery KS* which places a new current objective, " Obstacle Avoidance KS", on the **Aim Level**. The event "Moving-to B" is not erased, but kept in the background. The system will attempt to treat it later, after having performed its new objective. This behaviour is typical of an autonomous mode. In teleoperative mode, inaccurate information will give control back to the user.

(d) The event "Obstacle Avoidance" then triggers an *Avoidance Method Selector* which specifies an adapted technique according to the context. For instance, a vision method may be selected and placed on the **Objective Level**. A *Parameters Specification KS* then writes an event "Vision Pattern Follow-up" on the **Action Level**. The **Interpretation level of the CPBB** receives an event triggering the corresponding *vision-based routine*. Concurrently, sensors' consultations allow to globally examine the robotic system evolution.

The above description gives an idea of the whole architecture and the interaction between modules. We now study the distributed control module but first review work on parallel blackboards.

3 Blackboard and Parallelism

3.1 Survey of Existing Parallel Blackboards

All work on parallel blackboard aims to improve performance by concurrent execution of KS. In order to do this, classical architecture is modified by convenient scheduling of KS, communication synchronisation and shared data access. Work on parallel implementations of blackboard based systems follows one of the approaches below:

1. *The classical architecture is preserved but concurrent execution of KS is allowed.*

In this class, we find systems aiming to extend classical existing blackboard systems, such as CAGE [12], [15].

Velthuijen [15] presents a system dedicated to robotic cells control. It is based on a BB1-like Blackboard shell and adapted to KS parallelism for robotics. Problems related to real time constraints occur, due to the insufficient performance of blackboard based independent control process [11].

CAGE is a generic blackboard system, extending to parallelism the serial system AGE. A base control cycle is intended to offer two kinds of parallelism : parallel solution exploration and KS pipe-line between abtraction levels.

2. *The classical architecture is preserved but the monitor is distributed.*

Transactional blackboards [5] preserve the blackboard as a centralised communication tool and join to it a hierarchical access manager. It is a coarse grain parallelism; each node (processor) is composed of several groups of KS and associated control. The complexity of this non-reactive structure and the low performance of its hierarchical goal directed control mean that this approach must be discarded for applications in mobile robotics.

CMU-Rover ([4]) uses a blackboard-based robot controller. A centralised monitor ensures blackboard management, but the scheduler has both a blackboard structure (master control) and expert processes (slave control).

3. *The classical architecture is preserved but scheduler and monitor are distributed.*

The real time expert system RT-1 [3] is composed of complex reasoning modules distributed on different processors. The reasoning modules are blackboard-based systems. Knowledge granularity in our context do not justify the use of a such real time orientated structure.

CODGER for CMU-NAVLAB is dedicated to mobile robotics (involving an autonomous vehicle on a known map) [7]. It consists of a set of large independent modules. When in operation, these modules write and read on a local database (local map) which is managed by local processes. Shared structure access routines are provided to each module as local interfaces. This blackboard structure which is quite different from the classical one, is called *whiteboard*.

4. *The blackboard structure is distributed and both scheduler and monitor are integrated to its distributed structure.*

POLIGON ([12]) is a development framework for parallel systems. Its main feature is the blackboard's tendency to suppress serial effects. Each hypothesis becomes an active node associated to a processor and is related to a concerned KS set. There is no longer a control unit since message passing integrate control to each node. The model used gives priority to data parallelism in addition to KS parallelism, dictates fine grain parallelism use and requires a massively parallel hardware such as Connection Machine.

The approaches we have reviewed are either not real time systems (1.), have specific hardware architectures (2.,3.), are systems with insufficent performance due to the blackboard control type used (1.,2.), or are too complex (3.,4.) and so inadequate for the kind of knowledge used in our robotics problem. This brief survey makes us study a new proposal for our specific mobile robotic control software.

3.2 Presentation of an original parallel blackboard architecture

The fourth approach has been chosen for the distributed control module and so *the blackboard structure, its scheduler and control are all distributed. Real time constraints of the context are satisfied through the choice of a procedural control.*
Our approach consists of a parallel blackboard data structure, a monitoring and scheduling control distribution and a parallel KS execution. The blackboard becomes reactive by the distribution of the control over the different abstraction levels to which KS are suitably attached. Each level receives messages from other levels or from other modules. On each level a control unit identifies the type of the events and triggers related KS. KS can be activated concurrently. When activation of an already active source is required, the control unit duplicates the code of the KS. No bottleneck will occur for the trigger action of KS since each source is supported by a new process. This specification is illustrated by figure 2.
Because of hard real time constraints, the sequential control cycle is replaced by level

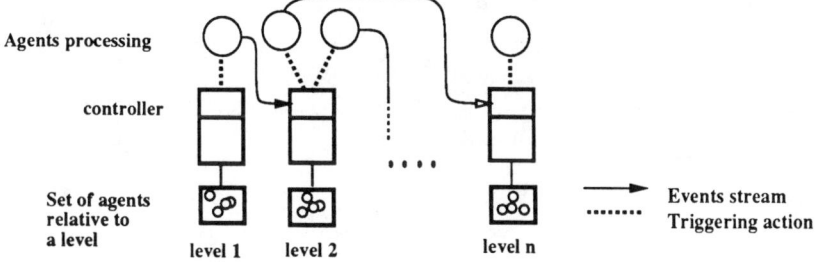

Figure 2: A parallel blackboard architecture

reactivity which is obtained from object paradigm and message passing. The classical technique of focussing [9] is then forced to level observation. Parallel KS activation operates according to Fennell and Lesser's study [6]. An attempt to modelling is presented in the next section, in order to clearly describe this parallel blackboard architecture.

4 Modelling

For the application considered here, we can restrict the above proposal to an unidirectional events stream. This assumption is justified by the function of the distributed control module: translation of symbolic requests into robot orders.

4.1 Modelling a level

Each level is modelled by a Petri net [13] as illustrated in figure 3. An input transition

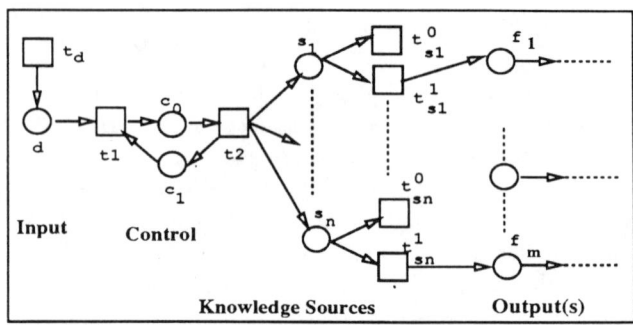

Figure 3: Petri net model for a level of the distributed control module

(t_d) represents the arrival of events on a level (place d). The controller is considered as a resource and only one message is processed at any given time. The place (c_0) is the controller freeness, while (c_1) is the controller processing. KS, however, are duplicated according to requirements and then places ($s_{j,\{j=1..n\}}$) can contain several tokens at once. A token in a place s_j will induce the firing of $t^0_{s_j}$ (no use of the agent) or the firing of $t^1_{s_j}$ (agent's s_j processing) as the case may be. Delays are associated with places c_0 (controller processing duration) and $s_{j,\{j=1..n\}}$ (agent processing duration). Finally, the average population in exit places $\{f_k\}_{\{k=1..m\}}$ will caracterise the event generation of this level.

The different types of event likely to arise at a particular level are known in advance. Thus, we can associate an average firing frequency to the input transition t_d. To each event type will correspond the triggering of one (or more) KS. Arrival probability is defined for each level and each event type, and an average firing frequency is naturally associated to $t^1_{s_j,\{j=1..n\}}$.

Notation

- $\bar{F}_{t_d}$: average frequency of arrivals on the level.
- $\bar{F}_{t_1}$: average firing frequency of transition t_1 (i.e. controller frequency processing).
- S: KS set (card(S) = n), $\mathcal{P}$ set of S's subsets.
- $\pi(e)$: probability of a message being of type e.
- $\Pi : \mathcal{P}(S) \longrightarrow [0...1]$: probability function of triggering of a set of KS.
- T_{c_0}: duration of message processing by the controller.
- T_{s_j}: duration of processing by an agent j.
- $M(s_j)$: average number of tokens in place s_j (i.e. average number of j duplication).
- $\bar{F}_{s_j}$: average frequency of request (triggering action) for agent j.
- e_i: number of events generated by agent j.
- $\bar{O}$: events stream of a level output (number of events generated per time unit).

4.1.1 Average utilisation of the controller

Clearly, we have

$$\bar{F}_{t_1} = \bar{F}_{t_2} = Min(\bar{F}_{t_d}, \frac{1}{T_{c_0}}) \qquad (1)$$

Indeed, the firing of t_1 depends on the presence of one token in c_1 and d. The tokens arrive at c_1 with a maximal frequency $\frac{1}{T(c_0)}$ and in d with an average frequency $\bar{F}_{t_d}$. The execution time of the controller, which performs a choice operation, may reasonably be assumed to be less than the time between two arrivals ($\frac{1}{\bar{F}_{t_2}} \geq T(c_0)$) and so events are processed when they arrive. In the case $\frac{1}{\bar{F}_{t_2}} \leq T(c_0)$, the utilisation rate of the controller is maximal and so the c_i marking increases, causing the controller to become overworked. This situation is not acceptable for real time applications.

4.1.2 Average duplication of agents

The controller triggers the set of agents with regard to the current context and the events which have arrived. Its average frequency of triggering is $\bar{F}_{t_2}$. In a simplified way, we consider that the arrival probability associated to each message e can be assimilated to the probability $\Pi(\{s_i\}_{i=1..k})$ of an agent set triggering action and so $\pi(e) = \Pi(\{s_i\}_{i=1..k})$ where $\{s_i\}_{i=1..k}$ being the set of the k agents wich must be triggered when a message e arrives. Then, the average frequency of the triggering agent j is :

$$\bar{F}_{s_j} = \left[\sum_{X \in \mathcal{P}(S)/s_j \in X} \Pi(X) \right] \cdot \bar{F}_{t_2} \qquad (2)$$

Each agent j has an associated processing duration, T_{s_j} (tokens corresponding to $t^1_{s_j}$ firing will stay in place s_j for this time). For all agents j, the average number of duplications $M(s_j)$ (corresponding to the average marking of place s_j) is:

$$\bar{M}(s_j) = T_{s_j} \cdot \bar{F}_{s_j} \qquad (3)$$

4.1.3 Output events stream

Each agent generates one or more events. These events have their own pre-determined destination (i.e. a further level). We notice that for the agent j, its output stream corresponds to its average request frequency ($\bar{F}_{s_j}$). Formally, a number of events, e_j, is associated to each agent j (number of tokens produced by transition $t^1_{s_j}$). Thus, we can determine the output stream level, which we shall call $\bar{O}$, is given by

$$\bar{O} = \sum_{j \in S} e_j \cdot \bar{F}_{s_j} \qquad (4)$$

4.2 Modelling the whole module

The global model of the distributed control module is obtained by connecting the models relative to each level. A simplified way is to consider each level as a closed box, by which we mean a self-contained process having an input stream and an output stream. These datas are obtained from the previous equations.

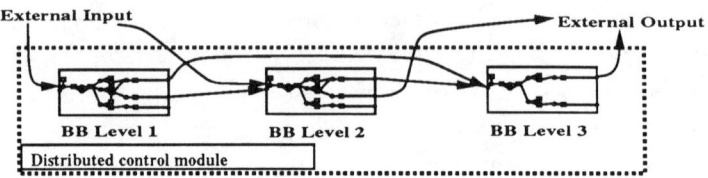

Figure 4: Distributed control module model

Notation
- L: the set of levels of which the distributed control module is composed. The assumption of an unidirectional events stream allows the elements of L to be ordered.
- S_i: the agents set of level i.
- D_i: the set of levels likely to receive an event from level i.
- $\bar{O}_{i|k}$: the events stream from level i to level k.
- $\bar{E}_k$: the events stream from the environment to level k.
- $\bar{I}_k$: input stream of events for level k.

From these definitions along with equation 4, we can determine the input event stream of level k and thus the average frequency of arrivals for this level (denoted by $\bar{F}_{t_d}$):

$$\forall k = 1..\#L, \quad \bar{I}_k = \bar{E}_k + \sum_{i \prec k} \bar{O}_{l|k} = \bar{E}_k + \sum_{(l \prec k) and (D_l \ni k)} \left[\sum_{i/i \in S_l} 1 \cdot \bar{F}_{s_i} \right] \quad (5)$$

The viability of the complete module is known from the duration of the controller process execution and from the arrival frequency of external events. An adapted host hardware architecture can be easily chosen since the model allows the evaluation of the average duplication of agents (i.e. concurrent processes). This evaluation supplies very useful information on each level's load.

5 Conclusion

An experimantal prototype is developed in relation to an existing project (I3S, ISIA, INRIA, Cie ECA). This project aims to develop a reconfigurable mobile robot. This robot must support gradual modes of operation from being pure teleoperated to being autonomous. The software multi-shape blackboard architecture insures that these modes are integrated. The inherent modularity of Blackboard systems means that the whole control architecture may be easily adapted so that it can be applied to robotics devices. A multi-robots extension can even be envisaged.

Model obtention allows, in addition to a formal architecture specification, a feasibility evaluation for a given hardware configuration. The model's generalisation to a proposed complete parallel blackboard architecture (i.e. without events stream assumptions) is now being studied. This work will be an interesting contribution to distributed problem solving theory. This issue will be treated in a forthcoming paper.

References

[1] M.Arnoux and M.Thomas. A blackboard system to assist telerobotics tasks. In *Proc. of Int. Workshop on IROS'89*, Tsukuba, Sept. 1989.

[2] R.Brooks. A layered intelligent control system for a mobile robot. In *Proc. of the Third Robotics Reseach Int. Symp.* O. Faugeras and G. Giralt Ed., 1986.

[3] R.Dodhiawala, N.Sridharan, P.Raulefs, and C.Pickering. Real-time AI systems: a definition and an architecture. In *Proc. of the IJCAI-89*, 1989.

[4] A.Elfes and S.Talukdar. A distributed control system for the CMU ROVER. In *Proc. of AAAI-83*, pp.830–833, 1983.

[5] J.Ensor and J.Gabbe. Transactional blackboards. In *Proc. of IJCAI-85*, pp.340–344, Los Angeles, Aug. 1985.

[6] R.Fennell and V.Lesser. Parallelism in solving: A case study of Hearsay-II. *IEEE Transaction on Computers*, C-26:98–111, February 1977.

[7] Y.Goto, K.Matzukaki, I.Kweou, and T.Obatake. Cmu sidewalk navigation system: A blackboard based outdoor navigation system using sensor fusion with colored-range images. In *Proc. Fall Joint Computer Conf.*, Dallas, nov. 1986.

[8] S.Harmon, D.Gage, W.Aviles and G.Bianchini. Coordination of intelligent subsystems in complex robots. In *Proc. of 1st Conf. on AI Applications*, pp.64–69, december 1984.

[9] B.Hayes-Roth. A blackboard architecture for control. *Artificial Intelligence*, 26(3),pp.251–321, 1985.

[10] H.Laasri and B.Maitre. Organisation du contrôle dans les architectures de blackboard: Etude, evaluation, comparaison. *Revue d'Intelligence Artificielle*, 3(1):105–146, 1989.

[11] P.Nii, N.Aiello, and J.Rice. Frameworks for concurrent problem solving; a report on CAGE and POLIGON. In R.Engelmore and T. Morgan, editor, *Blackboard Systems*, chapt.25, pp.475–501. Addison Wesley, 1988.

[12] J.Peterson. *Petri Net Theory and Modeling of Systems*. Prentice-Hall Inc., Englewood Cliffs, 1981.

[13] H.Velthuijen, B.Lippolt, and J.Vonk. A parallel blackboard system for robot control. In *Proc. of Scandinavian Conf. on AI*. H. Jaakola & S. Limnaima Ed., 1987.

**PART 5
ROBOT PROGRAMMING AND SENSORY DATA
PROCESSING**

Graphical Robot Programming: Requirements and Existing Systems

George Nikoleris, PhD
Dept. of Production and Materials Engineering
Lund University

Abstract

The use of a graphical system for the programming of industrial robots offers a number of advantages. Graphical systems employ an interface that is much easier to use. The resulting motion trajectories can be simulated and tested for collisions or singular points. Contemporary graphical systems use the geometry of the objects that are to be handled or processed in order to generate robot motion trajectories. Robot motions are however constrained by the process requirements which should be one of the main sources of the generation of robot motion trajectories. In this article the methodology of the generation of the robot motion trajectories in three commercial systems is examined as well as the requirements of process-related robot programming with special emphasis on high-quality arc welding.

1 A Short Introduction to Robot Programming

Since the introduction of the first teach-and-repeat manipulators in the early 1960's, industrial robots have been used in several manufacturing operations, such as spot and arc-welding, spray painting, deburring and automatic assembly. These operations are defined by programs that instruct the system controller to move the end-effector of the robot along described paths and to take several peripheral actions. Examples of such actions are the control of the process equipment in use, the activation of devices like fixtures and clamps, interaction with the environment through sensors and finally communication with other computer controlled equipment.

Robot programming is still a major research topic in robotics. The industrial robot is an essential manufacturing unit in the factory of the future, where high flexibility and short set-up times are necessary. Thus, the methods for producing robot programs must be as flexible and effective as the industrial robot itself. Different languages have been developed in order to make robot programming easier and to help programmers exploit the inherent versatility and flexibility of the industrial robots.

The difficulties in programming an industrial robot originate in the dual nature of the robot program: a logical structure with controlling statements that has to be developed at the same time as spatial relations or motions are defined. Experience from both the programming of numerically controlled machines and the industrial robots indicates that two-dimensional relations and geometry is easily taken care of by NC-operators. However, modelling and manipulation of three-dimensional bodies set a limit for many operators, and significantly increase the cost of the programming system.

1.1 User categories

The design of robot languages has been strongly influenced by the potential users and the application area of the robotic system. The users of robot programming languages can be either *operators, end users* or *application developers*.

Robot operators usually describe the operations that are to be carried out by the robot. Operators normally have little programming skill and make best use of high-level systems that do not include operations like robot modelling or sensor system specification. End users have good process knowledge as well as some programming knowledge and can use more complex systems.

Application developers add specific facilities in order to increase the ease of use for a particular class of end users or robot operators, while decreasing the generality of the system. Application developers combine greater programming skill with a deeper knowledge of the application area. The systems that are best used by application developers must allow for both low-level system-specific facilities and the programming environment of a high-level language. This mix of users and the way they work has been of great importance to the design of robot programming languages.

1.2 Robot languages

Robot languages are divided into three categories : *textual programming languages, simulation systems* and *implicit programming*. Textual languages focus on the description of robot motions as relations between objects, the interface of multiple sensor systems and the execution of peripheral actions using a formal programming language. Simulation systems are used to validate and to verify the generated programs as well as to choose an optimal robot configuration. Finally, implicit robot programming takes a top-down approach to the logical structure of the robot programs, using high-level statements that specify the execution of tasks rather than explicit operations.

Most of the research in implicit robot programming is devoted to the development of high-level assembly operations. In this type of programming assembly tasks are specified by the spatial relations of the objects to be assembled prior to and after the assembly operation. Less attention has been paid to the programming of process robots where the logical structure of the robot program is in essence quite simple. The robot is programmed to move along a path while the process equipment is activated with the proper process parameters. Leadthrough programming has proved to be sufficient in such applications. However, the tasks that a robot program must perform may increase significantly if the industrial robot becomes a part of an integrated manufacturing unit. These tasks include general administrative functions, coordinated operation between several manipulators and integration with external sensor systems. Complexity increases sufficiently even if only small units are integrated, as in the case of a welding robot and a positioner. In this example, in order to define relatively simple motions, such as circular welds on an arbitrary plane, the operator has to teach the controller a large number of poses. Other programming aspects that increase the complexity of process-related programs are the optimal choice of operation sequences and process parameters as well as the use and the programming of process-related sensors. Research in this area requires specialist knowledge of the processes involved and the use of artificial intelligence techniques.

2 Graphical Robot Programming Systems

The description of three-dimensional object relations and movements becomes much easier if the objects are graphically represented by a programming system. Such systems allow the modelling of objects in the working area of the robot and the simulation of movements. Although off-line programming of complete installations is the desired result, graphical systems are often used in order to test and to present different production solutions without actually creating the final robot programs.

The following steps are common for all graphical programming systems

- generation of models for the robots and their working environment. Establishment of hierarchical relations between the different objects
- creation of programs for the industrial robot (logic and motions)
- simulation of motions and interaction with the environment
- optimisation by testing different robot types
- generation and downloading of code to the industrial robot
- calibration and adjustment of the generated program

2.1 Cell and Robot Calibration

Off-line programming is still limited by two accuracy problems. The method requires that the workcell is accurately represented by the computer model and that the accuracy of the robot is the same or at least of the same magnitude as its repeatability.

Robots are often working with devices that are not part of their own mechanical system, like positioners, tables etc. In order to generate working robot programs, the exact location of these devices must be represented very accurately in the programming system. Programming systems must therefore include a calibration module that can be used in order to adjust the positions, orientations and, eventually, the shapes of the modelled objects so that they are in agreement with the real world.

Calibration of robots is more complicated. Differences in the mechanical structure and misalignments of electrical components have to be measured and taken into consideration during the inverse kinematics solution by the programming system or the controller [5]. The customization of the inverse kinematics for each individual robot is naturally a task for the robot manufacturer and should be an integral part of the robot controller. In this case, specification of position and orientation by the programming system should be at the object-level.

A special problem is the combination of robots and moving devices like positioners. The positioner must be calibrated both locally in order to eliminate joint misallignment or other irregularities, and in relation to the robot.

2.2 GRASP, B.Y.G. Systems

GRASP[2] is a language-based system; all commands can be typed in the input. The system accepts combinations of commands that can be used to perform more than one operation at a time :

$ TARG TRANSLATE Z 500 ROTATE GL 30 TARG X

GRASP also provides a *menu* mode which is a mouse-menu interface.

GRASP operates on solid models.[1] The models are generated either as simple primitives (cuboids, cylinders and prisms) or using translational and rotational sweeps of planar shapes.

GRASP allows a very general description of mechanisms. Both revolute and prismatic joints along arbitrary axes can be modelled. Closed-loop mechanisms can be implemented as special FORTRAN routines. However, BYG recommends the user to contact their support department if that kind of modelling is necessary. External devices can be modelled as auxiliary axes; an apporach common to many robot controllers.

The description of a motion in GRASP is a two step procedure: The user has to define the *type* (point-to-point, linear or circular) and the *geometry* of the motion. The type of the motion becomes a PATH while the geometry and related information becomes a TRACK in the system.

A TRACK is very similar to the final robot program except that it may also include information for the simulation phase. After the creation of a TRACK, robot motion is defined by a number of poses. In order to simulate real motion the user has to

---

[1] An extra module allows the display of B-spline curves and surfaces that are imported from a CAD/CAM-system.

create a PROCESS, an animation of the defined motion. In this last phase the user can check the program for collisions or make a presentation film.

Special process-oriented modules make it possible to define movements using check surfaces and to specify process related information such as welding current, voltage and travel speed.

3 ROBCAD, Tecnomatix Technologies

ROBCAD [7] functions in almost the same way as GRASP. The system works with solids while supporting the import of surfaces from other systems. ROBCAD accommodates two categories of kinematic components. A *robot* that incorporates a base frame, a tool frame and can be driven using inverse kinematics as well as a *device* that is a mechanism that can only be driven in direct mode.

Robot programs can be created either in the controller's own language or in a general language supported by ROBCAD. Basic motion is defined by diskrete poses. A welding module makes possible the definition of motion by lines and orientation as well as the specification of the related process parameters.

4 CimStation, SILMA

CimStation [6] is controlled either from a menu interface or a Pascal-like language called SIL. SIL is decribed in [6] as "an interactive language implemented in C for portability and efficiency, based on LISP for advanced programming capability, and made to look like Pascal so that it is easy to use."

CimStation uses boundary representation for geometrical modelling and a hierarchical structure in order to establish object relations. Temporary connections can also be specified using *affixments* between objects. The display is controlled by an object called *lens* that can be moved around or be *affixed* to moving parts. Motion is represented by *paths* that can be specified as a sequence of poses, by use of the geometrical models in the system or by SIL routines.

In the CimStation environment, robots are programmed in SIL, a very powerful robot language. A special feature of the SIL language is the use of non-unique function names (polymorphism), which results in homogeneous and compact robot instructions.[2] In SIL, programs are considered to be a type of data, so that functions can operate on programs.

5 AUTOMATOS, IVF

AUTOMATOS [4] is developed by the IVF, The Swedish Institute of Production Eng-

---

[2]Function statements as jmove, move, dmove etc can be replaced by one unique name *move*

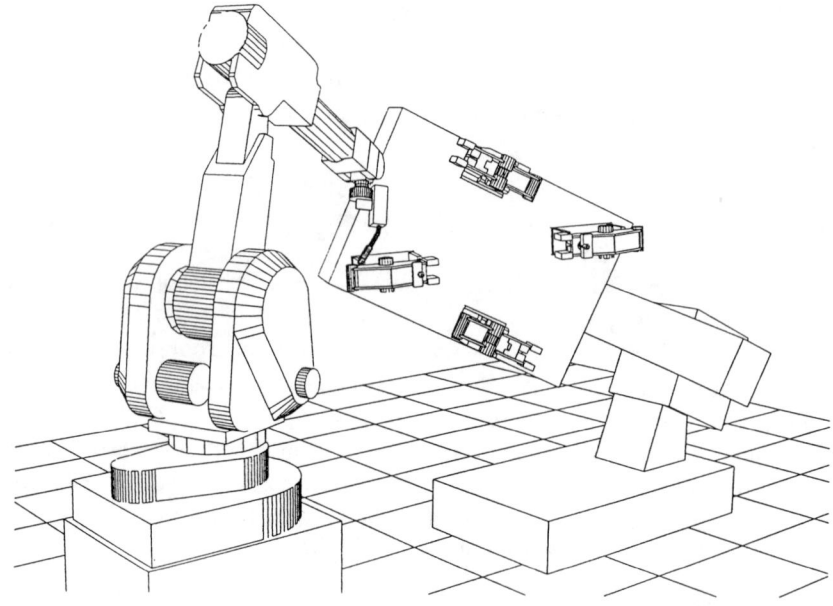

Figure 1 *Welding robot and positioner modelled in* AUTOMATOS

ineering Research. AUTOMATOS does not have the extensive display and simulation functions of the previous systems and should be regarded as a system that provides graphical support to textual off-line programming.

AUTOMATOS is an add-on to AutoCAD (Autodesc); it is written in AutoLisp and makes use of the modelling and displaying functions in AutoCAD. AUTOMATOS is used to program ABB-robots only.

6 Standardization Attempts

Development of high-level robot-independent programming languages would become much easier if the interface to the robot was standardized. A first step is the development of an intermediate code for robots. This approach is similar to the standardization of an intermediate code for the programming of numerically controlled machine tools, the Cutter Location Data (CLDATA) which has become an internationally accepted standard.

In Germany, attempts have been made to establish a similar intermediate code for the off-line programming of industrial robots. The proposal made was to extend the existing CLDATA interface to encompass robot programming. The new standard

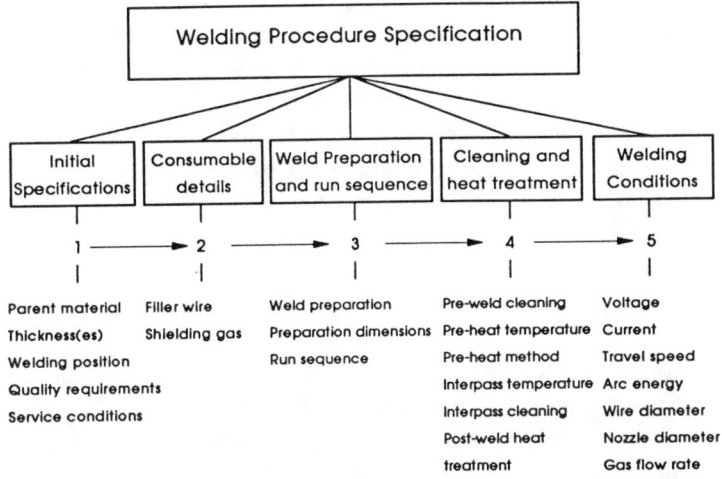

Figure 2 *Structure of a welding procedure specification*

(IRDATA) consists of textual records of variable length that specify poses, complex trajectories, tools, robots, sensors, boolean and arithmetic functions and general programming instructions. The standard also includes a number of reserved records, which are not standardized and are reserved for special applications. An ISO proposal for a new intermediate code (ICR) is currently being prepared [1].

Another solution is the use of a standard robot language. Several standardization proposals have been made and these include SLIM, Standard Language for Industrial Manipulators, a BASIC-like Japanese proposal and IRL, Industrial Robot Language, which is a more structured Pascal-like language under development in Germany.

7 Graphical Programming : Requirements

Present off-line programming systems do not satisfy the requirements of a modern production department [3]. The specification of welding procedures for a robotic cell necessitates the use of an augmented programming method. The desired production result and the constraints imposed by the welding process must be taken into account. The programming requirements of welding robots are thus twofold: the kinematic specification of the necessary motions and the choice of the welding parameters. The choice of welding parameters may be made implicitly in a manner similar to the programming methods presented previously. Thus, the programmer should be able to define the material type and the results of the welding process and let the system derive the necessary process parameters such as current, voltage, welding speed etc. Robotic and mechanized welding implies that two important variables in manual welding, the arc length and the welding speed are controlled according to machine settings. This results in consistent quality and better weld appearance but increases the number of free process variables and necessitates a better joint fit-up.

In high-quality arc welding it is desirable that weldments are aligned in such a way that all welding is performed in the flat position. This is possible if suitable positioners are used. An important requirement for a graphical programming system is therefore the ability to calculate positioner angles so that all welding is performed in the flat position.

8 Conclusions

The three major systems that were described in this article provide efficient simulations of workcells allowing for the testing of different manipulators. The created programs are based on the static robot models; cell calibration is also possible. However, none of the systems covers difficult programming situations like the simultaneous movement of a positioner and a robot, or the orientation of an object by a positioner. Robots are programmed in the object-level, even though process programming is highly desirable.

Off-line programming will become a much more used method if the systems function on a high, process-programming level, if geometrically difficult programming is possible and if robot manufacturers improve the accuracy of their robots

References

[1] ISO/DP 10562-1. ICR : Intermediate Code for Robots. Technical Report ISO/TC 184/SC 2, 1987.

[2] B.Y.G. Systems Ltd. *GRASP Reference Manual*, 1989.

[3] V. Gorbachev, V. Lipping, and A. Dryomov. Perestroika and off-line welding. *Industrial Robot*, 18(1):24–27, 1991.

[4] B. Gustafsson. A system for the off-line programming of abb-robots. Master's thesis, Dept. of Production Eng., CTH, 1989.

[5] S. Hayati. Robot arm geometric link parameter estimation. In *Proc. of the 22nd IEEE Conf. on Decision and Control*, December 1983.

[6] Silma Inc. *CimStation User's Manual*, January 1990.

[7] Technomatix Technologies,. *ROBCAD Technical Description*, Sepetember 1990.

Practical Error Compensation for Use In Off-Line Programming of Robots

S. Albright and K. Schröer
Fraunhofer-Institute for Production Systems
and Design Technology, IPK
Pascal Strasse 8-9, D-1000 Berlin 10

1 Introduction

As tasks of robots become more sophisticated, the process of robot task-teaching and re-teaching becomes longer and more tedious. For complex tasks, re-teaching can require that a robot be out of production for several weeks. To circumvent this process, a task can be programmed off-line as a series of robot poses which define an efficient trajectory and enable the robot to complete the task. These poses are then entered into the robot controller which directs the robot from one pose to the next. How accurately the robot can attain the mathematically generated poses must be considered when using an off-line generated routine.

For taught routines, the robot´s accuracy is defined by its ability to repeatedly reach a taught pose. Pose repeatabilities of today´s robots are usually better than 1 mm while absolute pose accuracy (i.e. precision which a robot can reach a numerically given position and orientation relative to an external frame) are found to be at least an order of magnitude worse.

Robot pose accuracy issues have been investigated by several researchers [2-7]. These investigations include the development of various robot calibration techniques used to determine the kinematic and sometimes mechanical features of the static robot by identifying (through measurements) model parameters. Absolute pose accuracy of a robot is improved using the identified parameters in an error compensation procedure making off-line programming a useful tool.

The accuracy of a compensation routine can be hindered when not adequately identifying the tool-mounting flange frame in terms of the robot. With some exceptions [6], the need for proper robot tool-mounting flange identification has not been recognized by researchers. This identification is important in order to accurately describe tool location in terms of the robot. A technique for flange identification is discussed in the following sections.

2 Review of IPK Calibration

A light review of the calibration procedure is included here as a precursor to describing how the tool flange is defined. Details of this procedure are found in [5].

2.1 Robot Model and Measurement

The calibration procedure identifies both kinematic and mechanical features of a robot. The kinematic model is based on the Denavit-Hartenberg (DH) parameter notation [1]. Any combination of rotary and prismatic joints can be modeled. However, for consecutive joints which are parallel, the extended DH-model proposed by Hayati [3] is used because its parameters for this configuration are continuous with small changes in joint geometry while the normal DH parameters are not. This is important since parameters are identified numerically.

Using elementary transformations R_u to symbolize rotation about coordinate axis u and T_u to symbolize translation along coordinate axis u, three consecutive coordinate frames are described:

a) When first joint is rotary and next is (near) orthogonal; $R_z T_z T_x R_x T_z$
b) When first joint is rotary and next is (near) parallel; $R_z T_x R_x R_y T_z$
c) When first joint is prismatic; $T_z R_x R_y T_z$

Because the parameter of the last elementary transformation in each of the above is redundant, it is not included in the numerical identification and thus retains its orignal given value.

Defining the transformations from the robot base frame to the first joint frame and from the last joint frame to the measurement object (also refered to as tool-center-point,TCP), completes the kinematic description. The transformation from the robot base to first joint frames is constant and is modeled either as a) or b) depending on the relative orientation of the base frame's z-axis and the first joint axis. The transformation from the last joint to the TCP is described using three parameters since only position measurements are made of the TCP. This transformation is written $R_z T_z T_x$.

Once the kinematic model is defined, the mechanical model is constructed. The robot characteristics which are modeled and identified are:

- Gear elasticity; modeled assuming linear elasticity is used.
- Gear transmission and coupling effects; important since joint encoder values are used as system input.
- Gear eccentricity; modeled by a first order Fourier approximation.
- Gear backlash; important for diagnostic purposes.
- Link elasticity; modeled as a cantilevered beam using linear beam theory.

Given all model parameters and joint encoder values, the series of elementary transformations from robot base frame to TCP frame is defined by a continuously differentiable function describing the stationary robot configuration.

The parameters are identified using 100 to 200 theodolite measurements of a lit ceramic ball-target fixed to the robot's flange. A Leica theodolite measuring system is used which has a position-measurement accuracy of ±0.05 mm.

Measured and predicted target positions are used in a numerical least-squares calculation to identify the actual robot parameters. The results of three calibrations conducted at IPK have been evaluated by measuring robot poses other than those used for calibration. In all three cases, robot accuracy was improved to the limits of repeatability (approximately ± 0.5 mm) disregarding possible errors in the tool-mounting flange.

2.2 Parameter Identification for Compensation

Error compensation of an off-line generated pose is made using calibrated robot model results. As shown in Figure 1, the target poses are converted from cartesian space to joint-encoder space using the calibrated robot model in a kinematic inverse calculation.

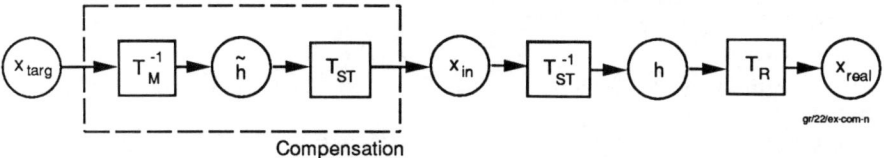

Figure 1. Off-line error compensation: X_{targ}, target pose; T_M, calibrated robot model; h, $\tilde{h}$, joint encoder values; T_{ST}, control model; X_{in}, compensated pose; T_R, function describing actual robot; X_{real}, actual pose.

Using the controller model in a forward calculation, the joint-encoder space poses are converted back to cartesian space. These two conversions comprise the compensation routine. The compensated poses are entered into the robot controller which converts the poses back joint-encoder space poses and sends them to the robot. If the calibrated model is an accurate representation of the actual robot, then the off-line generated target poses and actual robot poses will not differ. This is a practical form of compensation since target poses are corrected outside of the controller and do not disturb controller algorithms.

Off-line generated target poses and calibrated robot data must be compatible to make compensation possible. The target poses generated by an off-line program are likely to describe a tool frame with respect to a work-cell frame. In contrast, the calibrated robot is defined from its base frame F_b to a coordinate frame fixed in the measured target F_{mt} as seen in Figure 2.

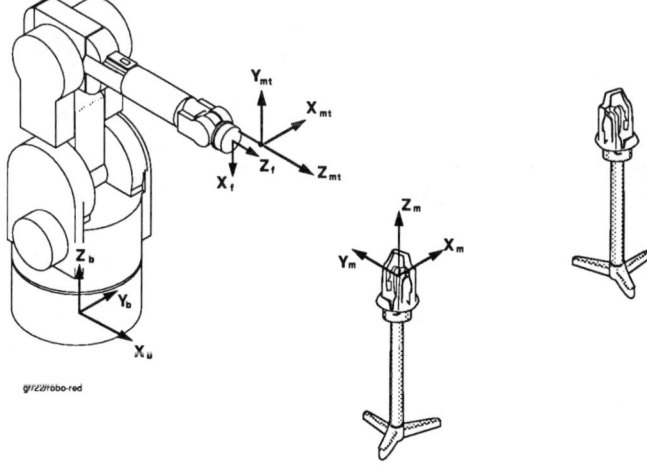

Figure 2. Measurement-system and robot coordinate frames defined by identified parameters.

In order to use the calibrated robot data to compensate a target pose, both robot model and target pose must describe the same single robot frame with respect to a single reference frame. Since various tools are likley to be mounted to any one robot in more than one work-cell environment, the obvious description to chose is one that defines the robot's tool-mounting flange frame, termed here as F_f, with respect to the robot base frame.

To identify parameters from frame F_b to F_f requires:

a) Knowledge of the precise position of measured-target frame F_{mt} in flange frame F_f before calibration.
b) Identification of parameters describing the robot from frame F_b to frame F_f.

It is assumed that the transformation of the selected tool (end effector) with respect to its mounting flange frame is known. The transformation from the work-cell to the robot base can be determined by explicit measurement or by using the calibrated robot as a measurement tool. Section 3 addresses the above requirements a and b.

3 Parameter Identification from Robot Base to Tool-Mounting Flange

Redefinition of the calibrated robot such that the flange is defined in terms of the robot base means some changes to the calibration procedure. As presented by Spur and Schröer, the calibrated robot is defined from the measurement system reference frame F_m, through the robot-base frame F_b, to the frame F_{mt} attached to the center of the measured target as shown above in Figure 2.

The robot base frame is arbitrarily chosen–typically following manufacturer's guidelines–and separating out the parameters defining the transform from frame F_m to frame F_b is a simple matter. However, the position of frame F_{mt} is defined in terms of the last joint coordinate frame using three parameters (i.e. $R_z T_z T_x$) and frame F_f is not specifically identified. To identify this frame, the target frame F_{mt} must be measured separately with respect to its flange frame before calibration. Parameters following the last joint frame can then be defined to describe frame F_f. Because of its nature, the flange-frame parametric description must differ from those described in section 1.1.

The kinematic model describing the relationship between succeeding joint frames uses 4 to 5 parameters for each joint relationship. The orientation of a joint frame is defined according to the direction of the joint axis and its relationship to the preceding joint axis. In contrast, the frame of the tool-mounting flange is defined by robot manufacturer conventions which are often determined by fittings on the flange. Therefore, it must be assumed that the last joint frame and the flange frame have a general relationship to one another. A complete description of two frames in general relative pose necessitates 6 parameters.

Assuming a robot of n joints, the transformation from joint n to frame F_f is written here as

$$A_n = R_z T_z T_x R_x R_y R_z \qquad (1)$$

The first three elements of A_n define the position of the flange frame and the last three, the orientation. A cylindrical position transformation allows expression of rotary joint motion.

The tool-mounting-flange frame can be identified only if the position and orientation of the flange frame are measurable. Using a single target for measurement allows only position identification and not orientaion identification. If it can be assumed that the last joint axis (i.e. last joint frame z-axis) coincides with flange frame coordinate axis defining the flange normal (i.e. flange frame z-axis) and that the angle between the last joint frame x-axis and flange frame x-axis is precisely known, then a single target is sufficient. However, these assumptions are likely to introduce tool positioning errors whose magnitude depend on tool size and robot manufacturing precision. In order to include the identification of the flange frame orientation, at least three targets must be available for measurement.

3.1 Measurement Object

Because the added precision gained by measuring flange frame orientation may not always be practical, two measurement-object concepts are introduced here. One includes the assumptions which make a single target sufficient and the other assumes a general relationship between last joint frame and flange frame requiring at least three targets. In both cases, the target centers must be precisely measured (approximately ±0.02 mm) with respect to the predefined measurement-object flange frame.

For clarification purposes, three flange frames are defined. One frame is fixed in the flange at the end of the robot arm, frame F_{fr}, and an other in the flange of the measurement object, frame F_{fm}, as seen in Figure 3. When these two flanges are mounted to one another, it is assumed that F_{fr} and F_{fm} are coincident. The orientation of these two flange frames is dictated by a third flange frame F_{ft} fixed in the mounting flange of the work-tool. The orientation of F_{ft} is typically defined by a fitting on the tool flange.

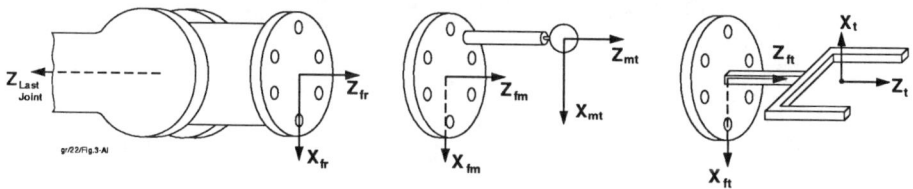

Figure 3. Conceptual illustration of flange frames.

The single target measurement object currently used in IPK calibration procedures is adequate only for flange position identification. Position of the ball-target's center must be precisely measured (e.g. using coordinate measuring machine) to determine its location with respect to the flange frame F_{fm}.

For a multi-target measurement object, four targets are desirable because of the chance that a theodolite's view of one target will be blocked by another during the measurement procedure. As in Figure 4, the targets are configured so their centers coincide with the corners of a tetrahedron and hence, the distances between each target are equal. This distance and the accuracy of the measurement system determine the orientation measurement accuracy.

Orientation measurement accuracy influences the precision of tool-mounting flange identification which in turn effects tool positioning accuracy, depending on the size of the tool. For small robots which can carry only light payloads, the tool is likely to be small and vise versa for large robots. Therefore, the size of the target assembly should be dictated primarily by the size of the robot. As an example, when wanting to limit the introduction of tool positioning errors to 0.10 mm, the distance between targets should range from 60 mm to 180 mm for tools sizes ranging from 100 mm to 300 mm (measured from tool frame to tool-center-point).

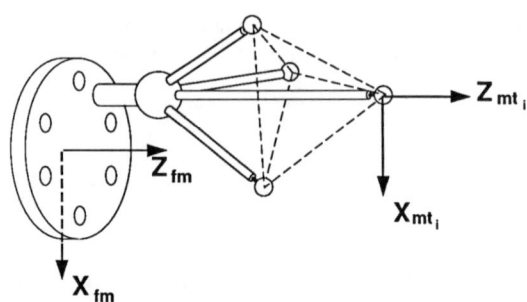

Figure 4. Measurement object with 4 targets.

3.2 Parameter Identification

Extending the modeling concept of Spur and Schröer, the robot is described to its end-flange frame F_{fr}–the last six parameters being those included in the transformation of equation (1). For the case where flange orientation is not identified and only a single target is used, only the first three parameters of A_n defining position are necessary. Formulation of the parameter identification problem is as follows.

Given

m:	Total number of identified parameters.
p:	Model parameters vector describing robot from measurement system to robot end flange.
h:	Joint encoder values vector.
T (p, h):	Nominal homogeneous transformation defining frame F_{fr} in terms of frame F_m.
$M_i(h)$:	Measured-target position "i" for robot pose h in frame F_m.

For a given h and using model parameters p, a measured-target position "i" in frame F_m can also be described as

$$t_i(p,h) = T(p,h) \cdot b_i \qquad (2)$$

where vector b_i defines a target position in frame F_{fm} and is embedded in 4-dimensional projective space described by

$$b_i = [x_i, \ y_i, \ z_i, \ 1]^T. \qquad (3)$$

For k measured poses, the following are defined.

$$\overline{h} = [h_1, ..., h_k]^T$$ (4)

$$\overline{t}(p,\overline{h}) = [t_1(p,h_1), t_2(p,h_1), ..., t_3(p,h_k), t_4(p,h_k)]^T$$ (5)

$$\overline{M}(\overline{h}) = [M_1(h_1), M_2(h_1), ..., M_3(h_k), M_4(h_k)]^T$$ (6)

The solution to the non-linear least-squares problem is the parameter vector p* where

$$\|\overline{t}(p^*,\overline{h}) - \overline{M}(\overline{h})\| = \min\|\overline{t}(p,\overline{h}) - \overline{M}(\overline{h})\|, \text{ for all } p.$$ (7)

The non-linear problem is linearized using a first order Taylor Series which approximates the actual frame F_{fr} pose relative to frame F_m.

$$T(p + \Delta p, h) = T(p, h) + D_p T(p, h) \cdot \Delta p.$$ (8)

Multiplying by vector b_i, the actual measured-target position in frame F_m can be approximated as

$$T(p + \Delta p, h) \cdot b_i = (T(p, h) + D_p T(p, h) \cdot \Delta p) \cdot b_i.$$ (9)

$D_p T(p, h)$ is the Jacobi matrix of $T(p, h)$. The measured position is equated to the above linearly approximated position to get

$$\overline{M}(\overline{h}) - \overline{t}(p,\overline{h}) = D_p \overline{t}(p,\overline{h}) \cdot \Delta p$$ (10)

where the Jacobian is written

$$D_p \overline{t}(p,\overline{h}) = \begin{bmatrix} \frac{\partial}{\partial p_1} T(p,h_1) \cdot b_1, & ..., & \frac{\partial}{\partial p_m} T(p,h_1) \cdot b_1 \\ \frac{\partial}{\partial p_1} T(p,h_1) \cdot b_2, & ..., & \frac{\partial}{\partial p_m} T(p,h_1) \cdot b_2 \\ \vdots & & \vdots \\ \frac{\partial}{\partial p_1} T(p,h_k) \cdot b_4 & ..., & \frac{\partial}{\partial p_m} T(p,h_k) \cdot b_4 \end{bmatrix}$$ (11)

ignoring the fourth row in the result of each matrix multiplication. Vector Δp is written

$$\Delta p = [\Delta p_1, ..., \Delta p_m]^T$$ (12)

A least-squares procedure is then used to solve for Δp to get $p^* = p + \Delta p$.

The relationship between frame F_{fr} and the last joint axis of a robot can cause problems in the numerical solution of equation (10). Because the last joint axis and the frame F_{fr} z-axis are likely to lie approximately along the same line, the rotation parameter in the first element R_z of expression (1) can vary from 0 to π with small variations in the z-axis position of frame F_{fr}. This causes discontinuity of the model transformation T(p,h) as a function of the model paramters p. The problem is overcome by identifying an intermediate flange frame F_{fr}' which is shifted from frame F_{fr} in the xy-plane.

In this case, F_{fr}' is arbitrarily defined by a 100-mm shift in x so that

$$^{fr'}\mathbf{T}_{fr} = \begin{bmatrix} 1 & 0 & 0 & -100 \\ 0 & 1 & 0 & 0 \\ 0 & 0 & 1 & 0 \\ 0 & 0 & 0 & 1 \end{bmatrix} \tag{13}$$

When shifting frame F_{fr} 100 mm to F_{fr}', a new vector $\mathbf{b}_i'$ must be used which is written

$$\mathbf{b}_i' = [x_i - 100,\ y_i,\ z_i,\ 1]^T. \tag{14}$$

Replacing vector **b** with **b**′ in the above numerical problem results in a solution vector p* which identifies frame F_{fr}'. Assuming F_{fr}' is defined with respect to the last joint frame by $\mathbf{A}_n'$, the parameters defining F_{fr} can be extracted from the equation

$$\mathbf{A}_n = \mathbf{A}_n'\ ^{fr'}\mathbf{T}_{fr}. \tag{15}$$

4 Conclusion

Compensating errors in robot trajectories generated by off-line programs can be accomplished outside of the robot control so not to disturb control algorithms. The parameters of the robot must be properly identified in order to make compensation possible.

Presented in this paper is a technique which identifies robot parameters from the robot base frame to its tool-mounting flange frame using a single set of position measurements. Required is a measurement object having at least 3 targets enabling the measurement of flange orientation. A measurement object with four targets whose relative positions locate the corners of a tetrahedron is proposed. Size of the tetrahedron is important for accurate orientation measurement.

The numerics of the identification routine are adapted to include known data of the measurement object. This is accomplished through an extension of the Spur and Schröer calibration procedure.

5 References

[1] Denavit, J. and Hartenberg, R.S., "A Kinematic Notation for Lower-Pair Mechanisms Based on Matrices," Journal of Applied Mechanics, June 1955, vol. 77, pp. 215-221.

[2] Everett, L.J. and Suryohadiprojo, A.H., "A Study of Kinematic Models for Forward Calibration of Manipulators," Proc. of IEEE Int'l. Conf. on Robotics and Automation, 1988, pp. 798-800.

[3] Hayati, S.A., and Mirmirani, M. "Improving the Absolute Positioning Accuracy of Robot Manipulators," Journal of Robotic Systems, 1985, vol. 2(4), pp. 397-413.

[4] Judd, R.P. and Knasinski, A.B., "A Technique to Calibrate Industrial Robots With Experimental Verification," IEEE Transactions on Robotics and Automation, Feb. 1990, vol. 6, no. 1, pp. 20-30.

[5] Spur, G. and Schöer, K., "Kalibrierung von Industrierobotern," Vorschubantriebe in der Fertigungstechnik, eds.: G. Pritschow, G. Spur, M. HWeck. München: Hanser, 1989, pp. 129-149.

[6] Veitschegger, W.K. and Wu, C-h., "A Method For Calibrating and Compensating Robot Kinematic Errors," Proc. of IEEE Int'l. Conf. on Robotics and Automation, 1987, vol. 1, pp. 39-44.

[7] Ziegert, J., "Robot Calibration Using Local Pose Measurements," International Journal of Robotics and Automation, 1990, vol. 5, no. 2, pp. 68-76.

Intelligent Programming of Force-Constrained Cooperating Robots

G. Duelen, H. Münch, Y. Zhang
Fraunhofer-Institute for Production Systems
and Design Technology (IPK)
Pascalstr. 8-9, D-1000 Berlin 10, Germany

Abstract
In this paper, concepts and algorithms for the coordinated optimal motion planning and force distribution for force-constrained cooperating robots are presented. The developed algorithms are characterized by the possibility to take various task specific requirements into consideration and by its adaptability to any kinematic and any assignment in the production cell through corresponding modelling, without change of the solution structure. This implies, on one hand, free programmability, and on the other hand, the integration of all task-related constraints and requirements.

1 Introduction

Due to the increased industrial requirements, there is great interest in using multiple robot arms to cooperatively manipulate an object, process a workpiece or perform part mating operations. The introduction of such systems will allow the performance of more sophisticated tasks than those currently undertaken using single robots. Cooperating robot systems can be principally devided into two categories, namely contact-free and force-constrained cooperating robots. The coordinated continuous path control of contact-free cooperating robots has been studied in /1/. The main interest centered around synthesizing the optimal motion trajectory for two cooperating robots in a common continuous path mode while satisfying task requirements, and avoiding collisions among the robots and the environment. In contrast to contact-free cooperating robots, force-constrained cooperating robots are both kinematically and dynamically coupled. In this case, coordinated planning and control of the motion and force behaviours are necessary. With respect to characteristic properties, relevant application areas for force-constrained cooperating robot systems can be divided into three basic classes.

Case 1: Two robots transporting a workpiece. For such tasks, the position and orientation difference between the end effectors of the involved robots has to be kept constant. Typical examples are the transport of liquids, of heavy or large workpieces with low bending resistances which are beyond the carrying capacity of one robot.

Case 2: Two robots performing a common processing task. These tasks require relative motion and contact forces between the tool and workpiece carrying robot endeffectors. In this case, the cooperating robots work in a common continuous path mode with defined contact force. Typical examples are the common processing of workpieces with a complex geometry, e.g. deburring or polishing of complex waved surfaces which are difficult for a single arm to approach.

Case 3: Two robots performing an assembly task. These tasks include primarily transitions between contact-free and force-constrained motions, which arise e.g. during the common execution of a mating task.

In this paper, the cases 1 and 2 will be discussed. In the following, the term cooperating robots refers to force-contrained cooperating robots, the term position refers to position and orientation, the term force refers to force and torque, and the upper index "*" indicates the optimal values, unless otherwise cited. The realization of cooperating motion of industrial robots is achieved almost exclusively in two sequential phases, the planning phase and the execution phase, due to the problem complexity. In the planning phase, the nominal trajectories of the motion and the force behaviours for each involved robot are generated based on the production task requirements. In the execution phase, the approach of the nominal motion and force behaviours generated in the planning phase is realized.

For a cooperating robot system, the degrees of freedom are in general less than the total number of the joints. Such systems are also kinematically redundant. In addition to this, since generally every joint of industrial robots is installed with an actuator, the number of the actuators will be greater than the degrees of freedom. As a result, there are in general infinitively many solutions for the nominal motion and force behaviours of the individual robots for a given task. A number of publications on the task planning and control of cooperating robots have appeared recently /2...5/. Orin and Oh /2/ have studied the control of force distribution for closed-chain robotic mechanisms. A weighted combination of energy consumption and load balancing was selected as the criterion. The linear programming technique was used to obtain a solution and the method was used to compute joint torques to drive a hexapod locomotion vehicle. Nakamura et. al. /3/ focused on the object dynamics and contact stability. The forces required to move the object are distributed in such a way that the internal forces (such as internal grasping forces) are minimized while satisfying the static frictional constraints. The manipulator dynamics were not considered in /3/. Zheng and Luh /4/ have addressed the optimal load distribution for two robots handling a single object. A solution was proposed for the optimal load distribution with minimum exerted forces on the object. These papers are concerned only with the aspect of force distribution for cooperating robots handling a single object. Bruhm /5/ has studied the coordinated motion planning and force distribution for cooperating robots handling a single object. Since the motion trajectory of the object was assumed to be prespecified, the nominal motion trajectories of the involved robots were generated by direct decomposition. The redundancy in the force distribution was used to optimize a object-based criterion, such as minimization of forces exerted on the object. However, the manipulator dynamics were not considered /5/.

In this paper, concepts and algorithms for the coordinated optimal motion planning and force distribution for cooperating robots performing different types of tasks are presented. The nominal motion and force behaviours of the individual robots are determined in such a way that redundant degrees of freedom within the system may be used to optimize a task-related criterion while satisfying kinematic and dynamic constraints. Based on the task specific requirements different optimization criteria are discussed. This paper is organized as follows. Section 2 describes briefly the kinematic and dynamic models of cooperating robots. In section 3, the algorithms for the coordinated optimal motion planning and force distribution are detailed. Conclusions and remarks follow in section 4.

2 Kinematic and Dynamic Modelling of Cooperating Robots

2.1 Kinematic and dynamic modelling of cooperating robots transporting a workpiece

Kinematic relation

Consider two robots, each with six degrees of freedom (d.o.f.), transporting a workpiece (Fig. 1) which is beyond the carrying capacity of one robot. To move the workpiece, two robots must grasp it at two specified points G1 and G2. It is assumed that the end-effectors furnish tight grips so that there are no relative motions among the end-effectors and the workpiece. Thus a closed kinematic chain is formed.

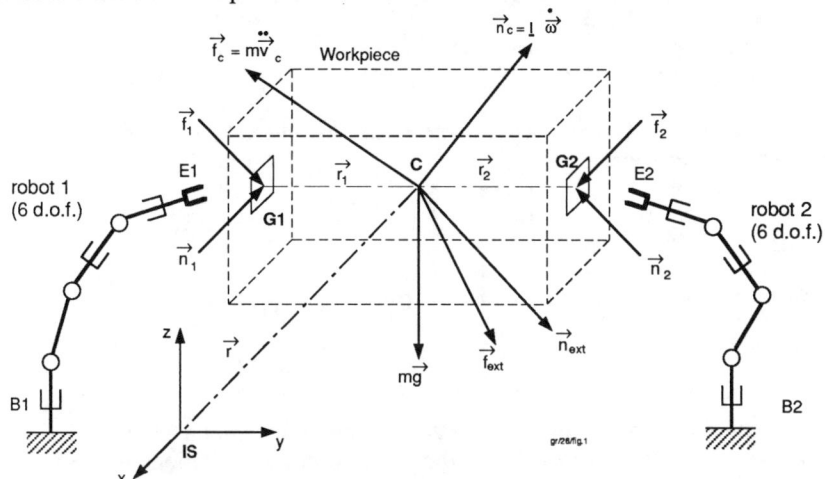

Fig. 1: Two industrial robots transporting a workpiece

The homogeneous transformation matrix T_B^E describes generally the position and orientation of the frame E with reference to the coordinate system B. Using the Denavit-Hartenberg convention /6/ and the definition of the homogeneous transformation matrix, the position constraint equations can be represented by the matrix equation:

$$T_{IS}^{B1} T_{B1}^{E1}(Q_1) T_{E1}^{C} = T_{IS}^{B2} T_{B2}^{E2}(Q_2) T_{E2}^{C} \tag{1}$$

Q_1 and Q_2 are the joint coordinate vectors of both robots. The notations for the coordinate systems are indicated in Fig. 1. Gripping point coordinates G1 and G2 on the workpiece are so defined, that they are identical to the end-effector coordinate systems E1 and E2 during transportation.

There are 12 scalar equations in Eq. (1), but only six equations are independent because the position and orientation of each frame can be uniquely defined by six parameters. Eq. (1) can be also represented by six scalar equations which are highly nonlinear in joint displacements and are difficult to solve. However, these nonlinear equations can be linearized for infinitesimal joint increments to yield the following linear constraint equation:

$$\underbrace{[\underline{J}_{C1} \ -\underline{J}_{C2}]}_{\underline{J}} \underbrace{\begin{bmatrix} \Delta Q_1 \\ \Delta Q_2 \end{bmatrix}}_{\Delta Q} = \underline{0} \tag{2}$$

J_{C1} and J_{C2} describe the Jacobian matrices of robots 1 and 2 with C as the object coordinates and IS as the reference coordinates.

Motion equation of the workpiece

As shown in Fig. 1, the workpiece is assumed to be rigidly held by the end-effectors of the cooperating robots. Under the influence of the forces exerted by the end-effectors, the workpiece moves along the desired motion trajectory. Using the Newton and Euler equations allows the derivation of the motion equation of the workpiece:

$$\sum_i \vec{f}_i + m\vec{g} + \vec{f}_{ext} = m\ddot{\vec{r}} \qquad (3)$$

$$\sum_i (\vec{n}_i + \vec{r}_i \times \vec{f}_i) + \vec{n}_{ext} = \underline{I}\dot{\vec{\omega}} + \vec{\omega} \times \underline{I}\vec{\omega} \qquad (4)$$

Equations (3) and (4) may be expressed with reference to the workpiece coordinate system C and combined to yield

$$\underbrace{\begin{bmatrix} \underline{R}_C^{G1} & 0 & \underline{R}_C^{G2} & 0 \\ \vec{r}_1^C \otimes \underline{R}_C^{G1} & \underline{R}_C^{G1} & \vec{r}_2^C \otimes \underline{R}_C^{G2} & \underline{R}_C^{G2} \end{bmatrix}}_{\underline{B}(6 \times 12)} \underbrace{\begin{bmatrix} \vec{f}_1 \\ \vec{n}_1 \\ \vec{f}_2 \\ \vec{n}_2 \end{bmatrix}}_{F_E(12 \times 1)} = \underbrace{\begin{bmatrix} m\underline{R}_C^{IS} \ddot{\vec{r}}^{IS} - \vec{f}_{ext}^C - m\underline{R}_C^{IS} \vec{g}^{IS} \\ \underline{I}^C \dot{\vec{\omega}}^C + \vec{\omega}^C \otimes (\underline{I}^C \vec{\omega}^C) - \vec{n}_{ext}^C \end{bmatrix}}_{F_L(6 \times 1)} \qquad (5)$$

where $\underline{R}$=rotation matrix, f and n=force and torque, v and ω=linear and angular velocity, m and $\underline{I}$=mass and inertia moment of the workpiece, g=gravitational force, $\vec{r}$=position vector, $\vec{r}\otimes$=notation for the vector product, the upper or lower indices indicate the coordinates that corresponding vectors or matrices refer to.

$\underline{B}$ is a (6x12) matrix and depends only on the workpiece's geometry. The vector F_E contains the forces exerted on the workpiece by the end-effectors. F_L is a force vector which is dependent on the motion trajectory of the workpiece, and on the external force f_{ext} and torque n_{ext} exerted by the environment.

Manipulator Dynamics

The motion equation of the cooperating robots can be obtained by extending and combining the dynamics of the individual robots, as described by:

$$U_i = \underline{H}_i(Q_i)\ddot{Q}_i + \underline{h}_i(Q_i, \dot{Q}_i) + G_i(Q_i) + \underline{J}_{Ei}^T F_{Ei} \qquad i = 1, 2 \qquad (6)$$

where U_i = joint driving torques, $\underline{H}_i$=inertia matrix, $\underline{h}_i$=Coriolis, centrifugal, and frictional force vector, G_i=gravitational force vctor, $\underline{J}_{Ei}$=Jacobian.

2.2 Kinematic and dynamic modelling of cooperating robots performing a common processing task

Kinematic relation

We consider now two robots performing a common processing task, such as deburring or polishing a workpiece with complex waved surface which is for a single robot difficult to approach. Refer to Fig. 2, the robot on the left carries the tool and the robot on the right carries the workpiece. Both robots are described by solide lines. Additionally, two virtual kinematic chains, described by dashed lines, for modelling and handling of restrictions and tolerance areas are introduced /1/. The magnified detail in Fig. 2 shows the actual trajectory segment defined with reference to the end-effector frame E2, and the virtual kinematic chain v1 modelling the current position and orientation tolerances as well as restrictions related to the actual trajectory segment. The other kinematic chain v2 represents a model of the inertial system related restriction and tolerance areas. Both virtual kinematic chains are represented by three orthogonal prismatic and a spherical joint without any displacements or link lengths.

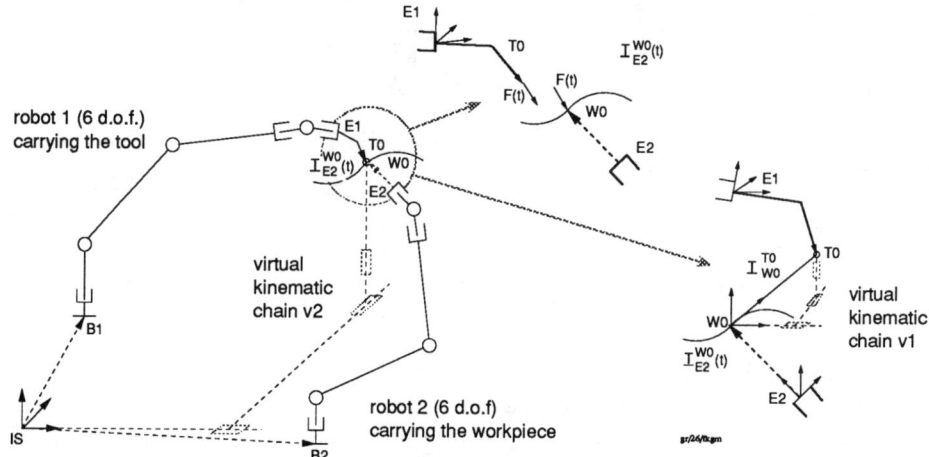

Fig. 2: Two robots performing a common processing task

As indicated in Fig. 2, the desired motion trajectory of the tool center point TO is defined with reference to the frame E2 as $\underline{T}_{E2}^{TO}(t)$. Then the position and orientation constraint can be represented by the matrix equations:

$$\underline{T}_{IS}^{B1}\underline{T}_{B1}^{E1}(Q1)\underline{T}_{E1}^{TO} = \underline{T}_{IS}^{B2}\underline{T}_{B2}^{E2}(Q2)\underline{T}_{E2}^{WO}(t)\underline{T}_{WO}^{TO}(Qv1) \tag{7}$$

$$\underline{T}_{IS}^{TO}(Q2,Qv1,t) = \underline{T}_{IS}^{TO}(Qv2) \tag{8}$$

Each of the matrix equations (7) and (8) can be represented by six scalar equations which are highly nonlinear in joint displacements and are difficult to solve. However, these nonlinear equations can be linearized for infinitisimal joint increments and combined to yield the accumulated linear equation system (9).

$$\underbrace{\begin{bmatrix} \Delta X \\ \Delta \tilde{X} \end{bmatrix}}_{\Delta X_{ac}} = \underbrace{\begin{bmatrix} \underline{J}_{E1}(Q1) & -\underline{J}_{E2}(Q2) & -\underline{J}_{v1}(Qv1) & \underline{0} \\ \underline{0} & \underline{J}_{E2}(Q2) & \underline{J}_{v1}(Qv1) & -\underline{J}_{v2}(Qv2) \end{bmatrix}}_{\underline{J}_{ac}} \underbrace{\begin{bmatrix} \Delta Q1 \\ \Delta Q2 \\ \Delta Qv1 \\ \Delta Qv2 \end{bmatrix}}_{\Delta Q_{ac}} \tag{9}$$

To solve the underdetermined equation system (9) the accumulated Jacobian matrix $\underline{J}_{ac}$ has to be determined. The Jacobians of the tool- and workpiece carrying robots can be evaluated based on the corresponding DH-representation /7/. The determination of the Jacobians of the virtual kinematic chains is quite simple /1/.

Manipulator Dynamics

Refer to Fig. 2, the contact force between the tool and the workpiece F(t), exerted at the contact point, is defined with reference to the frame E2. In order to derive the manipulator dynamics, F(t) is transformed into the end-effector frames E1 and E2. Then the dynamic equations of the manipulators can be represented by:

$$U_i = \underline{H}_i(Q_i)\ddot{Q}_i + \underline{h}_i(Q_i,\dot{Q}_i) + G_i(Q_i) + \underline{J}_{Ei}^T \underline{\mathfrak{R}}_i F(t) \qquad i=1,2 \qquad (10)$$

where $\underline{\mathfrak{R}}_i$ is (6x6) matrix which transforms the contact force F(t) into the end-effector frame Ei (i=1,2). These are defined by:

$$\underline{\mathfrak{R}}_1 = \begin{bmatrix} \underline{R}_{B1}^{E2} & \underline{0} \\ \underline{R}_{B1}^{E1}\tilde{\gamma}_1 \otimes \underline{R}_{B1}^{E2} & \underline{R}_{B1}^{E2} \end{bmatrix} \qquad \underline{\mathfrak{R}}_2 = \begin{bmatrix} \underline{R}_{B2}^{E2} & \underline{0} \\ \underline{R}_{B2}^{E2}\tilde{\gamma}_2 \otimes \underline{R}_{B2}^{E2} & \underline{R}_{B2}^{E2} \end{bmatrix} \qquad (11)$$

$\tilde{\gamma}_1$ and $\tilde{\gamma}_2$ are the position vectors from the origins of the frame E1 and E2 to the contact point with reference to E1 and E2, respectively.

3 Optimal Motion Planning and Force Distribution for Cooperating Robots

3.1 Optimal Motion Planning and Force Distribution for Cooperating Robots Transporting a Workpiece

3.1.1 Optimal Motion Planning

- *The motion trajectory of the workpiece is prespecified*

In the case that the motion trajectory of the workpiece is predefined with reference to the inertial system as $\underline{T}_{IS}^C(t)$, the nominal motion trajectories of the involved non-redundant robots can be generated by direct decomposition as described by:

$$\underline{T}_{B1}^{E1}(t) = (\underline{T}_{IS}^{B1})^{-1}\underline{T}_{IS}^C(t)\underline{T}_C^{G1}, \quad \underline{T}_{B2}^{E2}(t) = (\underline{T}_{IS}^{B2})^{-1}\underline{T}_{IS}^C(t)\underline{T}_C^{G2} \qquad (12)$$

$\underline{T}_{IS}^{B1}$ and $\underline{T}_{IS}^{B2}$ depend only on the assignment of the robots in the production cell, while $\underline{T}_C^{G1}$ and $\underline{T}_C^{G2}$ depend on the geometry of the workpiece. These matrices can be also calculated a priori, so that this solution is well suitable for on-line realization.

- *Only the beginning and end positions of the workpiece are prespecified.*

In this case only the beginning and end configurations of the closed kinematic chain are given, while the connecting trajectory is underdetermined. Motion synthesis implies the selection of the optimal trajectory under consideration of the robot and task specific constraints. The constraint relation between the joint increments is described by the linear equation system (2). Using the linear programming technique /8/ and choosing the weighted quadratic norm of the joint increments $\Delta Q^T \underline{M} \Delta Q$ as the optimization criterion, we can obtain the following local optimal solution:

$$\Delta Q = (\underline{J}^+\underline{J} - \underline{E})\nabla f(Q) \quad \text{with} \quad \underline{J}^+ = \underline{M}^{-1}\underline{J}^T(\underline{J}\,\underline{M}^{-1}\underline{J}^T)^{-1} \qquad (13)$$

where $\underline{J}^+$ is the Pseudo Inverse matrix, $\underline{M}$ is a symmetric, positive semi-definite weighting matrix, and $\underline{E}$ is a unit matrix. The vector $\nabla f(Q)$ presents the gradient of the function f(Q). Because the matrix $(\underline{J}^+\underline{J} - \underline{E})$ describes an orthogonal space of $\underline{J}$, the component $(\underline{J}^+\underline{J} - \underline{E})\nabla f(Q)$ produces also no relative motion between the end-effectors. However, this component forces, among other things, the decrease of f(Q). The goal of the considered task lies in the optimal transportation of the workpiece from the beginning position to the end position with minimization of the chosen criterion. Therefore, it is suitable to choose f(Q) as a proper function of the position difference between the actual position $\underline{X}_C(t)$ and the given end position $\underline{X}_C(t_e)$ of the workpiece, where t and t_e are the actual and the end time. If f(Q) is chosen as the quadratic norm of the position difference, then $\nabla f(Q)$ can be calculated by:

$$\nabla f(\underline{Q}) = \begin{bmatrix} 2(\underline{J}_{C1})^T (\underline{X}_C(t) - \underline{X}_C(t_e)) \\ \underline{0}(6 \times 1) \end{bmatrix} \tag{14}$$

3.1.2 Optimal Force Distribution

Eq. (5) shows that the distribution of the forces required to move the workpiece among the individual robots, is generally a mathematically underdetermined problem. According to the task specific requirements, the optimization criteria for the force distribution can be usually specified by one of the following aspects:

- *Minimization of the exerted forces on the workpiece*

When transporting a large workpiece with low bending resistance, excessive stress on the workpiece has to be avoided. In this case, the weighted quadratic norm of the forces exerted on the workpiece by the end-effectors $(F_E)^T M\ F_E$ is chosen as the criterion. The force constraint is specified by Eq. (5). Using the linear programming thechnique gives the local optimal solution:

$$\underline{F}_E^* = \underline{B}_M^+ \underline{F}_L \text{ with } \underline{B}_M^+ = \underline{M}^{-1} \underline{B}^T (\underline{B}\,\underline{M}^{-1}\underline{B}^T)^{-1} \tag{15}$$

It can be seen from Eq. (5) and (15) that the nominal forces of the end-effectors of the involved robots are equal to zero, if no external forces and gravitatioal forces are present. This means that the force distribution according to Eq.(15) eliminates all static internal forces in the workpiece. This solution does not take the kinematic and dynamic behaviour of the involved robots into consideraton and provides also the optimal force distribution in the task space. The matrix $\underline{B}_M$ depends only on the geometry of the workpiece, and can be calculated a priori. This solution requires also a relatively low computation effort and is therefore well suitable for on-line calculation of the force distribution.

- *Minimization of the joint torques*

When transporting a heavy workpiece, special attention must be paid to the stress of the joint driving mechanism in order to avoid over stressing any particular joint. In this case, the weighted quadratic norm of the joint torques, which result from the end-effector forces, $\overline{U}^T \overline{M}\, \overline{U}$ is chosen as the criterion. In this case, the local optimal solution is represented by:

$$\underline{F}_E^* = \underline{B}_D^+ \underline{F}_L \text{ with } \underline{B}_D^+ = \underline{D}^{-1} \underline{B}^T (\underline{B}\,\underline{D}^{-1}\underline{B}^T)^{-1} \text{ and } \underline{D} = \underline{J}_E\,\underline{M}\,(\underline{J}_E)^T \tag{16}$$

In contrast to Eq. (15), Eq. (16) provides the optimal force distribution in the joint space. The relation between the end-efector forces and the joint torques depends on the Jacobians. The matrix $\underline{B}_D^+$ depends also on the workpiece's geometry and the actual configuration of the kinematic chain. The calculation of $\underline{B}_D^+$ requires, among other things, the determination of the Jacobians. The computation effort is therefore larger than that required for the solution Eq. (15). This solution takes the robot kinematics into consideration, while the robot dynamics are ignored.

- *Minimization of the energy consumption*

In the application of space robots the minimization of the energy consumption plays a very important role. The driving torques, determined by Eq. (6), consist of two parts, where the first part is used for the motion of the involved robots and the other for the compensation of the end-effector forces. Once the nominal motion trajectories of the individual robots are determined, the joint driving torques can be formulated as a function of the end-effector forces. The incremental energy consumption of the cooperating robots in the time interval (t_k, t_{k+1}) can be represented by

$$\Delta e^{(k)} = 0.5 (F_E)^T \underline{A} \, F_E + \underline{b}^T F_E + c \tag{17}$$

where the matrix $A(\cdot)$, the vector $b(\cdot)$ and the scalar $c(\cdot)$ can be evaluted from the dynamic model /9/. The corresponding local optimal solution is represented by:

$$F_E^* = \underline{B}_A^+ F_L + (\underline{B}_A^+ \underline{B} - \underline{E}) \underline{A}^{-1} \underline{b} \quad \text{with} \quad \underline{B}_A^+ = \underline{A}^{-1} \underline{B}^T (\underline{B} \underline{A}^{-1} \underline{B}^T)^{-1} \tag{18}$$

Since this solution takes the complete dynamics of the cooperating robot system into consideration, the computation effort is much higher than the solutions described by equation (15) or (16).

3.2 Optimal motion planning and force distribution for cooperating robots performing a common processing task

3.2.1 Force distribution

The desired contact forces between the workpiece and the tool, exerted at the actual contact point, are specified with reference to the coordinate system E2, which is fixed to the workpiece. The contact forces can be uniquely transformed, respectively, to both end-effector frames E1 and E2, according to:

$$F_{E1} = \underline{\mathfrak{R}}_1 F(t) \, , \qquad F_{E2} = \underline{\mathfrak{R}}_2 F(t) \tag{19}$$

$\underline{\mathfrak{R}}_1$ and $\underline{\mathfrak{R}}_2$ are defined by Eq. (11). Once the forces at the end-effector are determined, the corresponding joint torques can be obtained by using the transpose of the Jacobians. In this case, the force distribution is dependent on the desired motion trajectory and the actual joint displacements. A well suitable force distribution among the joints requires also efficient motion behaviour of both robots.

3.2.2 Optimal motion planning

Like the desired contact forces, the desired relative motion between the workpiece and the tool is specified with reference to E2. The nominal motion of the robots in relation to the inertial system is generally underdetermined. This kinematic redundancy can be used to optimize the motion behaviour of the cooperating robots and to satisfy the task dependent constraints. To realize a more efficient motion synthesis the dynamics have to be taken into account. Especially in the case of two robots performing a common prcessing task, the desired contact forces play an important role in optimized motion synthesis. In the following, basic mathematic relations for a local energy optimal motion synthesis are outlined.

The constraint among the joint incremets are described by the linear equation (9). The incremental energy consumption of two contact-free cooperating robots was derived in /10/. Taking into account the term of contact forces in the dynamic model a similar representation for the incremental energy consumption of two force-constrained cooperating robots can be expressed by

$$\Delta e = 0.5 \Delta Q^T \underline{A} \, \Delta Q + \underline{b}^T \Delta Q + c \tag{20}$$

where $\underline{A}(.)$, $\underline{b}(.)$ and $c(.)$ are different from that of Eq.(17) and the determination is based on the complete dynamic model /9/. The corresponding local optimal solution is:

$$\Delta Q_{ac}^* = \underline{J}_{ac}^+ \Delta \underline{X}_{ac} + (\underline{J}_{ac}^+ \underline{J}_{ac} - \underline{E}) \underline{A}^{-1} \underline{b} \quad \text{with} \quad \underline{J}_{ac}^+ = \underline{A}^{-1} \underline{J}_{ac}^T (\underline{J}_{ac} \underline{A}^{-1} \underline{J}_{ac}^T)^{-1} \tag{21}$$

It is to point out that the solution Eq. (21) is only local optimal and the whole energy consumption for a prespecified task is additionally dependent on the chosen starting configuration. However, the second term in Eq. (21) can be used in searching near optimal starting configuration. Based on this starting configuration a generally satisfactory motion behaviour can be generated. Alternatively, by using Pontryagin's minimum principle a global optimal solution can be formulated. But this will lead to the solution of highly nonlinear differential equations with the two-point boundary-value problems, which make the solution difficult for practical implementation.

4 Conclusion and Remarks

The algorithms described above were successfully proved by simulation on typical cooperating robot systems with two six-jointed industrial robots using exemplary tasks. These methods are characterized by the possibility to take various production task related requirements and constraints into consideration. The optimization criteria can be chosen based on the task-related requirements. It can be seen that the algorithms produce the optimal motion and force behaviours which satisfy the highest task-related requirements.

ACKNOWLEDGEMENT: The algorithms presented in this paper are a part of the task planning and control system for force-constrained cooperating robots developed at the IPK, Berlin. This work was supported in part by the Deutsche Forschungsgemeinschaft. The autors alone are responsible for the content.

REFERENCES

1. G. Duelen, U. Kirchhoff, J. Held and H. Münch: "Concept and Algorithms for the Coordinated Path Control of Two Robots", Int. Symp. on Robot Manipulators: Modelling, Control and Education, Albuquerque, New Mexico, 12-14 Nov. 1986.
2. D. E. Orin and S. Y. Oh: "Control of Force Distribution in Robotic Mechanisms Containing Closed Kinematic Chains": Trans. of ASME J. of Dynamic Systems, Measurement, and Control, Vol. 102, June 1981, pp. 134 - 141.
3. Y. Nakamura, K. Nagai and T. Yoshikawa: "Dynamic and Stability in Coordination of Multiple Robotic Mechanisms", Int. J. of Robotics Res., MIT, Vol, 8, No. 2, April 1989, pp. 44 - 61,
4. Y. F. Zheng and J. Y. S. Luh: "Optimal Load Distribution for Two Industrial Robots Handling a Single Object", Proc. 1988, IEEE Int. Conf. on Rob. and Auto., pp. 344 - 349.
5. H. Bruhm: "Coordinated Motion Planning and Optimal Force Distribution for Robots With Multiple Cooperating Arms", Int. Symp. on Industrial Robots, 1985.
6. J. Denavit and R. S. Hartenberg: "A Kinematic Notation for Lower Pair Mechanics based on matrices. ASME Tran., J. Appl. Mech., 22 June 1955, pp. 215 - 221.
7. M. Vukobratovic, N. Kircanski: Scientific Fundamentals of Robotics , Vol. 3, Berlin, Heidelberg, New York, Springer-Verlag 1985.
8. T. L. Boullin, P. L. Odell: Generalized Inverse Matrices, Wiley-Interscience, 1971.
9. H. Münch, Y. Zhang: "Dynamic and Control of Force-Constrained Cooperating Robots", Research Report #DU-102, Deutsche Forschungsgemeinschaft, Sept. 1990.
10. G. Duelen, U. Kirchhoff, J. Held, and H. Münch, "Concept and Algorithms for the Coordinated Optimized Path Control of Two Robots", 25th IEEE CDC Athen 1986.

SENSING STRATEGIES GENERATION FOR MONITORING ROBOT ASSEMBLY PROGRAMS *

Vincenzo Caglioti, Massimo Danieli, Domenico Sorrenti
Dipartimento di Elettronica, Politecnico di Milano, Italy

Abstract

A method is presented for the generation of sensing strategies aimed at detecting errors in the execution of robot programs. The method employs sensor models, in which both the measurement accuracy, and the probability of misinterpretation are taken into account. A sensor simulator evaluates the attitude of each available sensor to test the relevant workcell variables. The sensing strategy generation problem reduces to a weighted set-covering problem which can efficiently be solved by a Branch and Bound algorithm.

1. Introduction

During the execution of robot programs errors frequently occur. These errors could be detected (or even avoided) by adding appropriate sensor detections to the program. Since these detections are not strictly related to the robot task, it would be desirable to automatically generate the appropriate *monitoring* instructions to be added to an assigned program [2]. In addition, if a robot has to operate in the presence of not completely known environment conditions, the robot task can be specified by a sensorized program. In this case, it is not sufficient to verify that the actual evolution coincides with one of the correct (expected) ones, but it is also necessary to check that the actual evolution is compatible with the environment conditions as they are perceived by the sensor instructions of the program.

This paper illustrates a method to plan sensing strategies aimed at assessing the correctness of the execution of a robot program.

Since the correct evolution associated to a robot program depends on the environment conditions, the execution flow may follow different paths. A system for the monitoring of the execution of sensory programs has been overviewed in [1]. In this paper we concentrate on the generation of monitoring strategies relative to a single execution flow (i.e., a single workcell evolution). An evolution is represented as the sequence of states, which the workcell assumes in correspondence to the end

---

* Work partially supported with grants of IBM-SEMEA and C.N.R., PFR 2, Alpi

of the execution of each program instruction. Each state is represented by the *state variables*. In manipulation tasks, the state variables describe the positions of the objects, which are present in the workcell, together with their contacts and constraints.

The *correctness condition* associated to the program, is organized as a sequence of *nodes*, which correspond to the end of the execution of the program instructions.

The monitoring system plans and executes sensing strategies, in order both to acquire information about the actual evolution of the workcell, and to detect errors.

In addition to the correctness condition the input to the monitoring system includes the CAD models of the robot and of the objects to be manipulated, and the models of the input devices and of the sensors. The sensor models are exploited in order to evaluate both the accuracy and the *correctness factor* of the detections. The correctness factor of a sensor detection of a physical variable is intended as the probability to correctly associate the measured value to the physical variable itself. This factor is relevant, especially when dealing with complex sensors (such as vision ones).

In Sec. 2, the problem of the program monitoring is discussed, and reduced to a weighted set covering problem. In Sec. 3, an example of the monitoring of a simple assembly program is presented, and in Sec. 4 some conclusive considerations are outlined.

2. Monitoring Robot Programs

At the end of each step of the program execution, the monitoring system has to verify the part of the correctness condition concerning the node associated to the execution step. Thus it must be verified that the actual workcell state matches the expected state.

A sensor simulation module analyzes the expected state of the workcell, in order to produce an expected sensor output. This information is then used to plan the verification of the correctness condition: in particular, the state variables that are to be measured are first determined, and then the most appropriate sensor detections are selected according to a criterion based on accuracy and correctness factor. Both these coefficients are produced by the simulation of the sensor detections. Among the sensor detections minimizing the criterion, a further selection takes place which is based on the time cost of the global strategies, which include the candidate sensor detections.

2.1 Simulation of Sensor Primitives

In each node, the data acquired by means of the following sensor primitives are simulated.
1. A *locate* primitive, which determines position and orientation of a modeled object basing on the linear edges extracted from a single image.
2. A *match* primitive, which compares the expected line-drawing of the scene image with the actual one.
3. A *photocell* primitive, which yields a boolean output variable, corresponding to the presence of an object between the emitter and the detector of the photocell.

The simulation of both the *locate* and the *match* primitives involves the construction of the line-drawing of the image of the expected state.

In each node, the simulated sensor data must be associated to the related physical variables. A *test* is defined as a pair $t = (v, N)$, intended as the fact that the physical variable v can be tested at the state corresponding to the node N. The association between a set D of sensor data and a physical variable v at the node N is then represented by a *sensor-test pair* (STP): (D, t). This association states that the test t can be performed by analyzing the data D. Notice that each test may appear in zero, one, or more STPs, according to the presence of appropriate sensors for the test.

The accuracy coefficient of STPs related to the *locate* primitive is estimated by propagating the expected uncertainty on the visible line features to the object position parameters. The correctness factor related to the localization of a given object O is related to the complexity of the (expected) scene: it is evaluated basing on the number of objects, other than the object O to be localized, such that there exists a spatial position compatible with the expected scene. Accuracy and correctness factor associated to STPs related to the *match* primitive are defined as for the *locate* primitive. Notice that the accuracy and the correctness factor depend on the state in which the detection takes place.

Each STP is initially associated to a structure called *sensor detection* (SD) which contains a call descriptor to the sensor primitive producing the data D. The SD structure also contains an estimate of the time cost of the sensor activation.

2.2 Verification of the Correctness Condition

Only the following physical variables are considered for the verification: the presence of the objects inside the cell (or in appropriate devices such as, e.g., grippers or feeders); the position of the objects; the presence and correct placement of the physical edges of the objects.

Some terminology related to the verification of the physical variables is now introduced. A *critical event* for the physical variable v is an event, related to a transition between nodes N_i and N_{i+1}, that can (even incidentally) cause the modification of the value of v. For instance the insertion of a peg into a holed base is a critical event for any physical variable relative to the holed base, since the base can incidentally be moved. A *non critical sequence* S_v for v is either a sequence of nodes between its creation and a critical event for v or between two consecutive critical events for v. S_v constitutes a connected part of the node sequence describing the program execution. Each variable can be tested once during each *non critical sequence* since its value should not change by the way. The most appropriate test of the variable v has to be selected, within the *non critical sequence*, out of the set of the tests for which at least one STP exists. The generation of a sensing strategy is subdivided in two phases.

The first phase of the planning activity consists in determining the best STP out of the set of the STPs related to the same physical variable; the STPs are evaluated with the following criterion, based on both accuracy and correctness:

$$J = K_\sigma \cdot \sigma^2 + K_\gamma \cdot (1 - \gamma), \quad \sigma^2 < \bar{\sigma}^2, \quad \bar{\gamma} \leq \gamma \leq 1,$$

where σ^2 and γ indicate the estimated accuracy of the STP and, respectively, its correctness factor. $\bar{\sigma}^2$ and $\bar{\gamma}$ are thresholds on the above coefficient, K_σ and K_γ express the relative weights of the accuracy and correctness.

The STP which minimizes J is not necessarily unique: a sensor detection can provide the same information (with the same accuracy and correctness factor) about a physical variable at different nodes in the same non critical sequence.

An *augmented* set S of sensor detections is constructed by combining SD relative to optimal (i.e., minimizing J) STPs. Suppose, e.g., that two optimal STPs correspond to the localization of two objects A and B in a same node. Then a new SD *(locate (A, B))* can be constructed, starting from the two SD *(locate (A))* and *(locate(B))* relative to the two optimal STPs. The new SD is associated to both STPs, and its time cost is given by combining the costs of the composing SDs.

At this point the planning problem reduces to a *weighted set-covering* problem [3], where the set of the physical variables to be verified in the node sequence represents the set to be covered, and the SDs of S represent the covering elements. Each SD covers one or more physical variables. The weight of each SD is represented by the estimate of its time-cost.

Consider a $n \times |S|$ binary matrix $\mathbf{E}$, where n is the number of the physical variables to be verified in the node sequence covered by at least one SD of S. The element $e_{i,j}$ is set to 1 iff the j-th element SD_j of S covers the physical variable v_i: therefore the column $\mathbf{E}_j$ represents the covering of SD_j. A solution, i.e. a sensing strategy, can then be represented by a $|S| \times 1$ binary vector $\mathbf{x}$, whose j-th component coincides with 1 if and only if SD_j is in the solution (i.e., in the sensing strategy). Since a feasible solution must cover all the physical variables, it can be expressed by the following inequalities:

$$\forall i \in \{1 \ldots N\} \quad \sum_{j=1}^{|S|} e_{i,j} \cdot x_j \geq 1$$

Therefore $\mathbf{x}$ is a feasible solution if and only if for each variable v_i at least one sensor detection SD_j, which covers v_i, is in the solution $\mathbf{x}$.

If c_j is the time cost of SD_j, the solution time cost is given by:

$$C(\mathbf{x}) = \sum_{j=1}^{|S|} c_j \cdot x_j$$

Therefore the problem can be formulated as a binary linear programming problem:

$$\min C(\mathbf{x}) = \sum_{j=1}^{|S|} c_j \cdot x_j, \quad \text{where } \forall i \in \{1 \ldots N\} \quad \sum_{j=1}^{|S|} e_{i,j} \cdot x_j \geq 1$$

We solved the problem by means of a specialization for the zero-one problem of the Branch and Bound algorithm [4]. The optimal strategy therefore consists of the

SDs, that covers all the physical variables in the node sequence with a minimum total time cost.

3. An Example

The experimental workcell is equipped with a four degrees of freedom SCARA IBM-7547 robot and a conveyor, which carries objects to the workcell. Because of the particular geometry of the robot, the position and orientation of its end-effector (a two-finger gripper) is given in terms of the values of x, y, z and $roll$ parameters with respect to the base reference frame (angles are expressed in degree). Also the object position is given in terms of the values of x, y, z coordinates of the object frame origin with respect to the base frame, while the object orientation is defined giving only the $roll$ angle. In addition, the workcell is equipped with the following sensors. A photocell, mounted on the robot gripper, indicated by $photo$-1; a fixed camera, indicated by tv-1, which observes the objects carried by the conveyor. A mobile camera (tv-2), mounted on the robot gripper. On both these cameras the primitive $locate$ is implemented.

Two objects are involved in this example. The first object (O1), which belongs to the OBJ1 class, is a holed parallelepiped with dimensions $(11 \times 7 \times 3)$. The hole is centered on the center of the upper face of the parallelepiped and its dimensions are $(3 \times 1 \times 1.5)$. The second object (O2), belonging to the OBJ2 class, is a parallelepiped with dimensions $(1 \times 3 \times 5)$.

In the initial state of the program execution, only O2 is present in the workcell. Its initial position is (70 70 −50 0). The object O1 is introduced into the workcell by the conveyor at the initial position (−55 −2.5 −50 5.01).

The robot program consists first in grasping O2 in its initial position and then in inserting it into the hole of O1. The AML/E program is now presented:

```
HOME: NEW PT(103,0,0,0);    - - the parking position
O2VERT: NEW PT(70.5,71.5,-30,90);    - - a point along the vertical of O2
O1VERT: NEW PT(-49.826,1.467,-30,5.01);    - - a point along the vertical of O1
pmove(O2VERT);    - - move to the vertical of O2
zmove(-46);    - - moves toward the grasping position of O2
grasp;    - - grasps O2
zmove(-30);
pmove(O1VERT);    - - a position on the vertical of O1
zmove(-44.5);    - - insert O2 into O1
release;    - - open hand
zmove(-30);
pmove(HOME);    - - return to parking position
:END.
```

The node sequence representing the correctness condition of this program, is reported in Fig. 1. For instance, the node $N8$ corresponds to the state after which the object $O2$ has been inserted into $O1$ and released by the gripper, and

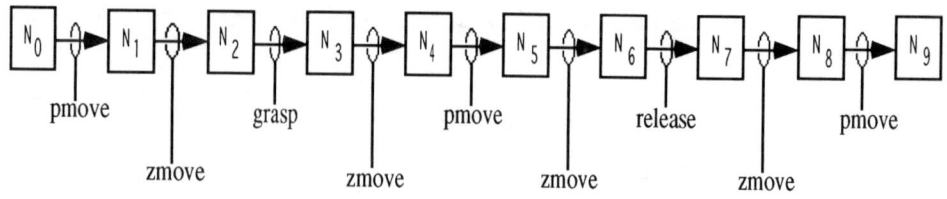

Fig. 1 The node sequence

the gripper has been lifted. This node is characterized by the following assertions (among others):

(next-nodes (N9))
(position gripper N8 (-49.827 1.467 -30.0 5.01))
(status gripper N8 open)
(holds gripper N8 nil)
(position O1 N8 (-55 -2.5 -50 5.01))
(position O2 N8 (-51.365 1.834 -48.5 -84.99))
(impeded-directions conveyor O1 N8 ((0 0 -1)))
(impeded-directions O1 O2 N8 ((1 0 0) (-1 0 0) (0 1 0) (0 -1 0) (0 0 -1)))
(contact conveyor O1 N8 ((face1 face10)))
(contact O1 O2 N8 ((face3 face4) (face5 face6) (face2 face2) (face1 face1) (face4 face5)))

The node sequence is analyzed, in order to look for the physical variables to be verified and to find the corresponding *non critical sequences*. For instance, at node N0 the physical variable $x_{O2} = 70$ with a related *non critical sequence* (N0 N1 N2) is detected. The *non critical sequence* for x_{O2} ends at node N2, since in N3 the object O2 is grasped by the robot. After the analysis has been completed, the sensor simulation is executed along the node sequence.

As an example, the simulation results of the activation of the fixed camera (i.e., *tv-1*) at the node N8 are reported in Fig. 2. The table below reports the accuracy σ^2 and the correctness factor γ relative to the STPs involving the position parameters of the objects O1 and O2 and the data produced by the (*locate*) primitive localization of the objects in the scene by means of the *tv-1* camera, as calculated basing on the expected line drawing. The value of the criterion J is also reported, where $K_{\sigma_X} = K_{\sigma_Y} = 1.33 cm^{-2}$, $K_{\sigma_Z} = 0.66 cm^{-2}$, $K_{\sigma_\phi} = 2.424 \cdot 10^{-5}$, $K_\gamma = 1.11$:

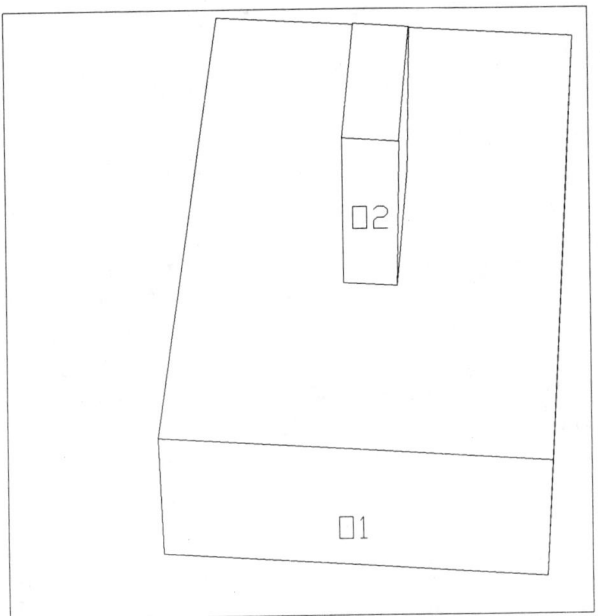

Fig. 2 Simulated view of node N8 from camera *tv-1*

variable	σ^2	γ	J
x_{O1}	0.5583	0.5	1.230
y_{O1}	0.0092	0.5	0.569
z_{O1}	0.3791	0.5	0.808
ϕ_{O1}	$5.821 \cdot 10^{-7}$	0.5	0.556
x_{O2}	0.122	0.25	0.996
ϕ_{O2}	$8.735 \cdot 10^{-4}$	0.25	1.063

The results concerning the y and z coordinates of the object $O2$ are neglected, since their accuracy exceeds a given threshold.

The plan generated in correspondence to the node sequence is composed by the following sensor instructions, in which the spatial positions are indicated by the roll, pitch, and yaw angles, and by the x, y, z coordinates referred to the frames attached to the cameras *tv1* and *tv2*:

(:read (photo1) :correct (value 0) :node N0)
(:read (tv2 (LOCATE OBJ2)) :correct (value (OBJ2 (1.571 0.785 3.141 -1.5 1.788 26.49))) :node N1)
(:read (photo1) :correct (value 1) :node N3)
(:read (photo1) :correct (value 1) :node N4)

```
(:read (tv1 (LOCATE OBJ1)) :correct (value (OBJ1 (3.018 -0.782 -3.055 4.246 -2.5
72.464))) :node N4)
(:read (photo1) :correct (value 1) :node N5)
(:read (tv2 (LOCATE OBJ1)) :correct (value (OBJ1 (0 0 2.356 -5.587 5.318 22.996)))
:node N5)
(:read (photo1) :correct (value 1) :node N6)
(:read (tv1 (LOCATE OBJ1)) :correct (value (OBJ1 (3.018 -0.782 -3.055 4.246 -2.5
72.464))) :node N6)
(:read (tv2 (LOCATE OBJ2)) :correct (value (OBJ2 (1.548 0.785 3.126 -1.586 1.430
24.733))) :node N8)
(:read (tv1 (LOCATE OBJ1 OBJ2)) :correct (value ((OBJ1 (3.018 -0.782 -3.055 4.246
-2.5 72.464)) (OBJ2 (-1.644 -0.073 2.359 0.614 1.833 73.975)))) :node N9)
(:read (photo1) :correct (value 0) :node N9) ,
```

where the keyword *:read* indicates the sensor that must be activated; the keyword *:correct* is followed by the predicates to be satisfied by the sensor data. The keyword *:node* indicates the node to which the instruction is associated. An instruction execution implies that the data are acquired, and then they are evaluated basing on the correctness predicates.

4. Conclusions

A method has been presented for generating sensing strategies aimed at monitoring the execution of robot programs. The suitable sensory detections for the measurement of the relevant state variables are characterized both with respect to their accuracy and correctness factor. This characterization involves a sensor simulation. The appropriate sensing strategy is selected, among a set of candidate ones. The selection is performed first by evaluating a criterion, based on the above features, and then by a time cost driven selection of the best sensing plan. This last step is performed by means of a Branch and Bound technique. The presented method is currently being tested on assembly programs executed by an IBM-7547 robot equipped with two cameras and a photocell sensor. Our future work includes the introduction of redundant sensing strategies, and the real time monitoring of the transitions between consecutive steps of the program execution.

References

[1] V.Caglioti, M.Danieli, D.Sorrenti - "Monitoring the Execution of Sensory Robot Programs" in Proc. 5th ICAR, (1991)
[2] R. E. Smith, M. Gini - "Reliable Real Time Robot Operation Employing Intelligent Forward Recovery" - Journal of Robotic Systems, vol. 3, pp. 281-300, (1986)
[3] M. Danieli - "A Monitoring System for robot sensory Programs" - (in italian), Master Thesis, Politecnico di Milano, (1991), Milano (Italy)
[4] H.M. Salkin, K. Mathur - "Foundations of Integer Programming" - North Holland, (1989)

INTEGRATION OF A CONSTRAINT SCHEME IN A PROGRAMMING LANGUAGE FOR MULTI-ROBOTS

D. DUHAUT, E. MONACELLI
L.R.P., Laboratoire de Robotique de Paris
UPMC - ENSAM - CNRS(URA 1305)
tour 66, Universite P. et M. Curie,
4, Place Jussieu 75252 Paris Cedex 05
FRANCE   Fax: 44 27 62 14
E-Mail : ddu@ccr.jussieu.fr

Overview

This paper describes a structure of dynamic management for robotic processes at task level programming. An interpreted language IAda, based on Ada, has been developed including two programming levels : the first is to perform concurrent applications with tasks or procedures written on Ada, and the second is to control the execution of the previous.

1. Introduction

The object of this paper is to propose a structure for task level programming of robotic applications [LO 83].
This structure must allow the manipulation and the cooperation of different agents, (which are independent processes associated with the physical resources of complex robotic systems), in order to perform certain applications.
Because the difficulties of modeling robots, the process, which must control the robot and its environment, often must function in the presence of uncertainty. Thus, performing a robotic application require a programming structure which can check temporal or space informations to compensate for uncertainty.
Consequently, the problem for a task level programming is two fold: enable the description of distributed applications, and check the execution.
The often solution is to use a supervisor process, which is more or less intelligent, concurrently with the task level program to validate its execution. A problem inherent in these systems is that they are not able to closely follow the execution.
In an other hand, reactive programming [ES 90] includes execution and verification in the same design. Giving the control to the processes, the program only evolves at event occurrences.
Based on an combinaison of reactive and supervision programming, we have constructed an interpreted concurrent language, IAda (Interpreter on Ada), including a constraint scheme. IAda manipulates agents with classical constructors and with the constraint scheme can assess dynamically its activity: for instance, a constraint relation on force or position for a manipulator displacement .
Because IAda is an interpreted language we can modify that constraint declaration (and its associated treatment) during the execution of the task level program without any problems.
An IAda program can be viewed as an interpreted Petri nets, with sequential and independent phases of task activations or procedure executions, improved of constraint declarations. We choose that structure because it allows a closely description of an robotic application [JO 90], and moreover, it provide facility on elaboration phase of the program, either on specification or validation [TU 90].
IAda have been written on Ada, as robot primitives, on account of task capability and portability of this high level language [CO 89].
In this paper, we propose after a description of IAda, a simulation of an application with mobile robots to show flexible capabilities of such programming.

2. IAda description

The IAda language contains a set of instructions to activate or disactivate Ada compiled primitives, and includes control schemes to build a program.
The manipulated primitives type are :
- Ada tasks and their entries,
- Ada procedures and their parameters,
- global variables.

2.1. Ada primitives manipulation

on subroutines
      EXECUTE (<subroutine>,<variables>);
         -- which computes the subroutine with argument cross over
on tasks
      INSTALL (<task>,<variables>);
         -- the task is placed in the scheduler

      ACTIVE (<task>);
      SUSPEND (<task>);
      ABORT (<task>);
         -- activates, suspend or abort a task by the primitive state variable (§.)

classical control schemes

      LOOP ... EXIT WHEN <test> ... END LOOP;
      IF <test> THEN ... ELSE ... END IF;

where <test> is either a boolean expression, or terminate(task) or an <exception_name>.

2.2. Sequence controls

To allow flexible programming for robotic applications, IAda includes independent phases with classical control schemes. Thus, an IAda program is decomposed in concurrent sequences with possibilities of interactions between their executions

sequence commands :

      SUSPEND_SEQUENCE (<reference>)
         -- used to freeze the execution of a sequence from an other sequence
      ACTIVE_SEQUENCE (<reference>)
         -- used to start the execution of a sequence from an other sequence
      RESET_SEQUENCE (<reference>)
         -- used to reset the execution of a sequence from an other sequence

Moreover, that kind of control could be used in constraint declaration, allowing to suspend a main sequence and active an other one to avoid a problem.

3. Constraint schemes

With sharing their area executions, concurrent applications could involve problems. Indeed, it may occur interactions due to dead lock situations or manipulation errors. In order to control that kind of problems, we introduce a control level in the IAda programming.

We add a constraint scheme :

$$\{ \text{<test>} \Rightarrow \text{<IAda\_program>} \}$$

as soon as the condition expressed by <test> becomes true, on the execution of previous IAda instructions, the <IAda_program> becomes a concurrent sequence added on the other process and executed in priority .
Note that this condition treatment can be a process activation, a task or sequence suspension, a correction on primitive variables, or at last, an operator call to solve non determinist programming.

A use of this scheme is, for example, the handling of a parcel by two manipulators with constraints on effector relative positions during the execution.
Assume that Move_robotA and Move_robotB are respectively the names of motion of a manipulator.

```
ACTIVE ( Move_robotA (Pos_actual_A, Pos_future_A));
ACTIVE ( Move_robotB (Pos_actual_B, Pos_future_B));
{ Coordinate (Pos_actual_A, Pos_actual_B)
=>
EXECUTE ( Readjust (Pos_actual_A,Pos_actual_B) ) }
```

where *Pos_actual_* and *Pos_future_* are global variables and *ACTIVE*, an IAda activation command;

where the function *Coordinate* expressed a properties to be verified during the displacement, and *Readjust*, the treatment called by the IAda command, *EXECUTE*.

In the constraint scheme, we distinguish three kinds of <test> :
1. <test> is an Ada exception raising to IAda level. For instance, an unexpected obstacle during the robot motion, unable to be treated at this level, can raise an IAda exception.
2. <test> is a properties to be verified (see the previous example). Coordinate could be a sample <test> like [(Pos_actual_A - Pos_actual_B) > Epsilon], or an expert program call (some explicit supervisor);
3. <test> is an Ada task terminating execution. We have to control that some task are still alive during the process.

Like it is shown in the previous example, constraint properties could depend on particularly variables of an agent. Those variables, called pertinent variables, describe the agent behavior their choice must be effective in order to limit the tested data flow.
A pertinent variable declaration is made according writing rules (see §4) to allows a visibility since IAda level.

In order to control condition states (see §6), we have introduced two IAda commands:
ACTIVE_CONDITION (<reference>)
SUSPEND_CONDITION (<reference>)
-- used to active or suspend the evaluation of a constraint scheme

The control of pertinent variables, declared on the primitive interface, is done in time of their affectation. In order to not interfere with the agent action, the verification phase must take a processor constant duration, as shorter as possible.
Considering that the verification duration cannot be constraint a priori,
- the verification is performed in three phases:
. a verification request is executed in constant duration by the agent,
. the verification in fact, is executed (on an other processor when it possible),

. if necessary, the raise of an IAda exception , which is treated by an unit control.
- the declared property is not necessary true during the execution of the next instructions, because of the delay between the request and its evaluation.
- the verification request must be done after variable assignments to perform Ada exceptions, before starting complex verifications by IAda exception treatments.

4. Rules on writing of primitives

The construction of the IAda language needs to write primitives in regard of few rules.

We present here the most significants:
- The pertinent variables, on which constraints can be expressed, are called <variable>_VIEW (to bind with the notion of task variable visibility). We associate to each of pertinent variables a <variable>_ABLE which allows to disable its visibility and thus, stops the flow up of new values (see §3).
- To allow the treatment of exceptions, we must define exception names, which can be raised to IAda level.
- Furthermore, special rules are imposed to allow a neat treatment into primitives. A <variable>_STATE is included on each primitive and expresses the access to the <variable>_STATE of the task. This <variable>_STATE permit suspend or abort commands.

5. IAda organization

IAda structure is composed of two parts :
- the interpreter, depending on manipulated primitives, obtained by a rewriting algorithm [DU 89],
- the IAda manager, which control the IAda execution by three tasks:
    . the sequential manager controls the sequence activities of the IAda program,
    . the condition manager and exception manager controls the treatment of constraints.

The value carry up is made for exception manager by rendez-vous processing. Indeed, it issued of the particularly of primitive exception raised, which involves a stop because of the importance of this problems, like a motion out of the robot limits. In contrary, it made by buffer structure for the sequential manager and the condition manager; because the evaluation request is done without stopping process computing.

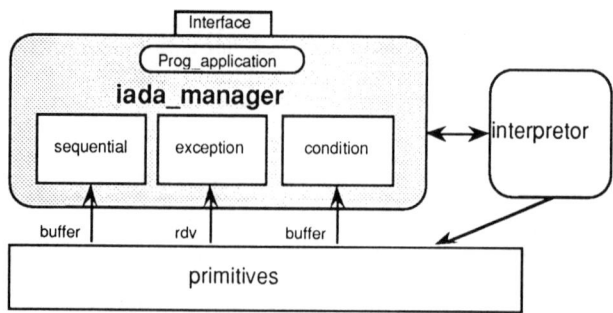

6. Example of IAda program

Let's take the example of the "princess escape", which demonstrate the flexibility of IAda program. This example has to be considered in the largest study of traffic control of multiple robot vehicles [GR 88]. There are three mobile robots able of linear motion and equiped with sensors. The first Robot is placed in a room, "prisoner", waiting for escape that the third robot action the open door button.

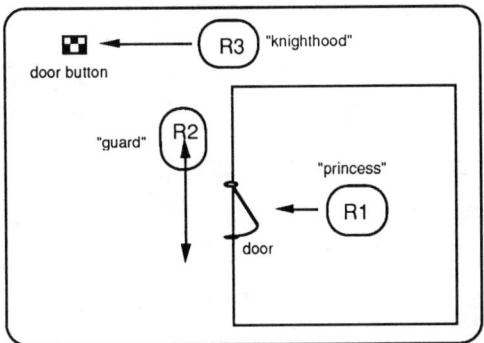

But like in all tales, there is a guard robot, the second, in front of the door. If the first robot is catched, during its escaping, by the second one then it must go back to its gaol. The translation of this problem in IAda is direct. It is decomposed first, in three sequences describing the task to each robot, and in part, the declaration of catched constraint between the robot one and two.
Here is the schema of a blocking situation

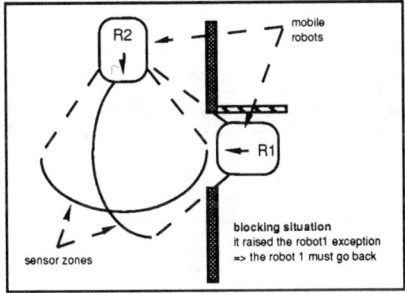

By security, any blocking situation raise an IAda exception which can be treated as same as any condition.

6.1. IAda program

To facilitate the exit of robot one (the princess), we introduce a condition expression to keep a minimal the distance between robot one and two (the jailer). When the distance between robot one and robot two is under a certain value then robot one stops to let the robot two pass. If the robot two detects somebody (exception is raised) an the robot one gets back into the jail.

```
sequence 1 :
install (ROBOT3(+50.0,+80.0))
active (ROBOT3(-64.0,+80.0))
execute (OPEN_DOOR)
```
{ move of robot 3 to the button and execution of the procedure open

```
sequence 2 :
install (ROBOT2(-21.0,-21.0))
loop
 active (ROBOT2(-21.0,-65.0))
 active (ROBOT2(-21.0,+65.0))
end loop
```
{ loop of robot 2, express the guard round in front of the door

```
sequence 3 :
install (ROBOT1(+20.0,+00.0))
loop
 exit when term(ROBOT3)
end loop
 active (ROBOT1(-30.0,+00.0))
 active (ROBOT1(-40.0,+80.0))
```
{ approach of robot 1, then wait for open and escape

```
condition :
1.
{ Distance_mini (Pos_R1,Pos_R2)
           => suspend_sequence(3)
              active_condition(2)
}
2.
{ not (Distance_mini (Pos_R1,Pos_R2))
           => active_condition(1)
              active_sequence(3)
}
exception :
{ *ROBOT2_STOP
           => suspend_condition(2)
              active (ROBOT1(-05.0,+00.0))
              execute(Dest_R1:=-30.0,+00.0)
              active_condition(1)
              active_sequence(3)
}
```
{ if the distance is too short, we stop the princess

{ when the distance becomes large, the princess move again

{ a collision is detected, and then the princess steps back before restarts

{ ROBOT2_STOP is an primitive exception, simulate of extero-sensors around a robot

where :
    ROBOTx (pos_x_act, pos_y_act, pos_x_end, pos_y_end) :
      motion primitives of robot x (move from an actual position to an end position).
    OPEN_DOOR : a procedure to open the gaol door.
    Distance_mini () : an evaluated obstacle distance function.
    term() : boolean function, it express a terminate task.

Moreover, we can easily activate a fourth robot without interfere with computing, during or after the execution.

7. Conclusion

The implementation of IAda have been made with Ada on SUN workstation with simulated primitives of robot motion (§9). IAda had demonstrated a good flexibility for the programming of robotic applications particularly when we compare it with classical independent supervisor.

Nevertheless, it appears that the number of tasks running at the same time could involve problems. Consequently, IAda requires sufficient computation capability.
We have started to experimented IAda for real time assembly phases in our laboratory, using as manipulator a PUMA 560. We employed several CPU boards, on VME-bus, based on Motorola 68020 to control robots and the different levels of IAda.
According to high concurrently material design of IAda, we are studying of a physical repartition algorithm for the different IAda processes.

8. References

[CO 89]   I. COX, N. GEHANI, "*Concurrent programming and robotics*", The international journal of robotics research, vol 8, No2, April 1989.

[DU 89]   D.DUHAUT "Ada et les différents niveaux de programmation en robotique", Congres ADA-FRANCE, Afcet, Paris December 1989

[ES 90]   B. ESPIAU, E. COSTE-MANIERE, "*A synchronous approach for control sequencing in robotics application*" IEEE Istanbul, august 1990.

[GR 88]   D. GROSSMAN, "*Traffic control of multiple robot vehicles*", IEEE journal of robotics and automation, vol 4, No5, October 1988.

[JO 90]   M. JOCKOVIC, M. VUKOBRATOVIC, Z. OGNJANOVIC, "*An approach to the modeling of the highest control level of flexible manufacturing cell*" Robotica, vol 8, pp 125-130, 1990.

[LO 83]   T. LOZANO-PEREZ, "*Robot Programming*", IEEE vol 71 July 1983

[TU 90]   S. TU, S. M. SHATZ, T. MURATA, "A*pplying Petri net reduction to support Ada tasking deadlock detection*" IEEE, 1990.

9. Exprimentation

A screen dump of the simulation of the "Princess escapes".

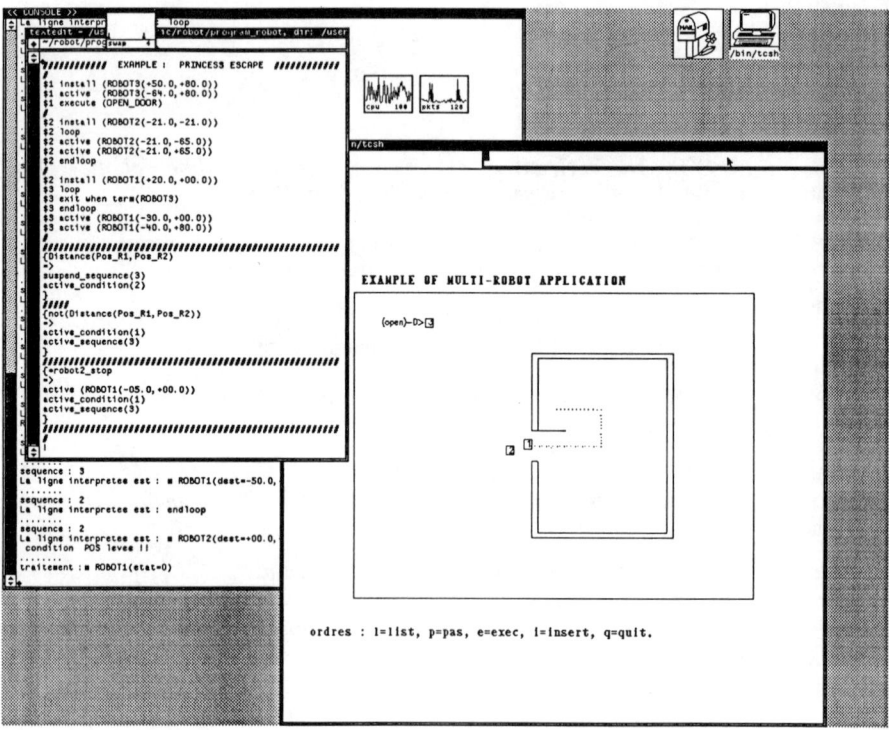

Software architecture and simulation tools for autonomous mobile robots[1]

G.D. van Albada, J.M. Lagerberg, B.J.A. Kröse
Department of Computer Systems
University of Amsterdam
Kruislaan 403, 1098 SJ Amsterdam

Abstract

In the development of software for experimental autonomous robotic vehicles various problems have to be solved.

Firstly, as the systems are experimental, frequent changes in the sensor configuration, the computer hardware and the control software must be easily incorporated. This requires a modular software design. Secondly, it is frequently necessary to test new software, new algorithms and even entirely new control paradigms. This is facilitated by the use of suitable simulation software. Thirdly, the software developed for one vehicle should be easily portable to vehicles of a somewhat different mechanical design. A common control level for all vehicles makes this possible.

In this paper we will address the solutions that we have adopted within the ESPRIT II project MARIE[2], starting with a presentation of the low level control structure adopted for our vehicle. The emphasis of our paper will be on the extensive simulation system that has been developed to support our software development efforts. Important aspects of this simulation system are the simulation of the response of the vehicle to the commands generated by the software, the simulation of the response of the sensors as the vehicle moves through the environment, and the user friendly graphical interface.

The simulation software allows the user to construct a complex environment and to find the output that would be produced by various types of sensors, such as ultrasonic range finders, as the robot drives through this environment. The graphical interface not only shows the progress of the vehicle, but also indicates how obstacles are seen by the sensor system.

Introduction

One of the central themes in the research effort at our institute is the achievement of autonomous behaviour by robotic systems. We study both stationary robot arms and robot

---

[1] This paper describes work done for the ESPRIT II project 2043: "Mobile Autonomous Robot in Industrial Environment" (MARIE)

[2] The other partners participating in the MARIE project are VOLMAC (prime), Robert Bosch GmbH, Framatome, Framentec, Hitec, IAI, Indecon, University of Strathclyde.

vehicles, both in the context of nationally funded projects and ESPRIT projects. In this paper we describe aspects of the software architecture and the simulation tools that we utilize in building an autonomous mobile robot. This approach was developed mainly in cooperation with other partners in the ESPRIT project MARIE.

The objective of the work described here is the achievement of a software environment and design philosophy that facilitate the rapid design, implementation and testing of sensor data processing and of high and low level control modules for mobile robots.

The first issue that we have to address is that of software flexibility. Various approaches to the problems of task execution, path planning, navigation and collision avoidance must be tested without completely redesigning or rebuilding the control software for the cart. This problem is generally solved by using a modular design, with well defined interfaces. The data representations within each module are hidden from the other modules.

The second issue is that of portability. To this date almost all mobile robots have a unique hardware and software architecture. Yet, it is desirable to build the control software in such a manner that it can easily be ported between various vehicles. The solution in this case is the design and implementation of an appropriate generic control layer or virtual machine having an "instruction set" that is independent of the underlying hardware.

The third issue concerns the initial testing of the software for the cart and predictions of the effects of modifications in the software and hardware. Furthermore, the effects of system failures etc. must be easily and safely testable. A suitable simulation environment greatly facilitates these procedures.

In this paper we will first describe the computing environment that we use. Next we will touch upon the general structure of our control software and describe the manner in which we strive to ensure modularity and portability. Subsequently, we will describe the simulation software packages that have been developed and our experience with the actual implementation and use of the software.

The computing environment

Both within the MARIE project and the Department of Computer Systems of the University of Amsterdam we strive for standardization of our computing equipment and operating systems. We use UNIX[3] workstations, mostly SUN Sparc stations, and a variety of dedicated experimental systems, linked together through ethernet.

The experimental systems are used for image acquisition, image processing, and robot control. On the hardware level, where possible, we have opted for VME and MC680x0 based processor boards as a standard, viz. Force 30ZBE. On the software level, we have chosen for standard C and a widely used real-time operating system, viz. VxWorks[4]. VxWorks is in many ways compatible with our UNIX environment, e.g. UNIX files can be accessed directly and Internet type sockets can be used in much the same way as in UNIX, including the use of select statements.

---

[3] UNIX is a registered trademark of AT&T Bell Laboratories.

[4] VxWorks is a trademark of Wind River Systems, Inc.

The computing hardware for the Marie vehicle is a case in point. It is based on a VME bus with one or more general purpose processors and one or more dedicated processors. We currently use a Force 30ZBE with VxWorks as the main processor, a MC68000 based system for the ultrasonic sensor system and dedicated PID controller hardware (four NS LM628 on a Philips PG3679 board – the "Philips Motion Controller" or PMC) for the low level control. A (removable) Ethernet connection provides the communication between the VME bus system and the SUN network.

Software design for the MARIE vehicle

As stated before, we attempt to realize as much as possible a modular software structure for our robot. The philosophy of our approach is to develop control and sensor data processing modules as stand-alone processes, which communicate via standardized communication channels, viz. sockets. In specifying the modules, we strive for a structure that makes the software easily portable to other mobile robots, having a different hardware and different sensor types. This puts requirements on the level of abstraction at the interfaces, which must be as high as possible. I.e., going up from the lower software levels to the higher levels, we strive to go from the specific to the general as quickly as possible.

Part of the structure of the currently implemented control software is illustrated in Figure 1.

As we go up from the PMC driver module, which implements the driver for the hardware PID controller, to the virtual cart, we can clearly demonstrate this increase in abstraction level.

The PMC driver module must know about the representation used in the NS LM628 for all the PID control parameters. This involves knowledge about clock frequencies, scaling factors etc. The interface for the PMC driver module has been designed such that calling modules can specify filter parameters, velocities and such in terms of encoder ticks (this cannot be helped here), seconds and an output to the motors/actuators normalized to the range from -1.0 to +1.0.

The next higher level, the virtual cart, will be discussed more extensively below. It completely hides the existence of the PID controller hardware and many other specific properties of the cart from the higher levels and accepts path specifications in SI units. Thus we have arrived in just two steps at an interface that can be used for virtually any cart.

The Virtual Cart

The virtual cart is a software layer that provides the developer of high-level software with a simple and consistent call interface that is, as far as possible, independent of the underlying hardware. It also provides certain basic safety features. The concept of a virtual cart plays an important role in achieving portability and data abstraction.

In designing the virtual cart interface, it was necessary to choose a suitable control paradigm. Basically, three different types of wheeled vehicles are in use (disregarding a plethora of special purpose vehicles). The most versatile, but probably least common of these is the vehicle that can move in any direction in any orientation. The other two types have one fewer degree of freedom. They are those with one or more drive wheels providing propulsion, and one or more steered wheels, determining the path curvature and those where the driving and steering are combined. The former ("bicycles") cannot in general turn in place, the latter ("wheelchairs") can. The most appropriate set of parameters to describe the behaviour of bicycles are speed and path curvature, for wheelchairs speed and turning speed. These differences are only important at low speeds and high path curvatures, where a singularity occurs in the conversion, otherwise both models are compatible.

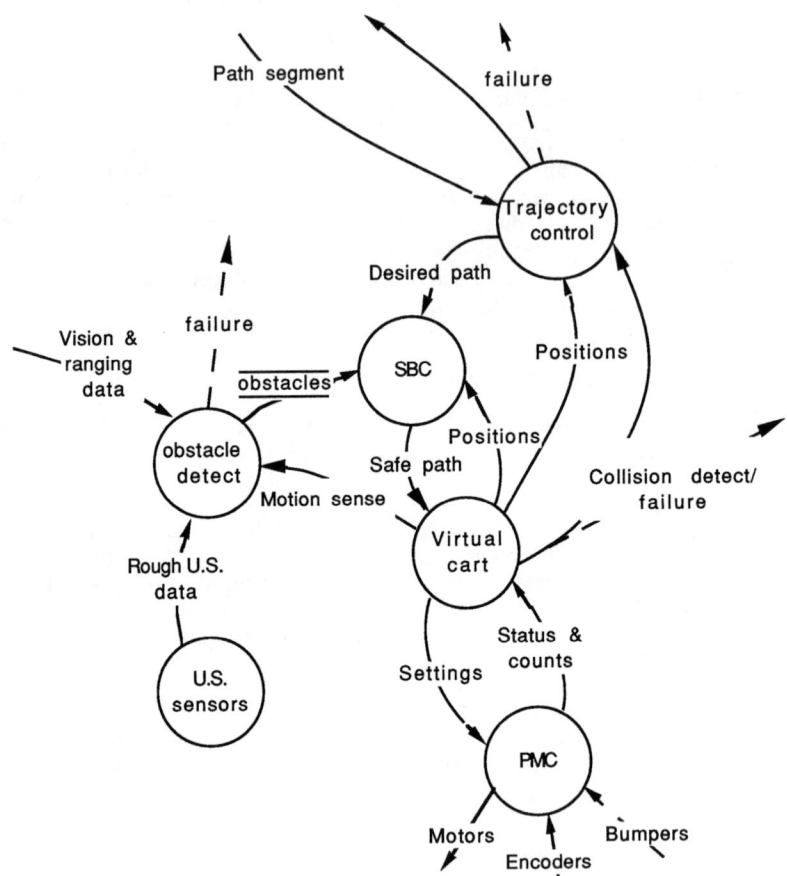

Figure 1. Part of the low-level software structure for the MARIE vehicle. The PMC module is the Philips Motion Controller board plus driver, described in this paper; the SBC module implements a sensor based collision avoidance strategy. The trajectory control and the SBC module are currently being tested in simulation in an environment provided by CARSIM and ASSIM.

We elected to implement both the bicycle and the wheelchair paradigms for the virtual cart. The virtual cart layer has been implemented and tested for a four wheeled vehicle that has two separately powered rear wheels and two steered front wheels coupled through a traditional trapezoidal system and controlled by a single electric motor. The cart thus has one more controlled axis than it has degrees of freedom. The virtual cart layer not only hides this redundancy from the user, but also hides details such as the wheelbase, encoder resolution etc. Commands to the virtual cart (bicycle paradigm) specify a path length relative to the current position of the cart, velocity, acceleration, path curvature and time derivative of the curvature all in SI units. Each new path specification immediately replaces the current specification.

Besides providing the desired abstractions, the virtual cart layer does more. Firstly, it significantly decreases the frequency at which higher level modules need to be executed. It does this by ensuring that a moderately complex path segment can be driven with good accuracy and by ensuring that those sensors that signal conditions that need to be reacted to on very short time scales are monitored, e.g. collision detectors attached to the bumpers of the cart. It also integrates the actual path as derived from the encoder readings to obtain estimated Cartesian positions (dead-reckoning).

Another important service of the virtual cart is that it provides some basic safety measures. The most evident measure is the monitoring of the collision detectors. If a collision is detected on one of the bumpers, the cart is prevented from driving further in that direction until the condition is reset by a higher software level and an exception is signalled. Similarly, if the desired and actual position of the cart differ too much, a stop is forced and an exception is signalled. A third measure is more subtle, and more easily circumvented - the higher level software must always specify a path length. This ensures that the cart will be stopped reasonably soon even if the higher level software crashes.

Development Environment

The software for the cart is developed and tested in simulation in a UNIX environment. UNIX is used because it provides a rich set of software development tools and is widely available. All the code has been written in C.

In the development stage new software modules are always tested in the simulation environment, after which they are implemented on the vehicle for final testing.

To support the development of software for experimental autonomous robots simulation is almost essential. New algorithms and new control paradigms can be developed and tested in a simulation environment before the hardware is even available. Debugging and testing of algorithms on a real machine is very time consuming, inefficient and in some situations even might be considered as dangerous, while off-line experiments can save a lot of time and allow the generation of measurement data in a very easy and flexible way. A simulation environment can give access to parameters which are not available in the real time application, like for example the position and orientation of the robot, or the voltages sent to the motors. Furthermore, a simulated environment can often be modified more easily than the real environment.

Yet, on the other hand, it must be realized that simulation can never entirely replace testing on the real system. Modelling of the exact mechanical behaviour of the cart is difficult at best, as is the simulation of the exact timing behaviour of a collection of processes running in a real-time environment. Also, the requirements of a simulation environment in the area of initialization, resets and user interfaces necessarily differ somewhat from the final application environment, often necessitating some code modifications when porting the tested software between the two environments. A simulation tool always implies a trade-off between realism and cost.

As our simulation tools are very much development tools indeed, they tend to be modified and extended as the development work on the vehicle's software progresses. Therefore, the work described in this paper must be seen as a snapshot of the status of a very useful and very intensively used set of tools but not as a final product.

The simulations are implemented on a graphical, UNIX-based workstation. We have two sets of simulation packages that we are currently integrating into a single environment. Some interaction is possible already. The first package - CARSIM - provides a simulation of the cart plus the low level control software. The second - ASSIM - provides sensor-simulation capabilities for a cart moving in a given environment.

The usefulness of simulation in the development of autonomous mobile robots systems has been reported by other authors; descriptions of other simulation packages for mobile robots can be found, e.g. in [1], [2], [3] and [4].

CARSIM

The simulation package for the vehicle consists of various sets of routines. One set of routines provides the actual simulation of the cart and the interface to the control software as provided by the Philips Motion Controller (PMC) driver. A second set of routines implements the simulation of the virtual cart. Furthermore two different user-interfaces are provided.

Besides providing essentially the same call interface as the actual PMC driver routines, the simulator routines in the first set also provide mechanisms that allow the state of the simulated vehicle to be interrogated and a routine to reset the simulation to its initial state.

The lowest level of the simulation software consists of computations of the dynamic and kinematic behaviour of the MARIE vehicle which has two independently driven rear wheels and two coupled, steerable front wheels controlled by one motor. In this part of the simulation the control outputs of the next level are translated, through a calculation of the mechanical response of the DC motors, to simulated sensor outputs (the axis encoders of the three motors) which are used to control the different motors. The dynamical model[5] used for the MARIE cart takes into account some non-linearities as they occur in a real cart, such as a static friction, but disregards other complexities, such as the change in effective mass as the steer is turned, the free play in the rear wheels and the elasticity of the cart. Some care has been taken to ensure that the simulated cart can behave differently from the control model in the virtual cart layer, but in a physically sensible way. E.g. various dimensions can be chosen slightly different, allowing the effect of such discrepancies to become apparent. The equations are solved to second order accuracy where possible. The correspondence between the simulated cart and the real cart is such that it is found that the control parameters that work well for the simulated cart also allow a satisfactory control of the real cart.

The cart simulator also generates information about the absolute position of the cart, the average control outputs to the vehicle motors and several other state variables that are not normally accessible in a real cart. These value can be interrogated by e.g. the user interface and the sensor simulation software. A separate monitoring program is available that can plot the attained absolute positions and generate a log file of various state quantities.

---

[5] This model was for a large part designed and built by G.M. van der Molen of the University of Strathclyde.

A second set of routines extends the interface level for the vehicle simulation up to the level of the virtual cart. Extending the simulation to this level provides a uniform call interface, valid for all vehicles on which the virtual cart layer has been implemented. These routines differ only in certain control aspects from the real-time version.

Two sets of user-interface routines are provided. The first provides only some very basic capabilities to call all the various routines in the PMC driver and the virtual cart, but provides extensive status reporting facilities and thus allows the user to examine the effects of various calls in the various simulated control layers in detail. The primary purpose of this user-interface is to provide detailed testing and debugging facilities.

The second user-interface is more suited to the study of the interaction of the simulated vehicle plus control software with higher level control routines. It provides extensive graphical display facilities and user interaction. It also provides an interface to a sensor simulation package "ASSIM". The combination of this user-interface with the vehicle simulator and various layers of control software is referred to as "CARSIM".

ASSIM

The "Amsterdam Sensor Simulator"[6] is a software package which simulates sensor data from various sensors mounted on a (simulated) mobile robot. One or more robots can be positioned in a 2D environment. When an event in the environment occurs (the robot on which the sensors are mounted, or one of the other robots moves), the simulator recalculates all sensor data. All sensors that are mounted on a robot are thus kept on their correct simulated output continuously.

The current version contains a simulation of a laser range finder, an ultrasonic range finder, an ideal range sensor and a collision detector. For navigation on a grid, a grid-line detector and a transponder detector are implemented.

The environment is represented as a set of 2D polygons, each with a starting height and end height. Surface properties (as for example reflectivity for sound or for electromagnetic waves) can be set by the user. Data about the environment as well as data about the robots (shape etc.) can be stored in a simple data base.

For the generation of the sensor data, a model of the (physical) properties of the sensor is used. For the ultrasonic sensor we have taken into account the intensity profile of the transmitter, the attenuation by beam divergence, absorption in the air and by reflection, specular reflections and (spurious) signals because of multiple reflections.

The simulation package can be used stand-alone, where the user has the choice of three different user interfaces:

a) When running the simulator on a Sun workstation, the graphical user interface can be used. Different maps of the environment can be loaded using the buttons. Robots can be moved either by using the mouse or by entering the desired position and orientation

---

[6] ASSIM was originally developed for the SPIN project 0710.133: "Planning methods and simulation for a semi--automatic vehicle," in collaboration with Industrial Contractors Holland BV.

in a special window. Sensors can be mounted or unmounted with buttons. Every sensor has a graphical representation of the data in a separate window.

b) In the non-graphical mode, a command line interpreter can be used to communicate with the program.

c) The user can write his own initial setup and sensor configuration and create routines for particular reactions on mouse or keyboard events.

A more extensive, but slightly outdated description of ASSIM can be found in a paper by Kröse [5].

Currently ASSIM is integrated with CARSIM in a single simulation environment. The environment and sensor configuration are still read from the database, but the position and orientation of the vehicle is provided by CARSIM. The simulated sensor data is available for other modules, and can also be represented graphically in a window on the screen. In this set-up, sensor based control modules (SBC in Fig. 1) can be tested in simulation before implementation on the actual vehicle.

Results

The modular software structure and extensive simulation tools have been tremendously useful in designing and testing our software, and also in the evaluation of new concepts. The availability of the virtual cart paradigm has greatly facilitated the porting of the trajectory controller software from the simulation environment to two different carts, one at the University of Amsterdam, the other at Robert Bosch GmbH.

After the initial design of the PMC driver interface, the simulation software for this driver was built first. Building this simulator was quite helpful in verifying the validity of the interface design and also in improving our understanding of the actual operation of the real system. Part of the simulation routines, especially unit conversions, was directly useful for the actual driver.

The next step was the implementation of the simulation routines for the cart and the basic user interface. The first simulated version of the virtual cart was built and tested in this environment.

Work on the graphic interface and an early, simplified version of the cart simulator and control software (not incorporating either the PMC or the virtual cart) proceeded simultaneously. After the first tests on the virtual cart, a merged version was constructed.

The current simulation package still suffers somewhat from the different origins of its components. We are currently working on integration of ASSIM and CARSIM, while simultaneously providing facilities to implement the various control layers as separate UNIX processes, using slightly modified call libraries. One of the major issues in this case is the implementation of a suitable "event" or time manager to simulate the progress of time for all the simulated processes in a manner consistent with the inherently concurrent nature of the processes in the real-time system.

In porting our software to the real vehicle, we found that, as the level of abstraction increased, the differences between the simulated and the real software became progressively smaller. The PMC driver and its simulator have some common code, but differ greatly in most respects. The algorithmic structure of the virtual cart is essentially the same for the simulation and the real vehicle and most of the code is shared. It was necessary, however, to modify the control structure of the virtual cart, in order to ensure that it could meet the real-time requirements of the actual cart. Initial experience with the trajectory controller shows that the differences between the simulated and real version are even smaller.

We also found that the most damaging remaining errors occurred in the PMC driver software, which had not been tested in simulation. The virtual cart, though it had been tested with the simulator of the Amsterdam cart, and was first tested on the Bosch vehicle, did contain fewer damaging errors, as most errors had been found in simulation.

Acknowledgements

Most of the work described in this paper was done for the ESPRIT II project 2043 "MARIE". The authors acknowledge the support of the ESPRIT programme of the European Communities and the constructive interaction with other partners in the consortium. In particular, we wish to thank G.M. van der Molen of the University of Strathclyde for his contributions to CARSIM and M. Bergman of the University of Amsterdam for his work on the graphical interface.

References

[1]   J. Meyer, "An emulation system for programmable sensory robots," IBM J. Res. Develop. (25) 6, 1981, 955-962

[2]   J. Raczkowsky, K.H. Mittenbuehler, "Simulation of cameras in robot applications," IEEE Computer Graphics and Applications, January 1989, 16-25

[3]   P. Adolphs, P.Léonard, J. Amelung, M. Augustyniak, A. Bletz, "SAMOS, a flexible simulation program for autonomous mobile systems," in "Intelligent Autonomous Systems 2," ed. T. Kanade, F.C.A. Groen, L.O. Hertzberger, 1989, Stichting International Congress of Intelligent Autonomous Systems, ISBN 90-800410-1-7, 630-640

[4]   T. Knieriemen, E. von Puttkamer, R. Trieb, " 3d7 - A 3D simulation environment for autonomous system design," in "Intelligent Autonomous Systems 2," ed. T. Kanade, F.C.A. Groen, L.O. Hertzberger, 1989, Stichting International Congress of Intelligent Autonomous Systems, ISBN 90-800410-1-7, 434-440

[5]   B.J.A. Kröse, E. Dondorp, "A sensor simulation system for mobile robots," in "Intelligent Autonomous Systems 2," ed. T. Kanade, F.C.A. Groen, L.O. Hertzberger, 1989, Stichting International Congress of Intelligent Autonomous Systems, ISBN 90-800410-1-7, 641-649

A Logical Framework of Sensor/Data Fusion

Mieczyslaw M. Kokar
Kyriakos P. Zavoleas

Department of Industrial Engineering and Information Systems
Northeastern University
Boston, Massachusetts

kokar@northeastern.edu
kyriakos@nueng.coe.northeastern.edu

Abstract

A sensor/data fusion system must integrate reasoning about both quantitative information (raw sensory inputs) and symbolic information (theories about the world perceived through sensors). In this paper we outline a logical framework for sensor/data fusion. In order to be classified as "fusion", the system must include at least two theories about the world and two data structures for storing and manipulating the data (in logical terms these are called *models*). We first introduce some formal requirements for fusion to be a logically sound process. Then we show that fusion can be designed either as a monotonic or a nonmonotonic process. We then use an example to illustrate our points. We believe that understanding of general principles of sensor/data fusion is necessary for designing fusion systems. This issue becomes even more important when these design decisions are shifted to the area of responsibility of an intelligent controller.

1 Introduction

A dynamic system in interaction with its environment is a subject of study from various viewpoints: *What information does the system collect from the environment, and by what means? What is the system's knowledge, and how is it actually stored? What are the system's goals? What are the system's data and knowledge processing mechanisms, and how do these mechanisms interact with the knowledge stored in the system knowledge base? How does the system reach decisions about its reactions to the environment? How are actions produced?* In the intelligent control paradigm all these questions can be classified into three groups related to the three elements of the perception-inference-action loop [7]. In this paper we deal with sensory input data interpretation for situation assessment, which is related mainly to the perception element of this loop. The information produced in the perception phase will naturally be the basis for generating system's actions.

During the last few years a substantial progress in signal/data processing theory has been accomplished. In many situations data from a single source has proved to be insufficient for deriving meaningful conclusions. One of the observations that led to such a conclusion is that the information content of any single sensor is limited, dependent on sensor type. Also, there are a number of uncertainties associated with data aquisition and processing that propagate errors in the process of transforming sensor data into decisions and actions. The utilization of muptiple sensors seems to be promising for the improvement of a signal processing system's performance. One of the problems it creates is increased needs for computer processing power, but recent progress in the area of microprocessors can compensate for this increase.

The high popularity of this approach has created an illusion that there exists a scientific discipline dealing with the integration of information from multiple sources, called "sensor fusion". Although a number of verbal descriptions can be found in the literature ([1], [8]), a formal definition for Sensor/Data Fusion has not been attemped. Thomopoulos [8] gives a systematic overview of sensor integration and data fusion and provides the following defintion of sensor fusion:

> "Let an observed phenomenon or a process that is being monitored, the sensors that are used to collect data and possibly produce inference, and a fusion center, all be considered as

one system. Sensor fusion is then the process of integrating raw and processed data into some form of meaningful inference that can be used intelligently to improve the performance of the system, measured in any convenient and quantifiable way, beyond the level that any one of the components of the system separately or any subset of the system components partially combined could achieve."

Although this definition covers most of the aspects that can be intuitively understood as "fusion", since it is verbal, it is open to various interpretations. Even if fusion is to be implemented by a programmer developing a situation recognition system, this definition may create serious misunderstandings. Such phrases as "intelligent use", "meaningful inference", or "improved performance" may have significantly different meanings. If the goal is to either generate a formal specification for a sensor/data fusion system or implement a system that "reasons" about fusion (automatically selects and/or synthesises fusion algorithms) then the need for a more formal treatment of the subject of fusion becomes a necessity.

In this paper we are making a step towards a formal theory of sensor/data fusion. First of all we provide a formal-logic-based interpretation for the notion of fusion. The immediate advantage of this interpretation is that the distinction between "data fusion" and "information fusion" becomes apparent. Data fusion corresponds to combining models of theories (quantitative data representing raw sensory inputs). Information fusion corresponds to generating new theories using other existing theories (symbolic information – theories about the world perceived through sensors). Fusion, in order to be logically sound, must include both of these activities at the same time.

One of the issues investigated in this paper is the issue of nonmonotonicity in fusion. By nonmonotonicity we mean such a reasoning process in which assertions made by the system based upon data from any single sensor can be invalidated after all this information is fused. Although we did not see a formal analysis of monotonicity in the fusion literature, it seems that there does not exist an agreement on whether fusion should be a monotonic or a nonmonotonic process. In this paper we show that the assumption of monotonicity cannot be accepted as a rule. Through simple examples we show that conclusions derived from any single theory may become invalid in the combined theory; likewise, statements that evaluate to false in one or both of the original theories may become true in the fused theory. In our opinion, non-monotonicity is inherent in fusion processes; one should consider non-monotonicity as a policy of fusion design, rather than a "bug" that must be eliminated.

2 A Logical Framework for Fusion

The logical framework for fusion is represented in Figure 1. In this figure W represents the world measured by two sensors through the corresponding measurement functions Z_1 and Z_2. Sensory data is stored in data structures M_1 and M_2. In logical terms these are models for two theories T_1 and T_2 respectively. The two theories are interpreted in the models through mappings I_1 and I_2.

We presume that these four components (two theories and two models) contain all knowledge that the system has about the world before fusion. In logical terms, fusion is creating a new (fused) model M_F using models M_1 and M_2, and a new (fused) theory T_F using the information contained in respective theories T_1 and T_2 associated with the two models. For this description to be complete we need to consider one more theory (in [4] it is called *metatheory*), which "controls" the fusion process.

As we stated in the introduction, our interest lies in transferring fusion decisions to an intelligent controller which exhibits a high level of autonomy, we will call such a controller an *intelligent agent*. To provide a more specific definition of an intelligent agent for fusion we need to introduce some formal notation.

2.1 Notation and Definitions

For the sake of self containement of this paper we give some defintions of formal logical systems. We generally follow the notation and terminology introduced by Lyndon [6].

Definition 1 *Let L be a formal language, A - a set of axioms (sentences presumed to be true), C - rules of inference (rules of deriving true conclusions given true premises), T - theorems (sentences that*

are shown true by application of the rules of inference C to the axioms A), I - rules of interpretation (mappings assigning objects in the domain of interest of the system S), and W - a world frame.
The quintuple
$$S =< L, A, T, C, I >$$
is called a formal system if:
(a) A is a minimal set such that $A \vdash_C T$,
(b) $W \models_I T$,
where the operator $\vdash_C$ is the provability operator for the given collection of inference rules C and $\models_I$ stands for the semantic satisfaction in the given interpretation I.

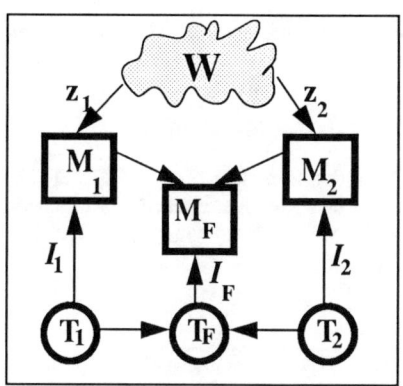 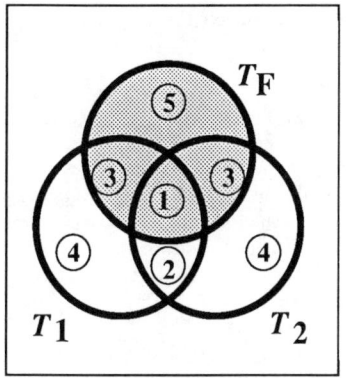

FIGURE 1
Logical Framework for Fusion

FIGURE 2
Fusing Theorems

Referring to Figure 1, we have three formal systems S_1, S_2 and S_F whose parts (theorems and interpretations) $T_1, T_2, T_F, I_2, I_2, I_F$ are explicitly represented in the figure. Since in a formal system axioms are parts of theorems ($A_1 \subset T_1, A_2 \subset T_2, A_F \subset T_F$) they are also captured in Figure 1. Languages L_1, L_2 and L_F are not explicitly represented and neither are rules of inference C_1, C_2 and C_F.

In formal logic, a *model* (M) is considered to be an abstract algebra with a non-empty set D called *domain*, and a set of operators $o_j \in O$, each of some rank $n(o_j)$, index[Bed by a set J

$$M =< D, O >.$$

These operators are represented in the agent's theories as either relational or functional symbols. In our case, a model contains data which were either obtained through the agent's sensors, or through the agent's reasoning process. In the latter case, the agent interprets the sensory data, applies the reasoning rules of appropriate theories, and then applies the interpretation rules to the derived sentences. It is thus presumed that the agent's model at any instance of time contains a collection of facts whose actual source is the current situation of the environment the agent is in.

To be able to fuse two theories there must be a common reference base for the information contained in the theories. In the fusion literature the process of association of information of one data set with the information of another data set is called *registration* [2]. This justifies the introduction of the notion of *relation* among theories.

Definition 2 *Two formal systems S_1 and S_2 are related ($R(S_x, S_y)$) if*

$$I_1(T_1) \cap I_2(T_2) \neq \emptyset.$$

This means that for two systems to be related they must have some common reference frame in their data sets. Although this fact is not explicitly represented in Figure 1, it is presumed that the two models, M_1 and M_2 are not disjoint.

Definition 3 *A meta-system mS (or meta-theory) is such a formal system whose interpetation contain other theories as objects.*

Meta-theories can contain partial-order relation among object theories.

Having defined all the necessary elements we can define a framework called *Intelligent Agen Framework (IAF)* in which we can carry out our logical analysis of fusion.

Definition 4 *An* Intelligent Agent Framework *(IAF) is a collection of formal systems, meta-systems and relations among the systems*
$$IAF = < S_x, mS_y, R_z >.$$

More discussion on the Intelligent Agent Framework can be found in [5], where it was initially defined.

3 Fusion of Theories and Their Models

Taken together with their interpretations I_1 and I_2, the two formal systems S_1, S_2 and their models M_1, M_2 fulfill the requirement of *logical adequacy* [6], i.e., any sentences p_1 and p_2 that can be derived within the formal systems S_1 and S_2 respectively can be semantically justified in their models Formally, it can be stated as:

$$T_1 \vdash_{C_1} p_1 \quad \text{iff} \quad M_1 \models_{S_1} p_1,$$
$$T_2 \vdash_{C_2} p_2 \quad \text{iff} \quad M_2 \models_{S_2} p_2.$$

Without such an requirement reasoning about the data stored in models would not guarantee *logica soundness* [6]: from correct presumptions we could derive false conclusions. It seems natural to require that the fusion process should preserve this kind of requirement, too. In other words, we propose that the fusion process fulfills the constraint of logical soundness of the resulting information. This gives rise to the following attempt of defining fusion in logical terms.

Definition 5 *Fusion is a process of combining two formal systems S_1, S_2 and their models M_1, M_2 into a new formal system S_F and a new model M_F in such a way that the following adequacy condition is fulfilled for any sentence p_F of theory S_F:*

$$T_F \vdash_{C_F} p_F \quad \text{iff} \quad M_F \models_{S_F} p_F.$$

This definition seems to be acceptable from the logical point of view since it stresses logical soundness of the result of fusion; in other words, the restriction it implies is that, for every sentence p_F in the fused system S_F, there must be some interpretation I_F, which is supported by the data in the fused model M_F. Also, this definition is appealing from the application point of view because it covers fusion of both data and information, which has been a topic of many discussions in the sensor/data fusion community. It is, however, reasonable to ask whether the definition of fusion should be more specific, more restrictive. In its current form it gives much of flexibility for ways of combining models and theories. The question is, therefore, whether we can add some more logical restrictions to this definition. One of the directions for searching for additional logical constraints – the monotonicity constraint – is investigated in this paper.

Perhaps it is worthwhile to mention that in our definition we are not trying to cover technical aspects of fusion which are captured by other definitions, as the one stated above. Our goal here is to investigate only logical aspects of fusion. We believe that our logical definition can be easily combined with other, more specific definitions in the literature.

4 Should Fusion Be Monotonic?

In considering possible restrictions to the definition of fusion we first tried to understand the issue of whether the fusion process should be required to be monotonic or not. We were not able to answer this question uniquely by studying the fusion literature and consulting practitioners in the field. To begin the analysis of the issue of monotonicity we need to accept some interpretation of our definition in terms of components of formal systems presented earlier in this paper. Although our definition gives strict requirements on the consistency of the resulting representation of fused information, it does not tell us how to combine particular elements of two formal systems – languages, theorems, inference rules, and interpretations – into one fused representation. To be able to discuss monotonicity we will propose an interpretation of the fusion operations, which perhaps is not the most general, but is sufficient for the sake of our goal.

First of all, we presume that sensory data from two sensors stored in two data structures M_1 and M_2 are fused into M_F through some *model merging* operation

$$M_F = M_1 \nabla M_2.$$

We do not formally define the specifications of a model merging operation; this is part of our ongoing research. Instead, we give an example of such an operation in the next section of this paper. Roughly, it is an operation in which two data structures, not necessarily of the same type, are combined into one. The necessary requirement for this operation is that appropriate associations between data points $d_1 \in M_1, d_2 \in M_2$, representing the same object in world W (Figure 1) are preserved

$$Z_1^{-1}(d_1) = Z_2^{-1}(d_2) \Rightarrow d_1 \xrightarrow{\nabla} d_F \wedge d_2 \xrightarrow{\nabla} d_F, \quad d_F \in M_F$$

After this kind of fusion, all the operations that were defined for any single model are still valid for at least the same model, or possibly also for the fused model. In addition to this some new operations may become admissible. For instance, one of the models may contain a two-dimensional array of pixels representing vision input, and the other model may represent pixels generated by a range sensor (such a system is described in [3]). The fused model may be stored as a three dimensional array with all old operations applicable to both sub-arrays, and in addition to this, with an additional legal operation of difference between values associated with respective pixels.

The question of how to fuse formal systems S_1 and S_2 can be restated in terms of their components: how to fuse languages, inference rules theorems and axioms, and finally, rules of interpretation.

Fusion of inference rules is straightforward if we limit ourselves to standard inference rules used in first-order logic:

$$C_F = C_1 = C_2.$$

Rules of interpretation I_F must include I'_1 and I'_2, where I'_i are syntactically modified versions of I_i so that they can interpret sentences of S_F in the fused model M_F. In addition to this, they must include interpretations for the new operations that are applicable to the fused model.

Fusing two languages L_1 and L_2 is similar to fusing models. We need to preserve consistency of naming objects in the world. Since we presume that the issue of consistency between the world and the models has been resolved, all we need to worry about is the consistency of assigning names to those elements of the model that have names in both languages,

$$L_F(I'^{-1}_1(M_F)) = L_F(I'^{-1}_2(M_F)) = L'_F.$$

The rest of these languages, we denote them as L''_1 and L''_2 ($L''_i = L_i - L''_F$) can be combined through the union operation. Thus the fused language is represented as

$$L_F = L'_F \cup L''_1 \cup L''_2.$$

Finally, we need to propose a method for fusing theorems T_i. This also will include axioms, since axioms are part of theorems ($A_i \subset T_i$). In proposing rules for fusing theorems, one is tempted to consider only such methods that guarantee that all the theorems T_1 and T_2 of the systems to be fused are preserved in the fused system ($T_1 \cup T_2 \subset T_F$). This would result in a monotonic fusion procedure. But we think that the question of whether fusion should be monotonic or not is an empirical question, and not a logical one. We state our claim in the following way.

510

Claim. *The fact that T_1 and T_2 are collections of theorems of two formal systems S_1 and S_2 does not imply that all these theorems are included in the theorems T_F of the resulting fused formal system S_F, i.e., not necessarily it is the case that $(T_1 \cup T_2) \subset T_F$.*

To prove such a claim it should be sufficient to show counterexamples of its negation. We go one step further; we also classify all possible relationships that can hold between theorems T_1, T_2 and T_F. They are represented symbolically using a Venn diagramm in Figure 2. Particular areas in the diagram represent the following situations:

1. Theorems supported by both formal systems are also supported by the fused formal system.
2. Theorems supported by both formal systems are false in the fused formal system.
3. Theorems supported by only one of the formal systems are true in the fused system.
4. Theorems supported by only one of the formal systems are false in the fused system.
5. Sentences that are false in both formal systems are true in the fused formal system.

An example where these situations appear is presented in the following section.

5 Example

In this section we present an example of a world seen by two sensors. Sensory data is stored in two model data structures. Reasoning about the world is done using two theories associated with two sensors. Although the world is somewhat artificial, we feel that its simplicity will allow us to make our claims more explicit.

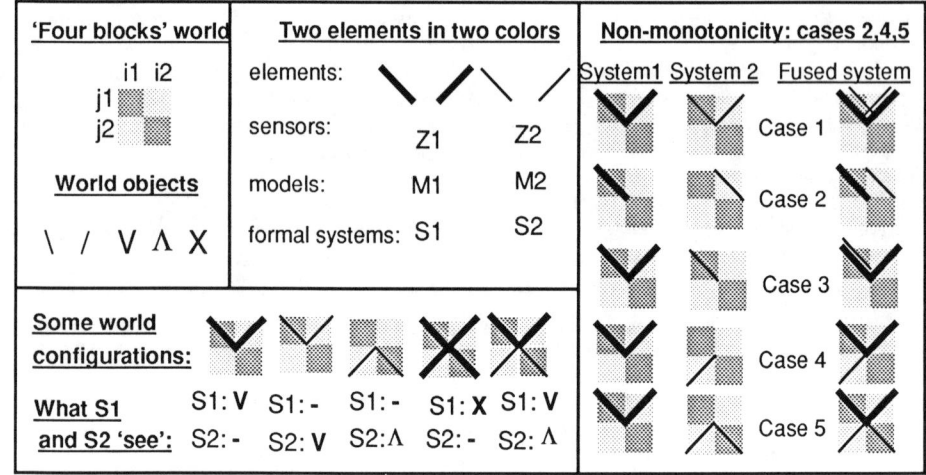

FIGURE 3: Example

Consider a world W consisting of four blocks as in Figure 3. Each of the blocks can have one of two possible elements, \, / (or be empty, $\emptyset$). The two elements (\, and /) can be in one of two colors; we represent one of the colors using bold face symbols (\, /). Suppose that the two sensors are able to recognize these elements, and that each of the sensors is sensitive to only one of the colors. Each sensor can then identify $3^4 = 81$ possible instances out of $6^4 = 1,296$ combinations possible in the world. It is therefore obvious that any single sensor has lower resolution than two sensors combined, which means that fusion is justifiable.

5.1 Models

Domains D_1 and D_2 for models M_1 and M_2 are:

$$D_i = G \cup P_i \cup \{0,1\},$$

$$G = G_x \times G_y, \quad G_x = \{i1, i2\}, G_y = \{j1, j2\},$$

where G represents objects of the world (blocks), $P_1 = \{\emptyset, \backslash, /\}$ and $P_2 = \{\emptyset, \backslash, /\}$ represent elements of the objects, and the two extra elements '0' and '1', rerpesent interpretations of T and F — logical "true" and "false" [6].

The operators applicable to these two domains are:
l : $G \rightarrow \{0,1\}$, r : $G \rightarrow \{0,1\}$, u : $G \rightarrow \{0,1\}$, d : $G \rightarrow \{0,1\}$, b : $G \rightarrow \{0,1\}$, f : $G \rightarrow \{0,1\}$

The first four opearators return '1' if the position of the block that contains the element referred to is *left, right, up, down* respectively. The remaining two operators return '1' if the element referred to is a '\' (back slash) or '/' (forward slash) respectively. In additon to this there is one more operator, c : $G \times G \rightarrow \{0,1\}$, which returns '1' if the two elements referred to can be interpreted as "connected" (e.g., like in a 'V' letter).

5.2 Theories

We use the same letters typed in italics to represent constants describing objects and predicates: $l(), r(), u(), d(), b(), f(), c()$, T (for "true"), F (for "false"). We also have at least three theoretical terms, 'V()', 'Λ()' and 'X()', which represent compound objects in the world that are of interest to us; another predicate of interest is the number of elements present in the world denoted by the predicate $elts()$. Our language includes logical constants $\vee, \wedge, \Rightarrow, \bar{}$ (for logical "not") etc. In the following we present a sample set of theorems in each of the two theories. x_i denotes an element occrring in T_i, for $i = 1, 2$.

$$\left. \begin{array}{l} l(x_i) \wedge b(x_i) \wedge r(y_i) \wedge f(y_i) \wedge \\ c(x_i, z_i) \Rightarrow z_i = y_i \wedge \\ c(y_i, w_i) \Rightarrow w_i = x_i \end{array} \right\} \Rightarrow V(x_i, y_i)$$

$$\left. \begin{array}{l} l(x_i) \wedge f(x_i) \wedge r(y_i) \wedge b(y_i) \wedge \\ c(x_i, z_i) \Rightarrow z_i = y_i \wedge \\ c(y_i, w_i) \Rightarrow w_i = x_i \end{array} \right\} \Rightarrow \Lambda(x_i, y_i)$$

$$\left. \begin{array}{l} l(x_i) \wedge u(x_i) \wedge b(x_i) \wedge r(y_i) \wedge u(y_i) \wedge f(y_i) \wedge \\ l(x'_i) \wedge d(x'_i) \wedge f(x'_i) \wedge r(y'_i) \wedge d(y'_i) \wedge b(y'_i) \end{array} \right\} \Rightarrow X(x_i, y_i, x'_i, y'_i)$$

5.3 Model fusion

Fusion of models and theories should be done according to some rational rules. In logical terms, these rules would constitute a theory of fusion. In the framework of IAF, they would be stored in a metatheory mS.

A new data structure is needed for the fused model. In this example the 'merging' of the two data structures of each individual model is straightforward; the fused model contains four blocks, and each one of them contains information about both of the colors that are measured. Thus only the word size of the original data structure needs to be augmented; since each sensor may assign one out of three possible values to each block, measurements of two sensors may lead to nine (3×3) possible values. Operators can be easily extended to accomodate the new domain, but they are semantically the same.

With such a model merging, no information is lost; however, no substantial processing is performed, apart from assigning data correspondence between the two separate models. The correspondence assessment constitutes the model fusion process.

5.4 Theory fusion

In this case the unification of all terms of the two theories is natural, since both of them describe the same objects in the world, use the same sets of elements, and the same sets of predicates. The most important extensions to be made are the following:
– The connectivity predicate c is now true, even if the elements are of different color.

— Our interest focuses on the pattern shape, not on the structure of colors; thus, all of the following are considered equivalent: V(x_1, y_1), V(x_2, y_2), V(x_1, y_2), V(x_2, y_1), V(x_F, y_2), V(x_1, y_F), V(x_F, y_F), for they all identify the pattern of a V. Denoted as $x_F(y_F)$ is an element which is detected by both systems i.e. it has both of the measurable colors.

Providing a good theory of fusion that a computer system could use to fuse two theories (formal systems) automatically is not an easy task; the question that arises is how to fuse theorems. In our fused theory, the predicates taken from both individual theories should be used, in order to identify the pattern in the world. We thus admit all theorems of the individual theories for identifying V's, Λ's and X's, allowing the elements to be of any color, not necessarily the same. However, patterns identified from the individual theories may not hold in the fused theory, and this is mainly because of the extension introduced in the connectivity predicate c.

The following cases express the idea that the theory of fusion does not need to be based on the principle of monotonicity, i.e., that the fused theory does not need to preserve truthfulness of the constituent theories. Besides case 1, where both theories consent in their conclusions, and case 3, where a conclusion of one of the theories is accepted after fusion without any conflict, we show examples of nonmonotonic situations (cases 2, 4 and 5), where conclusions in the individual theories cannot hold after fusion is performed.

Case 1. Both individually reached conclusions agree that the elements in the world correspond to a V pattern; no contradiction occurs, and the conculusion holds in the fused theory as well.

$$V(x_1, y_1) \wedge u(x_1) \wedge V(x_2, y_2) \wedge u(x_2) \Rightarrow V(x_F, y_F)$$

Case 2. Arguments supported in T_1 and in T_2 are refuted in T_F. In this example, each individual system detects one single element in the world; however, fusion concludes that it is not the same element detected by both systems, and updates the number of elements to two.

$$u(x_1) \wedge l(x_1) \wedge b(x_1) \wedge elts_1(1) \wedge u(x_2) \wedge r(x_2) \wedge b(x_2) \wedge elts_2(1) \Rightarrow elts_F(2)$$

Case 3. An argument that holds in one theory, T_1, still holds in the fused theory T_F; information from T_2 neither provides new clues nor produces conflicts.

$$V(x_1, y_1) \wedge u(x_1) \wedge u(x_2) \wedge l(x_2) \wedge b(x_2) \Rightarrow V(x_F, y_1)$$

Case 4. An argument that holds in one theory, T_1, is refuted in T_F, in the light of evidence from the other theory, T_2.

$$V(x_1, y_1) \wedge u(x_1) \wedge c(x_1, y_2) \wedge d(y_2) \Rightarrow {\sim}V(x_1, y_1)$$

Case 5. Arguments from T_1 and T_2 lead to the generation of a new argument that holds in T_F, even it is not supported by any of the individual theories.

$$V(x_1, y_1) \wedge u(x_1) \wedge \Lambda(x_2, y_2) \wedge d(x_2) \Rightarrow {\sim}V(x_1, y_1) \wedge {\sim}\Lambda(x_2, y_2) \wedge X(x_1, y_1, x_2, y_2)$$

5.5 Discussion

The natural question that one can raise at this point is whether fusion *can* be implemented as a monotonic process. One could argue that if we reduce the number and extent of conclusions derived in the individual theories, and the range of possible scenarios, then we might escape the need for refuting theorems in the process of fusion, and the fusion process could be monotonic. In our opinion, such an approach has at least two disadvantages. First, such a fusion system would generate wrong conclusions should the scenario it encouters be different than the ones envisioned by the system designer. Second, it would be vulnerable to sensor errors and failures. In real sensor systems, measurement errors cannot be fully eliminated. If the fusion process is monotonic, it becomes useless when one of the sensors fails. On the other hand, when the fusion system is flexible, it is able to recover from erroneous conclusions after gathering more data from other working sensors.

Consequently, we believe that the fusion system should be a reactive system; it should always try to reach and use decisions based upon the currently available information, even if they are to be invalidated later due to new evidence. The process of refining and adjusting derived conclusions should be inherent in the sensor/data fusion systems.

6 Conclusions

In this paper we investigated the basics of a logical framework that may contribute to formulation of a theory for fusion. From the perspective of formal logic, we distinguished between fusion of models and fusion of theories. Maintaining consistency between fusion of models and fusion of theories is necessary for interpretation of the fusion results. If a theory of fusion is to be used by a computer system to fuse theories there must be a general theory of fusion that can be applicable to many situations. Then an automatic reasoning system can apply it whenever some applicability preconditions are met.

The main question investigated in this paper is whether a theory of fusion should be monotonic, i.e., whether it should preserve all theorems of the theories being fused, or at least those that are supported by at least one of the theories. We have shown that the answer is not positive. Our conclusion then is that we should continue search for other principles that might serve as axioms of a fusion theory.

A complete theory of fusion will be invaluable not only for developing automatic fusion systems, but also for designers of fusion systems. It will help the designers in setting constraints and specifications, and would preclude waisting efforts due to wrong design decisions. Moreover, more experience needs to be aquired about the concept and nature of fusion. Since, as we showed in this paper, purely theoretical approaches to fusion may lead to erroneous theories, a combination of theoretical and experimental research is the best approach towards the development of an acceptable theory of sensor/data fusion. Overall, the logical framework proposed in this paper constitutes a contribution to a general theory of fusion.

7 Acknowledgements

The research reported here was supported in part by the Center of Electromagnetics Research, Northeastern University, and the "Alexandros S. Onassis" Public Benefit Foundation. We would like to thank Spyros Reveliotis for interesting comments and discussions.

References

[1] Kenneth L. Carlsen. Navy's fusion requirements. In *Proceedings DSF-87*, pages 41–79, 1987.

[2] Martin P. Dana. Registration: A prerequisite for multiple sensor tracking. In Yankov Bar-Shalom, editor, *Multitarget-Multisensor Tracking: Advanced Applications*. Artech House Inc., 1990.

[3] M. Hebert, T. Kanade, and I. Kweon. 3-d vision techniques for autonomous vehicles. In R.C. Jain and A.K. Jain, editors, *Analysis and Interpretation of Range Images*. Springer-Verlag, 1990.

[4] M. M. Kokar and W. Zadrozny. A logical structure of a learning agent. In Z. W. Ras, editor, *Methodologies for Intelligent Systems, 4*, pages 305–312. North-Holland, 1989.

[5] M. M. Kokar and W. Zadrozny. Specifying the knowledge level of a learning agent. In *Proceedings of ISMIS'90*, 1990.

[6] R. G. Lyndon. *Notes on Logic*. D. Van Nostrand Company, Inc., 1966.

[7] H. E. Stephanou, A. Meystel, and J. Y. S. Luh. Intelligent control: From perception to action. In *Proceedings of the IEEE International Symposium on Intelligent Control*, 1988.

[8] Stelios C.A. Thomopoulos. Sensor integration and data fusion. In *Proceedings: Sensor Fusion II: Human and Machine Strategies*, pages 178–191. SPIE, 1989.

MAINTAINING WORLD MODEL CONSISTENCY BY MATCHING PARAMETRIZED OBJECT MODELS TO 3D MEASUREMENTS

Markku Järviluoma[1] & Sakari Pieskä[2] & Tapio Heikkilä[2]
[1] University of Oulu/Systems Engineering Laboratory
Linnanmaa, SF-90570 Oulu, Finland
[2] Technical Research Centre of Finland/Electronics Laboratory
P.O.Box 200, SF-90571 Oulu, Finland

ABSTRACT

A robot control system must maintain knowledge about its surrounding world to be able to intelligent, goal-oriented behavior. Maintaining the consistency of the world model is important to assure reliable operations. Verification of the position and orientation of objects is the main task in the consistency maintenance process. This verification involves determining values for transformation parameters describing the six degree-of-freedom transformation of the object in relation to the sensor and manipulating systems. Our solution in determining this relationship is matching parametrized object models to measurements by using Newton's iterative method. In this paper we present one example case: matching a paper roll model to 3D vision sensor data. We also shortly present our research project called the 'Machine of the Future', during which this case has been studied.

INTRODUCTION

Maintaining world model consistency is essential research issue in several works on model-based robot control or on model-based vision. Newton's method has been used for tasks required in model updating, e.g. in camera calibration and matching 3D objects to 2D camera images; see review in [Lowe 1989]. In addition Boult and Gross [Boult & Gross 1987] have used Newton's method for depth information analysis represented by superquadricks. Sakane et al. [Sakane et al. 1987] has used it for visual control with 2D correspondences and Heikkilä [Heikkilä 1990] for 3D point/plane correspondences. Different approaches to verification of the world model based on sensor data processing include e.g. use of recursive Kalman filtering algorithms (see e.g. [Faugeras et al. 1987] and utilization of Dempster-Shafer formalism and its derivates (see e.g. [Safranek et al. 1990]).

World modelling and sensing strategy in our research:

The goals in our national research project "Machine of the

Future" include development of an integrated intelligent control method which can adapt to unexpected situations by dynamic planning and efficient usage of sensor data. The control method is based on hierarchically organized Planning-Executing-Monitoring (PEM)-triplets [Heikkilä 1990]. In the initial phase of the project (1989), the basic ideas of the control scheme were demonstrated with a simulator of an indoor mobile robot. In the ongoing laboratory phase (1990 - 91) experiments are made with industrial robots equipped with range and force sensors. Next year (1992 - 93) the research expands to an outdoor application, where the test equipment include a large paper roll loading manipulator equipped with a sophisticated gripping device and appropriate sensors. The consistency maintenance of the world model will be very important especially in the outdoor application because the work environment changes quite often and improper manipulation can easily cause damages to the transferred objects.

There are two kinds of world models in our control systems. The Static World Model (SWM) contains all permanent and known information, which does not change during the execution of tasks. When the system is started, the Dynamic World Model (DWM) is created and all the information needed to accomplish the task is copied from the SWM. After the given task is completed the SWM is updated from DWM to reflect the changes in the environment and in the system itself. Object model matching, which is described in details in this paper, is the most important part in maintaining the consistency of world models.

In our project the robot or manipulator system has a following sensor selection:

- a global range sensor for identifying and coarse matching of objects,
- a local range sensor for accurate matching of objects during the manipulation phase,
- a force/torque sensor for verifying gripping and for force guided placing operations, and
- a set of limit switches as safety sensors.

The sensing strategy is based on PEM-triplet based sensory operation planning and distribution of sensor information as presented in Figure 1. [Heikkilä 1990, p.41]. One goal in our sensing strategy is to avoid computational burden caused by frequently performed updates of models with sensor data. Therefore, the planner always chooses only the most essential part of sensor information to be used in the model updating.

In the control system the sensing related operations are distributed into several hierarchy levels and they may also be operated parallel due to real-time requirements. The sensing strategy for matching object models to 3D measurements will include the following phases:

- decision making when inconsistency between the model and the real world has increased so that it requires sensory operations,
- planning of required sensory operations; in the global range sensor's case this includes planning of measurement shapes, which the single

point measurements form,
- execution of the actual sensing,
- filtering of raw sensor data and decision of disregarding nonsense measurements,
- matching parametrized object models to measurements, and
- estimating the uncertainty of the matching.

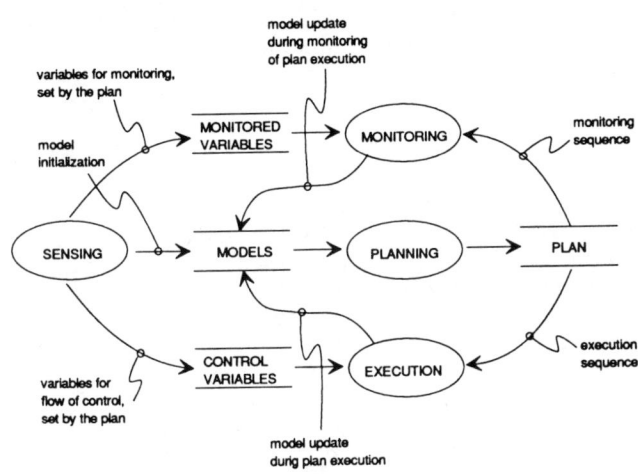

Figure 1. Distribution of sensor information in PEM-based control.

The sensor we have used in testing the algorithms in practice is a 3D vision sensor (a global range sensor) which can be used either as a 3D image scanner or a programmable coordinate meter [Moring et al. 1989]. This sensor consists of a pulsed time-of-flight laser range finder with servo-controlled pan and tilt movements. We used it as a programmable coordinate meter in our tests, so that the measurements sequence was planned before measurements. In our test we used matrix- and circular-type measurement sequences.

POSE AND POSITION ESTIMATION OF REGULAR OBJECTS

Our approach to pose and position estimation is to fit regular surface models to measured 3D points by using iteratively either least squares or Bayesian estimation. We also have included estimation of some other model parameters, such as the radius of a cylinder surface, in the same iteration procedure. The iteration procedure is descriebed in details in [Järviluoma 1991] and only the basic principles are reviewed below.

The model of the pose and position of the object is expressed as a homogeneous transformation matrix, which transforms 3D points from object's own coordinate frame to the common world's frame as $\underline{p}_w = \underline{M}_o \underline{p}_o$, where $\underline{p}_w$ and $\underline{p}_o$ is the same 3D point expressed as a homogeneous vector in world's and object's frames and $\underline{M}_o$ is a homogeneous transformation matrix. The pose and position corrections are made by rotating object's frame around its z-, y- and x-axes (in that order) and moving its origin in the the world's frame. Hence, our pose and position correction consists of six parameters: three rotation angles and three translation increments.

The correction method is based on finding these translations and rotations, and possibly additive corrections to other model parameters, by solving a linearized group of equations obtained for the distances of the measured points from the model surfaces. This group of equations can be expressed in the form

$$\underline{e} = -\underline{J}_\theta \underline{\theta} - \underline{J}_\varphi \underline{\varphi} + \underline{J}_m \delta\underline{m}$$

where $\underline{e}$ is a column matrix containing the distances of all measured points from the model surface, $\underline{\theta}$ is a column matrix containing the required translation and rotation corrections to the pose and position of the model, $\underline{\varphi}$ is a column matrix containing additional corrections to other model parameters, $\delta\underline{m}$ is a column vector containing the errors of all measured points, $\underline{J}_\theta$ and $\underline{J}_\varphi$ are Jacobian matrixes containing the partial derivatives $\partial \underline{e}_i/\partial \underline{\theta}$ and $\partial \underline{e}_i/\partial \underline{\varphi}$, $i=1,\ldots,N$, respectively and $\underline{J}_m$ is a block diagonal matrix containing the partial derivatives $\partial \underline{e}_i/\partial \delta\underline{m}_i$, $i=1,\ldots,N$, as the diagonal blocks with $\underline{e}_i$:s being the distances of the individual measured points from the model surface (possibly vector valued), $\underline{m}_i$:s being the individual measured points (vectors), $\delta\underline{m}_i$:s being the errors of individual measured points (vectors) and N being the number of measured points.

If the errors $\delta\underline{m}$ are zero mean, the corrections $\underline{\theta}$ and $\underline{\varphi}$ can be estimated using the least squares method. If the errors are Gaussian, the Bayesian a posteriori estimates can also be computed. In Bayesian estimation the corrections $\underline{\theta}$ and $\underline{\varphi}$ are handled as Gaussian random variables, which have the a priori means of zero and a priori covariances $\underline{P}_\theta$ and $\underline{P}_\varphi$. These covariances describe the uncertainty of the model before corrections. Bayesian estimation gives a posteriori estimates for $\underline{\theta}$ and $\underline{\varphi}$ and also for their covariances; hence it gives also an estimate for the uncertainty of the corrected model.

Because of linearization, the result of these estimated corrections are not accurate. The accuracy can be increased by repeating the estimation with the same measurements. The corrections $\underline{\theta}$ and $\underline{\varphi}$ should be carried out after each estimation (hence, their a priori estimates are zero in each estimation), but the covariances should be updated only after the last estimation. This iteration is halted after the absolute values of $\underline{\theta}$ and $\underline{\varphi}$ are close enough to zero.

In Figure 2 is presented an example of the convergence of the estimation of the pose, position and radius of a paper roll model. In this case the object consists of one cylinder surface and two plane surfaces and there has been 9 measurements from the cylinder surface and 3 measurements from the bottom plane surface (which is the same as the floor surface).

In this case one of the columns of the Jacobian matrix $\underline{J}_\theta$

contains only zeros. This column corresponds to the pose correction 'rotation around the axle of the cylinder' and it should be left out of the Jacobian and this correction should be kept zero. Hence, the model of the paper roll can freely rotate around its axle and the degree of freedom in pose and position estimation decreases from the original six to five.

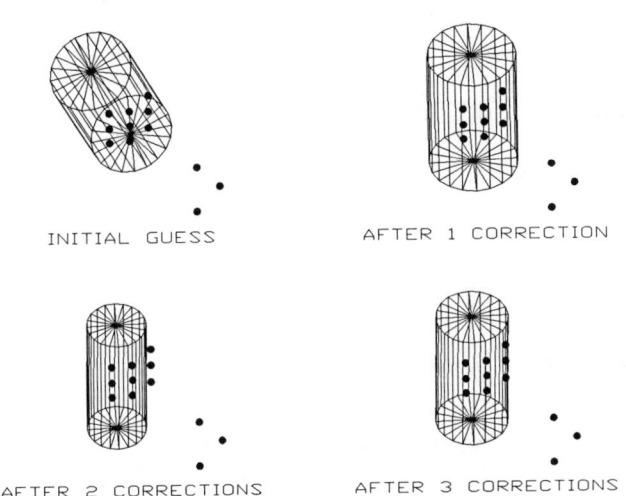

Figure 2. Convergence of the iteration in paper roll model correction. The black dots are point measurements from the cylinder surface and the floor obtained with the range sensor.

In this case the convergence of the iteration was found to be very good with both least squares and Bayesian estimation, provided, that the pose and position errors of the initial a priori model was less than about 45 degrees and less than about the radius of the roll and that the measurements were appropriately choosen. The choice of measurements should fullfil the rule: the number of linearly independent rows in the Jacobian matrix should be at least the same as the number of correction parameters in θ and φ together. In practice this means, that the number of 3D point to be measured should be high enough and that they should be placed on the surfaces as far apart in all three dimensions as possible.

The correspondence between the resulting updated model covariance and the actual uncertainty of the corrected model was tested with simulations. The model covariance estimate was computed using a 'true' model and error free measurements. The 'true' model was corrected by taking the correction as a sample from a zero mean normal distribution with this covariance matrix. The 'true' model was also corrected by adding random errors from a zero mean normal distribution with known covariance to the error free measurements and using Bayesian estimation. In Figures 3 and 4 are presented the results of one set of one thousand such simulations as the histograms of the coordinates of the origin and the components of the direction vector of the z-axis of the model's frame (which in this case coincides with the rotation axle of the cylinder surface). It can be seen that the distributions of these parameters are nearly equal in both cases.

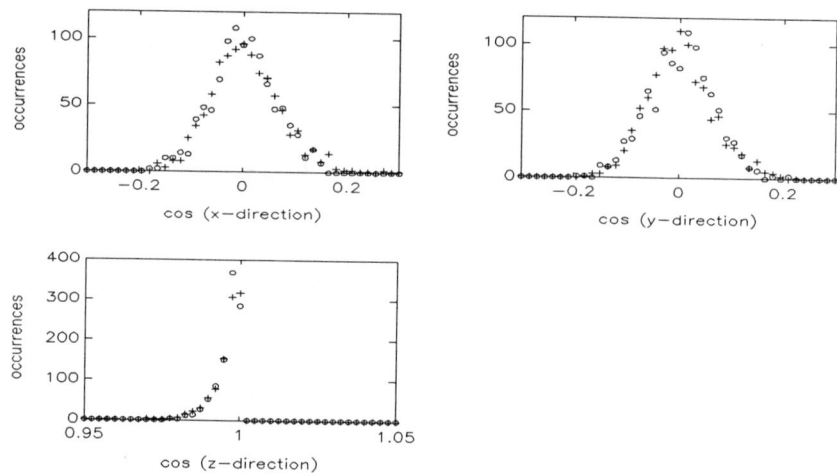

Figure 3. Histograms of the elements of the z-direction of the model's frame from 1000 corrections. The '+':s are from the correction algorithm and the 'o':s are computed with the updated model covariance.

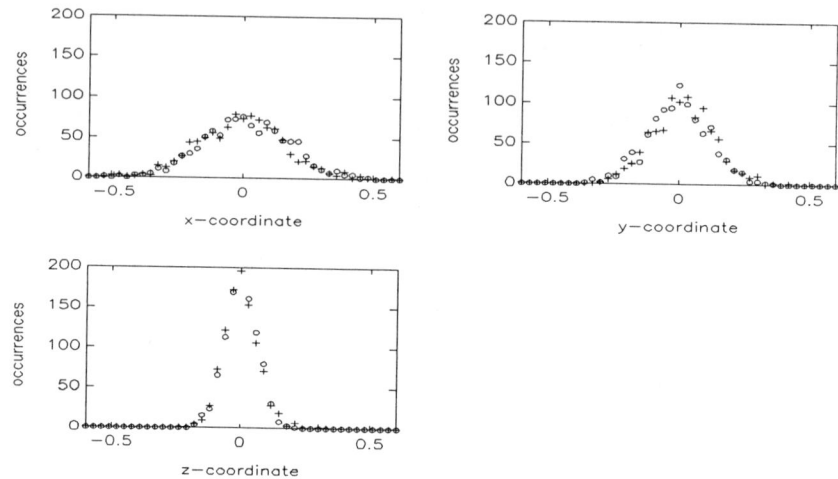

Figure 4. Histograms of the coordinates of the origin of the model's frame from 1000 corrections. The '+':s are from the correction algorithm and the 'o':s are computed with the updated model covariance.

The planning of measurements is very important also from the point of view of the accuracy of the result. For example in Figure 5 is presented the dependency of the inaccuracy of the result of cylinder surface model correction on the size of six different eight point measurement shapes. The size of the shapes is expressed as a sector angle inside which the shape has been fitted on the real cylinder surface and the shapes 1-6 are presented in Figure 6. The inaccuracy of the result is expressed as the trace of the covariancematrix obtained from Bayesian estimation, which is the same as the sum of the variances of final correction parameters. As can be noted, the best layouts for the measured points are in this case the cross and circular shapes.

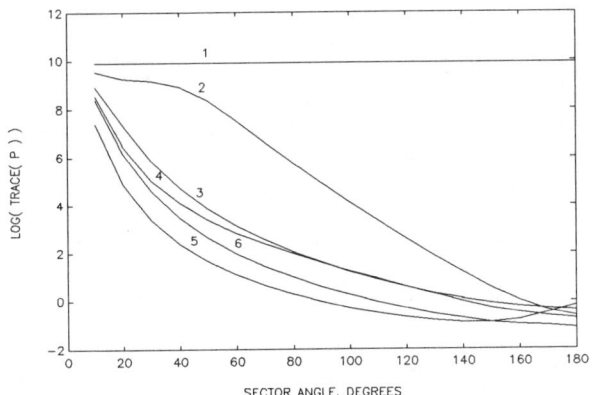

Figure 5. The inaccuracy of cylinder surface correction as a function of the size of some simple shapes of measured points. The shapes 1-6 are presented in figure 6.

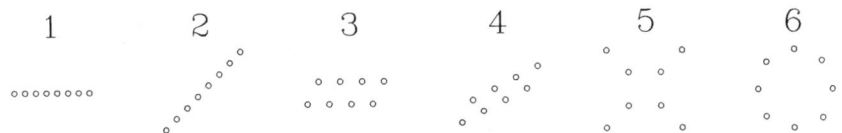

Figure 6. Shapes of measured points on a cylinder surface, which are used in figure 5. The orientation of the cylinder with respect to these is vertical.

CONCLUSIONS

The described method for the correction of the pose and position of regular object models according to 3D measurements has shown very good behaviour, provided, that the initial errors are within reasonable limits. The use of Bayesian estimation gives also good estimates for the uncertainty of the result, provided, that the initial model and measurement uncertainties are known well enough. Apart from the initial errors, the convergence of the iteration is also dependent on the number and layout of the measurements made. If the measurement errors are significant, as they usually are, the design of the measurements shape plays an important role also from the point of view of the accuracy of the result.

ACKNOWLEDGEMENTS

The authors would like to thank the following organizations for financing the project *'Machine of the Future'*: the *Technology Development Centre of Finland* (TEKES), *Auramo Cargo Systems*, *Finnish National Road Administration*, *Lännen Engineering*, *Normet*, *RoBosys*, *Steveco*, *Team SAGA*, *TH-Engineering*, *VTT* and the *University of Oulu*.

REFERENCES

Boult, T.E. & Gross, A.D. 1987. Recovery of superquadrics from depth information. Spatial Reasoning and Multi-Sensor Fusion Workshop, St. Charles, IL, USA, 5-7 October 1987. Los Altos, California, Morgan Kaufmann Publishers, Inc. Pp. 128-137.

Faugeras, O.D. & Lustman, F. & Toscani, G. 1987. Motion and Structure from Point and Line Matches. Proceedings of the First International Conference on Computer Vision, London, 8-11 June 1987. Washington D.C., IEEE Computer Society Press. Pp. 25-34.

Heikkilä, T. 1990. A Model-Based Approach to High-Level Robot Control with Visual Guidance. Espoo, Finland, Technical Research Centre of Finland, Publications 71. 77 p. + app. 58 p.

Järviluoma, M. 1991. Object Model Correction with Iterative Estimation. University of Oulu, Systems Engineering Laboratory, Report C 6. 41 p. + app. 7 p.

Lowe, D.G. 1989. Fitting Parameterized 3-D Models to Images. The 1989 Stockholm Workshop on Computational Vision. Report from Computational Vision and Active Reception Laboratory. 22 p.

Moring, I. & Myllylä, R. & Honkanen, E. & Kaisto, I. & Kostamovaara, J. & Mäkynen, A. & Manninen, M. 1989. New 3D-Vision Sensor for Shape Measurement Applications. Optics, Illumination, and Image Sensing for Machine Vision IV, Philadephia, Pennsylvania, USA, 8-10 November 1989. Washington, USA, the Society of Optical Instrumentation Engineers. Pp. 232-242.

Safranek, R.J. & Gottschlich, S. & Kak, A.C. 1990. Evidence Accumulation Using Binary Frames of Discernment for Verification Vision. IEEE Transactions on Robotics and Automation, vol. 6, no. 4. Pp. 405-417.

Sakane, S. & Sato, T. & Kakikura, M. 1987. Model-Based Planning of Visual Sensors Using a Hand-Eye Action Simulator HEAVEN. Third International Conference on Advanced Robotics, Versailles, 13-15 October 1987. Kemston, Bedford, IFS Publications Ltd. Pp. 163-174.

PART 6
SOPHISTICATED ROBOTIC SYSTEMS AND APPLICATIONS

A MODEL OF MANNED ROBOTIC SYSTEMS

P.H. Wewerinke
Department of Applied Mathematics, Systems and Control Group
and Mechatronics Research Centre Twente,
University of Twente
P.O. Box 217, 7500 AE Enschede,
The Netherlands

Abstract

In this paper a modelling approach is presented to describe complex manned robotic systems. The robotic system is modelled as a (highly) nonlinear, possibly time-varying dynamic system including any time delays in terms of optimal estimation-, control- and decision theory. The role of the human operator(s) is modelled varying from supervisor of the automated (part of the) system to controller in terms of the various functions involved to perform goal-oriented tasks.

It may be expected that the model is capable of answering questions related to reliability and efficiency, design alternatives, function allocation, automation, etc.

1. INTRODUCTION

Robotic systems are more and more applied in many areas. Examples of industrial operations are part assembly, material transfer, repair of parts and inspections. In addition, robotic systems play an important role in many teleoperations, e.g. in space applications (space stations, serviceable sattelites, material processing platforms) and operations in a risky or unaccessible environment.

Autonomous systems can meet the safety, reliability and especially economic requirements for specified tasks, but many operations involve the interacting contribution of both human operator(s) and robotic system(s). This concerns especially complex, non-standard operations in an unstructured environment. One example is shared compliant control, especially in applications with telemanipulators where time delays are considerable.
The role of the human operator(s) may vary from direct controller to supervisor of the automated (part of the) system. This depends on the goals to be achieved and the related functions to be fulfilled.

In the next section manned robotic systems are discussed in more specific terms. One approach to design and analyze a manned robotic system is based on mathematical models of this complex man-machine system. This is contained in section 3. The paper is concluded with some remarks about how the model can be utilized.

2. MANNED ROBOTIC SYSTEMS

In general the task to be performed with a manned robotic system can be described in terms of the various components involved.
The first important aspect of the task are the goals to be achieved under given boundary conditions. Realistic operations may involve a complex goal hierarchy (interrelated, or even conflicting goals, subgoals, procedures, etc.). These goals dictate the tasks to be performed. The complexity of the task hierarchy will correspond with the complexity of the goal hierarchy. The defined task will be affected by the operational environment.
Next, the functions can be derived to perform the defined task. The motives to fulfill these functions originate from the goals to be achieved. The human operator (HO) will perform these functions utilizing the available resourses, being separate items or elements of the system. These functions are the result of a function allocation procedure to the man and the machine, taking into account the human capabilities and limitations and the possibilities of the system.

Finally, the result of HO functions are actions, taken on the basis of drives (derived from his motives, etc.). Simple stimulus response behavior takes place at this level only. The actions affect the system resulting in a certain system behavior. Based on performance criteria and measures of this behavior, total system behavior can be evaluated with respect to the goals to be achieved.

More specifically, a manned robotic system can be analyzed or designed based on the assumption that a goal-oriented operation can be defined, e.g. controlling the robot from A to B, with given constraints (due to robot dynamics, control limits, environment, etc.) in a given (disturbance) environment. Furthermore, the robotic system is described as a (highly) nonlinear, possibly time-varying dynamic system including any transport or communication time delays.
The role of the HO may vary from supervisor of the automated (part of the) system to controller, or combinations. This involves HO functions such as perception of sytem outputs provided by (e.g.) displays and /or the visual scene, information processing to assess the task- and system variables of interest, decision making involved in monitoring the autonomous (sub)system and involved in intermittent control, and control. In this context, control is used in a broad sense including planning and compensatory actions.

There are various ways to analyze a manned robotic system.
One approach utilizes a variety of system performance and HO measures obtained in an experimental situation. The drawback of an experimental approach is that it involves per se comlex man-in-the-loop simulation. This implies that experiments mainly serve to verify system

performance rather than to predict the performance of new systems. This limits their utility, especially early in the design stage.
An alternative method to describe manned robotic system behavior is the use of mathematical models. The main advantage of mathematical models is that they provide a precise (quantitative) formulation of task aspects such as goals and system dymanics and basic concepts of HO functioning. Therefore, models potentially have a predictive capability, rendering a basis for selecting design alternatives. Furthermore, models provide a powerful tool to analyze operational problems of existing robotic systems systematically.
In the next section a model structure is presented describing the foregoing manned robotic system.

3. MODEL STRUCTURE

3.1 General

In this section a model of the manned robotic systems discussed in section 2 is presented. For details the reader is referred to Ref.1. Basically, the task considered is to control the robotic system from an initial state X_0 at time $k = 0$ to a final state X_n at time $k = N$ following a given desired trajectory $X_d(k)$.
Strictly speaking, the time aspect involved in this formulation is absent in most of the control tasks of interest. For example, the task is to arrive at a certain point in the state space, irrespective of time, but following a given sequence (e.g. a spatial trajectory). This is known in control theory as a space constrained control problem. Theoretically this is a difficult problem, certainly for robotic (i.e. nonlinear) systems. Thus, it is attractive to prescribe to desired trajectory for each time step and solve for the corresponding optimal control at each time step. This seems for many control tasks a meaningful approach.

As indicated before, the role of the HO may vary from supervisor of the automated (part of the) robotic system to controller. In this paper the manual control task will be assumed. The human functions involved include supervising. So the supervisor task is a special case of the manual control task. This will be discussed in the following.

The following elements are included in the model as indicated in the block diagram of the manned robotic system model shown in Fig. 3.1. The HO perceives (e.g. via a TV-camera/monitor system) the outputs of the system state (characteristic system variables such as position, velocity, attitude, etc.), which is perturbed by external disturbances. The HO utilizes these (inaccurate) observations of the state to estimate the state recursively. Based on this estimate and the task specifications a control input is selected and executed, resulting in a closed loop manual control.
In addition, the future desired state is observed (e.g. from pictorial information) and estimated. Based on this estimate and the system dynamics a pre-programmed control is selected. The resulting open loop

control is combined with the feedback control to compensate for deviations from the planned trajectory due to random disturbances.

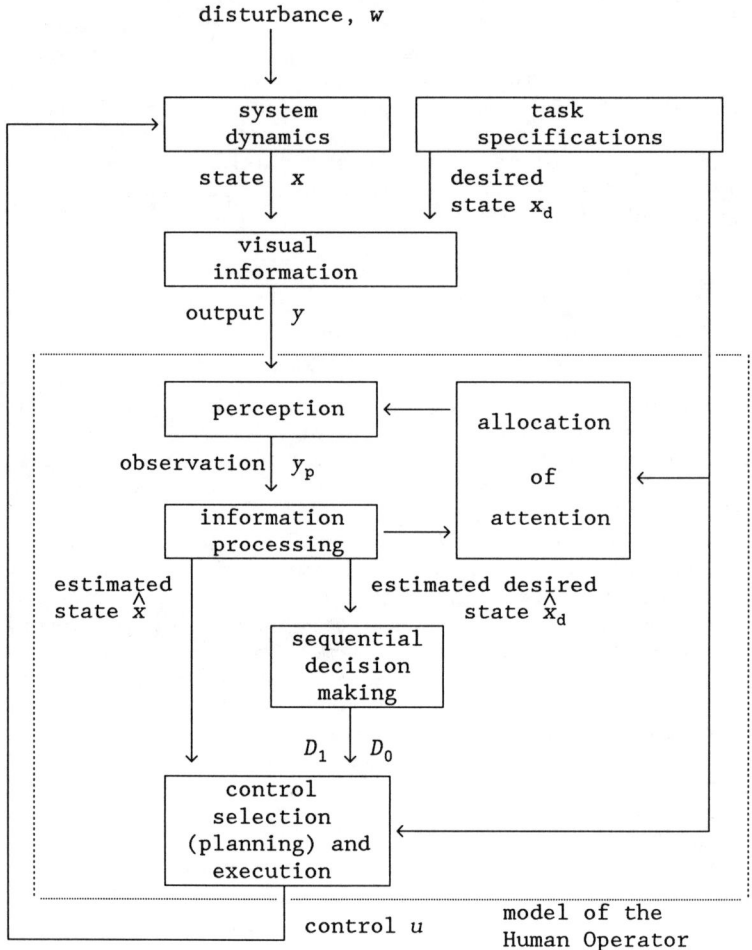

Fig. 3.1   Block diagram of the Manned Robotic System model

3.2 System dynamics

A robotic system may be represented as a nonlinear, time-varying dynamic system by

$$X(k) = f(X(k-1), U(k-1), W(k-1), k) \qquad (1a)$$

$$Y(k) = g(X(k), U(k), k) \qquad (1b)$$

where $X(k)$ is the n-dimensional state vector at time k, f is the n-dimensional vector function, U is the r-dimensional control vector, W is the l-dimensional disturbance vector, Y is the m-dimensional system output vector and g is the m-dimensional vector function.

The standard procedure is followed to describe the nonlinear system behavior X in terms of a state reference X_0 and a "small" perturbation x around this reference; thus $X = X_0 + x$, $U = U_0 + u$, etc. This linearization scheme yields a time varying reference model

$$X_0(k) = f(X_0(k-1), U_0(k-1), W_0(k-1), k) \qquad (2a)$$

$$Y_0(k) = g(X_0(k), U_0(k), k) \qquad (2b)$$

and a time-varying **linear** system description

$$x(k) = \Phi x(k-1) + \Psi u(k-1) + \Gamma w(k-1) \qquad (3a)$$

$$y(k) = Hx(k) \qquad (3b)$$

with $\Phi = \Phi(k, k-1)$ being the Jacobian matrix of f with respect to X, etc.; w is assumed to be a zero mean, Gaussian, purely random sequence, with covariance matrix R, representing disturbances and other system uncertainty.

This linearization scheme holds for relatively small x, u and w. This fact dictates the update rate of the reference and, therefore, of the various Jacobian matrices.

3.3 Control task

The task considered is to control the system state x of eq (3a) over some fixed interval of time $[0, N]$, so as to follow the desired state x_d, by realizing a control sequence $\{u(k), k = 0, 1, \ldots, N-1\}$ that minimizes the performance index

$$J_N(u) = E\{ \sum_{i=1}^{N} (x(i) - x_d(i))'Q_x(i)(x(i) - x_d(i))$$

$$+ u'(i-1)Q_u(i-1)u(i-1) \} \qquad (4)$$

where Q_x and Q_u are weighting matrices.
The solution of this optimal control problem is discussed in the following subsection dealing with human control behavior.

3.4 Human observer and controller

The model of the HO comprises various functional aspects, which are shown in the block diagram of Fig. 3.1.

Perception

The perceptual model describes how the system outputs y are related to the perceived variables y_p. It is assumed that the HO perceives these system outputs with a certain inaccuracy and with a certain time delay

$$y_p(k) = y(k-i) + v(k-i) \tag{5}$$

Here v is an independent, Gaussian, purely random observation noise sequence, representing the various sourses of human randomness (unpredictable in other than a statistical sense). Each element v_j, therefore, is specified by its covariance V_j. This covariance is functionally related to the signal level, human attention allocation and threshold phenomena.

The lumped time delay i can be associated with the HO's internal time delays related to perceptual, central processing, and neuromotor pathways. For systems with relatively large time constants, these delays can be neglected. Also system-related delays, for instance the communication or transport delays of remotely controlled systems may be modelled as a lumped equivalent "perceptual" time delay involved in eq(5).

Information processing

Based on the perceived data y_p up to time k (corresponding to data y up to k-i) and the known (learned, thus assuming that the HO is well-trained) dynamics of the system, the best (minimum variance) estimate $\hat{x}$ of the system state x can be made corresponding to time k - i. The resulting Kalman-Bucy filter equations are given by

$$\hat{x}(k-i) = \hat{x}(k-i/k-i-1) + K(k-i)n(k) \tag{6}$$

with

$$\hat{x}(k-i/k-i-1) = \Phi\hat{x}(k-i-1) + \Psi u(k-i-1) \tag{7}$$

$$K(k-i) = P(k-i)H'V^{-1}(k-i) \tag{8}$$

and

$$n(k) = y_p(k) - H\hat{x}(k-i/(k-i-1)). \tag{9}$$

Hence $\hat{x}(k-i/(k-i-1))$ indicates the estimate of x at time k - i based on the data y up to time k - i - 1, K represents the optimal trade-off between system uncertainty (in terms of the estimation error covariance P) and reliability of the data (in terms of the observation noise covariance V), and the innovation sequence n represents the new information.

The best estimate of x at time k, $\hat{x}(k)$, is obtained on the basis of $\hat{x}(k-i)$ and the known system dynamics by means of an optimal linear prediction process. The resulting prediction equation becomes

$$\hat{x}(k/k-i) = \Phi^i\hat{x}(k-i) + \sum_{\ell=0}^{i-1}\Phi^{i-1-\ell}\Psi u(k-i+\ell) \tag{10}$$

which corresponds to eq(7) for the one step prediction.

Perception and estimation of the future desired state x_d is described in a similar way. It is assumed that visual cues y_0 can be observed that are related to the difference between the present state and the future desired state. Estimation of the future desired state is depending on the assumed a priori knowledge that the HO may have about x_d. Three cases may be considered: no prior knowledge of x_d, only statistical knowledge of x_d and imperfect knowledge of x_d. The latter case results in the same filter equations as above. For the other cases the reader is referred to Ref. 1.

Sequential decision making

After the finite time interval for which the control task defined in subsection 3.3 has been performed, the decision has to be made about what to do next.
This amounts to the binary decision as to whether the system behaves according to the small perturbation model of eq(3), corresponding to a given state reference, or a systematic discrepancy between both necessitates a correcting action of the HO and an update of the system model. In the first case the HO continues to control this system steady-state. In the second case, the HO initiates another maneuver to track the systematic deviation (x_d) over some fixed interval of time, after which the HO updates his system model.
It might be necessary to update the sytem model more often. The extreme is to update the system model each time step. The reference is adjusted based on the estimated state (deviation from the old reference). This is known as the extended Kalman filter.

The comparison of system behavior as observed by the HO in terms of y_0 and the expected behavior on the basis of the present system model is made by the HO in terms of the innovation sequence n_0. A systematic deviation of the zero mean sequence x due to a change in the desired state (with respect to the present state reference) results in a non-zero mean innovation sequence (Ref. 1).
This can be tested by comparing (the log of) a generalized likelihood ratio with a threshold T according to

$$L(k) \underset{D_0}{\overset{D_1}{\gtrless}} T \qquad (11)$$

with

$$L(k) = L(k-1) + \frac{1}{2} \tilde{n}'_0(k) N_0^{-1}(k) \tilde{n}_0(k) \qquad (12)$$

and

$$T = \ln((1-P_M)/P_F) \qquad (13)$$

where $\tilde{n}_0$ is a moving average of n_0, N_0 is the covariance of n_0 and P_M and P_F are the decision error probabilities ("miss" and "false alarm").

Human control behavior

The control task discussed in subsection 3.3 is defined in terms of $J_N(u)$ of eq(4). Optimal human control corresponds to the minimal J_N. The resulting optimal control sequence $\{u(k), k = 0,1,\ldots,N-1\}$ is derived in Ref. 1. The result is given by

with
$$u(k) = S(k)\hat{x}(k) + S_m(k)z(k+1) \qquad (14)$$

$$S(k) = S_m(k)W(k+1)\Phi \qquad (15a)$$

$$S_m(k) = -(\Psi'W(k+1)\Psi + Q_u)^{-1}\Psi' \qquad (15b)$$

$$W(k) = Q_x(k) + \Phi'W(k+1)[\Phi + \Psi S(k)] \qquad (15c)$$

and

with
$$z(k) = [\Phi' + S'(k)\Psi']z(k+1) - Q_x(k)\hat{x}_d(k) \qquad (16)$$

$$z(N) = -Q_x(N)\hat{x}_d(N), \quad k = N-1,\ldots,0. \qquad (17)$$

Thus, the control is composed of two parts: a feedback control operating on the state estimate and a feedforward (open loop) control operating on the estimate of the desired state x_d and computed recursively backwards in time.

3.5 Constraints

So far, the estimation and control problem of the nonlinear system is solved by linearizing the nonlinear system around an estimated reference yielding a linear estimation and control problem. Control comprises permanent feedback control and intermittent open loop control based on sequential decision making.

Several constraints may play a role in a given task. Control is, in principle, constrained because of limited control authority, hardware limits, etc. Generally these constraints have to be included a priori in formulating and solving the optimal control problem.

Secondly, it may be necessary to consider (hard) constraints in the state space. Examples are the requirement to realize precisely a final destination of the system state (apart from stochastic effects), the limited space available to go from A to B (partly due to fixed obstacles), and limited space because of moving obstacles.

Both types of constraints can be handled in a similar way. Conceptually, the procedure is (as discussed in Ref. 2) to "adjoin" the constraints to the performance index J_N, given by eq(4), by a set of (socalled Lagrange) multipliers, which are chosen in such a way, that an optimal control is obtained (corresponding with an minimal J_N) given the constraints. Such optimization problems with constraints are conceptually straightforward. Numerical solutions, however, can require considerable computational cost.
This can be aggravated if the control problem is stochastic by nature. In that case the state has to be estimated requiring the solution of a

nonlinear estimation problem (e.g. based on an extended Kalman filter).

For example, the solution of the optimal control task inclusive collision avoidance of moving obstacles requires:
- estimation of own system state and the state of relevant obstacles;
- definition of the constraints that have to be met in terms of the estimated states (e.g. the estimated distance to obstacles);
- computation of the optimal control while meeting the constraints.

A more simple (engineering) approach to solve this and similar problems is given in Ref. 3.

4. CONCLUDING REMARKS

In this paper a model structure is presented of manned robotic systems in terms of optimal estimation-, control- and decision theory. The resulting model can be utilized to analyze, design and evaluate these systems by establishing the effect of task variables of interest on model outputs.

In case the model is used in a time simulation mode, the results are in terms of time histories (sequences) of interesting system- and HO-related variables. For the linearized model version, statistical measures can be obtained (i.e. ensemble mean values, covariances and probabilities) of all variables of interest. In the latter case the model provides a very cost-effective tool to assess the performance and reliability of manned robotic systems.

It may be expected that the model is capable of answering questions related to design alternatives, function allocation, automation, system reliability and efficiency, etc.

5. REFERENCES

1. Wewerinke, P.H.   Models of the human observer and controller of a dynamic system.
   Ph.D. thesis, University of Twente, the Netherlands, 1989.

2. Bryson, A.E. and   Applied optimal control.
   HO Y.C.   Halsted Press, 1975.

3. Hoogland, M.   Modeling vessel traffic (in Dutch).
   Thesis, University of Twente, the Netherlands, 1991.

Design, construction and performance of an anthropomorphic robot head

P Mowforth and D Wilson, The Turing Institute, UK

1 Introduction

There has been much recent interest in the idea of active perception [1, 2, 6]. By this we mean the use of sensors combined with control systems and motors to carry out a number of reflex behaviours. Examples of such reflexes at a simple level might include tasks such as automatic focus and aperture control in cameras. At a more sophisticated level this would also include optical vergence and tracking, Vestibulo-Ocular Reflexes (so as to maintain fixation during a head movement), and orientation to sound sources.

This paper reviews the various components of a project, codenamed RICHARD, to develop a motorised, anthropomorphic robot head featuring a variety of sensors including cameras and microphones. By anthropomorphic we mean that the physical dimensions, degrees of freedom and positions of sensors are in approximately the same locations as in a human head. Furthermore, that the overall goal for the project is to develop sensory reflexes for the robot head with a speed, precision and operation similar to that found in biological systems.

2 Mechanical design

To support a full range of reflexes, the head features six motorised degrees of freedom with azimuth and elevation control for the neck and each of the two eyes. Figure 1 shows the basic mechanical layout of the first prototype, RICHARD 1.

A pair of miniature, remote-head CCD cameras is used for the eyes. Each of these cameras is mounted on an angle bracket which can be rotated independently in azimuth and elevation by two separate motors fitted respectively to the base

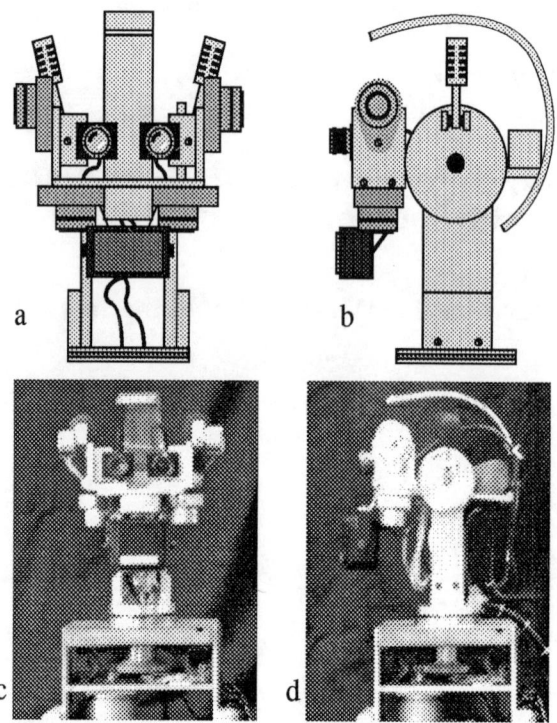

Figure 1: The first head prototype, Richard 1st, showing diagrams and photographs for front (a,c) and side (b,d) views.

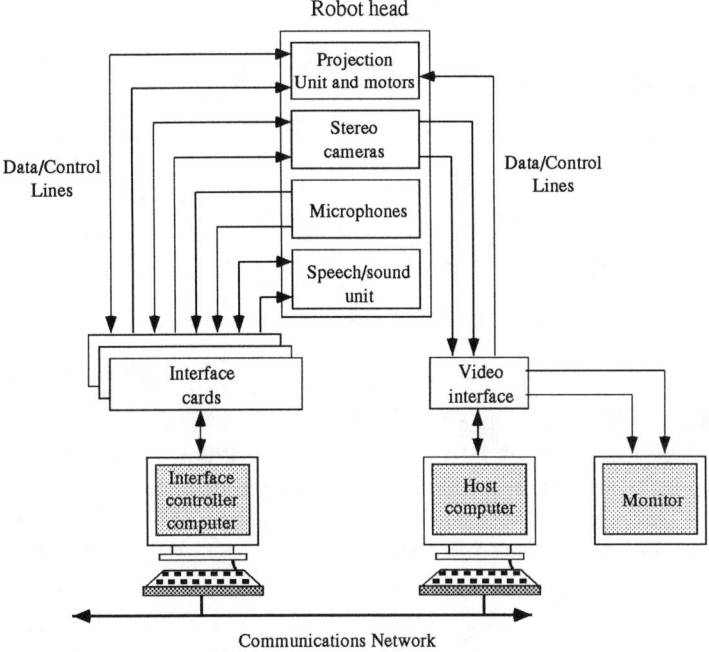

Figure 2: Computational architecture for the robot head.

and side of the bracket. The cameras are positioned with a baseline separation of approximately 6cm.

Azimuth control of the neck is available via a fifth motor fitted at the base of the head. This motor drives the entire head platform across a 4:1 gearing system. Elevation control relies on a motor mounted vertically within the neck structure which rotates the head about a central horizontal pivot using a drive-belt.

The head also supports non-visual sensory input. In particular, sound can be captured via a pair of microphones mounted in fixed positions, one at either side of the central head pivot. A speech synthesiser unit is also included which can handle both text-to-speech and synthesised tone generation. This unit is fitted directly below the gantry on which the cameras are housed.

3 Computational architecture

The robot head is a single component of an integrated computational architecture. Figure 2 identifies the other principal components of this architecture which includes interface electronics to acquire the incoming sensor data, electronics to drive the speech/tone generator unit, control systems for each of the actuators and a host computer with software support to coordinate the various reflex behaviours.

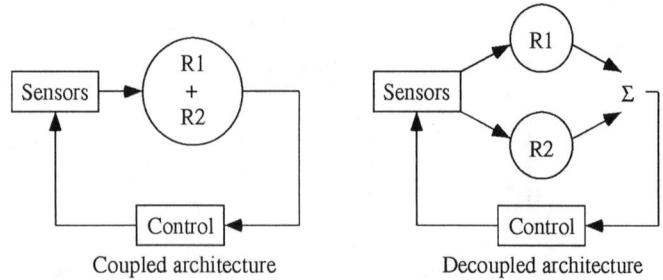

Figure 3: Diagramatic representations for coupled and decoupled control architectures.

All the interface electronics for the head are inserted in an interface controller computer together with the necessary software drivers. This computer effectively forms a device driver for the robot head. The current prototype makes use of a PC for this task and to ensure that a wide range of such computers may be used, all of the electronics have been designed to be fully compatible with the PC ISA (AT) bus standard. To facilitate any future maintenance or upgrade of the electronics, a modular design has been followed with separate cards for each of the interfaces for the neck motors, eye motors, speech unit and microphones.

The only interface which is not inserted in the PC interface controller computer is the video interface. Instead, this consists of a framegrabber unit which is resident in the host and an associated display monitor. Stereo images of the scene viewed by the cameras can be captured via the framegrabber for subsequent processing and displayed to the user on the monitor.

Software routines to implement each of the reflex behaviours execute under the control of a software workbench which is installed on the host computer. This workbench supports a library of simple processing modules, each of which performs a simple self-contained function. Complex sequences to implement each of the desired reflex behaviours can be built by bringing together a suite of these modules in a concurrent processing pipeline.

Processing sequence for reflexes such as vergence or tracking will typically involve the acquisition of some sensory input (video and/or sound data), the processing of this data in some intelligent fashion followed by the generation of appropriate commands to reposition the eye or neck motors. In each case, the workbench will ensure that these movement commands are sent across a communications link to the PC interface controller computer which will, in turn, drive the motors. This scheme algorithms to be developed for reflex control which are completely independent of the underlying hardware, thus enhancing system evolution.

With our current system, a SUN Sparcstation is used as the host computer and is linked to the PC interface controller across an RS-232 connection. The bandwidth provided by this link is suitable for the simple data transfer protocol which we have implemented. If further research shows that we require greater throughput, an alternative link can be substituted since the workbench assures device independence.

Two alternatives approaches to the problem of integrating different reflex behaviours are illustrated in figure 3. In the first approach, the sensory input undergoes sensor fusion to produce some behaviour based jointly on the two reflexes, $R1$ and $R2$, i.e. the relexes are coupled. The alternative approach views each of the two reflexes as independent and discrete, i.e. decoupled. In this case, the reflexes can operate completely independently or require only some form of primitive integration where, for example, they share a common actuator.

The robot head workbench is geared towards the integration of sensory reflexes in a decoupled architecture. The video or sound input is fed directly to independent processing sequences, each of which implements a single reflex behaviour. Bringing together separate reflexes in this way as discrete independent automata ensures an easier path for system evolution in addition to offering increased robustness.

4 Vergence

This project is aimed at developing a range of reflex behaviours for the robot head. However, to describe the main thrust of our work, the remainder of this paper will look at the vergence reflex in particular.

Vergence control is necessary to bring images captured from two locations in space into approximate correspondence [3]. This may be used directly to provide information about range as well as providing a stereo process with favourable starting conditions. Stereo matching algorithms contain fusional limits for the maximum allowable disparity; the vergence mechanism simply attempts to maximise the amount of the image pair which can then be fused.

The algorithm design is strongly constrained by the requirement for fast execution time. To achieve this, a pyramid image architecture has been adopted. As can be seen from Figure 4, the pyramid architecture is similar to retinal tessellation both in terms of its foveal magnification and the total number of pixels it uses.

Figure 5 provides a summary of the software architecture for the vergence control algorithm. Following pyramid construction, , the image pair is blurred by a small sigma gaussian function at each pyramid level. This filtering is necessary in order to achieve left right matches at each level. Next, a small fixed size window is applied to the left and right images, again at each level. The window is able to move to the left or to the right. Correlation is now carried out between the image pair masked by these windows as a search operation through the pyramid.

The progressive focussing of the solution through a series of octave separated levels is effectively a coarse-to-fine fovea. The algorithm is based on the Multiple

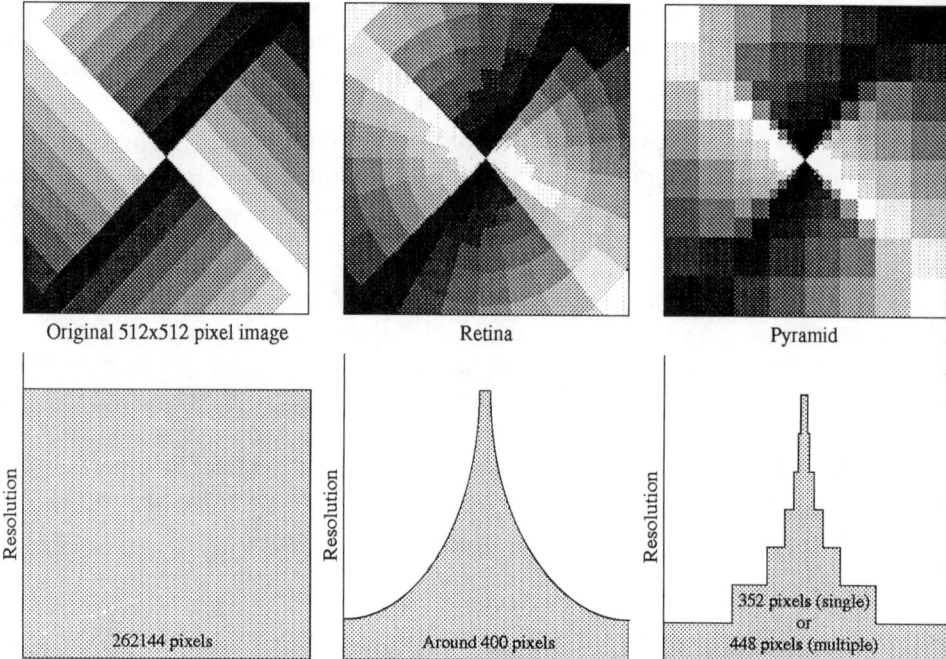

Figure 4: The original image can be seen tessellated onto both retinal and pyramidal datastructures.

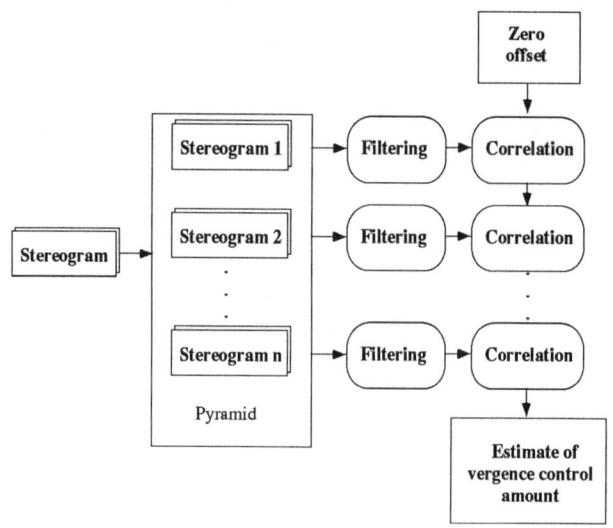

Figure 5: Software architecture of the vergence control algorithm.

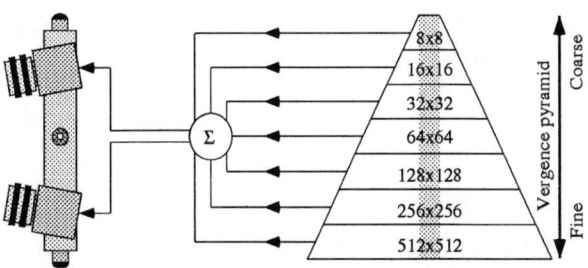

Figure 6: Hierachical control architecture for vergence and tracking.

Scale Signal Matching algorithm developed for matching in stereo and motion computation [4]. Like *MSSM*, the algorithm also produces a confidence value accompanying its estimate of vergence. On a Sun Sparcstation, vergence signals are produced in around 100mS. Further details of the algorithm and results for a range of test images are provided in [5].

Our current research is directed towards the construction of a hierarchical control system for both vergence and tracking in which each layer of the processing pyramid produces a control signal; see Figure 6. Preliminary work indicates that effective integration for the control of the system may develop naturally from the timing delays through the pyramid processing hierarchy.

5 Discussion

This paper provides some preliminary descriptions of an anthropomorphic robot head along with its associated electronics and software support. For the purposes of system maintainance, debugging and evolution, we have ensured that each of the various perceptual reflexes carried out by the system are independent and closed-loop.

For many of the reflexes, hierarchies of control become necessary. To accomplish this, we have attempted to use scale-based algorithms for which the hierarchies are *implicit*.

Future work will be directed towards the integration of each of the system reflexes. Early results suggest that the most effective approach here is to use the natural time constants for each reflex arc for the integration process.

Acknowlegements: This work has been supported by the Turing Institute and by the UK IED project *Active Stereo Probe*. In addition to the authors, those directly involved in the project include Britt Eiklid (DC servo motor control), Richard Fryer and John McDonald (active projection), Gordon Mair (engineering design), Paul Siebert (data representation), Simon Solbakken (vergence and tracking), Ketil

Undbekken (system electronics), Colin Urquhart (head modelling), Jin Zhengping (signal matching) and Tatjana Zrimec (reflex learning).

References

[1] J. Y. Aloimonos and A. Bandyopadhyay. Active vision. In *Proceedings of the first international conference on computer vision.*, pages 35–54, Washington, DC, June 1987. IEEE, IEEE Computer Society Press.

[2] J. Y. Aloimonos, I. Weiss, and A. Bandyopadhyay. Active vision. In *DARPA image understanding workshop*, pages 552–573, Los Angeles, LA, Feb 1987.

[3] R. H. S. Carpenter. *Movements of the eyes*. Pion, London, 1977.

[4] Z. Jin and P. Mowforth. A discrete approach to signal matching. Research Memo TIRM-89-036, The Turing Institute, Glasgow, UK, January 1989.

[5] P. Mowforth, P. Siebert, Z. P. Jin, and C. Urquhart. A head called richard. In BMVA, editor, *British Machine Vision Conference '90*, pages 361–366, Oxford, UK, 1990. BMVA.

[6] P. Mowforth and T. Zrimec. Learning of causality by a robot. In J. Hayes-Michie, D. Michie, and E. Tyugu, editors, *Machine Intelligence 12*, pages 225–240, Oxford, UK, 1991. Clarendon Press.

MULTI-FINGERED ROBOT HANDS

E.A. Al-Gallaf, A.J.Allen, K.Warwick
Department of Cybernetics, University of Reading,
Whiteknights, P.O.Box 225, Reading, Berks, RG6 2AY, U.K.

Abstract

The lack of truly dextrous end-effectors which exhibit fine manipulation under the control of multi-level strategy has certainly led to current limitations which prevent a much broader application of robots. This paper describes work which is specifically directed towards dextrous robot end-effectors, something which, in terms of a multi-fingered hand, involves not only motion of the fingers and the object, but also the exertion of forces in order to achieve manipulation requirements. Manipulation can be considered to be made of a number of separate issues, particulary in terms of multi-fingered hands. These are task planning, hand design, motion control, grasping, and sensing. The grasping issue can be further partitioned into the study of grasp taxonomy, type of contacts, the number of fingers fixed to the palm, grasp compliance and grasp stability. Some work has been carried out in specifically designed articulated hands, meanwhile an ongoing research is being carried out to design a four-finger robot hand to be presented.

Introduction

A number of researchers have focused on the design and control of many complex end-effectors in an effort to increase robot dexterity and adaptability. We would like to think of a robot manipulator as a machine that performs tasks just by adapting its task planning, however, the state of the art in robot hands is far away from this goal. Efforts have been directed towards multi-fingered robot hands, since parallel jaw grippers are characterized by some issues that have limited their potential for use in complex manufacturing manipulations, from these :

1) Parallel jaw grippers can assume a limited range of configurations, since having only one degree of freedom in a hand limits choice of grasping configurations, Salisbury[10].

2) It is unreasonable to expect large, massive links in a robot arm to be able to cover a large working volume and simultaneously have high bandwidth and sensitivity, Salisbury[10].

3) Multi-fingered hands provide for a unique sensing modality, the incorporation of position, force, tactile and proximity sensors that provide an opportunity for obtaining information about the physical attributes of objects and tasks.

4) The need of mechanical interfacing between the end-effector and the arm [eg.remote centre of compliance] when used for assembly tasks due to, (A) Robot oscillation, (B) Robot programming errors, (C) Unexpected environmental forces and displacement and (D) Part fixture errors.

5) Dexterous hands can be used in direct human control where they are utilised in hazardous locations.

In manufacturing environments, the use of the hand to obtain an effective grasp and manipulate the grasped object to achieve a given task, is an objective that has led the science of manipulation be divided into the following topics :
(1) Task planning.(2) Hand design.(3) Grasping issues.(4) Hand control.(5) Sensing capabilities.
Further more, each topic is subdivided into a number of issues some are related to each other depending in specific task requirements.

(1) Task Planning : Develops a sequence of homogeneous manipulations to execute a task. Since most tasks require a number of different grips, task planning provides the necessary transition procedures for grip changes. Changes must be done in real time without losing control of the object.

(2) Hand Design : The human hand is a prototype of a general purpose manipulator, do we have to begin the design to end with an anthropomorphic hand? How many fingers are needed, what size should each finger be? How many joints per finger? Do we need special hand designs to execute specific tasks? All these questions should be asked and considered while designing multi-fingered robot hands, Silver[6].

(3) Grasping Issues : Grasping of objects by a multi-fingered robot hand is much more complicated than the manipulation of an object rigidly attached to the end of a six-axis robotic arm for two reasons. (A) Kinematic relations between the finger joints motions and grasped object motion are complicated. (B) The hand has to firmly grasp the object while manipulating it.

The grasping issue is further classified into a number of topics that include : (1) Grasping taxonomies. (2) Grasp quality measure. (3) Type of contact. (4) Grasp compliance. (5) Grasp stability.

(4) Control System It is the control that deals with hardware such as motors, sensors, A-D converts ...etc. A control system should be built to manage a large number of actuators at the same time, it adapts its strategy according to a given task. How the control system interacts with previously considered issues is an important aim while planning hand control scheme.

(5) Sensing : Sensing was considered as a part of the hand mechanical design, something is true because it is recommended where various sensing elements should be fixed, whether they occupy a considerable volume, and how electrical conductors are led to them. Tactile sensing provides the control system with full up dated information of the interaction of the hand with the environment it is facing.

To link between manipulation science and designed hands, this paper has been divided into two main sections. The first section deals with general grasping issues, where as the second section gives a brief explanation of a currently designed four-finger hand at the Department of Cybernetics, University of Reading.

Section - 1 Grasping by Multi-fingered Robot Hands

One issue of manipulation science is grasping, here we shall concentrate on grasp taxonomy, grasp quality measure, grasp type of contact, grasp compliance, and grasp stability.

1.1 Grasping Taxonomies

Grasp selection depends on the hand orientation, object shape, and task to be performed. Grasp taxonomy is the study of grip choice and how grip choice is influenced by the performed task. Grasp study was confined to single-handed operations by machinists working with metal parts and hand tools. The machinists were observed and interviewed about their grasp choices and perceptions of their work. There are limitations of taxonomy as applied to manufacturing environment from these: (1) It is an incomplete system of investigation, since there are numerous every day grasps which are not included. (2) It was found in the study, machinists adopted numerous variations on the grasp partly in response due to personal preferences and differences in size and strength of their hand, Cutkosky[3].

1.2 Measure of Grasp Quality

While analytic approaches for manipulation is complex, a few assumptions have been made to be used while grasping an object between fingers of robot hand : (1) Linearization of kinematics. (2) Quasi-static analysis. (3) No sliding or rolling of the fingertips. (4) No over-constrained grasps. (5) And rigid body model with point contacts between the fingertips and the grasped object. Based on the above assumptions a number of quality measures have been developed and are summarized:

1) Connectivity : How many degrees of freedom are there between the grasped object and the hand ? How many independent parameters are needed to completely specify the position and orientation of the object.
2) Stability : Will the grasp return to its configuration after being disturbed by an external force or moment?
3) Compliance : What is the effective compliance of the grasped object with respect to the hand.
4) Grasp Isotropy : Does the grasp configuration permit the finger joints to accurately apply forces and moments to the object? For example, if one finger is in a singular configuration, it will be impossible to control force and motion in a particular direction.
5) Resistance to slipping : How large can the forces and moments on the object before the fingers will start to slip? The resistance to slipping depends on the configuration of the grasp, on the type of contacts, Cutkosky[2].

Grasp quality measure issues are important when considering a multi-fingered robot hand to accomplish a specific task. The common factors among these measures are: (1) All are functions of the contact conditions between the fingers and the grasped object. (2) Are functions of the actuation system, coupling among different joints and coupling among fingers. (3) Function of fingers location geometries. (4) Number of degrees of freedom of the hand itself.

1.3 Grasp type of contact

Type of contact defines the interaction between the fingertips and the object. Two approaches have been followed. Analysis based on screw theory where as the other is based on fingertip geometry. Grasp stability, manipulability, and compliance are functions of contact type since all depend on the grip transform. For example, a soft fingertip with friction would reduce the object DOF to two rather than a frictionless point contact with five DOF. Although soft finger model is complicated mathematically when considering the dimension of the grip transform, it does however, model a human finger.

1.4 Grasp Compliance

Salisbury[8] extended his theory of arm compliance by applying another approach based on grip transform he developed for articulated hands. The role played by the grip transform G in manipulation by robot hands is similar to the role played by the Jacobian matrix in manipulation by an open chain arm. The grip transform allows determination of contact wrenches to be exerted for grasping an object. From velocity point of view it allows determination of object velocity from sensed contact velocities when n wrenches are acting on an object. A (6xn) contact matrix is formed and designated by **W**.

$$W = [\overline{W}_1, \overline{W}_2, \overline{W}_3, \ldots \ldots \ldots, \overline{W}_n]$$

where **n** is the total number of screw coordinates of each wrench. Grip transform **G** is constructed by augmenting the (6xn) **W** matrix with (n-6) element basis vectors of internal forces accounting for internal forces between fingers, Salisbury[10]:
An external wrench acting on an object given by :

$$\overline{F} = [F_x, F_y, F_z, M_x, M_y, M_z, F_{12}, F_{13}, F_{23}]$$

is related to forces supported by the fingers **f**

$$\overline{f} = [f_1, f_2, f_3, \ldots \ldots f_n]$$

through the grip transform **G**

$$\overline{F} = [G]^{-T} \times \overline{f}$$

An object stiffness, **Kc** , accounts for object deviation from a prescribed position, so that restoring forces return it to its location once an external wrench is applied to the object.

$$K_c = dig \, [K_x, K_y, K_z, M_x, M_y, M_z]$$

The composite jacobian matrix is obtained by concatenating all fingers jacobian matrices diagonally :

$$J = dig[J_1, J_2, J_3, \ldots . J_n]$$

where **n** is the number of finger fixed to the palm.
Finally, finger joint space stiffness matrix used to adjust actuator constants in servo loop is given by , Salisbury[8]:

$$K_\Theta = J^T \, G^T \, K_c \, G \, J$$

$$\overline{\tau} = K_\Theta \times \overline{d\Theta}$$

where **dθ** is the change in joint space due to object deviation.
Grasp stiffness, or its inverse compliance, is a function of Servo stiffness, structural compliance, changes in contact geometry, coupling among different joints and fingers, and type of contact, Cutkosky[1].

1.5 Grasp Stability
Asada[4], introduced the concept of stability by centring a robot hand,

with elastic fingers and rollers on fingertips, on an object mass centre and checking for grasps that are stable with respect to rotation and translation. For the three dimensional case, Nguyen[7], made some assumptions for obtaining analytical conditions for which a grasp is in stable equilibrium. Using the potential function of the grasp Nguyen modeled contacts to be frictionless point contacts, hard finger contact, or soft finger contacts with friction. Fingers are modeled as virtual springs, each virtual spring is a set of independent linear and angular spring independently controlled. The main Nguyen objective was, to obtain from the potential function as analytical conditions for which a grasp is in equilibrium. The potential function is the sum of the potential functions due to all **n** finger wrenches [springs].

$$V = \frac{1}{2} \sum_{i=0}^{n} K_i \times \sigma_i^2$$

where K_i is the ith spring constant and σ_i : compression of a linear and angular springs due to object twist.
A grasp is in equilibrium when the gradient of its potential function is zero, when no compressions, all σ_{is} equal zero, and the grasp is stable.

Sometime later Salisbury ,Salisbury[8], used the screw theory to determine if a grasp is stable or not, he concluded a grasp is stable if the grasp matrix was full row rank. For the grip matrix to have full rank it is sufficient that none of the rows be linearly dependent on each other.
As a basic requirement any choice of a grasp must satisfy the stability criterion, therefore we could narrow the set of all grasps of an object to a smaller set of stable grasp. That has motivated the need to select an optimal grasp among a set of stable grasps. Task-oriented quality measures take account of the task requirements [eg. object twist and wrench directions] and the way a hand grasps an object in a modeled environment.

Section - 2   Reading Dextrous Hand, [RDH]

General purpose hands have been designed previously, for example, Okada hand Okada[9], Utah/MIT Jacobsen[5], and Stanford/JPL hand Salisbury[8]. However, we shall concentrate on a new effort to produce a tendon operated multi-degrees-of-freedom [MDOF] dextrous robotic hand as a research tool. The four finger (with each finger having three-degrees-of-freedom) hand has been designed at Reading Department of Cybernetics.

2.1 Configuration of the Hand The designed hand was considered to be anthropomorphic in size and geometry. Each finger has three DOF for a total of **12** DOF in the hand. Each finger DOF is actuated by a pair of tendons driven by a single actuator, requiring 12 actuators to operate the 12 DOF hand, resulting in a considerable reduction in the number of actuators and weight. The current version include three fingers and one thumb, with each digit containing three joints. The two distal joints of the fingers and thumb are capable of excursions from -70 to +70, whereas the first joint capable of +/-20 motions. The hand includes up to 36 low friction pulleys for the purpose of tendon routing. Figure 1 shows the kinematic structure of the hand, and frame assignments .

2.2: Actuation System: The actuation system for the **RDH** is based on permanent magnet d.c motors. Actuators are located at a distance from the hand and the forces transmitted through tendons. This has the advantages of reducing the mass and volume at the hand and permits the motors to be

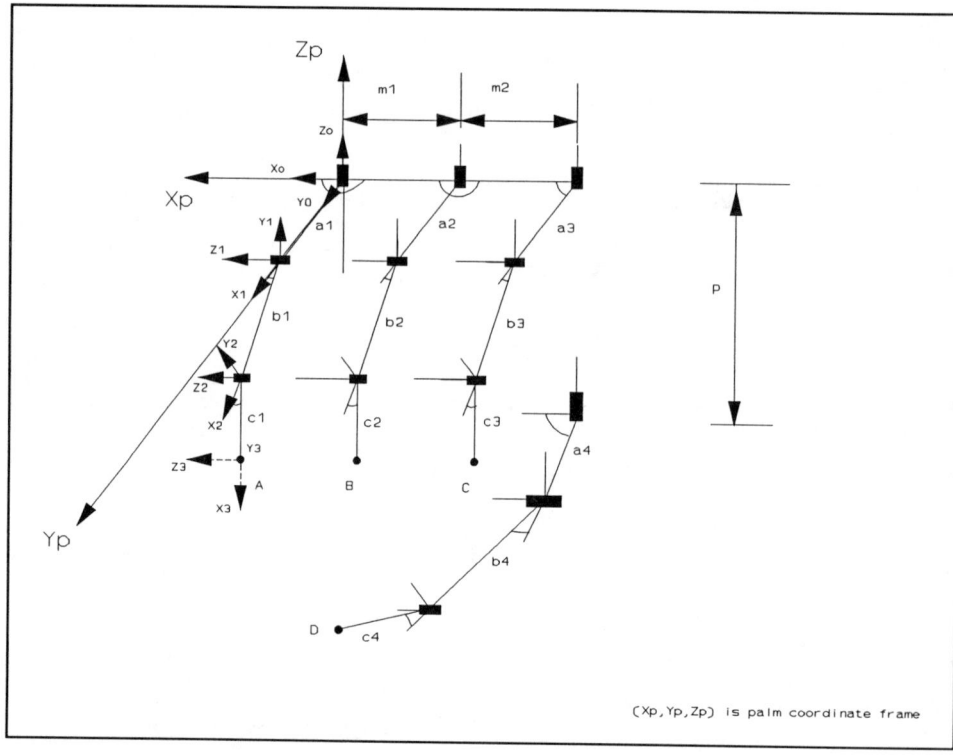

Figure 1 Some Kinematic parameters and frame assignments.

located on the robot's forearm where it will have less effect on the manipulator dynamic performance. The robot arm with the hand have 18 totally degrees of freedom.

2.3: Internal Sensors and Control Hardware: The RDH is equipped with 12 position sensors located near each driving actuator where joint angular positions are sensed indirectly. The hand is also equipped with force sensors near to fingertips to avoid force errors due to friction and tendon routing over the coupling pulleys. Each of the tendons pass over a small pulley mounted on a cantilever beam whose deflection is measured via a pair of strain gauges. Fingertips have been designed in such a way to allow tactile sensors to be mounted on the fingertips for future research. The controller hardware allows flexibility in the control algorithm by changing controller parameters accordingly. The digital control has been implemented using 12 dedicated motion processors mounted on a single board. Each motion processor allows the implementation of trajectory path by loading down path constants. Motion processors allow the specification of digital controller parameters based on a position version of a conventional digital PID controller. All software development is done on a SUN under UNIX environment communicating with an intermediate master processor that links higher level software with low-level digital servo system, as shown in figure 2. The master processor communicates with the 12 motion processors and reports unexpected performance by reading in real time status registers and loads them in a table of events. It also has direct communication with D/A and A/D converters and signalling of various error conditions.

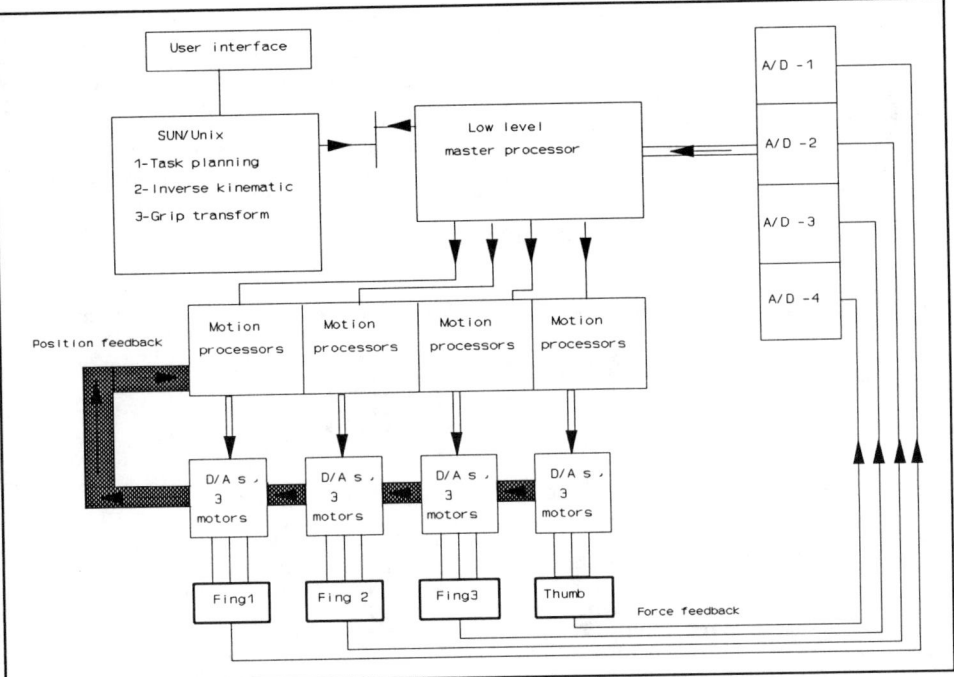

Figure 2 Hierarchical control architecture of RDH .

2.4: Task and Trajectory Structure: In order to make full advantage of the hand, it has been designed to be fitted as an end-effector of PUMA-500 robot manipulator. To perform a given task the user has to program the hand accordingly. The hand higher control provides a (4xn) array of records, where **n** is the number of sequence the hand has to perform. An inverse kinematic process is done either by an algebraic approach using **DENAVIT** and **HARTENBERG** scheme, Denavit[3], or using a geometric solution by decomposing the spatial geometry of the fingers into several plane geometry problems. Three vectors are in each record containing the acceleration, velocity, and position of a given joint. For a given position, a given acceleration , and a given path time, the time t_p required for a joint to reach a constant velocity with parabolic blends is found by:

$$t_b = \frac{t}{2} - \frac{\sqrt{\alpha^2 t^2 - 4\alpha(\theta_f - \theta_0)}}{2\alpha}$$

where **t** is the total travel time of the path, α is the desired acceleration, θ_f final position, and θ_0 initial position.
During the course of object manipulation, a multi-segment linear path with blends is used as a high level commands.

2.5: Fingertips and Tendon Coordination : In real time we need to make the construction of force closure by computational geometry as simple as possible. To overcome grasp uncertainties, the fingertips have been made of soft material (like rubber) since such contact allows a finite contact area to exert a 4 wrench - system . Salisbury has investigated such contact to show that soft finger contact consists of all zero pitch wrenches acting about all lines in the tangent plane, wrenches of infinite pitch acting about any line normal to the tangent plane and wrenches of any pitch acting about the contact normal, Salisbury[8]. Tendon management portion of the controller receives the inverse kinematics solution of cartesian fingertip as an input, corresponding to a finger joint space :

$$\overline{\Theta} = [\Theta_1 \; \Theta_2 \; \Theta_3]^t$$

and outputs Θ_a, the actuators angular displacements. Θ_a is determined by:

$$\begin{vmatrix} \Theta_1 a \\ \Theta_2 a \\ \Theta_3 a \end{vmatrix} = \begin{vmatrix} \rho_1 & 0 & 0 \\ \beta_1 & \rho_2 & 0 \\ \beta_2 & \beta_3 & \rho_3 \end{vmatrix} \times \begin{vmatrix} \Theta_1 \\ \Theta_2 \\ \Theta_3 \end{vmatrix}$$

where β_s and ρ_s are constants determined by coupling pulleys radii .

General Conclusion Manipulation science, which is a sub division of robotics, comprises task planning, control, grasping, hand design and sensing. In order to assist robot hands in grasping certain objects and develop robot hands, taxonomies have been used utilizing the physiological aspects of human hand. When a task to be accomplished stable grasp is not a sufficient condition, however, from different possible grasps we have to choose one much suitable for the task. For compliance control,we need much faster algorithms to be implemented in real time. Finally we have gone through one example of general purpose robot hands which is designed at Reading University.

References:-
[1]   Cutkosky, M. R. and Kao, I. 1989 . Computing And Controlling the Compliance of a Robotic Hand.IEEE T. of Robotics And Automation 5(2): 151-165.
[2]   Cutkosky, M. R. and Wright, P.K. 1986. Modelling manufacturing grips and correlation with the design of robotic hands. Pro.IEEE Int.Conf. on Robotics and Automation.1533-1539.
[3]   Denavit, J. and Hartenberg, R.S.,1955. A Kinematic Notation for Lower-Pair Mechanisms Based on Matrices. J of Applied Mechanics: 215-221.
[4]   Jacobsen, S.C., et al, 1986 (San Francisco, April). Design of the Utah/MIT dextrous hand.Proc. IEEE Conf. on Robotics and Automation. Washington: 1520-1532.
[5]   Hanafusa, H. and Asada, H. 1977(Tokyo). Stable prehension by a robot hand with elastic fingers.Proc. 7th Symp. on Industrial Robots : 361-368.
[6]   Mishra, B. and Silver, N.1989.Some Discussion of Static Gripping and its Stability.IEEE T. of Systems,Man,and Cybernetics 19(4): 783-796.
[7]   Nguyen, V.D. 1989.Constructing Stable Grasps.Int.J Robotics Research 8(1): 26-37.
[8]   Mason, M.T. and Salisbury, J.K., Jr. Robot Hands and the Mechanics of Manipulation. Cambridge,MA MIT Press,1985.
[9]   Okada, T. 1979. Object-handling system for manual industry.IEEE J. of Systems,Man,And Cybernetics 9(2): 79-89.
[10] Salisbury, J.K.,and Craig,J.J. 1982. Articulated hands:force control and kinematic issues. J. Robotics Research 1(1): 4-17.

MULTIDIRECTIONAL PNEUMATIC FORCE SENSOR FOR GRIPPERS

CAEN, R.* and COLIN, S.**
* Professor, Institut National des Sciences Appliquées de Toulouse - Département de Génie Mécanique. Avenue de Rangueuil, 31077 Toulouse Cedex, FRANCE.
** Researcher, Institut de Mécanique des Fluides. Allée du Professeur Camille Soula, 31400 Toulouse, FRANCE.

Abstract

The aim of this paper is to present an original multidirectional pneumatic force sensor based on the pneumatic potentiometer effect. It allows a gripper to maintain an object with the minimum necessary gripping force. The sensor is able to detect the forces due to the carried object (weight, inertia forces) which are directed in any direction in a plane perpendicular to the sensor axis. This detection in a whole plane considerably increases the possibilities of the I.M.F.T. Pneumatic Prehensor.

Key Words

Force, acceleration, pressure, sensing systems, prehensors, grippers, robotics.

Introduction

A pneumatic prehensor has been studied in the *Fluidics and Jets Dynamics* team of the *Institut de Mécanique des Fluides de Toulouse*, for a few years. It is able to self-evaluate the grasping force required for the various phases of the objects manipulation.
This prehensor has obtained an international patent [1] and some distinctions [2]. Two sensors can be used as shown in Fig. 1 : a contact sensor described in paper [3] and an unidirectional force sensor described in paper [4], the behavior of which is the following.
A determined unbalance between the two pneumatic potentiometers allows an initial gripping force when the mobile finger comes in touch with the object. Its value is about one Newton. As soon as the robot wrist moves up, the blade bends down, the pressure increases in the lower potentiometer and decreases in the upper one. Thus the acting force of the jack on the object increases gradually until the object takes off. This phase can be short (about 0.1 s).
This prehensor is performing but the unidirectional sensing limits its degrees of freedom number. For that purpose, a multidirectional force sensor which can take the place of the unidirectional one has been realized. It allows the prehensor to keep its tactile sense in a whole plane and so it increases considerably the gripping possibilities of the manipulators, by adding three degrees of freedom to the wrist.

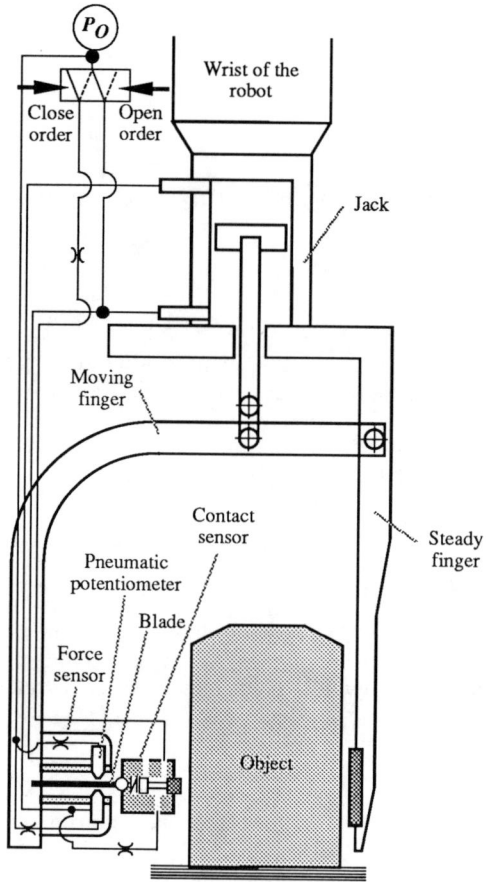

Figure 1

Behavior of the Multidirectional Pneumatic Force Sensor

Principle

The originality of the sensor is due to the association of a flexible circular rod with an obturator made of caoutchouc or soft polymer, both inside a cylindrical chamber with a circular section. Among the shapes of the obturator which can be envisaged, three examples are shown on Fig. 2. One criterion of choice could be the facility of realization of the sensor.

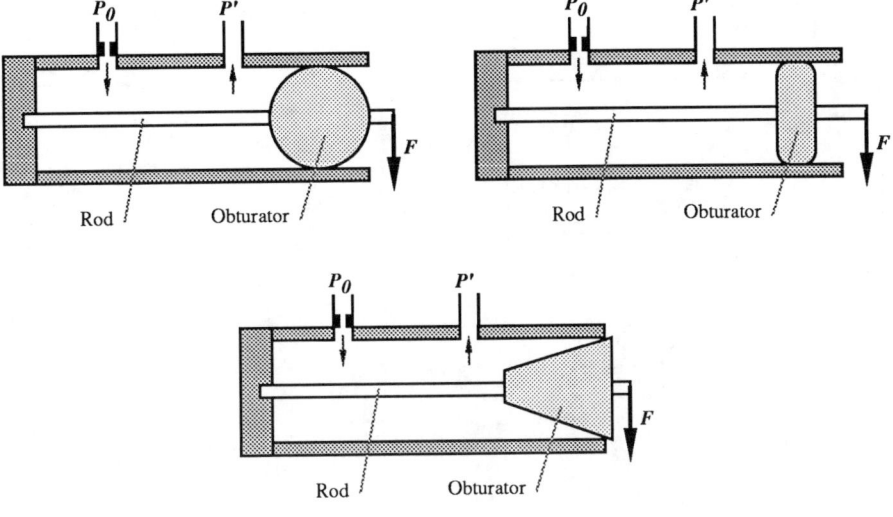

Figure 2

Each pressure is an absolute pressure. The supply pressure is called P_0, the output pressure P' and the atmospheric pressure P_a. If a force F is applied to the rod extremity along any direction in the plane perpendicular to the cylinder axis (Fig. 3), the flow rate q' increases and the output pressure P' decreases. The variation of P' depends on the geometrical and mechanical characteristics of the obturator and of the rod. The rod can be fixed to the bottom of the sensor chamber with a ball-and-socket joint or an embedding. The pressure P' also depends on the diameter d of restriction R (Fig. 4).

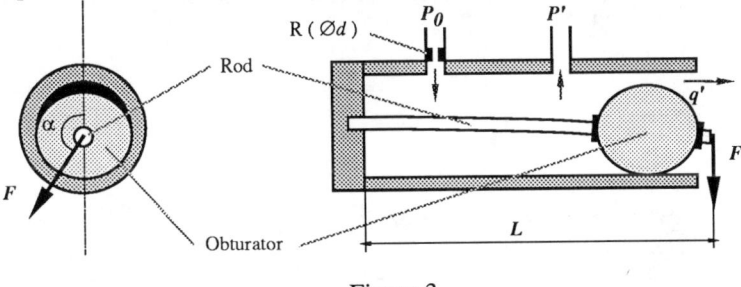

Figure 3

Static Behavior

The experimental influence of the diameter d of restriction R on the output pressure P' is shown in Fig. 4. These characteristic curves point out that the static behavior of the sensor is akin to the typical pneumatic potentiometer one.

Remark : P' may be different from P_0 for small values of F because the initial escape flow rate q' can have a value different from 0. Fig. 5 shows the evolution of the output pressure P' vs the force F for different directions of this force (marked by the angle α on Fig. 3).

These experimental points are very close, pointing out the homogeneous multidirectional behavior of the sensor.

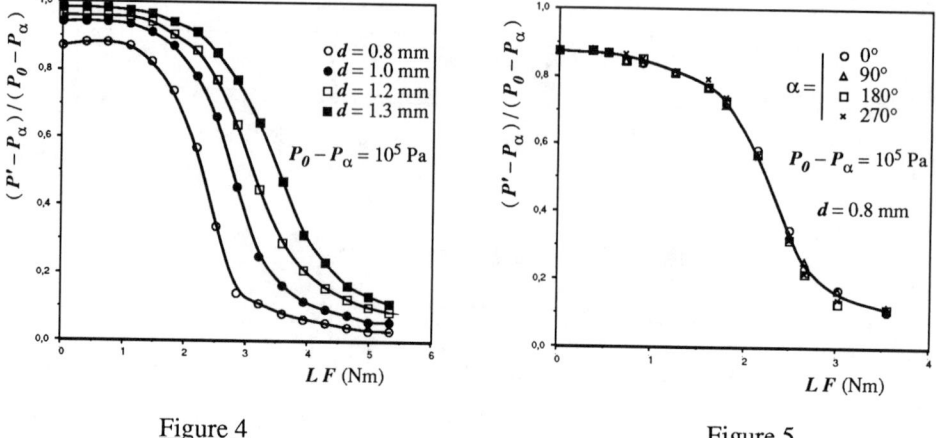

Figure 4

Figure 5

Dynamic Behavior

The purpose of this paragraph is to study very simply the influence of the principal parameters acting on the dynamic behavior of a system which associates the sensor and a jack. This system (Fig. 6) is in fact the basic device of a multidirectional force sensing prehensor. Several simplification hypothesis are made. For example, wave propagation is neglected, the average density of air is supposed to be constant and the bulk modulus in the jack chambers are assumed to be equal to the absolute pressures inside these chambers. These hypotheses can be justified in the case of short pneumatic lines and of low supply pressure ; they are in agreement with previous results [3] and [4].

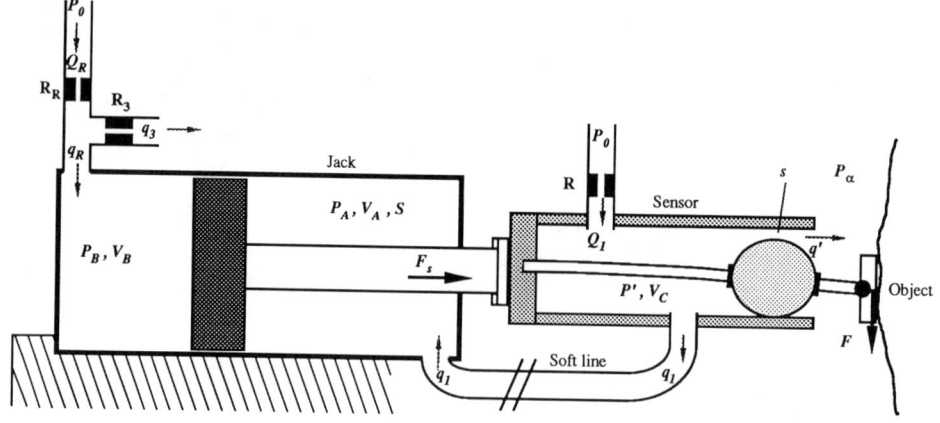

Figure 6

Transfer Function of the System

The subscript (0) is referring to the initial conditions. The following relations can be written:

$$P'_{(0)} = P_{A(0)} = P_{B(0)}$$
$$q_{1(0)} = q_{R(0)} = 0$$
$$Q_{1(0)} = q'_{(0)}$$
$$Q_{R(0)} = q_{3(0)}$$
$$q'_{(0)} = k\, s_{(0)} \sqrt{P'_{(0)} - P_\alpha}$$

where P', P_A, P_B and P_α are absolute pressures (Pa); Q_1, q_1 and q' are the flow rates in the pneumatic lines (m^{3}s^{-1}); s is the minimum section of the escape flow (m^2) and k a flow rate coefficient (m$^{3/2}$kg$^{-1/2}$).

The functioning of the system is studied after the sensor came in touch with the object. Then it may be assumed that the piston of the jack does not move any more since the object is supposed to keep its shape. Moreover, the parameters variations (Δ) are assumed to be small compared with the mean values. Under these hypotheses,

$$P' = P_A \Rightarrow \Delta P_A = \Delta P'$$
$$P_B \text{ is constant} \Rightarrow \Delta P_B = 0$$
$$\Delta Q_R = \Delta q_3 = \Delta q_R = 0$$

The dynamic equations are:

$$\frac{\Delta q'}{Q_{1(0)}} = \frac{\Delta s}{s_{(0)}} + \frac{\Delta P_A}{2(P_{A(0)} - P_\alpha)}$$

$$\Delta Q_1 = p\, V_C \frac{\Delta P_A}{P_{A(0)}} + \Delta q' + \Delta q_1$$

$$\frac{\Delta Q_1}{Q_{1(0)}} = -\frac{\Delta P_A}{2(P_0 - P_{A(0)})}$$

$$\Delta q_1 = p\, V_A \frac{\Delta P_A}{P_{A(0)}}$$

where p is the Laplace variable. The transfer function of the system is then obtained:

$$\frac{\Delta P_A}{\Delta s} = -\frac{\dfrac{1}{C_A\, s_{(0)}}}{1 + \dfrac{V_A + V_C}{Q_{1(0)}\, P_{A(0)}\, C_A} p}$$

with $C_A = \dfrac{1}{2}\left[\dfrac{1}{P_{A(0)} - P_\alpha} + \dfrac{1}{P_0 - P_{A(0)}}\right]$

The gripping force F_s and the pressure P_A are linked by:

$$\Delta F_s = -S\, \Delta P_A$$

where S is the useful section of the jack. The time constant of the system is then given by:

$$\tau = \frac{V_A + V_C}{Q_{1(0)} P_{A(0)} C_A}$$

Experimental verification

An experimental verification of the time constant τ has been made with different values of the restriction diameter d and consequently with different values of $Q_{1(0)}$, $P_{A(0)}$ and C_A. Fig. 7 is obtained from photographies on a screen of a memory oscilloscope.

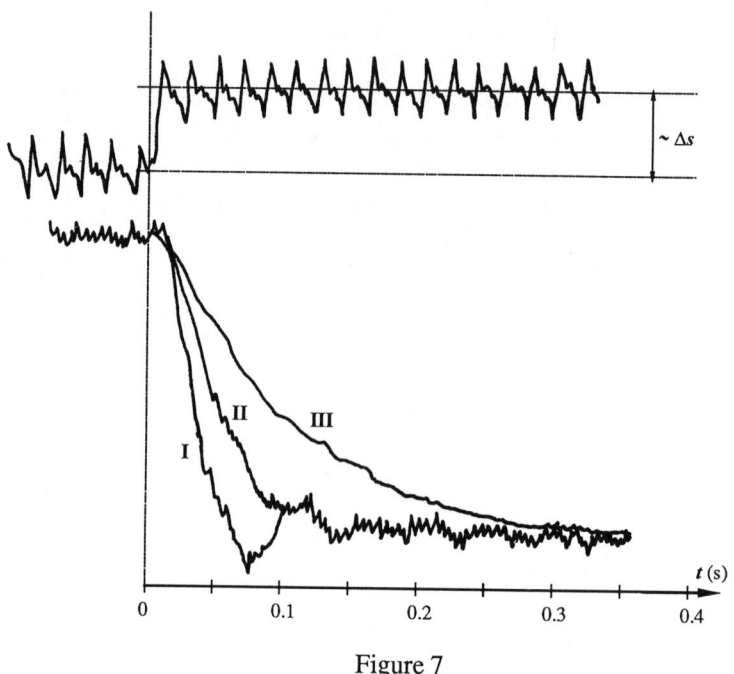

Figure 7

The theoretical results are :

I	$P_{A(0)} = 1.9 \; 10^5$ Pa	$Q_{1(0)} = 0.19 \; 10^{-3}$ m^{3}s^{-1}	for	$d = 1.3$ mm
II	$P_{A(0)} = 1.8 \; 10^5$ Pa	$Q_{1(0)} = 0.18 \; 10^{-3}$ m^{3}s^{-1}	for	$d = 1.0$ mm
III	$P_{A(0)} = 1.65 \; 10^5$ Pa	$Q_{1(0)} = 0.16 \; 10^{-3}$ m^{3}s^{-1}	for	$d = 0.8$ mm

$$V_C + V_A = 8.24 \; 10^{-5} \text{ m}^3$$

which lead to :

τ (I) = 0.04 s τ (II) = 0.08 s τ (III) = 0.14 s

The agreement between theoretical and experimental results could seem rather poor (about 30%). This discrepancy is largely ascribable to the experimental lack of accuracy and also to the very simplified theoretical model. It is obvious for example that the evolution **I** in Fig. 7 does not correspond to a first order response.

However, for an applied study like this one, it is very useful to bring to the fore the main parameters which play part in the dynamic behavior of the system. It is also of interest to get average values of the time constant and to compare them to these obtained with the unidirectional force sensor [1]. We notice that in the same conditions, these time constants are very close. This implies that the performances of the pneumatic prehensor are not reduced when it is equipped with the new sensor.

Application of the multidirectional force sensor on a gripper

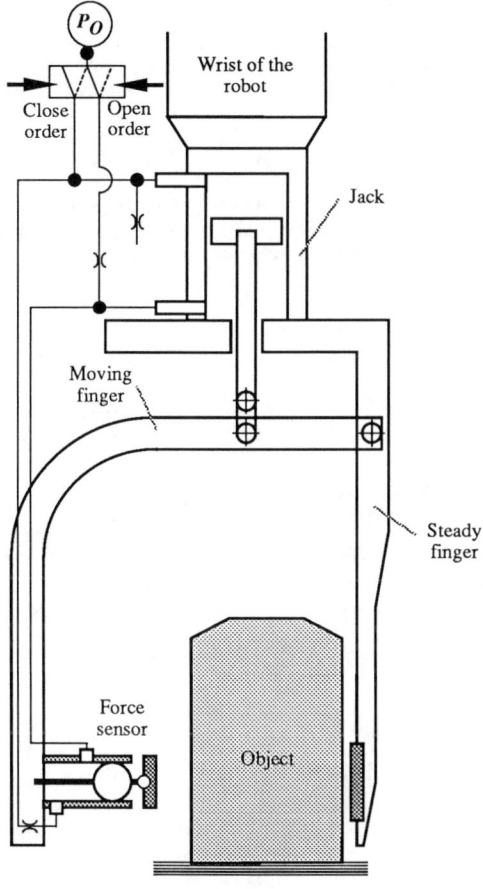

Figure 8

The multidirectional sensor is placed on a gripper according to the scheme of Fig. 8. A series of objects the weights of which vary from 2 to 23 N is manipulated. Slow and quick rotations of the arm which supports the prehensor are tested in the plane of the above figure. The gripping force appears as a convenient direct function of the objects weights.

Conclusion

The principle of the behavior of an original multidirectional force sensor, as well has the feasibility of a gripper equipped with this sensor, have been described in this paper. Such a gripper can have several applications in fragile objects carrying. However, some of the sensors presented in this paper have not a very good symmetrical behavior, owing to difficulties of manufacture.

The sensor sensibility could be improved for small values of F (see Fig. 4 and 5) by a convenient choice of materials for the rod and the obturator, or by the use of an intermediate fulcrum of the rod [1].

The next step of our work will lead to an optimization of the sensor, to its adaptation to the gripper of an industrial robot and to the study of several simulated real behaviors such as fruits picking for example.

Acknowledgements

The authors would like to thank Mr. J. Boscary for his contribution to this work and Messrs. P. Delmas and P. Fohanno for the realization of the test setups.
Financial support of the Conseil Régional Midi Pyrénées has been much appreciated.

References

[1] Patent CNRS PCT / FR 86 / 0054 5 May 1986 (Inv. CAEN, R. and FONADE, C.).

[2] First price of Laboratoires de Recherches Région Midi Pyrénées, December 1989 and Gold Medal in Salon International des Inventions et Techniques Nouvelles de Genève, April 1990.

[3] CAEN, R., DAT, J. and COLIN, S. "About tactile transducers used on a pneumatic prehensor with integral control". Proceedings of International Conference on Automation, Robotics and Computer Vision, 19-21 September 1990, pp. 313-317, Singapore.

[4] CAEN, R. and KHATTAB, A., "Pneumatic prehensor with tactile transducers". Journal of Theoretical and Applied Mechanics, Vol. 7, N° 6, 1988, pp. 875-897.

Grasping in an Unstructured Environment using a Coordinated Hand Arm Control

Sanjay Agrawal and Ruzena Bajcsy*

GRASP Laboratory
Department of Computer and Information Science
School of Engineering and Applied Science
University of Pennsylvania
Philadelphia, PA 19104, U.S.A.

Abstract

Abstract

Dextrous hands used in robotic systems require equally dextrous control architectures in order to perform the complex tasks that the hands were built for. Even with its multitude of degrees of freedom, the hand must closely interact with the arm on which it is mounted. Coordination with the arm takes place at the planning level, where desired motions for the hand and arm are synchronized. At the reflexive level, more dynamic interactions between the hand and arm are required, such that the arm is capable of using the force information from the hand, to modify its own trajectory. This coupling between mechanisms is not restricted to hands and arm, and so we need a multi layered control architecture through the actions of various mechanisms and active sensors are coordinated.

1 INTRODUCTION

Humans perform a variety of complex manipulatory tasks using their hands. In general the hand requires complimentary motions of the arm, which is termed as hand-arm coordination. The arm is responsible for repositioning and reorienting the hand and assists in the application of forces and torques upon the environment. We find both low level reflexes that control the hand-arm system, as well as high level task based motions that allow for dextrous and cooperative compliant motions. The human hand and arm coordinate their motions both at the planning stage, by preshaping while reaching, and while performing the task where feedback for arm is provided via the sensors in the hand, enabling the arm to move along a suitable trajectory. Even though the fingers possess multiple degrees of freedom, they have limited ranges of motions, and during active exploration, require the arm to provide the gross motion adjustments [9, 7]. Most previous work in the area of robot hands has focused on the abilities of the hand as an independent system addressing issues relating to the application of controlled forces and torques on an object that has been grasped within the hand [11, 16]. Systems where a hand is mounted on an arm have in general been confined to operate in static, structured environments requiring limited interaction between the hand and arm [13, 12, 1].

---

*Acknowledgments : This research was supported in part by Air Force AFOSR Grants 88-0244, 88-0296; Army/DAAL Grant 03-89-C-0031PRI; NSF Grants CISE/CDA 88-22719, IRI 89-06770; DARPA Grant N0014-88-0630 and DuPont Corporation.

The human hand is much more than a grasping mechanism. It serves to provide the sense of touch through which we explore and learn about the environment [10]. Similarly a robotic hand arm system needs to function as a probe, directing the exploration of a surface or object, as well as a gripper that can enclose a variety of shapes and sizes of objects. Manipulators that can behave both as controllable mobile sensors and configurable grasper have far reaching uses. Many hazardous environments require a dextrous manipulator that can function in an autonomous manner in locating and manipulating objects under uncertain conditions. Sensor based hand-arm systems can also be used, in manufacturing cells where the system has to deal with a variety of object shapes and uncertainty in their position and orientation.

Clearly a hand-arm robotic system requires coordination from the lowest level upwards to allow the hand and arm to perform cooperative motions. Cooperation takes place at the sensory level and at the planning level. The controller for each mechanical agent coordinates its activities with our agents via the sensory level and accepts commands from the planning level [2, 3]. The sensory level is responsible for transforming the feedback from other agents into a format acceptable to the receiving agent. This transformation of sensed data is a function of the current command being executed by the receiving agent. This coordination between agents happens at the servo level, allowing the agents to have a fast response time to any sensory feedback within the system. We use this system to coordinate the Penn hand and a robot arm in performing manipulation tasks.

2 SYSTEM ARCHITECTURE

The architecture is designed to allow for two kinds of communication. In order for a robot system to function in an autonomous and intelligent manner, the mechanical system must be able to react to sensory information in a reflexive manner, while the planner generates a new command for the mechanisms. This dual reaction in the system implies to separate channels of information flow, and two separate sets of computations. The agents controlling the mechanical system must have the ability to react to any external sensory feedback, and modify its current command accordingly. The planning level is required to generate a new plan, or decide that the existing plan is still valid. Plans constructed have intermediate goals that are defined in terms of desired sensory states. The plan must be able to distinguish between an acceptable sensory feedback in progressing towards the goal, as well as sets of feedback, that would require a replanning of the task solution.

We address these two requirements by providing dynamic planning and intelligent control agents. Mechanical and sensory agents that need to communicate in real time are not always physically coupled, nor is the raw information provided by one agent meaningful to the receiving agent. We specify a minimum connectivity between the system that would allow sensory information available to any sensory or mechanical agent, as well as all the layers within the system.

Conceptually the system has three separate layers: planning, sensory and control, each with its own functionality, and set of communication threads. Unlike Brooks [5], we do not construct the layers as a subsumption architectures, but distinguish them by the function they serve. At the lowest layer lie the mechanical and sensory agents. Each agent has a dedicated controller that is responsible for performing a given command, and a communication layer that interacts with the planning level by receiving new commands and sending back its current state. Another channel of communication also takes place with the sensory layer. Feedback from the agent is sent to the sensory layer, and data from other agents received from it. The command and data elements are used by the controller to drive its mechanical or sensory entity. The complexity of the control module in these agents depends on the entity the are controlling. We describe three of the controllers designed for the agents in the system implemented.

2.1 THE PLANNING LAYER

Within this layer of the system, two functions are performed; A task framework is created within which task is segmented into primitive motions. These primitive motions must be performed sequentially in order to complete the task. Primitives are conditionally linked together by a set of sensory condition vectors. If any of the condition vectors are satisfied, the current primitive can be terminated, and the next primitive instantiated.

The framework is implemented as a graph, where each node represents primitive, and each link is a sensory vector that must be satisfied, in order to move along that path. This graph structure allows for dynamic planning by providing multiple transitions from a single node. The initial task framework consists of nodes with exactly one edge leading from them.

This edge represents an expected sensory vector, on satisfying which, a transition can be made to the next primitive in the framework. A finite set of sensory vectors can be satisfied within any given primitive. We examine a small number of the more likely outcomes, instantiating the primitive associated with the eventual outcome. Once a new transition has been established, a new likely path to the goal state is constructed, by creating the necessary new nodes, which are linked by the expected sensory vector.

2.2 THE SENSORY LAYER

The sensory layer's primary function is to collect the feedback from each of the agents in the system. Each agent sends data to the sensory layer at the start of their respective cycles. Thus agents update the sensory layer at different rates. When a command is sent to an agent, the sensory layer is provided with a list of data, that the agent should be sent. This data list specifies the format in which the agent expects the data, and the rate at which the agent will expect to receive the data. Thus agents with a very low cycle time would be sent data only at some multiple of their cycle time. The data obtained from one agent must usually be transformed in order to be useful to the receiving agent. Position data is usually transformed by changing the frame of reference, while force data, is usually filtered and summed up to provide a single force/moment vector.

2.3 COMMAND QUEUES

A primitive consists of commands for the different agents in the system. These commands are either sent in parallel to the agents, or they have dependencies between each other.
The planner maintains a queue of commands for each agent in the system. When the primitive is instantiated, the queues of the involved agents get filled up, and the queue managers, responsible for maintaining the queues. In case of a dependency the manager decides if the dependency is satisfied, before sending the command to the agent.

If the planner decides that a primitive has completed its task, the remaining elements in the queue are discarded, and a new set of elements are replace them in the queue. The queue elements for each agent are structured based on the attributes of the commands sent to the agent.

2.4 ARM CONTROLLER

The arm on which the hand is mounted must couple its motions very closely with that of the hand. The arm controller receives commands from the planner in the form, of an absolute position, a relative position, or a vector along which to move. The arm controller is able to react to any forces on the hand, by receiving data from the hand, via the sensory level. The arm treats the feedback from the sensors on the hand as a force vector, along which it must comply. Along with the command to the arm, the sensory level is provided

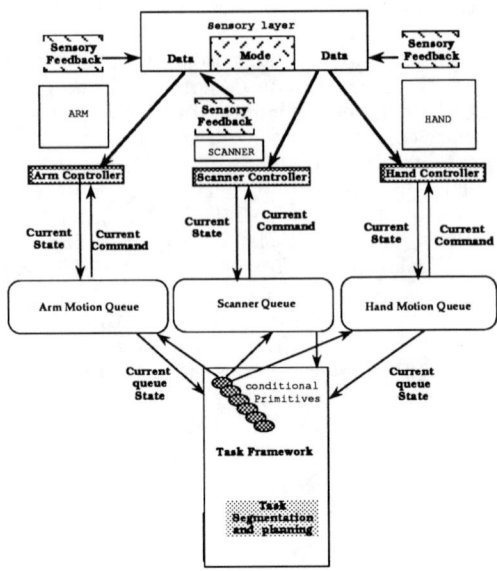

Figure 1: Layered Control Architecture

a mode by the planner. This mode is determines what forces are sent back to the arm. If the hand has grasped an object, the arm must not react to the grasping forces exerted by the hand, thus the sensor layer filters out such forces.

2.5 HAND CONTROLLER

The hand is a fairly complex mechanical device. It can be commanded into two ways; either by specifying desired forces for a set of sensors, a set of joint positions, a net desired grasping force, or desired pushing force can be provided. Thus the controller is designed with two layers. The lower layer is responsible for performing the force or position servo for the individual joints. At the higher layer, the net grasping or pushing forces are monitored, and the individual joints are set based on this desired force. The sensory layer can modify the net desired grasping force, or an individual desired joint position, based on information from other sensors in the system. This modification does not destroy the current command structure, but merely modifies the desired setpoint.

3 SYSTEM DESIGN

Conceptually the architecture provides three layers, where the control layer consists of independent subsystems that interact with each other through the sensory layer, and also interact with the planning layer. In implementing the system we found it desirable to mirror this conceptual architecture onto actual hardware. Thus each agent in the system, must be able to send and receive messages at two distinct communication rates. Communication with the sensory layer is a tightly bound time delay associated with it.

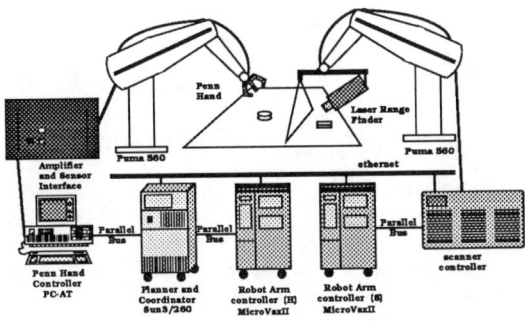

Figure 2: Distributed Environment

This requirement is due to the fact that the agent's controller needs to have all the sensory data in order to compute its next set point. To provide such a tightly bound time delay, each agent has a dedicated link to the sensory level. Communication between the agents and the planner are not as time critical, and thus bounded delays in the delivery of the command or status information can be tolerated.

The hardware used in our implementation consists of a Sun3 running SunOS which provides the platform for the planner and the sensory layer. The sensory layer resides within the kernel allowing the agents to communicate with the sensory layer over a dedicated parallel bus with small bounded delays. The planning layer lives as a user process, that can communicate with the agents via ethernet, using TCP packets, or via a serial link running at 9600 baud. The sensory layer and the planning layer communicate on the Sun using shared memory.

The agents for both the arms reside on MicroVAX II's that are each connected to a Unimate controller using a parallel bus. The control and sensory communication processes reside within the kernel. The Unimate controller provides an interrupt at the start of each cycle, causing the arm controller to receive data from the sensory level, and send its current position. Communication to the planning level is done as a user process over the ethernet.

3.1 PENN HAND CONTROLLER

The Control for the Penn hand is done using a 80286 based PC. All the hardware for controlling the hands resides in the PC's base aside from the power amplifiers and the tactile data signal conditioning hardware. The PC contains the encoder interface cards, a 16 channel A/D board, to read the output of the tactile pads and the potentiometers, and a 4 channel A/D used to send a voltage to the amplifiers which servo the four motors actuating the PENN hand.

The software for the controller is designed to provide an interrupt driven background loop which is used to servo the four motors, read the tactile sensors and the joint positions, and communicate and receive sensory data to and from the sensory level of the system. The foreground process receives commands from the planning level, and uses the current command to set force and position setpoints, to serve as a goal force or position for the background process to servo on.

Table 1: Scanner Parameters

Axis	X (width)	Y (length)	Z (height)
Imaging volume (mm)	135	164	172
Resolution (mm /pixel)	.23	.485	.672

3.2 THE MOBILE LASER RANGE SCANNER

The range image obtained from the scan is assumed to contain data corresponding to at most one object. If an object is detected, a superquadric representation of the object is attained by using the computational methods found in [8]. This parametric representation of the object is used to compute, the distance of the centroid of the object from the base of the arm carrying the hand. The orientation and the length along each axis are also computed and sent to the coordinator.

The Laser Range Imaging System consists of a laser stripe generator and SONY XC-39 camera which generates video signal of the images obtained under the illumination of the laser stripe. The processing unit [14] receives the continuous sequence of laser images and generates a range image of the scene in real time. the Laser range generator and the camera are mounted at the end of a Puma560 thus providing.

In operation, it moves linearly at a known constant velocity under robot control, scanning the desired workspace we are interested in. It can be shown that the position of the laser stripe as observed by the camera is a measure of the height of the nearest object intercepted by it. This video signal is sampled at a rate of $60Hz$ by the processing unit and the range image is produced in real time. Synchronization between scanning motion and image generation is ensured by the ability to send a triggering command along a serial line connecting the host computer controlling the robot and the processing unit. Since the size of an image is limited by the imaging volume of laser unit for a given resolution, multiple scans could be required to cover the relevant workspace. Having the scanner under manipulator control provides the flexibility of variable resolution in the Y direction, since the resolution in Y the scanning direction is a function of the velocity of the scanning motion. A limitation of the imaging system is that orthographic projection is assumed in the generation of the range image. Software compensation is employed to counteract errors of this kind, especially for tall objects.

The imaging volume of a single scan and the resolution of the range image are summarized in Table 1.

4 THE SENSORIZED PENN HAND

The PENN hand was designed and built at the GRASP lab at the University of Pennsylvania. The design justification and methodology is explained in [15]. Parts of the hands were redesigned in order to incorporate the tactile sensors [6], the potentiometers, and all the wiring that was required to send control signals and read the data from the sensors on the hand.

The Penn hand shown in figure 3 was designed to be a compact and lightweight mechanism, thus all the modifications were performed keeping these design criteria in mind. The Penn Hand is a compact mechanism weighing 3.8 pounds and actuated by four motors which are all located directly under the palm It has three fingers with two joints each. Two of the fingers can rotate around the palm to provide the flexibility of different grasping configurations. Each finger has one degree of freedom, with two joints that are coupled and driven by the same actuator. The sensors on the fingers were thus positioned so as to get the forces being sensed to to be complied with along the direction of motion. A single braided cable that is attached to the base of the hand and runs down to the

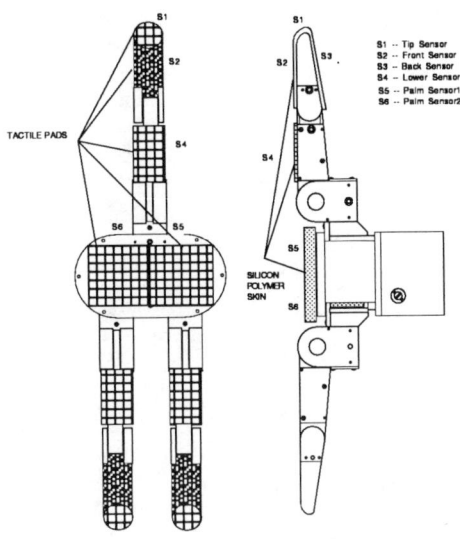

Figure 3: Sensorized Penn hand

controller. From the base of the hand, the cable branches along four paths, each hooked up to a single actuator. The cable hooked to each actuator, also carries the data lines to read all the sensors related to the control of that mechanism. The actuators, for two of the fingers that rotate around the palm, are located in the base of the fingers, and thus rotate with the fingers. This rotation requires the cables going to these actuators to be flexible and long enough to allow for travel around the palm.

The fingertip was designed to have multiple sensor sites, keeping in mind that all likely contact areas were covered by sensors. The design was constrained by the shape and size of the available sensors, as also by the initial design of the PENN hand. The tip of the finger was shaped anthropomorphically to provide contoured surfaces at all contact points, as well as allow for a single sensor pad to cover the tip. The tactile pads on the fingers and palm are built by Interlink Technology and they measure the applied pressure. There range of force measurements and other properties can be referenced in [6]. The sensor transduction incorporates FSR Technology. Each finger carries four tactile pads, one of which covers the inward facing side of the proximal link, while three pads entirely cover the distal link. These are arranged such that one pad covers the inward facing side, one pad covers the tip, and the third pad covers the back of the link. This arrangement of sensors is designed to decouple grasping forces from external contacts. Two large pads are fitted in the palm. All potential contact surfaces and sensors are entirely covered by a molded silicon elastomer [4] that provides compliance, friction and adhesion properties when grasping.

The potentiometers at the base of the fingers are used to allow calibration of the fingers, as well as allow the position of the lower joint to be determined independent of the upper joint, in case the upper joint disengaged via the breakaway mechanism [15].

5 EXPERIMENTS

In order to test the system a series of manipulation experiments were performed that allowed us to empirically verify the ability of the agents to communicate in a timely manner. A set of primitives were constructed in order to accomplish the tasks. These primitives do not form a complete set of possible motions, but were sufficient to allow a certain amount of replanning and error recovery for each of the tasks.

5.1 SURFACE FOLLOWING

195A unique surface tracking algorithm has been designed, where the fingertips maintain contact with the surface by adapting to small perturbations in the surface, and the arm reorients the hand to correct for large changes in the surface curvature. Four primitives are linked together to perform the experiment. The first orients the hand from its current position to one where the finger tips are placed equidistant from each other around the palm and point downwards. The second primitive moves the arm, till some part of the hand contacts a surface, or a position limit is reached. On contact the next primitive orients the hand till the three fingertips contact the surface. The last primitive moves the arm in a the required direction maintaining fingertip contact with the surface.

5.2 GRASPING STRATEGIES

Of the two grasping strategies developed for the hand-arm system, one uses the volumetric description and position of the object provided by the range finder to approach the object. While in free space the hand-arm system complies with any obstacles by modifying its trajectory on contact, and determining a new path towards the object. A precontact point at which the hand starts to approach the object along a straight line is the initial goal position. On reaching this point, the fingers are configured to match the shape of the object. The arm moves the hand till contact is made with the object. If contact is made with the palm, the fingers can enclose around the object till a balanced grasp force is achieved. If any of the fingers contact, the arm moves away opens the hand wider and reapproaches the object, till the palm contacts the object.

In the second strategy, the object is completely unknown, thus the surface tracking primitives are generated and used till a possible object is detected. Objects are detected when a sharp concave discontinuity is detected by the sensors at the back of the fingers. The object is examined by exploring it for potential edges, or protrusions about which to grasp.

6 CONCLUSION

The architecture that has been constructed provides a structure by which real time communication between cooperating agents, can be decoupled from the other communications within the system. In addition a standard interface between the different layers in the system and the agents has been provided. At present the visual system does not interact intelligently with the other agents in the architecture. We would like to be able to use the vision system, to continuously monitor the task, and allow the visual feedback aid in the selection of a new plan. Future work must be focused on building a large library of primitive motions such that a wider set of tasks can be tackled.

References

[1] P. Allen, P. Michelman, and K.S. Roberts. An integrated system for dextrous manipulation. In *Proceedings of IEEE International Conference on Robotics and Automation*, pages 612–61, Scottsdale AZ, 1989.

[2] M. A. Arbib, T. Iberall, and D. Lyons. *Coordinated Control Programs for Movements of the Hand*. Technical Report 83-25, COINS University of Massachusetts, Amherst, 1983.

[3] D. Brock and S. Chiu. Environment perception of an articulated robot hand using contact sensors. In *Proceedings of the ASME winter Annual Meeting: Symposium on Robotics and Manufacturing Automation*, pages 89–86, Miami Beach, Fl, 1985.

[4] Dow Corning. *Silastic MDX4-4210 Medical Grade Elastomer*. Technical Report 51-202G-88, Dow Corning Medical, 1988.

[5] P. Cudhea and R. Brooks. Coordinating multiple goals for a mobile robot. In *International Conference on Intelligent Autonomous Systems*, pages 168–174, 1986.

[6] Interlink Electronics. *The Force Sensing Resistor*. Technical Report, Interlink Electronics, 1989.

[7] J.M. Elliot and K.J. Connolly. A classification of manipulative hand movements. *Developmental Medicine and Child Neurology*, 26:283–296, 1984.

[8] A. Gupta and R. Bajcsy. Part description and segmentation using contour, surface and volumetric primitives. *SPIE Sensing and Reconstruction of Thre-Dimensional Objects and Scenes*, 1260, 1990.

[9] M. Jeannerod. *Attention and Performance*, chapter Intersegmental coordination during reaching at natural visual objects. Erlbaum, Hillsdale, NJ, 1981.

[10] R. L. Klatzky, S. Lederman, and C. Reed. Haptic integration of object properties: texture hardness and planar contour. *Journal of Experimental Psychology*, 15(1):45–57, 1989.

[11] Z. Li and S. Sastry. Task oriented optimal grasping by multifingered robot hands. *IEEE Journal of Robotics and Automation*, 4(1):32–44, 1988.

[12] K. Rao, G. Medioni, H. Liu, and G.A. Bekey. Robot hand-eye coordination: shape description and grasping. In *Proceedings of IEEE International Conference on Robotics and Automation*, pages 407–411, Philadelphia PA, 1988.

[13] S.A. Stansfield. *Visually guided Haptic Object Recognition*. PhD thesis, University of Pennsylvania, 1987.

[14] C.J. Tsikos and R. Bajcsy. *Redundant Multi-Modal Integration of Machine Vision and Programmable Mechanical Manipulation for Scene Segmentation*. Technical Report MS-CIS-88-41, GRASP Lab, University of Pennsylvania, 1988.

[15] N. Ulrich, R. Paul, and R. Bajcsy. A medium-complexity end effector. In *Proceedings of IEEE International Conference on Robotics and Automation*, Philadelphia, 1988.

[16] T. Yoshikawa and K. Nagai. Evaluation and determination of grasping forces for multi-fingered hands. In *Proceedings of IEEE International Conference on Robotics and Automation*, pages 765–771, Philadelphia PA, 1988.

Effective Integration of Sensors and Industrial Robots by Means of a Versatile Sensor Control Unit

Jürgen Wahrburg

ZESS (Center of Sensory Systems) and Institute of Control Engineering
University of Siegen, Germany

Abstract. This paper presents the basic features of a novel sensor control unit conceived as a programmable system to connect a robot controller with various intelligent sensors. The approach is adapted to the use of standard industrial robots shifting sensor related data processing tasks to external units. Based on a highly modular design the sensor control unit can be adapted to different robots and applications leaving the basic system architecture unchanged. It can be furnished with a powerful multiprocessor system to perform extensive signal processing operations, as coordinate transformations and filtering functions. A prototype has been connected to an ABB robot and tested successfully in an industrial environment.

INTRODUCTION

The automation of more complex manufacturing tasks by use of industrial robots demands the fulfilment of very narrow positional tolerances, particularly in the areas of assembly and machining, where the robots are in direct contact with other objects. In many applications the solution of these problems presupposes the availability of appropriate sensors which supply the robots with some information about their outer world. However, today the majority of robot installations in industrial automation is still characterized by purely position controlled applications. The potential sensor related enhancements are opposite to unsolved questions concerning the integration of robots and sensors which have prevented a wider use of sensor based robots so far. In order to contribute to the solution of this problem area a suitable system architecture is presented to connect sensors and standard industrial robots. It is based on the design of a novel sensor control unit conceived as a uniform platform which can be tailored for different sensor applications.

In the framework of this paper the term "sensor" will be focussed on those devices which - in contrast to simple binary switching sensors - provide one or more analog output signals, as for example seam finding or force-torque sensors. Special attention is given to the most challenging case of closed sensor control loops where sensor signals are used to perform a fine tuning of preprogrammed robot motions and must be processed on-line during the motion.

PRESENT STATE OF SENSOR-BASED ROBOTIC SYSTEMS

The standard robot control systems which are used in *industry* today must be regarded as closed systems offering only a few predetermined sensor functions which can hardly be modified or enlarged. As the features of most of the available sensors are fixed likewise, a lack of common interfaces arises as well in a physical as in a logical sense. For this reason a direct robot-sensor connection demands a lot of individual, application specific efforts, if it is possible at all. The task is made even more difficult due to the still fairly long cycle times of the present industrial robots. Deadtimes in the range of 50 to 100 msec can lead to stability problems in closed loop sensor systems and allow rather slow motion speeds only. As it is costly and expensive for robot as well as for sensor manufacturers to support many different products versions which are specifically tailored to certain applications, only very few complete solutions for sensor integration are available on the market today.

In the area of *research*, several valuable investigations in sensor-based robotic systems have already been published, see for example [1]...[3]. However, they often rely on self-designed robot controls or substantial changes of existing systems and thus impose difficulties for verification with standard industrial controllers. The proposed combinations of sensors and robots are often very closely adapted to the specific problems being discussed, or cannot be transferred to other applications due to further reasons. In general the research work has influenced the industrial scene only to a very small extent so far.

BASIC CONCEPTS TO IMPROVE SENSOR INTEGRATION

In many applications the use of industrial robots cannot be regarded as an isolated task, but the robot(s) are embedded in a flexible workcell together with further components, possibly as part of a CIM-controlled plant. The different levels of a sample configuration are illustrated by Fig. 1. Sensor integration in this context can start from two contrasting approaches. One the one hand, a top-down design at first describes sensor functions on a high level, that is in a systematic, more abstract way. However, the realization of such concepts may turn out to be difficult, because available industrial systems do not offer the presupposed features in lower levels.

The alternative approach is given by a bottom-up design which starts robot-sensor integration from the lowest level. The idea is that if satisfactory solutions have been found

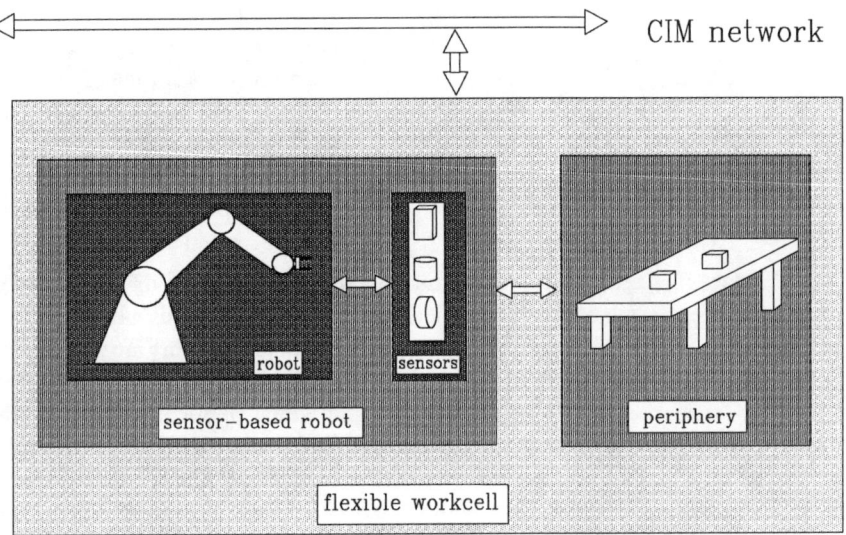

Fig. 1. Embedding of sensor-based robots in industrial automation.

in a lower level the combined components appear as one integrated system to the next upper level in the hierarchy, where problem solving can thus be based on the features of the integrated system without having to bother about details of the separate components in the lower level. For example, the design of a flexible workcell consisting of sensor-based robots and peripheral components will rely on well defined features of the sensor-based robots which have been obtained due to a successful integration of sensors and robots in the next lower level.

Within the framework of this bottom-up strategy our efforts to integrate sensors into robotic systems start out from the use of standard industrial robots and their associated controls, in order to circumvent as far as possible the difficulties which are responsible for the low acceptance of sensors in industrial systems and have been outlined in the previous section. The limited sensor functions of the robot controllers are accepted intentionally, although theoretically more sophisticated approaches might be desirable. As a consequence, all sensor related signal processing tasks which go beyond the built-in functions must be shifted to additional external units. The necessary hardware ressources in most cases cannot be provided entirely by the sensors, because these are subjected to spatial limitations, are not fitted with a sufficiently powerful electronics, or are not expandable by the user to match the demands of a specific application.

Instead of a direct connection we propose the introduction of a separate **sensor control unit (SCU)** as a connecting link between sensors and robot control. As illustrated by Fig. 2,

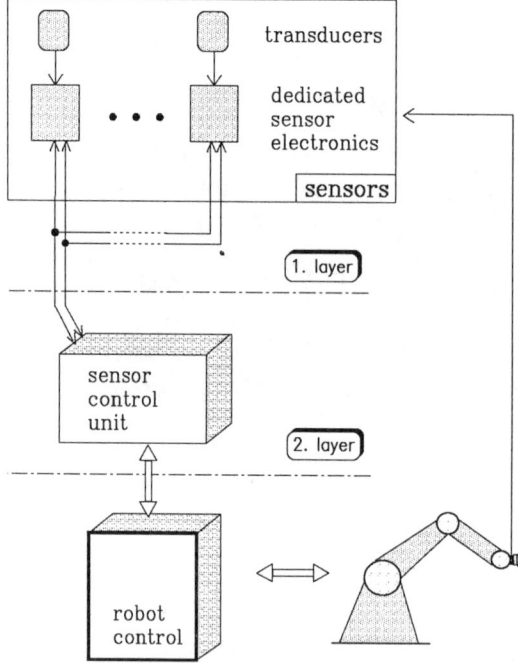

Fig. 2. Sensor control unit to connect sensors and robot control.

sensor signal processing is splitted into two layers by this way. In the first layer each sensor is furnished with its own dedicated electronics which is located as close to the transducers as possible and carries out the basic functions of signal amplification and adaptation. The second layer is formed by the SCU. It supplies a common bidirectional interface to the robot control and performs more general, computing-intensive tasks. A more detailed description is to follow in the next sections. The use of smart sensors is advantageous in several aspects. Firstly, an integrated electronics is less sensitive to electrical noise than a connection to an external unit, secondly, the SCU is not loaded with very specific functions of several sensors, and thirdly, a common serial link may be established to connect all sensors and the SCU offering a high speed bidirectional data transfer.

CONSIDERATIONS TO DESIGN A VERSATILE SENSOR CONTROL UNIT (SCU)

Main Objectives

Although at a first glance the introduction of the SCU may seem to suggest itself as a useful division of sensor related data processing, several points come out in detail which must be regarded carefully to achieve satisfactory solutions fulfilling industrial requirements. The main objectives to be taken into account for the design of the SCU are summarized in Tab. 1, supplemented by some explaining remarks.

Apart from the basic task to provide the necessary ressources for sensor signal processing and transmission, a high flexibility of the SCU is desirable. A modular design will enable the user to adapt the system to his specific demands in a convenient way without having to repeat all of the work once again in each new case. The SCU as an open, flexible part in sensor-based robotic systems relieves of the necessity to change robot or sensor characteristics, or to design problem-specific hardware enlargements.

Tab. 1. Features to be incorporated into the SCU design.

task / requirement	accompanying explanations
interface transducer	Information received from the sensors must be converted to match the requirements of a robot control as well in a physical sense, concerning appropriate analog and digital hardware interfaces, as in a logical sense, concerning a software to process sensor signals adapted to the sensor functions of the robot control.
performance of extensive arithmetic operations	Depending on a specific application the preprocessing of sensor signals can reach a significant extent in order to supply sensor information in a format that can be readily used by the robot control.
data fusion in multisensor systems temporal synchronization of robot and sensors	The SCU is the bus-master of the serial databus connecting it to all sensors and has to control sensor activities and the data transmission on this line. Merging of the data by the SCU lets several sensors appear as one versatile sensor system to the robot. In parallel the SCU must synchronize the data exchange to the robot control.
fast signal processing	The execution times of sensor signal processing should be as short as possible in order to not further increase the deadtime in closed sensor loops.
easy to use	Operating of the SCU should be possible on such a level that the same staff who program and operate industrial robots is also capable to use the SCU.

Implementation of a Prototype Design

The hardware selected to build up the SCU consists of a self-designed, modular system which is more flexible and, if the expenses for specific input-/output-enhancements are included, even cheaper than a readily available system, e.g. an industry PC-system. As illustrated by Fig. 3, a microcontroller has been chosen as central processing unit (CPU) due to its additional integrated features, including a serial interface controller which is used to

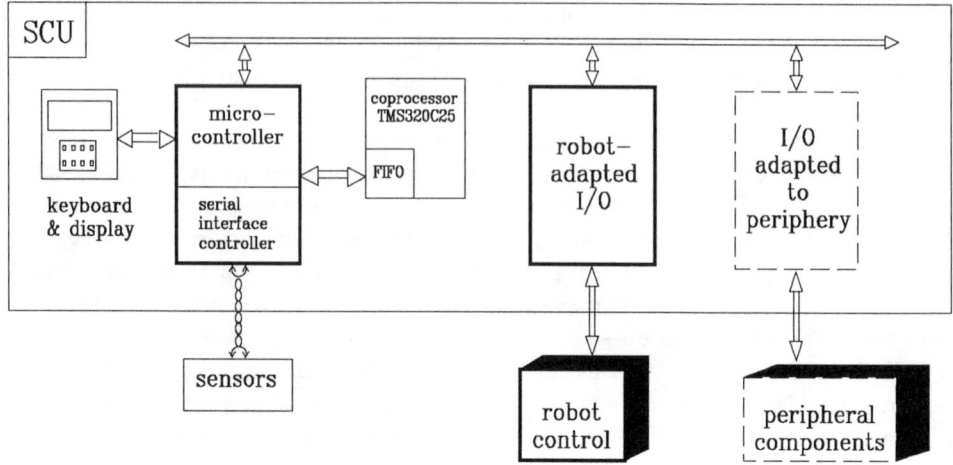

Fig. 3. Block diagram of the SCU hardware.

provide the serial link to the sensors. Presently an 8-bit type of the 8051 processor family is used, while a 16-bit 80C196 - based system is evaluated for future applications. Different I/O-boards which are connected to the CPU via a common system bus can be selected to establish an efficient connection to various types of robot controls.

It is important to point out that the connection between robot control and SCU is conceived to be bidirectional. All functions of the SCU and the sensors connected to it can thus be initiated dynamically by corresponding control commands and parameter values incorporated into the motion program of the robot. This will facilitate the temporal coordination of robot and sensor actions, and enables a user to handle the sensor system as an integral part of an enhanced robot, not being forced to familiarize himself with completely separate, new tools to program and use the sensors.

Some essential hardware features of the SCU can be listed as follows.

- *Sensor interface.* A multipoint serial connection according to RS-485 specifications is used, allowing baudrates of up to 500 kB.

- *Robot interface.* The I/O-boards have been designed to meet well accepted industrial standards, as for example decoupling via optocouplers and 24 V power supply in case of digital signals.

- *Options.* • A TMS 320C25 signalprocessor board may be furnished as a coprocessor to execute computing intensive tasks.

- Fast transients of sensor signals can be recorded in realtime into internal RAM storage of presently 60 KByte for a following detailed analysis.
- A keyboard and a 2-line LCD-display are available for local programming and monitoring functions.

Due to the powerful coprocessor the execution times even of extensive calculations, as for example coordinate transformations, filtering functions, and sophisticated control algorithms, are reduced to the microsecond range. The cycle times of the sensor system, including data processing within the sensor electronics and the SCU as well as data transmission, are one to two orders of magnitude smaller than the deadtimes of available robot controllers and therefore do not increase the stability problem in closed sensor loops, described e.g. in [4]. If in a certain application the time lag of the robot system cannot be circumvented, the SCU can also be used to control peripheral actuators in the environment of the robot which may have faster response times, as indicated by the dotted block in Fig. 3.

A FIRST TEST INSTALLATION

A prototype of the SCU is presently tested in an industrial application in combination with an ABB robot, type IRb 2000, and a novel force-torque sensor [5] which is under development at the swiss firm Kistler. A photo of the sensor without housing is given in Fig. 4, showing in the upper part the printed circuit board of its integrated electronics.

Fig. 4. Kistler force-torque sensor with integrated electronics.

The problem to be solved is a deburring task. The robot picks up parts consisting of brass casting and conducts them along a stationary grinding belt in order to remove their burrs, as shown in Fig. 5. All of the burrs must be removed completely, but on the other hand not too much material must be taken away. As the shape of the parts sligthly varies from piece to piece, the forces and torques detected by the sensor during the grinding process are used to perform a correction of the preprogrammed trajectory. The robot follows the contour of the actual part with a constant contact force.

First experiences with the integrated robot-sensor system show encouraging results. Present work focusses on the development of appropriate control algorithms to further reduce the deviations from the ideal trajectory during contour following and thereby

Fig. 5. Robot-sensor integration in a deburring task.

improve the uniformity of the product quality.

CONCLUSION

The introduction of the presented SCU offers a well suited approach to improve the performance of sensor-based robot systems. Using it as a uniform platform future work can concentrate on more specific aspects of a given automation task. The usefulness of the SCU will be increased by future robots, because they will have shorter cycle times to reduce the deadtime problem in closed sensor loops, but will not be equipped with sensor functions being as powerful and flexible as those of the SCU.

REFERENCES

1. Whitney, D.E., "Historical perspective and state of the art in robot force control." Proc. IEEE Conf. on Robotics and Automation, St. Louis, 1985, pp. 262-268.
2. Tzafestas, S.G., "Integrated sensor-based intelligent robot system." IEEE Control System Magazin, vol. 8, Apr. 1988, pp. 61-72.
3. Van Brussel, H., "Sensor based robots do work!", Proc. 21st Int. Symp. on Ind. Robots, Copenhagen, IFS Publications, Bedford 1990, pp. 5-25.
4. Wahrburg, J., "Control concepts for industrial robots equipped with multiple and redundant sensors," in: J.T. Tou, J.G. Balchen (Eds.), Highly Redundant Sensing in Robotics, Springer Verlag, Berlin, Heidelberg, New York, 1990, pp. 277-291.
5. Cavalloni, C., "A novel 6-component force-torque sensor for robotics." Proc. Int. Conf. on Advanced Mechatronics, Tokyo, May 1989.

IDENTIFICATION AND EVALUATION OF HYDRAULIC ACTUATOR MODELS FOR A TWO-LINK ROBOT MANIPULATOR

Jian-jun Zhou   Finn Conrad

Control Engineering Institute
Technical University of Denmark
DK-2800 Lyngby Denmark

ABSTRACT. This paper describes the practical application of system identification techniques to obtain a open-loop linear model of an electrohydraulic actuator which is used to drive a two-link robot manipulator. The obtained model is intended to provide a basis for control system design. The identified model has been evaluated, respectively, by applying spectral analysis and simulation. The results show that the identified models give a very good description of the dynamic behaviour of the hydraulic actuator.

INTRODUCTION

Although robot manipulators have been used in industry for a number of years, their capabilities reach far beyond their present level of application. At present, most industrial robot control systems still utilize decoupled controllers, e.g. PID controllers, which control joint angles separately through simple position loops. However, in view of the trends in industrial applications for faster robots, the applicability of these simple control strategies is limited.

A two-link hydraulic test robot has been designed and constructed at the Control Engineering Institute, Technical University of Denmark. One of the purposes of the test robot is used it to investigate digital control and design techniques of fast robots. In the recent years, many advanced control strategies have been proposed for controlling robots moving at high speed. Most of these control strategies are based on a deeper understanding of the robot dynamics and the incorporation of the robot dynamics in the control design.

Hydraulic robot manipulators are complex non-linear systems, and it is very difficult to obtain an accurate model based on physical principles especially for obtaining an accurate low-order linear model. However, application of system identification techniques provides an alternative way to obtain accurate robot models.

Identification experiments of the hydraulic test robot have been carried out. In the identification of the robot dynamics, the system is divided into two subsystems, i.e. the robot arm and the hydraulic actuators, whose models were identified, respectively. The arm dynamic parameters, i.e. inertial moments, and mass centres etc., of the robot were estimated based on measured piston positions, velocities and pressure differences across the pistons. The dynamic parameters of the actuator were identified based on the servovalve input voltage and measured piston position. Furthermore, the kinematic parameters of the robot were also estimated based on the measured piston position and the tool centre point position in Cartesian coordinates. This paper only presents and discusses the identification of the hydraulic actuator dynamics.

SYSTEM DESCRIPTION

The hydraulic test robot is shown in Fig. 1. Each link of the robot is driven by an electrohydraulic actuator. The actuator controlling the lower joint is fixed on the basis platform, and another one is fixed on the lower link. The hydraulic pistons are directly coupled to the links to reduce backlash. Two-stage super high frequency servovalves are chosen to control the cylinders.

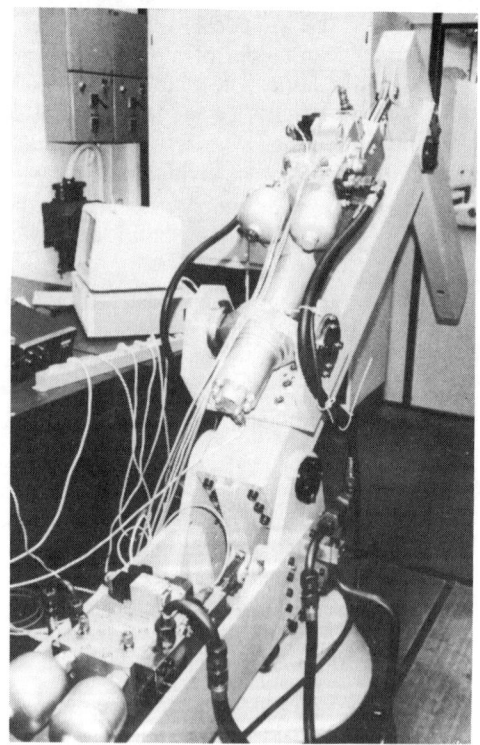

Figure 1   The hydraulic test robot

A digital control system based on an AT&T DSP32 signal processor is developed for collecting data, monitoring the system, and performing the mathematical calculations. A position transducer and a velocity transducer supply the displacement and velocity of the piston of each actuator to the digital controller. Two pressure transducers, one for each servovalve, transmit the pressure drops across the pistons to the digital controllers. A DT2841 card from Data Translation is chosen for the AD/DA conversion.

The actuator consists of a hydraulic two-stage servovalve of flow control type and a two rod cylinder. From the physical principles the dynamics of the actuator can be described by the following equations.

The dynamic response of the servovalve can be described by the second-order transfer function:

$$\frac{x_v}{u} = \frac{K_v}{s^2/\omega_s^2 + 2x_s s/\omega_s + 1} \tag{1}$$

where x_v is the displacement of the spool valve from null; u the signal voltage input to the amplifier; and K_v the total gain.

The static characteristic of the servovalve can be approximately described by the nonlinear flow equation:

$$Q_f = \begin{cases} c_v w x_v \sqrt{(P_s - P_f)/\rho} & for \quad x_v > 0 \\ c_v w x_v \sqrt{(P_s + P_f)/\rho} & for \quad x_v < 0 \end{cases} \tag{2}$$

where Q_f is the flow across the piston; c_v the discharge coefficient of the valve orifices; w the area gradient of the spool valve; P_s the supply pressure; P_f the pressure drop across the piston; and ρ the mass density of oil. Linearization of the flow equation gives a linear equation

$$Q_f = K_q x_v - K_c P_f \tag{3}$$

The continuity equation of the cylinder is:

$$Q_f = A_t \cdot \dot{y} + \frac{V_t}{4\beta} \cdot \dot{P}_f + c_{sl} \cdot P_f \tag{4}$$

where V_t is the total volume of the cylinder chambers; c_{sl} the total leakage coefficient of the piston; β the effective bulk modulus of the system; A_t the end area of the piston; and $\dot{y}$ the piston velocity.

The force equation of the piston is

$$A_t \cdot P_f = m_t \cdot \ddot{y} + B_t \cdot \dot{y} + F_L \tag{5}$$

where m_t is the total mass of the piston; B_t the viscous damping coefficient of the piston; and F_L the reaction from the link.

After Laplace transformed, equations (3-5) may be solved simultaneously to obtain

$$Y = \frac{\frac{K_q}{A_t} X_v - \frac{K_{ce}}{A_t^2}\left[1 + \frac{V_t}{4\beta_e K_{ce}} s\right] F_L}{\frac{V_t m_t}{4\beta_e A_t^2} s^3 + \left[\frac{K_{ce} m_t}{A_t^2} + \frac{B_t V_t}{4\beta_e A_t^2}\right]s^2 + \left[1 + \frac{B_t K_{ce}}{A_t^2} + \frac{K_t V_t}{4\beta_e A_t^2}\right]s + \frac{K_{ce} K_t}{A_t^2}} \tag{6}$$

where $K_{ce} = K_c + c_{sl}$. If there is no any internal spring loads ($K_t = 0$) and $B_t K_c / A_t^2$ is much smaller than unity, (6) reduces to

$$Y = \frac{\frac{K_q}{A_t}X_v - \frac{K_{ce}}{A_t^2}\left[1 + \frac{V_t}{4\beta_e K_{ce}}s\right]F_L}{s\left[\frac{1}{\omega_h^2}s^2 + \frac{2\xi_h}{\omega_h}s + 1\right]} \quad (7)$$

where

$$\omega_h = \sqrt{\frac{4\beta_e A_t^2}{V_t m_t}} \qquad \xi_h = \frac{K_{ce}}{A_t}\sqrt{\frac{\beta_e m_t}{V_t}} + \frac{B_t}{4A_t}\sqrt{\frac{V_t}{\beta_e m_t}}$$

If the piston has no load force ($F_L = 0$) corresponding to decoupling the piston from the link, (7) reduces to

$$\frac{Y}{X_v} = \frac{K_q/A_t}{s\left[\frac{1}{\omega_h^2}s^2 + \frac{2\xi_h}{\omega_h}s + 1\right]} \quad (8)$$

When the piston is coupled with the link, a similar equation as (8) can be obtained. It follows from (1) and (8) that open-loop system is of fifth-order.

IDENTIFICATION EXPERIMENTS

For a hydraulic actuator, identification of open-loop dynamics is not a straightforward matter, because the actuator can not be operated in open-loop. Therefore, the model only can be identified via closed-loop experiment. In our identification experiment, feedback control is performed by a proportional controller. The input of the servovalve is thus given by

$$u(t) = K_p(w(t) - y(t))$$

where K_p is controller gain; $w(t)$ is test signal; and $y(t)$ is measured piston position.

In industrial practice a pseudo random binary sequence (PRBS) is frequently used as test signal in order to improve the statistical information content of the measured data. Selection of level of the binary signal has big influence on the accuracy of the identified model. For the actuator in which the piston has a large moving rage, a big level will result in saturation of servovalve input signal. In result a saturation nonlinear element is artificially introduced into the system. Furthermore, the big level will also cause that the identified model has a low frequency bandwidth. But the model obtained by using a small level may only be valid for a certain working range if the system is nonlinear.

To avoid the problem we propose an alternative solution by using the following test signal

$$w(t) = w_{\sin}(t) + w_{PRBS}(t) \quad (9)$$

where $w_{\sin}(t) = a \cdot \sin(\omega t)$ is a low frequency *sine* signal which can drive the system moving slowly and covering most working range; and w_{PRBS} is PRBS signal with a small level which can guaranty that the system is fully excited. Because variation of the test signal between sequent discrete points is not very big in this case, it guaranties that the control signal is not saturated. Correspondingly, system output can be presented as

$$y(t) = y_{\sin}(t) + y_{PRBS}(t) \quad (10)$$

where $y_{\sin}(t)$ and $y_{PRBS}(t)$ are outputs corresponding to the inputs $w_{\sin}(t)$ and w_{PRBS}, respectively, where $y_{\sin}(t)$ can be measured from an alternative experiment. Identification calculation is carried out based on $w_{PRBS}(t)$ and $y_{PRBS}(t)$.

The experiment set up is shown in Fig. 2, where the actuator was fixed in horizontal and operated in closed-loop. The link was decoupled from the piston rod. So no external load is acting on the piston.

The identification experiments for the case the link was coupled with the piston rod were also carried out. In this case the identified model actually presents the dynamics of one-link robot.

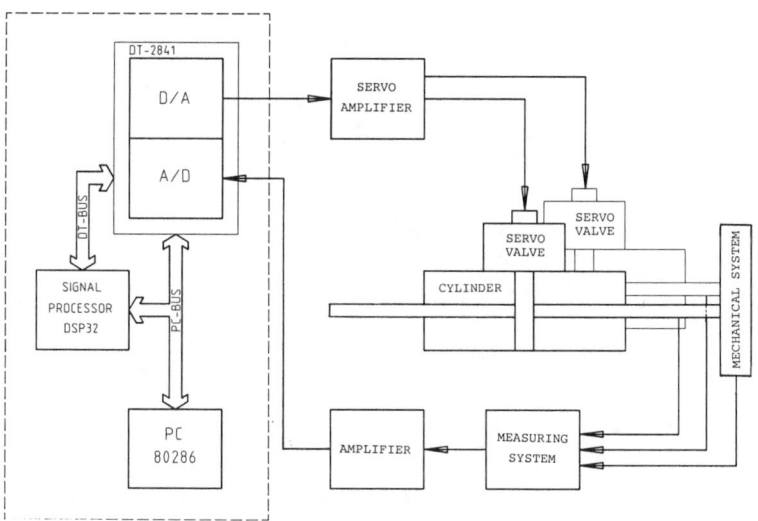

Figure 2   The experiment set up

IDENTIFICATION METHODS AND RESULTS

Because the two actuators are almost identical, only experiment results for the actuator driving the upper link are presented here. Fig. 3 shows a set of input-output data. The supply pressure was 210 bar and the sampling period was 10 ms.

Model Structure

The purpose of the identification here is to provide a basis for controller design. A simple model structure should be used without significantly reducing the linear dynamics of the system, so that control design and implementation can be achieved easily. The discrete-time linear model to be identified has the following structure (ARX model):

$$A(q^{-1})y(k) = B(q^{-1})u(k-d) + e(k) \qquad (11)$$

where d is delay time,

$$A(q^{-1}) = 1 + a_1 q^{-1} + \cdots + a_{n_a} q^{-n_a} \qquad B(q^{-1}) = b_0 + b_1 q^{-1} + \cdots + b_{n_b} q^{-n_b}$$

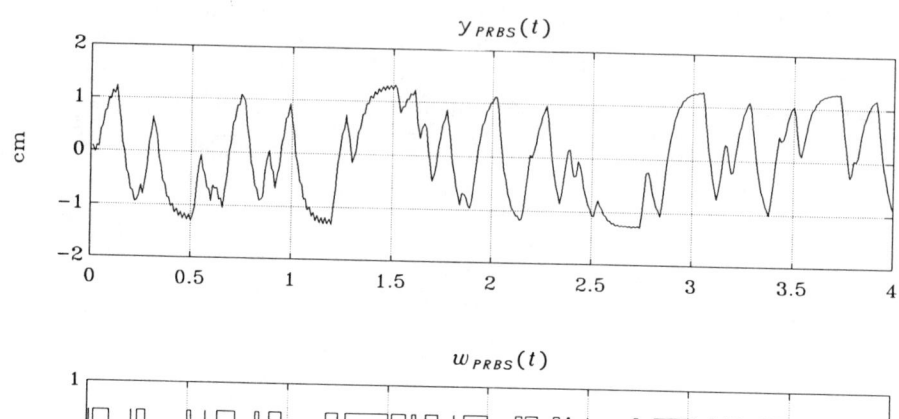

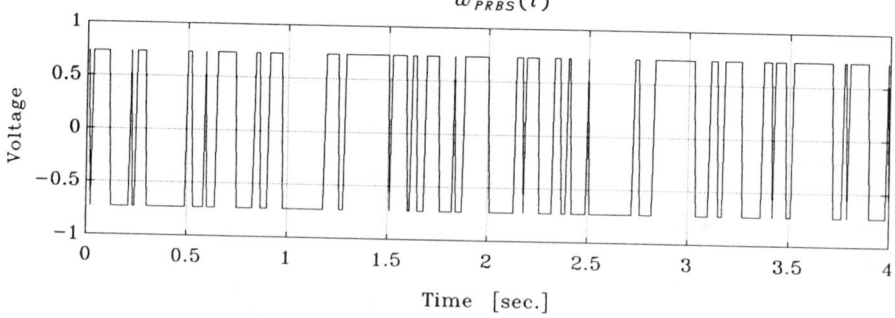

Figure 3 The measured input-output data

Identification Methods

Two approaches, direct and indirect identifications, may be considered for estimating the open-loop parameters of a system operating in closed-loop. The difference between the two approaches lies in which signal is used as input signal in identification computation. In direct identification, the open-loop model is directly estimated from the input-output data, $u_{PRBS}(t) (= K_p(w_{PRBS}(t) - y_{PRBS}(t)))$ and $y(t)$. In indirect identification, the closed-loop system can be regarded as a whole and its model can be estimated based on the input-output data, $w(t)$ and $y(t)$. The open-loop model may then be determined from the estimated closed-loop model using the knowledge of the controller. If we use $Q(s)$ and $G(s)$ to denote the closed-loop and open-loop transfer functions, respectively. The open-loop transfer function can be determined from

$$G(s) = \frac{Q(s)}{K_p(1 - Q(s))} \qquad (12)$$

The prediction error method (PEM) (Ljung 1987) was used to estimate linear model parameters. When the estimated model is an ARX model, PEM is identical with the least squares estimation method which has very good computational characteristics, accuracy, and reliability.

Identification Results

In general the Akaike Final Prediction Error Criterion (FPE) can be used as a reference for determining model order. FPE is defined as

$$FPE = \frac{1+n/N}{1-n/N} \cdot v$$

where n is the total number of estimated parameters, and N is the length of the data record, and v is the minimum value of the criterion function. From theoretical point of view, the model with the smallest value of FPE should be chosen. In our case, a fifth-order model gives the smallest FPE value. This is consistent with the theoretical analysis result in previous section. But we also find that the improved fit for a fifth order model is rather marginal compared with a third order model. By checking the pole-zero configuration for the fifth order model, we see that the two pole-zero pairs are likely to cancel, so that a third order model would be adequate.

The identified open-loop transfer function is

$$G(s) = \frac{Y(s)}{U(s)} = \frac{-3.325s^2 + 441.0s + 3231.9}{s^3 + 149.74s^2 + 987.0s - 66.2}$$

where there is one pole close to zero. It conforms that the system has indeed an integration. This also keeps when the piston is coupled with the piston. The result also shows that the system is of non-minimum phase.

In our identification experiments, the same results were obtained for indirect and direct identification approaches.

Model Evaluation

One way for evaluating the obtained model is to simulate it and compare the simulated model response with the measured system response. Fig. 4 shows comparison of computed response and measured response obtained from alternative experiment using another set of PRBS as inputs. As it can be seen the agreement is quit good.

We also compute the residuals of the estimated third order model and their autocorrelation function as well as their cross-correlation function with the input signal. The results show that the residuals are quite small compared to signal level of the output and they are reasonably well uncorrelated with the input and between themselves. We also compare the computed closed-loop frequency response with the measured closed-loop frequency response. The agreement is also quit good.

CONCLUSIONS

Models of an electrohydraulic actuator for a hydraulic test robot have been identified by applying system identification techniques. Direct comparisons of the simulated response using identified models with corresponding measured results from the system show the accurate model can be achieved by the identification. The obtained results establish a basis for the application of the controller design.

When the actuator is decoupled from the link, it can be pretty well described by a linear model. Direct and indirect identification methods give the same models. From the Akaike

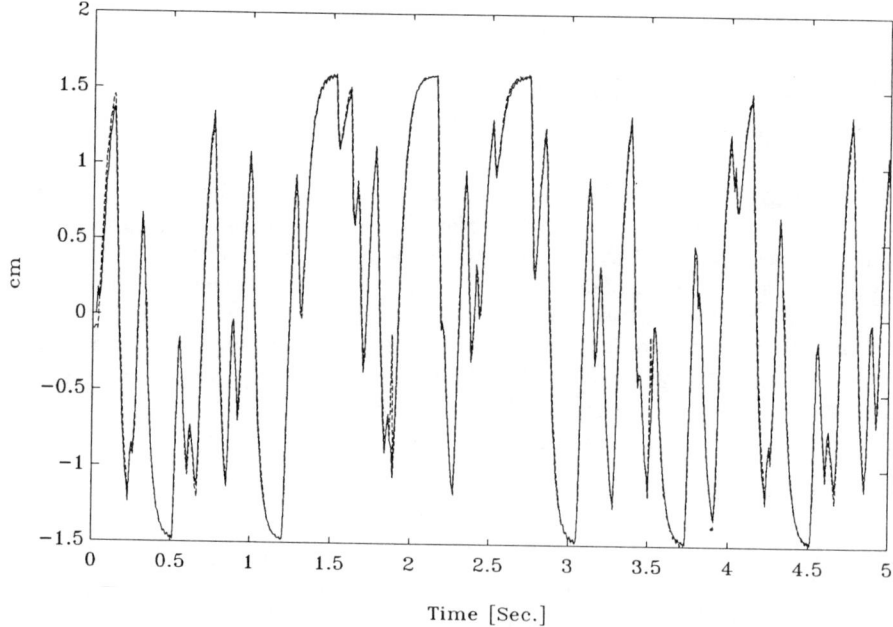

Figure 4 Comparison of computed and measured system responses

Final Prediction Error Criterion, the best fit of the system is a fifth order model. But a third order model also can give very good description of the system and it is especially adequate for the application of adaptive controller.

Identification results for the case that the actuator is coupled with the link show that the system can still be approximately described by a third order linear model even though it is not as good as the case that the piston is decoupled with the link. This result inspires that it may be possible to apply a optimal controller based on a linear model to the test robot.

REFERENCES

Conrad, F., P. H. Sørensen, E. Trostmann, J. J. Zhou: "Design and Digital Control of A Fast Hydraulic Test Robot Manipulator." Proc. of the 9th BHRA International Conference on Fluid Power, Cambridge, England, 25-27, April 1990.

Ljung, L. "System Identification: Theory for the User." Prentice-Hall, Inc. Englewood Cliffs, Mass., 1987.

Gustavsson, I., L. Ljung, and T. Soderstrom, "Identification of Processes in Closed Loop." Automatica, Vol. 13, pp. 59-75.

Zhou, J. J. "Adaptive Control and Applications to Hydraulic Robot Manipulators." Ph.D.-thesis. Control Engineering Institute, Technical University of Denmark, Lyngby, 1989.

A MODULAR ARCHITECTURE FOR CONTROLLING AUTONOMOUS AGENTS

Lykourgos Petropoulakis
Industrial Control Unit
Electronic & Electrical Engineering Department
University of Strathclyde
Marland House, 50 George Street
Glasgow G1 1QE, SCOTLAND
Tel: +44 41 552 4400 ext 2549
Fax: +44 41 553 1232
E-mail: akis%uk.ac.strath.icu@nsfnet-relay.uk

Abstract

Controlling robotic systems to perform tasks in real world environments has proved a difficult problem. This is the main reason why robots operate - almost exclusively - in well-structured domains performing repetitive tasks which may require accuracy but little, if any, intelligence. A lot of research effort has been devoted in an attempt to redress the balance and equip robots with sufficient initiative to perform complex tasks in an autonomous and robust way. In this paper a modular parallel processing architecture is proposed for robotic assembly, with emphasis placed on efficiency robustness, autonomy, and - in case of failure - graceful degradation. This is a new approach to the robotic control problem, and similar lines of research have produced interesting results. The paper explains the proposed approach, reviews some of the concepts presented by other researchers, and gives a brief account of the robotic research programme which has just started in the Industrial Control Unit.

1 INTRODUCTION

It can be argued that the architectural framework within which a system develops is critical because its structure ultimately affects the overall performance. This argument appears to be true for computer systems (compare the architecture and performance of distributive systems with that of mainframes). Drawing an analogy between computers and robotic systems it can be argued that the degree of intentionality and intelligence exhibited by a robotic mechanism depends - amongst other things - on its architectural control mechanism. Furthermore, it can be said that the classical sequence of *plan-execute-sense-evaluate-think-replan* – normally associated with intelligent activity – is inflexible and therefore limited in its scope. The inflexibility arises out of a highly structured framework that enforces a control sequence which is not always desirable in robotic applications. This argument appears to be particularly true in cases where reflex actions are needed. Thus, although a robot equipped with sufficient computational power can, in theory, take the time to analyse its environment, think out its actions, and act in the desired manner, practical considerations normally preclude a cycle of this kind. In the case of autonomous mobile systems for example, a fast reaction time is particularly important.

Robotic assembly is equally demanding and the same considerations apply. In the case of assembly systems, a typical sequence of events may be the formulation, modeling, and analysis of the assembly problem [6,12], followed by path planning [9] and the definition of the assembly strategy (the plan). This sequence is normally done off-line, and the plan is passed on to the on-line system for execution. During the execution stage, the environment is monitored, the information is processed, error recovery procedures are activated, a world model is updated, and finally, the planning strategy is reformulated and execution can continue. Such architectures suffer from a number of drawbacks: Firstly, the off-line, mathematical analysis of the assembly design, has proved to be computationally expensive and has been criticised [7] for being based on the *worst case* principle which, on combination can result in large overestimates of error. Secondly the technology to support on-line updating of 3D world models [2] is expensive, thus beyond the practical economics of most industrial assembly systems; the reliance on accurate models is unrealistically high, and error recovery processes based on world model considerations are computationally very intensive [8]. Moreover the robustness of such systems is questionable.

2 MODULAR METHODS

Believing that centralised architectures do not meet the demands of robotic applications, many researchers have directed their efforts into developing modular architectures [1, 4]. This is hardly surprising since modularity permits the implementation of parallel processing which is central to achieving the speed of reaction needed in robotic systems, enables the designer to formulate and test each section independently, and appears to provide a better basis for system expansion as this grows in size and/or complexity.

Some of these methodologies are based on the *subsumption* architecture which was initially applied to mobile systems [3] and has more recently been applied to manipulators [5, 14]. Typically these approaches incorporate the sensing action in the design of the modules, called *behaviours*, instead of reasoning about it at the symbolic level. It has been shown [5] that some manipulative functionality can be obtained without any representations, models or plans, but through a combination of hardware design, and suitable actions under the control of parallel processing. It is worth noting however, that the domain of application was not robotic assembly.

A slightly less radical approach applicable to robotic assemblies is presented in [10] where a virtual machine is used between the planning process and the assembly output. In this case, the planner constructs its plans in terms of the existing modules which have the task of translating planner designed part motions into robot actions. The result is a decoupled system (off-line/on-line) potentially applicable to other domains [11]. The on-line architecture is similar to [5]. The system behaviours are activated, and their actions are dictated by two factors: a) their design potential, and b) the sensors of the system. The outputs of the modules are designed to directly affect the environment in which the robot operates.

The modules are designed to operate in parallel, but unlike [5], communications between modules are not restricted to simple signals. However, there is no explicit representation of the environment and the sensor action is integrated into the modules. The latter is an implementation which is based on the belief that sensors have to be an integral part of the design and cannot be brought into the system as an afterthought. This approach however carries a penalty. Given that by design the sensors are at the very heart of each module, large system changes may be necessary when new sensors are introduced.

3 A ON-LINE HYBRID APPROACH

The methodology described in this paper shares a lot of common ground with some of the aforementioned strategies but, at the same time, differs from them in a number of ways. Our primary concern is the on-line implementation of an assembly process, rather than off-line task-planning. This is because we hold the view that the majority of problems associated with robotic assembly occur due to on-line uncertainty, and we aim to concentrate our efforts on this issue. It is firmly believed however, that, for reasons of efficiency and flexibility, the planning and execution stages of an assembly process must remain independent of one another, with the on-line system capable of coping with the varied conditions that can arise in an assembly cell. In turn, this requirement implies a flexible, fast-acting execution system with a number of in-built intelligent actions.

Our view is that, in order to support intelligent activity, it is essential for a robotic system to possess a certain degree of on-line reasoning. Moreover, in order to also retain flexibility, some of this reasoning process must be performed at a symbolic level. We therefore aim to develop a system which will support on-line higher-level cognitive processes but avoid the penalties that a classical architecture carries. In addition to this, and in view of the varied assembly environments and needs, it is felt that the architecture should be capable of supporting incremental expansion of the system both in terms of its cognitive powers and in terms of its manipulative capabilities.

We aim to accomplish this goal through the use of independent agents which we call *System Action Module(s)* (SAM). These agents provide the required system competences for completion of the assembly task. In this respect we identify two categories of System Action Modules; *configuration_dependent* (CSAM), and *task_oriented* (TSAM).

CSAMs are the most elementary multi-function independent modules that can exist in the system. Their design – hence their functionality – is dictated by hardware and environmental factors within the assembly domain (e.g. the robot, the assembly cell, the position and type of sensors, etc.). Naturally because of hardware dependencies, these agents may have limited capabilities. For example, a *grasp* module would be limited in the size of objects that it can handle depending on the design of the gripper. Similarly a *position_robot* agent for a robot with less than six degrees of freedom would be restricted.

The function of TSAMs is to translate the task requirements as expressed by the operator, or planning system, into viable robot and part motions that would ensure the correct completion of the assembly. They achieve this functionality by interacting with one another and by activating CSAMs as they are needed. In this way the *task_oriented* modules provide an interface between the high-level task requirements, and the low-level hardware-dependent modules of the system.

3.1 THE ARCHITECTURAL DESIGN

Starting with the CSAMs, the design of these agents is achieved by combining elements from four basic units.[1] The first unit is the *sensory_unit*, and it incorporates the sensors of the system. Initially there would be 5 different types of sensors in the unit; tactile, force, proximity, light sensitive sensors (photocells), and a vision system. The second unit is the unit of *primitives* and this contains rudimentary robot control functions such as *open_gripper*, *rotate_wrist*, *approach*. At present there are twenty such primitives which enable total control over the robot. The third unit is the *reactive_unit* and it contains low-level actions that are triggered automatically by the sensors. The degree of functionality of the elements in this unit may be influenced externally (this is explained later). The fourth unit is the *information_unit*. Upon initialisation of a process, this unit supplies the process with the necessary precompiled knowledge needed to assist its functionality (i.e. changes in the assembly cell, other existing processes, hardware alterations, etc.). By design, each process is made aware – through the information unit – of all other processes designed to complement its functions. This property is of unique importance to the architecture since it allows processes to be attached to or detached from a particular

---

[1] By units we mean a collection of processes or actions which have similar characteristics. These are the elements of the unit.

agent with the minimum of effort.

As an example of a CSAM agent consider the *grasp* competence. To construct it, a pressure sensor and an array of photocells (or simply a light beam) may be used, together with the primitives *rotate_wrist*, *open_gripper*, *close_gripper*, and the reactive element of *change_force*. The latter ensures that a tighter grip can be obtained if a component slips in the fingers, or that the grip can be eased. Another useful reactive element in this situation is *evaluate_slip*. This obtains the amount by which a component has changed position in the fingers (e.g. using the photocells or other sensors [13]). This information may be utilised by higher level competences if needed. The given example makes explicit the fact that one sensor (photocells) may activate more than one action (increase_force, evaluate_slip).

The behaviour of *primitive* and *reactive* processes can be controlled by information passed to the SAM upon activation of the module. In fact, processes within units can be enabled or disabled and can be requested to limit their outputs[2] (e.g. exert a limited amount of force, open gripper a specified amount). This enables higher-level processes to influence the activities of other processes according to knowledge, sensory information, and expectation. However, once such a request has been issued, the system is left free to carry out its instructions and outputs of lower-level competences cannot be suppressed in any way.

At the lowest level of control, *primitives* can be driven by elements in the *sensory_unit*. This allows actions to be attached to sensors. In this way, and as far as higher-level processes are concerned, a similar functionality can be obtained even if the sensor (or its position) changes. For example, the process of using vision to align the gripper in a part acquisition procedure, differs from that of using a contact sensor. However, if both processes result in correct alignment, then as far as the *grasp* module is concerned, the process of alignment was successful. Thus, having the alignment process attached to the corresponding element in the *sensory_unit* has the advantage of being able to replace sensors and still maintain the functionality of the system. This of course implies that each sensor invokes a different alignment procedure, but the cost of this is minimal when compared to the cost of redesigning a whole *grasp* module.

Basic TSAMs are constructed using CSAMs and two units: the *information_unit* (already examined), and the *reasoning_unit*. The elements within the *reasoning_unit* are meant to provide some means of analysing situations not easily resolved at lower-levels. To do so, these elements need to have some knowledge regarding the assembly system and its limitations (provided by the information unit) and they need to be able to reorganise and redirect the activities of other agents. For example, in order to make the most efficient use of a robot's working envelope whilst avoiding part motions which may violate the limits, some evaluation of the motion must be made on-line. A reasoning process (which may invoke a variety of sensors) may be triggered by a

---

[2]This feature is not valid for the *sensory_unit*.

control error code (e.g. cannot reach position), or through a sensor (e.g. detecting an obstacle). The reasoning power of the system can be increased progressively through the utilisation of additional elements and sensor capabilities (e.g. widen camera view to find part) as required.

As an example of a TSAM, consider the case where a part (e.g. a bolt) is acquired by the robot. The parameters passed on to the module by the off-line system would indicate the way the part must be held to be used in the assembly. Once this – and any additional information – is known, the TSAM must invoke the following activities: a) locate the part, b) determine its orientation c) evaluate how the part may be reoriented (if needed), d) activate CSAMs to position the robot and grasp the part.

Activity (a) could be achieved through another TSAM such as *locate_part*. This agent would use partial knowledge of the expected position of the part, drive the camera unit, use templates to match visual data, and interpret the result. Activities (b) and (c) require additional reasoning power (and incremental use of sensing capability) in order to evaluate additional regrasping actions.

3.2 THE COMMUNICATIONS SYSTEM

It is probably true to say that one of the most difficult problems in the presented architecture is that of getting processes to communicate with one another. However, it is believed that the communication protocol presented below will be adequate for the demands of our system.

Starting with the lower-level, when a System Action Module is activated, it also activates all processes which are immediately required and are part of the module. Each process is assigned to a processing unit (ideally on a one-to-one basis). One processing unit within each module is dedicated to managing communications. This operates on a first-come first-served principle, and keeps a record of the transmitting and receiving processes and signals. As each process, unit, and module are indexed, it is easy for the system to know the origin and destination of each signal. Communications are not restricted to simple "on/off" signals. Once a process "attracts the attention" of another process, then any type of information can be exchanged. The protocol between the two processes is determined by the information unit and is passed on to any process as part of its initialisation procedure. To cut down on unecessary "chattering" over the network, replies are provided only when requested.

There are two ways of activating a process: as a continuous process or as a temporary process. Continuous processes remain active until the module is deactivated. These are normally fast control loops and sensor monitoring processes. Temporary processes are engaged by other processes and are "attached" to the module of the parent process. The communications manager of the module allocates the process to a processing unit. Temporary processes perform a task, communicate the result, and then "die". In this way processing units can be reassigned. The communications manager is made aware of the free processing unit, and this is reassigned if and when is needed.

Higher-level competences operate in a similar fashion. Because higher-level modules can invoke more than one lower-level competence at the same time, it is possible for two different competences to share the same facilities (e.g. sensors, reasoning, units, etc.) for entirely different purposes. There are of course restrictions in order to avoid conflict (e.g. attempting to use a camera to monitor two different areas within the cell at the same time). For this reason *sensory* and *primitive* units cannot be accessed by more than one process at a time.

4 CONCLUSIONS

It is believed that the presented architecture provides an easier and more flexible way of programming and controlling robotic assembly processes, by virtue of the fact that the user need not be concerned with the underlying functionality of individual competences. As each agent is designed to be an autonomous and robust unit – within its design limitations – the user need only assess whether a given competence fulfills the requirements of the intended task. The methodology provides the feature of constructing functional systems for various domains out of relatively simple blocks. This is a particular important property if the architecture is to be utilised in various robotic applications. In addition it gives the user the opportunity to add new competences or modify existing ones and thus customise the system. It can therefore accommodate small or large, complex or simple, general or specialised domains.

The architecture permits activation or inhibition of individual processes according to requirements. This means that malfunctioning or redundant processes or sensors can be disabled effortlessly. At the same time, in a lot of cases, addition of new processes to the system can be achieved by the mere updating of the information unit and the minimum of reprogramming.

Our approach, like others, attempts to avoid explicit representation of models on the on-line system. However, it also recognises the need for some kind of in-built knowledge and representation and ensures that this can be accommodated as required (e.g. use of templates for vision system, knowledge of general location of parts). It is also worth noting that this approach does not attempt to implement a one-to-one relationship between modules and actions. It was felt that in an assembly domain this is an impractical requirement. We therefore opt for a situation that incorporates parallel processing to accommodate speed of reaction when this is needed, whilst permitting the system the luxury of some cognitive process when this is affordable or essential. More importantly however, it allows the user of the system to determine the percentages in this amalgamation.

An assembly cell is currently being constructed to implement the presented approach. It will be used to control two RTX-UMI robots and the aforementioned sensors. At the same time, the control system of the robots is being upgraded through the addition of sensors and additional functions. We hope to report soon on our first experimental results.

5 BIBLIOGRAPHY

[1] Arkin, R.C, "Motor schema based navigation for a mobile robot: An Approach to programming by behavior", Int. Con. on Robotics & Automation, IEEE Proc., 1987, pp 264-271.

[2] Ayache, N, and Faugeras, O.D, "Building a consistent 3D representation of a mobile robot's environment by combining multiple stereo views", Proc. IJCAI-87, pp 808-810, 1987.

[3] Brooks, R.A, "Achieving Artificial Intelligence through Building Robots", AI Memo 899, AI Lab., MIT, 1986.

[4] Brooks, R.A, and Connell, J.H, "Asynchronous distributed control system for a mobile robot", SPIE's Cambridge Symp. On Optical and Optoelectronic Eng., 26-31, Oct., 1986, Hyatt Regency Cambridge, Cambridge, MA.

[5] Connell, J.H, "A behavior-based arm controller", IEEE Trans. on Robotics and Automation, vol 5, No 6, Dec. 1989.

[6] Flemming, A, " Analysis of Uncertainties and Geometric Tolerances in Assembly of Parts", PhD Thesis, University of Edinburgh, Scotland, 1987.

[7] Latombe, J-C, "Automatic Synthesis of Robot Programs from CAD Specification", Robotics and Artificial Intelligence, eds Brady et al., NATO ASI Series F vol 11, 1984, 200-219.

[8] Lemmer, J.F, and Kanal, L.N, (eds), Uncertainty in Artificial Intelligence 2, Elsevier Science, Science Publishers B.V. (North-Holland), 1988.

[9] Lozano-Perez, T, "A simple motion planning algorithm for general manipulators", MIT AI Lab. Tech. Rep. AIM-896, Cambridge, MA, 1986.

[10] Malcolm, C, and Smithers T, "Symbol Grounding via a Hybrid Architecture in an Autonomous Assembly System", Proc. of Cost-13 Workshop, Representation and Learning in an Autonomous Agent, Nov 16-18, 1988, Lagos, Portugal. Also Edinburgh University DAI Res. Paper 420.

[11] Petropoulakis, L, and Malcolm, C, "Programming Autonomous Assembly Agents: Functionality and Robustness", Journal of Mechatronic Syst. Engin., Vol 1, pp 107-113, 1990. Also Edinburgh University DAI Res. Paper 468.

[12] Requicha A.A.G., and Tilove, R.B, "Toward a theory of geometric tolerancing",

Int. Journal of Robotic Research, Vol 2, No 4, 1983.

[13] Tomovic, R, and Stojiljkovic, Z, "Multifunctional Terminal Device with Adaptive Grasping Force", Automatica, Vol 1, No 6, elmsford, N. Y.: Pergamon Press, 1975 pp. 567-570.

[14] Yamauchi, B, "Real-Time Sensorimotor Control Using Independent Agents", Image Understanding and Machine Vision, Optical Society of America Technical Digest, June 1989.

Development of Intelligent Control for Robot Cells using Knowledge Based Simulation.

Zoe Doulgeri
Dept. of Electrical Eng., University of Thessaloniki, Thessaloniki, Greece
Giuseppe D'Alessandro
Dept. of Robotics and Automation, Tecnopolis CSATA, Valenzano/Bari, Italy

ABSTRACT

This paper describes the development and use of a knowledge based system for the simulation and control of robot cells. The system has been developed as an extension of the KEE-SIMKIT knowledge based simulation toolkit to allow the reuse of simulation modules for the development of system configuration and actual robot supervisory control policy. The operation of the system has been applied to the configuration and control problem of a real manufacturing and assembly robot cell with a given basic structure. Control strategies include the development of an expert robot scheduler.

1. INTRODUCTION

The widely recognised advantages of the use of automated systems like FMS, robot cells etc, are potential benefits which are only realised when the system is carefully designed and effectively controlled. An approach used extensively in every application phase of such systems, from design to implementation and operation, is discrete-event simulation. Discrete-event simulation is the only tool which can capture the necessary detail and complexity of such systems and it was therefore successfully used to predict system performance and evaluate different configuration and control alternatives. However, use of simulation tools have been limited by a number of characteristics. First, there is a need to employ a simulation expert to model, program and analyse and interpret simulation output and who has preferably some knowledge of the domain area. Second, there is a lack of necessary analytical tools to develop optimal solutions. Third, there is a weakness in handling qualitative factors. Last, the translation of the operating logic contained within the simulation to the actual control policy is not straightforward. In pursuing solutions to these problems, efforts are concentrated in developing more flexilbe and easy to use simulation tools. Knowledge Based Simulation systems (KBS) appear to be the most promising.
This paper presents work on the development of a KBS tool for the configuration and control problem of robot cells. The tool is intended for use by people of little simulation expertise. Thus, it is developed to allow first, rapid modelling and second development of decision logic in a way that it can easily be translated for use by the real system controller. The tool was developed as an extension of the KEE-SIMKIT toolkit and its operation is illustrated by describing its application to the design and control problem of a real manufacturing and assembly robot cell for the production of various types of a machine tool component.
The rest of the paper is organised as follows. Part 2 discusses knowledge based simulation. Part 3 describes the major features of the Kee-Simkit toolkit. Part 4 introduces

the knowledge-based system for the simulation and control of robot cells through a description of the extensions made to provide support for simulating robot activities. The remaining parts detail the initial application of the tool to the scheduling of robot tasks for a real manufacturing and assembly robot cell. The structure of the cell model developed is discussed and the ease of implementing different configuration and scheduling strategies is demonstrated. Scheduling strategies include the development of an expert robot scheduler which is intended for the actual implementation.

2. KNOWLEDGE-BASED SIMULATION SYSTEMS

Discrete-event simulation is a tool which has been extensively and succesfully used as a test bed to mimic the detailed operation of complex systems like FMS under different design alternatives and to test different operational decisions without disturbing the actual system [1].

The requirement for more flexible and easy to use simulation tools led researchers in concentrating their efforts in merging the technologies of simulation and artificial intelligence (AI) [2, 3, 4]. The emerged knowledge-based simulation systems incorporate the powerful representational techniques of AI research in modelling and simulation software. These systems are mostly built on top of hybrid AI development environments which typically support a range of knowledge representation mechanisms [5, 6, 7].Thus, economy of expression is encouraged and complex decisions are defined in the most appropriate manner. A lot of these systems utilise the object-oriented programming paradigm which promotes modularity and clarity of models [8]. By combining reasoning and simulation facilities such systems may not only provide assistance to the user for model construction, experimental design and result interpretation but they also enable him to attack an entire new class of complex problems.

3. THE KNOWLEDGE-BASED **SIMKIT** SIMULATION TOOLKIT

SIMKIT is an integrated set of general purpose simulation and modelling tools built on top of KEE, an AI software development system which provides tools for the construction of expert systems many of which are explicitly designed to facilitate modelling of domain knowledge. Thus, SIMKIT takes full advantage of the expressive representation, reasoning and interface tools that KEE provides.

KEE is a frame based system implemented in LISP, supporting object oriented programming, inheritance of two types (class/subclass and class/member) and rule-based reasoning, all integrated with high resolution graphics.

Object-oriented programming involves the definition of objects which store both data and procedures and which interact with one another by passing messages. Typically, an object class definition will consist of a set of slots storing attributes of the object and a set of methods defining how the object behaves on receipt of messages. Such behavior would usually modify slots or send further messages.

SIMKIT has extended KEE to provide first, general support for the rapid construction of specialised knowledge bases (KB) called libraries and models and their underlying KB representation through different editors and verifyers and second, extensions specifically designed to support discrete-event simulation, that is the representation of time dependent behavior and stochastic phenomena and the collection and analysis of data generated during simulation runs. Specifically, these extensions include a clock to represent the passage of time, a calendar to store future events, a simulator for the initialisation and running of a model, a library of random number generators and a library of data collectors [5].

Libraries constructed with SIMKIT will contain class level units representing object types, instances of which can be used in models to represent particular objects, and relation units to describe the types of relationships that can exist between objects in a model. An example of a library built with SIMKIT is QLIB a library dealing with general queueing problems which contains object classes of permanent components like servers and queues, transitory items which flow through servers, and the downstream relationship between components which indicates that one component is downstream of another [5]. The behavior of objects may be modelled using LISP methods within which rules can also be fired. Message passing to methods is scheduled for specific future simulated times. In QLIB for example, components comprise method slots for the basic events: generate.item, item.arrives, start.activity, complete.activity, item.departs, which define the changes in slots due to the occurence of these events and may schedule further events concerning the same or other related components for future simulated times. QLIB, in other words, takes the event scheduling simulation world view.

The KEE system's inheritance facilities enables propagation of attributes and behavior to all subclasses and instances of object classes when adding slot values and methods at the class level. SIMKIT libraries and models are KEE KBs and thus can be accessed, manipulated, inspected and edited using KEE facilities.

4. THE KNOWLEDGE-BASED SIMULATION SYSTEM FOR INTELLIGENT CONTROL OF ROBOT CELLS

The principal characteristic of a robot cell is the high dependancy on the robot of the flow of parts through the system's primary and secondary resources. Workstations may require robot service either for loading/unloading operations or for the execution of their main operation for example in assembly. Pallets and fixtures may also require robot service for their loading and unloading with parts. Thus, a great number of different configuration of robot activities exist in a physical robot cell and the complexity of interlinks has greatly complicated traditional simulation models of robot cells [9].

The objective of this work was to develop a knowledge based simulation tool which could be used for easy model construction of a robot cell, independent development of a control module for easy specification and modification of the decision elements, simulation of the cell's behavior and generation of the actual control policy for implementation on the robot cell controller.

<u>Model construction support- The RCL library</u>

A library called the Robot Cell Library (RCL) was built on top of KEE-SIMKIT which will ideally comprise all elemental model objects of the components of a robot cell. The designer would only then have to extract these elemental modules from the library in order to build the model of a specific cell.

The RCL keeps the ideas introduced in QLIB on the distinction of object classes to permanent components (servers and queues) and transitory items which are pushed downstream from server to server but extends these ideas to include robot served components and robot activities.

Robot served components comprise LISP methods which generate requests for the appropriate robot activity to enable part flow to the subsequent components. Requests belong to the class of transitory items which are deleted as soon as they are satisfied. Transitory items also include fixtures and pallets which are recycled in the system and parts, batches and assemblies which are removed from the system at the end of their processing.

The robot is modelled by a permanent component related to a set of robot activities. Robot activities are modelled by an object modelling the robot service which is associated with a queue, where generated requests are stored temporarily when the robot is busy and thus unable to satisfy them. The robot keeps track of the unsatisfied requests at an attribute slot. Thus, in a model constructed with RCL, robot activities are explicitly modelled as opposed to other activities which are carried out in servers. In this way modularity together with clarity and understanding of the model's operation is enhanced. In addition, the problem of controlling the robot can be well defined since the list of unsatisfied requests for various robot services represent the alternatives open to the robot. The robot controller should decide which request to satisfy based on some control policy.

Permanent components in RCL include batch and assembly stations. Batch stations model the collection of a number of items (eg. palletising station) which are intended to be processed as a batch for a part of their downstream path and are subsequently separated. In an assembly station assembled parts are deleted to generate a new item which will then proceed further down the manufacturing path.

The Robot Cell Control module

In general, the form of the Control Module assumes an interlinked hierarchy of rulesets which are used to choose which requests of those waiting is to be satisfied by the robot. The control module is defined in the library level so that reasoning is carried out by the rulesets within the system's KB. Thus, decisions are made on the basis of the global system state taking into account a wide range of factors. Implementation of alternative strategies is achieved by simply redefining the rulesets.

The decision making as to the behavior of the robot is carried out by the rulesets of the control module and is achieved by determining the *preferred.request* from those waiting. The times the preferred request must be specified, i.e the times of the firing of the rulesets, are the times the robot becomes available.

5. THE MANUFACTURING AND ASSEMBLY ROBOT CELL

The System

Figure 1 shows the basic cell structure. The cell consists of the manufacturing area and the assembly line. A track connects the two areas on which the robot is allowed to move to serve both areas. The cell is able to manufacture two kind of components, B and L which are then joined into preassemblies consisting of one B and two L. Two of these preassemblies are assembled to produce the intermediate component M. Ten M components are joined to produce the final product C.

Pallets holding raw and manufactured components are stored and circulating in a pallet ring connecting the manufacturing and assembly area. Each pallet holds both B (a batch of 8 items) and L (a batch of 16 items) components. Apart from the pallet ring, hardware elements of the manufacturing area comprise: machine tools associated with an automatic pallet changer (APC), working on either a B or an L batch of componets loaded on a dedicated fixture, a washing station, two hardening stations for each kind of component associated with a hardening table and a load/unload station of one fixture capacity.

The robot is responsible for the direct and indirect feeding of all the workstations in the manufacturing area. It feeds the machine tools and the washing station through the load/unload station by loading raw components from a pallet to an empty fixture which will then proceed to the machine tools through the APC and by unloading a fixture of manufactured components directly to the washing station. It also feeds the hardening stations through the hardening table on which it unloads the washed components. The robot is also responsible for loading hardened components from the table to the pallet to enable the transfer of the pallet to the assembly area.

Hardware elements of the assembly area comprise the robot served area of the assembly line consisting of a vertical and horizontal press and the assembly stations further down the line on which secondary assembly operations are performed to the final product C transfered by a conveyor belt. The robot assembles components to produce the M intermediate component with the help of the two presses.

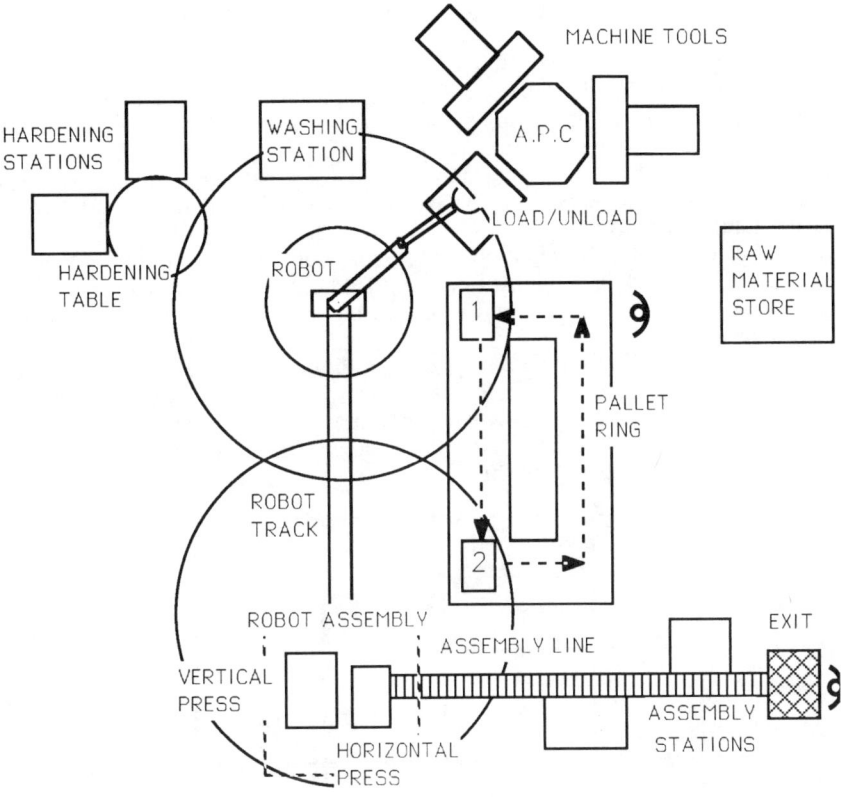

Figure 1: A schematic representation of the manufacturing and assembly robot cell

The Cell Model

In creating object definitions for all the major entities comprising the physical cell, multiple inheritance was used where possible to record similarities between objects.

Thus, the *load_unload_station* and the *washing_station* are robot served stations whilst machine tools, hardening stations and secondary assembly stations are simple server type objects. The *hardening_table_buffer* models the part of the hardening table which holds the hardened parts and it is a robot served queue whereas the *washed_table_buffer* is a simple queue type object.

The critical ring position 1 is modelled by *pallet_stand* which is a batch station whereas ring position 2 is modelled by *assembly_stand* which is a robot served queue (figure1). A server models the movement of the pallet from the one position to the other (*ring_station*).

Requests are issued from the robot served components to the following robot activities: *pallet_to_fixture* and *fixture_to_wash* from the *load_unload_station*, *wash_to_table* from the *washing_station*, *table_to_pallet* from the *hardened_table_buffer* and *robot_assembly* from the *assembly_stand*.

As the RCL does not provide directly for robot movements, the robot track movement was modelled indireclty. Specifically, each robot activity was provided with a *location_required* slot which may be either the manufacturing or the assembly end of the robot track. In order to start a robot activity, a robot move to the required location might be necessary.

The Configuration Problem

Since the basic structure is already determined, the design task for this particular cell is largely a configuration problem determining the quantities of the standard units required to meet a particular set of production requirements.

The estimated work load of the cell's workstations is highly unbalanced, the greatest load put on the machine tools. Thus, to examine the system operation for increasing production requirements the number of machine tools was increased from one to three. The system operation was also examined for the assembly line running for two and one shifts for every three manufacturing shifts. This operational requirement stems from the system's anticipated ability to work unmanned in the manufacturing area whereas the assembly line requires the presence of operators.

The necessary changes involving the above system configurations were easily performed with the developed software tool and table 1 summarises some results concerning the system performance for 1- 3 machine tools where the machine and robot utilisations, the daily production output and the required fixture and pallet quantities are reported for the case of the assembly line running for one shift.

no. of M.T	Utilisations		daily production	fixtures	pallets
	machine	robot			
1	99.4	23	3.6	2	7
2	M1 99.5 M2 98.7	48	7.6	4	13
3	M1 95.8 M2 98.7 M3 96.0	72	11.6	4	20

Table 1: System performance for different configurations

The Control Problem

This section demonstrates the ease of implementing scheduling strategies for robot tasks by describing the rule set definitions being developed for the manufacturing and assembly robot cell. When the robot becomes available a selection from the list of unsatisfied requests for robot activities which are issued by the robot served components can be made according to:

1) some heuristic dispatching rule based for example on either the request generation time FIFO, LIFO etc, or on the request processing time SPT,LPT etc, or on the request type or last on a combination of these,
2) some expert robot scheduler

The term expert is applied to the robot scheduler in this paper to mean that the logic by which decisions are made is too complex to be described coherently by a single heuristic rule. An expert scheduler bases its decision on a wider range of information describing the system state and in this way tries to outperform dispatching rules [10, 11, 12].

It is envisaged that a fully developed expert robot scheduler for the manufacturing and assembly robot cell would operate in two levels:
1) A higher decision level which, by observing system behavior, will determine the values of parameters used by the lower level
2) the lower level decision logic which consists of a range of heuristic dispatching rules implemented by rule sets which are used to choose the *preferred_request* to give to the robot as a command.
Decisions are solely determined by the state of the cell. They are therefore independent of whether this state is derived by the simulation model or by communication with the real devices comprising the cell.
According to the lower level decision logic requests are initially prioritised according to the robot activity type they are associated with. Within types requests are answered according to a simple priority rule. Each request is subsequently associated with an urgency level which is determined by examining the upstream and downstream components of the cell. In other words, there is a low urgency for satisfying a request in the current time if there are not any items pushing to proceed to the robot served component which has issued the request under consideration and/or the downstream component is at the moment not starving. The examination of the urgency conditions involves comparison with limit values which express quantitavely the pushing and starving terms. Priorities among types, within types and limit values are determined by the higher decision level.

An initial implementation of this logic favors requests concerning services to the machine tools in an attempt to facilitate item flow from what it was proved to be the cell's bottleneck. Rules are weighted according to the above objective giving the greatest weight to requests concerning the *fixture_to_wash* and *pallet_to_fixture* activities. Within type, requests are answered in a fifo order. The initial type priority might not be respected if urgency levels favor another request type.
Results of simulation runs for a simple priority rule and the above lower level heuristic rules showed that the expert heuristic rule achieves better results particularly when the system consists of more than one machine tools and it is running for one assembly shift every three manufacturing shifts. Table 2 summarises performance results of the two machine tool system for a period of five working days.

Control	Utilisation		Robot	Total production
	M1	M2		
priority rule	94.7	95.1	46.1	38
expert decision logic	99.2	99.1	49.2	40

Table 2: System performance in five working days for two control alternatives

Three shifts per day are assumed for the manufacturing area and one shift per day for the assembly area. A fifo priority rule is used. Improved performance under the expert decision logic is attributed to the fact that the robot in this case can more intelligently allocate its time among the various requests.

The higher-level may act by changing request type priority if under some other load or operational conditions another workstation is idenified as the bottleneck. Limit values may also change if excessive delays and/or workstation starving is observed.

5. CONCLUSIONS AND FUTURE WORK

This paper has outlined the operation of a knowledged based simulation and control tool for robot cells which may be applied to the design and control problem of such systems. The modularity and flexibility of the overall approach has been demonstrated by showing the implementation of the tool to the configuration and control problem of an actual manufacturing and assembly robot cell for the production of machine tool components.

Future work on the application of the tool will involve the full development and trial of the expert robot scheduler in a intermediate generality level so that it may be implemented to other robot cells with the least of modifications.

REFERENCES

1. P. J. O' Grady and U. Menon, A concise review of Flexible Manufacturing Systems and FMS literature, *Computers in Industry*, Vol.7, 155-167, 1986
2. M. S. Elzas, Relations between Artificial Intelligence Environments and modelling and simulation support systems, in *Modelling and Simulation Methodology in the Artificial Intelligence Era,* Ed. M. S. Elzas, T. I. Oren and B. P. Zeigler, Elsevier Science Publishers B. V., North-Holland, 1986
3. R. M. O' Keefe and J. W. Roach, Artificial Intelligence approaches to simulation, *Journal of the Operational Research Society*, Vol. 38, 713-722, 1987
4. R. E. Shannon, Knowledge-based Simulation techniques for manufacturing, *International Journal of Production Research*, Vol. 26, 953- 973, 1988
5. D. Silverman, M. Stelzner, SIMKIT Knwoledge-based Simulation Tools, in *Knowledge Systems' Paradigms*, Ed. M. S. Elzas, T.I. Orin and B. P. Zeigler, Elsevier Science Publishers B. V., North-Holland, 1989
6. Y. V. R. Reddy, M. S. Fox, N. H. Husain and M. Mc Roberts, The knowledge-based simulation systems, *IEEE Software*, Vol. 3, 26-37, 1986
7. D. J. Mc Arthur, P. Klahr and S. Norain, ROSS: an object oriented language for constructing simulation, in *Expert Systems Techniques, Tools and Applications,* Ed P. F. Kahr and D. Waterman, Addison-Wesley, Mass. 70-94, 1986
8. B. J. Cox, Object-oriented Programming: An Evolutionary Approach, Addison-Wesley, 1986
9. B. H. Claybourn, Scheduling robots in flexible manufacturing cells, *CME,* Vol.30, No. 5, 36-40, 1983
10. D. Ben-Arieh and C. l. Moodie, Knowledge-based Routing and Sequencing for Discrete Part Production, *Journal of Manufacturing Systems*, Vol. 6 No. 4, 287-297, 1987
11. P. Rogers, D. J. Williams, P. S. Wesley and J. N. Clare, On-line scheduling of machining cells using knowledge-based simulation, *Proc. 4th International Conference Simulation in Manufacturing,* 151-163, 1988
12. S-Y. D. Yu and R. A. Wysk, Mulri-pass Expert Control System_ A control/Scheduling Structure for Flexible Manufacturing Cells, *Journal of Manufacturing Systems,*Vol.7, No. 2, 107-120, 1988

RECENT ADVANCES IN ROBOT GRINDING

A. IKONOMOPOULOS, L. DRITSAS
ZENON SA
Egialias 48, 151 25 Athens

ABSTRACT
Robot finishing is a very challenging application because of the related numerous unsolved technical problems. Among these, finishing process control and robot programming are the most critical.
ESPRIT Project 2640 was launched two and a half years ago with main objective to develop a new robot grinding cell based on advanced solutions to the above problems. In addition, the development and integration of a vision based quality inspection station has traced a completely new approach to robot finishing. The achievements of the project with respect to robot programming, process modelling and quality inspection will be presented in this paper.

1. Introduction
1.1 Fettling and Deburring
The concepts of Fettlig, Deburring, Polishing and Buffing are frequently encountered in the robot application lieterature. It is important to distinguish between these two types of finishing processes. The casting-related finishing is typically referred to as "fettling" or "cleaning of castings". Fettling (i.e. cutting through thick sections of cast metals) is distinguished by the large forces occuring (up to 400N), the powerful tools needed (45 hp) and the requirements for large rigid robots. On the other hand, deburring (i.e. edge breaking, chamfering, grinding and brushing) is distinguished by smaller forces (4 ÷ 45 N) and tools (1/3 hp) and the requirement of smaller dexterons robots with quick response for sudden path change [7].

1.2 Polishing and Buffing
Polishing is an abrading operation employed for the removal or smoothing out of grinding lines, scratches, pits, mold marks, parting lines, tool marks, stretcher strains etc. Polishing follows a machining or a grinding operation and precedes buffing. The operation is usually done in successive stages from coarser to finer, and may be performed with a belt to which an abrasive is bonded. The belt rides on a contact wheel.
Polishing must be distinguished from grinding in that only the surface finish is altered with a minimum material removal. Both rigid and flexible grinding disks, as well as sanding belts are currently in use for robot guided grinding and polishing. Most applications involve the workpiece handling concept where the workpiece is held by the robot and carried to a stationary grinding or polishing unit. In many cases special appliances are constructed by users themselves for a particular applications. Some

equipment producers (such as Bosch, Hilti, Asea) offer specialized devices. Surface polishing is normally made with more compliant grinding disks or belts.

Buffing is a process for producing smooth, reflective, scratch free surface on workpieces by bringing them into contact with revolving cloth or sisal buffing wheels (buffs) charged with a suitable compound. The action of the wheels, together with that of the compound, either cuts or flows the metal, thus removing minor defects and imparting a smooth, lustrous finish.

1.3 Philosophies of Robot Grinding

At the present time, three approaches are used for robotic deburring. All three of them were attempted in the EP2640 namely:
1. the compliant approach
2. the fine-tuned robot
3. sensor-based-programming
4. force feedback

1. *The compliant approach* is used to accommodate robot inaccuracies and part variations. It relies on programming "tricks" and special design tool holders. Teach pendant programming on actual parts is used.

   Active Compliance (Force-Motion Relation) means that forces of motions occur directly as a result of providing energy to actuators (necessary for smooth metal removal in deburring tasks).

   Passive Compliance means that forces or motions occur directly as a result of the deformation of elastic bodies. The inputs of the actuators driving these bodies are not altered as a result of the interaction forces [10].

2. *The fine-tuned Robot*

   This approach is based on robot programming "tricks" [7, ch.7 and 9] and on highly accurate robots. In practice, because of part variations, the fine-tuned approach is not normally recommended, unless some well designed compliance exists either on the robot or on the tool. The main reasons ar that each time a maintenance is performed or an accident happens, all points in the programs must be retaught.

3. *Sensor-based-programming*

   Procedures using sensors in robot programming are summarized according to implementation requirements as follows:
   - Sensor-aided teach-in,
   - Sensor-aided playback,
   - Automatic sensor-based path generation, and
   - Sensor-aided off-line programming

   Each procedure is extensively discussed in [7].

4. *Force Feedback*

   Manipulation using force control is different from the usual movements of the robot arm. There are two basic tracks in pursuit of the issue of how a computer-controlled arm can use F/T sensor information to modify position [10]. The **logic branching feedback** method and the **continuous feedback** method.

- The essence of the logic branching is a set of discrete moves terminated by a discrete event, such as contact. It is primarily scalar in nature and consists of statements in a computer-controlled language:

MOVE in X-direction UNTIL FORCE EXCEEDS FMAX
TWIST about Z UNTIL TORQUE EXCEEDS TMAX
The current trend to create rule-based force control systems belongs to this track of thought.

- Continuous force control, on the other hand, is based on <u>multiaxis F/T information</u> combined with <u>multiaxis coordinated response</u>. It has been applied to assembly and edge-following tasks (e.g. deburring) and was based on a feedback matrix (with constant entries) that converted a sensed F/T into a "response-velocity" vector which was combined with the original velocity command.

Despite the diversity of approaches to the use of F/T sensors in the continuous feedback method of robot motions, we can categorize all of them into two classes: hybrid control and impedance control. A description of hybrid control can be found in [14].

Reference [10] presents application of the "adaptive-passive" compliance and the "active accommodation" approaches to assembly and contour tracking applications.

2. Modeling of the Grinding Process

The main objective of Process Modeling (P.M.) is to permit an easy, fast and accurate reprograming (possibly off-line) of the Robot Path for different taps. The main input variables entering the finishing process are the normal and tangential forces, the wheel speed, the cutting power, the feedrate, the grit size, the workpiece and backing material properties. The main output variables are the volumetric metal removal rate (MRR), the surface finishing, the location of the contact patch, the disc wear parameters and the temperature. Since it is difficult to measure directly any of the output parameters, some modeling is necessary.

A fairly extensive literature search showed that the team at Charles Stark Draper Laboratory (MIT) have done a significant progress in this direction and have already applied their (static and dynamic) modeling to real world Robot Grinding Systems.

The first and still one of the most significant (experimental) result was the one by Ivers [1] that related the Volumetric Metal Removal Rate (Q mm^3/S) (M.R.R.) with the Mechanical Power (P) delivered to the contact patch. The "linear" relationship

$$Q = C1 * P + C2 \tag{1}$$

C1, C2 = constants

was proved to be valid for a great range of Normal Contact Forces (**Fn**), Feed Rates (**Vf**), grit sizes and backing materials.

The power (P) delivered to the contact path is given by:

$$P = /Ft * Vc/ \tag{2}$$

where **Ft** is he tangential cutting force and **Vc** ("Cutting velocity") is the average velocity of the grits that are cutting, relative to the workpiece.

Denoting by **Vg** the grit velocity relative to the grinder (Vg = Ω*Rc where Rc is the average radius to the center of the contact patch and Ω the angular speed).

In general Vc = Vf + Vg
In the case of Vf<<Vg we can assume Vc = Vg and hence

$$P = /Ft * \Omega * Rc/ \tag{3}$$

For real experiments, Ft is measured via a force-sensor and Ω through a

tachometer. From (1) and (3)
Q = C1/Ft * Ω * Rc/ + Co   (4)

Kurfess [2] developed and verified a <u>dynamic model</u> for the grinding process. There the grinding system is again an electric grinder with a flexible abrasive disk, mounted on a (stiff) vertical milling machine with one degree of freedom: rotation about the pivot point.

The first order nonlinear differential equation for the angle of the grinding disk with respect to the horizontal (Θ), was numerically integrated for various grinding conditions. The simulation results compare well with the actual results (for typical mean cut depths of 0.1mm the simulator predicted the final height within a standard deviation of 0.01mm).

3. The EP2640 Project
3.1 Description of the Project's Goals

The project is performed by a multinational consortium composed by ZENON SA (Coordinator), FRAUNHOFER INSTITUTE (IPK), JOYCE LOEBL, ENOSA and EPFL and its goal is to develop a robot based system for the surface finishing of castings by application of the new developments in Information Technologies.

A robot based finishing cell has been designed and includes the following:
- 6 axis industrial robot,
- adaptable grinding tool system,
- transport system, specialized pallets and grippers,
- path programming system,
- process control system,
- inspection unit and
- cell management unit.

The main main ideas and developments on robot cell layout are presented below.

A graphical cell-layout representing the arrangement of functional components within a given space aids in the actual placement of the equipment in the prototype cell. The selection of the appropriate tool system and the right positioning are related to the workpiece accessibility and the ability to finish all selected workpiece surfaces.

3.2 Process Modeling

Based on the modeling concepts presented in section II, a power transducer for the three-phase AC grinding motor was installed, based on the assumption that within certain range of forces and velocities, the measured electrical power of the motor (Pm) can be "linearly" related to he contact-patch power (P).

P = a1 * Pm + a2   (5)

a1,a2 = constants

Experiments have been designed to verify the ranges within which equation (5) holds true.

Equation (5) models the fact that the electrical power delivered to the motor (input power) is converted into "noload" or "rotational" losses (friction, core losses, windage, bearing)-which is independent of the load, and "loading and heat" losses which depend on the load in a presumably linear way [8].

Assuming that the angular velocity Ω of he wheel is constant and independent of the normal force F_n, within the range of the typically applied forces for finishing, the previous process-model results can be used for the shaping of tap surfaces by "actively" correcting Fn.

The output of the power transducer is sampled with a sampling rate of 50 Hz and the values are saved is a text file for further processing.

A routine for the finishing of the flat surface of the tap was programmed the Hitachi robot. The programmed depth of cut is 1mm and the feed rate provided by the robot is either zero or very low (10 mm/s).

Another interesting set of experiments investigated the system response (in both time and frequency domains) of the grinding root and a hydraulic X-Y table carrying the grinding machine. The main result was the confirmation of the initial speculation that industrial robots and their controllers do not have dynamic response fast enough to meet the needs of closed-loop force control in terms of the economics of the system. Instead the use of the hydraulically actuated X-Y table for high speed correction motion has been adopted.

3.3 Robot Programming

The execution planning of the finishing process requires four steps:
- division of the surface into patches,
- tool selection,
- creation of tasks and
- tasks sequencing.

Two approaches for path generation have been developed: the first aims to automatically generate patches on the basis of tool characterization and part geometry while the second aims to provide the user with an interactive CAD tool for making surface divisions rely on the judgment and skill of an experienced user.

Once a surface patching and a tool selection have been completed, a robot program must be generated where individual subprograms are dedicated to particular patches. The robot motion in a subprogram is based on a series of robot positions and interpolation between them.

The procedure of specifying robotic movements is based on the interpolation between static positions. Positions definition is performed in the following manner. Axis systems (frames) are defined in both the environment and robot sets. Matching of these two axis systems (one directly on top of the other) results in a unique robot position (all 6 axis constrained). These positions, or individual contact cases, are then combined by interpolation (joint or linear) to form the robot motion needed.

The axis system belonging to the robot are situated on the surface of the workpiece. One axis of this frame points, in a direction normal to the surface. A second axis points along the surface tangent and is oriented to achieve the proper feed characteristics. The third axis is chosen in a way that the complete frame makes a right-handed coordinate system. A procedure to automatically place the origins of these frames on a path surface is based on the analyzed curvature. These frames are generated by first defining a curve on the workpiece surface. The curve is then divided into segments according to the curvature. The endpoints of these segments form the origins of the alignment frames. The criteria for segment division attempts to compute enough origins so that the maximum deviation between the curve and a linearly interpolated path between origins is relatively small. That is, the deviation is small enough to be within an acceptable polishing tolerance range or compensated by the compliance of the tool or a micro-manipulator system.

The finishing of different surface patches on the workpiece may involve a set of vastly different configurations of the robot. This may be the case even with adjacent patches. An acceptable task sequence minimizes the amount of effort needed to re-configure the robot between subprograms. One quality requirement for the finished workpiece, is smooth transitions on the surface without stop marks. Polishing tools characterized by sharp edges are more sensitive to inaccurate position than finishing tools with large surfaces. Thus, a strategy for defining sequences is based on curvature as well as on the robot kinematics related to patch accessibility.

Sensor-aided teach-in remains the most practical approach for this project. Using this approach, the robot programmer can assess CAD results and provide expert knowledge required to generate appropriate paths. He is also capable of fast reaction and hence immediate adaptation. The only remaining question is which sensors will be most useful for the programming tasks.

As mentioned previously, the finishing task requires contact between the tool surface. The section on manual programming practices describes how the programmer attempts to assess the contact force between workpiece and tool system by roughly gauging the amount of force needed to move the polishing belt past the workpiece. Force-Torque sensors have been repeatedly suggested as measuring systems in both the teach-in and finishing phases. Interest in EP 2640 was concentrated primarily on implementations based on 6 degree-of-freedom force torque sensors. Although the quality of measurements from such devices has greatly improved, such sensors lack the ability to identify the actual point of contact. Without this information it is impossible to make a valid interpretation of the measured forces and torques. The advantage of such sensors, however, is the ability to mount the sensor between the robot and the robot-held workpiece. Thus it is possible to also make force measurements during actual finishing.

A more realistic solution concentrates, however, only on the programming tasks. The use of sensors should give the robot programmer information that would not normally be available during normal programming. Using profiled wheels (convex and concave), it becomes imperative that the programmer has the ability to assess the quality of contact between the workpiece and the tool.

3.4 Optical Inspection

The inspection cell is designed to work in line with the grinding cell. The casting arrives on a pallet to the grinding cell which after grinding transfers the pallet to the inspection cell. Both grinding and inspection cells are controlled by a task manager who commands the cells and collects information from the process.

The inspection cell includes a PUMA 560 robot for handling the part, a vision system from Joyce Loebl for the inspection and the necessary mechanical infrastructure for handling and lighting. The robot will handle the tap and will present to the camera according to a path, planned in a way permitting to see all the points of the tap. The inspection window being 15x15 mm, the image resolution reaches 0.06 mm.

This work has been articulated around four main directions:
- image acquisition
- defect detection algorithms
- dedicated hardware architecture and
- inspection path planning

Image acquisition from metal surfaces presents some inconvenience because of the strong reflectance. The problem has been faced by designing a lighting unit permitting to acquire reliable pictures.

The defects detection problem was faced by thorough investigations along three directions, namely, mathematical morphology algorithms, adapting thresholding and edge detection algorithms.

In order to face the time constraints of the inspection, a special hardware architecture for parallel processing of inspection frames was conceived and developed.

The inspection path is designed and programmed manually. However, theoretical studies, are also made in parallel for developing an automated path planner.

The software under development will include the following functions:
- input model data in the form of surface patches from AUTOCAD
- calculate robot positions and orientations for inspection
- calculate the window sizes for visual analysis
- produce data in the form of robot paths which may then be imported into the robot simulation package (GRASP) for robot configuration checking.

4. Other Approaches

4.1 The Work at IPK

The work by Hsieh's team by in IPK [11], uses a sensor integrated robot system based on a laser scanner to acquire workpiece geometry and a feed path planning module to detect burrs and generate the feed paths automatically. A universal process model based on a frame concept with homogeneous transformation matrices has been applied for implementation on a microcomputer thus coupling robot control and sensorial devices.

4.2 The Active End Effector Approach

An active end effector was developed at McMaster University to improve precision in robotic deburring, and is described in [12]. The design objectives are obtained from a dynamic analysis of the combined system of robot arm, end effector and deburring process dynamics. The unit allows robot-independent positioning, over a range of ±15mm in two orthogonal directions with an accuracy of better than 0.01mm and a bandwidth of 20Hz. A combination of d.c. servo motors with linear ball screws achieves both high precision and a large mechanical advantage. A high quality chamfer is obtained by performing position adjustments normal to the part edge for constant cutting force. An extended discrete PID controller based on an ARMAX plant model is designed and simulated off-line prior to real-time implementation. In real-time force control tests, the active end effector system improves the precision of a PUMA 560 robot by an order of magnitude over the open-loop case.

4.3 Impedance Control Approach

Kazerooni [13] has developed a two DOF active end effector which surpasses the limitations of passive designs. The impedance in the normal direction is used to obtain a precise chamfer, while the impedance tangential to the edge is used to control the rate of material removal. With a passive design the impedances in the normal and tangential directions are coupled by the tool mass. With active impedance control the impedances in the normal and tangential directions can be selected independently for improved performance.

An X-Y servocontrolled table is used to move the rigidly mounted tool, while the part is held by the robot.

4.4 The Work at CSDL

The process modeling at Charles Stark Draper Laboratory was later integrated into a Robotic Grinding system adopting the "Tool - on - Robot" approach, started with the work of Tate [3] on closed Loop Force Control, continued with Todtenkopf's work [4] and culminated in Kurfess's and Brown's Ph.D thesis [5]. The implementation of force feedback is based on the capability of the PUMA 560 controller to modify robot trajectories in real time ("Path Control Interface").

Todtenkopf [4] investigated and verified the effectiveness of controlling surface shape by controlling power applied to the workpiece. Force Control has been chosen as the independent control variable and a structured light - vision provided an accurate measurement of weld bead volume in order to determine desired MRR's and feedback information for lateral control.

A simple planning scheme ("...maximum power where the cross sectional area is maximum and minimal power where it is minimum... until... the maximum deviation in area can be eliminated in one grinding pass") is used to generate the desired power trajectories which can be translated into nominal force trajectories (F_n) along the weld bead, basically using eq (3). This nominal force profile along with real time alterations, which reflect variations in the force - speed characteristic of the grinder, constitute the input to be tracked by the force controller.

References

1.to 5. D.E. Whitney et al, "Intelligent Robot Grinding System", CSDL Report R01999
6. Zajac, C.J. "Analysis and Implementation of Robotic Deburring Systems", Proceedings of Robots 13, Maryland, 1989 (ABB Robotics)
7. Gillespie L.K "Robotic Deburring Handbook " (SME, Dearborn, 1987)
8. Pherson G., "An Introduction to Electrical Machines and Transformers" (ch. 4) John Wiley and Sons, 1981
9. EP2640, 1st, 2nd, 3rd, 4th and 5th Interim reports
10. H.Van Brussel "Sensor Based Robots do Work", 21st ISIR, October 1990
11. G. Seliger, L-H. Hsieh "Sensor Aided Programming and Movement Adaptation for Robot-Guided Deburring of Castings" (to be published)
12. G.M. Bone, M.A. Elbestawi "Active End Effector Control of a Low Precision Robot in Deburring", Robot. & CIM, Vol.8,No.2, pp.87-96, 1991
13. H. Kazerooni, "Automated Robotic Deburring Using Impedance Control", IEEE Control Systems Magazine, Feb.1987, pp.21
14. J. Craig, "Introduction to Robotics, Mechanics and Control" Addision Welsey 1989

A low cost robot based integrated manufacturing system for the garment industry

I.Gibson, P.Bowden, P.M.Taylor, A.J.Wilkinson
Robotics Research Unit
University of Hull, UK

1. Introduction

As the rest of the world becomes more competitive so the garment manufacturing industry in the UK suffers from the effects of low cost imports. High levels of employment and high wage figures in the UK and a noisy, undesirable work environment results in the garment industry suffering from a high employment turnover. Companies in the UK are therefore being forced to consider either drops in overall output, shifts to higher cost/quality markets, or increases in the level of productivity through automation.

A number of systems are being set up to attempt to fully automate the garment manufacturing process [1,2]. At Hull University, the Robotics Research Unit, in collaboration with the UK manufacturer Corah under SERC, ACME funded projects, has been investigating problems associated with garment manufacture for around ten years [3,4]. Working in particular on knitted fabric, systems have been developed to make possible the integration of robots and sensors for garment manufacture. Findings from previous projects have aided the construction of a multiple robot system for the assembly of a complete garment [5]. At all times, the underlying philosophy behind these projects have required that attention to low costs of this system be maintained.

2. Garment assembly system

The garments made by this system are items of underwear in the form of mens and ladies briefs. The mens brief is the more complex design, requiring 9 separate sewing operations to complete (see figure 1). The ladies design is known as a single, rear concealed gusset structure, requiring 7 operations to complete. The garments are made from interlock knitted fabric, a highly flexible material commonly used in such products.

The system layout can be seen in figure 2. The garments are constructed from individual pieces of cut fabric, each style requiring 4 such pieces, a front piece, a back piece, and 2 gusset pieces. These pieces of fabric are taken form cut stacks and are the starting point for the automated garment assembly line. A finger mechanism, used in conjunction with a high pressure air jet separates the top piece from a fabric

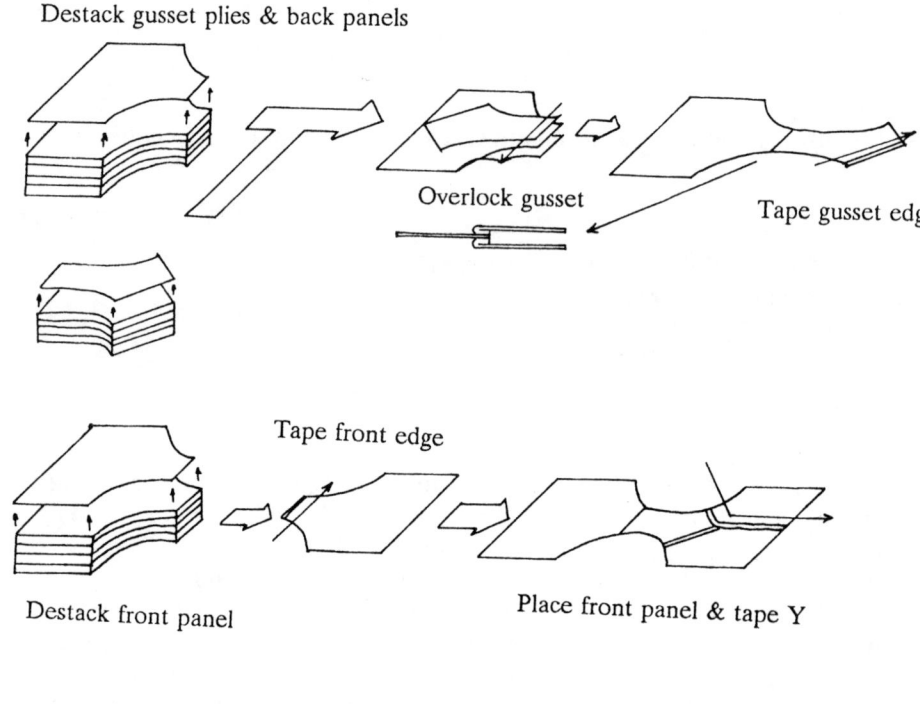
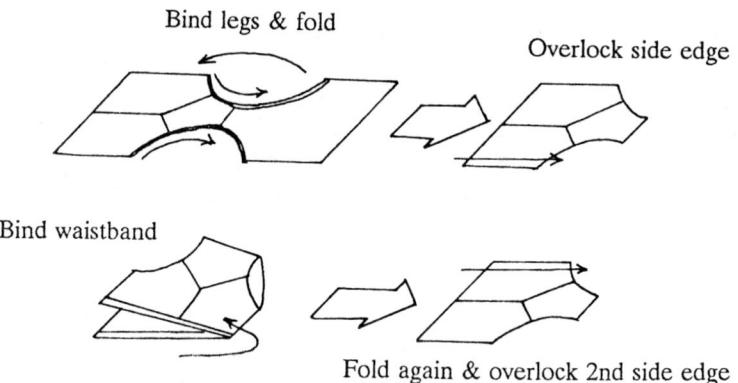

Figure 1 Construction of mens design

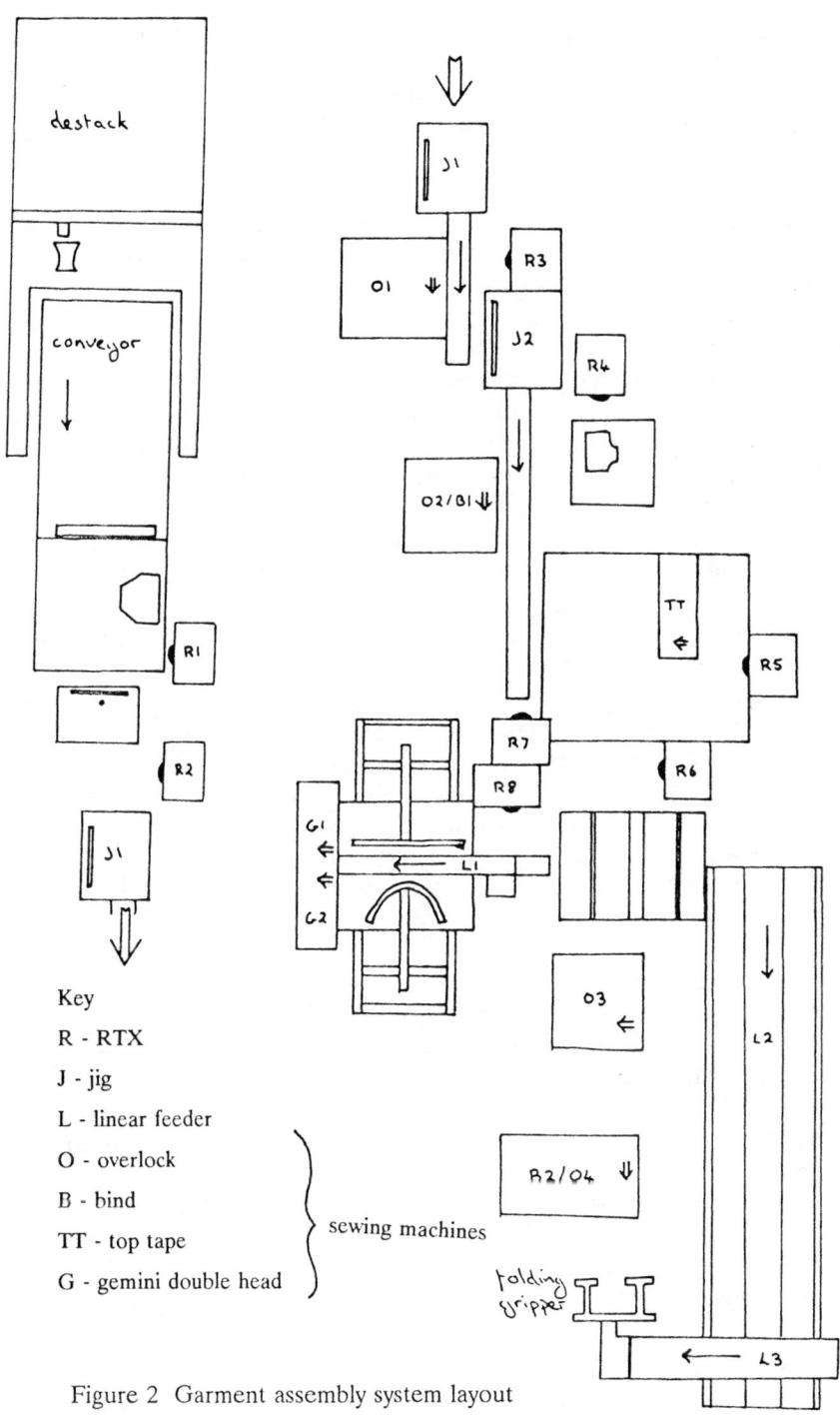

Figure 2 Garment assembly system layout

stack and places it on a conveyor [6]. All the pieces can be separated in this fashion and are collected from the conveyor by a robot prior to the first sewing operation. A robot mounted gripper is used in conjunction with a sensory vibrating table to align the 3 piece assembly (the rear panel sandwiched between 2 gusset pieces) along the edge to be sewn.

The second sewing operation is performed on the other side of this subassembly. By this time the ladies design will have had a front fabric panel added to it. The mens design requires the substructure to be added to the front panel after the second sewing operation. Prior to this the front panel will have had tape added to one edge. The mens design requires the extra operation of top taping to the front panel [7] to complete the classical shaped front.

The operations required to complete, or 'close', the garment are the same for both mens and ladies styles. Tape and/or elastic is sewn on to the leg arches of the open net structure. The garment is then folded to match the opposing edges of the structure to form side seams. Between the sewing of the first and last side seam, the structure is unfolded so that a waistband can be added.

Robots, equipped with appropriate grippers and sensory devices are used to handle between sewing stages. With one exception, linear feed mechanisms are used to guide the fabric through the sewing machines. Where concave curved external edges are to be sewn, the fabric can be straightened using a specially designed mechanism [8], utilising a property of knitted fabric to advantage. The exception is the top tape sewing of the mens design [7]. This operation requires at least 3 degrees of freedom to perform, and is therefore best suited to a robot. The final three closing operations are performed using a gripper attached to a 2-axis linear feed mechanism. This reduces the requirement for complex handling of a folded object between sewing stages. The operations can be performed in this fashion because no extra fabric pieces need to be added to the structure.

The entire system was constructed with specific attention to cost. There are a number of facets to the system that contribute to a low cost final solution. These facets are:-

    The use of low cost robotics
    The devolution of the assembly process into simple, individual robot tasks
    The ability of the robotic handling equipment to interface with existing sewing machine equipment
    The adoption of an assembly line principle, employing a JIT philosophy
    The design of a flexible system allowing for ease of reconfiguration

3. Low cost robots

The robot arm used throughout the system was a UMI RTX. In comparison with other robots this SCARA configuration has an impressive work envelope and compact design. A PC is used to control the actions of the RTX. The robot however is extremely slow (approx. 200mm/s point to point, 50 mm/s straight line interpolated movement) and inaccurate (approx. 0.5-1mm repeatability). This low performance is mirrored by the low cost (approx. 1/5th the cost of a comparatively sized Unimation Puma). The system uses 8 RTX robots for a variety of handling tasks. It would be difficult to visualize a working system based on only 2 or 3 Unimation Pumas. The inaccuracy of the RTX is not a problem in this particular case. The garments in question have overall dimensional tolerances of 5-10mm. The design of the garment and the construction techniques allow inaccuracies of 1mm to go unnoticed.

An improved controller for the RTX, designed and developed at Hull [9], will increase the robots velocity by up to 3 times with improved repeatability. This nonlinear control technique also provides smoother straight line motion. A replacement kit for the existing controller, which communicates with the PC via a parallel interface, will cost only a few hundred pounds.

The final completion stages of the system show an increasing use of linear feed mechanisms. This adoption of modular robotic principles can be applied to the earlier assembly stages once the development is complete and where less than 4 degrees of freedom are required. At present 4 RTX robots are used in this fashion. This would further reduce costs and increase speed due to the reduced control requirements.

4. Robot tasks and sensing

The large number of robots utilised in this system seems like an extravagant use of equipment. A better solution may be to reduce the number of robots and increase the complexity of handling operations. This however would be a false economy. Greater flexibility in gripping function could lead to greater cost, particularly with respect to turret or interchangeable grippers. Also, by using the same manipulator to perform multiple tasks then some degree of parallelism is lost and scheduling of operations to optimise throughput is more difficult. An ideal situation would be a sequence of operations that execute in the same time. Time would not be wasted waiting for a previous operation to complete. In order to maintain this parallelism the system at Hull was broken down into a sequence of operations [10]. Where possible each operation is executed by an individual robot. The more complex folding operations would require extra manipulation and sensing to resolve a complex fabric bundle. Such operations are not split up for this reason.

Wherever possible the sensing systems have been kept to a minimum. Vision sensors are in general out of the question in their traditional form. Low cost DRAM cameras [11] are used for calibration purposes, allowing robots to be placed in conjunction with other pieces of assembly equipment without the need for complex and costly jigging. Position sensing of fabric panels is performed using retro-reflective optical sensors. Since some knowledge of the shape and general position of the fabric is known, position servoing of robots can be performed. In some instances these sensors are fixed, usually in the bed of the jig platform but they have also been used in a gripper mounted form. In each case the sensors are primarily intended to minimize any knock on effects of process inaccuracies resulting from the cutting or sewing of fabric.

5. Interface with existing sewing equipment

Industrial sewing machines are much more complex and robust compared with their domestic counterparts. In the Hull system, the cost of a sewing machine is comparable to the anticipated cost of the robotic equipment used to serve it. The automated assembly system is designed to be highly flexible, dealing with variations in garment design that may require different sewing machines. For instance, although the sewing operation is the same, some designs may require elastic to be added whilst others require binding material. The most economical way of effecting these changeovers is to make use of existing sewing machines that can be utilised elsewhere in manual or batch production areas. There is a distinct advantage in having sewing equipment that can be used by manual operators as well as in conjunction with robotic equipment.

For full robotic integration, the sewing operation, feeding, trimming, and thread cutting must all be performed automatically. Industrial sewing machines and their controllers can interface easily with part detecting sensors, pneumatic thread trimmers, and stitch counting facilities supplied as standard. The feed rate can be monitored by tapping into the control loop at the optical shaft encoder, once again a standard feature on an industrial sewing machine. The Quick Rotan controller and motor was adopted throughout the system for convenience. This common control system has a 4 bit digital speed select which can be correlated with the linear feed controller. By using this system, a number of common makes and types of sewing machine are compatible with the robotic equipment.

No matter how reliable a sewing machine is, there will always be routine maintenance required with the need to react to thread breakages and needle wear. The flexibility of the system is such that problems of this nature can be dealt with in two ways. Firstly, the sewing machine requiring work could be physically removed from the system. A substitute machine could then be positioned in its place. There may be a number of instances where this is not possible or desirable, the time required to service the machine being significantly lengthened by the time spent repositioning. It is possible therefore to work on the machine whilst it is still in its designated position. To avoid the system operation coming to a complete halt whilst this work is being carried out, a machine operator can interface with the system. This is done by using a simple push button module which coordinates the completion of non-automated phases with the computer information and control system.

The possibility of equipment being moved around the system emphasises the importance in the use of the Dynamic RAM calibration cameras. The cameras cost around £100 and can be multiplexed to a single central processing unit. An LED barcode is attached to each piece of equipment, identifying it to the robot and providing alignment data for the robot coordinate frame.

6. Assembly line principle

As manufacturers become more competitive, adoption of the 'Just In Time' (JIT) philosophy is becoming more evident. In terms of the manufacturing principles, two factors are paramount in order to obtain a quick response, product flexibility and batch sizes.

An assembly system must be flexible enough to cope with changes in customer requirements. This can be achieved by either designing the system to handle a variety of materials and components or allowing for a rapid system reconfiguration.

As was mentioned in section 5, the Hull system was designed to cope with variations in part size without having to stop production. Variations in style that involve the use of different fabrics or machines may require some reconfiguration. Outside of the two garment types already mentioned, there is at present no flexibility to deal with others.

Provided the flexibility to cope with change is implemented, another key advantage of JIT is the reduction in batch sizes. The saving in storage and management of work in progress (WIP) is obvious. The ultimate aim is to minimize this WIP so that there are no batches at all and each assembly operation hands directly on to the next. The pioneering work on this was done on car manufacture but the principles are also illustrated by the Hull garment assembly system. Each operation, when completed, has a robot handling device that passes the partly finished garment on to the next operation. Other robotic devices are used to add further components, with no stacking or bundling required. The sensors, handling technology, and therefore the cycle times are all kept to a minimum. As has already been mentioned, cycle times comparable with manual techniques (240s for the Hull system, compared with 80s for manual assembly) have been achieved with modest robot and sewing speeds and with non-optimized sensory interaction.

It can be seen therefore that by using robotic operations combined to form a JIT system is a very economic way to make garments. For optimum use of resources (in this case sewing machine equipment) a combination of JIT and batch manufacture is usually appropriate. The batch system is used as a background task, making products that are less sensitive to changes in customer requirements. If robots are going to be used for either assembly line or batch manufacture then a highly flexible network and man machine interface, like the one used at Hull [12], is essential.

7. Overall system

There are two major areas that restrict the system performance. The first relates to the design criteria of the major product. Whilst it is safe to assume that the system can be reconfigured to cope with slight design variations like the gusset structure, it is not possible to assume that the system can deal with different products, like shirts. These would require radical system changes and attention to scale.

The second restriction is fabric type. Most of the assembly operations and handling sequences rely on the properties of knitted fabric in order to function. Woven and lace fabrics require different handling mechanisms that would not be adequate for knitted fabrics and vice versa. Similarly, assembly procedures would require different considerations (e.g. tailoring). However, it is unusual to find these particular underwear products not being made form single or double knit fabrics.

8. Conclusions

The Hull complete garment assembly system was designed with close attention to cost on a number of levels. System cost was kept to a minimum by using affordable technology in the form of low cost robotics. Care was also given to the avoidance of high cost sensing systems (e.g. vision systems) wherever possible. By using an assembly line configuration, parallel operations could be carried out, optimising machine usage and throughput. The cost of reconfigurating the system was also kept low by using sensory based calibration techniques and a non-hardware limited network for software based control.

The system was developed using cost effective techniques that could be applied to other product areas. Some of the solutions, like the use of low accuracy robots, are more specific to the problem of garment manufacture. Other solutions, like the use of the DRAM calibration devices and the devolution of complex handling tasks, can be applied to a number of different manufacturing problems.

References

[1] Aisaka N., "Technical development in apparel industry", JTN (Shirley Institute), vol. 390, pp 47-51, 1987.

[2] Gershon D., Porat I., "Vision servo control of a robotic sewing system", Proc. IEEE Int. Conf. on Robotics, 1988.

[3] Kemp D.R., Taylor P.M., Taylor G.E., "Adaptive sensory gripper for fabric handling", Proc. 4th IASTED symp. on Robotics and Automation, Amsterdam, Holland, 1984.

[4] Taylor P.M., Koudis S.G., "Automated handling of fabrics", Science Progress, Oxford, vol. 71, pp351-363.

[5] Gibson I., Taylor P.M., Wilkinson A.J., Palmer G.S., Gunner M.B., "Complete garment assembly using robots", NATO ASI Conf. on Expert Systems and Robotics, Corfu, Greece, July 1991.

[6] Gunner M.B., Taylor P.M., "Placing fabric onto moving surfaces", Proc. Int. Clothing Conf. on Textile Objective Measurement and Automation in Garment Manufacture, Bradford, July 1990.

[7] Gibson I., Taylor P.M., Wilkinson A.J., "Robotics applied to the top taping of mens Y-front briefs", Proc. Int. Cloting Conf. on Textile Objective Measurement and Automation in Garment Manufacture, Bradford, July 1990.

[8] Palmer G.S., Wilkinson A.J., Taylor P.M., "The design of an integrated robotic garment assembly cell", Proc. 6th Nat. Conf. on Production Research, Glasgow, 1990.

[9] Gilbert J.M., "Nonlinear control of an industrial robot", Ph.D. thesis, Univ. of Hull, UK 1989.

[10] Taylor P.M., Wilkinson A.J., Gibson I., Palmer G.S., Gunner M.B., "The automation of complex handling tasks requiring out-of-plane manipulation", '91 ICAR, 5th Int. Conf. on Advanced Robotics in Unstructured Environments, Pisa, Italy, June 1991.

[11] Palmer G.S., "An RTX assembly cell calibration system prototype", Univ. of Hull internal report 76/89, 1989.

[12] Taylor P.M., Wilkinson A.J., Gunner M.B., Sawyer A.J., Gibson I., "Integration of a flexible cell for garment assembly", Proc. 28th Int. MATADOR Conf. on CIM, FMS and Robotics, pp 131-138, 1990.

VISION FOR ROBOT GUIDANCE IN AUTOMATED BUTCHERY

G Purnell and K Khodabandehloo.
Advanced Manufacturing and Automation Research Centre (AMARC).
University of Bristol,
Queens Building, University Walk,
Bristol, UK. BS8 1TR.

Abstract.

The visual methodology employed to derive cutting information for automated beef forequarter deboning is described. Machine vision assessment of carcasses is made, and forequarter size/shape descriptor points for robotic meat cutting are defined. These points are either matched against a database thus inferring cut data from previous experience or used as input to vision algorithms to produce cutting information directly. The visual methodology for locating descriptor points has been proven for a large number of carcasses. The method for subsequent derivation of cutting data has yet to be proven conclusively.

Introduction.

Tasks within the meat industry tend to be repetitive, laborious and labour intensive. Couple this with a hazardous and unpleasant environment and it is plain to see why the industry has a problem with high staff turnover rates. The potential benefits from the introduction of robotic systems would be high, not only is the human operative removed from the harsh environment, the costs associated with training replacement staff are reduced. However, the majority of robotic systems currently utilised in manufacturing industry do not fulfil the requirements for use in meat processing plant[Marchant, 1985].

The Agriculture and Food Research Council provided funds for a three year link research project between the University of Bristol and the Institute of Food Research Bristol into the application of robots in the meat industry. In particular the task of deboning of beef forequarters (FQs) for the production of processing meat for inclusion in pies, ready meals etc would be considered where cut meat appearance is unimportant[AFRC, 1987].

Analysis of existing butchery techniques suggested that force feedback and vision are the major senses utilised by the human butcher. Hence the main operation principle for the automated system would be force feedback control (FFC) of the cutting device allowing it to 'feel' its way along the bones. In order to get the cutter to a point on the bone where FFC could commence, some form of initial positional information would also be required. A combination of visual information and geometric knowledge of forequarters would be employed for this prior positioning.

Overview of Butchery System Operation.

From an early project stage the vital system components were identified(Purnell, 1990) and these are depicted in Figure 1.

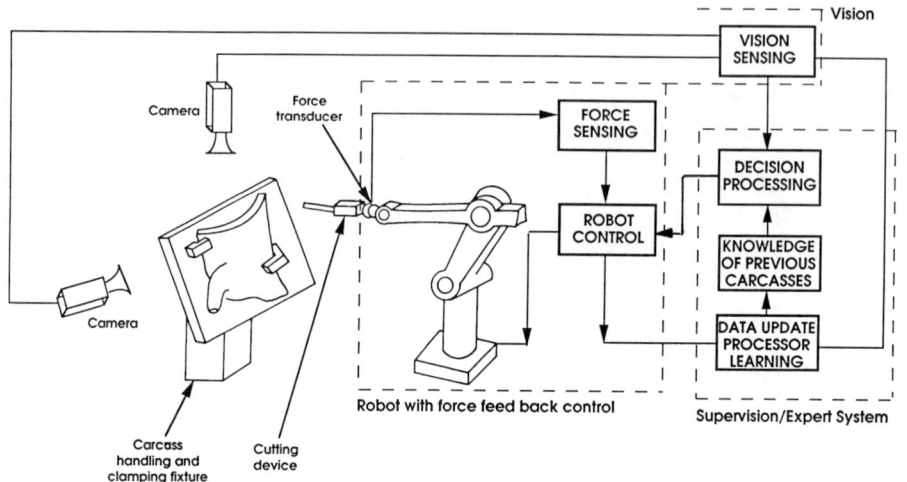

ROBOTIC MEAT CUTTING SYSTEM GENERAL CONFIGURATION FIG 1

Each forequarter (FQ) input to the butchery system would be clamped to a support fixture and presented in the correct orientation to the robotic manipulator. The carcass would then be viewed with the vision system and a set of visually identifiable points indicative of the FQ size and shape would be found. These descriptor points would then be compared to other sets of descriptor points held in the knowledge base for previously processed FQs. Along with the descriptor points, the FQ database would contain the actual cut paths followed with force feedback control (FFC) and also the cut start and end points obtained when each FQ was processed in the past. By searching the database for a match of descriptor points, cut start points can be inferred for the current FQ. The cut paths and end points also inferred by a descriptor point match can be used to monitor progress of the cuts and signal if significant deviation from the path used previously occurs. Should no match of descriptor points be forthcoming then cut start points and initial directions will be defined directly from the descriptor points.

Once the cut start point has been obtained this will be used to guide the robot to position the cutter at the approximate starting location. The blade will then be driven into the meat and made to track along the bone profile utilising FFC. Continual comparison of the current cutter position with the cut path contained in the FQ database would be used to trigger error recovery routines should the discrepancy become too large. Cutting will proceed until the expected cut end point is neared where additional force sensing for end of cut will be initiated. Having completed the cut, the start point for the next cut in the cutting sequence will be approached and the cutting process repeated until all the meat is removed from the FQ. A block diagram describing the interaction of the database and vision with the rest of the system is shown in Figure 2.

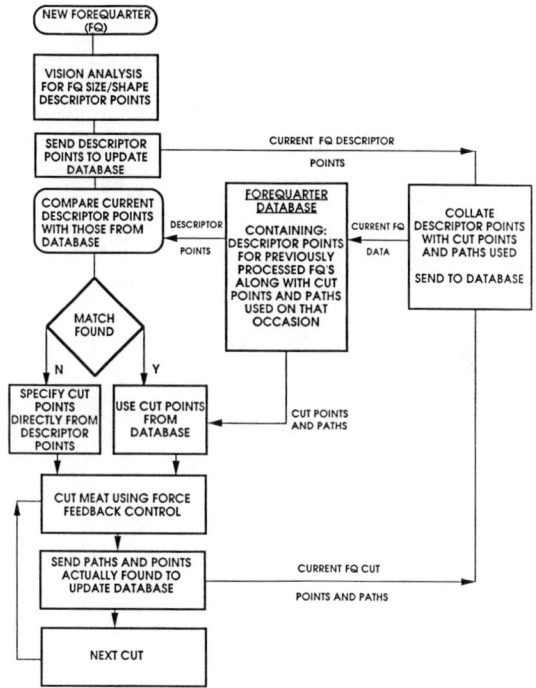

AUTOMATED BUTCHERY OPERATION SCHEME FIG 2

The full system operation relies on the use of the FQ database. However, this database cannot be created until the system cutting hardware is capable of processing complete FQs. This is currently not the case and processing is restricted to the no match route (see Figure 2). Thus techniques of producing cut point data directly from descriptor points are required. First, this would enable an initial database to be established for use in the final system and secondly it provides the methodology for cutting point generation in a no descriptor point match situation.

Cutting Scheme and Measurement Points.

In order to remove meat from FQs it is necessary to establish a generic cutting scheme, common to all carcasses. Existing schemes were examined and found to require excessive handling and manipulation for robotic use, consequently a scheme specific to FFC cutting was developed where all cuts lie along the surface of bones (Figure 3). In parallel to the development of the cutting scheme the variability of FQs was assessed by anatomically defining a number of measurement points that describe the overall carcass shape and relate to the cutting scheme (Figure 4). The intention has been to locate these measurement points visually and use their positions for matching against the FQ database (i.e. use the measurement points as descriptor points). If no match was found then the inherent relationship of the measurement points to the cutting paths would be used to define the cut instructions.

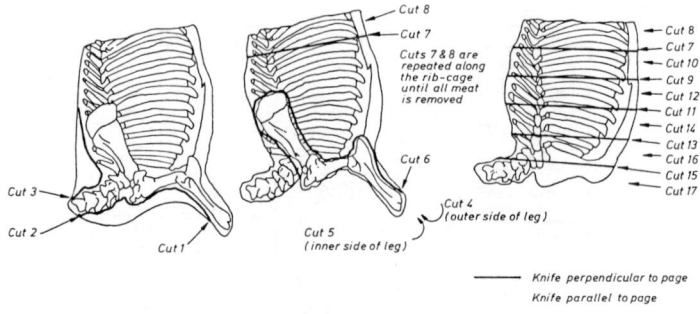

FIG.3 STANDARD FOREQUARTER CUTTING SCHEME

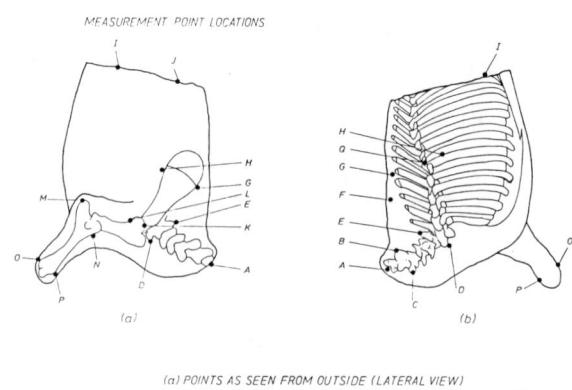

(a) POINTS AS SEEN FROM OUTSIDE (LATERAL VIEW)
(b) POINTS AS SEEN FROM INSIDE (MEDIAL VIEW)   FIG 4.

Measurement Point Collection Via Machine Vision.

The measurement point definitions were examined (see Figure 4) and it was obvious that the positions of points beneath the surface of the carcass (G, H, D etc) could not be found with vision. The remaining points were categorised into groups defined on the estimated difficulty of recognition and a series of trials instigated to verify the groupings made.

Using the Automatix AV3, 64 grey scale, 383 x 255 resolution machine available, the feasibility of visual location for each of the measurement points (see Figure 4) was investigated in turn. Points on the leg end (O, P) were readily found as the leg feature is easy to locate. Other points such as M, N, and F could be located once they were highlighted by the lighting system. Since these points lie on or close to edges or sharp curves of the FQ surface it is possible with strategically placed cameras and lighting to enable the visual thresholding function to obtain the location of the point. This technique worked for most of the carcasses tested but since separate cameras and lighting set-ups would be required for each measurement point it is not a practical solution. Reliable identification of some of the points was not possible, notably where the position varied greatly between carcasses (eg. E, F) or where visual highlighting with structured lighting was not feasible (eg. B, C, Q).

Given the limited success of these trials alternative methods for cut point generation were considered - a technique of using visually identifiable FQ features in place of the anatomically defined measurement points has proved more workable. The measurement point definitions would still be used for FQ measurement and the subsequent analysis of size/shape variations but the descriptor points for matching purposes would become visually defined.

Visual Determination of the Cutting Scheme Points.

Working from photographs of FQs, vision algorithms were produced that located the cut start point, initial directions and end point for each cut. After algorithms had been completed for several cuts, it was noted that certain points were common in the derivation of cut information for many of the individual cuts. These common points in the cut definitions were obvious choices for the visually identifiable descriptor points. The vision algorithms then subdivided into two parts:
i). Derivation of salient common points.
ii). Subsequent processing for each cut data.

Each incoming FQ would be visually appraised to determine the salient points, which would be used for matching to the FQ database. If no match was forthcoming the second set of algorithms would be executed, producing cut points and initial directions directly from descriptor (salient) point data.

3D Measurement and Scaling.

The above methodology specifies cutting points and initial directions, however the position definitions produced are in pixel co-ordinates. The positions need to be converted into the robot co-ordinate system to enable cutting to take place. It is a simple matter to correct for non-linearity in pixel position and calibrate for measurements in a single plane parallel to the camera CCD element. However a FQ is not planar and 3D measurement is required - this introduces the problems of perspective.

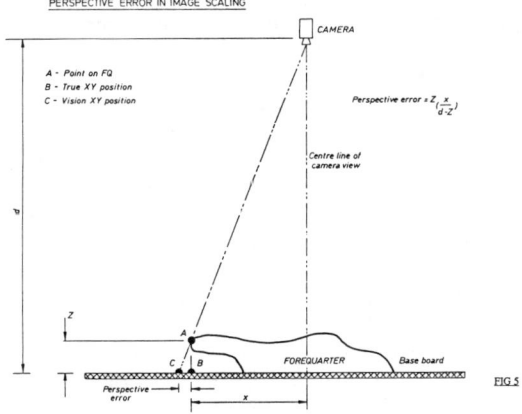

FIG 5

As the point of interest moves towards the camera, it appears to move toward the edge of the image screen. The discrepancy between this apparent position and the true position is the perspective error (Figure 5). The magnitude of the perspective error is given by:

$$e = Z.x / (d-Z)$$

Where:-
- $e =$ Perspective error.
- $Z =$ Perpendicular distance from the baseboard (calibrated plane).
- $x =$ Distance from centre of image.
- $d =$ Camera - Baseboard separation.

Thus if Z (distance from calibrated plane) can be determined, corrections for perspective can be made.

The rig for measuring FQs would consists of 3 cameras; the main camera is be mounted with the CCD element parallel to the FQ clamping plate. This provides data from the lateral side of the FQ. Cameras 2 and 3 are mounted to view parallel to the plane of the base board giving the neck (dorsal) view and leg (ventral) views respectively. For the lateral view the base board forms the calibrated plane and for the other cameras the calibration planes are defined in space at known XY positions perpendicular to the baseboard surface. The remaining problem is the determination of Z distance.

Z could be measured by one of the auxiliary cameras but this view would also be subject to perspective error. To circumvent this problem an iterative method using both views could be used, but this may prove time consuming. Another possibility may be to use statistical data from the FQ measurement phase of the project to estimate Z height at the required position, but this would take no account of the actual FQ seen. Although none of these methods alone is very effective, a hybrid method combining elements of the possibilities has evolved which has been used to produce Z distances.

The hybrid scheme assumes the point of interest has been found in both the lateral (LA) and one other view, the offset from the LA plane is approximated from the FQ measurement data. This enables the LA pixel position to be corrected for pixel shape and perspective. The corrected position is then converted into millimetres and transformed into the robotic co-ordinate system. From this robotic XY position and knowledge of the dorsal (DO) or ventral (VE) calibration plane position the point offset from the auxiliary view calibration plane can be found. Using this value, and the point position in the DO or VE view, corrections for auxiliary camera perspective are made and a robotic Z position, perpendicular to the baseboard, can be defined. If required, this Z value can be compared to the original estimate and an iterative cycle instigated.

Derivation of Vision Algorithms for Cut Point Determination.

Initially the algorithms used photographs of FQs as input data, and later a one third scale model of a beef side was used which permitted investigation into the use of structured lighting. This model had to be returned before the completion of the algorithms and a full size model was constructed by moulding from a real FQ. With each change of input data type the vision test rig had to be rebuilt and recalibrated to

cope with the increase in size. The final modification to work with the full size moulded model not only allowed development work to proceed with a true shape FQ but the rig also had the capability of testing the algorithms against real FQs.
The final algorithm for visual cut point derivation has 3 parts:-

Part 1: This is a series of set-up functions describing camera relative positions and areas of interest in each view. The scaling constants for each view are also input to the system. Datum and calibration features are recognised in all views and the positions determined. The transformations required to map corrected pixel points into the robotic co-ordinate frame are then defined.

Part 2: In this part, each of the camera images of the FQ is visually processed to obtain the salient (descriptor) points in that view.

Part 3: Here, the salient feature positions extracted previously in part 2 are manipulated to produce cut points. However, some additional image processing is necessary.

The separation of image processing allows for the majority of the repetitive image processing to be performed in the same section thus reducing the programming load. Since the salient points are defined in part 2, if a match occurs to the FQ database the algorithms for visual determination of cutting points (part 3) need not be executed.

Testing was accomplished in two sections, with the salient point identification scheme evaluated separately to the cutting point deviation algorithms.

Testing of Salient Point Extraction Algorithms.

A sonic digitiser producing XYZ data with accuracy of up to +/- 1mm was used to collect edge information for 40 of the FQs processed in the measurement phase. Using a 3D CAD package the expected views for each of the cameras were synthesised. This permitted the methodology of salient point extraction to be examined. In general the scheme proposed worked for all FQs although a few points were misplaced. However, in most cases the algorithms would recover after such an error and subsequent salient points would be located correctly even though a misplaced point was used in their generation.

Testing of Cutting Point Derivation Algorithms.

The cutting point positions were established manually on a real FQ and the positions measured with the sonic digitiser. Each FQ was then placed into the vision rig and the cutting point positions determined using the vision algorithms. A comparison of the two sets of cutting point positions was then made. To date, the cutting point derivation routines have only been tested on 5 FQs, to complete the analysis more carcasses will need to be processed.

The visually derived cutting points are related to the robotic datum and the manually collected points to the measurement points datum at the front edge of the 5th vertebra. Thus the point positions could not be compared directly, and in order to relate both sets of data, some form of normalisation was required. The straight line distance from cut start to end point and the angle between this and the initial direction vector were

calculated for each cut or section of cut. A comparison of the visual and digitised measurements was made to assess the algorithm accuracy.
The correlation of results between methods for cut 1 was good, as it was for the majority of cut 6.1 values. The remainder of the values however bore little resemblance between methods. Although FFC can accommodate a small errors the discrepancies seen in the comparison were unacceptable. The possible reasons for the difference are many, the vision rig may not be calibrated correctly, or there may be some inconsistency in the positioning of the probe when taking the digitised data. Another factor which may affect the results was the field of view selected for each of the cameras on the rig. Further testing will be necessary to determine the cause of the discrepancies.

Conclusions.

A generic cutting scheme and a system of carcass measurement specifically adapted for robotic deboning of a beef forequarter has been defined. A methodology for the derivation of cutting data for each carcass has been proposed that involves the comparison to a database of previous experience. Algorithms for synthesis of cutting information from a visual appraisal of each carcass have been implemented and tested. The algorithms for collection of comparison points have been successfully proven, however tests on the routines for the derivation of cutting data proved inconclusive. The vision system has been linked to a robot arm and successful deboning cutting trials have been made. Work continues to fully verify the operation of the algorithms and implement the complete beef forequarter deboning system.

Acknowledgements.

The authors wish to thank the Agricultural and Food Research Council (AFRC) for funding the research, and the Institute of Food Research - Bristol for provision of facilities, knowledge and advice. Thanks must also be given to Terry Gorman, Neil Maddock, Paul Drewery and John Byles for their input and support.

References.

AFRC.
    (Agricultural and Food Research Council).
    Link Research Grant Application (LRG114). March 1987.

Marchant. J A
    The use of Robotics in Agricultural and Food Industries.
    National Institute of Agricultural Engineering, Silsoe, Bedford, UK. 19th November 1985.

Purnell G, Maddock N A, and Khodabandehloo K.
    Robot Deboning for Beef Forequarters.
    Robotica (1990) volume 8, p303-310.

USE OF ROBOTS IN SURGICAL PROCEDURES

P N Brett and K Khodabandehloo,
Advanced Manufacturing and Automation Research Centre,
Faculty of Engineering, University of Bristol, Queen's Building,
University Walk, BRISTOL, BS8 1TR, UNITED KINGDOM.
Tel: ++44 272 303240
Fax: ++44 272 251154

SUMMARY

Future implementation of a surgery assistant robot will require development of dedicated manipulators, end-effectors and tools to perform fundamental surgical tasks. Currently the success of surgical procedures depends very much upon the skill of the surgeon to react to variations in the working region. Extensive use is made of tactile and vision feedback. Surgery assistant robots will need to reflect this capability and may be used to reduce the required skill necessary to perform procedures and, as a result, increase their availability.

This paper discusses safety considerations in view of manipulator configuration and the needs of a robotic system for manipulating surgical tools.

INTRODUCTION

There are a growing number of examples of robot systems where integration of the manipulator, sensory and expert system technologies have increased the level of system autonomy to cope automatically with variation in working conditions. Examples include the sheep shearing robot [1] where the cutter is able to react to variations in sheep geometry and a moving surface, robot systems for fruit picking [2] and poultry packaging [3] involving the identification of objects of varying size and geometry to guide a manipulator to pick and place, a robot for playing a game of snooker [4]; there are others. This has demonstrated increased scope for potential applications of robot technology.

Surgery is one such application area. Difficulties are imposed by the complex working environment, extreme safety issues and limitations imposed by the current state of technology. Whilst it is accepted that the capabilities of the surgeon cannot be replaced in full there are applications where benefits can be realised in the near future. These examples offer some containment of the difficulties of recognition, location of the working point and in dealing with non-rigid tissues. It is important that these tasks are identified and demonstrated successfully for acceptance of the introduction of robotics in surgery.

The fundamental benefits offered are:-

- Repeatable tool position and trajectory
- Steady motion
- Ability to react rapidly to changes in force level
- Remote operation
- Ability to remain poised in a fixed position

Such qualities can lead to more effective treatment and reduced risk of infection in the case of contagious diseases by reducing human contact and in some cases reduced risks to the patient by offering the potential of less invasive surgery.

There are examples of robots used without physical contact with patients. In these cases the non-invasive action of the robot has avoided some of the difficulties over safety [5] and the possibility of litigation whilst utilising some of the benefits. More recent investigations are approaching human contact using a robotic device [6]. Such a device is being developed to be used for transurethral resection of the prostrate [7] and research is on-going for a robotic system to assist in stereotactic procedures in France. An automated retraction system is undergoing research in Canada [8].

There are a number of areas of technology that need to be considered before a suitably autonomous device can be developed, not least the role and evaluation of 'Expert Systems' for high level system control. This paper discusses manipulator configuration with respect to safety and examines requirements for robot hardware in contact with the patient.

SYSTEM CONFIGURATION

It is not considered reasonable that a robot system should operate without continuous back up from a surgeon who would be in a position to define and check the step by step progress of the task and to override the system in case of unforeseen difficulties.

The robotic device would be driven to an acceptable start under position control by the surgeon and, upon instruction, perform the defined task prompting the surgeon at each stage. The system will need to react to given situations in real time by comparison of behaviour with reference models of characteristic behaviour of the medium during cutting or handling. This data can assist in the identification of material properties or to anticipate undesirable behaviour requiring remedial action.

An appropriate system configuration is shown in the figure 1. The system is rich in sensors. At the tool or end-effector torque and force measurements would be combined with vision data. The expert system provides High Level Control in relation to progress and task description. By comparison with behavioural reference models the planning of manipulator motion can be achieved. The low level controller of joint motion will need to react to tool force levels as well as velocity and position.

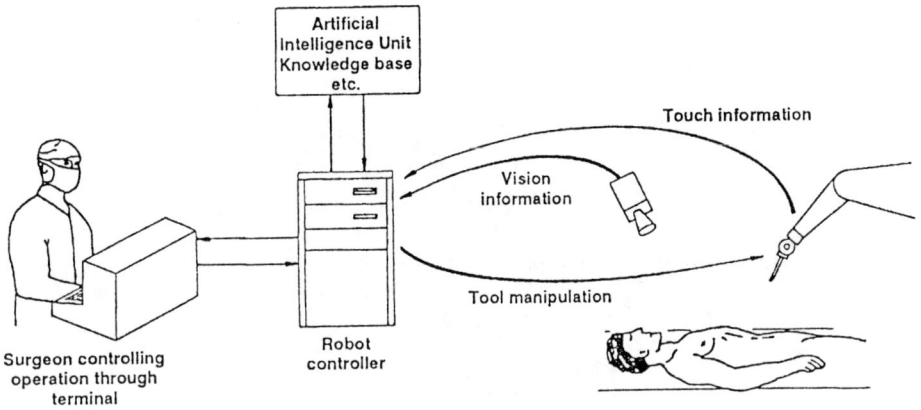

SYSTEM CONFIGURATION

FIGURE 1

MANIPULATOR REQUIREMENTS

The nature of the profession is such that descriptions of techniques tend to be qualitative rather than quantitative. Therefore considerable cooperation is required between physicians and engineers to produce technical specifications for a robot and the tools, reflecting manipulation requirements and acceptable force levels.

Manipulator Design

In the case of the manipulator the requirements are driven by the required access and manipulation of tools, accuracy of position and force level control.

High safety standards must be observed as the robot system is to operate in close proximity with the patient and the team of theatre staff. A potential hazard is a "run away" joint. The risk can be reduced by selecting a suitable manipulator configuration.

The following criteria are considered important:-

- At the working site, tool motion to be achieved by moving a minimum number of joints.

- Tool motion to be achieved by decoupled joint motion at the end-effector.

- The facility for tool and end-effector position to be deduced simply by visual inspection of joint positions must be available.

- By manual override, the tool and end-effector to be extracted and withdrawn from the working site safely by moving one joint at a time.

- The work space of the manipulator to be discernible by theatre staff. This is best achieved if the work space is internal to the structure of the manipulator.

- The patient to be outside of the work space available to the manipulator when moving the end-effector to the working site.

- Manipulator structure to enable reasonable access to the patient.

To satisfy such constraints manipulator usage is unlikely to be general and more likely dedicated to a particular procedure or, at best, to working on a region of the body. One such configuration is shown in figure 2.

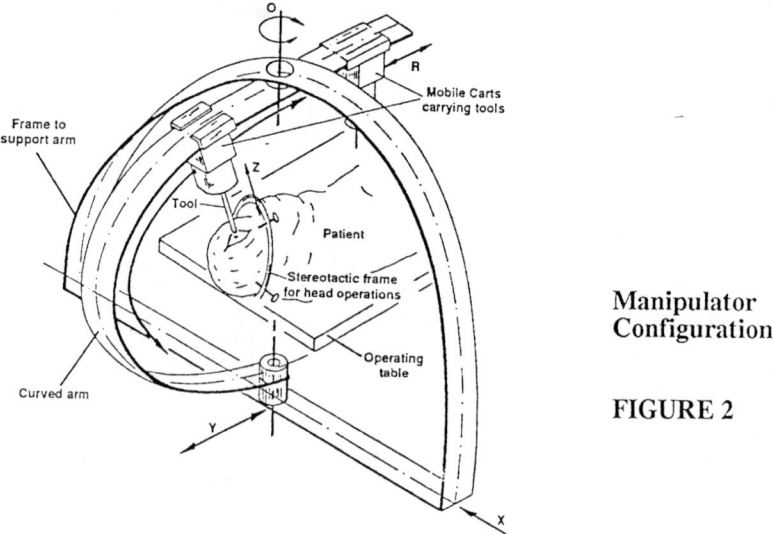

Manipulator Configuration

FIGURE 2

A configuration of this sort is suited to working on an area of the head. The design features three linear drives, a rotary axis and a curved rail on which carriages supporting specialised end-effectors can be moved. Many procedures require two working points of contact in the manipulation of tissue or the use of tools. Safety considerations are reflected in the method of operating the manipulator. Initially drives x and y are moved to align the manipulator with the operating table. In some procedures the position of the patient needs to be assured with respect to a reference position by supporting the patient suitably. The end-effector would then be moved to the correct position and the manipulator joints then locked in this position. Tool motion can then be controlled by drives activated at the end-effector.

The movement of the joints of the manipulator can be restricted to avoid contact with the patient on moving the manipulator into position and as the robot cannot move outside of its well defined work envelope, given by the curved members, there is reduced risk to theatre staff.

End-effector motion should be achieved by decoupled joint motion such that end-effectors can be removed easily manually and quickly without risk to the patient.

Similar arrangements utilising different shapes to the curved rail could be applied to other parts of the body offering the same benefits with regard to safety.

TOOLING, TECHNIQUES AND REQUIREMENTS

A variety of fundamental hand-held tools are available for carrying out primary functions, for example.

>   Scalpels
>   Scissors
>   Retractors
>   Diathermy Probes

These are available in a variety of sizes and forms. Where possible tools of a robotic solution should be of similar form to existing tools as these are in ready supply. When applied by surgeons it is often the case that manipulation of tools is in response to tactile feedback. In order to determine typical working conditions measurements of force levels and characteristics associated with tool use have been made on cadavers. Force levels and required positional accuracy define strength and stiffness specifications for manipulator, tools and sensors. Characteristics associated with changes in force with respect to time or displacement can be used to anticipate a particular condition.

Cutting using a scalpel
Slide incisions and stab incisions are the most common cutting actions. In the case of stab incisions the motion and applied pressure are in the same direction and perpendicular to the surface of the skin. Although accurate incision length and direction can be achieved, depth control is poor due to the required change in cutting force as the bursting strength of the tissue is exceeded. When cutting through to the surface of a bone the abrupt change in force can be detected simply, however incisions into the first layer of a series of tissue layers will prove more difficult. Cutting force will depend upon blade size and condition, tissue type, depth of penetration and velocity. Detection of the transition between layers will require a sensitive sensor although depth can be anticipated from previous knowledge. Often a surgeon may resort to scraping using a sub-bursting force with the motion of the blade parallel to the surface and perpendicular to the blade where thin tissue layers are to be separated.

In slide incisions the motion of the blade is parallel to the surface of the skin. Examples of force levels in stab and slide incisions were obtained from measurements on the dura of a dead pig. This is the tough flesh like layer on the head below the skull. Typical results are shown in figures 3 and 4 using a No. 15 blade and a cutting velocity of 1.27 mm/s. Maximum force is found in the slide incision of the beginning and end of the cutting stroke.

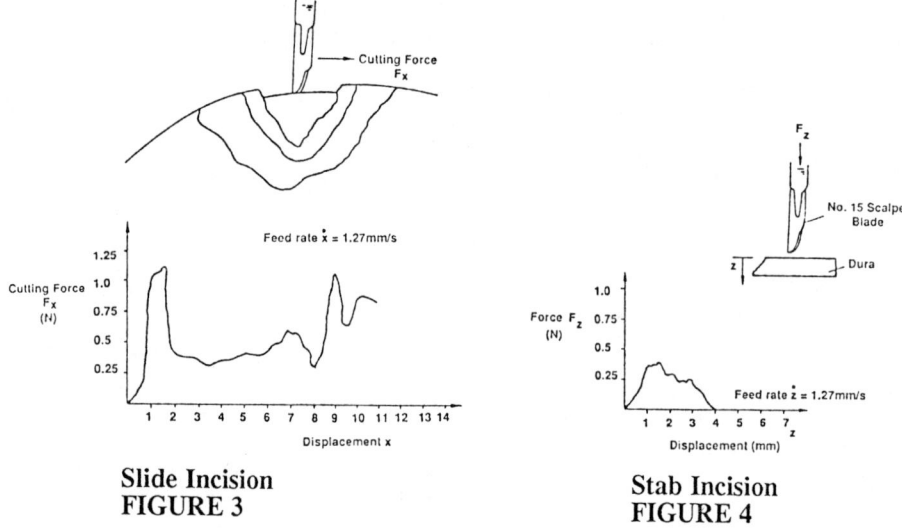

Slide Incision
FIGURE 3

Stab Incision
FIGURE 4

Frequent blade changes would be required, maintaining sterility and sharpness and the scalpel end-effector should accept the existing standard blades, the end-effector will need to resolve force components along orthegonal x, y and z axes.

When a slide incision is made to a large area of unrestrained skin, tension should be applied by stretching the surface. This avoids jagged cuts produced by rippling the flesh. Control of the depth of cut using vision is possible by separating the edges of the skin soon after the blade has passed as the base of the cut soon fills with blood.

It is necessary to define parameters such as the start point, length, direction, contour, depth and cross-section configuration of the incision. Skin marking pens could be used to indicate the desired path of the blade.

Other Cutting Tools
Scissors: These have the advantage over a scalpel of cutting flacid tissue that cannot be held under tension. Normal scissor cuts or push cuts can be made. Push cuts enable rapid, accurate cuts in sheet-type tissues in the direction of the grain.

A mechanism could be designed to reproduce the cutting action of a pair of scissors but the design would be greatly eased if a shearing or guillotine action could be used instead.

An end-effector of such types would require force sensors to monitor the blade closing force and a power supply for mechanism actuation.

Diathermy Devices: These devices employ a high frequency electric current and are used to penetrate tissue providing for coagulation of blood. A clean incision is made using such tools without bleeding. This is an advantage for an automated system as the working area is drier and easier for visual inspection. The main disadvantage is poor mechanical feedback for judging depth.

Saws and Drills: When using these tools there is often a need anticipate penetration by changes in force level as in some cases the thickness of the bone is not known prior to cutting and the cutting region may not be easily visible, particularly in the case of some drilling applications.

INVESTIGATIONS ON EXAMPLE ROBOTIC APPLICATIONS IN SURGERY

Stapedotomy

This procedure is carried out to restore hearing loss resulting from fixation of the stapes, a small bone of the middle ear. A hole is bored through the footplate of the stapes, penetrating the region of the inner ear. A prosthesis is attached to the moving parts of the middle ear and this disturbs fluid in the inner ear in response to vibration at the eardrum.

The robotic device shown in figure 5 is able to demonstrate automated tool control in drilling on cadavers. The device demonstrates:-

- Detection of drill-bone contact
- Detection of the far surface of the stapes before the drill breaks through (an unknown distance)
- Drilling under force control
- Drilling under position control
- Repeatable drilling trajectory

This successful research study is the result of close collaboration between the Robotics Research Group at Bristol University and the Department of Otolaryngology of the hospitals of Bristol, UK.

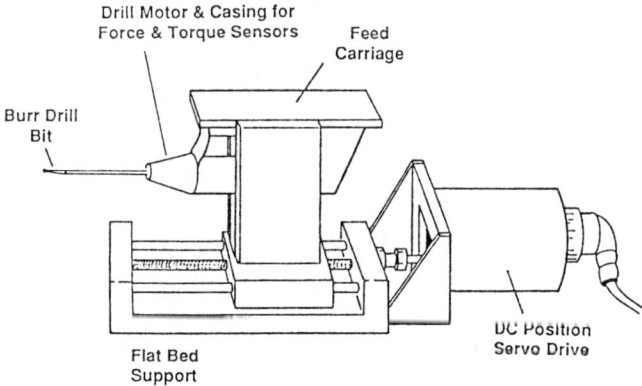

Bristol Demonstration Device for Drilling in Stapedotomy
FIGURE 5

Surface Vein Harvesting

Removal of surface veins is a common procedure. In the case of the cardiac by-pass operation, veins are removed in good condition to be reused. Often the saphenous vein is removed from the leg.

A technique that can be used is the 'loop method'. This requires only small incisions in the leg compared with more traditional methods involving an incision over the length of the leg. The 'loop method' is attractive as it results in reduced post operative pain and speedier patient recovery, however it requires considerable skill as it is important to detect tributaries to the main vein by a rise in force on the tool. The loop is threaded over the vein tracking along the vein separating it from surrounding layers as the loop passes along as shown in figure 6. On detecting a tributary an incision is made at the position of the loop ligation of the tributaries occurs and the free end of the vein is extracted through the incision. The loop is extracted to be rethreaded onto the vein to continue the process.

Laboratory equipment using an industrial robot supporting a loop tool with force feedback automatically demonstrated tracking of a simulated vein and detection of tributaries. This is a potential area for robotic devices and intelligent sensor devices for surgical application below the head.

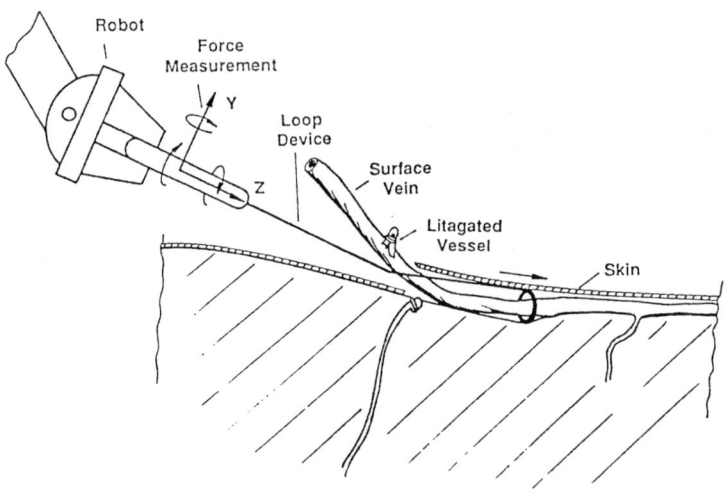

Application of the 'Loop Method' in Surface Vein Harvesting
FIGURE 6

CONCLUSIONS

Surgery is a complex working environment for robot systems and will require further development and integration of the enabling technologies to produce a working solution. In view of strict safety requirements and operating features, manipulator configuration is likely to be dedicated to working on particular areas of the body. End-effectors should where possible accommodate standard tools. Requirements and potential solutions for the manipulation and use of some primary tools are described. It is important to examine current practices carefully to determine sensing strategies. In most cases the application of surgical tools will be controlled by vision and force feedback data. There are requirements to distinguish layers of flesh or bone from cutting forces and this is normally achieved by the surgeon using previous experience. The capabilities of the system must reflect this. A knowledge base as part of the expert system is a likely solution for a high level controller.

ACKNOWLEDGEMENTS

The authors wish to thank Mr M Eastwood and the following staff of the Bristol Hospitals for their support: Mr M Griffiths, Mr D Baldwin, Mr J Blanshard, Mr Dhasmana and Mr J Hutter.

REFERENCES

1. Trevelyan, J P. 'Sheep handling and manipulation for automated shearing', The 19th Int. Symp. and Exposition of Robots, Sydney, Australia, Nov. 1988.

2. Marchant, J. 'The use of robots in the agricultural and food industries', Nat. Inst. of Agricultural Engineering, Silsoe, Bedford, UK, 1988.

3. Khodabandehloo, K. 'Robotic handling and packaging of poultry products', Robotica, 1990.

4. Rennell, I J and Khodabandehloo, K. 'Development of skilled robots : a new approach in robotics', 20th ISIR Tokyo, 1989.

5. Edwards, R. 'Robotics in Medicine : Safety Aspects', Conf. on Robotics in Medicine, IMechE HQ, London, June 1990.

6. Finlay, P A. 'SARAH: An advanced robot for assisting in precision surgery', Conf. on Robotics in Medicine, IMechE HQ, London, June 1990.

7. Davies, B L, Hibbard, R D, Timoney, A, and Wickham, J. 'A surgeon robot for prostatectomies', IARP, 2nd Workshop on Medical and Healthcare Robotics, Newcastle, UK, Sept. 1989.

8. McEwan, J A, Bunams, C A, Auchinleck, G F and Breault, M J. 'Development and Initial Clinical Evaluation of pre-robotic and robotic retraction systems for surgery', IARP, 2nd Workshop on Medical and Healthcare Robotics, Newcastle, UK, Sept. 1989.

AUTHOR INDEX

AGRAWAL S.	559	DRITSAS L.	603
ALAGAR V.	27	DUELEN G.	217, 469
ALBRIGHT S.	459	DUHAUT D.	487
ANDRE P.	71	ERGUEN T.	151
ARIMOTO S.	99	FANGHELLA P.	11
ARLABOSSE F.	351	FICOLA A.	79
BAJCSY R.	559	FOGAÇA P.	383
BAGHI A.	193	FONTAINE J.	313
BALAFOUTIS C.	45	FRAPPIER G.	391
BARRAL G.	367	FRAU J.	321
BECQUET M.	37	GALLETI C.	11
BENMOUNAH A.	279	GAUSSENS E.	351
BLAZEVIC P.	313	GENÇEV S.	151
BOWDEN P.	611	GIBSON I.	611
BRETT P.	627	GOUVIANAKIS N.	407
BUI T.	19	GRIFFIN M.	167
CAEN R.	551	HATZIVASILIOU F.	261
CAGLIOTI V.	431, 479	HEIKKILA T.	515
CHAOUIYA C.	439	HO P.	245
CHRISTOU N.	407	HODGSON S.	125
COLIN S.	551	HONDERD G.	61,225,235,415
CONRAD F.	577	IKONOMOPOULOS A.	603
D'ALESSANDRO G.	595	INNOCENTI G.	3
DANIELLI M.	479	ISTEFANOPOULOS Y.	151
De CAMPOS A.	**383**	JÄRVILUOMA M.	515
De HAAS J.	61	JONKIND W.	61,235
De KEYSER R.	159	KELEMEN M.	193
De LAPLACE S.	313	KHODABANDEHLOO K.	627,637
DESAI P.	109	KHOSLA P.	305
DIMITRIADIS B.	407	KOKAR M.	505
DOULGERI Z.	595	KOUMBOULIS F.	183

KROESE B.	495	PARTHENIS K.	407
La CAVA M.	79	PATEL R.	45
LAGERBERG J.	495	PÉREZ ORIA J.	175
LARRE A.	321	PERIYASAMY K.	27
LEMARQUAND P.	391	PETRIDIS V.	297
Le PAGE P.	351	PETROPOULAKIS L.	585
LIGEOIS A.	271	PIESKA S.	55
LOSCO L.	71	PIOTROWSKI L.	375
MALIK R.	423	PONS N.	313
MATOS M.	383	PRASAD S.	423
MAUBOUSSIN A.	287	PRUSKI A.	359
MENG A.	253	PURNELL G.	619
MERTZIOS V.	19	RABI J.	313
MILLAN J.	37	RENDERS J.-M.	37
MITCHELL R.	167	RIESWIJK T.	61,225,235
MITROUCHEV R.	71	SCARLATOS S.	19
MOIGNARD C.	271	SCHAEFFER V.	287
MONACELLI E.	487	SCHALKWIJK P.	225
MONTSENY E.	321	SCHMIDT G.	207
MOUMEN K.	359	SCHROER K.	459
MOWFORTH P.	535	SENTIEIRO J.	343
MUELLER P.	87	SEQUEIRA J.	343
MUENCH H.	469	SINHA P.	245,279
MURACA P.	79	SIRKS M.	235
NANIWA T.	99	SONG Y.	159
NEUBAUER W.	207	SORRENTI D.	479
NIEVERGELD A.	117	STOTEN D.	125
NIKOLERIS G.	451	SWIDER K.	53
NOVAKOVIC B.	135	TAHBOUB K.	87
NTAFOS S.	253	TAILLARD J.-P.	71
OCCELLO M.	439	TAM R.	143
OLIVER G.	321	TAYLOR P.	611
PAPANIKOLOPOULOS N.	305	THEVENON J.	351
PARASKEVOPOULOS P.	183	THOMAS M.-C.	439
PARENTI-CASTELLI V.	3	THOMOPOULOS S.	143

TSIBOUKIS T.	297
TSOUKALAS M.	253
TZAFESTAS S.	261
TZIERAKIS K.	183
Van ALBADA G.	495
Van de VEN H.	117
Van den BOGAERT T.	391
Van der MOLEN G.	399
Van TURENNOUT P.	415
WAHRBURG J.	569
WARWICK K.	367,543
WERSHOFEN K.	333
WEWERINKE P.	525
WILKINSON A.	611
WILLNOW C.	217
WILSON D.	535
ZAVOLEAS K.	505
ZHANG Y.	469
ZHOU J.-J.	577

International Series on
MICROPROCESSOR-BASED AND INTELLIGENT SYSTEMS ENGINEERING

Editor: Professor S. G. Tzafestas, *National Technical University, Athens, Greece*

1. S.G. Tzafestas (ed.): *Microprocessors in Signal Processing, Measurement and Control*. 1983   ISBN 90-277-1497-5
2. G. Conte and D. Del Corso (eds.): *Multi-Microprocessor Systems for Real-Time Applications*. 1985   ISBN 90-277-2054-1
3. C.J. Georgopoulos: *Interface Fundamentals in Microprocessor-Controlled Systems*. 1985   ISBN 90-277-2127-0
4. N.K. Sinha (ed.): *Microprocessor-Based Control Systems*. 1986   ISBN 90-277-2287-0
5. S.G. Tzafestas and J.K. Pal (eds.): *Real Time Microcomputer Control of Industrial Processes*. 1990   ISBN 0-7923-0779-8
6. S.G. Tzafestas (ed.): *Microprocessors in Robotic and Manufacturing Systems*. 1991   ISBN 0-7923-0780-1
7. N.K. Sinha and G.P. Rao (eds.): *Identification of Continuous-Time Systems*. Methodology and Computer Implementation. 1991   ISBN 0-7923-1336-4
8. G.A. Perdikaris: *Computer Controlled Systems*. Theory and Applications. 1991   ISBN 0-7923-1422-0
9. S.G. Tzafestas (ed.): *Engineering Systems with Intelligence*. Concepts, Tools and Applications. 1991   ISBN 0-7923-1500-6
10. S.G. Tzafestas (ed.): *Robotic Systems*. Advanced Techniques and Applications. 1992   ISBN 0-7923-1749-1

KLUWER ACADEMIC PUBLISHERS – DORDRECHT / BOSTON / LONDON